THE WATCH BOOK

COMPENDIUM

GISBERT L. BRUNNER
CHRISTIAN PFEIFFER-BELLI

teNeues

Contents

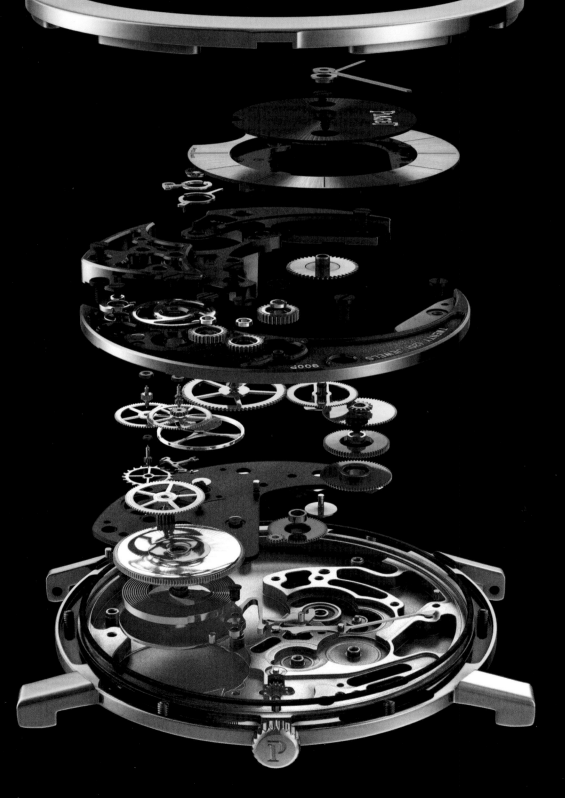

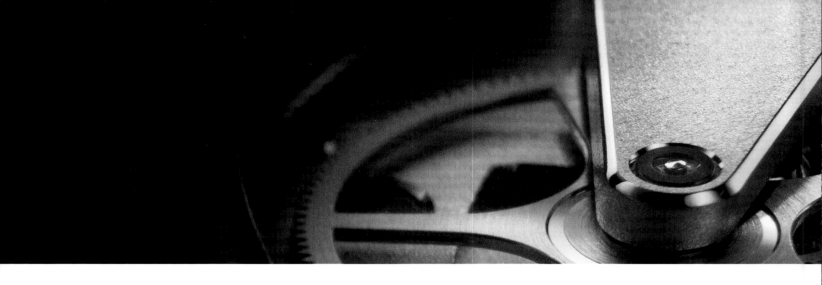

The History of the Wristwatch

This book is devoted to wristwatches. But every smartphone can show the time nowadays, so the role of a wristwatch must also be to attract admiring gazes as a piece of jewelry that can be seen from a distance. If you believe style consultants, then a wristwatch—alongside a wedding ring and a pair of cufflinks—is the only appropriate jewelry for a man. Quietly ticking on its wearer's wrist, a watch signifies the importance its owner ascribes to precious time.

To delve more deeply into the history of timepieces for the wrist, we must return to the era of Napoleon Bonaparte. One evening some time before his coronation as French emperor, the First Consul and his entourage were traveling by coach to the Théâtre-Français in Paris. The horses suddenly bolted in the Rue Saint-Honoré and the noble carriage toppled in front of Marie-Étienne Nitot's little shop. The jeweler and his staff hurried to the fallen vehicle, helped the theatergoers to their feet, and invited them into the shop. After the coach had been righted, Bonaparte thanked his hosts and promised not to forget their kindness. Nitot recalled this pledge when he heard about the upcoming coronation, which was to occur on December 2, 1804. Hoping to be commissioned to craft the crown jewels, he and the gemstone merchant Salomon Halphen entered the lion's den in the Tuileries Palace, where they convincingly presented their offer to the future emperor. Pressed for time, Napoleon agreed. His visitors explained that they lacked liquidity to purchase the precious materials, so their patron gave them a prepayment of 2.5 million francs. The sum on the jeweler's invoice was ultimately six times greater, but Napoleon and his wife were wholly satisfied. It went without saying that an appropriate gift was needed two years later, when Josephine's son from her first marriage wed a daughter of Bavaria's King Maximilian I. The court jeweler created a pair of sumptuous bracelets that Auguste Amalia of Bavaria could wear on her left and right wrists. One of these bangles included a manually adjustable calendar, the other a little watch movement. It thus seems legitimate to describe Marie-Étienne Nitot as the inventor of the first wristwatch in the truest sense of the word.

But a history of this genre of timepiece would be incomplete if it failed to mention England's Queen Elizabeth I, who was delighted to receive a ticking present on the occasion of the reinstitution of the Reformation in 1571. Her favorite, the Earl of Leicester, had arranged to have a small portable clock redesigned so she could wear it on her wrist. The French intellectual Blaise Pascal is believed to have affixed his pocket-watch to his forearm in the mid 17th century. Mothers and nursemaids did likewise because they felt that their timepieces were better protected there from children's unpredictable little hands than if they were worn around the neck as pendants or pinned to clothing as brooches.

Henri-Louis Jaquet-Droz merits our attention too. This Geneva-based watchmaker's ledgers record that he made an ornamental timepiece for the wrist in 1790. The successors of the ingenious watchmaker Abraham-Louis Breguet tempted their prosperous clientele with similar items between 1831 and 1838. Geneva's elite Patek, Philippe & Co. manufactory fabricated a golden bracelet with a baguette-shaped movement in 1868, but five years would pass before it found its buyer. A large diamond had to be ostentatiously flipped upward before its dial could be seen.

Ten years later, a small series of wristwatches were made in Vienna in 1878. According to their chronicler, these were intended "for gentlemen to wear for their convenience." The first series of wristwatches with chain bracelets were probably made by Girard-Perregaux in 1880: Germany's Kaiser Wilhelm II ordered them as equipment for naval officers. Constant Girard-Perregaux later also offered his creations for sale in the USA, albeit with only moderate success because men there didn't yet trust this genre of timepiece. But American women tourists in Luzern felt differently around 1886: these ladies were enthusiastic about golden and silver wristwatches with so-called "scissor flex" wristbands.

Jewelers in other metropolises likewise had ample reason to be grateful to the fair sex. Clothing fashions succeeded one another at a progressively faster pace so traditional pendant watches were only limitedly useful, but the new wristwatches went perfectly with every outfit. Despite brisk sales to women, one dealer viewed

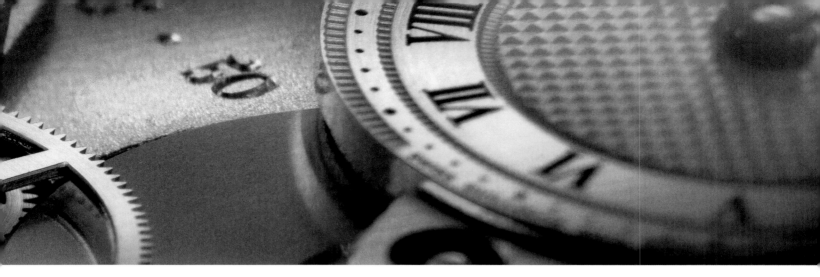

the preference for wristwatches as "an aberration of female taste" because the wrist is "surely the least appropriate place to affix a timepiece."

Professor Hermann Bock agreed in Hamburg in 1917: "We can only hope for the speedy disappearance of the foolish fashion of wearing a watch on the most active part of the body: namely, the wrist." Bruno Hillmann criticized the female gender in 1925. In his book *Die Armbanduhr: ihr Wesen und ihre Behandlung bei der Reparatur* (*The Wristwatch: Its Characteristics and Repair*), this watchmaker from Berlin condemned "a very special kind of watch that's as difficult to handle with watchmaker's tools on a workbench as it is with a writing pen at a desk." He hoped that with the "continually increasing masculinization of the female sex, the "gent's vest will triumph among ladies" and "the bell will toll to save us from the tyranny of the wristwatch." But the wish of this artisan cum author didn't come true. Even worse: more and more men succumbed to the appeal of the wristwatch. A watch for the wrist couldn't be as bad as some claimed because this genre of timepiece had survived the rigors of the Second Boer War from 1899 to 1902 on the wrists of British soldiers. Positive experiences during the First World War confirmed this view. A soldier in the heat of battle could instantly read the time simply by rotating his forearm. Grids above the delicate crystals protected the latter from shattering. Radium on dials and hands showed the necessary information in the dark. And chronographs with telemeter scales made it simple to calculate the distance to the enemy lines based on the different speeds of light and sound.

The Parisian jeweler and designer Louis Cartier had proven in 1904 that wristwatches needn't necessarily be miniaturized versions of pocket-watches. The "Santos" that he created for the pioneering Brazilian aviator and bon vivant Alberto Santos-Dumont decisively contributed to the emancipation of wristwatch design. Innovative English armored vehicles inspired Cartier to create his rectangular "Tank" watch on September 15, 1916. In gratitude for their courageous military service, Louis Cartier presented the first specimens of this wristwatch to General John Joseph Pershing (the supreme commander of American troops in France) and high-ranking officers.

Hans Wilsdorf similarly numbers among the recognized pioneers in the history of the wristwatch. Among his other contributions, Rolex's founder deserves credit for the first wristwatch chronometer, the first genuinely watertight case and a rotor-based automatic winding mechanism for wristwatches.

The unique and ongoing success that began in the 1930s, coupled with the continued decline of the pocket-watch, can be credited to small but indispensable components such as shock absorbers, as well as an extraordinarily wide spectrum of forms and cleverly engineered additional functions. Thanks to these and other achievements, the ticking wristwatch could evolve into a practical mass-produced item for daily use, a cultural asset par excellence and a reliable companion in the conquest of outer space and the world's oceans.

The first flourishing of mechanical wristwatches ended in the 1970s, when oscillating quartzes violently ousted traditional mechanical movements from watches' cases. Hard pressed to compete with these new multifunctional electronic watches, time-honored timekeeping mechanisms had little to offer except proven reliability, longevity, and undeniably loveable charm. Fortunately, this depressing phase didn't last very long. Collectors soon discovered the lasting values of the allegedly "old-fashioned" watch. This trend initiated a spectacular renaissance for mechanical timekeeping. Never before in the more than seven centuries since the invention of geared clockworks have there been so many interesting new developments and complications. This situation notwithstanding, ticking microcosms must now once again defend their niche on the wrist. The latest assailants are smartwatches and their manufacturers. But chronometric luxury and long-lasting value cannot be found among these newfangled inventions, so it would seem that we can reconfirm the correctness of an opinion aptly expressed by the American psychologist Robert Levine: "The ticking of a mechanical watch is the heartbeat of human culture." ◦

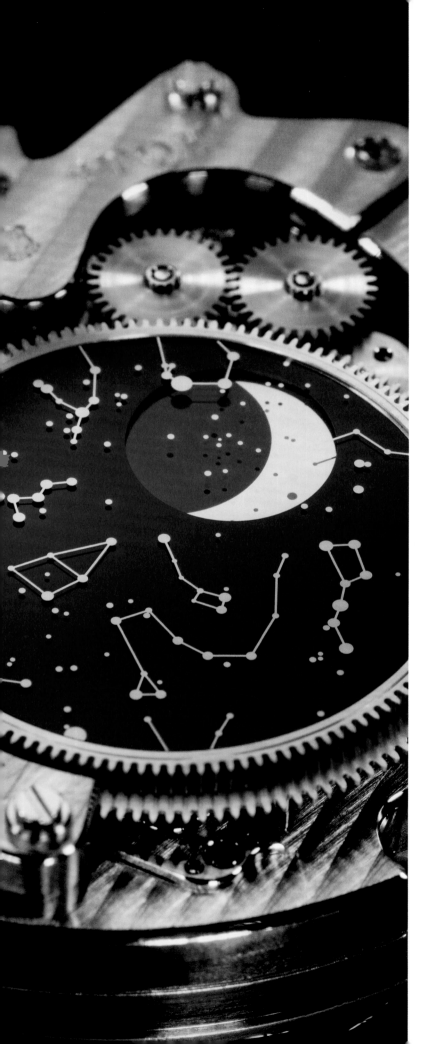

Geschichte der Armbanduhr

Dieses Buch ist Armbanduhren gewidmet, die heutzutage, da jedes Smartphone ganz nebenbei die Uhrzeit anzeigt, auch als weithin sichtbare Schmuckstücke punkten müssen. Übrigens die einzig legitimen bei Männern – neben Ehering und Manschettenknöpfen –, wenn man maßgeblichen Stilberatern glaubt. Ganz nebenbei lassen sie, tunlichst tickend, auch noch erkennen, welche Bedeutung ihre Besitzerin oder ihr Besitzer dem kostbarsten Gut der Menschheit beimisst.

Die Beschäftigung mit der Genese des ans Handgelenk geschnallten Zeitmessers verlangt eine Rückblende in die Ära von Napoleon Bonaparte. Eines Abends, noch vor seiner Krönung zum französischen Kaiser, ließ sich der Erste Konsul samt Entourage zum Pariser Théâtre-Français kutschieren. In der Rue Saint-Honoré gingen die Pferde durch, und das noble Gefährt stürzte vor dem kleinen Geschäft des Marie-Étienne Nitot um. Der Juwelier und seine Mitarbeiter eilten zu Hilfe und baten die verdutzte Abendgesellschaft herein. Nach dem Aufrichten der Kutsche ließ Bonaparte seinen Gastgeber wissen, dass er dessen Hilfsbereitschaft nicht vergessen werde. Und genau darauf besann sich Nitot, als er von den Krönungsfeierlichkeiten am 2. Dezember 1804 erfuhr. In der Hoffnung, die Kronjuwelen fertigen zu dürfen, wagte er sich gemeinsam mit dem Edelsteinhändler Salomon Halphen in die Höhle des Löwen. Dort, im Tuilerienpalast, trug das Duo dem künftigen Herrscher sein Anliegen so überzeugend vor, dass Napoleon mit Verweis auf seine knappe Zeit einwilligte. Nachdem die Bittsteller ihre mangelnde Liquidität zum Materialeinkauf gebeichtet hatten, erhielten sie 2,5 Millionen Francs Vorschuss. Obwohl sich die Juweliersrechnung letztendlich auf den sechsfachen Betrag belief, zeigten sich Napoleon und seine Frau höchst zufrieden. Zwei Jahre später, als sich Josephines Sohn aus erster Ehe mit einer Tochter des bayerischen Königs Maximilian I. vermählte, brauchte es natürlich ein adäquates Geschenk. Der Hofjuwelier kreierte ein Paar wertvoller Armbänder, die Auguste Amalia von Bayern am linken und rechten Handgelenk tragen konnte. Eines mit manuell schaltbarem Kalendarium, das andere mit kleinem Uhrwerk. Somit scheint es legitim, Marie-Étienne Nitot als Urheber der ersten Armbanduhr im wahrsten Wortsinn zu bezeichnen.

Die lange Geschichte dieses Typus Zeitmesser wäre unvollständig geschildert, fände nicht auch Königin Elisabeth I. von England gebührende Erwähnung. Anlässlich der Wiedereinführung der Reformation durfte sich die Monarchin schon 1571 über ein tickendes Präsent freuen. Ihr Günstling, der Graf von Leicester, hatte eine kleine tragbare Uhr so umgestalten lassen, dass sie Halt am Handgelenk fand. Mitte des 17. Jahrhunderts soll auch der französische Gelehrte Blaise Pascal seine Taschenuhr am Unterarm befestigt haben. Gleiches taten Mütter und Kindermädchen, die ihre wertvollen Zeitmesser an dieser Stelle besser vor dem Griff unberechenbarer Kinderhände geschützt sahen als an einer Halskette oder einer Brosche.

Bleibt Henri-Louis Jaquet-Droz. Im Jahr 1790 soll der Genfer Uhrmacher den Rechnungsbüchern zufolge einen schmückenden Zeitmesser fürs Handgelenk hergestellt haben. Zwischen 1831 und 1838 bedienten die Nachfolger des genialen Uhrmachers Abraham-Louis Breguet einige ihrer ausgesprochen wohlhabenden Kunden mit Ähnlichem. Die Genfer Nobelmanufaktur Patek, Philippe & Co. fertigte 1868 ein feines Goldarmband mit baguetteförmigem Uhrwerk. Bis zum Verkauf dieses erlesenen Stücks zogen allerdings fünf Jahre durch die eidgenössischen Lande. Das Ablesen der Zeit bedingte eine ostentative Geste, weil Frau zunächst einen großen funkelnden Brillanten hochklappen musste.

Zehn Jahre später, 1878, entstand in Wien eine Kleinserie von Armbanduhren, welche den Chronisten zufolge „von Herren zur eigenen Bequemlichkeit" getragen wurden. Die aller Wahrscheinlichkeit nach ersten Serien-Armbanduhren mit Kettenbändern stammten 1880 von Girard-Perregaux. Der deutsche Kaiser Wilhelm II. hatte sie zur Ausrüstung von Marineoffizieren geordert. Anschließend bot Constant Girard-Perregaux seine Kreationen auch in den Vereinigten Staaten von Amerika an. Offenbar mit mäßigem Erfolg, denn die dortigen Männer trauten ihnen nicht. Ganz im Gegensatz zu amerikanischen Touristinnen, welche gegen 1886 in Luzern Gefallen an goldenen und silbernen Armbanduhren mit sogenannten Scherenbändern fanden.

Auch andernorts hatten Juweliere allen Grund, den Vertreterinnen des zarten Geschlechts ihre Dankbarkeit zu bezeugen. Bedingt durch die immer rascher wechselnde Kleidermode ließen sich traditionelle Anhängeuhren nur noch eingeschränkt verwenden. Die neuen Armbanduhren passten hingegen zu jedem Outfit. Trotz guter Geschäfte mit Frauen betrachtete ein Fachhändler die Vorliebe für die Armbanduhr als „eine Verirrung des weiblichen Geschmacks", denn das Handgelenk sei „sicherlich der unpassendste Ort zur Befestigung einer Uhr". Ähnlich äußerte sich 1917 der Hamburger Professor Hermann Bock: „Die Modenarrheit, die Uhr an der unruhigsten Körperstelle, im Armbande, zu tragen, verschwindet hoffentlich bald wieder." Und 1925 ging Bruno Hillmann mit den Frauen der Schöpfung ins Gericht. In seinem Buch *Die Armbanduhr: ihr Wesen und ihre Behandlung bei der Reparatur* wetterte der Berliner Uhrmacher gegen „eine ganz besondere Uhrenabart, die so schwierig wie am Werktisch auch mit der Feder am Schreibtisch zu behandeln ist". Außerdem äußerte er die Hoffnung, dass durch die „stetig zunehmende Vermännlichung des weiblichen Geschlechts", mit der auch die „Herrenweste bei den Damen Trumpf" werde, „endlich die Erlösungsstunde von der Tyrannei der Armbanduhr" schlage. Der Wunsch des schreibenden Handwerkers ging, wie sich bald schon zeigen sollte, natürlich nicht in Erfüllung. Es sollte sogar noch deutlich schlimmer kommen, denn mehr und mehr Männer verfielen dem Reiz der Armbanduhr. So schlecht konnte schließlich nicht sein, was sich während des Zweiten Burenkriegs von 1899 bis 1902 unter höchst strapaziösen Bedingungen an den Handgelenken britischer Soldaten bewährt hatte. Während des Ersten Weltkriegs

bestätigten sich die positiven Erfahrungen. Im Eifer des Gefechts reichte zum Ablesen der Zeit ein kurzes Drehen des Unterarms. Schutzgitter bewahrten die delikaten Kristallgläser vor Bruch. Radiumzifferblätter und -zeiger lieferten im Dunkeln die nötigen Informationen. Und Chronographen mit Telemeterskala ermöglichten sogar unkomplizierte Entfernungsmessungen zur gegnerischen Front durch die unterschiedliche Ausbreitungsgeschwindigkeit von Licht und Schall.

Dass Armbanduhren nicht zwangsläufig verkleinerte Derivate von Taschenuhren sein mussten, hatte der Pariser Juwelier und Designer Louis Cartier schon 1904 bewiesen. Seine „Santos" für den brasilianischen Flugpionier und Lebemann Alberto Santos-Dumont leistete einen entscheidenden Beitrag zur gestalterischen Emanzipation. Am 15. September 1916 inspirierten ihn die neuen englischen Kampfpanzer zur rechteckigen „Tank". Die ersten Exemplare verschenkte Louis Cartier aus Dankbarkeit für ihre mutigen Dienste an General John Joseph Pershing, den Oberbefehlshaber der amerikanischen Truppen in Frankreich, und an andere hohe Offiziere.

Zu den anerkannten Pionieren in der Geschichte der Armbanduhr zählt auch Hans Wilsdorf. Dem Rolex-Gründer sind erste Armbandchronometer, uneingeschränkt wasserdichte Gehäuse, der Rotorselbstaufzug für Armbanduhren und weitere Großtaten zu verdanken.

Die einzigartige Erfolgsbilanz ab den 1930er Jahren und die zunehmende Verdrängung der Taschenuhr gründen sich ferner auf kleine, aber unverzichtbare Bauteile wie die Stoßsicherung, auf unbändige Formenvielfalt und ausgeklügelte Zusatzfunktionen. Dank dieser und anderer Leistungen konnte sich die tickende Armbanduhr zum alltagstauglichen Massenprodukt, zu einem Kulturgut par excellence und zur zuverlässigen Begleiterin bei der Eroberung des Weltraums und der Meere entwickeln.

Der erste Höhenflug endete in den 1970er Jahren, als Schwingquarze die überlieferte Mechanik vehement aus den Gehäusen verdrängten. Der multifunktionalen Elektronik hatte das Traditionelle außer erwiesener Zuverlässigkeit, Langlebigkeit und einem liebenswerten Charme kaum etwas entgegenzusetzen. Doch die depressive Phase währte nicht lange. Sammler entdeckten die nachhaltigen Werte des vermeintlich Unmodernen. Und dieser Trend initiierte eine spektakuläre Renaissance der Mechanik. Niemals in der mehr als 700-jährigen Uhrengeschichte gab es so viele interessante Neuentwicklungen und Komplikationen. Ungeachtet dessen müssen die tickenden Mikrokosmen ihren Platz am Handgelenk aktuell erneut verteidigen. Angreifer sind Smartwatches und ihre Produzenten. Weil man chronometrischen Luxus und Werterhalt vergebens sucht, könnte sich letzten Endes einmal mehr die weise Erkenntnis des amerikanischen Psychologen Robert Levine bestätigen: „Das Ticken der mechanischen Uhr ist der Herzschlag der menschlichen Kultur." ○

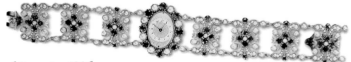

First wristwatch, made by Marie-Étienne Nitot for Auguste Amalia of Bavaria, 1806
Erste Armbanduhr, gefertigt von Marie-Étienne Nitot für Auguste Amalia von Bayern, 1806
Première montre-bracelet de Marie-Étienne Nitot pour Augusta-Amélie de Bavière, 1806

Histoire des montres-bracelets

Ce livre est un hommage à des montres-bracelets qui, du fait que tout Smartphone indique également l'heure, doivent aujourd'hui ostensiblement tenir lieu de parures. Si l'on en croit les conseillers en style faisant autorité, ce sont les seules admises chez l'homme, avec l'alliance et les boutons de manchette. Elles révèlent aussi en passant autant que possible par leur tic-tac l'importance que son ou sa propriétaire accorde au bien le plus précieux de l'humanité.

Pour aborder la genèse des garde-temps fixés au poignet, un retour en arrière sur l'ère napoléonienne s'impose. Un soir, alors qu'il n'a pas encore été couronné empereur des Français, le premier consul et sa suite se rendent à la Comédie-Française en calèche. Dans la rue Saint-Honoré, les chevaux s'emballent et le noble équipage se renverse devant la boutique de Marie-Étienne Nitot. Le joaillier et ses employés se précipitent au secours des passagers sous le choc et les prient d'entrer dans la boutique. Une fois la calèche redressée, Bonaparte informe ses hôtes qu'il n'oubliera pas leur sollicitude. Et c'est précisément de cette promesse dont se souvient Nitot lorsqu'il prend connaissance de la célébration prochaine du sacre le 2 décembre 1804. Espérant se voir confier la réalisation des joyaux de la couronne, il s'aventure dans l'antre du lion avec le joaillier Salomon Halphen. Au palais des Tuileries, les deux hommes présentent avec tant de conviction leur projet à Napoléon pressé par le temps que le futur souverain accède à leur requête. Ayant confessé leur manque de liquidités pour l'achat de matériaux, les requérants se voient offrir une avance de 2,5 millions de francs. Bien que la facture du joaillier ait fini par atteindre six fois cette somme, Napoléon et son épouse se montrent pleinement satisfaits. Deux ans plus tard, lorsque le fils d'un premier mariage de Joséphine épouse une des filles de Maximilien Ier, roi de Bavière, il faut bien sûr un cadeau approprié. Le joaillier à la cour crée une paire de bracelets précieux, un pour chacun des poignets d'Augusta-Amélie de Bavière. L'un est équipé d'un calendrier manuel, l'autre d'un petit mouvement d'horlogerie. On peut ainsi légitimement dire que Marie-Étienne Nitot a lancé la première montre-bracelet au sens propre du terme.

On retracerait de manière incomplète la longue histoire de ce type de garde-temps si l'on n'accordait à Élisabeth Ire, reine d'Angleterre, l'importance qui lui revient. À l'occasion du retour à la Réforme, elle peut se réjouir de recevoir une montre en présent dès 1571. Son favori, le comte de Leicester, a fait adapter une montre portable de sorte qu'elle puisse trouver place au poignet. Au milieu du XVIIe siècle, le savant français Blaise Pascal aurait fixé sa montre de poche à son avant-bras. C'est aussi ce que font les mères et bonnes d'enfants de l'époque qui trouvent leur précieux garde-temps ainsi mieux protégé des gestes imprévisibles des enfants qu'il ne le serait sur un collier ou une broche.

Apparaît ensuite Henri-Louis Jaquet-Droz. D'après ses livres de comptes, cet horloger genevois aurait fabriqué en 1790 un garde-temps à porter au poignet en guise de parure. Puis, de 1831 à 1838, les successeurs du génial horloger Abraham-Louis Breguet auraient livré à des clients particulièrement fortunés ce type de bijou. En 1868,
la manufacture genevoise de haute horlogerie Patek, Philippe & Co. fabrique un bracelet raffiné en or dont le boîtier rectangulaire abrite un mouvement baguette. Mais il faudra tout de même cinq ans pour que cette pièce raffinée se vende. Pour lire l'heure, sa propriétaire doit tout d'abord faire un geste ostentatoire, à savoir relever un gros brillant étincelant.

Dix ans plus tard, en 1878, est créée à Vienne une petite série de montres-bracelets « que les hommes portent pour leur confort personnel », affirment les chroniqueurs de l'époque. Les premières montres à bracelet chaîne fabriquées en série datent de 1880 et sont l'œuvre de Girard-Perregaux. L'empereur allemand Guillaume II en commande pour ses officiers de marine. Constant Girard-Perregaux propose ensuite ses créations aux États-Unis, apparemment avec un succès modéré : dans ce pays, les hommes ne leur font pas confiance. Il en va tout autrement des touristes américaines qui, vers 1886, apprécient dans la ville de Lucerne les montres bracelets en or et en argent à bracelets « accordéons ».

Ailleurs, des joailliers ont aussi tout lieu d'être reconnaissants envers la gent féminine. La mode vestimentaire évoluant toujours plus rapidement, les montres pendentifs traditionnelles sont de plus en plus difficiles à porter. Les montres-bracelets au contraire s'accordent à chaque tenue. Malgré les bonnes affaires que lui font faire les femmes, un revendeur considère la préférence accordée à la montre-bracelet comme « un fourvoiement du goût féminin », le poignet étant pour lui « sûrement l'endroit le moins adapté pour fixer une montre ». En 1917, le professeur hambourgeois Hermann Bock le rejoint : « Espérons que la mode aberrante qui consiste à placer la montre à l'endroit le plus mobile du corps, sur un bracelet, disparaisse bientôt. » En 1925, l'horloger Bruno Hillmann chapitre dûment ces dames. Dans son livre *La montre-bracelet : ses caractéristiques et sa réparation*, cet horloger berlinois peste contre « une variété particulière de montres, aussi difficiles à manier à l'établi qu'au bureau pour écrire à la plume ». Il indique en outre espérer que « la masculinisation sans cesse croissante de la gent féminine », par laquelle « les vestons d'hommes deviendraient également un atout féminin », sonne enfin « la dernière heure de la tyrannie exercée par la montre-bracelet ». Bien entendu, le souhait de l'artisan écrivain ne se réalise pas, comme on le constate bientôt. Les choses s'aggravent même considérablement car de plus en plus d'hommes succombent eux aussi au charme de la montre-bracelet. Après tout, les montres qui ont montré leur valeur aux poignets des soldats britanniques dans des conditions très éprouvantes durant la Seconde Guerre des Boers (1899–1902) ne peuvent être aussi mauvaises que cela. Ces expériences positives se vérifient durant la Première Guerre mondiale. Dans le feu de l'action, il suffit pour lire l'heure de tourner légèrement l'avant-bras. Un grillage protège du bris les fragiles verres en cristal. Cadrans et aiguilles au radium fournissent les informations requises même dans l'obscurité. Et les chronographes à échelle télémétrique permettent de mesurer facilement la distance avec le front ennemi grâce aux vitesses de diffusion différentes de la lumière et du son.

Les montres-bracelets ne sont pas forcément des dérivés miniatures des montres de poche, comme le démontre le joailler et créateur parisien Louis Cartier dès 1904. Avec la « Santos » dessinée pour le pionnier de l'aviation et bon vivant Alberto Santos-Dumont, il contribue de manière décisive à l'évolution du design des garde-temps. Le 15 septembre 1916, il s'inspire des nouveaux chars d'assaut de l'armée britannique pour créer le modèle rectangulaire « Tank ». En 1918, il remet les premiers exemplaires de ce modèle au général John Joseph Pershing, commandant en chef des troupes américaines en France, et à d'autres hauts gradés de l'armée américaine, pour les récompenser de leur bravoure au combat.

Hans Wilsdorf, le fondateur de Rolex, est lui aussi un pionnier reconnu de l'histoire de la montre-bracelet. On lui doit nombre d'avancées décisives, notamment les premiers chronomètres-bracelets, les boîtiers totalement étanches et le système de remontage à rotor pour montres-bracelets automatiques.

Le succès exceptionnel des montres-bracelets à partir des années 1930 au détriment des montres de poche s'appuie également sur des composants aussi minuscules qu'indispensables, tels les pare-chocs, ainsi que sur une extrême diversité des formes et d'ingénieuses fonctions auxiliaires. Grâce à ces réalisations et à bien d'autres encore, la montre-bracelet est devenue un produit de grande diffusion du quotidien, un bien culturel par excellence et une alliée fiable dans la conquête spatiale et maritime.

Cette période de grands succès prend fin dans les années 1970, lorsque le quartz chasse avec vigueur des boîtiers la traditionnelle mécanique. Celle-ci n'a que peu d'atouts à opposer à l'électronique multifonctionnelle, si ce n'est sa fiabilité éprouvée, sa longévité et son charme désuet. Mais cette phase de recul n'est que de courte durée. Les collectionneurs découvrent en effet des valeurs sûres dans ce qui est prétendument démodé, une tendance qui déclenche une renaissance spectaculaire des mouvements mécaniques. Jamais, dans les plus de 700 ans d'histoire des montres, on n'assiste à l'éclosion d'autant évolutions intéressantes et de complications. Malgré tout, ces microcosmes mécaniques doivent une fois encore défendre leur place à nos poignets en s'affirmant face aux montres intelligentes et à leurs fabricants. Le luxe et la préservation des valeurs en matière de mesure du temps étant de vaines quêtes, la sage conclusion du psychologue américain Robert Levine pourrait une fois de plus se confirmer : « Le tic-tac d'une montre mécanique est le pouls de la civilisation. » □

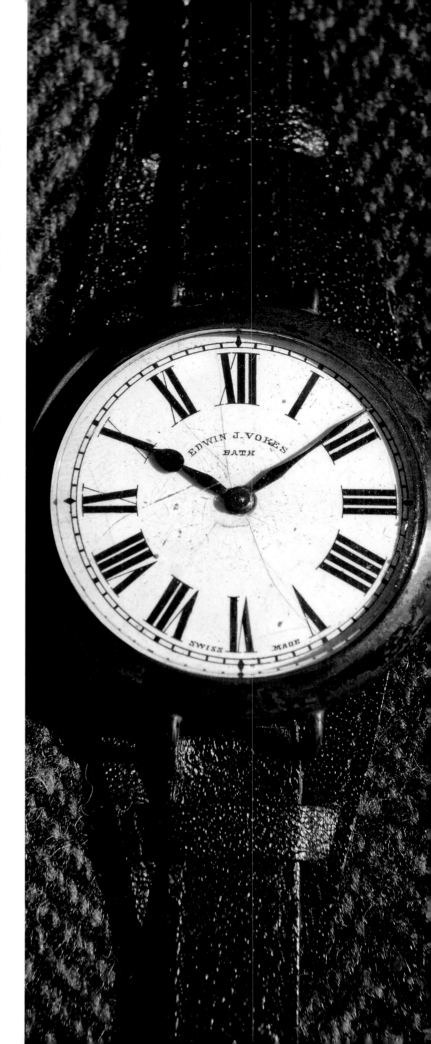

Omega wristwatch from the Second Boer War, 1899–1902
Armbanduhr von Omega aus dem Zweiten Burenkrieg, 1899–1902
Montre-bracelet Omega datant de la Seconde Guerre des Boers,
1899–1902

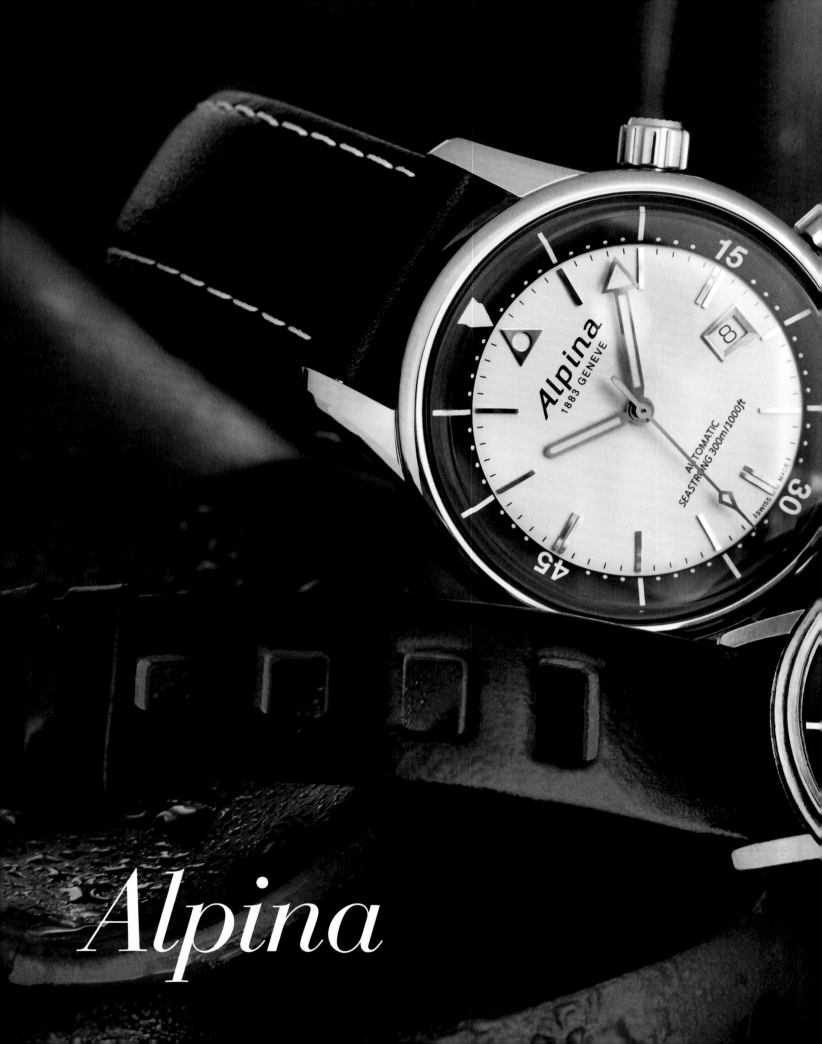

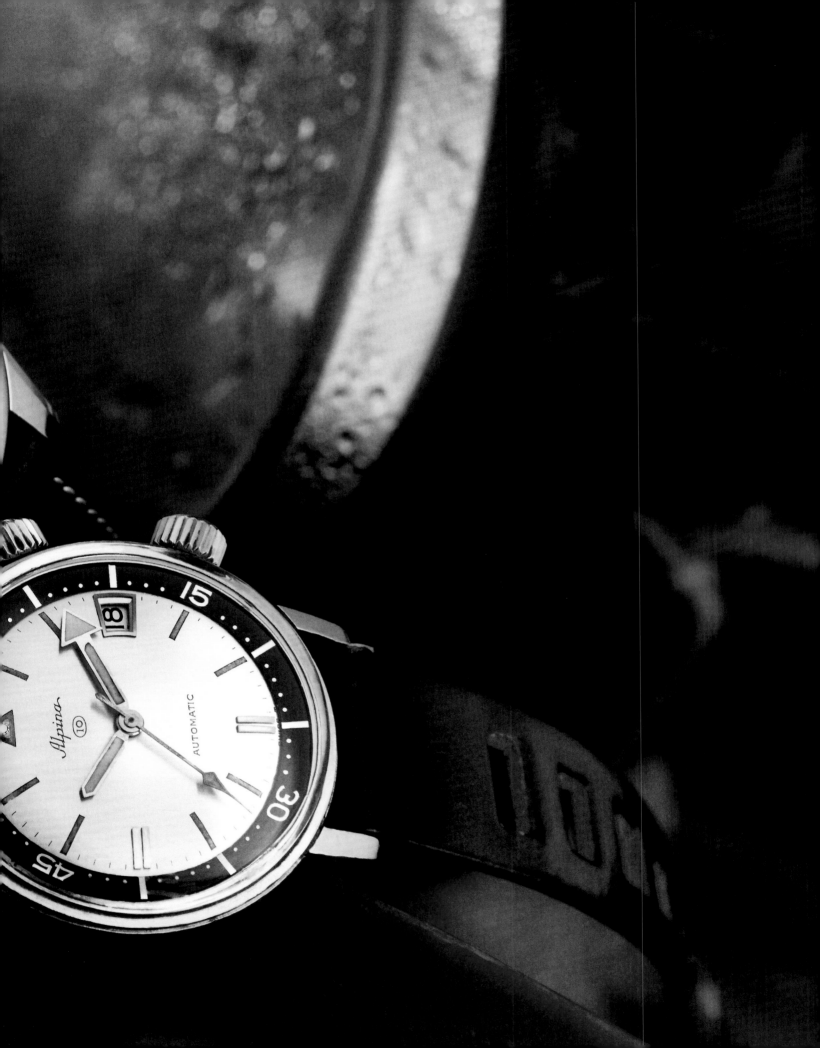

From a Cooperative Association to an Innovative Watch Manufactory

It's difficult to say what might have happened to Alpina without Peter C. Stas and wife Aletta. Blind speculations are fortunately unnecessary because at the beginning of the 21st century these two Dutch entrepreneurs, who had already boldly established Frédérique Constant and shepherded it to success, discovered the languishing Swiss traditional brand with the sporty touch. Alpina perfectly complemented the Stas' first brand Frédérique Constant, so the purchase took place relatively quickly in 2002. When the first collection in Alpina's new era debuted at the watch fair in Basel one year later, it was obvious that the elements of its design were clearly derived from Alpina's long tradition. But the new owners soon discovered that this reference to the brand's past didn't guarantee a successful future. After 17 lost years, Alpina's renaissance was tantamount to a complete new beginning. The rather uninspiring products from the epoch between 1985 and 2002 simply weren't worth mentioning, but collectors knew and appreciated Alpina timepieces from earlier epochs. Points of reference for the future could be derived from the brand's 120-year history, which began in 1883, when Gottlieb Hauser founded a cooperative association with a long German name that can be translated as the "Association of Swiss Watchmakers." His objectives were clearly defined: on the one hand, he wanted to foster collaboration among reliable manufacturers of high-quality timepieces and accessories; on the other hand, he hoped that group orders would lead to lucrative conditions for the association's members. The principle functioned well, both in Switzerland and in neighboring countries. After a contract was signed with the Straub & Cie. watch manufacturer, the association's headquarters were relocated to Bienne in 1890. Six years later, a change in the statutes resulted in a new name: "Union Horlogère, Schweizerische Uhrmachergenossenschaft, Association horlogère Suisse."

In the course of decades, orders placed by the association were profitable for various companies. Some of them were well known, others remained in the background. They included: J. Straub & Co., Bienne (complete timepieces), Favre of Geneva (complete timepieces), Kurth Frères of Grenchen (complete timepieces—later Certina), Duret & Colonnaz of Geneva (ébauches), Huguenin-Robert (cases), Schwob Frères & Co. of La Chaux-de-Fonds (Cyma—complete timepieces), Robert Frères of Villeret (complete timepieces—Minerva), Ali Jeanrenaud (pendants for pocket watches), and Numa Nicolet & Fils (dials). Harassing fire from uninvolved manufacturers and wholesalers caused occasional discomforts, but couldn't lastingly interfere with the successes. A second special caliber ticked inside many of the pocket watches around the turn of the 20th century. When the prosperous business celebrated its 25th anniversary in 1908, the name "Alpina," for which the Union Horlogère had received trademark protection in 1901, was registered as an independent watch brand. With this step, the association participated in the trend for short, catchy and thus successful brand names. The appellation kept its good reputation because the name "Alpina" was used only on selected watches with high-quality inner lives. A red triangle with a stylized dial and the signature served as a readily recognizable logo.

The attempt to establish a foothold in Germany, and especially at Glashütte in Saxony, began in 1909. The fine products made by the Präcisions-Uhrenfabrik Alpina Glashütte i. S. targeted the chronometric luxury segment. To achieve this goal, ébauches from Geneva underwent the necessary modifications and upgrading in the Müglitz Valley. Precisely this was a thorn in the eye of A. Lange & Söhne, which initiated a lawsuit that culminated in the termination of the competing company on July 17, 1922. The Alpina Gruen Gilde SA, which was cofounded with the American entrepreneur Dietrich Gruen in 1929, likewise had only a short lifespan: the self-proclaimed "largest syndicate of watchmakers of all times" ended its activities in 1937. Among the highlights of this era are the rectangular "Doctor's Watch" with the same baguette caliber from Aegler SA that was also encased in Rolex's legendary "Prince." But the intended win-win situation ultimately didn't arise. The subsequently founded joint-stock company, which was named Alpina Union Horlogère SA, was again headquartered in Bienne.

Alpina unveiled its steel "Block Uhr" as the first sport watch in 1933 and equipped it with a special crown that received patent protection in 1934. The "Alpina 4" was very successful on international markets in 1938 and afterwards. The number "4" in its name referred to four quality characteristics: antimagnetic; watertight "Geneva" case; Incabloc shock absorption; and the use of rustproof stainless steel for the sturdy case. The era of self-winding watches began at Alpina in 1944 with Caliber 582, which is also known as "P82."

Sanctions imposed after World War Two by the victorious Allies prohibited the use of the Alpina brand name in Germany, partly because seamen in the German navy had worn watches with this signature. Alpina was renamed Dugena. When the Quartz Tsunami arrived in the early 1970s, Alpina was nearly powerless to oppose it. Businesspeople from Cologne, who lacked the necessary affinity for watches, became the new owners. The rest of the sad story is well known.

Gradual but steady improvements began under the prudent aegis of Peter and Aletta Stas. Alpina currently produces circa 20,000 wristwatches per year and plans call for annual production to increase to 75,000 watches in the medium term. The price segment: affordable.

With an eye toward its history, Alpina feels equally at home on dry land, in the air and under water. The collection of pilot's watches, which premiered under the name "Startimer Pilot" in 2011, has been judiciously expanded. Diver's watches grew increasingly important too: the "Seastrong Diver Heritage" that debuted in 2016 recalls an eye-catching predecessor from the 1960s and the "Seastrong Diver 300" aptly embodies contemporary styling while remaining watertight to 300 meters. The electronic "Horological Smartwatch" with analog time display is an ideal partner for a smartphone. The "Alpiner 4" line has a sporty touch, while the "Alpiner Heritage Manufacture KM-710" and other models perfectly embody the retro look. As in the past, Alpina presently encases two types of movements: one variety is purchased from suppliers

Pilot's chronograph,
hand-wound movement, 1930s

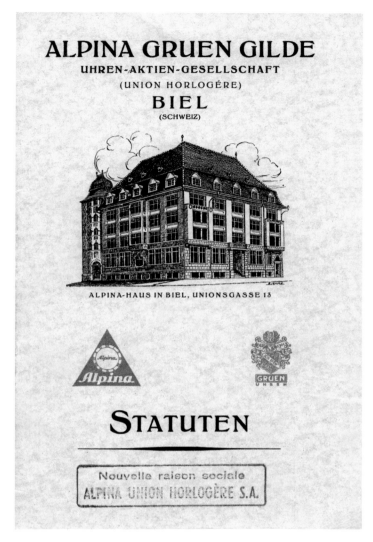

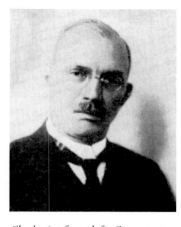

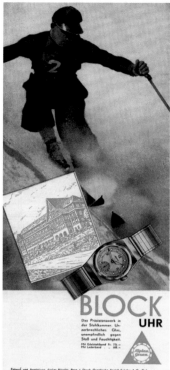

such as Eta or Sellita; the other is exclusively fabricated at Alpina's own manufactory in Geneva. The rotor of Caliber AL-710 is obviously similar to the oscillating weight in successful Caliber 582 from the 1940s. The AL-710 also provides power for Alpina's top-of-the-line model. Chief design engineer Pim Koeslag developed a unique chronograph module for the "Alpiner 4 Manufacture Flyback Chronograph." Koeslag's practicality and intelligence are evident in the fact that the cadrature requires only 96 components. A star-shaped construction replaces a conventional column-wheel. The coupling recalls elements of Edouard Heuer's oscillating pinion. The model's name refers to the flyback mechanism in Caliber AL-760.

Peter C. Stas has never regretted the purchase: "Alpina is a small child in a significantly larger business. We'll take pains to ensure that it continues to grow." ○

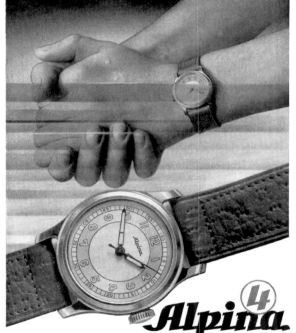

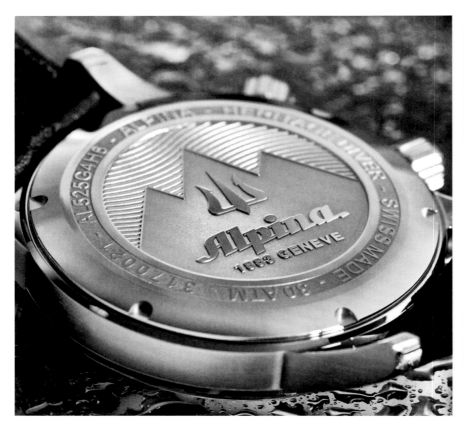

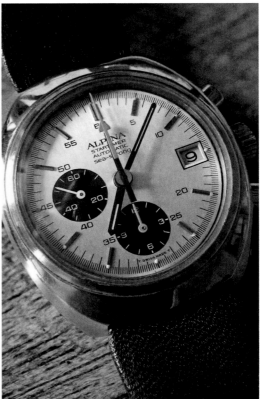

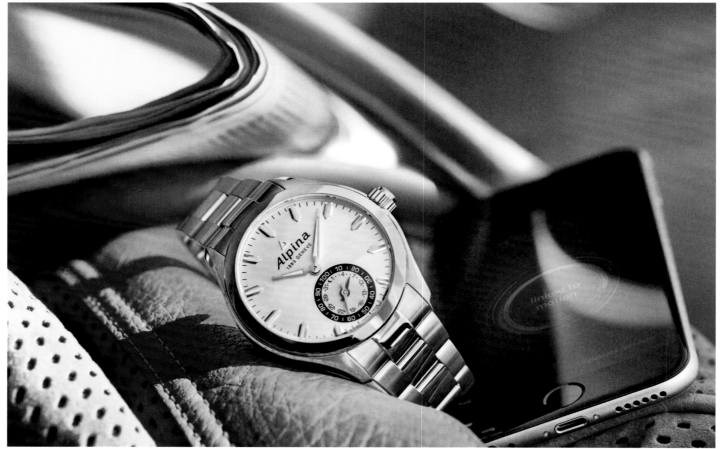

Clockwise from top left: Seastrong Diver Heritage, 2016 ◦ Seastrong Chronograph, 1960s ◦ Horological Smartwatch, 2015

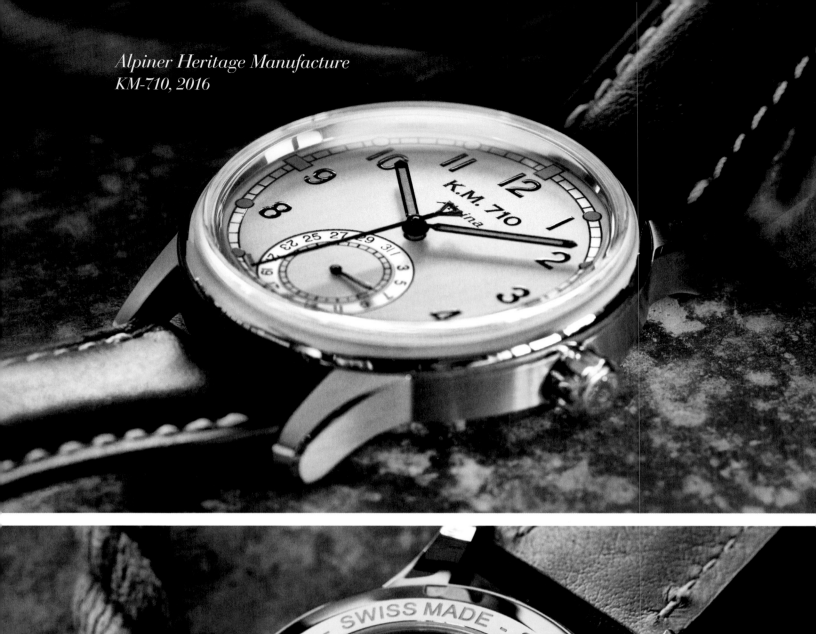

*Alpiner Heritage Manufacture
KM-710, 2016*

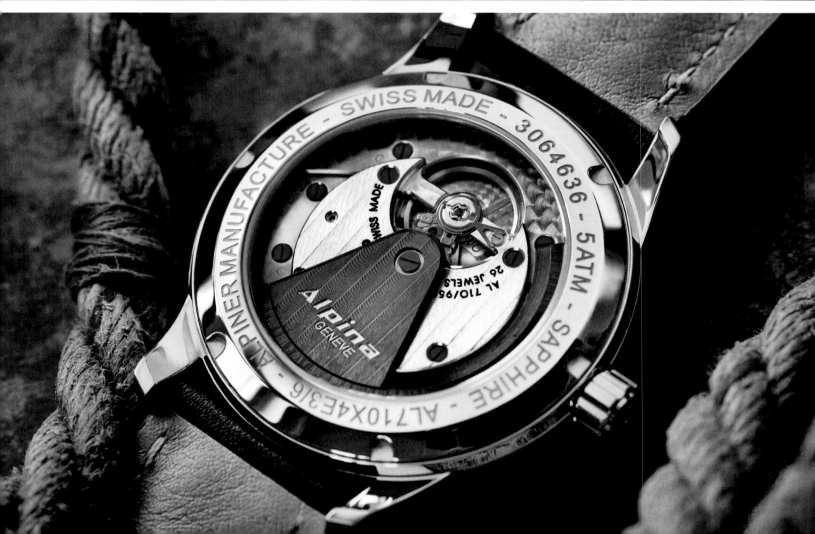

Von einer Genossenschaft zur innovativen Uhrenmanufaktur

Was ohne Peter C. Stas und seine Frau Aletta aus Alpina geworden wäre, lässt sich schwer sagen. Ein Stochern im Nebel ist zum Glück auch gar nicht nötig, denn zu Beginn des 21. Jahrhunderts entdeckte das holländische Unternehmerpaar, dem zuvor schon ein Gründungsabenteuer namens Frédérique Constant geglückt war, die dümpelnde Schweizer Traditionsmarke mit sportlichem Anstrich. Weil Alpina die Erstmarke Frédérique Constant perfekt ergänzte, kam der Kauf 2002 relativ zügig zustande. Bereits ein Jahr später war die erste Neuzeit-Kollektion während der Basler Uhrenmesse präsent. Unübersehbar die an das lange Erbe zurückreichenden Gestaltungselemente. Was jedoch kein Erfolgsgarant war, wie die neuen Eigentümer anschließend feststellen mussten. Wegen 17 verlorener Jahre kam die Alpina-Renaissance einem kompletten Neubeginn gleich. Die wenig erbaulichen Produkte aus der Epoche zwischen 1985 und 2002 waren nämlich nicht der Rede wert gewesen. Sammler kannten und schätzten die Zeitmesser aus früheren Epochen. Anhaltspunkte für die Zukunft ergaben sich aus der 120-jährigen Geschichte. Selbige reicht zurück bis ins Jahr 1883. Am Anfang stand die Vereinigung der Schweizer Uhrmacher, gegründet von Gottlieb Hauser. Seine Ziele waren klar definiert. Zum einen ging es um die Zusammenarbeit mit zuverlässigen Fabrikanten qualitativ hochwertiger Uhren und Zubehörteile. Zum anderen sollten Gruppenaufträge zu interessanten Konditionen für die Mitglieder führen. Das Prinzip funktionierte, und zwar auch in den Nachbarländern. 1890 bedingte ein Vertrag mit dem Uhrenfabrikanten Straub & Cie. die Verlegung der genossenschaftlichen Zentrale nach Biel. 1896 führte eine Statutenänderung zum neuen Namen Union Horlogère, Schweizerische Uhrmachergenossenschaft, Association horlogère Suisse.

Von den Aufträgen der Genossenschaft profitierten im Laufe der Jahrzehnte teils bekannte, teils im Hintergrund wirkende Firmen: J. Straub & Co., Biel (Fertiguhren), Favre, Genf (Fertiguhren), Kurth Frères, Grenchen (Fertiguhren – später Certina), Duret & Colonnaz, Genf (Rohwerke), Huguenin-Robert (Gehäuse), Schwob Frères & Co., La Chaux-de-Fonds (Cyma – Fertiguhren), Robert Frères, Villeret (Fertiguhren – Minerva), die Pendant- und Bügelfabrik von Ali Jeanrenaud und die Zifferblattfabrik von Numa Nicolet & Fils. Störfeuer nicht involvierter Fabrikanten und Großhändler brachten gelegentlich Unruhe, konnten die Erfolge auf Dauer aber nicht stören. Um die Wende zum 20. Jahrhundert tickte in etlichen der Taschenuhren schon ein zweites Spezialkaliber. 1908 zelebrierte das prosperierende Unternehmen sein 25. Jubiläum. Aus diesem Anlass erfolgte die Eintragung des von der Union Horlogère seit 1901 für hochwertige Taschenuhren geschützten Namens „Alpina" als eigenständige Uhrenmarke. Mit diesem Schritt folgte die Genossenschaft dem Trend zu kurzen, einprägsamen und deshalb erfolgreichen Signaturen. Banalisierung des Namens vermied die Beschränkung auf ausgesuchte Uhren mit hochwertigem Innenleben. Als Logo mit hohem Wiedererkennungswert diente ein rotes Dreieck mit stilisiertem Zifferblatt und Namenszug.

Der Versuch, in Deutschland und dort speziell im sächsischen Glashütte Fuß zu fassen, startete 1909. Die feinen Produkte der Präcisions-Uhrenfabrik Alpina Glashütte i. S. sollten das chronometrische Luxussegment besetzen. Zu diesem Zweck erfuhren Genfer Rohwerke im Tal der Müglitz die gebotene Modifikation und Aufwertung. Genau das kam speziell bei A. Lange & Söhne gar nicht gut an. Ein jahrelanger Rechtsstreit führte am 17. Juli 1922 zur Löschung der Firma. Auch die 1929 zusammen mit dem amerikanischen Unternehmer Dietrich Gruen gegründete Alpina Gruen Gilde SA hatte nur ein begrenztes Leben. 1937 endeten die Aktivitäten der selbsternannten „größten Interessengemeinschaft von Uhrmachern aller Zeiten". Zu den Höhepunkten dieser Ära gehörte die rechteckige „Doctor's Watch" mit Baguette-Kaliber der Aegler SA. Selbige fand man auch in der legendären Rolex „Prince". Von der intendierten Win-win-Situation war am Ende aber keine Rede mehr. Die Folge-Aktiengesellschaft mit Namen Alpina Union Horlogère SA residierte weiterhin in Biel.

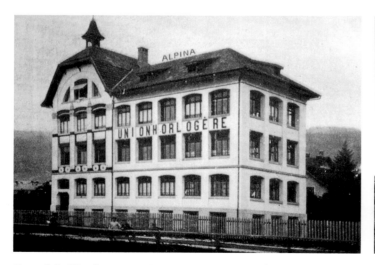
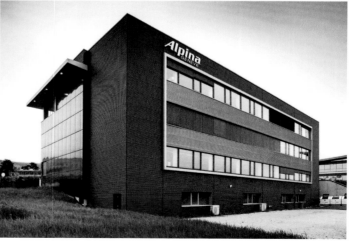

From left: Headquarters, Bienne, 1890 ◦ Manufacture building, Plan-les-Ouates, 2006 ◦ Von links: Hauptsitz in Biel, 1890 ◦ Manufakturgebäude, Plan-les-Ouates, 2006 ◦ De gauche à droite : Siège biennois, 1890 ◦ Bâtiment de la manufacture, Plan-les-Ouates, 2006

Bereits 1933 hatte Alpina die stählerne „Block Uhr" als erste Sportuhr mit der 1934 patentierten Spezialkrone vorgestellt. Ab 1938 agierte die Sportuhr „Alpina 4" höchst erfolgreich auf internationalen Märkten. Die Ziffer „4" wies dabei auf vier Qualitätsmerkmale hin: 1. antimagnetisch, 2. wasserdichtes „Geneva"-Gehäuse, 3. Incabloc-Stoßsicherung und 4. Verwendung von rostfreiem Edelstahl für die belastbare Schale. Mit dem Kaliber 582, auch P82 genannt, brach 1944 das Automatikzeitalter an.

Nach dem Zweiten Weltkrieg gipfelten Sanktionen der Alliierten darin, die Nutzung des Markennamens Alpina in Deutschland zu untersagen, weil u. a. Soldaten der deutschen Kriegsmarine auf Uhren mit dieser Signatur geblickt hatten. Aus Alpina wurde Dugena. Anfang der 1970er Jahre stand Alpina der ungestümen Quarz-Welle beinahe machtlos gegenüber. Kölner Unternehmer ohne ausgeprägte Uhren-Affinität wurden neue Eigentümer. Der Rest ist bereits bekannt.

Unter umsichtiger Ägide von Peter und Aletta Stas ging es langsam, aber kontinuierlich bergauf. Inzwischen produziert Alpina jährlich etwa 20 000 Armbanduhren. Mittelfristig sollen es rund 75 000 Exemplare werden. Preissegment: bezahlbar.

Mit Blick auf die Geschichte fühlt sich Alpina auf der Erde, in der Luft und unter Wasser zu Hause. Die 2011 eingeführte Flieger-uhren-Kollektion erfuhr unter dem Namen „Startimer Pilot" eine gezielte Ausweitung. Taucheruhren erlangten im Laufe der zurück-liegenden Jahre zunehmende Bedeutung. An ein markantes Vorbild aus den 1960er Jahren erinnert die 2016 vorgestellte „Seastrong Diver Heritage". Gegenwärtiges Designempfinden repräsentiert der bis 300 Meter wasserdichte „Seastrong Diver 300". Partner des Smartphones ist die elektronische „Horological Smart-watch" mit analoger Zeitanzeige. Sportlichen Touch verstrahlt die Linie „Alpiner 4", perfekter Retrolook zeichnet unter anderem die „Alpiner Heritage Manufacture KM-710" aus. Apropos: Wie

früher verbaut Alpina auch heute zwei Uhrwerk-Typen: zugekaufte von Eta oder Sellita und exklusive aus eigener Genfer Manufaktur. Der Rotor des AL-710 weist unübersehbare Ähnlichkeiten mit der Schwungmasse des erfolgreichen 582 aus den 1940er Jahren auf. Dieses Kaliber dient auch dem aktuellen Spitzenprodukt „Alpiner 4 Manufacture Flyback Chronograph" als Motor. Hierfür hat Chefkonstrukteur Pim Koeslag ein einzigartiges Chronographen-modul entwickelt. Von nutzbringender Intelligenz zeugt die Tatsache, dass die Kadratur mit gerade einmal 96 Komponenten funktio-niert. An die Stelle des klassischen Schaltrads tritt ein sternförmiges Gebilde. Die Kupplung lässt Elemente des Schwingtriebs von Edouard Heuer erkennen. Schließlich weist der Modellname auf die Temposchaltung im Kaliber hin.

Bereut hat Peter C. Stas den Kauf übrigens nie. „Bei Alpina handelt es sich um ein Kleinkind in einem deutlich größeren Geschäft. Wir werden dafür sorgen, dass es stetig wächst." ◦

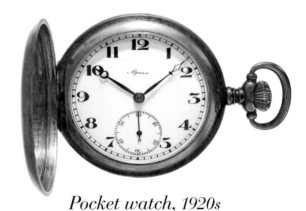

Pocket watch, 1920s

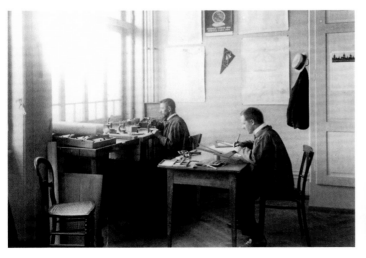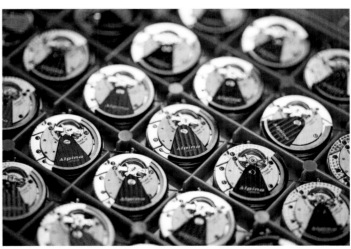

From left: Headquarters, Bienne, 1890 ◦ Quality control of manufacture movements ◦ *Von links: Im Bieler Hauptsitz, 1890 ◦ Manufaktur-werke in der Qualitätskontrolle* ◦ *De gauche à droite : Au siège biennois, 1890 ◦ Mouvements de manufacture au contrôle qualité*

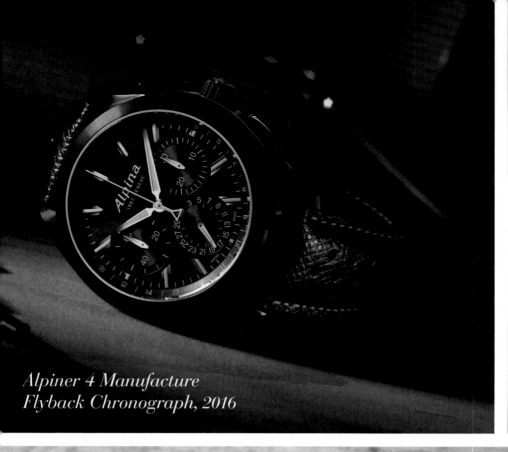

Alpiner 4 Manufacture
Flyback Chronograph, 2016

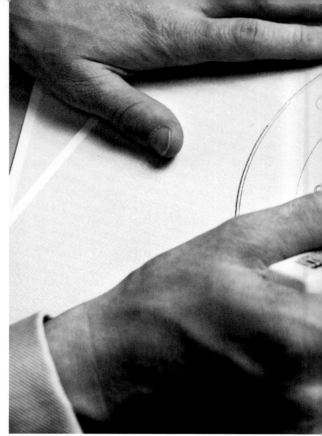

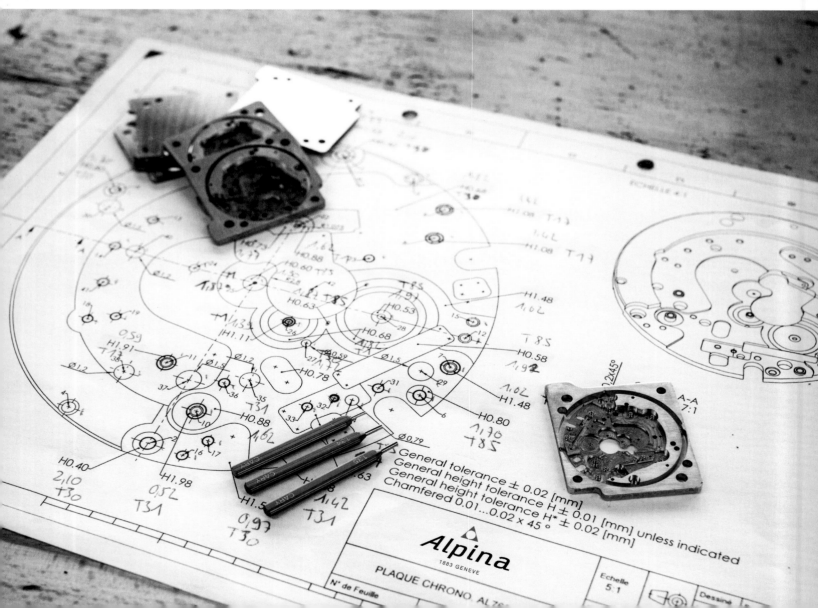

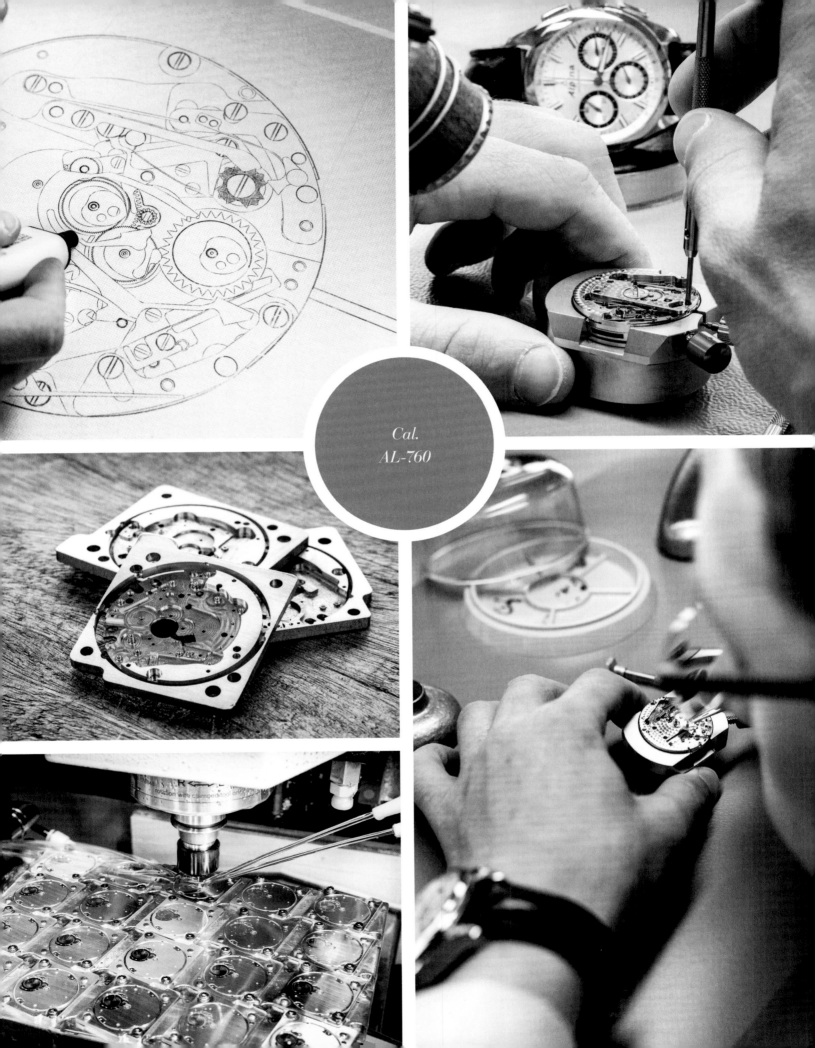

Cal.
AL-760

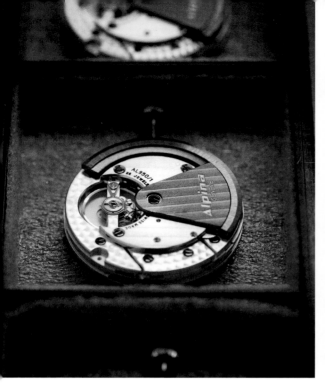

Une coopérative devenue manufacture horlogère innovante

Il est difficile de dire ce qu'il serait advenu des montres sportives Alpina sans Peter C. Stas et son épouse Aletta. Heureusement, nul est besoin de se répandre en conjectures car ce couple d'entrepreneurs hollandais, qui a déjà créé avec bonheur la maison Frédérique Constant, découvre au début du XXIᵉ siècle cette marque suisse de tradition, alors qu'elle commence à tanguer. Alpina complétant à merveille la marque principale Frédérique Constant, le rachat en 2002 est mené rondement. Seulement un an plus tard, Alpina présente sa première collection de cette ère nouvelle au Salon mondial de l'horlogerie à Bâle (Baselworld). Si l'on reconnaît d'emblée l'esthétique héritière d'une longue tradition, le succès n'est hélas pas garanti, comme les nouveaux propriétaires doivent le constater au lendemain de la manifestation. En raison de 17 années perdues, faire renaître Alpina revient à repartir de zéro.

Les produits médiocres sortis entre 1985 et 2002 ne présentant guère d'intérêt, les collectionneurs apprécient plutôt les garde-temps plus anciens. Aussi faut-il rechercher des orientations pour l'avenir dans l'histoire de l'entreprise vieille de 120 ans, qui remonte à la création, en 1883, de l'Union horlogère (Vereinigung der Schweizer Uhrmacher) par Gottlieb Hauser. Ses objectifs sont clairement définis : d'une part travailler en partenariat avec des fabricants fiables de composants de montres et d'accessoires de haute qualité ; d'autre part faire des achats groupés pour obtenir des conditions intéressantes. Ce principe fonctionne jusque dans les pays limitrophes. En 1890, pour honorer un contrat avec le fabricant de montres Straub & Cie, la coopérative transfère son administration centrale à Bienne. Six ans plus tard, en 1896, elle modifie ses statuts et prend le nom d'Union Horlogère, Schweizerische Uhrmachergenossenschaft, Association horlogère Suisse.

Les bénéficiaires des commandes passées à la coopérative au fil des décennies sont des entreprises connues à des degrés divers : J. Straub & Co. à Bienne (montres complètes), Favre à Genève (montres complètes), Kurth Frères à Granges (montres complètes – deviendra Certina), Duret & Colonnaz à Genève (ébauches), Huguenin-Robert (boîtiers), Schwob Frères & Co. à La Chaux-de-Fonds (à l'origine de Cyma – montres complètes), Robert Frères, à Villeret (montres complètes – deviendra Minerva), le fabricant de pendants et bélières Ali Jeanrenaud et la fabrique de cadrans Numa Nicolet & Fils. Des attaques de la part de fabricants et de grossistes externes n'apportent que des perturbations occasionnelles, sans entraver durablement la réussite de l'association. Au tournant du XXᵉ siècle, un second calibre spécifique anime désormais nombre des montres de poche. En 1908, l'association florissante fête ses 25 ans d'existence. À cette occasion, elle procède à l'enregistrement à titre de marque autonome du nom « Alpina », déjà protégé en relation avec des montres de poche haut de gamme. La tendance est alors en effet aux griffes courtes et faciles à mémoriser, clés de la réussite. Pour éviter une dilution de la marque Alpina, son emploi est limité à des montres de qualité supérieure, animées par des calibres haut de gamme. Facilement reconnaissable, le logo représente un cadran stylisé enserré dans un triangle rouge, le tout surmontant le nom de la marque.

En 1909, Alpina tente de s'implanter en Allemagne, précisément à Glashütte, en Saxe, avec pour objectif la conquête du segment des garde-temps de luxe par les produits haut de gamme de la fabrique de montres de précision Alpina Glashütte. À cette fin, Alpina apporte à des ébauches genevoises les modifications et valorisations nécessaires dans la vallée de la Müglitz. Voilà qui est fait pour déplaire, à A. Lange & Söhne tout particulièrement. Au terme d'un an de procès, l'entreprise est radiée du registre

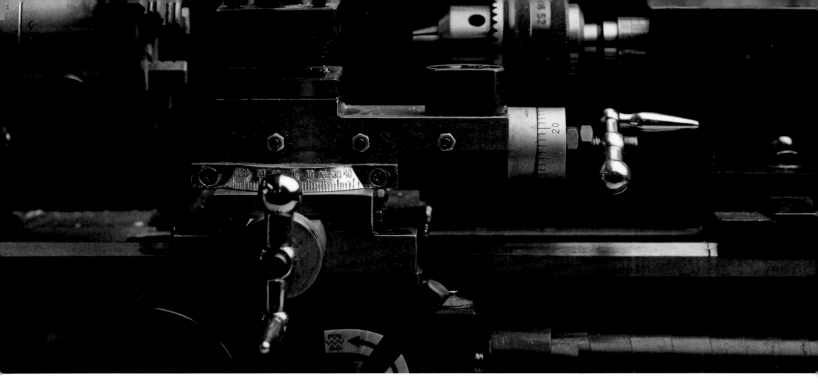

From left: Manufacture calibers ○ Blue screws ○ Old tooling machine ○ Von links: Manufakturkaliber ○ Blaue Schrauben ○
Alte Fertigungsmaschine ○ De gauche à droite : Calibres de manufacture ○ Vis bleuies ○ Machine-outil ancienne

des sociétés le 17 juillet 1922. Alpina Gruen Gilde SA, cofondée en 1929 avec l'entrepreneur américain Dietrich Gruen, ne sera pas non plus promise à un bel avenir : l'association qui se dit être le « plus grand groupement d'intérêts d'horlogers de tous les temps » cesse ses activités en 1937. Parmi les créations marquantes de cette ère figure la « Doctor's Watch » rectangulaire à mouvement baguette produit par Aegler SA, qui équipe aussi la légendaire Rolex « Prince ». La situation gagnant-gagnant escomptée au départ est restée un vœu pieux. La société anonyme qui prend la succession sous le nom d'Alpina Union Horlogère SA maintient le siège à Bienne.

Succédant à la première montre sportive présentée en 1933, la « Block Uhr » acier équipée de la couronne spéciale brevetée en 1934, la montre sportive « Alpina 4 » lancée en 1938 remporte un succès foudroyant sur les marchés internationaux. Le chiffre « 4 » fait référence aux quatre caractéristiques : 1. Antimagnétisme, 2. Boîtier étanche « Geneva », 3. Système antichoc Incabloc et 4. Acier inoxydable pour le boîtier résistant. En 1944, le calibre 582, aussi appelé P82, inaugure l'ère des mouvements automatiques. Au lendemain de la Seconde Guerre mondiale, les Alliés interdisent, à titre de sanction suprême, l'exploitation du nom de marque Alpina en Allemagne, notamment parce que la marine allemande avait lu l'heure sur des montres revêtues de cette signature. Alpina devient alors Dugena. Au début des années 1970, la marque est presque impuissante face à l'ouragan du quartz. Des entrepreneurs originaires de Cologne sans grande affinité pour l'horlogerie la rachètent. On connaît la suite de l'histoire.

Sous l'égide éclairée de Peter et Aletta Stas, l'entreprise remonte la pente, lentement mais sûrement. Alpina produit désormais quelque 20 000 montres-bracelets par an, un chiffre qui devrait atteindre à moyen terme les 75 000 exemplaires, et ce à des prix abordables.

Si l'on considère le passé, on peut dire qu'Alpina est actuellement dans son élément sur terre, sur mer et dans les airs. Lancée en 2011, la collection de montres de pilotes « Startimer Pilot » a été étoffée de manière ciblée. Par ailleurs, les montres de plongée ont gagné en importance au cours des dernières années : la « Seastrong Diver Heritage » présentée en 2016 n'est pas sans rappeler un modèle phare des années 1960, tandis que la « Seastrong Diver 300 » étanche jusqu'à 300 mètres de profondeur incarne la sensibilité contemporaine en matière de design. Au chapitre des montres connectées, la collection « Horological Smartwatch » est à affichage analogique.

Si l'allure sportive caractérise la collection « Alpiner 4 », le look résolument rétro est représenté entre autres par le modèle « Alpiner Heritage Manufacture KM-170 ». À noter : comme autrefois, Alpina recourt à deux types de mouvements différents : des calibres ETA ou Sellita, d'une part, et des produits exclusifs mis au point en interne à Genève, d'autre part. Le rotor du mouvement AL-710 a manifestement de nombreux points communs avec la masse oscillante du célèbre calibre 582 conçu dans les années 1940. C'est également ce calibre qui anime le produit phare actuel, le chronographe « Alpiner 4 Manufacture Flyback Chronograph », pour lequel le directeur technique Pim Koeslag a développé un module de chronographe inédit. Innovation aussi intelligente qu'utile, sa cadrature fonctionne avec tout juste 96 composants. La conventionnelle roue à colonnes est remplacée par une pièce en forme d'étoile. On reconnaît dans l'embrayage des éléments du pignon oscillant d'Édouard Heuer. Pour finir, le nom du modèle renvoie à la fonction retour-en-vol intégrée au calibre AL-760. Peter C. Stas n'a jamais regretté son achat : « Alpina est comme un jeune enfant perdu dans la cour des grands. Nous allons veiller à la faire grandir constamment. » ○

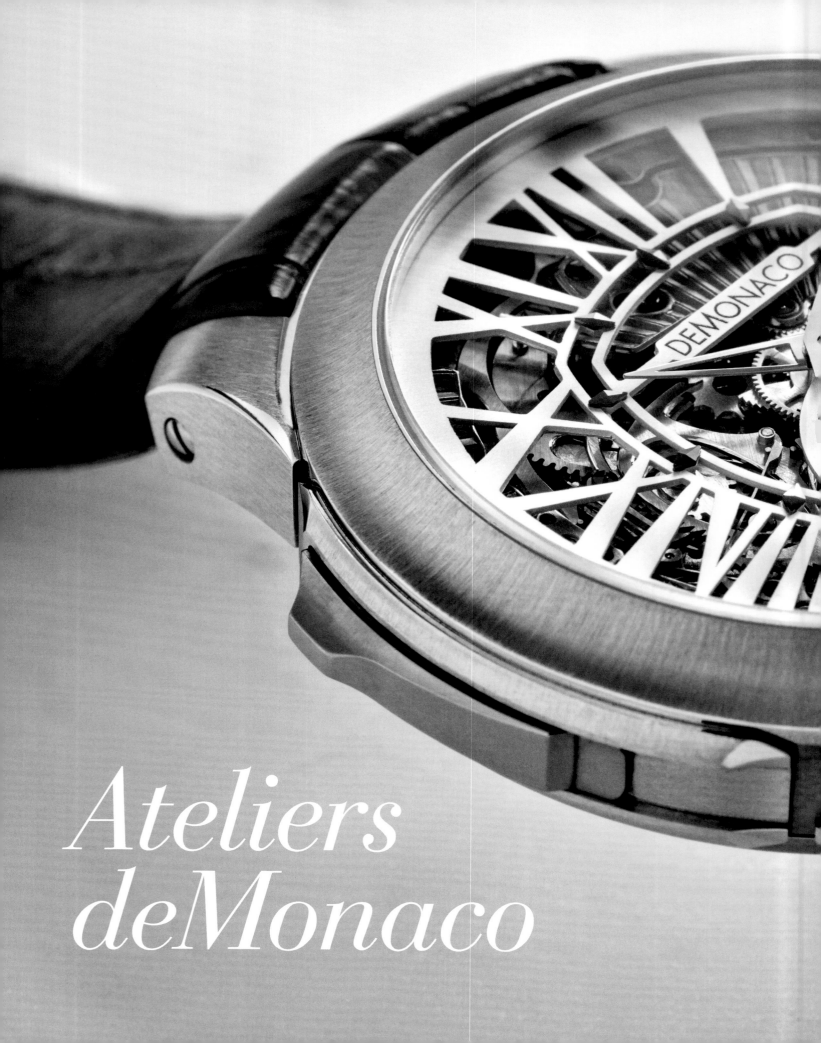

Ateliers
deMonaco

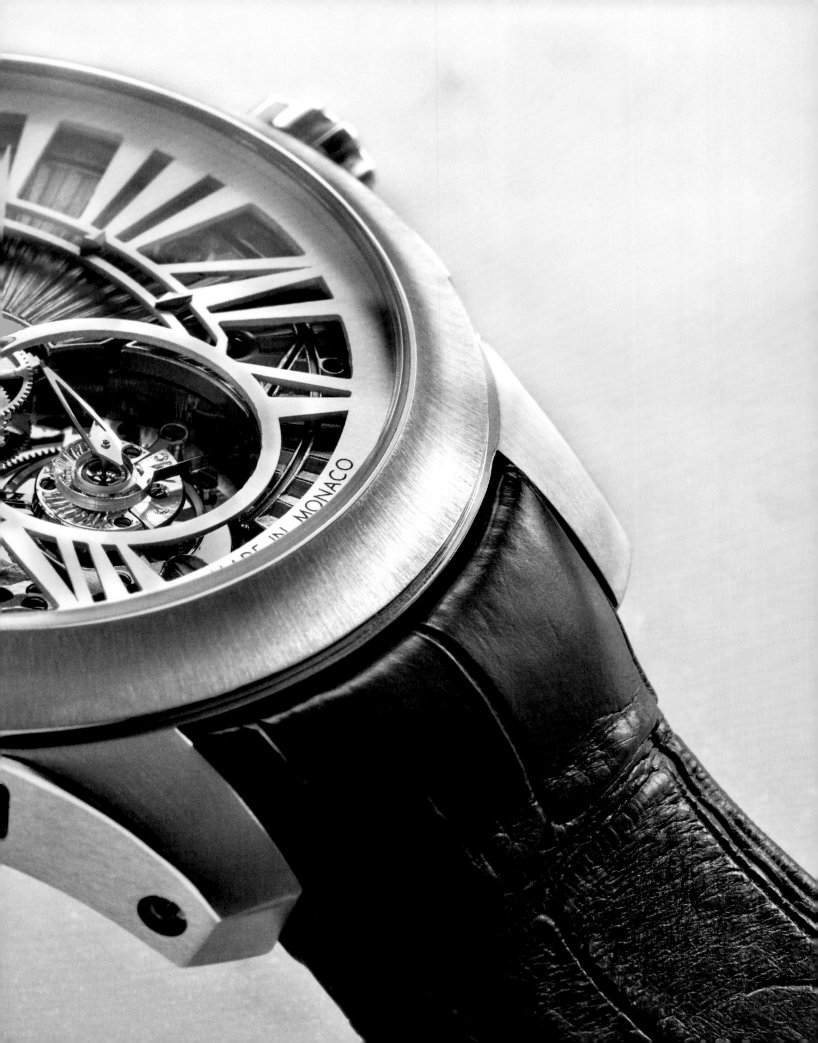

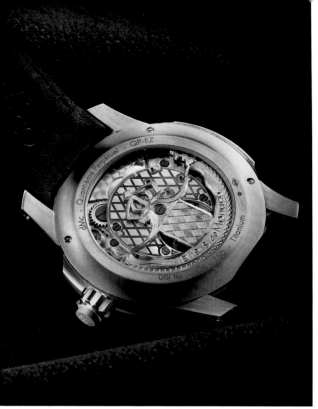
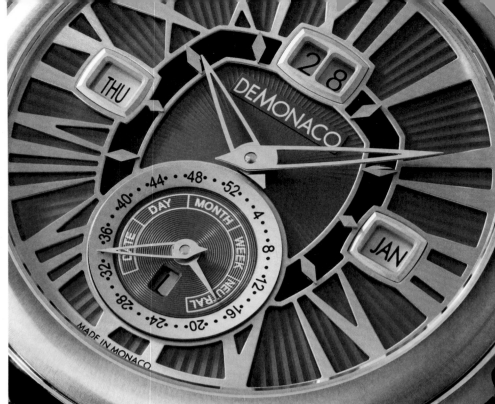

dMc-QP-RR-WG, Collection: Quantième Perpétuel Classique, in white gold, 2010

Horological Luxury from the Côte d'Azur

English

Blue ocean, gambling casinos, fascinating automobiles and an extraordinary Grand Prix are a few of the images that come to mind when one thinks about the little principality on the Côte d'Azur. Watches in Monaco primarily grace the wrists of beautiful and wealthy people or sparkle in the showcases of noble jewelry stores. But watch manufacturing was unknown here until 2009, when Peter Stas and the watchmaker Pim Koeslag opened a small but very fine watch atelier embodying a third pillar alongside Alpina and Frédérique Constant. With strictly limited series, useful complications and the utmost in the watchmaker's art, timepieces bearing deMonaco's signature were destined to play in the uppermost chronometric league. And they've done precisely that ever since 2009. Due to the extremely high proportion of manual craftsmanship, the artisans at Ateliers deMonaco annually produce only a homeopathically small number of exclusively mechanical timepieces. Nothing is left to chance, neither in the design nor in the craftsmanly fabrication of the cases, dials and hands. Each detail prioritizes meticulousness. But the special value of these watches is in their painstakingly finished and decorated movements. Every caliber assembled at Ateliers deMonaco was also developed and manufactured here, and each caliber has unique horological appeal. Already during the initial design and engineering phase, Pim Koeslag takes pains to implement small but ultimately decisive differences. For example, the patented "Quantième Perpétuel" not only has indicators which jump ahead punctually at midnight, but also hosts a convenient setting and correcting system, appropriately named "EaZy adjust." Turning the extracted crown lets the wearer select the display that he wants to adjust: date, day of the week, calendar week, or month. The preselected indicator can then be reset by pushing the crown. The ultra-complicated "Tourbillon Répétition Minute" collection embodies the high art of watchmaking. This collection's name aptly expresses its message: the one-minute tourbillon compensates for the negative effects that gravity exerts on the accuracy of the rate of a watch in a vertical position. Borne beneath a transparent sapphire bridge, the tourbillon's cage rotates once per minute around its own axis. The minute-repeater mechanism is nearly imperceptible to the eye, but considerably more complex. When the wearer wants to hear the time, he operates the little slide on the left-hand flank of the case. The mechanism responds by first chiming the current hour, then ringing the number of quarter hours since the preceding full hour and finally tolling the number of minutes that have elapsed since the previous quarter hour. For example, 32 tones resound at 12:59— a dozen low-pitched chimes, three duos of low and high chimes, and finally fourteen high-pitched chimes. A silent centrifugal-force regulator ensures that the sequence of strikes against the gongs occurs regularly, i.e., neither too quickly nor too slowly. Approximately three strikes can be heard during each two-second interval. The audible performance at one minute before one o'clock accordingly lasts between 20 and 25 seconds. An aficionado who would like to observe the hammers striking the two meticulously tuned gongs must first slip the watch off his wrist. Instead of showing its tourbillon, the movement now reveals the lovingly crafted decorations on its back side. The remarkable sound quality is partly due to the well-conceived design of the case, for which titanium and gold are used. Whether the case is angular or round, each shape assures an equally thrilling sonic experience. ◦

Uhrmacherischer Luxus von der Côte d'Azur

Blaues Meer, Spielcasinos, faszinierende Autos oder einen Ausnahme-Grand-Prix assoziiert man sehr spontan mit dem kleinen Fürstentum an der Côte d'Azur. Das Thema Uhren spielt sich in Monaco primär an den Handgelenken der Schönen und Reichen oder bei den zahlreichen Nobeljuwelieren ab. Uhrenfertigung war bis 2009 hingegen ein Fremdwort. Dann kamen Peter Stas und der Uhrmacher Pim Koeslag, um ein kleines, aber sehr feines Uhren-Atelier zu eröffnen. Und zwar als dritte Säule neben Alpina und Frédérique Constant. Durch winzige Stückzahlen, hilfreiche Komplikationen und allerhöchste Uhrmacherkunst sollten deMonaco signierte Zeitmesser in der obersten chronometrischen Liga spielen. Und das tun sie seit besagtem Jahr. Wegen des extrem hohen Anteils an zeitraubender Handarbeit entsteht bei Ateliers deMonaco jedes Jahr nur ein homöopathisches Quantum ausnahmslos konventionell tickender Zeitmesser. Schon beim Design sowie der handwerklichen Ausführung von Gehäusen, Zifferblättern und Zeigern bleibt nichts dem Zufall überlassen. Gestalterische Sorgfalt bis ins letzte Detail wird groß geschrieben. Der besondere Wert dieser Armbanduhren ist jedoch in den mit größter Sorgfalt finissierten Uhrwerken zu suchen. Jedes der in Monaco fertiggestellten Kaliber entstammt ausnahmslos eigener Entwicklung und Fertigung. Darüber hinaus besitzt jedes Uhrwerk aber auch seinen besonderen uhrmacherischen Reiz. Schon bei der Konstruktion achtet Pim Koeslag auf die Implementierung kleiner, letzten Endes aber entscheidender Unterschiede. In diesem Sinn besitzt der patentierte „Quantième Perpétuel" nicht nur pünktlich um Mitternacht springende Anzeigen, sondern auch ein komfortables, „EaZy adjust" genanntes Einstell- und Korrektursystem. Indem man an der gezogenen Krone dreht, lässt sich auswählen, was verändert werden soll: Datum, Wochentag, Kalenderwoche oder Monat. Danach lässt sich das Gewünschte per Druck auf die Krone verändern. Ganz hohe uhrmacherische Schule, weil besonders kompliziert, verkörpert die Kollektion „Tourbillon Répétition Minute". Ihr Name ist Botschaft. Der Kompensation negativer Auswirkungen auf die Ganggenauigkeit in senkrechter Position dient das Minutentourbillon. Sein Käfig dreht sich unter einer transparenten Saphirbrücke jede Minute einmal um seine Achse. Optisch kaum wahrnehmbar, aber deutlich komplizierter tritt die Minutenrepetition in Erscheinung. Wer hören möchte, wie spät es gerade ist, muss den kleinen Schieber im linken Gehäuserand betätigen. Danach ertönt zuerst die Zahl der aktuellen Stunde, gefolgt vom Quantum der vollen Viertelstunden. Als Drittes gibt der Mechanismus noch die Zahl der seit der letzten Viertelstunde verstrichenen Minuten wieder. Um 12:59 Uhr sind also summa summarum 32 Töne zu hören: zwölf tiefe, drei Mal tief und hoch sowie abschließend noch 14 hohe. Für gleichförmigen, nicht zu schnellen, aber auch nicht zu langsamen Ablauf der Schlagfolge sorgt ein geräuschloser Fliehkraftregler. Während zwei Sekunden sind etwa drei Schläge zu hören. Ergo dauert das Hörspiel kurz vor eins 20 bis 25 Sekunden. Wer sehen möchte, wie die Hämmer dabei gegen ein Paar sorgfältig gestimmter Tonfedern schlagen, muss die Uhr vom Handgelenk nehmen. Anstelle des Tourbillons zeigt sich nun die Rückseite mit ihrer liebevollen Dekoration. Die bemerkenswerte Klangqualität ist aber auch der durchdachten Gehäusekonstruktion unter Verwendung von Titan und Gold zu verdanken. Egal ob eckig oder rund: In jeder Form liefert sie berauschende Klangerlebnisse. ○

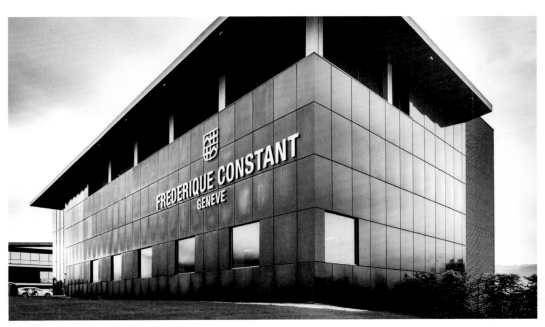

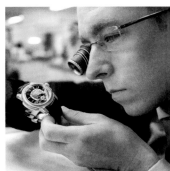

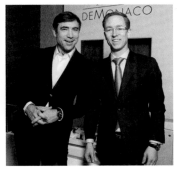

Top: *Manufacture building, 2015* ○ *Oben: Manufakturgebäude, 2015* ○ *Ci-dessus : Bâtiment de la manufacture, 2015* ○ *Right, from top: Pim Koeslag* ○ *Manuel Da Silva Matos and Pim Koeslag*

Horlogerie haut de gamme sur la Côte d'Azur

Français

Mer bleue, casinos, belles voitures ou Grand Prix d'exception, telles sont les images que l'on associe très spontanément à cette petite principauté de la Côte d'Azur. Au chapitre des montres, l'essentiel se joue aux poignets des stars et célébrités ou chez les nombreux grands joailliers. La production horlogère ne date en effet que de 2009. Cette année-là, Peter Stas et l'horloger Pim Koeslag ouvrent un atelier horloger, petit mais haut de gamme, qui constitue aujourd'hui le troisième pilier de la production horlogère dans la région, aux côtés d'Alpina et de Frédérique Constant. Avec une production confidentielle, des complications utiles et un art horloger d'exception, les ateliers deMonaco entendent jouer dans la cour des grands de l'univers horloger. Et c'est bien ce qu'ils font depuis le début. Compte tenu de la proportion élevée de travail artisanal chronophage, il ne sort chaque année des ateliers deMonaco qu'une quantité infime de garde-temps, exclusivement mécaniques. Rien n'est laissé au hasard, jusque dans la conception et la réalisation artisanale des boîtiers, cadrans et aiguilles. Le souci de l'esthétique est le maître mot jusque dans le moindre détail. Mais c'est dans les mouvements à la finition soignée à l'extrême que réside la valeur particulière de ces montres-bracelets. Tous les mouvements produits à Monaco sont de conception et de fabrication locales. Chacun d'eux possède en outre un attrait particulier sur le plan horloger. Dès la conception en effet, Pim Koeslag s'efforce d'intégrer des différences infimes mais décisives. Ainsi, la complication « Quantieme Perpetual » brevetée propose non seulement des affichages sautants à minuit précis, mais aussi un système de réglage et de correction très pratique baptisé « EaZy ». Il suffit de tirer le bouton-poussoir puis de tourner la couronne pour choisir la fonction que l'on désire ajuster : date, jour de la semaine, semaine civile ou mois. On peut alors la modifier, à toute heure du jour ou de la nuit, en appuyant sur le bouton-poussoir. Le « Grand Tourbillon Minute Repeater » est digne des plus hautes écoles d'horlogerie par ses nombreuses complications. Le nom de ce garde-temps est explicite. Le tourbillon, qui tourne une fois par minute sous un pont de saphir transparent, compense les influences négatives de la position verticale sur l'exactitude de la marche. La répétition à minutes, pratiquement invisible, se manifeste de manière bien plus complexe. Pour écouter l'heure qu'il est, il faut actionner le petit poussoir situé dans la tranche gauche du boîtier. On entend alors sonner autant de fois que le nombre d'heures écoulées, puis autant de fois que le nombre de quarts d'heure révolus. Le mécanisme donne pour finir le nombre de minutes écoulées depuis le dernier quart d'heure révolu. Ainsi, pour 12 heures et 59 minutes, on entend au total 32 sons : douze sons graves, trois sons alternant le grave et l'aigu et enfin 14 sons aigus. Un régulateur centrifuge silencieux veille à ce que la suite de tons soit régulière, ni trop rapide ni trop lente. Comme l'on entend environ trois sons toutes les deux secondes, la sonnerie qui retentit peu avant 13 heures dure donc 20 à 25 secondes. Pour voir les marteaux frapper sur les timbres soigneusement accordés, il faut retirer la montre et la retourner. Au lieu du tourbillon, on voit la face arrière décorée avec raffinement. L'excellente qualité du son résulte entre autres de la conception soignée du boîtier, réalisé en titane et en or. Quelle que soit sa forme, anguleuse ou ronde, il produit des sons enchanteurs. ○

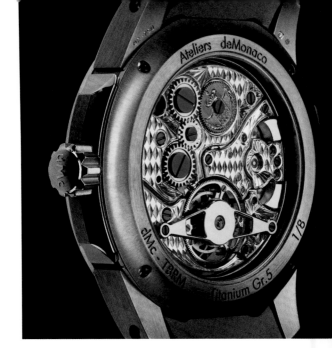

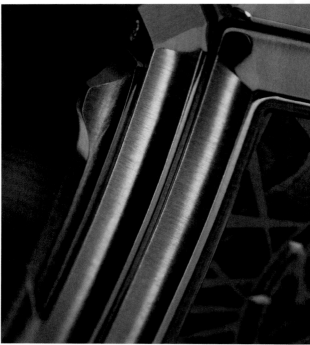

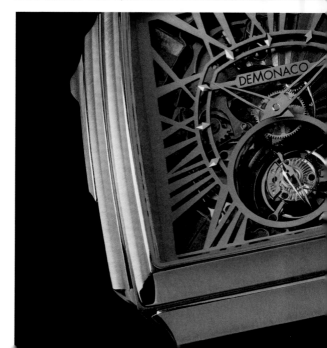

Top: dMc-TBRM-SR-GR, Collection: Tourbillon Répétition Minute, 2010
Middle and bottom: dMc-TBRM-SC-GR, Collection: Tourbillon Répétition Minute, 2010

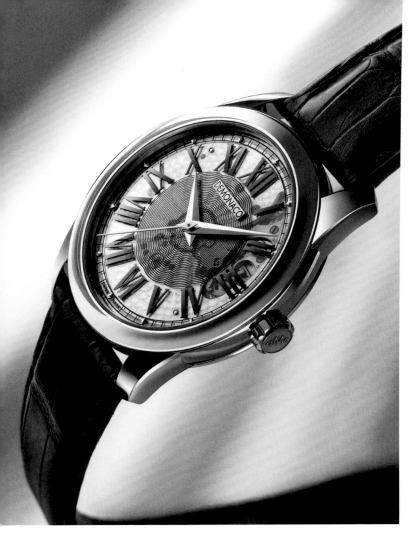

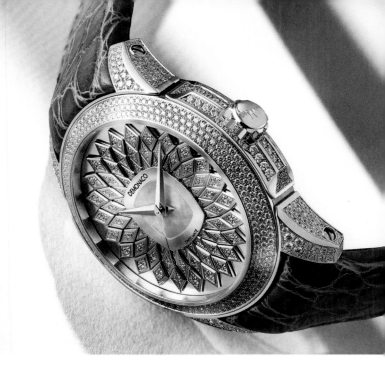

Clockwise from top left: dMc-PDG-SPH-WG, Collection: Poinçon de Genève, 2016 ○ dMc-MC-LSD-WGP, Collection: Ronde de Monte-Carlo, 2016 ○ dMc-TB-GP2-WG, Collection: Bespoke: Tourbillon, Grand Prix de Monaco 1966, 2016 ○ dMc-TB-OCBL-RL-TI, Collection: Tourbillon, Sub-collection: Oculus 1297, 2016

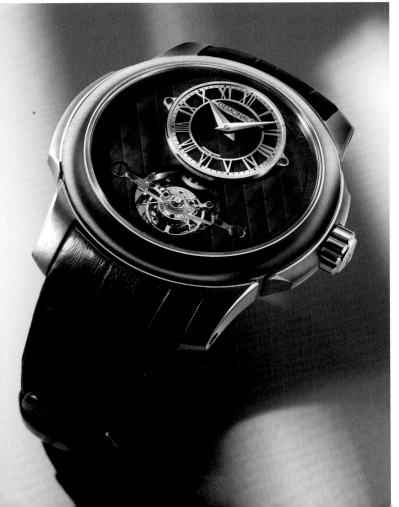

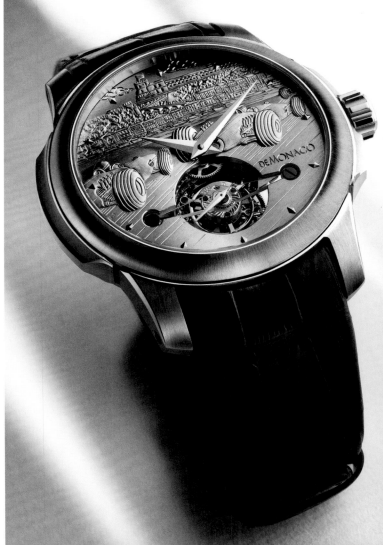

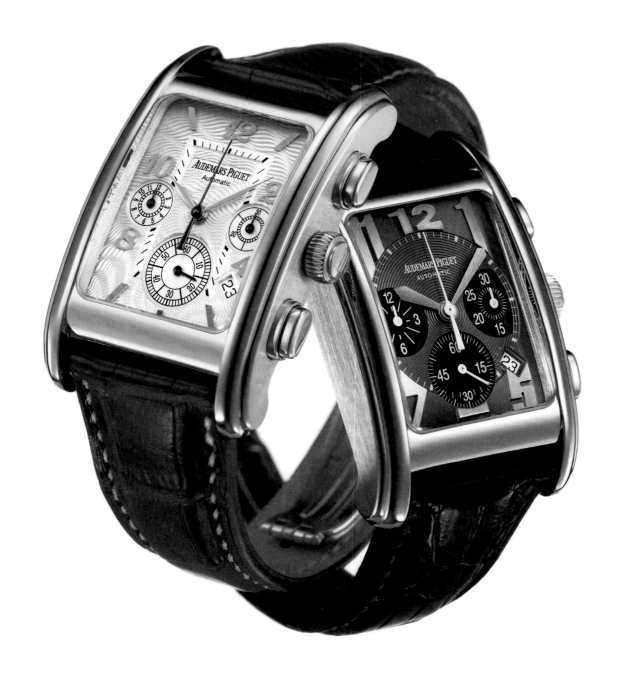

Audemars
Piguet

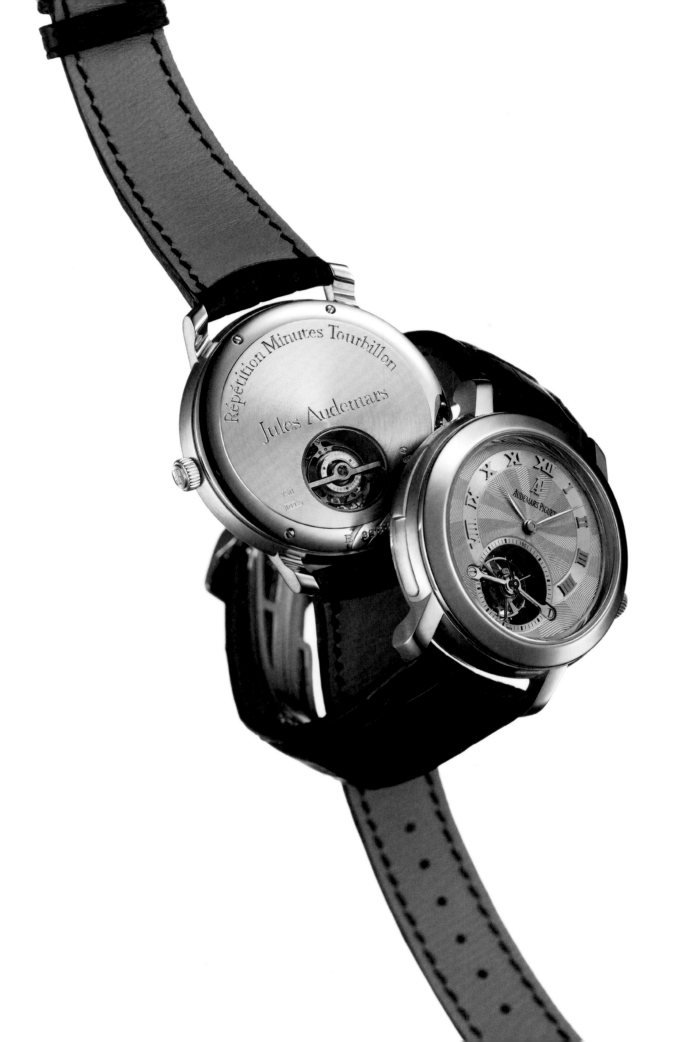

Owned by Its Founding Families since 1875

Nearly every great traditional brand in the watch industry has come under new ownership at least once in its history, but not Audemars Piguet. The spirit of the firm's founders has continued to live through their direct descendants ever since Jules-Louis Audemars joined forces with Edward-Auguste Piguet in the late 19th century. The family-owned manufactory writes new chapters in its history at precisely the same location where the two founding watchmakers first took the daring step into professional independence: namely, at Le Brassus in Vallée de Joux, which is affectionately nicknamed the "Valley of Tinkerers." A success story par excellence began here in 1875, when the region was suffering through an economic crisis and the young Jules-Louis Audemars, who was born in 1851, designed and built his first complicated movements, devoting all his passion and skill to this complex métier.

Well-filled order ledgers soon prompted him to seek competent support, which he found in the watchmaker Edward-Auguste Piguet, who was also familiar with business-related tasks. Their collaboration prompted them to found a jointly owned company. They signed a contract to establish Audemars Piguet & Cie, Manufacture d'Horlogerie in Le Brassus on December 17, 1881. The objective stated in the contract was to use the latest production methods to manufacture fine and complicated watches, e.g., repeater movements, calendar calibers, chronographs, etc. Jules-Louis Audemars served as technical director, while Edward-Auguste Piguet took care of administrative affairs. This time-honored division of labor continued after the founders' deaths: members of the Audemars family remained primarily responsible for technical concerns, while the Piguet family looked after commercial matters. Audemars Piguet & Cie employed a staff of ten people in 1889. Unlike many other businesses, these employees received salaries both summer and winter. The successful development continued into the early 20th century, so construction of a new building began directly alongside the traditional headquarters in 1907. This building, which has been enlarged and modernized several times, remains the home of Audemars Piguet today, while the older edifice serves as the company's museum.

The watchmaker Paul-Louis Audemars took over his father's posts as president of the administrative council and technical director in May 1917. After attending a school of watchmaking and business, Paul-Edward Piguet joined the firm in 1919 and continued to serve as its commercial director until 1962. Incidentally: the firm's directors have allowed the staff to share in the company's business success since 1912. Thanks to a consistent quality and product policy, Audemars Piguet has earned global renown and an illustrious clientele which includes Dent and Frodsham in London, Tiffany in New York and Paris, Cartier in Paris, Bulgari in Rome, Gübelin in Luzern, and Dürrstein in Glashütte and Dresden.

The disastrous crash of the stock market in New York on October 24, 1929 led to a long-lasting downturn. The global economic crisis triggered by the crash, coupled with protectionist measures implemented in many countries, caused a precipitous decline in the demand for luxury watches. In 1930, all watchmaking factories in Vallée de Joux felt compelled either to put workers on shorter hours or to temporarily close their facilities altogether. Some never reopened their gates; others survived with new owners. Survival at Audemars Piguet was assured by a strong sense of family coherence, which was also cultivated by the watchmaker Jacques-Louis Audemars, who joined the firm in 1933. As head of production, he brought new vitality into the languishing production at Audemars Piguet, which fabricated a mere 116 timepieces in 1935.

The situation had improved somewhat when Georges Golay joined the company on May 1, 1945, but the traces left by World War II were impossible to overlook. Acting on advice from this thoroughbred businessman, the directors commenced an operative restructuring. In its wake, the new signature "Audemars Piguet" was introduced onto the watches' dials and the suffix "& Cie" was eliminated. As general director from 1966 to 1987, Georges Golay formatively influenced the structure, philosophy, and appearance of Audemars Piguet to a greater degree than anyone before him. His name is associated with the launch in 1972 of the undisputed leader model "Royal Oak," which currently generates more than 70 percent of Audemars Piguet's revenues. Golay initiated a profound restructuring and reorganization in 1973. Workshops too underwent urgently needed expansion. Early in the 1980s, Golay established AP Technologies, a small but ultramodern factory for micromechanical and electronic specialties. Following Georges Golay's death in 1987, the company's business affairs have been entrusted without exception to external managers, but the important chair of the administrative council has always been occupied by a member of the founding family. The council is presently chaired by Jasmine Audemars, a great-granddaughter of the firm's founder. ○

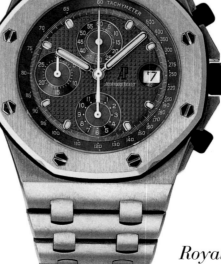

Royal Oak Offshore
Chronograph, 1994

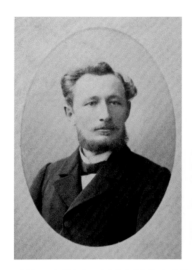

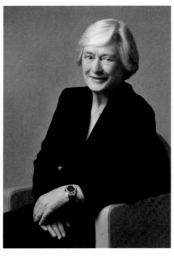

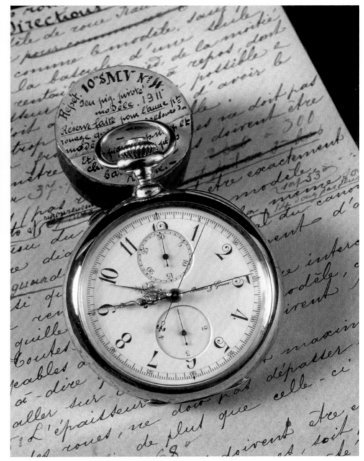

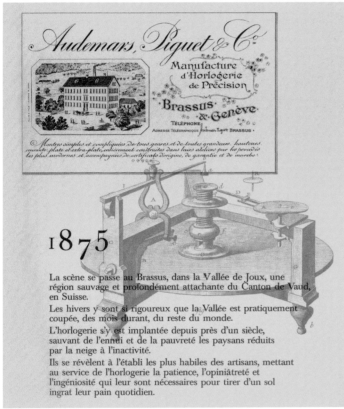

Audemars, Piguet & Co.
Manufacture
d'Horlogerie
de Précision

Brassus
& Genève

1875

La scène se passe au Brassus, dans la Vallée de Joux, une région sauvage et profondément attachante du Canton de Vaud, en Suisse.

Les hivers y sont si rigoureux que la Vallée est pratiquement coupée, des mois durant, du reste du monde.

L'horlogerie s'y est implantée depuis près d'un siècle, sauvant de l'ennui et de la pauvreté les paysans réduits par la neige à l'inactivité.

Ils se révèlent à l'établi les plus habiles des artisans, mettant au service de l'horlogerie la patience, l'opiniâtreté et l'ingéniosité qui leur sont nécessaires pour tirer d'un sol ingrat leur pain quotidien.

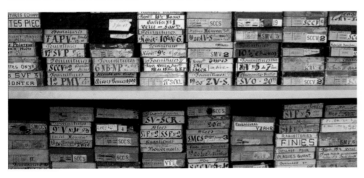

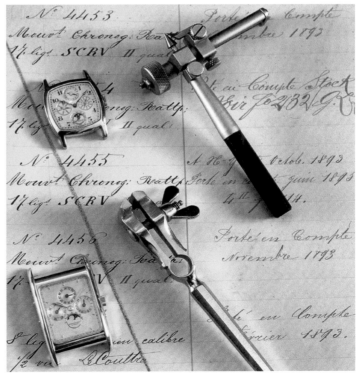

Clockwise from top left: Jules-Louis Audemars ◦ Jasmine Audemars ◦ Chronograph with 30-minute counter and jumping quarter-seconds ◦ Calendar wristwatches from 1925 and 1991 ◦ Boxes containing parts in the archive ◦ Von oben links im Uhrzeigersinn: Jules-Louis Audemars ◦ Jasmine Audemars ◦ Chronograph mit 30-Minuten-Zähler und springender Viertelsekunde ◦ Armbanduhren mit Kalendarium aus den Jahren 1925 und 1991 ◦ Fourniturenschachteln im Archiv ◦ Dans le sens horaire depuis la gauche : Jules-Louis Audemars ◦ Jasmine Audemars ◦ Chronographe au ¼ de seconde et compteur 30 minutes ◦ Montres-bracelets à calendriers des années 1925 et 1991 ◦ Boîtes de fournitures aux archives

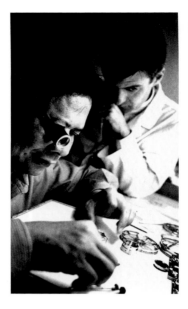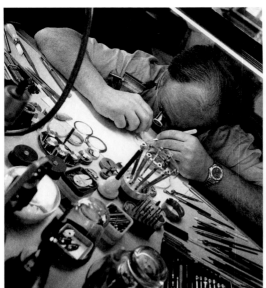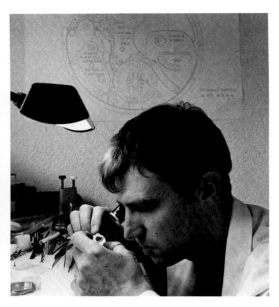

Seit 1875 mehrheitlich im Eigentum der Gründerfamilien

Deutsch

An diesem Faktum gibt es nichts zu rütteln: Nahezu alle großen Traditionsmarken der Uhrenindustrie haben im Laufe ihrer Geschichte mindestens einmal neue Eigentümer bekommen. Nicht so Audemars Piguet. Hier lebt, seit sich Ende des 19. Jahrhunderts Jules-Louis Audemars mit Edward-Auguste Piguet zusammentat, der Geist der Firmengründer in den direkten Nachkommen fort. Die Familienmanufaktur geht ihren Geschäften auch heute noch exakt dort nach, wo die beiden Uhrmacher einst den Start in die berufliche Selbstständigkeit wagten. Gemeint ist Le Brassus im Vallée de Joux, bekannt auch als „Tal der Tüftler". Dort begann 1875, als in der Region eine wirtschaftliche Krisenstimmung herrschte, eine Erfolgsgeschichte par excellence.

Der junge Jules-Louis Audemars, Jahrgang 1851, konstruierte und fertigte erste komplizierte Uhrwerke – ein komplexes Metier, dem seine ganze Leidenschaft galt. Gut gefüllte Auftragsbücher verlangten bald nach kompetenter Hilfe. Die fand sich in Person des Uhrmachers Edward-Auguste Piguet, der auch in wirtschaftlichen Dingen versiert war. Durch die enge Kooperation reifte der Entschluss, ein gemeinsames Unternehmen ins Leben zu rufen. Die Unterzeichnung des Vertrags zur Gründung von Audemars Piguet & Cie, Manufacture d'Horlogerie in Le Brassus erfolgte am 17. Dezember 1881. Das darin erklärte Ziel: Herstellung feiner und komplizierter Uhren nach modernsten Fertigungsmethoden. Darunter verstand das Duo unter anderem Schlag- und Kalenderwerke oder auch Chronographen. Von Anfang an übernahm Jules-Louis Audemars die Rolle des technischen Leiters, während sich Edward-Auguste Piguet um die administrativen Belange kümmerte. Die bewährte Arbeitsteilung setzte sich auch nach dem Tod der Firmengründer fort: Mitglieder der Familie Audemars zeichneten primär für technische, jene der Familie Piguet für kaufmännische Belange verantwortlich. 1889 beschäftigte Audemars Piguet & Cie schon insgesamt zehn Mitarbeiter. Die standen, was keineswegs üblich war, Sommer wie Winter auf der Gehaltsliste.

Weil sich die erfreuliche Entwicklung auch im frühen 20. Jahrhundert fortsetzte, entstand ab 1907 ein Neubau unmittelbar neben dem Stammsitz. Dieses Gebäude, zwischenzeitlich mehrfach vergrößert und modernisiert, beherbergt Audemars Piguet bis heute. Und das alte Haus dient dem Unternehmen als Museum.

Im Mai 1917 übernahm der Uhrmacher Paul-Louis Audemars als Verwaltungsratspräsident und technischer Direktor die Rolle seines Vaters. Paul-Edward Piguet trat 1919 nach dem Besuch von Uhrmacher- und Handelsschule in die Firma ein, um die Aufgaben des kaufmännischen Direktors bis 1962 wahrzunehmen. Schon ab 1912 hatte die Firmenleitung ihre Mitarbeiter übrigens angemessen am geschäftlichen Erfolg beteiligt. Dank konsequenter Qualitäts- und Produktpolitik konnte Audemars Piguet weltweit hohes Ansehen und damit bedeutende Kunden gewinnen. Dazu zählten die Häuser Dent und Frodsham in London, Tiffany in New York und Paris, Cartier in Paris, Bulgari in Rom, Gübelin in Luzern sowie Dürrstein in Glashütte und Dresden.

Der geschichtsträchtige New Yorker Börsencrash am 24. Oktober 1929 leitete einen nachhaltigen Bremsvorgang ein, denn die dadurch ausgelöste Weltwirtschaftskrise und die protektionistischen Maßnahmen vieler Länder ließen das Verlangen nach Luxusuhren gegen null tendieren. 1930 sahen sich sämtliche Uhrenbetriebe im Vallée de Joux gezwungen, entweder Kurzarbeit anzumelden oder die Fabrik zumindest zeitweise ganz zu schließen. Manche öffneten ihre Türen nie mehr, andere machten mit neuen Eigentümern weiter. Bei Audemars Piguet sicherte der ernst genommene Familiensinn das Überleben. Selbigen zelebrierte auch der Uhrmacher Jacques-Louis Audemars, der 1933 zum Unternehmen stieß. Als Fabrikationschef sollte er neuen Schwung in die darniederliegende Produktion bringen, denn 1935 entstanden bei Audemars Piguet gerade einmal 116 Zeitmesser.

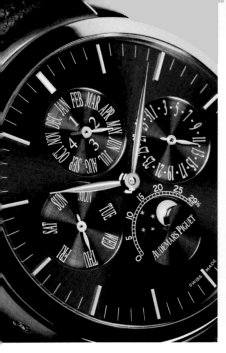
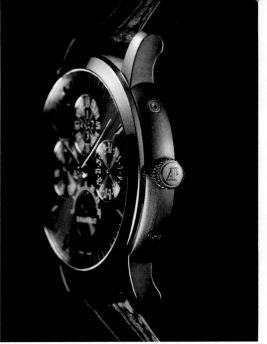
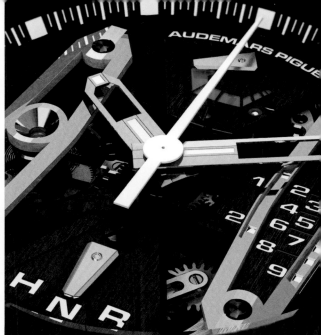

Top left and middle: Jules Audemars Perpetual Calendar, limited, 2008

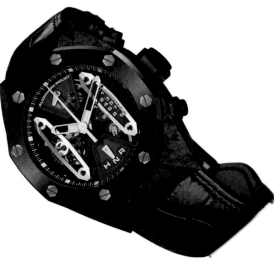

Als am 1. Mai 1945 Georges Golay anheuerte, hatte sich die Situation zwar gebessert, aber die Spuren des Zweiten Weltkriegs waren unübersehbar. Auf Anraten des Vollblutkaufmanns leitete die Direktion eine betriebliche Umstrukturierung in die Wege. Im Zuge dessen führte man die neue Zifferblattsignatur „Audemars Piguet" ein. Der Zusatz „& Cie" fiel weg. Von 1966 bis 1987 prägte Georges Golay als Generaldirektor Struktur, Philosophie und Erscheinungsbild wie kein anderer zuvor. Mit seinem Namen verknüpft ist 1972 die Einführung des unangefochtenen Leadermodells „Royal Oak" und seiner Derivate, die bei Audemars Piguet mehr als 70 Prozent des Umsatzes erzielen. 1973 initiierte Golay eine tiefgreifende Restruk-turierung und Reorganisation. Die Werkstätten erfuhren die dringend gebotene Ausweitung. Zudem rief er Anfang der 80er Jahre eine kleine, aber hochmoderne Fabrik für mikromechanische und elektronische Spezialitäten ins Leben, AP Technologies. Seit dem Tod von Georges Golay im Jahr 1987 steuerten ausnahmslos externe Manager die operativen Geschicke. Der wichtige Vorsitz des Verwaltungsrats lag jedoch stets in den Händen der Gründerfamilie. Aktuell steht Jasmine Audemars, die Urenkelin des Firmengründers, dem Gremium vor. ○

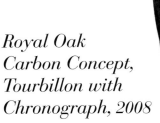

Royal Oak
Carbon Concept,
Tourbillon with
Chronograph, 2008

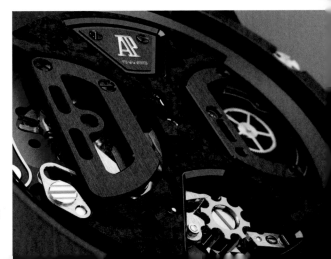

Right: Calibre 2895

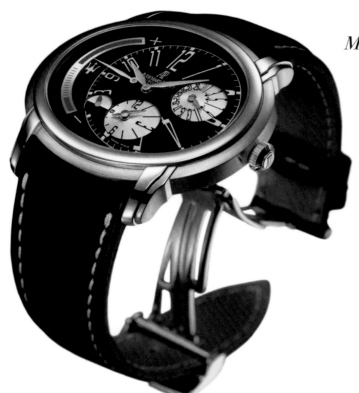

Millenary Maserati Dual Time, 2004

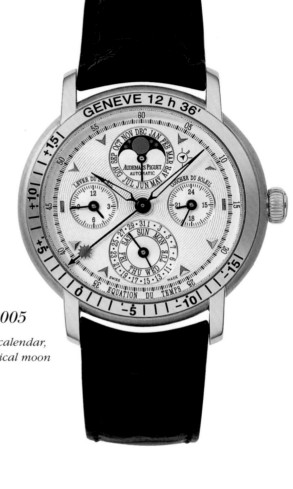

Jules Audemars Equation of Time, 2005

with sunrise and sunset times, perpetual calendar,
and astronomical moon

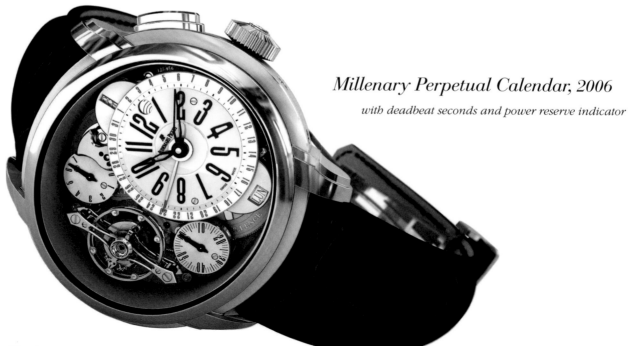

Millenary Perpetual Calendar, 2006

with deadbeat seconds and power reserve indicator

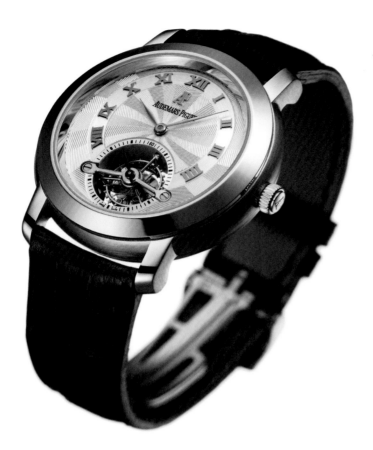

Jules Audemars Minute Repeater
Tourbillon, 2005

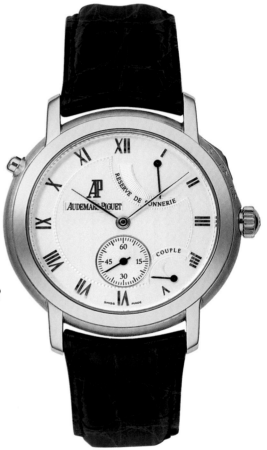

Grande et Petite Sonnerie
Minute Repeater, 2002

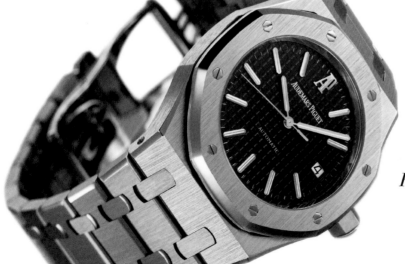

Royal Oak, 2005

Majoritairement propriété des familles fondatrices depuis 1875

C'est un fait indéniable : presque toutes les grandes marques de tradition de l'industrie horlogère ont changé de propriétaire au moins une fois dans leur histoire. Audemars Piguet fait exception à la règle. Depuis l'association de Jules-Louis Audemars et d'Edward-Auguste Piguet à la fin du XIXe siècle, l'esprit des fondateurs de la société survit chez leurs descendants directs. La manufacture familiale poursuit aujourd'hui encore ses activités sur le lieu même qui a vu jadis les deux horlogers débuter leur aventure professionnelle indépendante. Ce lieu, Le Brassus, est situé dans la vallée de Joux, le « Berceau de la haute horlogerie ». C'est là que débute en 1875, alors que la crise économique sévit dans la région, l'histoire d'une réussite par excellence.

Né en 1851, le jeune Jules-Louis Audemar conçoit et fabrique les premières montres à complications – une activité complexe, à laquelle il se voue corps et âme. Les carnets de commande étant bien remplis, le concours de quelqu'un du métier s'avère vite nécessaire. Cette aide, Jules-Louis la trouve en la personne d'Edward-Auguste Piguet, par ailleurs expert en matière commerciale. Collaborant étroitement, ils décident de créer leur propre entreprise. Le contrat portant création d'Audemars Piguet & Cie, Manufacture d'Horlogerie au Brassus, est signé le 17 décembre 1881. L'objectif déclaré est de fabriquer des montres raffinées et compliquées suivant les méthodes les plus modernes. Pour les deux partenaires, ces complications sont des sonneries et des calendriers, ainsi que des chronographes. Dès le début, Jules-Louis Audemars assume le rôle de directeur technique, tandis qu'Edward-Auguste Piguet s'occupe des affaires administratives. Cette répartition des tâches efficace perdure après le décès des deux fondateurs : les membres des familles Audemars et Piguet prennent surtout la responsabilité des aspects techniques pour l'une et commerciaux pour l'autre. En 1889, Audemars Piguet & Cie emploie déjà dix personnes à l'année, ce qui est à l'époque loin d'être habituel. Cette évolution réjouissante se poursuivant au début du XXe siècle, un nouveau bâtiment est créé en 1907 tout près du siège. Cet édifice, plusieurs fois modernisé et agrandi par la suite, abrite encore aujourd'hui Audemars Piguet. Le bâtiment d'origine sert de musée à la manufacture.

En mai 1917, l'horloger Paul-Louis Audemars succède à son père aux postes de président du conseil d'administration et de directeur technique. Après des études d'horlogerie et de commerce, Paul-Edward Piguet rejoint en 1919 la manufacture, où il assume les fonctions de directeur commercial jusqu'en 1962. Dès 1912, la direction fait par ailleurs dûment participer ses collaborateurs au succès commercial de la compagnie. Une politique qualité et produits résolue permet à Audemars Piguet de jouir d'une grande réputation internationale et d'acquérir ainsi des clients importants. Parmi eux figurent les maisons Dent et Frodsham à Londres, Tiffany à New York et Paris, Cartier à Paris, Bulgari à Rome, Gübelin à Lucerne, ainsi que Dürrstein à Glashütte et Dresde.

Le krach historique de Wall Street, le 24 octobre 1929, entraîne un ralentissement durable de l'activité : en raison de la crise économique qui s'ensuit et des mesures protectionnistes prises par de nombreux pays, la demande en montres de luxe devient quasiment nulle. En 1930, toutes les manufactures horlogères de la vallée de Joux se voient contraintes de fonctionner en horaires réduits ou de fermer, au moins temporairement. Si certaines resteront fermées, d'autres reprendront leur activité avec d'autres propriétaires. Audemars Piguet doit sa survie à un solide esprit de famille. C'est à lui que le célèbre horloger Jacques-Louis Audemars sacrifie lorsqu'il rejoint la manufacture en 1933. En sa qualité de chef de fabrication, il a pour mission de revivifier la production en crise : en 1935, Audemars Piguet produit tout juste 116 garde-temps.

Lorsque Georges Golay est engagé au 1er mai 1945, la situation s'est améliorée, mais les traces de la Seconde Guerre mondiale restent visibles. Sur les conseils de ce commercial dans l'âme, la direction lance une restructuration. C'est à l'issue de cette dernière qu'est introduite la nouvelle signature « Audemars Piguet » sur le cadran. Le suffixe « & Cie » disparaît. Directeur général de 1966 à 1987, Georges Golay façonne la structure, la philosophie et l'image de la manufacture comme nul autre avant lui. Son nom est associé en 1972 à l'apparition du modèle phare incontesté « Royal Oak » et de ses dérivés, qui entrent pour plus de 70 pour cent dans le chiffre d'affaires d'Audemars Piguet. En 1973, Georges Golay restructure et réorganise en profondeur la manufacture. Les ateliers connaissent l'extension qui s'imposait tant. En outre, il crée au début des années 1980 une usine de petite taille mais ultramoderne, spécialisée dans la micromécanique et l'électronique, AP Technologies. Depuis sa mort en 1987, ce sont presque toujours des personnes extérieures à la famille qui gèrent les affaires courantes. Le poste crucial de président du conseil d'administration reste quant à lui entre les mains de la famille fondatrice. Aujourd'hui, cette assemblée est présidée par Jasmine Audemars, arrière-petite-fille du cofondateur de la marque. ○

Royal Oak City of Sails Chronograph, 30th Anniversary Ed., 2008

Top left: *Royal Oak Concept, 2002* ○ Top right: *Royal Oak City of Sails Chronograph, 2008*
Bottom: *skeletonized movement of a Royal Oak* ○ Unten: *Skelettiertes Werk einer Royal Oak*
En bas : *Mouvement squelette d'une Royal Oak*

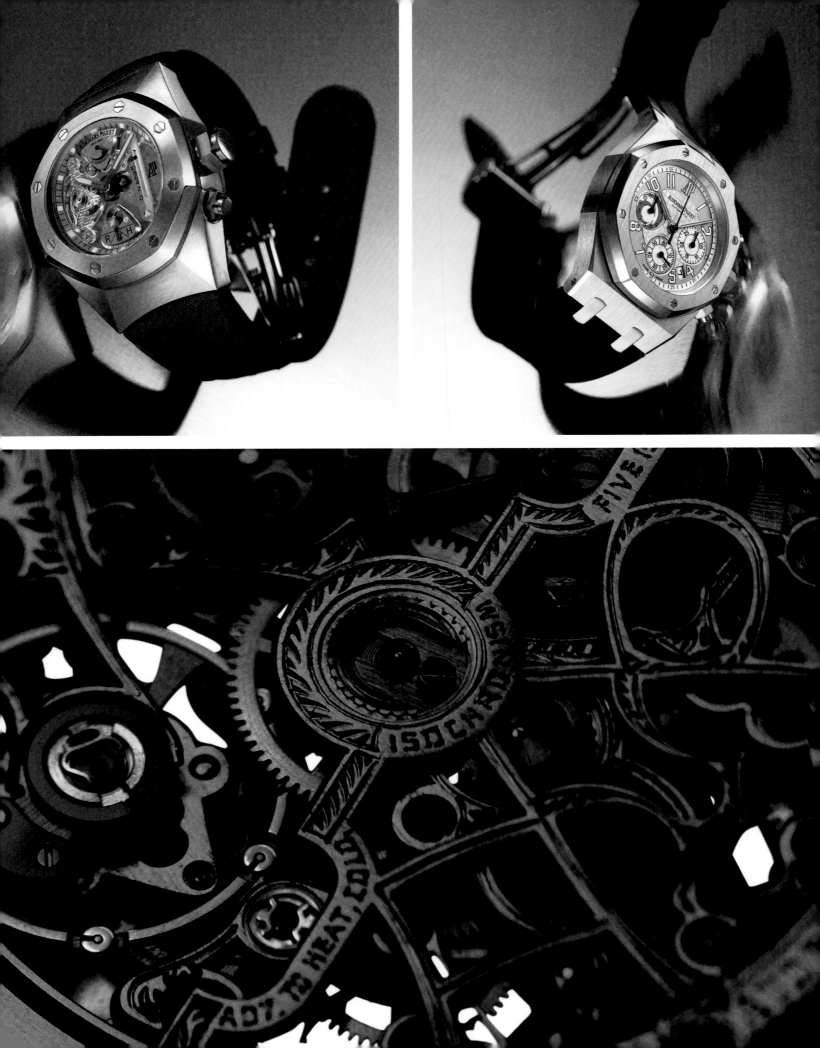

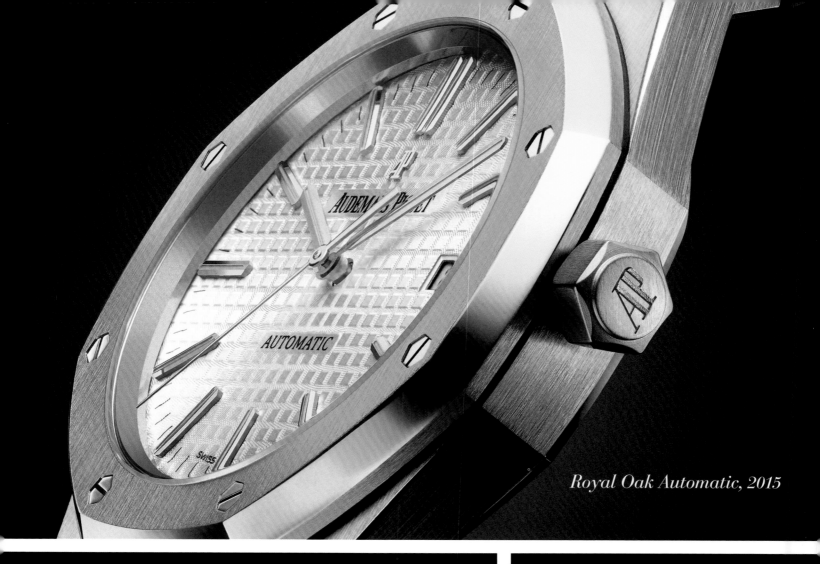

Royal Oak Automatic, 2015

Millenary Quadriennium, 2015

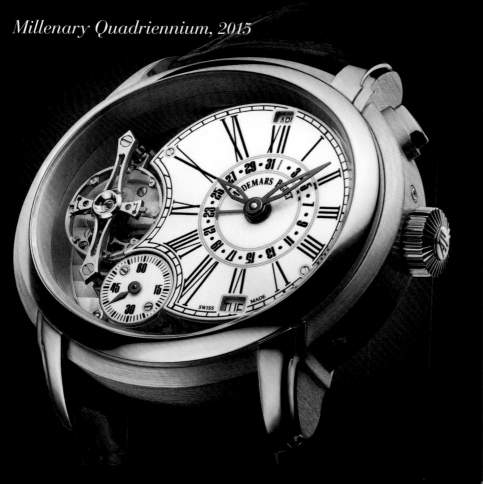

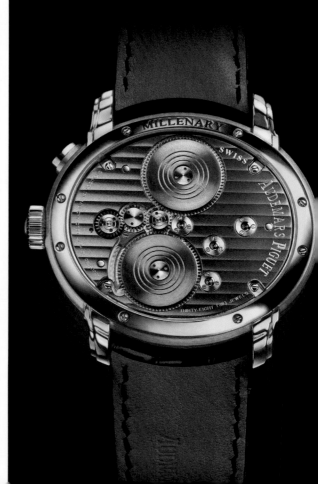

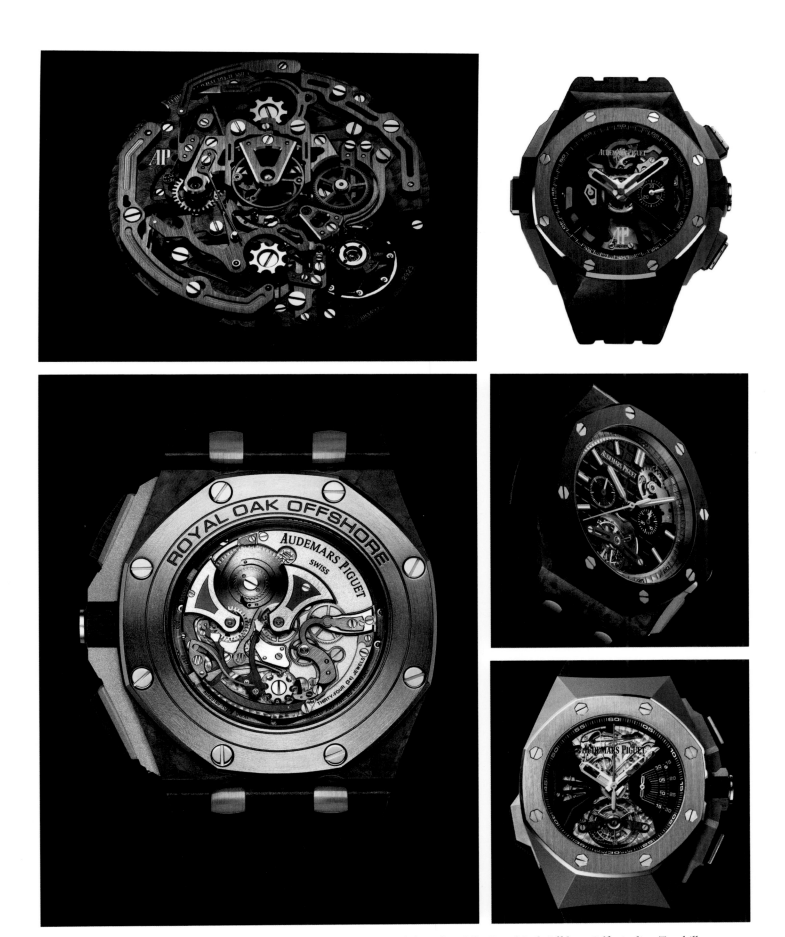

Top: Royal Oak Concept Laptimer Michael Schumacher, 2015 ◦ Bottom left and middle: Royal Oak Offshore Self-winding Tourbillon Chronograph, 2015 ◦ Bottom right: Royal Oak Concept RD#1, 2015

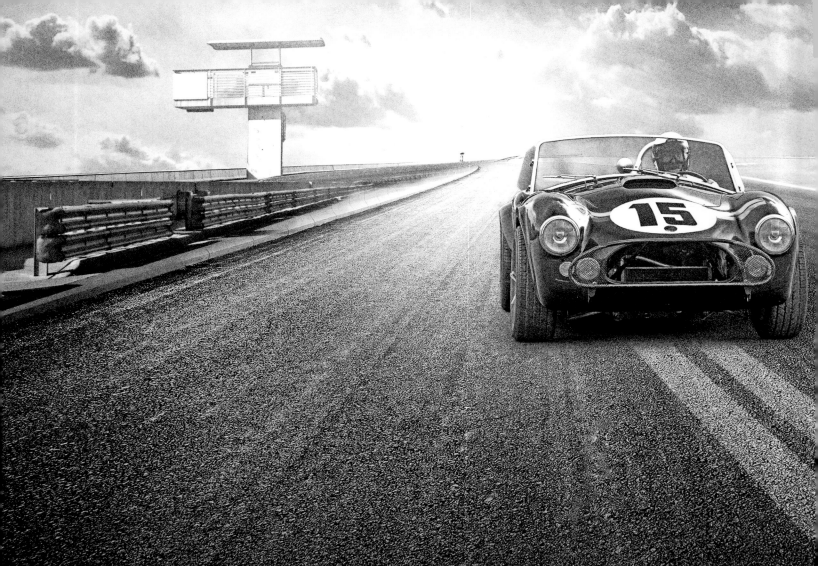

Baume & Mercier

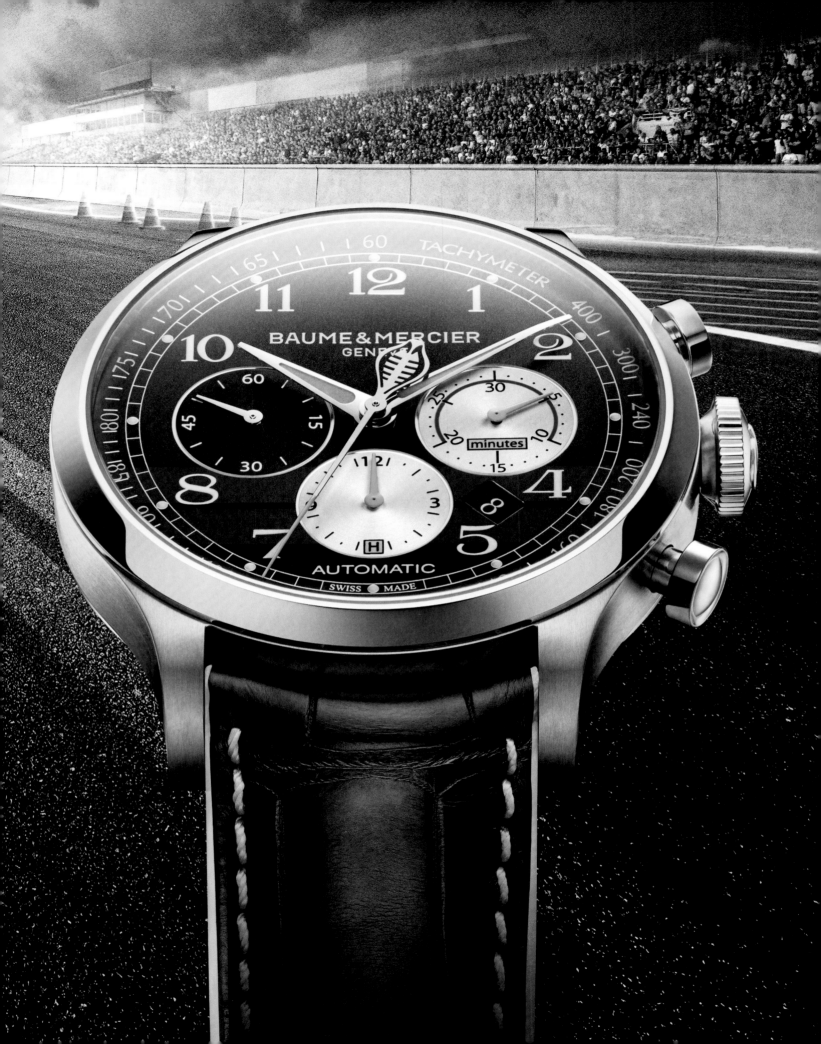

Fine Watches since 1830

Cultivate tradition, but renew it regularly: this credo runs like Ariadne's thread through the history of the Baume & Mercier watch brand, which was founded in 1830 at Les Bois, a little town in Switzerland's Jura region. These rather modest beginnings are linked with Joseph-Célestin and Louis-Victor Baume. Soon after the two brothers had established "Société Baume Frères," they became aware of the importance of international markets. "Baume Bros." was accordingly founded as a subsidiary in London in 1844. Not long afterwards, watches named "Waterloo," "Diviko" and "Sirdar" sparked enthusiasm among buyers in Australia and New Zealand. Innovation and an unconditional striving for quality and precision earned the duo no fewer than six medals at world expositions. A noteworthy pinnacle in the firm's history was reached in 1892, when a very fine tourbillon, which the famous Albert Pellaton-Favre had fabricated and regulated for the Baume Brothers, was awarded 91.9 of 100 theoretically possible points in the 44-day chronometer tests conducted at Kew Observatory in London. The importance of this sensational grade can be better appreciated when one considers the fact that 80 of 100 points suffice to earn the internationally recognized "Class A Certificate" with the qualifier "especially good." It's no wonder that this pocket tourbillon, which bore the number 103018, remained unbeaten for the next ten years. The second part of the brand's renowned name was added after the arrival of Paul Mercier in 1912. William Baume, a descendant of the firm's founder, was on a sales trip with a valise full of new products when he met the ambitious watchmaker and jeweler at Haas, a watch

and jewelry shop in Geneva. Originally named Tcherednitchenko, Mercier had earned an excellent reputation thanks to his savvy business practices and outstanding craftsmanly skills. The two men gradually realized that if they joined forces, they would be better equipped to cope with the big challenges that the future held in store for them. On November 26, 1918, their mutual amity and appreciation prompted them to sign a contract establishing the firm of Baume & Mercier, which would be headquartered in Geneva. After extremely successful careers, William Baume retired in 1937, followed by Paul Mercier several months later. The jeweler Constantin de Gorski began following in their great footsteps. The industry experienced a strong upswing after World War Two due to pent-up demand and the enormous popularity of decorative and precise wristwatches. Baume & Mercier primarily scored points with classical gents' watches, sporty chronographs and valuable ladies' watches. The Piaget Family became majority shareholders in 1965. Cartier Monde SA has made decisions for the label since 1988 because Christian and Yves Piaget sold 60 percent of Piaget Holding SA and Baume & Mercier SA to the French luxury corporation. The most recent chapter in the firm's history began in 1993, when Piaget and Baume & Mercier were wholly purchased by Vendôme Group, which watch lovers nowadays know as Richemont SA.

The course is presently steered by Alain Zimmermann, who learned—and learned to love—all aspects of the chronometric métier during the many years he spent at IWC in Schaffhausen. Avoiding both strict adherence to tradition and rash

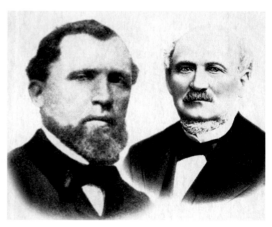

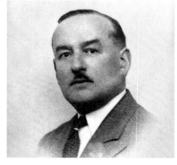

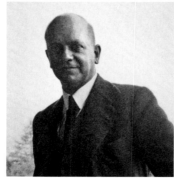

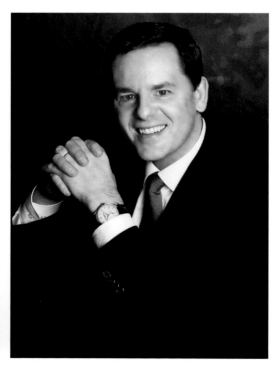

Top: Tourbillon pocket watch awarded a "Class A Certificate," 1892 ∘ Oben: Mit dem „Class A Certificate" ausgezeichnetes Tourbillon, 1892 ∘ En haut : Tourbillon récompensé par un « certificat de classe A », 1892 ∘ Clockwise from top left: Louis-Victor and Joseph-Célestin Baume ∘ Paul Mercier ∘ Alain Zimmermann ∘ William Baume

Pocket watch, 1940

Yellow gold key-winding watch with Poinçon de Genève *quality hallmark*
Uhr aus Gelbgold mit Schlüsselaufzug und Genfer Siegel
Montre en or jaune à remontage à clef, avec Poinçon de Genève, gage de qualité

stylistic experimentation, Zimmerman pursues a brand philosophy that embodies a healthy blend of clear, unadulterated design and the preservation of traditional values. Manufactory status is not presently planned for Baume & Mercier because this would propel its products far above their established price level. Commonly available and time-tested basic movements from Eta and Sellita ensure the quality and precision for which Baume & Mercier has been known worldwide since the 19th century. The undisputed bestseller is the "Classima" line, which was launched to celebrate the brand's 175th birthday in 2005. As its name states, "Classima" features gents and ladies watches that embody classicism in each of their details. For example, classicism is evident in the hands, which culminate in tips that extend all the way to their accompanying indices on the dial. The purest retro look is represented by "Hampton" and "Clifton": the rectangular shape of their cases aptly expresses the aristocratic lifestyle of high society on the American East Coast. Equally nostalgic but with an entirely different design, "Clifton" has a circular case and a combination of dial and hands that expressively conjures the currently trendy look of the 1950s. Distinguished by eye-catching horns, this watch line is offered by Baume & Mercier in many different variants ranging from a sleekly simple self-winding watch with a small second-hand, through models with full calendar and moon-phase displays, to a tourbillon wristwatch that recalls a prizewinning 19th-century pocket watch. (The tourbillon, which was patented in 1801, compensates for the ill effects that the Earth's gravity exerts on the accuracy of the rate of a mechanical watch in a vertical position.) The 30 specimens of the gold "Clifton" that debuted in 2013 with exclusive hand-wound Caliber P591 from ValFleurier are sold out, as are the same number of "Clifton" pocket watches, which premiered in 2015. The repeater movement inside the latter chimes precious time, which

it can audibly announce at five-minute intervals, if desired. Alain Zimmermann consoles aficionados who couldn't get their hands on the objects of their desire: "Other complicated timepieces, which similarly prove Baume & Mercier's time-honored horological competence, are already in the planning stages." Members of the fairer sex will surely appreciate the new "Promesse" or might opt instead for the "Linea," which debuted in 1987 and features a clever quick-change system for its wristbands. Without the need for a tool, a simple manipulation enables the sophisticated ladies who wear this watch to give it an impossible-to-overlook appearance. Last but not least, "Capeland" is a line of sportily elegant watches that pay homage to the popularity of the chronograph in various versions. Strictly limited and thus avidly coveted, these models were developed in a partnership, cultivated since 2015, with the American sportcar legend Shelby Cobra. Each model embodies the genetic code and indomitable spirit of a fascinating automotive mythos. This exciting theme, which is closely linked with the ingenious Carroll Shelby, obviously offers ample potential for Baume & Mercier. Despite the tremendous success of the limited-edition "Capeland Shelby® Cobra" collection, or perhaps because of its warm reception, Alain Zimmermann takes pains to avoid acting rashly. New models will be created judiciously and marketed only in limited quantities. With this in mind, we can look forward to wonderful surprises from Baume & Mercier in coming decades. ○

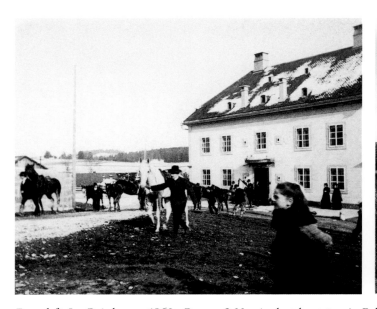 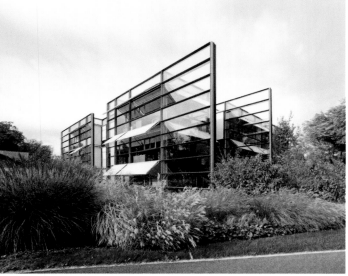

From left: Les Bois house, 1860 ○ Baume & Mercier headquarters in Bellevue, 2016 ○ Von links: Haus in Les Bois, 1860 ○ Hauptsitz von Baume & Mercier in Bellevue, 2016 ○ De gauche à droite : Le bâtiment aux Bois, 1860 ○ Le siège de Baume & Mercier à Bellevue, 2016

Pocket watch, 1892

Mechanical hand-wound movement, one-minute tourbillon, detent escapement chronometer, small seconds, 18-carat yellow gold case ○ *Mechanisches Handaufzugswerk, Ein-Minuten-Tourbillon, Chronometerhemmung, kleine Sekunde, Gehäuse aus 18-karätigem Gelbgold* ○ *Mouvement à remontage manuel. Tourbillon minute, chronomètre à échappement à détente, petites secondes. Boîtier or jaune 18 carats*

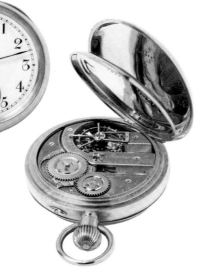

Rectangular wristwatch, 1940

Inspiration for the Hampton collection
Inspiration für die Kollektion Hampton
La source d'inspiration pour la collection Hampton

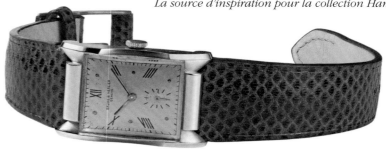

Rectangular wristwatch, 1920

Yellow gold case, manually wound movement
Gehäuse aus Gelbgold. Handaufzugswerk
Boîtier or jaune, mouvement à remontage manuel

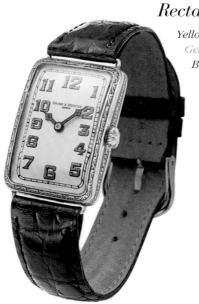

Wristwatch, 1920

Diamond-studded platinum case,
manually wound movement
Diamantenbesetztes Platingehäuse,
Handaufzugswerk
Boîtier platine serti de diamants,
mouvement à remontage manuel

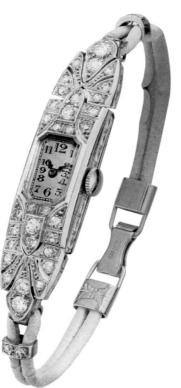

Chronograph, 1948

Inspiration for the Capeland collection
Inspiration für die Kollektion Capeland
La source d'inspiration pour la collection Capeland

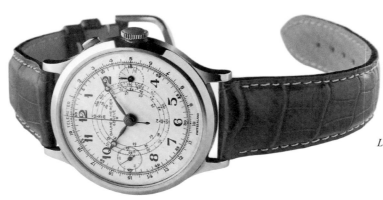

Round wristwatch, 1965

Inspiration for the Classima collection
Inspiration für die Kollektion Classima
La source d'inspiration pour la collection Classima

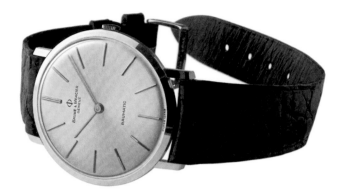

Tronosonic, 1971

With day and date apertures,
yellow gold case, electronic
tuning-fork movement
*Mit Tages- und Datumsanzeige,
Gehäuse aus Gelbgold, elektronisches Stimmgabeluhrwerk*
À deux guichets dateurs,
boîtier or jaune, mouvement
électronique à diapason

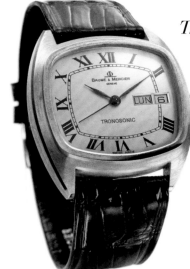

Galaxie, 1972

Diamond-studded white gold case and bracelet,
cabochon-cut sapphire
*Diamantenbesetztes Gehäuse und Armband aus Weißgold,
Saphir im Cabochonschliff*
Boîtier et bracelet or blanc sertis de diamants,
saphir taille cabochon

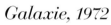

Catwalk, 1999

Diamond-studded steel case
Diamantenbesetzter Stahl
Acier serti de diamants

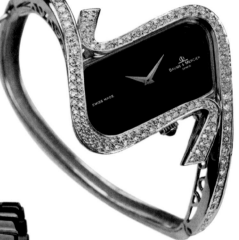

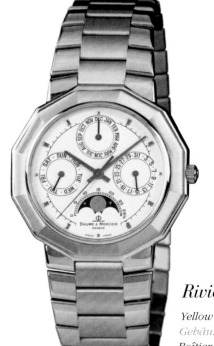

Riviera, 1987

Yellow gold case, chronograph, complete calendar function
Gehäuse aus Gelbgold, Chronograph mit komplettem Kalender
Boîtier or jaune, chronographe à calendrier complet

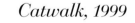

Feine Uhren seit 1830

Clifton 1830 Pocket Watch
Five-Minute Repeater, 2015

Das Überlieferte pflegen und sich trotzdem in schöner Regelmäßigkeit erneuern: Dieses Credo zieht sich wie ein roter Faden durch die Biographie der 1830 in Les Bois, einer kleinen Ortschaft im Schweizer Jura, gegründeten Uhrenmarke Baume & Mercier. Die eher bescheidenen Anfänge verknüpfen sich mit Joseph-Célestin und Louis-Victor Baume. Schon bald nachdem es die Société Baume Frères ins Leben gerufen hatte, erkannte das Brüderpaar die Bedeutung internationaler Märkte. 1844 wurde mit Baume Bros. eine Londoner Niederlassung etabliert. Danach begeisterten Uhren mit den Signaturen „Waterloo", „Diviko" und „Sirdar" Menschen in Australien und Neuseeland. Innovation sowie das unbedingte Streben nach Qualität und Präzision trugen dem Duo nicht weniger als sechs Medaillen bei Weltausstellungen ein. Einen bemerkenswerten Höhepunkt in der Firmengeschichte brachte 1892. In diesem Jahr erzielte ein hochfeines Tourbillon, welches der berühmte Albert Pellaton-Favre für die Gebrüder Baume gefertigt und reguliert hatte, bei den Chronometerprüfungen am britischen Observatorium in Kew bei London nach 44 Tagen die 91,9 von 100 theoretisch möglichen Punkten. Die Bedeutung dieses sensationellen Werts ergibt sich aus der Tatsache, dass bereits 80 von 100 Punkten die Zuerkennung des international anerkannten „Class A Certificate" mit dem Vermerk „besonders gut" nach sich zogen. Kein Wunder also, dass dieses Taschen-Tourbillon mit der Nummer 103018 zehn Jahre lang unübertroffen blieb. Den zweiten Teil des renommierten Markennamens brachte das Jahr 1912 in Person von Paul Mercier. William Baume, ein Nachfahre der Firmengründer, hatte sich einmal mehr mit neuen Produkten auf Reisen begeben. Den ambitionierten Uhrmacher und Juwelier lernte er im Genfer Uhren- und Juweliergeschäft Haas kennen. Durch umsichtige Geschäftsführung und außergewöhnliche handwerkliche Fertigkeiten hatte sich der geborene Tcherednitchenko in der Branche einen vorzüglichen Ruf erworben. Zug um Zug gelangten beide Seiten zur Erkenntnis, dass man die großen Herausforderungen der Zukunft gemeinsam besser bewältigen könne. Am 26. November 1918 fanden gegenseitige Sympathie und Wertschätzung ihren Niederschlag in der Unterzeichnung des Vertrags zur Gründung von Baume & Mercier mit Firmensitz in Genf. Nach überaus erfolgreichem Wirken zogen sich im Jahr 1937 zuerst William Baume und dann auch Paul Mercier aus dem operativen Geschäft zurück. In ihre großen Fußstapfen trat der Juwelier Constantin de Gorski. Großer Nachholbedarf und die enorme Beliebtheit schmückender und präziser Armbanduhren bescherten der Branche nach dem Zweiten Weltkrieg einen bedeutenden Aufschwung. Baume & Mercier punktete vor allem durch klassische Herrenuhren, sportliche Chronographen und wertvolle Damenuhren. 1965 gelangte die Familie Piaget durch den Erwerb der Aktienmehrheit ins Haus.

Ab 1988 hatte die Cartier Monde SA das Sagen, weil Christian und Yves Piaget 60 Prozent der Piaget Holding SA und der Baume & Mercier SA an den französischen Luxusmulti veräußerten. Das vorläufig letzte Kapitel in der Firmengeschichte begann 1993, als Piaget und Baume & Mercier komplett ins Eigentum der Vendôme-Gruppe übergingen, welche Uhrenliebhaber heute als Richemont SA kennen.

Den gegenwärtigen Kurs steuert Alain Zimmermann, welcher das chronometrische Metier während vieler Jahre bei der Schaffhauser IWC in all seinen Facetten kennen und lieben gelernt hat. Eisernes Festhalten an der Tradition ist ihm ebenso fremd wie ungestümes gestalterisches Experimentieren. Vielmehr zielt die Marken-Philosophie auf einen gesunden Mix aus klarem, unverfälschtem Design unter Beibehaltung tradierter Werte. Eine eigene Manufaktur ist derzeit kein Thema, denn dadurch würde Baume & Mercier das angestammte Preissegment verlassen. Andererseits gewährleisten speziell die gleichermaßen gängigen wie bewährten Basis-Uhrwerke von Eta und Sellita jene Qualität und Präzision, für die das in der ganzen Welt vertretene Unternehmen seit dem 19. Jahrhundert bekannt ist. Als unangefochtener Bestseller gilt die 2005 anlässlich des 175. Geburtstags lancierte „Classima"-Linie. Damen und Herren bietet sie, wie der Name unschwer erkennen lässt, klassische, bis ins letzte Detail durchdachte Optik. Sie zeigt sich zum Beispiel bei den Zeigern, deren Spitze, wie es sich gehört, bis an die zugehörige Zifferblatt-Indexierung reicht. Retrolook in Reinkultur repräsentieren „Hampton" und „Clifton". Durch ihre rechteckige Form bringt erstere den distinguierten Lebensstil der gehobenen amerikanischen Ostküsten-Gesellschaft in vorzüglicher Weise zum Ausdruck. Das nicht minder nostalgisch wirkende, gestalterisch jedoch völlig andersartige Pendant heißt „Clifton". Im Rund des Gehäuses und in der Zifferblatt-Zeiger-Kombination leben die angesagten 1950er Jahre ausdrucksstark fort. Diese Uhrenlinie mit markanten Bandanstößen offeriert Baume & Mercier in ganz unterschiedlichen Ausführungen, angefangen bei der schlichten Automatik mit kleinem Sekundenzeiger über Modelle mit Vollkalendarium und Mondphasenanzeige bis hin zu einem Tourbillon, welches an die preisgekrönte Taschenuhr aus dem 19. Jahrhundert erinnert. Wirkungsvoll kompensiert der 1801 patentierte Drehgang die unliebsamen Auswirkungen der Erdanziehungskraft auf die Ganggenauigkeit mechanischer Uhren in senkrechter Position. Die nur 30 Exemplare der 2013 vorgestellten Gold-„Clifton" mit dem exklusiven Handaufzugskaliber P591 von ValFleurier sind ebenso ausverkauft wie genauso viele „Clifton"-Taschenuhren von 2015. Deren Repetitionsschlagwerk verkündet die kostbare Zeit, wenn gewünscht, alle fünf Minuten. Für zu kurz Gekommene hat Alain Zimmermann einen Trost parat: „Weitere komplizierte

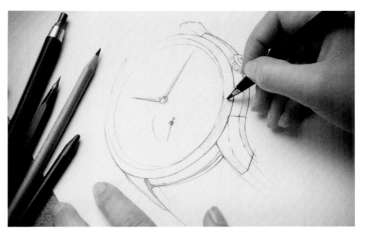

Zeitmesser, welche die jahrzehntelange uhrmacherische Kompetenz von Baume & Mercier unter Beweis stellen, sind bereits in Planung." Das zarte Geschlecht kommt bei der neuen „Promesse" zu seinem Recht. Oder bei der bereits 1987 aus der Taufe gehobenen „Linea" mit ausgeklügeltem Bandwechsel-System. Mit wenigen Handgriffen und ohne jegliches Werkzeug kann Frau von Welt ihrer Armbanduhr einen unübersehbar anderen Auftritt verleihen. Bleibt schließlich „Capeland", eine sportlich-elegante Uhrenlinie, die dem beliebten Chronographen in unterschiedlichsten Ausführungen huldigt. Streng limitiert und deshalb besonders begehrt sind jene Modelle, welche aus der seit 2015 gepflegten Partnerschaft mit der amerikanischen Sportwagen-Legende Shelby Cobra erwachsen. In allen Modellen finden sich der genetische Code und der Spirit eines faszinierenden automobilen Mythos. Dieses bewegende Thema, welches sich mit dem nachgerade genialen Carroll Shelby verknüpft, birgt für Baume & Mercier jede Menge Potenzial. Trotz oder gerade wegen des riesigen Erfolgs der limitierten „Capeland Shelby® Cobra"-Kollektion möchte Alain Zimmermann aber nichts überstürzen. Neuigkeiten werden mit Bedacht kreiert und stets nur in begrenzten Quantitäten auf den Markt gelangen. In jedem Fall ist Baume & Mercier auch in den kommenden Jahrzehnten für große Überraschungen gut. ○

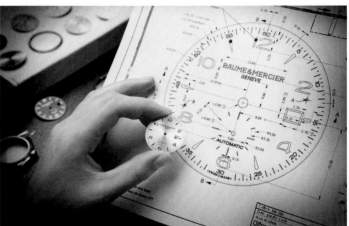

Clifton 1892
Flying Tourbillon,
2014

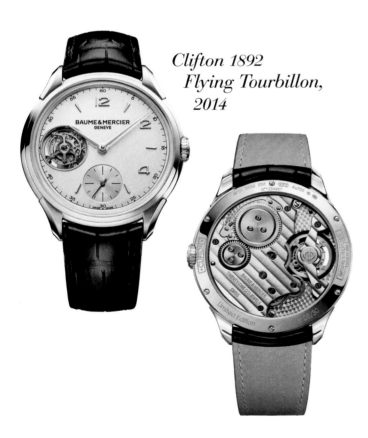

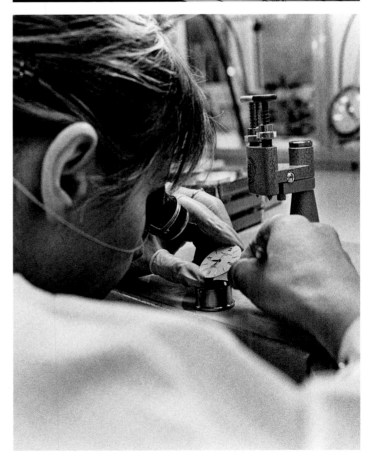

*Top and middle: Design studio, Clifton watch ○ Bottom: Les Brenets watch workshops ○ Oben und Mitte: Designstudio, Modell Clifton ○ Unten: Uhrenatelier, Les Brenets ○ **En haut et au milieu : Atelier de conception, la Clifton ○ En bas : Les ateliers horlogers aux Brenets***

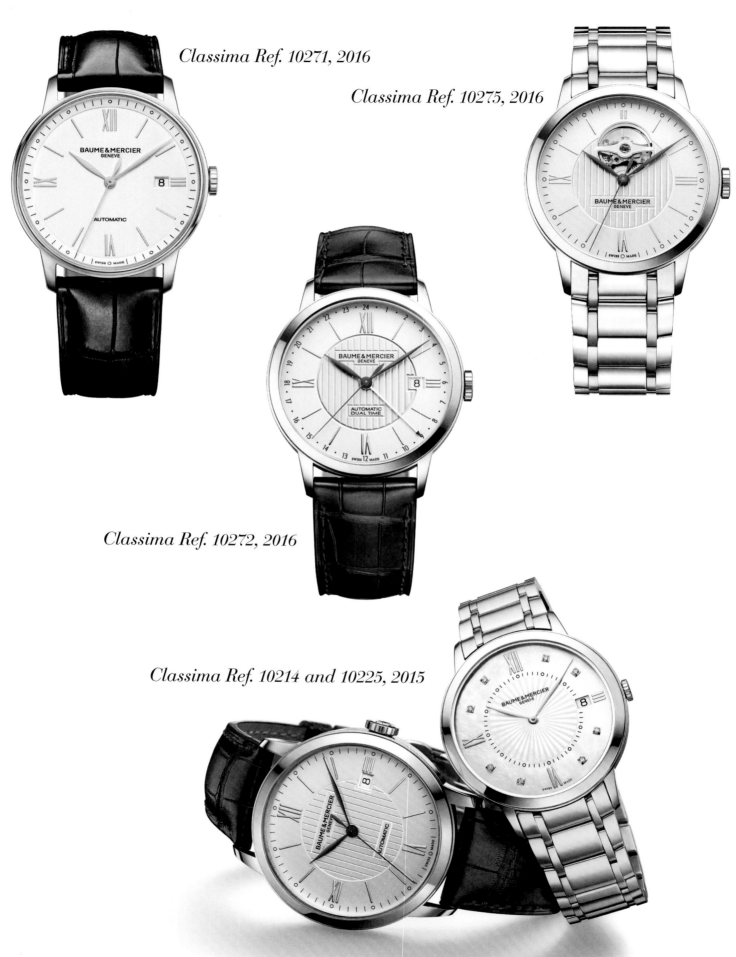

Classima Ref. 10271, 2016

Classima Ref. 10275, 2016

Classima Ref. 10272, 2016

Classima Ref. 10214 and 10225, 2015

Clifton Ref. 10058, 2013

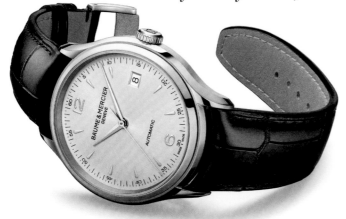

Clifton Ref. 10052, 2013

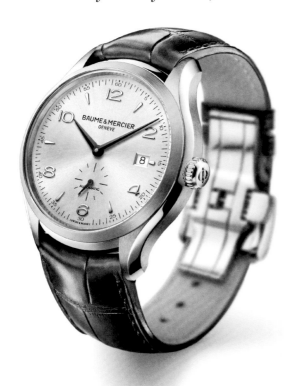

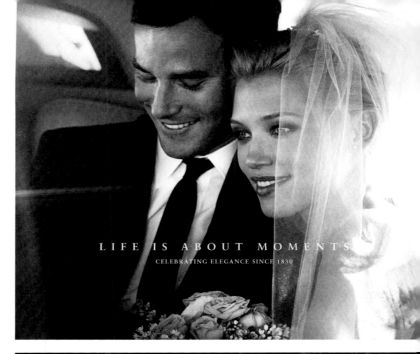

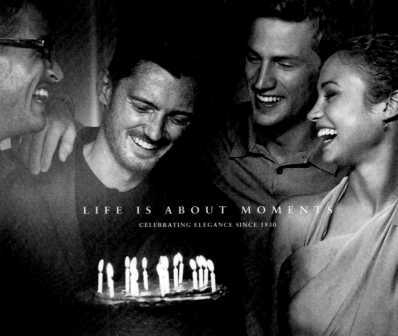

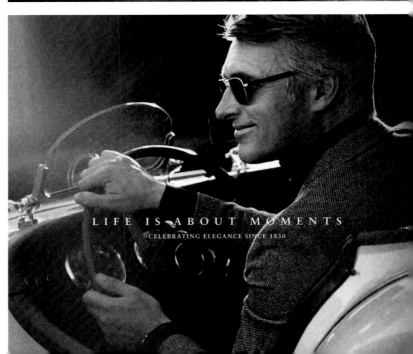

"Life is about moments" advertising campaign by Peter Lindbergh ◇
Werbekampagne „Life is about moments", fotografiert von
Peter Lindbergh ◇ « Life is about moments » – Campagne publicitaire
signée Peter Lindbergh

Des montres de haute précision depuis 1830

Français

Rester fidèle à la tradition tout en se renouvelant avec une belle régularité, ce crédo fait figure de fil conducteur dans l'histoire de la marque horlogère Baume & Mercier, qui remonte à 1830. Tout commence modestement aux Bois, petite localité du Jura suisse, par la création du comptoir horloger Frères Baume par Joseph-Célestin et Louis-Victor Baume. Saisissant rapidement l'importance des marchés internationaux, ceux-ci ouvrent en 1844 une succursale londonienne, Baume Bros. Les modèles « Waterloo », « Diviko » et « Sirdar » partent alors à la conquête de l'Australie et de la Nouvelle-Zélande. L'innovation et la quête sans concession de la qualité et de la précision rapportent aux deux frères pas moins de six médailles dans le cadre d'expositions universelles. 1892 est un temps particulièrement fort dans l'histoire de l'entreprise : un tourbillon de haute précision, réalisé et réglé par le célèbre Albert Pellaton-Favre pour les frères Baume, obtient 91,9 points d'un maximum théorique de 100 points à l'issue de 44 jours de tests chronométriques effectués à l'observatoire de Kew, près de Londres. Ce score est sensationnel dans la mesure où 80 points correspondent déjà à une distinction reconnue dans le monde entier, le « certificat de classe A » assorti de la mention « très bien ». Rien d'étonnant donc à ce que ce tourbillon de poche numéro 103018 soit resté inégalé pendant dix ans. La seconde partie du nom de cette marque renommée est apportée en 1912 par Paul Mercier. Au cours de son énième voyage d'affaires pour présenter de nouveaux produits, William Baume, descendant des fondateurs, fait la connaissance de cet ambitieux horloger-joaillier dans l'horlogerie-joaillerie genevoise Haas. Gestionnaire avisé et artisan hors pair, Tcherednitchenko, dit Paul Mercier, s'est vite fait une excellente réputation dans la branche. De fil en aiguille, les deux hommes parviennent à la conclusion qu'ensemble il leur serait plus facile de relever les grands défis de l'avenir. Le 26 novembre 1918, ces sympathie et estime mutuelles se traduisent par la signature du contrat de fondation de Baume & Mercier, entreprise sise à Genève. Les artisans de son succès, William Baume et Paul Mercier, se retirent des affaires en 1937. Le joailler Constantin de Gorski prend leur difficile relève. Après la Seconde Guerre mondiale, la fin des privations et l'engouement pour les montres-bracelets à la fois esthétiques et précises sont à l'origine d'une reprise significative dans le secteur. Baume & Mercier doit son succès avant tout aux montres pour hommes, aux chronographes de sport et aux montres-bijoux pour femmes. Par une prise de participation majoritaire, la famille Piaget met un pied dans la maison en 1965. À partir de 1988, Cartier Monde SA préside aux destinées de l'entreprise, Christian et Yves Piaget ayant cédé à la multinationale du luxe française 60 pour cent de leurs actions dans Piaget Holding SA et Baume & Mercier. Le dernier chapitre en date dans l'histoire de la manufacture débute en 1993, quand Piaget et Baume & Mercier sont absorbées par le groupe Vendôme, que les passionnés de montres connaissent de nos jours sous le nom de Richemont SA.

Les rênes de la maison sont tenues à l'heure actuelle par Alain Zimmermann qui, au cours de ses nombreuses années d'activité chez IWC, à Schaffhouse, a appréhendé la mesure chronométrique sous toutes ses facettes et appris à l'aimer. N'étant pas homme à perpétuer la tradition coûte que coûte, ni à se lancer tête baissée dans le design d'avant-garde, il veille à un juste équilibre entre un design épuré et authentique, d'une part, et le respect des valeurs traditionnelles, d'autre part. Une manufacture maison, qui signifierait un changement de gamme de prix pour Baume & Mercier, n'est pas à l'ordre du jour. De plus, la qualité et la précision qui font la réputation de cette entreprise représentée dans le monde entier tiennent justement à des calibres à la fois courants et éprouvés sortis des ateliers d'Eta et de Sellita. Lancée en 2005 à l'occasion du 175e anniversaire de la maison, la ligne « Classima » est un best-seller incontesté. Comme son nom l'indique, elle propose aux hommes et aux femmes un design classique, étudié jusque dans le moindre détail. Les aiguilles, par exemple, effleurent comme il sied, l'index correspondant sur le cadran. Les amateurs trouvent la quintessence du style rétro dans les lignes « Hampton » et « Clifton ». Par sa forme rectangulaire, la première est la parfaite expression du style de vie distingué des Américains de la côte Est. Entouré d'une aura non moins nostalgique, son pendant « Clifton » a un design totalement différent : le boîtier rond et le couple cadran-aiguilles perpétuent de manière expressive le style années 1950, qui connaît un regain. Baume & Mercier décline cette ligne dotée de cornes protubérantes au niveau de l'extrémité du bracelet dans des variantes très diverses, du modèle à mouvement automatique et petite trotteuse jusqu'au tourbillon rappelant la montre de poche du XIXe siècle largement primée, en passant par des modèles à calendrier complet et phases de lune. Mécanisme breveté en 1801, le tourbillon compense efficacement les incidences négatives de la gravité sur la précision de marche des montres mécaniques dans la position verticale.

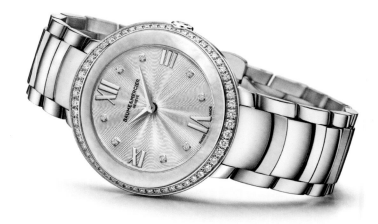

Promesse Ref. 10199, 2015

Produite à seulement 30 exemplaires, la « Clifton » or dotée de l'exceptionnel calibre à remontage manuel P591 de ValFleurier est épuisée, tout comme bien des montres de poche « Clifton » sorties en 2015, toutes dotées d'une sonnerie à répétition qui, si le propriétaire de la montre le souhaite, retentit toutes les cinq minutes. En guise de consolation pour tous ceux qui sont restés sur leur faim, Alain Zimmermann annonce : « Nous avons à l'étude d'autres garde-temps à complications témoignant du savoir-faire horloger qui est celui de Baume & Mercier depuis des décennies. » La gent féminine trouve son compte dans le nouveau modèle « Promesse » ou dans la collection « Linea », lancée en 1987, aux bracelets astucieusement interchangeables. En quelques manipulations et sans outillage spécifique, les femmes du monde changent l'aspect de leur montre-bracelet du tout au tout. Pour finir, la collection à la fois sportive et élégante « Capeland » rend hommage dans ses versions les plus diverses au chronographe très prisé. Les modèles issus du partenariat entretenu depuis 2015 avec le constructeur américain Shelby, auteur de la légendaire voiture de course Cobra, sont produits en édition limitée, ce qui en fait des produits très recherchés. Tous les modèles portent en eux l'ADN et l'âme d'un grand mythe automobile. Le thème du mouvement, lié à un homme vraiment génial du nom de Carroll Shelby, recèle des possibilités infinies pour Baume & Mercier. Malgré l'immense succès de la collection limitée « Capeland Shelby® Cobra » ou peut-être justement pour cette raison, Alain Zimmermann ne veut pas précipiter les choses. Les nouveautés sont créées avec circonspection et toujours produites en séries limitées. Quoi qu'il en soit, dans les décennies à venir, Baume & Mercier ne finira pas de nous surprendre. ○

Linea, 2016

A system lets the wearer easily remove one summery colored bracelet and insert another

System, bei dem die Trägerin das Armband im Handumdrehen selbständig wechseln kann, Sommerfarben

Système de bracelets interchangeables par la propriétaire au gré des saisons, coloris d'été

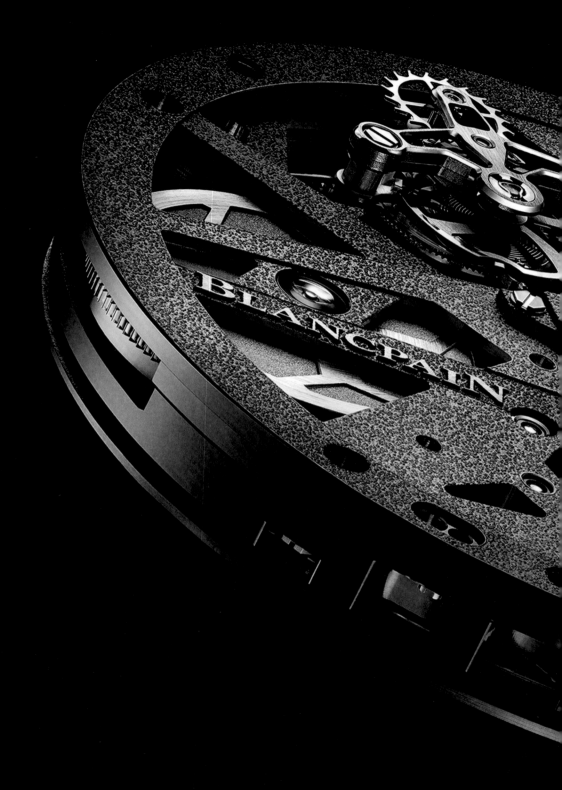

Blancpain

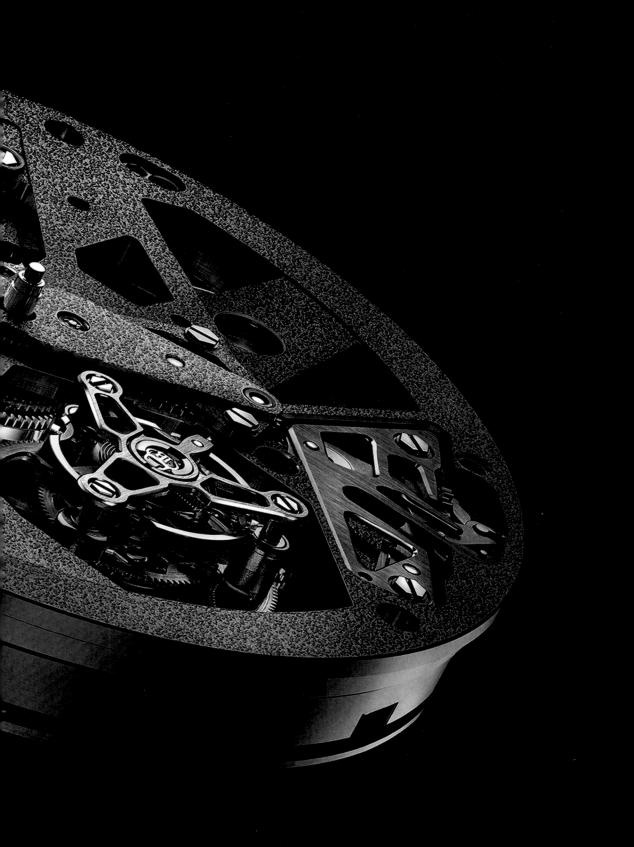

Mechanical Watchmaking Since 1735

English

Anyone who would like to explore Blancpain and its history must gaze far into the past. This watch brand's origins can be traced to 1735, when a certain Jehan-Jacques Blancpain lived and worked in Villeret, a little town in the Jura region of western Switzerland. Like many of his colleagues, this young watchmaker initially occupied himself with the fabrication of components. Many years would pass before complete pocket watches joined the product portfolio. The transformation from an artisanal operation to a small but partially mechanized watch factory named "Blancpain" occurred in 1815, but this brand's name was seldom heard among aficionados until the 20th century, e.g., with the fabrication in 1930 of the "Rolls," a rectangular wristwatch with unconventional self-winding inner workings. The aptly named "Fifty Fathoms" diver's watch set sail in 1953. Its special feature: a dive-time bezel that clicks authoritatively into place and is safeguarded against inadvertent repositioning. It's not surprising that Jacques-Yves Cousteau and his team consulted this wristwatch while filming the documentary *The Silent World*. The fabrication of tiny hand-wound movements was prioritized in the early 1950s. These include rectangular Caliber R 59, which has a total volume of less than 500 cubic millimeters. Circular Caliber R 550 is even smaller: a mere 425 cubic millimeters. It debuted with its crown positioned on its back under the name "Ladybird" in 1956. In 1983, when the renaissance of the mechanical wristwatch was gradually gaining momentum, Blancpain came under the aegis of Jean-Claude Biver and Jacques Piguet (the owner of the Frédéric Piguet ébauche manufactory) and began its modern era by unveiling a wristwatch with a moon-phase display. Two years later, Blancpain surprised aficionados with the debut of an extra-slim minute repeater wristwatch encasing a newly developed movement with a diameter of 20.3 millimeters and a low height of just 3.2 millimeters.

The world's smallest chronograph with date display, automatic winding and optionally also with split second-hand premiered in 1987. It was followed in 1990 by a tourbillon with an eight-day movement, a date display and a power-reserve indicator. This same year also brought the inarguable crowning touch to the fireworks of complications with the presentation of the "1735," which concatenates 740 components in its movement and was the world's most complicated wristwatch at the time of its debut. Another important year in this company's long history was 1992, when the Swatch Group took the traditional label under its capacious wings. "Léman," an alarm wristwatch that can also show the time in other time zones, was released in 2003. Blancpain shared automatic Caliber 1241 with Breguet, a fellow member of the Swatch Group, but couldn't yet claim the honor of being a full-fledged manufacture because Frédéric Piguet continued to supply the encased movements. This also applies to the "Carrousel Volant Une Minute," which debuted in 2008 with a "flying" (i.e., cantilevered) one-minute tourbillon. This mechanism embodies a remarkable evolutionary phase in a device invented in the late 19th century by the Danish watchmaker Bahne Bonniksen. Self-winding Caliber 225 combines 262 components and automatically winds itself. Under the aegis of Marc A. Hayek (the grandson of the great Nicolas G. Hayek), the movement manufacturer Frédéric Piguet merged with Blancpain to form a dyed-in-the-wool watch manufacture in 2010. In 2016, this manufactory presents a new model in the "Fifty Fathoms Bathyscaphe" family, which traces its ancestry to the 1950s. For the first time in Blancpain's history, gray plasma ceramic is used as the material for the watch's case and rotatable bezel. The indices are made of liquid metal, a deformation-resistant alloy. Self-winding Caliber 1315 with silicon hairspring and a five-day power reserve ticks inside the lightweight and scratchproof case, which is watertight to 30 bar. ○

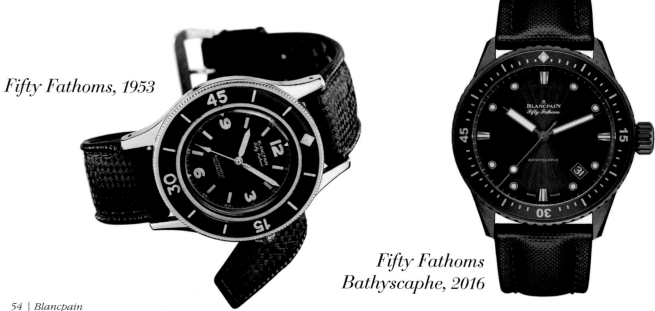

Fifty Fathoms, 1953

Fifty Fathoms Bathyscaphe, 2016

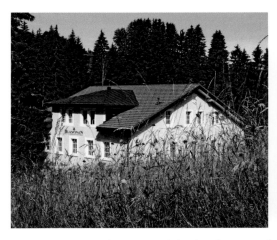
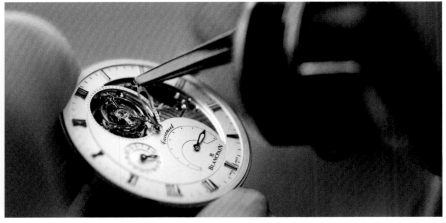

Opposite page: Cal. R 550 Ladybird ○ This page, from left: The manufactory in Le Brassus ○ Manually inserting a hand ○ Diese Seite, von links: Manufaktur, Le Brassus ○ Manuelle Montage eines Zeigers ○ De gauche à droite : La manufacture du Brassus ○ Montage manuel d'une aiguille

Mechanische Uhrmacherei seit 1735

Wer sich mit Blancpain und der Geschichte dieser Uhrenmarke beschäftigen möchte, muss sehr weit zurückblicken. Die Ursprünge liegen nämlich im Jahr 1735. Damals wohnte und arbeitete Jehan-Jacques Blancpain in Villeret, einer kleinen Ortschaft im Westschweizer Jura. Zunächst beschäftigte sich der junge Uhrmacher wie viele seiner Kollegen auch mit der Herstellung von Komponenten. Komplette Taschenuhren ließen noch auf sich warten. Die Umwandlung des Handwerksbetriebs in eine kleine, auf mechanisierter Basis arbeitende Uhrenfabrik namens Blancpain erfolgte 1815. Von sich reden machte Blancpain freilich erst im 20. Jahrhundert. Zum Beispiel 1930 durch die Herstellung der „Rolls", einer rechteckigen Armbanduhr mit ungewöhnlichem Selbstaufzugs-Innenleben. Die Vorstellung der Taucher-Armbanduhr namens „Fifty Fathoms" erfolgte 1953. Ihre Besonderheit: eine rastende, gegen unbeabsichtigtes Verstellen gesicherte Tauchzeit-Lünette. Kein Wunder, dass Jacques-Yves Cousteau und seine Mannschaft bei den Dreharbeiten zum Dokumentarfilm *Die schweigende Welt* auf diesen Zeitmesser blickten. Die frühen 1950er Jahre standen im Zeichen der Konstruktion winziger Handaufzugswerke. Ein Volumen von weniger als 500 Kubikmillimetern wies das rechteckige Kaliber R 59 auf. Das runde R 550 brachte es auf lediglich 425 Kubikmillimeter. Letzteres mit rückwärtig positionierter Krone debütierte 1956 unter dem Namen „Ladybird". 1983, als die Renaissance der mechanischen Armbanduhr zaghaft einsetzte, trat Blancpain unter der Ägide von Jean-Claude Biver und Jacques Piguet, Inhaber der Rohwerkemanufaktur Frédéric Piguet, mit einer Mondphasen-Armbanduhr in die Neuzeit seiner Existenz. Zwei Jahre später überraschte Blancpain mit einer extraflachen Armbanduhr mit Minutenrepetition, deren neuentwickeltes Werk bei 20,3 Millimetern Durchmesser lediglich 3,2 Millimeter hoch baute. Der seinerzeit weltweit kleinste Chronograph mit Datumsanzeige, automatischem Aufzug und – auf Wunsch – auch Schleppzeiger-Mechanismus erschien 1987 auf der Bildfläche. 1990 folgte ein Tourbillon mit

Acht-Tage-Werk, Datums- und Gangreserveindikation. Als unangefochtene Krönung des Komplikationen-Feuerwerks kann die im gleichen Jahr vorgestellte „1735" gelten, welche, zusammengefügt aus 740 Werkskomponenten, damals als weltweit komplizierteste Armbanduhr gelten durfte. Ein ebenfalls wichtiges Jahr in der langen Firmengeschichte war 1992. Hier brachte die Swatch Group das Traditionsunternehmen unter sein Dach. 2003 erschien „Léman", ein Armbandwecker mit Zeitzonen-Dispositiv. Das Automatikkaliber 1241 teilte sich Blancpain mit der Schwester Breguet. Von einer Manufaktur konnte man indessen noch nicht sprechen, denn die verbauten Uhrwerke stammten weiterhin von Frédéric Piguet. Diese Feststellung gilt auch für das 2008 vorgestellte „Carrousel Volant Une Minute" mit fliegendem Minutenkarussell, dessen Mechanik eine bemerkenswerte Evolutionsstufe der Erfindung des dänischen Uhrmachers Bahne Bonniksen aus dem späten 19. Jahrhundert verkörpert. Das Kaliber 225 mit Selbstaufzug und 100 Stunden Gangautonomie besteht aus 262 Komponenten. Unter der Leitung von Marc A. Hayek, dem Enkel des großen Nicolas G. Hayek, erfolgte 2010 die Verschmelzung des Werkefabrikanten Frédéric Piguet mit Blancpain zu einer waschechten Uhrenmanufaktur. 2016 wartet das Unternehmen mit einem neuen Modell der Taucheruhrenlinie „Fifty Fathoms Bathyscaphe" auf, die auf die 1950er Jahre zurückgeht. Erstmals in ihrer Geschichte besteht das Gehäuse aus grauer Plasmakeramik. Ihre Drehlünette ist aus dem gleichen Material gefertigt. Für die Indexe findet Liquidmetal, eine verformungsstabile Legierung, Verwendung. Die leichte, kratzfeste Schale des bis 30 Bar wasserdichten Modells schützt das Automatikkaliber 1315 mit Silizium-Unruhspirale und fünf Tagen Gangautonomie. ○

Rolls, 1930

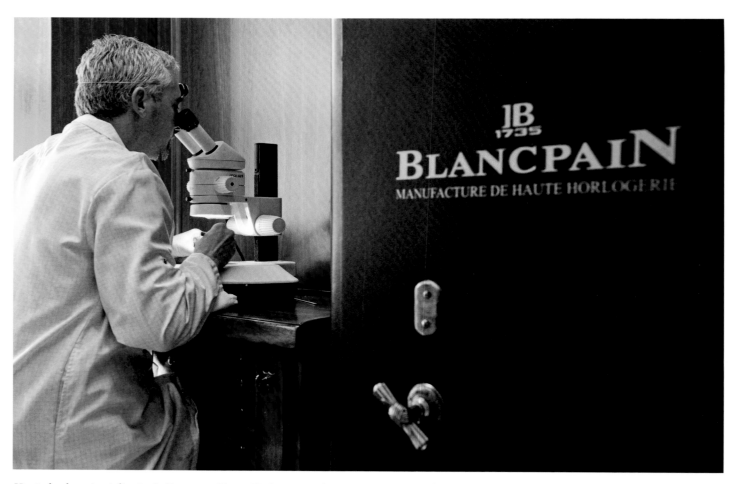

Haute horlogerie atelier in Le Brassus ○ *Haute-Horlogerie-Atelier in Le Brassus* ○ *Atelier de haute horlogerie au Brassus*

Horlogerie mécanique depuis 1735

Français

Les origines de la marque horlogère Blancpain remontent à une époque très lointaine, à savoir à l'année 1735. Jehan-Jacques Blancpain vit et travaille alors à Villeret, petit village du Jura suisse. Comme nombre de ses confrères, le jeune horloger fabrique également des composants. Les montres de poche complètes ne viendront que bien plus tard. C'est en 1815 que la petite entreprise artisanale devient une fabrique horlogère mécanisée répondant au nom de Blancpain. Il lui faudra cependant attendre le XXᵉ siècle pour se faire connaître, notamment en 1930 par le biais de son modèle « Rolls », une montre-bracelet rectangulaire équipée d'un mouvement à remontage automatique inédit. Lancée en 1953, la montre-bracelet de plongée « Fifty Fathoms » se distingue par sa lunette unidirectionnelle crantée, qui la préserve de tout dérèglement intempestif. Rien d'étonnant à ce que Jacques-Yves Cousteau et son équipe aient confié à ce garde-temps le décompte de leur temps d'immersion pendant le tournage du documentaire *Le Monde du silence*. Le début des années 1950 est placé sous le signe des mouvements à remontage manuel miniatures : le calibre carré R 59 tient dans un volume inférieur à 500 millimètres cubes, son homologue rond R 550 dans seulement 425 millimètres cubes. Ce dernier fait ses débuts en 1956 dans le modèle « Ladybird », à couronne déportée

sur le fond du boîtier. En 1983, tandis que la renaissance de la montre mécanique en est encore à ses balbutiements, Blancpain entame un nouveau chapitre de son histoire avec une montre-bracelet à phases de lune, sous l'égide de Jean-Claude Biver et de Jacques Piguet, propriétaire de la manufacture d'ébauches Frédéric Piguet. Deux ans plus tard, Blancpain crée la surprise avec une montre-bracelet ultraplate à répétition minutes, pourvue d'un tout nouveau mouvement d'une hauteur réduite à 3,2 millimètres pour un diamètre de 20,3 millimètres. Ce qui est à l'époque le plus petit chronographe au monde avec date, remontage automatique et rattrapante (en option) voit le jour en 1987. Il est suivi en 1990 par un tourbillon à indication de la date et de la réserve de marche, et qui offre 8 jours d'autonomie. Le bouquet final incontesté de ce feu d'artifice de complications est sans doute la « 1735 » présentée la même année : assemblée à partir de 740 composants, elle est considérée à l'époque comme la montre-bracelet la plus compliquée au monde. 1992 représente une autre année clé dans la longue histoire cette entreprise de tradition, qui passe dans le giron du Swatch Group. 2003 voit la sortie d'une montre-bracelet réveil dotée d'une fonction GMT, la « Léman ». Blancpain partage le calibre automatique 1241 avec son entreprise sœur Breguet.

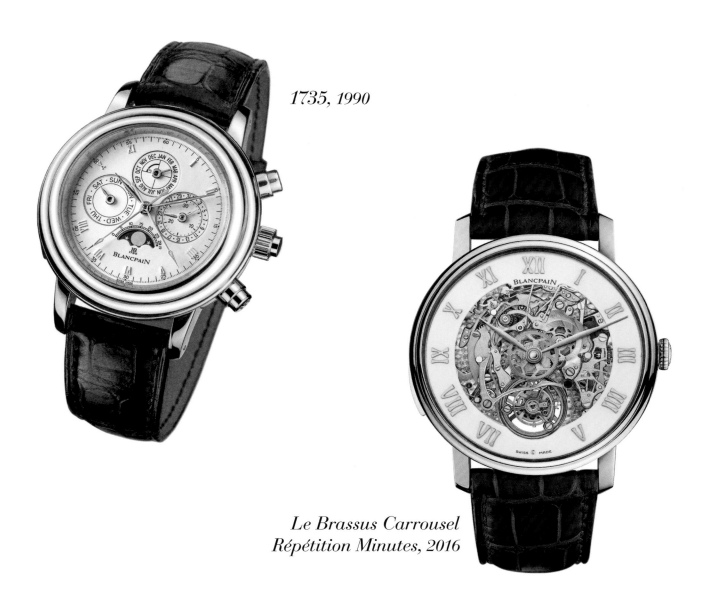

1735, 1990

*Le Brassus Carrousel
Répétition Minutes, 2016*

La société ne peut cependant toujours pas prétendre au titre de manufacture car elle continue d'équiper ses garde-temps de mouvements Frédéric Piguet. Ce sera encore le cas en 2008 du « Carrousel Volant Une Minute » qui, effectuant une rotation complète en une minute, incarne un perfectionnement remarquable de l'invention de l'horloger danois Bahne Bonniksen à la fin du XIXᵉ siècle. Son calibre 225 à remontage automatique affichant 100 heures de réserve de marche est constitué de 262 composants. Sous la direction de Marc A. Hayek, petit-fils du célèbre Nicolas G. Hayek, Blancpain fusionne en 2010 avec le fabricant de mouvements Frédéric Piguet et devient une manufacture horlogère à part entière. En 2016, l'entreprise sort une nouvelle montre de plongée, la « Fifty Fathoms Bathyscaphe ». Pour la première fois dans l'histoire cette collection qui remonte aux années 1950, la boîte et la lunette sont toutes deux en céramique plasma grise. Les index sont en Liquidmetal, un alliage à l'épreuve des déformations. Une boîte légère, résistante aux éraflures et étanche jusqu'à une pression de 30 bars protège le calibre automatique 1315 équipé d'un spiral en silicium et offrant cinq jours d'autonomie. ◦

Cal. 1735

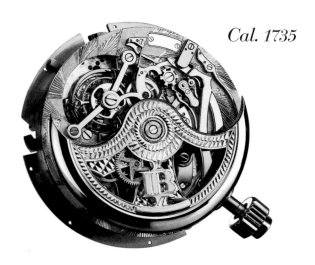

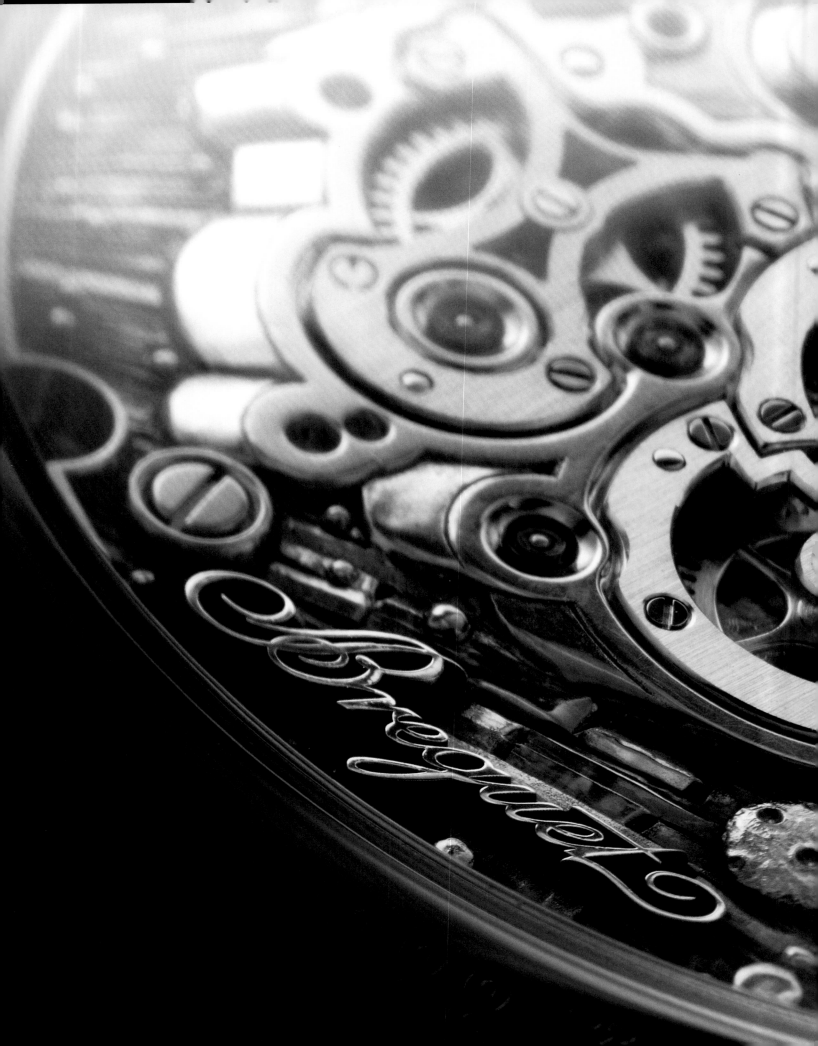

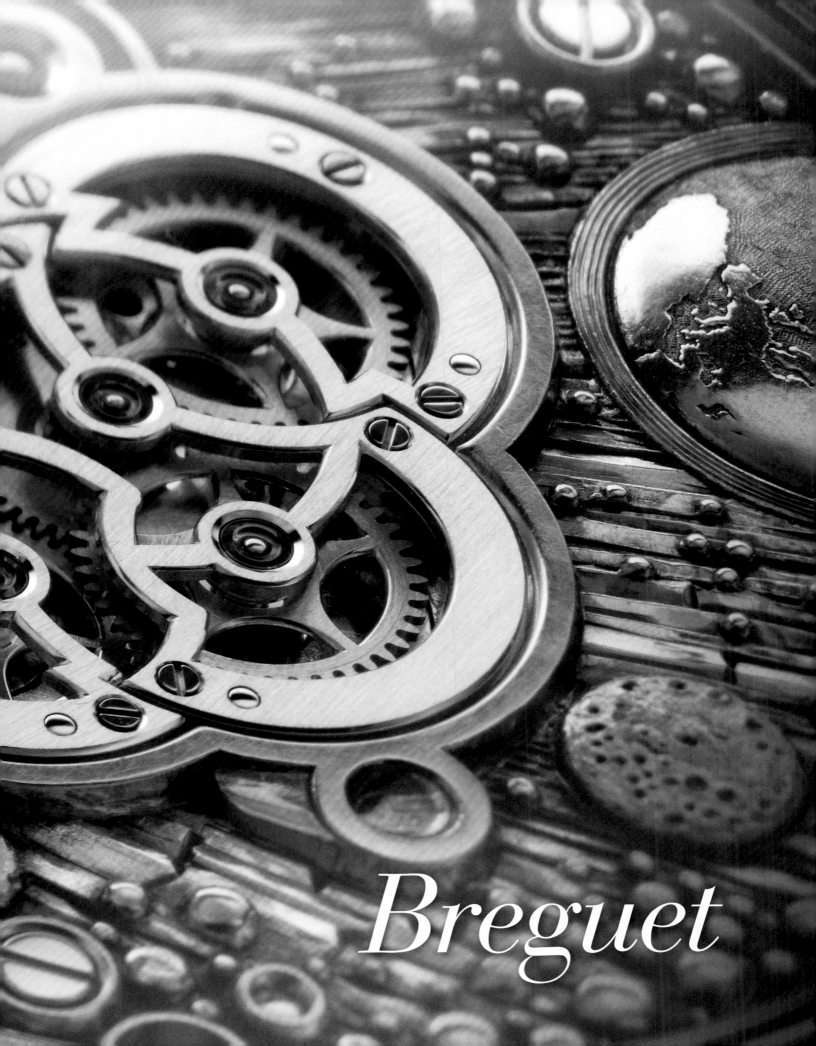

Breguet

Masterful Watchmaking since 1775

Abraham-Louis Breguet could justifiably be described as perhaps the most important watchmaker of all times. It would be no exaggeration to claim that this ingenious inventor, who was born in Neuchâtel in 1747, compressed two centuries of the watchmaker's art into a mere 50 years. After having learned the watchmaker's craft in Versailles, he studied mathematics alongside his career as a watchmaker. He married a woman from a prosperous Parisian bourgeois family and then opened his own shop, which catered to an illustrious clientele. Breguet's mathematics professor had facilitated the young man's access to the imperial court, so King Louis XVI immediately ordered several watches. This commission and the quality of the delivered timepieces attracted other noble clients, e.g., the Duke of Orleans, Queen Marie-Antoinette, the Queen of Tuscany, Prince Talleyrand, the Prince of Wales, the King of Spain, Czar Alexander, King George III of England and Hanover, Alexander von Humboldt, and King Friedrich Wilhelm II of Prussia.

Breguet's inventiveness and his ability to transform creative ideas into horological realities were inexhaustible. The watch world can thank this grand master for inventing: the so-called "Parachute," a shock-absorption system to protect the slender pivots of the balance's staff; the "Perpétuelle," a self-winding pocket-watch with a circa 60-hour power reserve; a precursor of today's chronographs; a gong for watches with repeater striking mechanisms; the "Montre Souscription," a partially prepaid and thus less costly serially produced watch that prevented idleness in his workshops; the "Pendule Sympathique," into which a pocket-watch could be inserted overnight, where the timepiece would be automatically wound and precisely set; and a balance-spring with an upwardly arcing terminal curvature, now usually known as the "Breguet balance-spring." His most important invention, which he patented in 1801, is the tourbillon, which compensates for the ill effects exerted by gravity on the accuracy of the rate of mechanical watches.

The words "Breguet et fils" appeared on some of his watches in 1801, confirming that his son Antoine-Louis had joined the business. Abraham-Louis Breguet surrendered the firm's leadership to his son in 1807, but he continued to work in the background. The situation at Breguet declined steadily after Abraham-Louis Breguet's death in 1823. Not until 1832, a year in which Antoine-Louis Breguet sold fewer than 50 watches, did the business's director finally acknowledge the extraordinary horological talents of his son Louis-François-Clément. Under the name Breguet Neveu & Cie, the founder's grandson re-established the business's erstwhile reputation.

For example, he instituted the parallel production of standardized and individualized watches. He also achieved a successful comeback for the "Pendule Sympathique." But despite all these successes, Abraham-Louis Breguet's grandson increasingly turned his back on classical watchmaking. His real passions were the modern telegraph, technical apparatuses, and scientist instruments. His son Antoine, who was born in 1851, shared these interests. This is why the watch department was sold to the experienced workshop master Edward Brown on May 8, 1870. Four generations of the Brown family led the traditional brand through its next 100 years. Like its founder, they primarily targeted the international upper class as customers for their elite products. Georges Brown, who directed Breguet from 1927 to 1970, also expanded the label's spectrum of aeronautic instruments, which include the legendary "Type XX" wristwatch chronograph with flyback function for naval pilots and the French air force.

After a long career, Georges Brown sold the venerable business to the brothers Jacques and Pierre Chaumet, who were heirs to an old Parisian jeweler dynasty. This episode in the firm's history likewise came to its end in 1987, when the brand was sold to the Investors Corporation, which relocated the label's headquarters to the country where Abraham-Louis Breguet was born. Breguet watches have been manufactured since 1987 in the remote Vallée de Joux, where complications for mechanical movements have traditionally been fabricated. Breguet came under the aegis of the Swatch Group in 1999. Its late president Nicolas G. Hayek fully agreed with the opinion which the great collector Sir David Lionel Salomons had expressed in 1921: "For someone who understands mechanisms, a Breguet watch is truly a painting."

In this sense, the Breguet brand and manufactory could continue in their founder's style and with his innovative dynamism. The spirit of Abraham-Louis Breguet still thrives in the successful "Tradition" watch line. And with a friction-free bearing for a balance's staff in the force field of a permanent magnet, the Breguet brand again proved that mechanical watchmaking can also offer delightful surprises in the 21st century. ◦

Ring watch with alarm, 1836
Ringuhr mit Wecker, 1836
Montre-bague dotée d'une fonction réveil, 1836

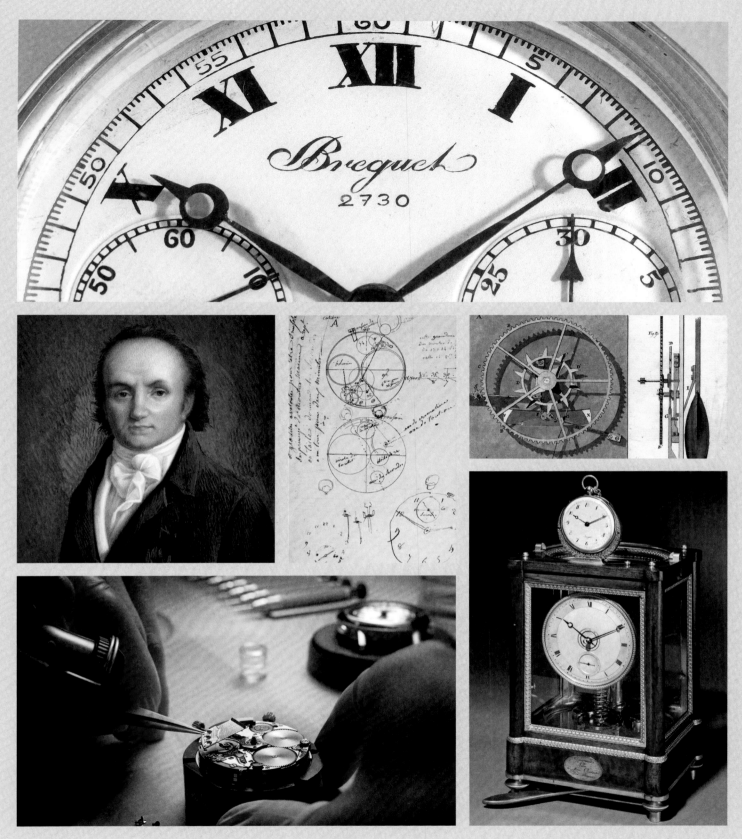

Top: *chronograph with 30-minute counter, N° 2730, 1930s* ○ Middle: *Abraham-Louis Breguet (1747–1823)* ○ *Diagrams from his sketchbook* ○ *Watercolor for patent of the "escapement with constant force," 1798* ○ Bottom: *the assembly process* ○ *Pendule Sympathique with pocketwatch N° 721 from 1814, owned by Queen Elizabeth II* ○ Oben: *Chronograph mit 30-Minuten-Zähler, N° 2730, 1930er Jahre* ○ Mitte: *Abraham-Louis Breguet (1747–1823)* ○ *Zeichnungen aus seinem Skizzenbuch* ○ *Aquarell zum Patent der „Hemmung mit konstanter Kraft",* *1798* ○ Unten: *Bei der Montage* ○ *Pendule Sympathique mit der Taschenuhr N° 721 von 1814, aus dem Besitz von Königin Elisabeth II.* ○ En haut : *Chronographe à compteur 30 minutes, N° 2730, années 1930* ○ Au milieu : *Abraham-Louis Breguet (1747–1823)* ○ *Dessins de son cahier d'atelier* ○ *Dessin aquarellé à l'occasion du brevet de l'« échappement à force constante », 1798* ○ En bas : *Assemblage* ○ *Pendule sympathique avec la montre de poche N° 721 de 1814, propriété de la reine Élisabeth II*

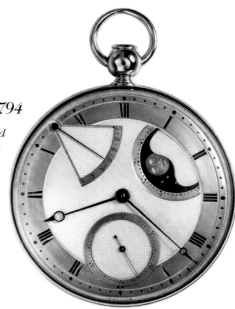

Tourbillon N° 1188, 1808

for the Ottoman market
für den osmanischen Markt
pour le marché ottoman

Breguet N° 1105, 1794

with automatic movement and
quarter-hour tapping repeater
*mit automatischem Werk und
Viertelstunden-Klopfrepetition*
avec mouvement automatique,
à répétition des quarts à toc

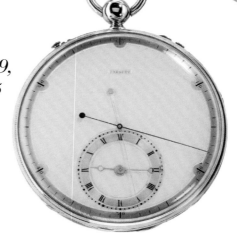

Breguet N° 4009,
observation chronometer, 1825

with chronograph rattrapante
mit Chronograph-Rattrapante
avec chronographe à double seconde

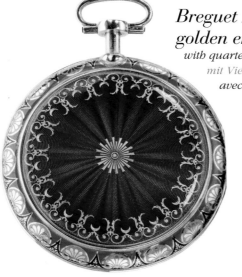

Breguet N° 2919,
golden enamel watch, 1804

with quarter-hour repeater for the Turkish market
mit Viertelrepetition für den türkischen Markt
avec répétition des quarts, pour le marché turc

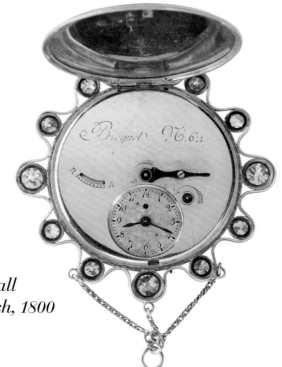

Breguet N° 611, small
medallion tact watch, 1800

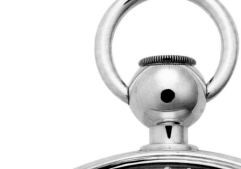

Breguet N° 2585, 1811

with 7½-minute repeater and engraved, enameled map of nine Italian provinces

mit 7½-Minuten-Repetition und gravierter, emaillierter Landkarte von neun italienischen Provinzen

avec répétition des demi-quarts, représentant au dos la carte gravée du Piémont et du Milanais, nouvelles provinces de l'Italie

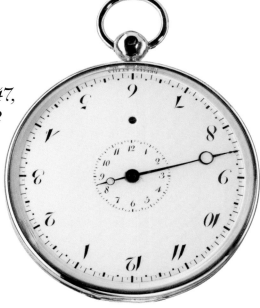

Breguet N° 947, subscription watch, 1802

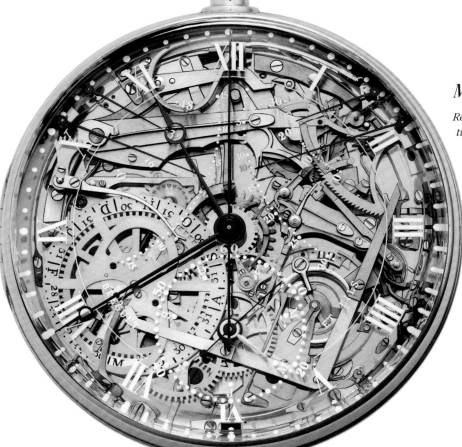

Marie-Antoinette, N° 1160

Replica (2005–2008) of the most complicated timepiece ever built by Breguet. The original is kept in the L.A. Mayer Institute for Islamic Art in Jerusalem.

Neuinterpretation (2005–2008) der kompliziertesten Uhr, die Breguet jemals gebaut hat. Das Original liegt im L.A. Mayer Institute for Islamic Art in Jerusalem.

Réplique (2005–2008) de la montre la plus compliquée jamais construite par Breguet. L'original est conservé au L.A. Mayer Memorial Institute for Islamic Art de Jérusalem.

Meisterliche Uhrmacherei seit 1775

Wer Abraham-Louis Breguet den wohl bedeutendsten Uhrmacher aller Zeiten nennt, liegt mit Sicherheit nicht falsch. Auch die Behauptung, dass der 1747 in Neuchâtel geborene Erfinder die Entwicklung von zwei Jahrhunderten Uhrmacherkunst in nur 50 Jahren vollzogen habe, trifft den Kern der Dinge. Seine Ausbildung zum Uhrmacher absolvierte Breguet in Versailles. Später studierte er neben dem Beruf auch noch Mathematik. Auf die Heirat einer gutsituierten Pariser Bürgerstochter folgte die Eröffnung eines eigenen Geschäfts mit renommierter Klientel: Breguets Mathematikprofessor hatte ihm den Weg zum kaiserlichen Hof geebnet. Ludwig XVI. orderte gleich mehrere Uhren. Das und die gelieferte Qualität brachten weitere hochrangige Kunden ins Haus wie den Herzog von Orleans, Königin Marie-Antoinette, die Königin der Toskana, Fürst Talleyrand, den Prinzen von Wales, den König von Spanien, Zar Alexander, König Georg III. von England und Hannover, Alexander von Humboldt oder König Friedrich Wilhelm II. von Preußen.

Schier unerschöpflich waren Breguets Einfallsreichtum und seine Fähigkeit, das kreative Gedankengut uhrmacherisch zu realisieren. Dem großen Meister seines Fachs verdankt die Uhrenwelt unter anderem die Stoßsicherung für die dünnen Zapfen der Unruhwelle, „Parachute" (Fallschirm) genannt; die „Perpétuelle", eine Taschenuhr mit Selbstaufzug und rund 60 Stunden Gangautonomie; einen Vorläufer des heutigen Chronographen; die Tonfeder für Uhren mit Repetitionsschlagwerk; die „Montre Souscription", eine teilweise vorausbezahlte und deshalb preisgünstigere Serienuhr zur kontinuierlicheren Auslastung der Werkstätten; die „Pendule Sympathique", welche die abendlich eingelegte Taschenuhr über Nacht selbsttätig aufzieht und exakt stellt; die Unruhspirale mit hochgebogener Endkurve, heute vielfach Breguet-Spirale genannt. Die wohl bedeutendste Erfindung ist und bleibt jedoch zweifellos das 1801 patentierte Tourbillon zur Kompensation negativer Schwerkrafteinflüsse auf Ganggenauigkeit mechanischer Präzisionsuhren.

Im gleichen Jahr verwies die Signatur „Breguet et fils" auf manchen Uhren auf den Eintritt des Sohnes Antoine-Louis. Ab 1807 lag die Firmenleitung zwar komplett in seinen Händen, aber der Vater wirkte im Hintergrund weiterhin tatkräftig mit. Nach dem Tod von Abraham-Louis im Jahr 1823 ging es bei Breguet kontinuierlich bergab. Erst 1832, als Antoine-Louis Breguet keine 50 Uhren mehr verkaufte, erkannte und akzeptierte er die außergewöhnliche uhrmacherische Begabung seines eigenen Sohnes Louis-François-Clément. Unter dem Namen Breguet Neveu & Cie gelang es dem Enkel des Firmengründers, das einstige Renommee wiederherzustellen. Er führte beispielsweise die parallele Produktion standardisierter und individueller Uhren ein. Und er bescherte der „sympathischen Pendule" ein erfolgreiches Comeback. Doch trotz aller Erfolge kehrte der Enkel der klassischen Uhrmacherei zunehmend den Rücken. Seine eigentliche Leidenschaft galt modernen Telegrafen, technischen Apparaten und wissenschaftlichen Instrumenten. Diese Interessen teilte auch sein 1851 geborener Sohn Antoine. Daher erhielt die Uhrensparte am 8. Mai 1870 mit dem versierten Werkstattmeister Edward Brown einen neuen Eigentümer. Während der folgenden 100 Jahre übten sich nacheinander vier Generationen dieses Namens in der Leitung des Traditionsunternehmens. Wie schon der Gründer wandten sie sich mit ihren hochkarätigen Produkten hauptsächlich an die internationale Upperclass. Daneben weitete Georges Brown, der von 1927 bis 1970 die Breguet-Geschicke lenkte, das Spektrum an aeronautischen Instrumenten konsequent aus. Dazu gehörte auch der legendäre Armband-Chronograph „Type XX" mit Flyback-Funktion für die Marineflieger und die französische Luftwaffe.

Am Ende eines langen Berufslebens verkaufte Georges Brown das alteingesessene Unternehmen an das Pariser Brüderpaar Jacques und Pierre Chaumet, Erben einer alten Juweliersdynastie. 1987 endete auch diese Episode in der Firmengeschichte. Fortan hieß die Eigentümerin Investors Corporation. Sie verlegte den Sitz ins Geburtsland

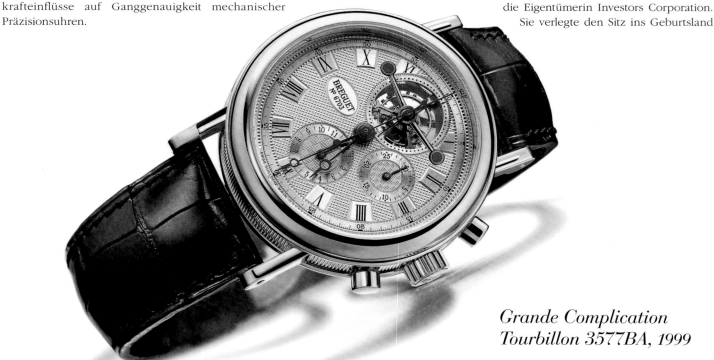

Grande Complication
Tourbillon 3577BA, 1999

des Abraham-Louis Breguet, in das abgeschiedene Vallée de Joux, wo traditionsgemäß Komplikationen für mechanische Uhrwerke entstehen. Im Herbst 1999 gelangte Breguet unter das Dach der Swatch Group. Ihr inzwischen verstorbener Präsident Nicolas G. Hayek konnte bestens nachempfinden, was der große Sammler Sir David Lionel Salomons 1921 formuliert hatte: „Für jemanden, der sich auf Mechanik versteht, ist eine Breguet-Uhr tatsächlich ein Gemälde."

In diesem Sinne konnten Marke und Manufaktur Breguet an die einstige Gestaltungs- und Innovationskraft anknüpfen. In der erfolgreichen Uhrenlinie „Tradition" lebt der Geist von Abraham-Louis Breguet ausdrucksstark fort. Und durch die reibungslose Lagerung der Unruhwelle im Kraftfeld eines Permanentmagneten hat Breguet bewiesen, dass die mechanische Uhrmacherei auch im 21. Jahrhundert für echte Überraschungen gut ist. ○

Breguet N° 2516, platinum wristwatch, 1933

with perpetual calendar and moon phases
mit ewigem Kalender und Mondphasen
avec quantième perpétuel et phase lunaire

Breguet N° 2926, platinum self-winding watch, 1933

with date and power-reserve indicator
mit Datum und Gangreserveanzeige
avec date et affichage de la rèserve de marche

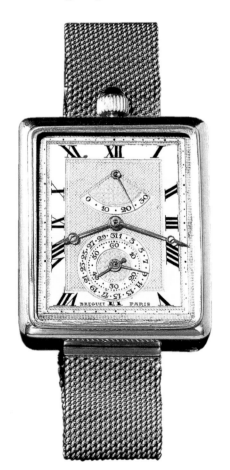

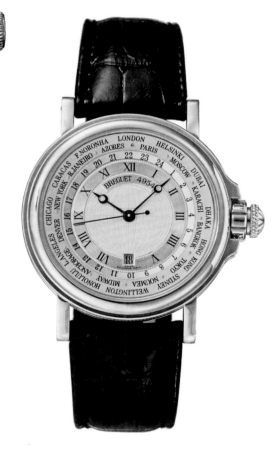

Marine Hora Mundi 3700, 1997

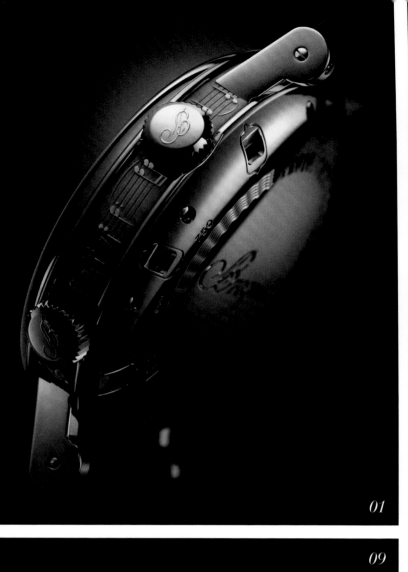

01

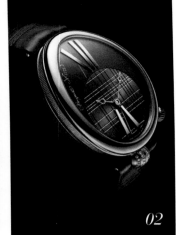

02

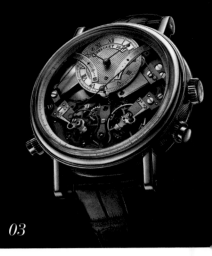

03

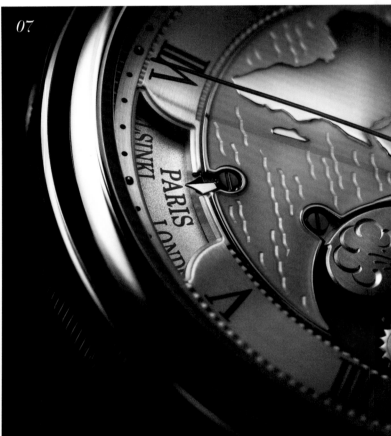

07

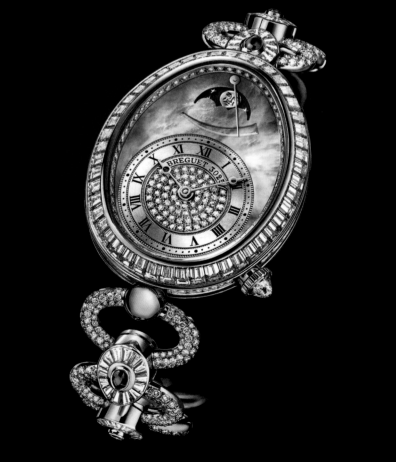

09

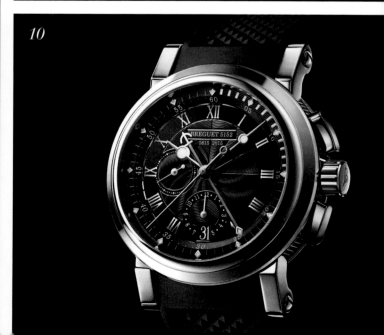

10

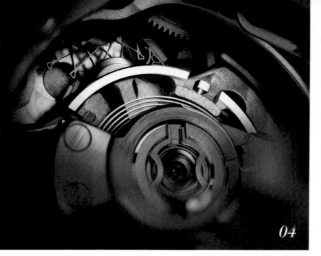

04

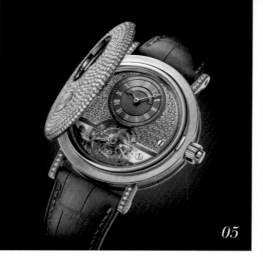

05

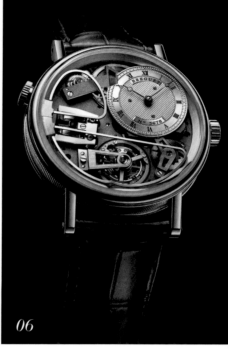

06

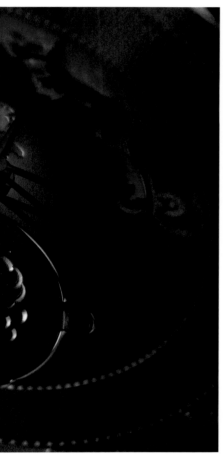

08

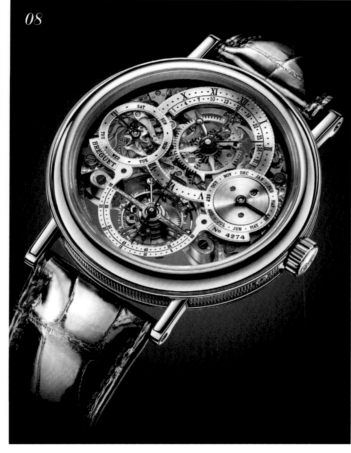

01. Grande Complication
    Réveil Musical 7800BA, 2010
02. Reine de Naples Mini 8968, 2015
03./04. Tradition 7077, 2014
05. Classique Grandes Complications
    Tourbillon, 2005
06. Tradition Minute Repeater
    Tourbillon, 2015
07. Hora Mundi (universal time), 2011
08. Classique Grandes Complications,
    2005
09. Reine de Naples, 2015
10. Marine Chronograph, 2015
11. Tradition 7097, 2015
12. Reine de Naples Mini 9808, 2015

11

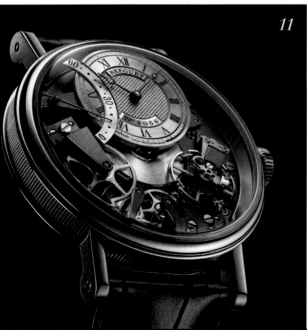

12

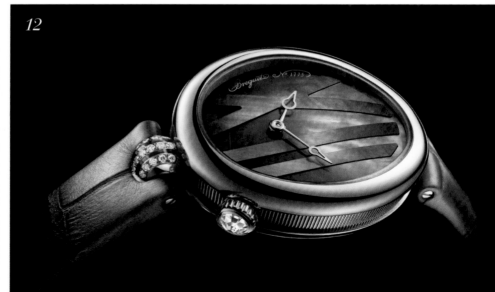

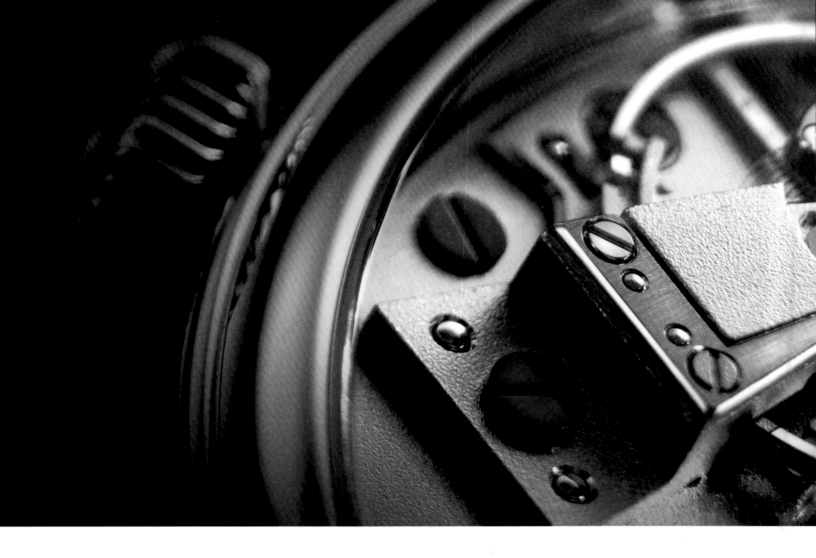

Horlogerie d'exception depuis 1775

Français

Ceux qui voient en Abraham-Louis Breguet le maître horloger le plus marquant de tous les temps n'ont certainement pas tort. Et ceux qui affirment que ce inventeur né à Neuchâtel en 1747 n'a mis que 50 ans à faire progresser l'art horloger de l'équivalent de deux siècles ne sont pas loin de la vérité. Breguet suit une formation d'horloger à Versailles. Plus tard, parallèlement à son activité professionnelle, il étudiera les mathématiques. Après son mariage avec la fille d'un bourgeois fortuné, il crée sa propre affaire avec une clientèle renommée, son professeur de mathématiques l'ayant introduit à la cour de France. Louis XVI lui commandera plusieurs montres à la fois. Ce fait et la qualité de ses réalisations lui valent d'autres clients de haut rang, notamment le duc d'Orléans, la reine Marie-Antoinette, la reine de Toscane, le prince Talleyrand, le prince de Galles, le roi d'Espagne, le tsar Alexandre, le roi George III d'Angleterre et de Hanovre, Alexander von Humboldt et le roi de Prusse, Frédéric-Guillaume II.

L'ingéniosité de Breguet et sa capacité à convertir ses idées créatives dans le domaine horloger sont quasiment inépuisables. Le monde de l'horlogerie doit à ce grand maître de la spécialité de multiples inventions : le système protégeant des chocs le délicat pivot de l'axe de balancier, appelé « pare-chute » ; la « Perpétuelle », une montre de poche à remontage automatique disposant d'une soixantaine d'heures de réserve de marche ; un précurseur de nos chronographes actuels ; le timbre des montres à répétition ; la « montre de souscription », une montre de série, et donc plus économique, prépayée en partie, un concept imaginé pour assurer la continuité du travail dans les ateliers ; la « pendule sympathique », qui remonte et règle automatiquement pendant la nuit la montre de poche que l'on y a déposée le soir ; le spiral à spire conique, aujourd'hui souvent appelé spiral Breguet. L'invention sans conteste la plus importante demeure le tourbillon. Breveté en 1801, ce régulateur compense l'influence de la gravité sur la régularité des mouvements d'horlogerie des montres mécaniques de précision.

La même année, la signature « Breguet et fils » sur certaines montres témoigne de l'arrivée dans l'entreprise du fils d'Abraham-Louis, Antoine-Louis. Si celui-ci assure seul la direction dès 1807, son père continue d'intervenir activement en coulisses. Après le décès du père en 1823, la manufacture ne cesse de décliner. Il faut attendre 1832, alors qu'il ne vend même plus 50 montres par an, pour qu'Antoine-Louis Breguet reconnaisse et accepte les talents

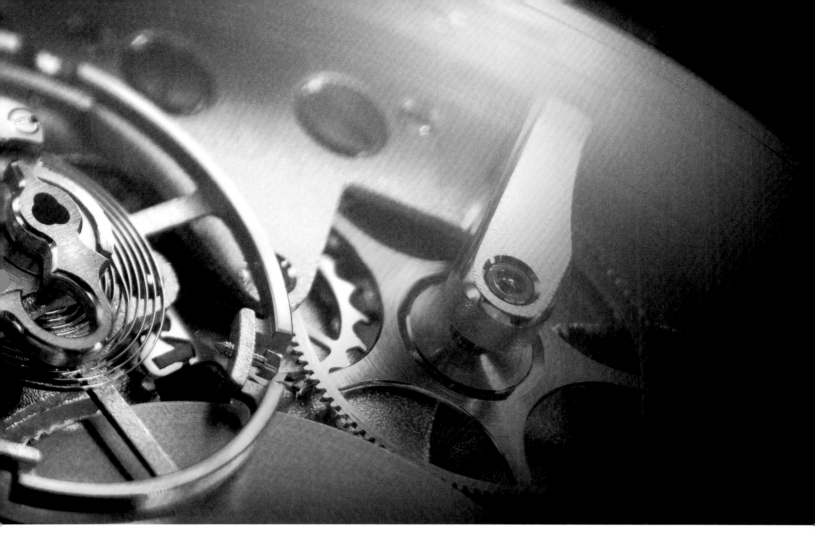

*La Tradition. The balance with screws along the inside for finely adjusting the rate, parachute for shock absorption, and the famous Breguet balance-spring ○ La Tradition. Die Unruh mit innenliegenden Schrauben zur Regulierung des Gangs, Parachute (Stoßsicherung) und der berühmten Breguet-Spirale ○ **La Tradition.** Balancier avec vis à l'intérieur pour le réglage, parachute (protection contre les chocs) et célèbre spiral Breguet*

d'horloger exceptionnels de son propre fils, Louis-François-Clément. Sous le nom Breguet Neveu & Cie, le petit-fils du fondateur réussit à redorer le blason de la manufacture, introduisant par exemple la production en parallèle de montres de série et de montres personnalisées. En outre, il signe un retour réussi pour la pendule sympathique. Malgré tous ses succès, il se détourne de plus en plus de l'horlogerie traditionnelle. Il se passionne en réalité pour les télégraphes modernes, les accessoires techniques et les instruments scientifiques, des centres d'intérêt que partage son fils Antoine, né en 1851. Aussi, la branche horlogère est-elle cédée le 8 mai 1870 à Edward Brown, maître d'atelier expérimenté. Au cours des 100 années qui suivent, quatre générations de la famille Brown se succèdent à la direction de cette entreprise de tradition. Comme son fondateur, elles s'adressent avec leurs réalisations de prestige essentiellement à la haute société internationale. Parallèlement, Georges Brown, qui préside aux destinées de Breguet de 1927 à 1970, élargit résolument la gamme des produits proposés aux instruments destinés à l'aéronautique. Parmi eux figure la légendaire montre-bracelet chronographe « Type XX » avec fonction de retour-en-vol pour l'Aéronavale et l'armée de l'Air française.

Au terme d'une longue carrière, Georges Brown vend l'entreprise solidement implantée aux frères Jacques et Pierre Chaumet, héritiers d'une longue dynastie de joailliers. En 1987, cet épisode dans l'histoire de la société prend aussi fin : la manufacture est reprise par Investors Corporation. Son siège est transféré dans la région qui a vu naître Abraham-Louis Breguet, la paisible vallée de Joux, où l'on fabrique dans le respect de la tradition des complications pour des mouvements mécaniques. À l'automne 1999, Breguet entre dans le giron de Swatch Group. Nicolas G. Hayek, son président, décédé depuis, a mieux que tout autre ressenti ce que le grand collectionneur Sir David Lionel Salomons avait déclaré en 1921, à savoir que pour qui s'y connaissait en mécanique, une montre Breguet avait tout d'un tableau de maître.

C'est dans ce sens que la marque et la manufacture Breguet ont pu renouer avec la capacité de création et d'innovation de jadis. L'esprit d'Abraham-Louis Breguet se perpétue avec beaucoup de force dans la ligne de montres à succès « Tradition ». En plaçant l'axe de balancier dans le champ de force d'un aimant permanent, ce qui permet d'éliminer les frottements, Breguet a démontré que l'horlogerie mécanique pouvait encore réserver de réelles surprises au XXIe siècle. ○

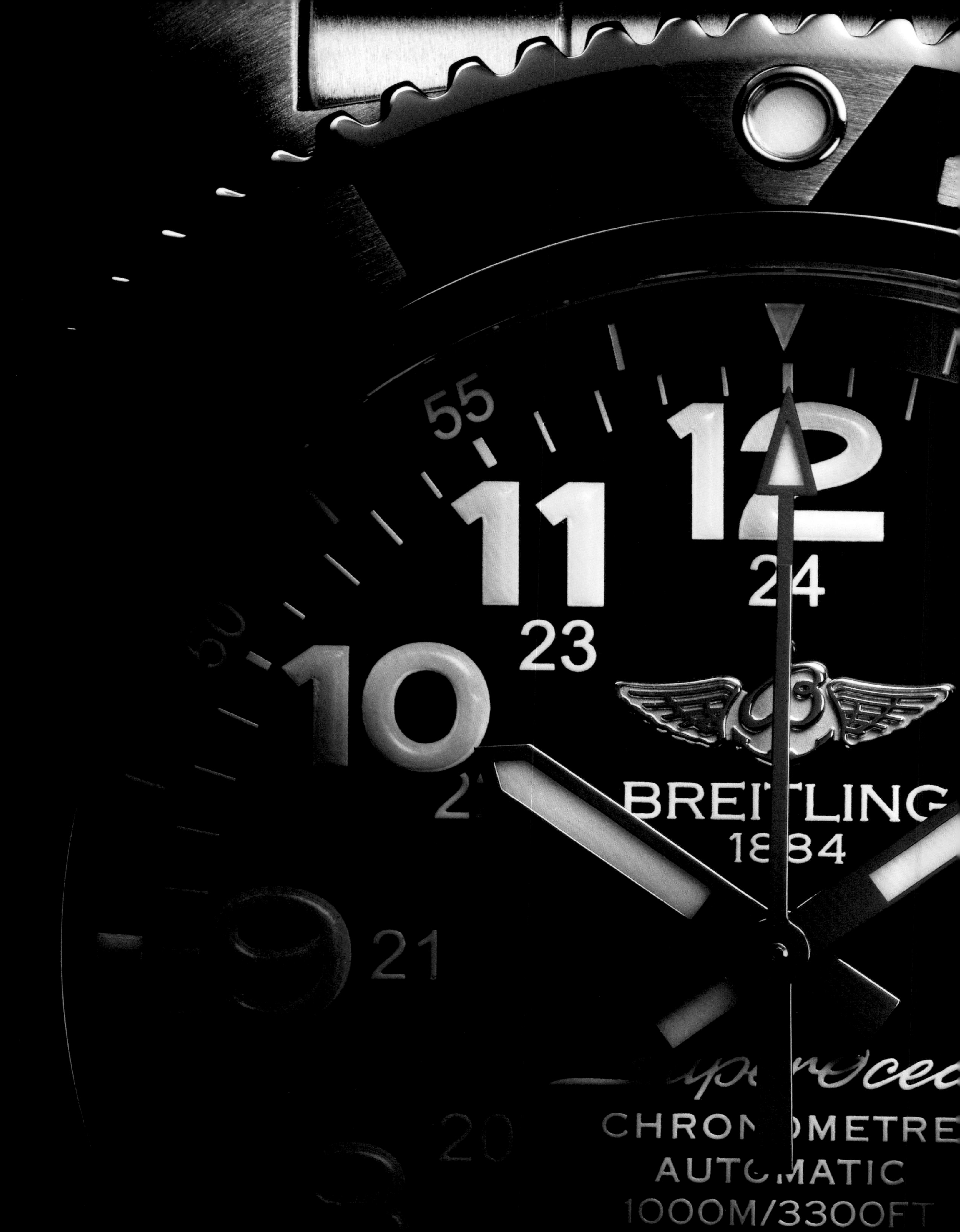
BREITLING
1884
CHRONOMETRE
AUTOMATIC
1000M/3300FT

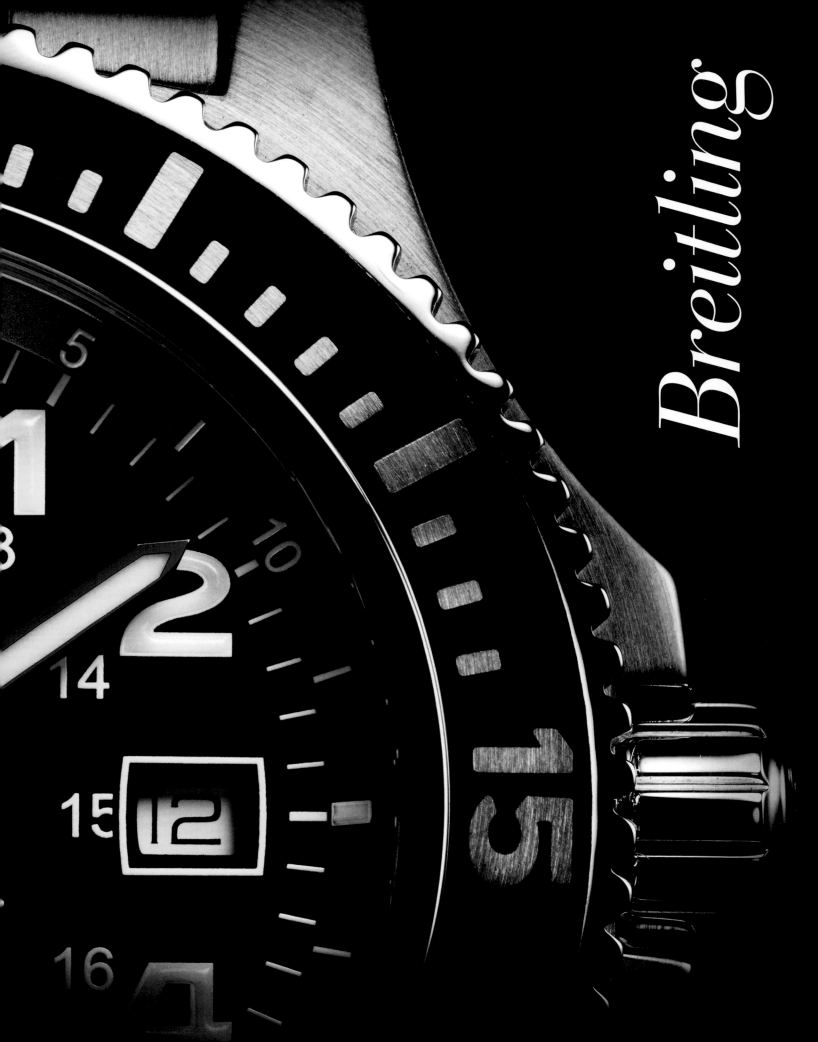

The Long Path toward Becoming a Chronograph Manufactory

Watch lovers almost always think of chronographs when they hear the name Breitling. This is by no means accidental: It's deeply rooted in the brand's history. Léon Breitling put his name on the dials of his first counter chronographs in 1884. Intelligent measuring tools for watchmakers also bore his signature, thus attracting gazes to Breitling. The firm's founder still worked at this time in his Swiss hometown of Saint-Imier. He established the G. Léon Breitling SA, Montbrillant Watch Manufactory in 1892 at La Chaux-de-Fonds in the Jura region, where his watchmakers focused on the task of measuring brief intervals. Léon Breitling had bet his money on the right racehorse!

His son Gaston joined the family business in 1914. Gaston Breitling had inherited his passion for chronographs from his father, who died that same year. Gaston unveiled one of the first wristwatch chronographs with a button alongside its crown in 1915. A patented special model named "Vitesse" helped policemen measure the speeds of motorized vehicles. In accord with the current trend, the ambitious businessman prioritized the development and production of wristwatch chronographs. His death in 1927 left a gap that was difficult to fill. Not until 1932 would the helm finally come into the competent hands of Willy Breitling, a third-generation family member with a thorough managerial and technical education. With him at its head, the Breitling firm debuted an impressive series of innovations. A patent was granted in 1934 to protect wristwatch chronographs with two separate buttons: one push-piece started and stopped the stopwatch, the other triggered its hands to return to their zero positions. A chronograph for airplane cockpits followed in 1936. A model with an innovative 12-hour counter based on a Valjoux ébauche was introduced in 1938. The legendary "Chronomat" with circular slide rule followed in 1941. The "Navitimer" with optimized calculating capabilities debuted in 1952. And a Breitling wristwatch accompanied the American astronaut Scott Carpenter into outer space in 1962.

Breitling numbered among the international pioneers of self-winding chronographs in 1969. The family business marched into a new era with the "Chrono-Matic" model. This watch wound itself via a microrotor, so its seldom-needed crown could be repositioned to the left-hand flank of the case.

Then came the mid 1970s, when the turbulent waves of the Quartz Revolution threatened to drown the new movement and the brand that had created it. Twenty-four employees lost their jobs in 1978. Breitling completely suspended operations one year later. As a grand seignior of the old school, the ailing Willy Breitling opted for an orderly retreat. He sold the well-established names "Breitling" and "Navitimer" to the engineer Ernest Schneider. After it was registered as Breitling Montres SA on November 30, 1982 and its headquarters were relocated to Grenchen in the Aare Valley, the business rose and soared to unprecedented heights. Under the aegis of Ernest's son Théodore, all movements bearing the traditional signature first had to pass the official chronometer tests before they were encased. Punctually in time for the brand's 125th anniversary, Breitling presented the self-winding chronograph B01 and thus attained the status of a full-fledged manufactory. The spectrum of Breitling's movements has grown with each passing year—and no end is in sight. ○

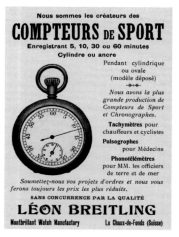

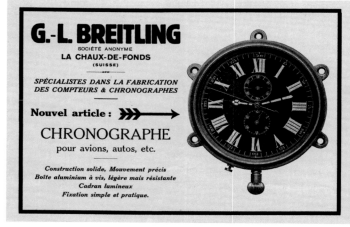

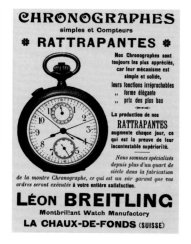

Top: Scott Carpenter in Breitling's catalogue from 1993 ○ Middle: Transocean Chronograph, 1915 ○ Bottom: advertisements from 1910, 1931, and 1912 ○ Oben: Scott Carpenter im Breitlingkatalog 1993 ○ Mitte: Transocean Chronograph, 1915 ○ Unten: Inserate von 1910, 1931 und 1912 ○ En haut : Scott Carpenter dans le catalogue Breitling de 1993 ○ Au milieu : Transocean Chronograph, 1915 ○ En bas : Petites annonces de 1910, 1931 et 1912

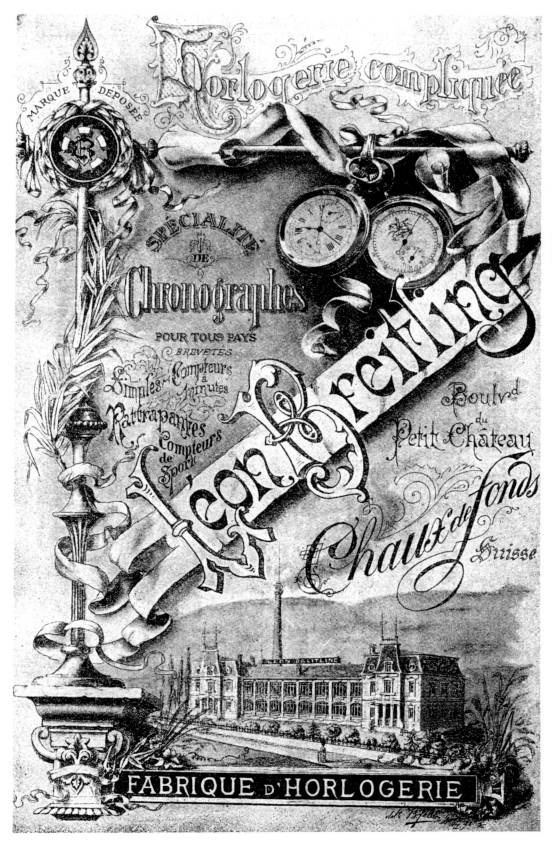

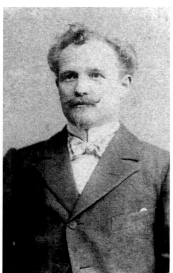

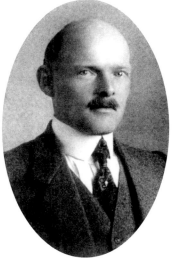

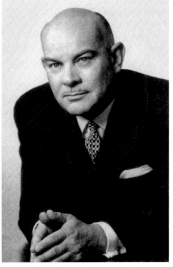

Advertisement from 1894 ◦ *Anzeige aus dem Jahr 1894* ◦ *Affiche de 1894*

Léon, Gaston, and Willy Breitling

Clockwise from top left: advertising from 1968 with Raquel Welch wearing a Co-Pilot, from the film Fathom ◦ *Quadra Chronograph Premier, 1946 ◦ Léon Breitling established his workshop in the leftward part of the building in 1884 ◦ Advertisement, 1940* ◦ *Von oben links im Uhrzeigersinn: Anzeige mit Raquel Welch mit einer Co-Pilot, 1968, aus dem Film* Fathom ◦ *Quadra Chronograph Premier, 1946 ◦ Im linken Gebäudeteil gründete Léon Breitling 1884 seine Werkstatt ◦ Werbung, 1940* ◦ *Dans le sens horaire depuis la gauche : Publicité pour le film* Une fille nommée Fathom, *représentant Raquel Welch avec une Co-Pilot, 1968 ◦ Quadra Chronograph Premier, 1946 ◦ En 1884, Léon Breitling crée son atelier dans l'aile gauche de ce bâtiment ◦ Annonce, 1940*

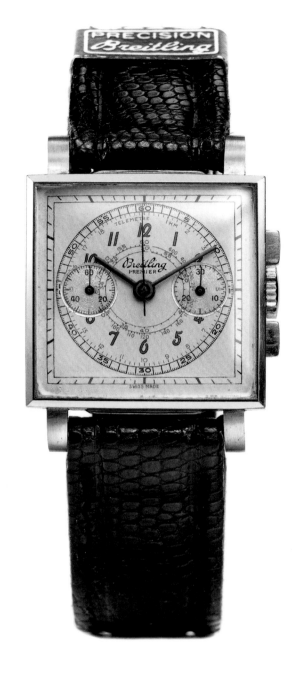

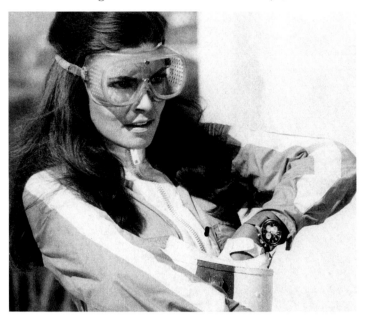

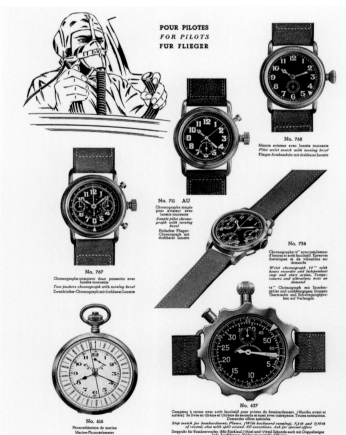

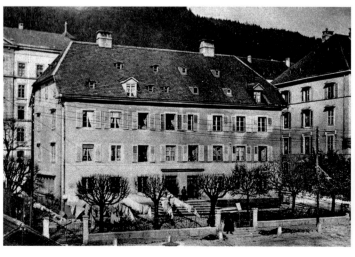

Chronomatic, 1965

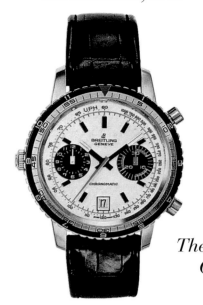

Aerospace, 1985

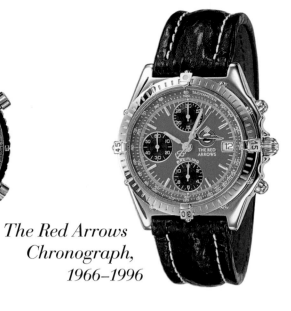

*The Red Arrows
Chronograph,
1966–1996*

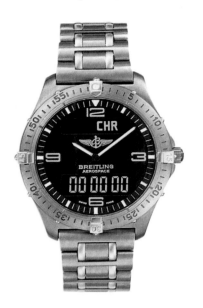

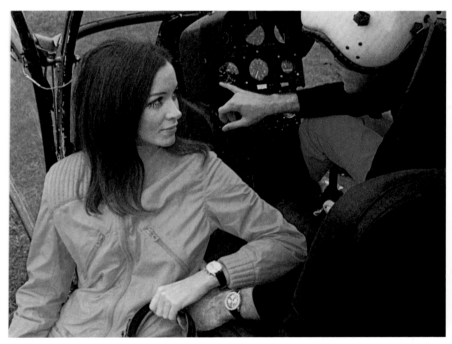

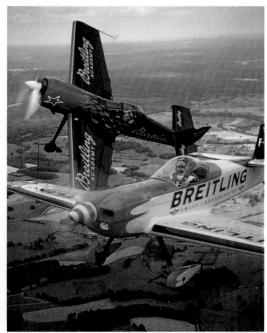

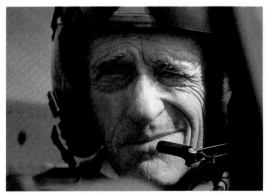

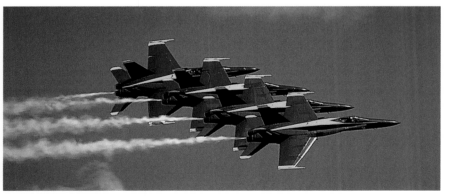

Bottom left: Ernest Schneider († 2015) bought the Breitling brand in 1979 and established Montres Breitling SA in 1982 ◦
Unten links: Ernest Schneider († 2015) kaufte 1979 die Marke Breitling und gründete 1982 die Montres Breitling S.A. ◦
En bas à gauche : Ernest Schneider († 2015) acquiert en 1979 la marque Breitling et fonde en 1982 la société Montres Breitling SA

B55 Connected, 2015

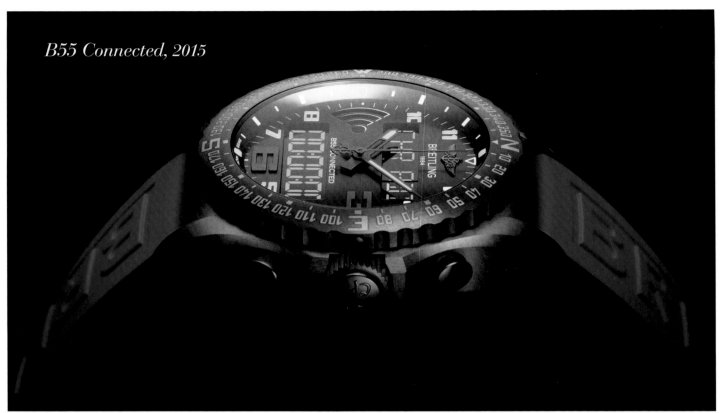

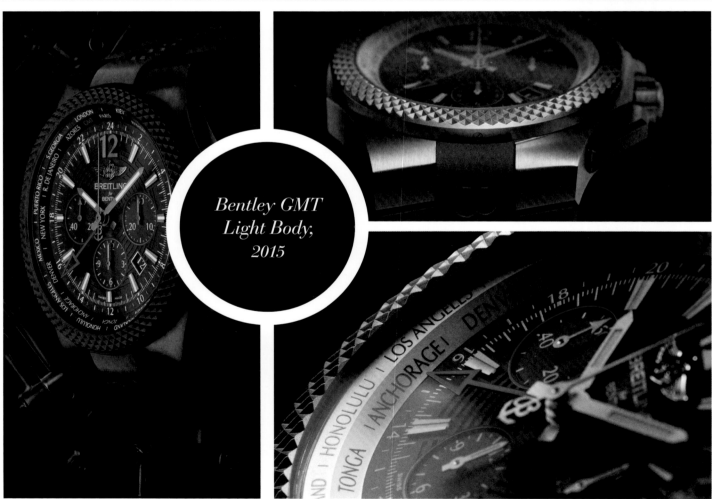

Bentley GMT Light Body, 2015

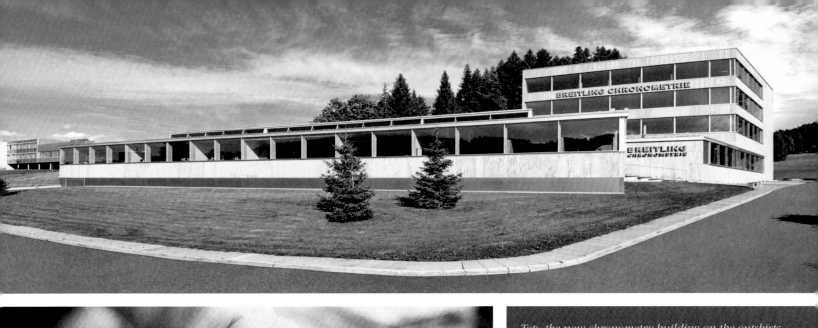

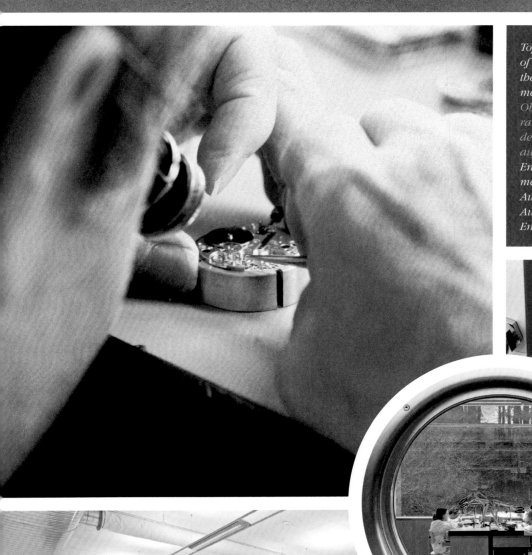

Top: the new chronometry building on the outskirts of La Chaux-de-Fonds ○ Middle from left: assembling the new caliber 01 ○ Testing the self-winding movement ○ Bottom: a view inside the manufactory ○

Oben: Das neue Chronometriegebäude am Stadtrand von La Chaux-de-Fonds ○ Mitte von links: Bei der Montage des neuen Kalibers 01 ○ Der Automat auf dem Prüfstand ○ Unten: Blick in die Manufaktur ○

En haut : Le nouveau bâtiment de Breitling Chronométrie à la périphérie de La Chaux-de-Fonds ○ Au milieu : Assemblage du nouveau calibre 01 ○ Au milieu à droite : Test du remontoir automatique ○ En bas : Vue de l'intérieur de la manufacture

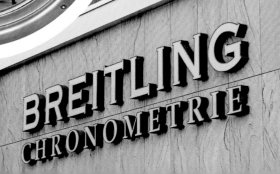

Der lange Weg zur Chronographenmanufaktur

Deutsch

Mit dem Namen Breitling verknüpfen Uhrenliebhaber beinahe zwangsläufig Chronographen. Diese Assoziation kommt nicht von ungefähr, sondern liegt in der Historie der Marke begründet. Man schrieb das Jahr 1884, als Léon Breitling erste Zählerchronographen mit seiner Signatur versah. Darüber hinaus lenkten intelligente Messwerkzeuge für Uhrmacher die Blicke auf Breitling. Zu diesem Zeitpunkt wirkte der Gründer noch in seinem Schweizer Heimatort Saint-Imier. Acht Jahre später, 1892 also, rief er dann in der Jura-Metropole La Chaux-de-Fonds die G. Léon Breitling S.A., Montbrillant Watch Manufactory ins Leben, welche der Kurzzeitmessung in den folgenden Jahrzehnten ein besonderes Augenmerk widmete. Damit setzte der Fachmann aufs richtige Pferd.

1914 startete der Sohn Gaston seine berufliche Karriere im Familienbetrieb. Seine Passion für Chronographen hatte er vom Vater geerbt, der im gleichen Jahr das Zeitliche segnete. 1915 präsentierte Gaston Breitling einen der ersten Armbandchronographen mit Drücker neben der Krone. Ein patentiertes Spezialmodell namens

„Vitesse" unterstützte die Polizei bei Geschwindigkeitskontrollen. Trendgerecht forcierte der ambitionierte Unternehmer die Entwicklung und Fabrikation von Armbandchronographen. Kein Wunder, dass sein früher Tod im Jahr 1927 eine schwer zu füllende Lücke riss. Erst 1932 konnte Willy Breitling als Repräsentant der dritten Generation das Ruder nach gründlicher kaufmännischer und technischer Ausbildung übernehmen. Dann jedoch ging es bei den Innovationen Schlag auf Schlag. Ein Patent von 1934 bezog sich auf Armbandchronographen mit zwei separaten Drückern:

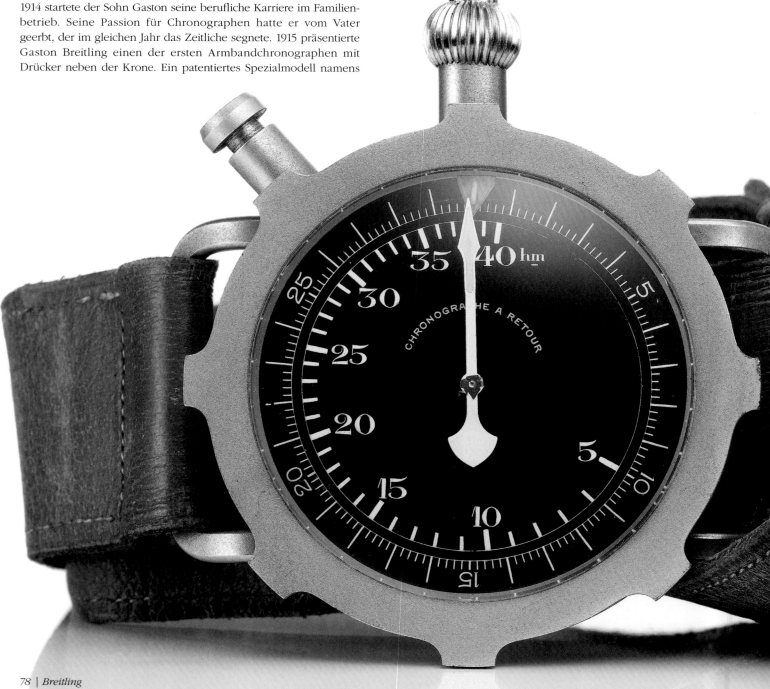

einer für Start und Stopp, der andere zum Nullstellen. 1936 kam ein Chronograph für Flugzeugcockpits heraus, 1938 ein Modell mit innovativem 12-Stunden-Zähler auf der Basis eines Valjoux-Rohwerks, 1941 der legendäre „Chronomat" mit Rechenscheibe, 1952 der „Navitimer" mit optimierten Rechenmöglichkeiten, und 1962 gab es für eine Breitling schließlich einen Ausflug ins Weltall am Handgelenk des amerikanischen Astronauten Scott Carpenter.

1969 gehörte Breitling zu den internationalen Pionieren des Chronographen mit automatischem Aufzug. Mit dem Modell „Chrono-Matic" marschierte das Familienunternehmen in eine neue Ära. Die wegen des Selbstaufzugs durch einen Mikrorotor nur noch selten benötigte Krone befand sich nun logischerweise in der linken Gehäuseflanke.

Mitte der 70er Jahre zog die ungestüme Quarzwelle das neue Uhrwerk und auch die Marke aber zunehmend nach unten.

1978 verloren 24 Mitarbeiterinnen und Mitarbeiter ihren Job. Im August des Folgejahres stellte Breitling den Betrieb komplett ein. Als Grandseigneur alter Schule wählte der gesundheitlich angeschlagene Willy Breitling den geordneten Rückzug. Er verkaufte die eingeführten Namen „Breitling" und „Navitimer" an den Ingenieur Ernest Schneider. Nach Registrierung der Breitling Montres S.A. am 30.11.1982 sowie der Verlegung des Firmensitzes nach Grenchen im Aaretal stieg das Unternehmen in ungeahnte Höhen auf. Unter der Ägide von Ernests Sohn Théodore mussten alle Uhrwerke mit der traditionsreichen Signatur vor dem Einbau ins Gehäuse die offizielle Chronometerprüfung bestehen. Pünktlich zum 125. Firmenjubiläum im Jahr 2009 präsentierte sich Breitling mit der Vorstellung des Automatikchronographen B01 als echte Manufaktur. Seitdem ist die Palette hauseigener Uhrwerke Jahr für Jahr gewachsen. Und ein Ende dieser Aktivitäten ist nicht abzusehen. ○

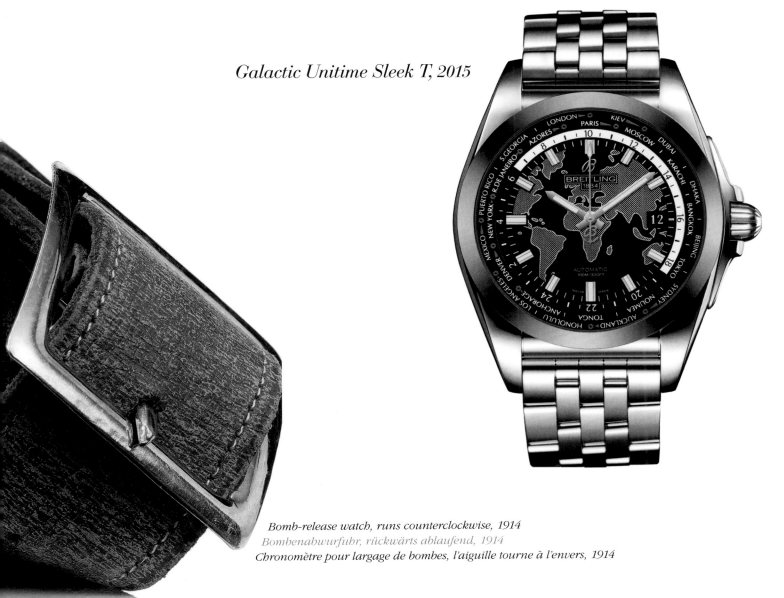

Galactic Unitime Sleek T, 2015

Bomb-release watch, runs counterclockwise, 1914
Bombenabwurfuhr, rückwärts ablaufend, 1914
Chronomètre pour largage de bombes, l'aiguille tourne à l'envers, 1914

Le long parcours jusqu'à la manufacture de chronographes

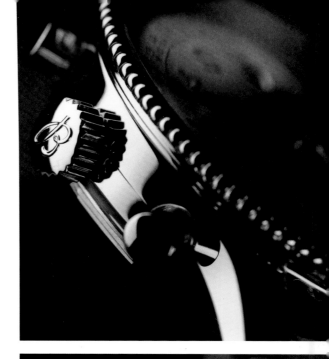

Les amateurs de montres associent presque immanquablement le nom de Breitling aux chronographes. Cette association n'est pas un hasard, elle est inscrite dans l'histoire de la marque. C'est en 1884 que Léon Breitling appose sa signature sur ses premiers chronographes à compteur. Il attire en outre l'attention sur lui par des instruments de mesure intelligents pour horlogers. À cette époque, le fondateur de la marque est encore établi à Saint-Imier, la ville suisse qui l'a vu naître. Huit ans plus tard, en 1892, il crée dans la métropole jurassienne de La Chaux-de-Fonds la G. Léon Breitling SA, Montbrillant Watch Manufactory, qui s'intéressera dans les décennies qui suivent tout particulièrement au chronométrage. C'est un choix judicieux que fait ce maître horloger.

En 1914, son fils Gaston débute sa carrière dans l'entreprise familiale. Il tient cette passion des chronographes de son père, lequel disparaît au cours de la même année. En 1915, Gaston Breitling présente l'un des premiers chronographes-bracelets à poussoir positionné près de la couronne. Un modèle spécial breveté, baptisé « Vitesse » est utilisé par la gendarmerie pour effectuer les contrôles de vitesse. Répondant à la tendance de l'époque, cet entrepreneur ambitieux accélère la conception et la fabrication de chronographes-bracelets. Rien d'étonnant à ce que son décès, en 1927, laisse un vide difficile à combler. Il faut attendre 1932 pour que Willy Breitling, représentant de la troisième génération, reprenne le flambeau après s'être donné une solide formation commerciale et technique. Les innovations se succèdent alors à un rythme soutenu. Un brevet de 1934 couronne un chronographe-bracelet à deux poussoirs : l'un pour les fonctions de « mise en marche/arrêt », l'autre pour la « remise à zéro ». 1936 voit naître un chronographe de bord destiné aux cockpits d'avion, et 1938 la sortie d'un modèle innovant à compteur 12 heures intégrant une ébauche de Valjoux. En 1941 sort le légendaire « Chronomat », doté d'une règle à calcul circulaire, en 1952 le « Navitimer », avec des capacités de calcul optimisées. En 1962, une Breitling s'envole pour l'espace au poignet du spationaute américain Scott Carpenter.

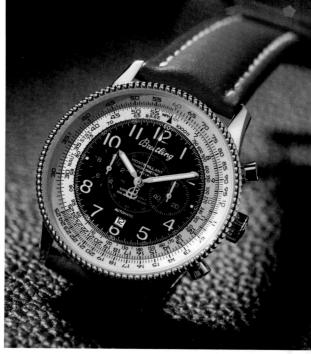

En 1969, Breitling rejoint les pionniers internationaux des chronographes à remontage automatique. Avec le modèle « Chrono-Matic », l'entreprise familiale entre triomphalement dans une ère nouvelle. La roue de couronne, rendue de moins en moins nécessaire par le mouvement automatique à microrotor, est alors logiquement située dans le flanc gauche du boîtier.

Au milieu des années 1970, la déferlante du quartz tire ce nouveau mouvement mais aussi la marque de plus en plus vers le bas. En 1978, 24 employées et employés sont remerciés. En août de l'année suivante, la société Breitling ferme complètement ses portes. En grand seigneur de la vieille école, Willy Breitling, dont la santé décline, organise la retraite en bon ordre. Il vend les noms déposés « Breitling » et « Navitimer » à l'ingénieur Ernest Schneider. Après le dépôt des statuts de la Breitling Montres SA le 30 novembre 1982 et le transfert du siège de la société à Granges, dans la vallée de l'Aar, l'entreprise renaît de ses cendres et atteint des sommets insoupçonnés. Sous l'impulsion de Théodore, le fils d'Ernest, tous les mouvements portant la célèbre signature seront soumis, avant assemblage, au test du COSC (Contrôle officiel suisse des chronomètres). En 2009, année du 125e anniversaire de la société, Breitling se pose en véritable manufacture en présentant le chronographe automatique B01. Depuis, la palette des mouvements produits en interne s'étoffe d'une année sur l'autre. Et cette activité créatrice ne devrait pas s'arrêter de si tôt. ○

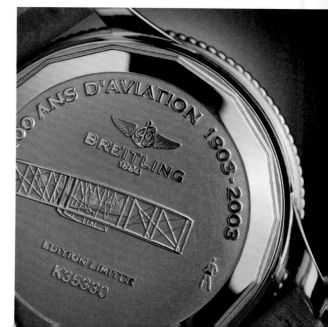

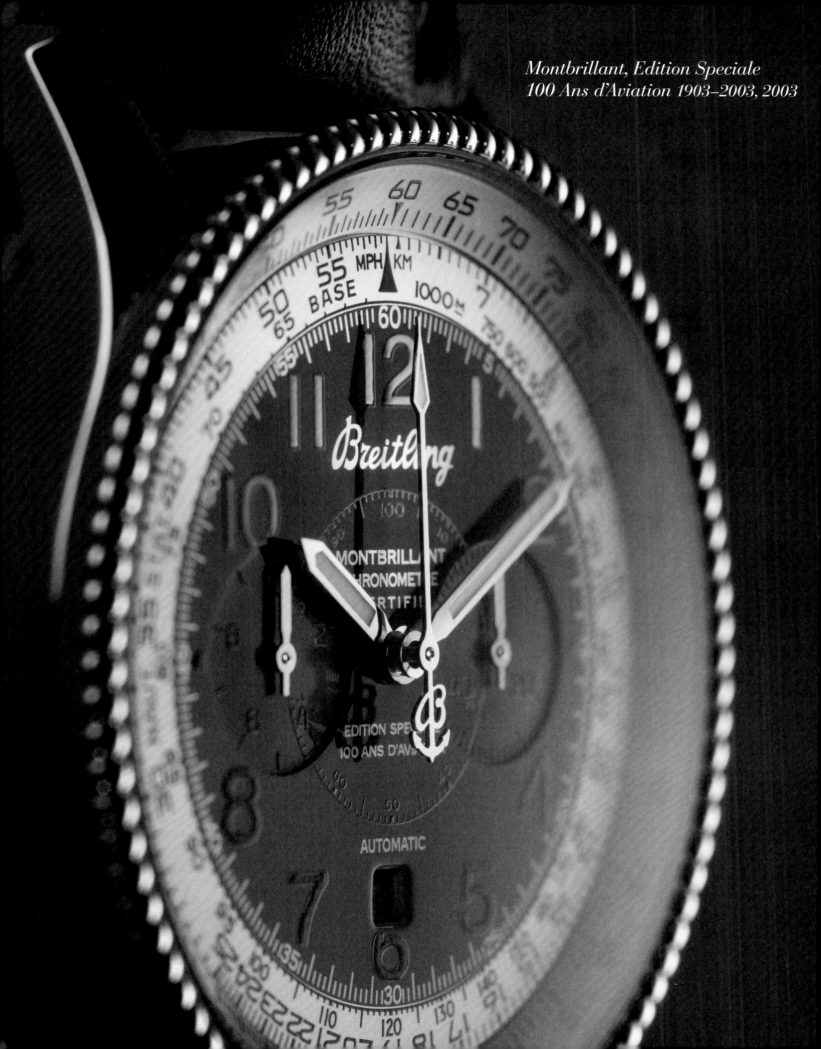

Montbrillant, Edition Speciale
100 Ans d'Aviation 1903–2003, 2003

Carl F. Bucherer

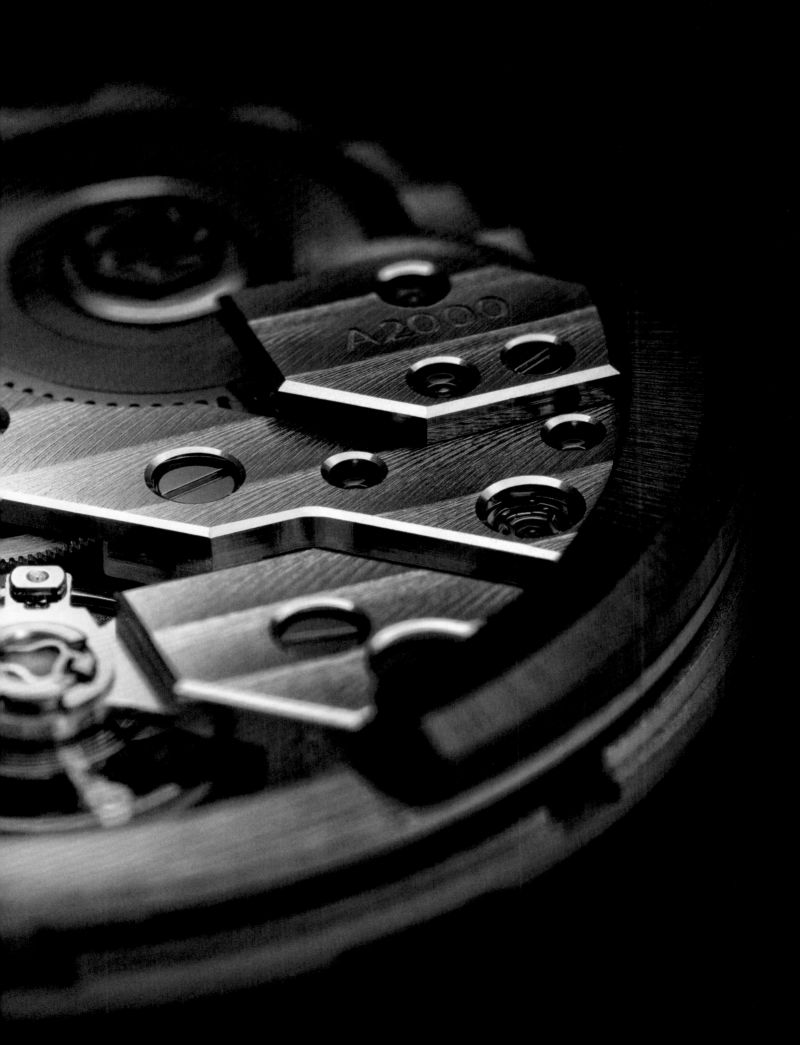

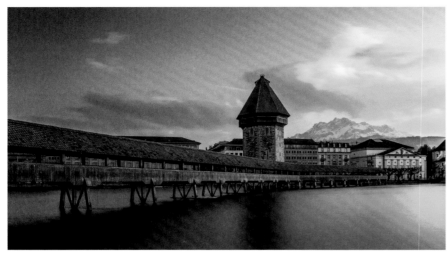
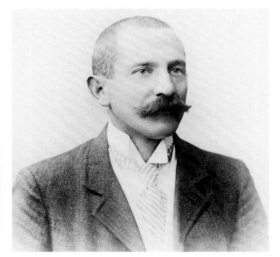

From left: Chapel Bridge, Lucerne ○ Carl Friedrich Bucherer ○ Von links: Kapellbrücke, Luzern ○ Carl Friedrich Bucherer ○ De gauche à droite : Pont de la Chapelle, Lucerne ○ Carl Friedrich Bucherer

From Watch Dealer to Watch Manufactory

English

Carl Friedrich Bucherer opened his first specialized store at Falkenplatz in Lucerne in 1888. He interpreted his clientele's sometimes unusual wishes as commands which it was his privilege to carry out. His reputation soon spread beyond the canton's frontiers. He opened additional stores, which contributed their fair shares to the flourishing of his business ventures. Carl F. Bucherer began offering watches bearing his signature in 1919. His standards were quite high, so he wasn't content to sell private-label goods that had been anonymously manufactured for him. This prompted the savvy businessman to open his own manufacturing facility in the little town of Cortébert in the Jura region in 1919. The workshop wasn't a full-fledged manufacture, but a so-called *établissage*, where purchased components were assembled into ticking products, some of which were delivered together with official chronometer certificates. After an intermezzo with affordably priced watches that simply bore Bucherer's signature on their dials, the traditional label returned in 2001. With this step, the firm's current owner Jörg Bucherer pays due respect to his successful grandfather. Carl F. Bucherer's customers can now choose among watches in five different lines: "Adamavi," "Alacria" (for ladies), "Manero," "Pathos" (for ladies), and "Patravi." Additional functions make the watches in the last-mentioned line into the collection's leading models. The

beginning is marked by a steel automatic chronograph with outsize date display below the "12." Bucherer offered a similar self-winding model encasing the Venus 210 column-wheel caliber in the 1940s. People with wanderlust could opt for the "Patravi" chronograph in 2004. This "time-writer" has an additional 24-hour hand that can be reset in hourly increments via the crown. Time-zone watches weren't entirely new for Carl F. Bucherer: a "Worldtimer" with universal time indication and encasing a hand-wound Derby 7510 caliber as its basic movement joined the collection in 1960. Carl F. Bucherer presented the exclusively developed "Patravi TravelTec GMT" at Baselworld in 2005. Alongside its chronograph, this wristwatch with automatic Caliber CFB 1901 also includes a cleverly designed function to simultaneously show the time in several different time zones. The central 12-hour counter can be reset either forward or backward via the crown. The date in a window in the dial automatically changes to stay in synchrony with the time in a new time zone. An additional 24-hour hand continues to show the reference time or home time. For multifunctional mechanisms, Carl F. Bucherer collaborated with the specialists at Dubois Dépraz in the Vallée de Joux. Carl F. Bucherer joined the elite circle of genuine manufactures in 2008. It earned this distinction by acquiring both a little atelier for horological complications and automatic Caliber CFB A1000, which was developed in the Jura region. The oscillating weight that supplies energy to this slim (4.3-millimeters-tall) microcosm moves peripherally around the caliber. The rotor is borne on little rollers which are equipped with maintenance-free (ceramic) ball bearings. Another special feature is the so-called Central Dual Adjusting System (CDAS), a cunningly engineered fine-adjustment mechanism with a central controlling element. Optimized Caliber CFB A2000 followed in this movement's footsteps in 2016. Unlike its predecessor, a balance with a variable moment of inertia and a freely breathing hairspring serves as the rate-regulating organ in this caliber. The balance's frequency was increased from three to four hertz. CFB A2050 is the first member of this evolutionary phase. It ticks inside the "Manero Peripheral." ○

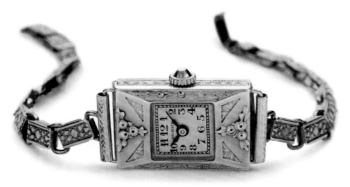

La Grande Dame, 1919

Vom Händler zur Uhrenmanufaktur

Deutsch

Die Kalender zeigten das Jahr 1888. Am Luzerner Falkenplatz eröffnete Carl Friedrich Bucherer sein erstes Fachgeschäft. Dort waren dem Kaufmann die mitunter recht ausgefallenen Wünsche seiner Kundschaft selbstverständlich Befehl. Bald schon reichte der Ruf über die Kantonsgrenzen hinaus. Weitere Läden förderten beständiges Wachstum. Ab 1919 offerierte Carl F. Bucherer auch Uhren mit eigener Signatur. Infolge seines Qualitätsanspruchs begnügte er sich nicht mit reinen Private-Label-Erzeugnissen, welche irgendwo anonym in seinem Auftrag hergestellt wurden. Stattdessen richtete der Unternehmer noch im gleichen Jahr eine eigene Fabrikationsstätte im Juraflecken Cortébert ein. Dort pflegte er keine eigene Manufaktur, sondern das, was man gemeinhin Etablissage nennt. Aus zugekauften Komponenten entstanden tickende Produkte, welche teilweise sogar mit offiziellem Chronometerzertifikat geliefert wurden. Nach einem Intermezzo mit preiswerten, lediglich mit der Signatur Bucherer versehenen Uhren kehrte 2001 das traditionelle Label zurück. Mit diesem Schritt erwies der gegenwärtige Firmeninhaber Jörg Bucherer seinem erfolgreichen Großvater die gebührende Referenz. Seinen Kundinnen und Kunden bietet Carl F. Bucherer derzeit Armbanduhren in fünf Linien an: „Adamavi", „Alacria" (Damen), „Manero", „Pathos" (Damen) und „Patravi". Letztere kann in punkto Zusatzfunktionen als Kollektions-Leader gelten. Den Anfang markierte ein stählerner Automatik-Chronograph mit Großdatum unterhalb der „12". Ohne Selbstaufzug hatte es so etwas bei Bucherer schon in den 1940er Jahren unter Verwendung des Schaltrad-Kalibers Venus 210 gegeben. Menschen mit Fernweh erhielten 2004 den Chronographen „Patravi GMT". Bei diesem Zeitschreiber ließ sich ein zusätzlicher 24-Stunden-Zeiger über die Krone in Stundenschritten verstellen. Gänzlich neu waren Zeitzonen-Armbanduhren für Carl F. Bucherer übrigens nicht. Schon 1960 fand sich ein „Worldtimer" mit Universalzeit-Indikation auf Basis des Handaufzugskalibers Derby 7510 in der Kollektion.

2005 präsentierte Carl F. Bucherer zur Baselworld die exklusiv entwickelte „Patravi TravelTec GMT". Neben dem Chronographen besitzt diese Armbanduhr mit dem Automatikkaliber CFB 1901 auch eine ausgeklügelte Funktion zur simultanen Darstellung mehrerer Zonenzeiten. Ihr zentraler 12-Stunden-Zeiger lässt sich per Krone beliebig vor- oder rückwärts verstellen. Das Fensterdatum vollzieht den Wechsel zu einer anderen Zonenzeit automatisch mit. Ein zusätzlicher 24-Stunden-Zeiger bewahrt die Referenz- oder Heimatzeit. Für die multifunktionale Mechanik kooperierte Carl F. Bucherer mit dem Spezialisten Dubois Dépraz aus dem Vallée de Joux. 2008 betrat Carl F. Bucherer den Kreis echter Manufakturen. Eintrittskarten waren der Erwerb eines kleinen Ateliers für uhrmacherische Komplikationen und das im Jurabogen entwickelte Automatikkaliber CFB A1000. Bei dem nur 4,3 Millimeter hohen Mikrokosmos bewegt sich die energiespendende Schwungmasse peripher rund ums Uhrwerk. Die Rotorlagerung erfolgt mit Hilfe kleiner Rollen, welche ihrerseits auch noch mit wartungsfreien (Keramik-)Kugellagern ausgestattet sind. Zu den Besonderheiten gehört ferner das sogenannte Central Dual Adjusting System (CDAS), eine ausgeklügelte Feinregulierung mit zentralem Steuerelement. In die Fußstapfen dieses Uhrwerks tritt 2016 das optimierte CFB A2000. Im Gegensatz zum Vorgänger dienen nun eine Unruh mit variablem Trägheitsmoment und eine frei schwingende Unruhspirale als Gangregler. Die Unruhfrequenz klettert von drei auf vier Hertz. CFB A2050 heißt das erste Mitglied der Evolutionsstufe. Es tickt in der „Manero Peripheral". ○

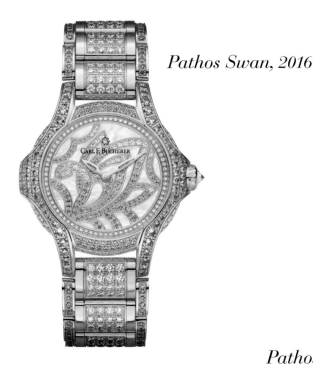

Pathos Swan, 2016

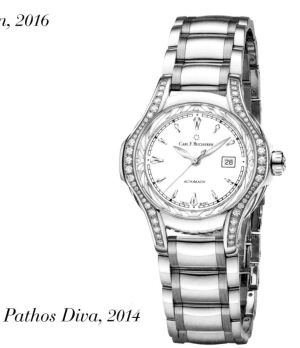

Pathos Diva, 2014

Du commerce à la manufacture de montres

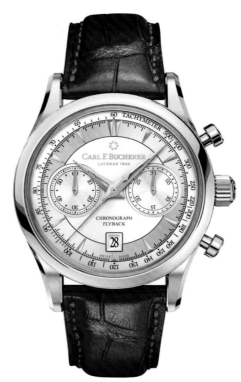

Manero Flyback, 2016

Nous sommes en 1888. Carl Friedrich Bucherer ouvre son premier magasin spécialisé à Lucerne, sur la Falkenplatz. Pour ce commerçant, les désirs parfois peu ordinaires de ses clients sont naturellement des ordres. Sa réputation s'étend par conséquent rapidement au-delà des frontières du canton. L'ouverture d'autres magasins favorise une croissance soutenue. À partir de 1919, Carl F. Bucherer propose également des montres de manufacture. Exigeant sur la qualité, il ne se contente pas de produits sous label privé fabriqués pour son compte, on ne sait où par un prestataire inconnu. L'entrepreneur fait construire la même année une usine à Cortébert, petite localité du Jura suisse. Cette usine n'est pas consacrée à la fabrication mais à ce que l'on appelle communément l'établissage, une activité qui consiste à assembler des composants achetés pour en faire des montres. Certaines seront même livrées avec un certificat de contrôle officiel suisse des chronomètres (COSC). Après un intermède avec des montres économiques simplement griffées Bucherer, le label d'origine est de retour en 2001. Jörg Bucherer rend ainsi dûment hommage à son brillant grand-père. Carl F. Bucherer propose à sa clientèle des montres-bracelets regroupées en cinq lignes : « Adamavi », « Alacria » (femmes), « Manero », « Pathos » (femmes) et « Patravi ». La dernière peut être considérée comme leader de la collection sur le plan des fonctions additionnelles. Le premier garde-temps proposé est un chronographe à remontage automatique et indication grande date à midi. Un modèle semblable était sorti chez Bucherer dès les années 1940, équipé du mouvement à roue à colonnes Venus 210. Sorti en 2004, le « Patravi GMT » est fait pour les gens ayant le mal du pays. Sur ce garde-temps, on peut régler par incrément d'une heure l'affichage 24 heures supplémentaire, à l'aide d'une couronne à poussoir. Les bracelets-montres avec indication de fuseaux horaires ne sont en fait pas une première pour Carl F. Bucherer. En 1960, on trouve déjà dans la collection le « Worldtimer », chronographe à remontage manuel et indication du temps universel fabriqué à partir d'un calibre Derby 7510. En 2005, Carl F. Bucherer présente au salon Baselworld le « Patravi TravelTec GMT », développé en exclusivité. Outre un chronographe, cette montre-bracelet, qui s'appuie sur le mouvement automatique CFB 1901, dispose d'un mécanisme astucieux permettant d'afficher simultanément plusieurs fuseaux horaires. L'aiguille centrale des heures peut être déplacée à volonté vers l'avant ou l'arrière grâce à la couronne à poussoir. Le quantième à guichet effectue automatiquement le changement de fuseau horaire. Une échelle 24 heures sur le réhaut conserve l'heure d'origine (ou de référence). Pour le mécanisme multifonctions, Carl F. Bucherer a collaboré avec l'entreprise spécialisée Dubois Dépraz, sise dans la Vallée de Joux. En 2008, Carl F. Bucherer entre dans le club des vraies manufactures grâce à l'acquisition d'un petit atelier spécialiste des complications horlogères et au développement du mouvement automatique CFB A1000 dans l'arc jurassien. Dans ce microcosme de seulement 4,3 millimètres de haut, la masse oscillante génératrice d'énergie (le rotor) tourne à la périphérie du mouvement. L'amortissement du rotor est assuré par de petits galets eux-mêmes équipés de roulements à billes en céramique n'exigeant aucune maintenance. Parmi les autres particularités du mouvement figure le CDAS (Central Dual Adjusting System), un système de réglage fin intelligent avec élément de commande central. Au mouvement CFB A1000 succède en 2016 une version optimisée, le CFB A2000. La régularité de marche est assurée en l'occurrence par un balancier à inertie variable et un spiral totalement libre. La fréquence d'oscillation du balancier passe de trois à quatre Hertz. Mouvement correspondant au stade d'évolution suivant, le CFB A2050 anime le « Manero Peripheral ». ○

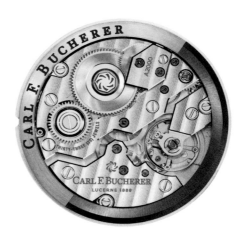

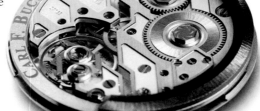

Right top: Cal. CFB A2000 ○ Right bottom: Cal. CFB A1000

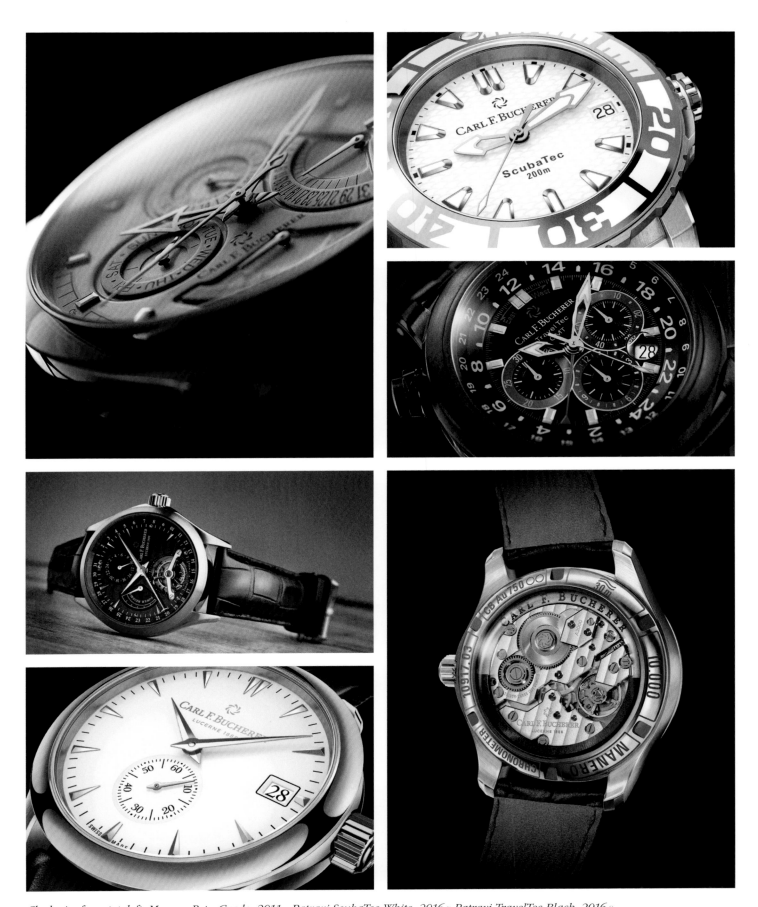

Clockwise from top left: Manero RetroGrade, 2011 ◦ Patravi ScubaTec White, 2016 ◦ Patravi TravelTec Black, 2016 ◦
Manero Peripheral, 2016 ◦ Manero Peripheral, 2016 ◦ Manero Tourbillon, 2015

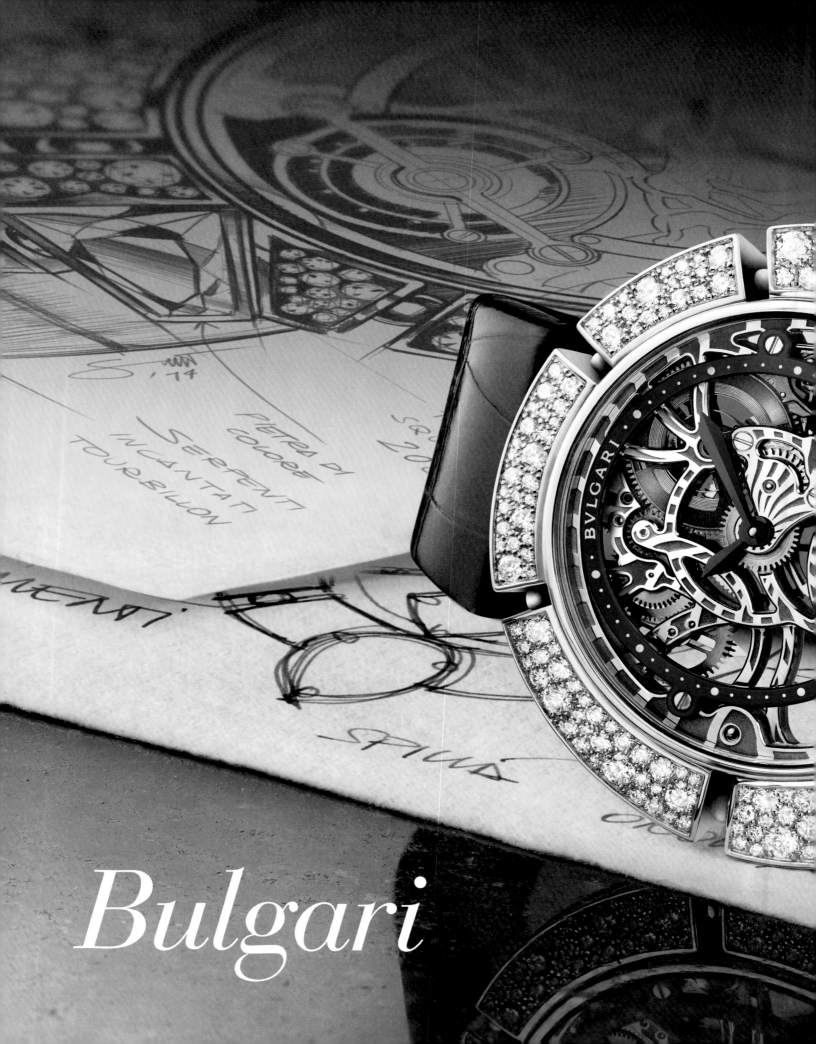

Bulgari

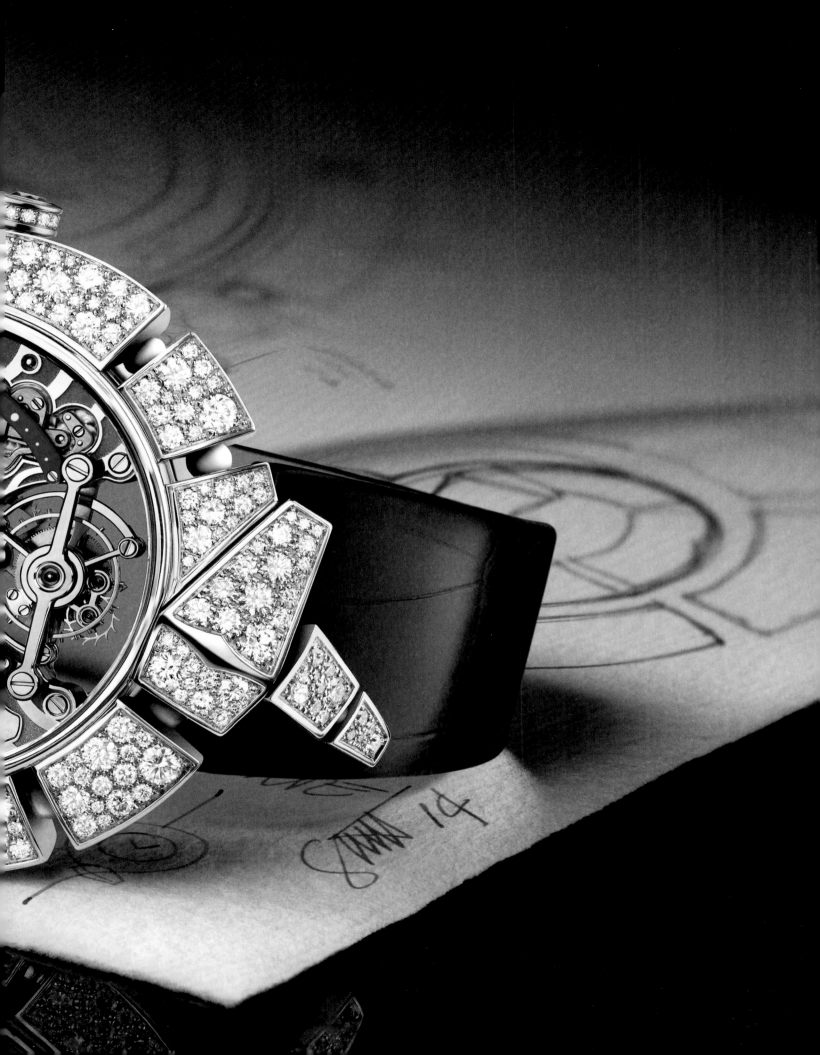

Italian Watch Luxury of Swiss Provenance

English

The name "Bulgari" is often associated with high-carat jewelry in the cosmos of luxury, but this label's watches all too often fly under the radar. The reasons for this may have to do with the history of the company and the biography of its founder Sotirios Boulgaris. After immigrating to Naples from Greece around 1880, he Italianized his name to "Sotirio Bulgari" and began earning a living in his new homeland as a silver merchant.

The official history of the Bulgari business began when Sotirio opened his first jewelry shop on Via Sistina in Rome in 1884. Relocation to the classier venue of no. 29 Via dei Condotti followed ten years later. Discriminating customers frequented the new shop, where they could chose among sparkling treasures, but timepieces weren't yet part of the firm's portfolio. Watches likewise remained absent when Sotirio moved his business again in 1905, reopening at no. 10 Via dei Condotti, near the famous Spanish Steps. Not until the early 1920s could upper-crust ladies select wristwatches with geometric platinum cases, linked bracelets and ample encrustations of diamonds. These timepieces were fabricated in France, but the words "Bulgari Roma" graced their dials.

After Sotirio died in 1932, his sons Giorgio and Costantino continued their father's passion for high-quality jewelry. Comprehensive renovation work coincided with the introduction of the distinctive BVLGARI logo in 1934. This insignia was also found on the slim pocket watches that Audemars Piguet delivered in the late 1930s. The subsequent decade witnessed Bulgari's creation of feminine wristwatches with flexible snakeskin wristbands. Swiss suppliers provided the movements and dials. Thanks to their uniquely ornamental look, these watches have remained en vogue decade after decade.

A new watch era, characterized by sustainability and innovative dynamism, began in 1975. Imaginative product designers rose to the challenge posed by the Quartz Revolution and conceived the luxurious "Bulgari Roma" with its unprecedented combination of a classically styled case and a liquid-crystal display. Following the example first set by Roman emperors, who demonstrated their power by having their names struck into coins, the words "Bvlgari" and "Roma" were engraved along the bezel. But this was only a modest beginning. The family business debuted the really big classic with a bright future in 1977. Along with a thoroughly optimized basic design, the "Bulgari Roma" also boasted the label's logo spelled out not once, but twice. Instant acceptance of the "Bulgari Bulgari" as a style icon practically demanded professionalization, so the label's watch-related activities were entrusted to Bulgari Time SA, which was established in Neuchâtel, Switzerland in 1982. Bulgari acquired ample competence in mechanical watchmaking by acquiring the little Gérald Genta and Daniel Roth manufactories at the turn of the millennium. The engineers, technicians and watchmakers in Vallée de Joux were and are thoroughly experienced in every conceivable facet of their métiers. Bulgari annually produced and sold more than 200,000 timepieces at this time. Sales of watches comprised 47 percent of total revenues, far outperforming the traditional jewelry division. After this acquisition, Bulgari started an unprecedented process leading toward a greater vertical range of manufacturing. The Romans acquired 50 percent of the shares of the dial maker Cadrans Designs SA in 2005. In October of this same year, they also bought a majority share of Prestige D'Or SA, a leading fabricator of steel and precious metal wristbands. The Bulgari Group became sole owner of both firms in 2009. The road to the brand's first basic self-winding movement was paved by a developmental agreement signed in 2007 with Leschot SA, which is likewise headquartered in Neuchâtel. Caliber BVL 168 debuted inside the case of the "Bulgari Sotirio" in 2010. The company is now no less fully integrated than the experienced case manufacturer Finger SA. The logical consequence of these activities followed in 2010 with the use of the unified signature "Bulgari" for all watches. Daniel Roth and Gérald Genta now belonged to the past. The retrospective ended in 2011, when the French luxury concern LVMH bought the traditional Italian company, which makes its watches exclusively in Switzerland.

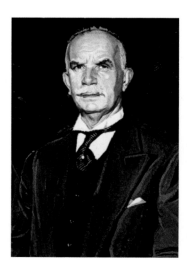

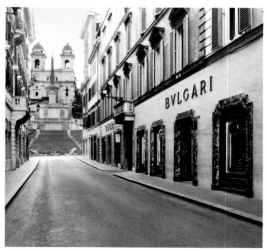

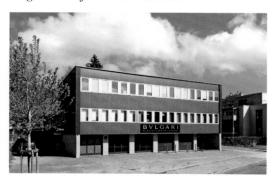

From left: Sotirio Bulgari ◦ Via dei Condotti, Italy ◦ Bulgari Manufacture de Cadrans

A person who wants to see what goes on at Bulgari's various production sites must first get behind the steering wheel of an automobile, drive to the secluded Vallée de Joux, and park the car near the railway station in Le Sentier, where Bulgari's experts cultivate the utmost in the watchmaker's art. Among their finest products is a 1.95-millimeter-tall one-minute tourbillon with a "flying" (i.e., cantilevered) rotating carriage. This minimal overall height defines only the carriage, inside of which the balance, hairspring and escapement continually turn their circles. To create the superlative movement known as "BVL 268," Bulgari eliminated the usual bearing jewels, as well as the jewels for the pivots of the balance's staff, and replaced them with seven miniature bearings. Craftspeople combine nearly 250 components for each movement. This device's little sister is Caliber BVL 128, a classical hand-wound movement with a height of just 2.23 millimeters: this extraordinary slimness qualifies as a horological complication. The third movement in the trio debuted at Baselworld in 2016. It can be triggered to audibly announce precious time to the nearest minute. Connoisseurs describe this complication as a "minute repeater." Without exaggeration, the "Octo Finissimo Minute Repeater" can be lauded as a unique achievement. The movement is just 3.12 millimeters tall; the titanium case is 6.85 millimeters thick, 40 millimeters in diameter, equipped with a transparent window in the back, and resists water pressure to five bar. A silent centrifugal-force regulator assures the equal length of intervals between consecutives chimes. As is true for all of Bulgari's first-rate chronometric products, the 362 components that comprise Caliber BVL 362 are meticulously finely processed by hand. A dexterous artisan requires up to 20 hours to put the finishing touches on the tourbillon cage in Caliber BVL 268. Patient craftsmanship culminates in the creation of superlative chronometric products that are much more than the sum of their parts. Blank components for the simpler "Solotempo" self-winding Calibers BVL 191 and BVL 193 are fabricated in the heights of the Jura region in western Switzerland. Man and machine collaborate along the modern assembly line where the calibers are decorated, assembled and finely adjusted. Automated machinery is used wherever machines can do more work —and above all better work—than a human hand. Cases and wristbands are built in the little town of Saignelégier in the Jura region; the dials are born in La Chaux-de-Fonds. Depending on the complexity, 100 or more work phases may be required for each of these little discs, which serve as the backgrounds for tirelessly orbiting hands. All of the threads conjoin at Neuchâtel, where Bulgari's headquarters are situated beside a picturesque lake. Design, planning and coordination of the products are undertaken in close consultation with the Italian headquarters. Qualified suppliers provide whatever Bulgari cannot produce on its own. Comprehensively trained colleagues in light-flooded ateliers work with the utmost conscientiousness to guarantee that each watch (regardless of whether its name is "Bulgari Bulgari," "Diagono," "Lucea," "Octo" or "Serpenti") upholds the manufactory's rigorous quality standards. After all, Bulgari has an illustrious reputation to defend. ○

Top: Cal. Finissimo BVL 128 SK
Bottom, from left: Lucea Il Giardino Paradiso, 2016 ○ Piccola Lucea, 2016

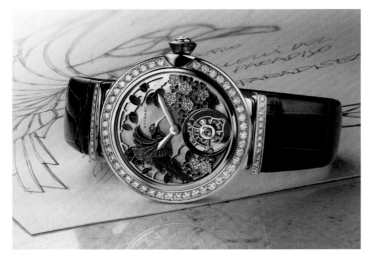

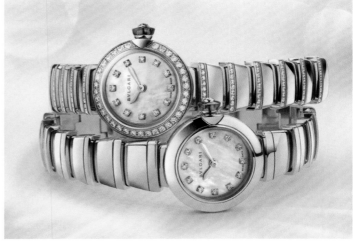

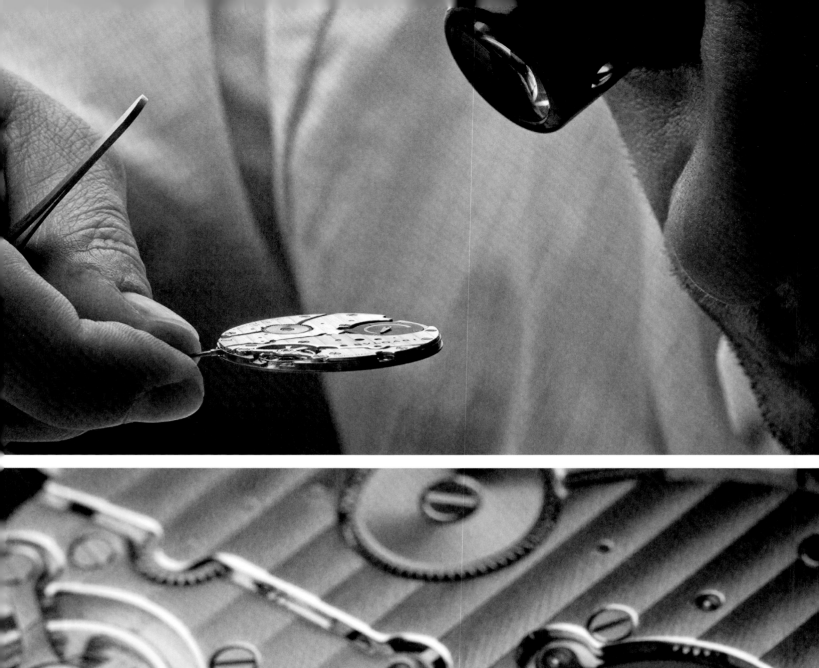
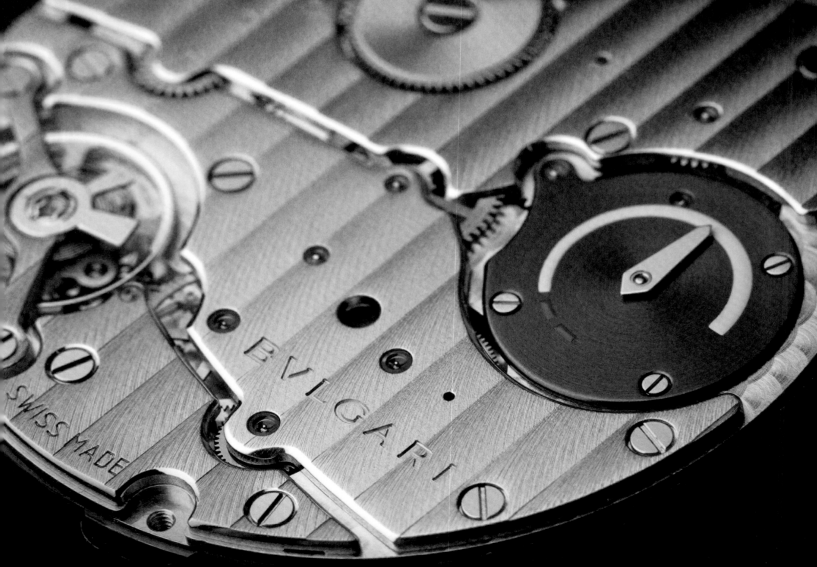

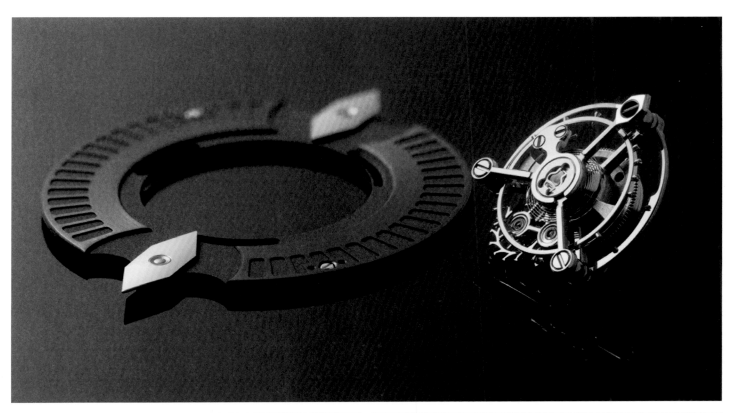

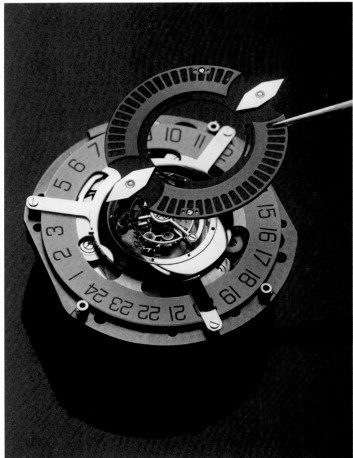

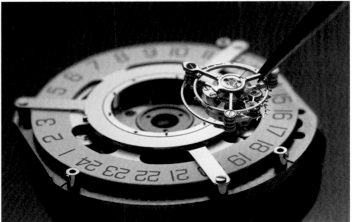

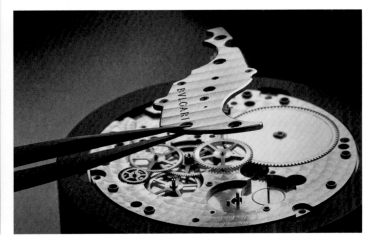

Movement manufacturing ○ *Montage des Uhrwerks* ○ *Réalisation de mouvements*

Italienischer Uhren-Luxus
eidgenössischer Provenienz

Deutsch

Im Luxuskosmos verknüpft sich der Name Bulgari oftmals mit hochkarätigem Schmuck. Uhren stehen bei manchen Zeitgenossen hingegen weniger im Rampenlicht. Die Gründe sind möglicherweise in der Unternehmensbiographie zu suchen und im Firmengründer Sotirios Boulgaris. Nach seiner Emigration begegnete man dem gebürtigen Griechen ab 1880 in Neapel. Und zwar nun unter dem italienischen Namen Sotirio Bulgari. Den Lebensunterhalt verdiente der Wahlitaliener mit dem Handel von Silberwaren.

Mit der Eröffnung eines ersten Schmuckgeschäfts in der Via Sistina in Rom nahm 1884 die offizielle Bulgari-Firmengeschichte ihren Lauf. Zehn Jahre später erfolgte der Umzug in die deutlich mondänere Via dei Condotti. Im Haus mit der Nummer 29 konnten sich anspruchsvolle Kundinnen an exquisiten Preziosen erfreuen. Uhren waren definitiv kein Thema. Auch nicht 1905, als Sotirio seine Aktivitäten in die Via dei Condotti 10 nahe der Spanischen Treppe verlegte. Erst zu Beginn der 1920er Jahre konnten sich die Damen der besseren Gesellschaft an Armbanduhren mit geometrischen Platingehäusen, Gliederbändern und reichlich Brillantbesatz erfreuen. Obwohl in Frankreich fabriziert, stand am Zifferblatt Bulgari Roma zu lesen.

Nach dem Tod des Firmengründers im Jahr 1932 propagierten die beiden Söhne Giorgio und Costantino ihre Passion für hochwertige Juwelen. Mit umfangreichen Renovierungsarbeiten ging 1934 die Einführung des markanten Logos BVLGARI einher. Selbiges fand sich auch auf flachen Taschenuhren, geliefert von Audemars Piguet in den späten 1930er Jahren. Im folgenden Jahrzehnt kreierte Bulgari feminine Armbanduhren mit flexiblem Schlangenband. Werke und Zifferblätter stammten von Schweizer Lieferanten. Wegen ihrer einzigartigen schmückenden Optik büßten diese Uhren über Jahrzehnte hinweg nichts an Aktualität ein.

Das von Nachhaltigkeit und Innovationskraft geprägte Uhren-Zeitalter begann indessen erst 1975. Einfallsreiche Produktgestalter hatten die Herausforderung der Quarz-Revolution angenommen und bei ihrer luxuriösen „Bulgari Roma" erstmals ein Flüssigkristall-Display mit klassischem Gehäusedesign kombiniert. Bei der Signatur folgten die Bulgari dem Brauch römischer Kaiser, Macht durch ihren Namenszug auf Münzen zu demonstrieren. In diesem Sinne fanden sich Bulgari und Roma auf dem Glasrand. Das war ein bescheidener Anfang. Den ganz großen Klassiker mit Zukunftspotenzial präsentierte das Familienunternehmen 1977. Zum gründlich optimierten Basisdesign der „Bulgari Roma" gesellte sich das signifikante Logo in gleich doppelter Ausführung. Die spontane Akzeptanz dieser Stilikone namens „Bulgari Bulgari" verlangte beinahe zwingend nach Professionalisierung. Seit 1982 zeichnet die Bulgari Time SA mit Sitz im eidgenössischen Neuenburg für das Uhrengeschäft verantwortlich. Jede Menge Kompetenz in Sachen Mechanik erwarb Bulgari zur Jahrtausendwende mit den kleinen Manufakturen Gérald Genta und Daniel Roth. Die Ingenieure, Techniker und Uhrmacher im Vallée de Joux waren und sind nämlich mit allen erdenklichen Wassern gewaschen. Zu diesem Zeitpunkt produzierte und verkaufte Bulgari jährlich mehr als 200 000 Zeitmesser. Damit stellte das Uhrenbusiness mit 47 Prozent

Umsatzanteil den tradierten Schmuckbereich deutlich in den Schatten. Nach dieser Akquise startete Bulgari einen beispiellosen Vertikalisierungsprozess. 2005 brachten die Römer 50 Prozent der Aktien des Zifferblattherstellers Cadrans Designs SA unter ihre Fittiche. Darüber hinaus kauften sie im Oktober des gleichen Jahres die Aktienmehrheit an der Prestige D'Or SA, einem führenden Fabrikanten von Armbändern aus Stahl und edlen Metallen. 2009 gehörten beide komplett zur Bulgari-Gruppe. Den Weg zum ersten eigenen Basis-Automatikkaliber ebnete ein 2007 geschlossenes Entwicklungsabkommen mit der ebenfalls in Neuchâtel beheimateten Leschot SA. Das BVL 168 debütierte 2010 in der „Bulgari Sotirio". Dieses Unternehmen ist mittlerweile ebenso vollständig integriert wie ein erfahrener Gehäusefabrikant namens Finger SA. Die logische Konsequenz aus diesen Aktivitäten bestand 2010 in der Verwendung einer einheitlichen Signatur für alle Uhren: Bulgari. Daniel Roth und Gérald Genta gehörten der Vergangenheit an. Der Rückblick endet 2011, als der französische Luxus-Multi LVMH das italienische Traditionsunternehmen kaufte, dessen uhrmacherische Aktivitäten ausnahmslos in der Schweiz über die Bühne gehen.

Wer erleben möchte, was Bulgari in seinen verschiedenen Produktionsstätten tut, muss sich hinter das Steuer eines Autos setzen. Im abgeschiedenen Vallée de Joux, konkret nahe dem Bahnhof von Le Sentier, übt sich Bulgari in höchster Uhrmacher-kunst. Zu den Spitzenprodukten gehört ein 1,95 Millimeter flaches Minutentourbillon mit fliegend gelagertem Drehgestell. Die minimale Gesamthöhe definiert allein der Käfig, in dem Unruh, Unruhspirale und Hemmungspartie beständig ihre Kreise drehen. Für diesen Superlativ namens BVL 268 musste Bulgari die üblichen Lagersteine einschließlich jener für die Zapfen des Ankerrads durch insgesamt sieben Miniatur-Kugellager ersetzen. Für jedes Uhrwerk benötigen die Handwerker knapp 250 Komponenten. Die „kleine Schwester" heißt BVL 128. Lediglich 2,23 Millimeter baut dieses klassische Handaufzugswerk hoch, was in der Uhrmacherei als Komplikation durchgeht. Das Dritte im Bunde gab während der Baselworld 2016 seinen Einstand. Auf Wunsch schlägt es die kostbare Zeit minutengenau. Kenner sprechen von einer Minuten-repetition. Bei nur 3,12 Millimetern Werks- und 6,85 Millimetern Gesamthöhe der limitierten „Octo Finissimo Minute Repeater" kann man ohne jede Übertreibung von einzigartiger Leistung sprechen. Das 40 Millimeter große Titangehäuse mit Sichtboden ist sogar wasserdicht bis zu fünf Bar Druck. Für den gleichförmigen Ablauf der Schlagfolge sorgt ein geräuschloser Fliehkraftregler. Die 362 Werkskomponenten des Kalibers BVL 362 werden, wie es sich für alle Bulgari-Produkte der chronometrischen Spitzen-klasse gehört, manuell aufs Sorgfältigste feinbearbeitet. Allein für die Finissage eines Tourbillonkäfigs im Kaliber BVL 268 benötigt ein kunstfertiger Handwerker bis zu 20 Stunden. Durch geduldige Handarbeit entstehen chronometrische Spitzenprodukte, welche deutlich mehr sind als die Summe ihrer vielen Teile. In den Höhen des Westschweizer Jura entstehen auch Roh-Komponenten für die einfacheren „Solotempo"-Automatikkaliber BVL 191 und BVL 193. Bei der Dekoration, Montage und Regulierung arbeiten Menschen und moderne Fertigungsstraßen Hand in Hand. Automaten kommen überall dort zum Zuge, wo sie mehr und vor allem Besseres leisten. Gehäuse und Bänder stammen aus dem Jurastädtchen Saignelégier, Zifferblätter aus La Chaux-de-Fonds. Je nach Komplexität verlangt eine der kleinen Scheiben, vor der sich die Zeiger unentwegt drehen, nach 100 Arbeitsschritten oder deutlich mehr. In Neuchâtel, wo Bulgari nahe dem malerischen See sein Uhren-Hauptquartier unterhält, laufen alle Fäden zusammen. Produktdesign, -planung und -steuerung erfolgen in enger Abstimmung mit der italienischen Zentrale. Was Bulgari nicht selber produzieren kann, steuern qualifizierte Zulieferer bei. In lichtdurchfluteten Ateliers geben umfassend geschulte Mitarbeiterinnen und Mitarbeiter ihr Bestes, damit ausnahmslos alle Uhren, egal ob sie beispielsweise „Bulgari Bulgari", „Diagono", „Lucea", „Octo" oder „Serpenti" heißen, den rigorosen Qualitätsstandards der Manufaktur entsprechen. Bulgari hat schließlich einen Ruf zu verlieren. ○

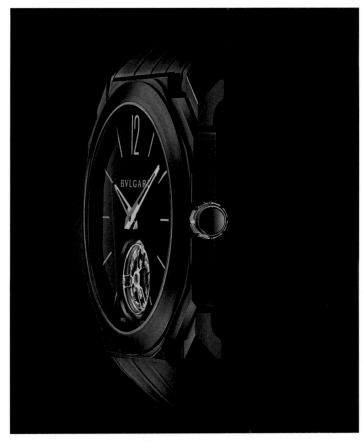

Clockwise from top right:
Octo Finissimo Ultranero Tourbillon, 2016 ○
Octo Finissimo Skeleton, 2016 ○ Octo Ultranero Solotempo, 2016

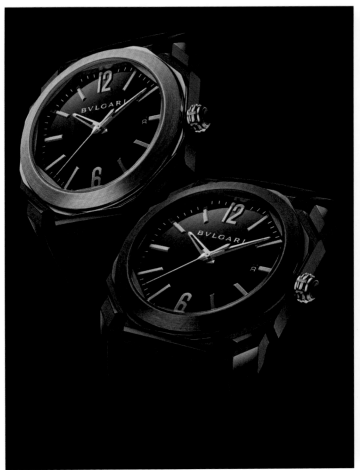

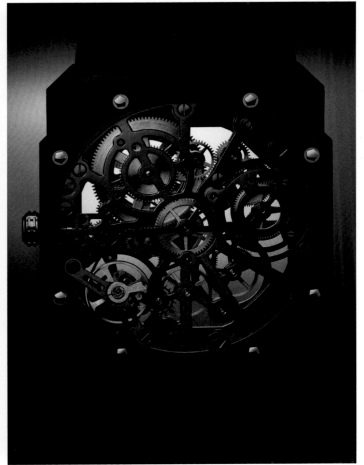

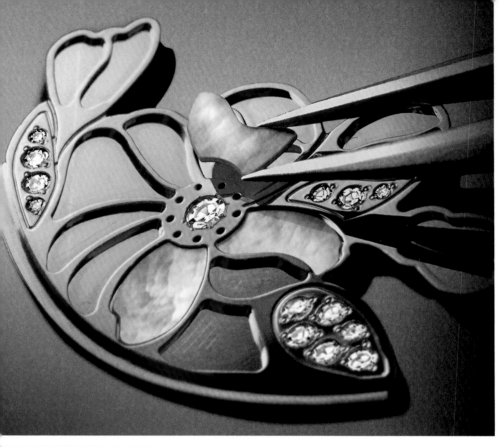

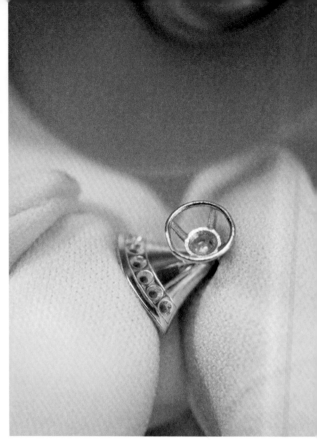

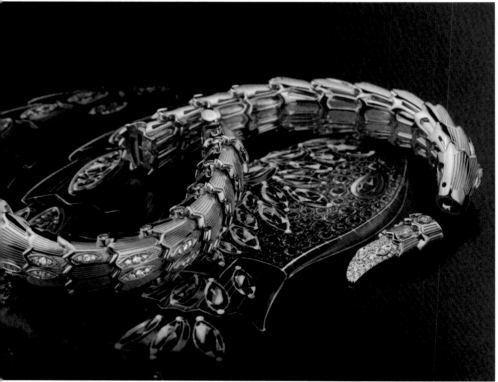

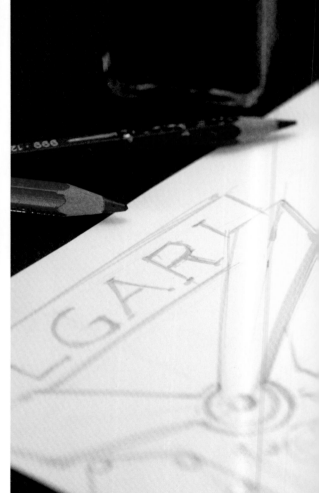

All of the high-jewelry watch creations are crafted in Italy and Switzerland:
the jewelry work is performed in Bulgari's specialized artisanal workshops in Italy;
the watch components are fabricated and assembled in Neuchâtel, Switzerland.
Alle Schmuckuhren werden in Italien und der Schweiz hergestellt. Die Juwelenarbeiten
stammen aus handwerklichen Betrieben, die sich auf Bulgari spezialisiert haben.
Die Uhrwerke werden in Neuchâtel in der Schweiz gefertigt.
La réalisation artisanale des montres de haute joaillerie se partage entre l'Italie et la Suisse.
Le travail de joaillerie est accompli en Italie, dans les ateliers artisanaux spécialisés de Bulgari,
tandis que la montre proprement dite est usinée et assemblée à Neuchâtel (Suisse).

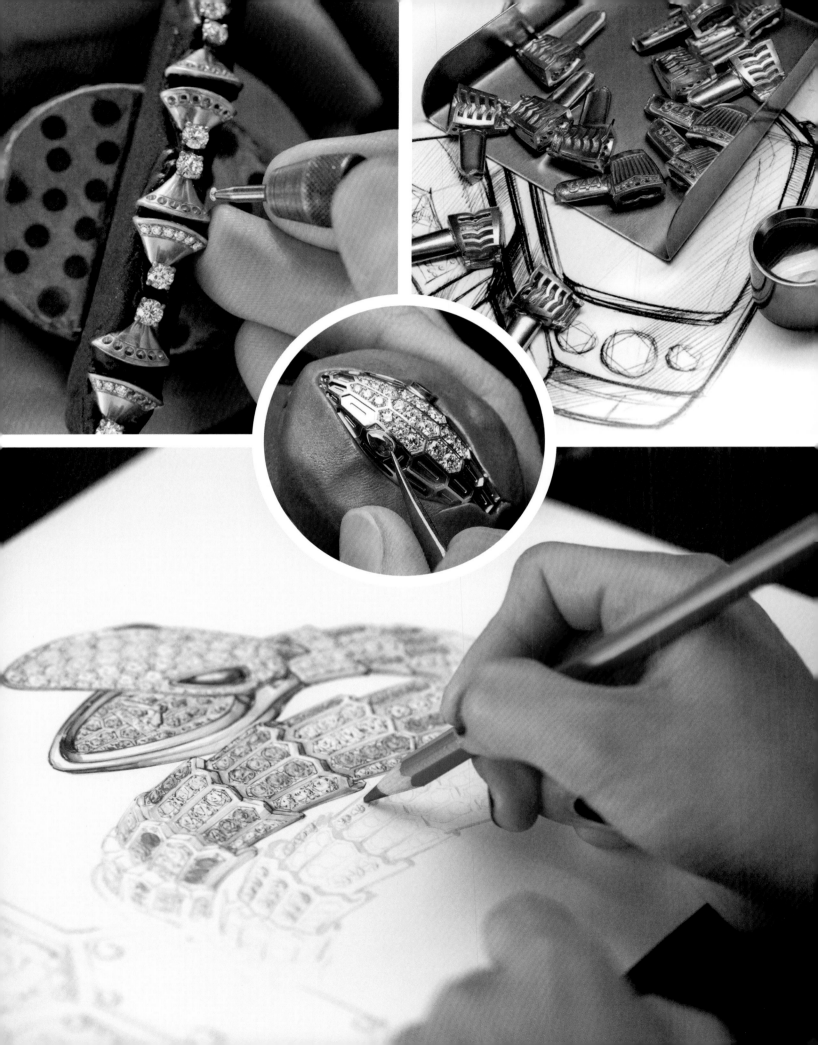

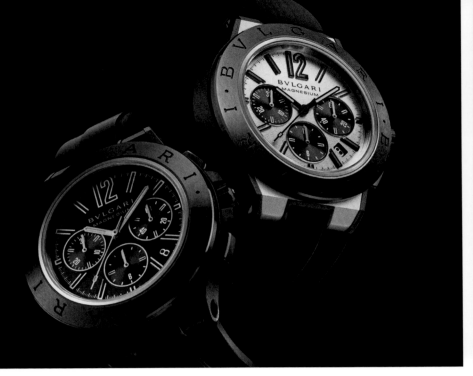
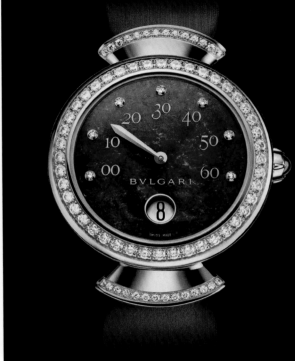

Luxe à l'italienne swiss made

Dans l'univers du luxe, on associe souvent le nom Bulgari à de somptueux bijoux, les montres de la marque ne bénéficiant pas d'une attention comparable de la part de certains de nos contemporains. Cela tient peut-être à l'histoire de l'entreprise et à la personnalité de son fondateur, Sotirios Boulgaris. Ce Grec qui a pris le chemin de l'émigration s'installe en 1880 à Naples, sous un nom italianisé, Sotirio Bulgari. Dans sa patrie d'adoption, il vend de l'orfèvrerie d'argent.

L'ouverture d'une première joaillerie via Sistina à Rome, en 1884, signe l'acte de naissance de la maison Bulgari. Dix ans plus tard, la boutique est transférée à une adresse bien plus chic, la via dei Condotti. Au numéro 29, les clientes difficiles sont comblées par des joyaux exquis. À l'époque, les montres ne sont aucunement d'actualité. Elles ne le sont pas davantage en 1905, lorsque Sotirio Bulgari transfère ses activités au numéro 10 de cette même rue, près de l'escalier de la Trinité-des-Monts. Ce n'est qu'au début des années 1920 que Bulgari propose des montres-bracelets pourvues de boîtiers géométriques en platine, de bracelets à maillons et de brillants à foison, pour la plus grande joie des femmes du monde. Bien qu'il s'agisse de produits fabriqués en France, le cadran porte la mention « Bulgari Roma ».

Après le décès du fondateur en 1932, ses deux fils Giorgio et Costantino laissent libre cours à leur passion pour les bijoux de luxe. En 1934, d'importants travaux de rénovation vont de pair avec l'introduction de BVLGARI, un logo marquant également apposé sur les montres de poche plates fournies par Audemars Piguet à la fin des années 1930. Au cours de la décennie suivante, Bulgari crée des montres-bracelets pour dames pourvues de mouvements et de cadrans achetés auprès de fournisseurs suisses ainsi que d'un bracelet « serpent » souple. Parce qu'elles ont des allures de bijoux, ces montres ne se démoderont pas au cours des décennies suivantes.

Cependant, l'ère horlogère placée sous le signe de la durabilité et de la capacité d'innovation ne débutera qu'en 1975. Relevant le défi de la révolution du quartz, d'ingénieux concepteurs associent pour la première fois un affichage à cristaux liquides à un boîtier au design classique dans leur luxueuse « Bulgari Roma ». Cette signature se rattache à une pratique des empereurs romains, qui affichaient leur puissance en apposant leur nom sur les monnaies. Bulgari et Rome sont ainsi réunis sur la lunette du cadran, un début modeste. C'est en 1977 que l'entreprise familiale présente son très grand classique à l'avenir prometteur. L'optimisation en profondeur du design de base de la « Bulgari Roma » s'accompagne de l'introduction du marquant logo, gravé en double. Face au succès rapide de cette icône du style répondant au nom « Bulgari Bulgari », une professionnalisation s'impose presque. À partir de 1982, toutes les activités horlogères sont entre les mains de Bulgari Time SA, société sise à Neuchâtel, en Suisse. Au début du XXIᵉ siècle, Bulgari se dote de compétences étendues en matière de mécanique en s'adjoignant les petites manufactures Gérald Genta et Daniel Roth. En effet, les ingénieurs, techniciens et horlogers de la vallée de Joux sont depuis toujours rompus à toutes les tâches. Bulgari produit alors plus de 200 000 garde-temps par an, soit 47 pour cent de son chiffre d'affaires. L'horlogerie éclipse la joaillerie, qui a constitué jusqu'alors le cœur de métier de Bulgari. Cet achat est le point de départ d'un processus de verticalisation sans précédent : en 2005, l'entreprise romaine prend sous son aile 50 pour cent des actions du fabricant de cadrans Cadrans Designs SA ; en octobre de la même année, elle prend une participation majoritaire dans Prestige d'Or SA, qui figure parmi les leaders sur le marché des bracelets en acier et en métaux précieux. En 2009, ces deux entreprises sont intégrées au groupe Bulgari. Un contrat de développement passé en 2007 avec l'entreprise neuchâteloise Leschot SA ouvre la voie à la conception du premier calibre automatique de base maison, le BVL 168, qui fait ses débuts dans la « Bulgari Sotirio ». Leschot SA

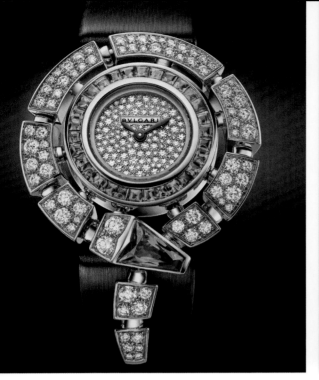
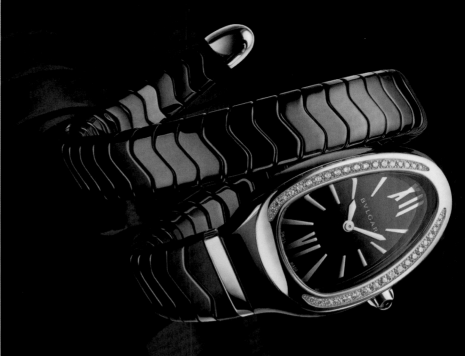

From left: Diagono Magnesium Chronograph, 2016 ◦ Diva's Dream, Retrograde Minutes and Jumping Hour, Heart of Ruby Dial, 2016 ◦ Serpenti Incantati Jewellery, 2016 ◦ Serpenti Tubogas Ceramic and Rose Gold, 2016

est désormais intégrée à Bulgari, au même titre qu'un fabricant de cadrans chevronné, Finger SA. Ces mesures débouchent logiquement en 2010 sur l'apposition d'une seule et même signature sur toutes les montres : Bulgari. C'en est fini de Daniel Roth et Gérald Genta. Notre rétrospective s'arrête en 2011, année du rachat par LVMH, multinationale française du luxe, de la maison de tradition italienne, dont toutes les activités horlogères sans exception se déroulent sur le sol suisse.

Qui a envie de voir Bulgari à l'œuvre dans ses différents ateliers de production doit se rendre dans la vallée de Joux. Près de la gare ferroviaire du Sentier, plus précisément, Bulgari s'adonne à l'art horloger au plus haut niveau. Parmi ses produits phares figure un tourbillon minute à cage mobile tournante affichant 1,95 millimètre d'épaisseur. Cette faible hauteur totale correspond à celle de la cage, dans laquelle le balancier, le spiral et l'échappement sont en constante rotation. Pour cette prouesse dénommée BVL 268, Bulgari a remplacé l'empierrage traditionnel, y compris celui des pignons de la roue d'ancre, par sept roulements à billes miniatures. L'assemblage d'un mouvement requiert près de 250 composants. BVL 128, son « cadet », est un mouvement à remontage manuel classique de seulement 2,23 millimètres de hauteur, une prouesse créditée du statut de complication horlogère. Le « benjamin » a fait son entrée dans le monde à l'occasion du Salon mondial de l'horlogerie à Bâle (Baselworld) 2016. Il sonne les minutes à la demande, une complication que les connaisseurs appellent la répétition minutes. Affichant une épaisseur de mouvement réduite à 3,12 millimètres pour une épaisseur totale de boîtier de 6,85 millimètres, le modèle « Octo Finissimo Minute Repeater » est, sans exagération aucune, une performance exceptionnelle. Le boîtier titane à fond en verre est même étanche jusqu'à une pression de 5 bars. Grâce à un régulateur centrifuge silencieux, la sonnerie des minutes s'égrène avec régularité. Les 362 composants

du mouvement du calibre BVL 362 sont le fruit d'un travail artisanal extrêmement soigné. Par ailleurs, le finissage d'une cage de tourbillon équipant un calibre BVL 268 représente à lui seul jusqu'à 20 heures de travail pour un artisan chevronné. Résultat d'un travail manuel minutieux, les garde-temps d'exception sont bien plus que la somme de leurs nombreux composants. Les hautes vallées du Jura, en Suisse occidentale, abritent également la fabrication de composants bruts pour des calibres automatiques plus simples, tels le BVL 191 de la montre « Solotempo » et le BVL 193. Leurs ornementation, assemblage et réglage résultent d'une étroite interaction entre des hommes et des lignes de production modernes. Des robots sont à tous les postes de production où ils sont plus performants que l'homme, quantitativement mais surtout qualitativement. Les boîtiers et les bracelets proviennent de la petite localité jurassienne Saignelégier, les cadrans de La Chaux-de-Fonds. Selon leur degré de complexité, ces composants servant d'arrière-plan au mouvement régulier des aiguilles requièrent jusqu'à une 100 d'étapes de travail, voire bien davantage. C'est à Neuchâtel, là où Bulgari a établi son quartier général horloger près du pittoresque lac, que sont prises toutes les décisions en matière de design, conception et gestion des produits, et ce en étroite concertation avec la maison mère en Italie. Bulgari achète auprès de fournisseurs qualifiés tout ce qui ne peut pas être produit en interne. Dans des ateliers inondés de lumière, des salariés dûment formés donnent le meilleur d'eux-mêmes pour que toutes les montres – qu'elles s'appellent « Bulgari Bulgari », « Diagono », « Lucea », « Octo » ou « Serpenti », pour ne citer que quelques noms – répondent aux exigences de qualité de la manufacture. Bulgari tient à sa réputation. ◦

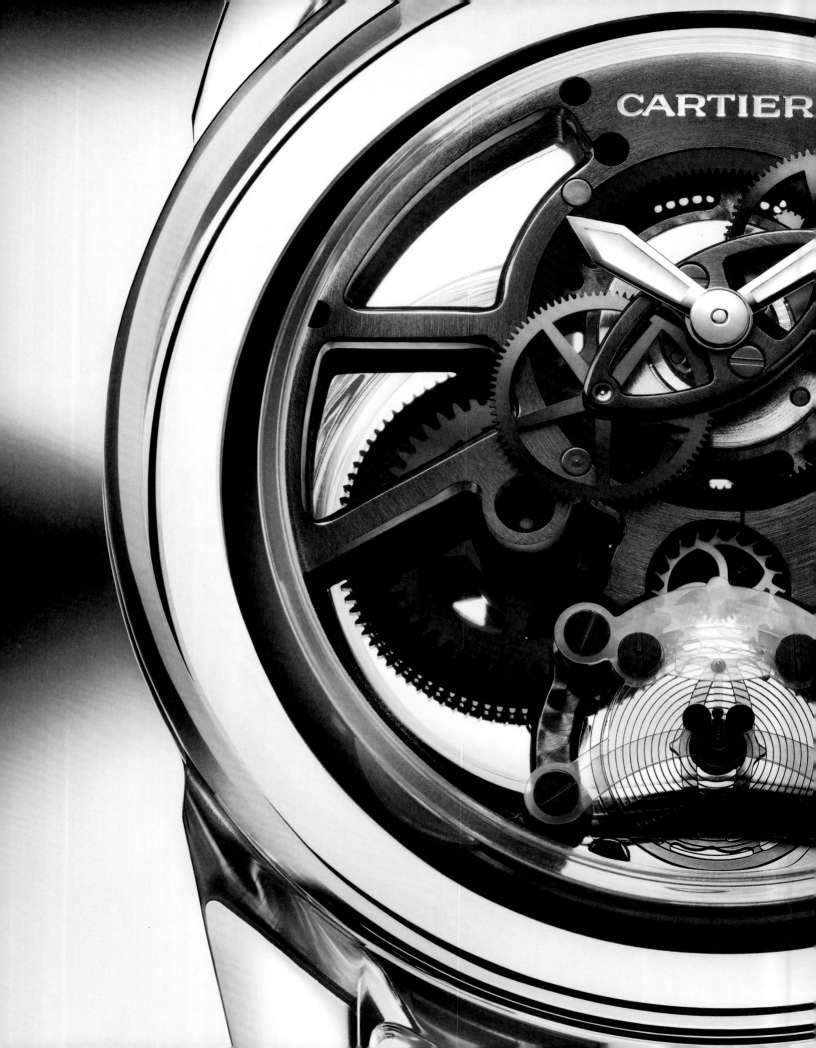

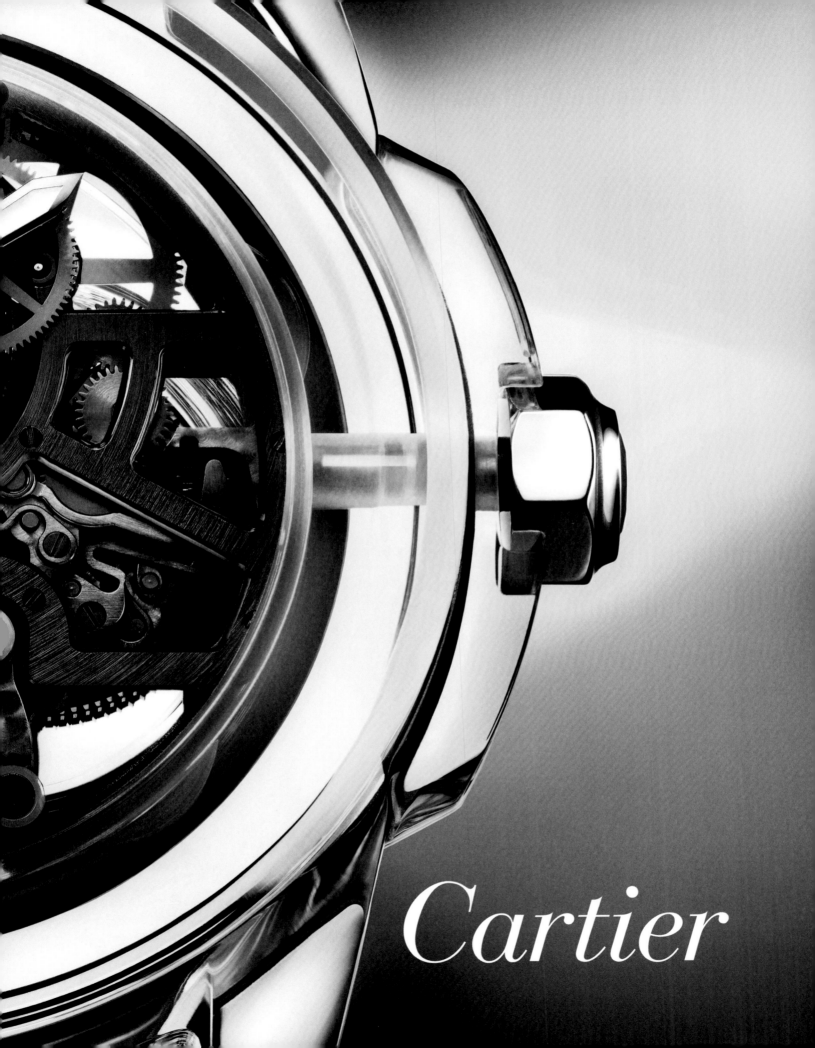

Cartier

From a Jeweler for Kings to a King among Watch Manufactories

English

Cartier is not an uncommon name in France and western Switzerland, but it earned global renown thanks to Louis-François Cartier, who was born the son of a Parisian maker of gunpowder horns in 1819. After apprenticing with Adolphe Picard, the young man took over his master's jewelry shop in 1847. Training and experience enabled the 28-year-old to know exactly what his noble clientele wanted. Chased or engraved gems and cameos were in fashion at this time, so the newcomer focused on these items for his first steps into professional independence.

Cartier's father initiated a second commercial cornerstone. Pierre Cartier encouraged his son to add ornamental watches made by external suppliers to his existing portfolio of art objects crafted from silver, bronze, or ivory. By 1892, his grandson Alfred Cartier, who had joined the up-and-coming business in 1874, had sold no fewer than 408 watches and châtelaines, i.e., chains for pocket-watches. Wristwatches probably joined Cartier's assortment in 1888. Alfred and his eldest son Louis established the firm of Alfred Cartier et Fils a decade later. Relocation into the Rue de la Paix, the world's first luxury shopping boulevard, followed in 1899. All of the leading jewelers and fashion houses were located here in the heart of debonair Paris. Louis Cartier further broadened his assortment of timepieces, which beautifully complemented his selection of exclusive jewelry. The long-cherished dream of the brand's own watches neared fulfillment in 1905. A contract with Edmond Jaeger, one of the foremost Parisian watchmakers, assured for Cartier an

exclusive collection of ladies' watches. Louis Cartier, who ranks among the world's most gifted watch designers, had joined forces with an outstanding horological technician. Their collaboration resulted in precious, attention-getting clocks and also enabled wristwatches to conquer men's wrists.

A significant contribution to their popularity was made by Cartier's friendship with the pioneering aviator and bon vivant Alberto Santos-Dumont. The Brazilian needed both hands to steer his flying machine, so he asked Cartier to create a special wristwatch to show precious time, which is especially precious at high altitudes. With the "Santos," which first became generally available for purchase in 1911, Louis Cartier emancipated the styling of this new star in the watch heavens. The same can be said of Cartier's rectangular "Tank," which he gave to General John Joseph Pershing and other high-ranking US military officers in 1918. Serial manufacturing of this watch began in 1919, but the extraordinary consequences of its development first became apparent in the 1930s, when rectangular watches in Art Deco style were en vogue around the world.

The golden age of the Cartier family ended with Louis Cartier's death on July 23, 1942. None of his heirs was able to even come close to continuing his great legacy. An epoch of disorientation began among the three Cartier branches in Paris, London, and New York, and this disunity negatively affected the house's success and creativity. Cartier still shone with the fading renown of its founding generation

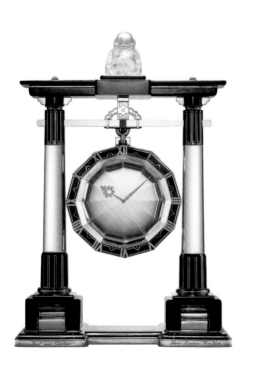
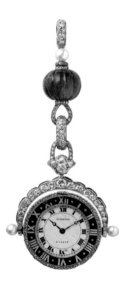
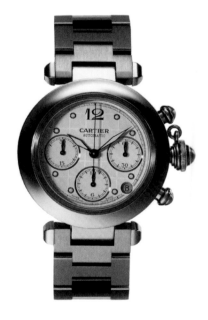

Top: Magnetic Table Clock, 1928 ○ Left: Large Portique Mystery Clock, 1923 ○ Middle: Pendant Watch, 1923 ○ Right: Pasha Chronograph, 1985

and members of high society still purchased its products because of the brand's high-sounding name, but genuine innovations were lacking. This was also the case among Cartier's watches. Alongside the brand's classics, during this transitional phase Cartier also offered watches made by famed Swiss manufacturers such as Audemars Piguet, Ernest Borel, Jaeger-LeCoultre, Movado, Patek Philippe, Piaget, or Vacheron & Constantin. A new beginning came in 1962 with the departure of family members Claude and Pierre Cartier, followed by the launch of the legendary cigarette lighter in 1968.

A new epoch on Rue de la Paix in Paris began in 1972, when a group of clever financiers headed by Joseph Kanoui acquired shares in Cartier Paris. Alain-Dominique Perrin, who had joined Briquet Cartier SA in 1969, took responsibility for the marketing. The French press was euphoric: "A new breeze of dynamism and youth now blows through number 13 Rue de la Paix." The "Must de Cartier" line of accessories began its steep upward trajectory in 1973. Watches with this name first became available in 1976. An important step toward becoming a global player came in 1979, when all worldwide participations were combined under the aegis of the Cartier Monde corporate group. Another step forward came in 1981 with the fusion of Cartier Joaillerie SA and Les Must de Cartier SA, along with the appointment of Alain-Dominique Perrin to the post of Cartier's president. Cartier Monde was acquired in 1993 by the Vendôme Luxury Group, which became the Richemont SA luxury holding in 2000.

A new era began at Cartier in 2009 with the development and production of its own watch movements in an impressive manufacturing building. Everything that has subsequently occurred at Cartier in La Chaux-de-Fonds can be subsumed under the concepts of vision, creativity, and quality. Built in 2000 and located halfway between La Chaux-de-Fonds and Le Locle, this building provides more than 320,000 square feet of floor space. The largest of Richemont's subsidiaries could successfully evolve here from a watch établisseur, which purchased its products from various suppliers, into a full-fledged manufacture with a steadily growing number of its own calibers. More than 30 exclusive movements in just six years' time bear impressive testimony to a remarkable horological feat. The most recent jewel in this crown appeared in 2015 with the "Rotonde de Cartier Grande Complication Squelette." This wristwatch combines a perpetual calendar, a minute repeater, and a "flying" tourbillon. Its automatic caliber 9406 MC concatenates 578 components and is the fruit of five years' developmental work. The prestigious Geneva Seal guarantees the outstanding quality and precision of this platinum wristwatch, which deviates from the astronomical norm by no more than 60 seconds per week. ○

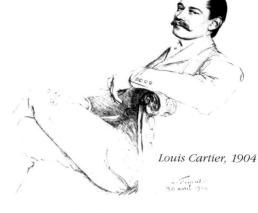

Louis Cartier, 1904

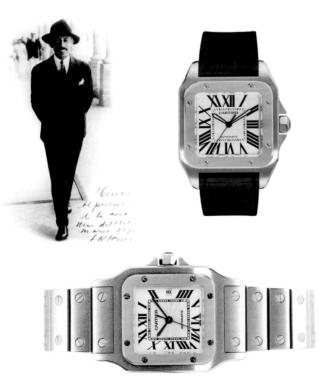

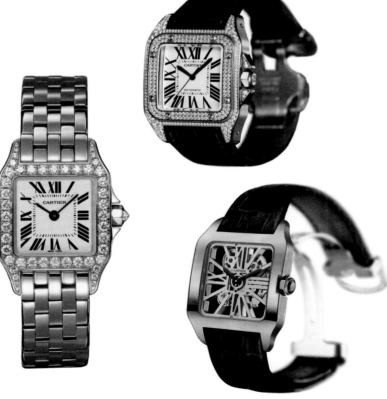

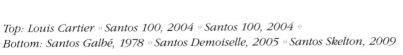

Top: Louis Cartier ○ Santos 100, 2004 ○ Santos 100, 2004 ○
Bottom: Santos Galbé, 1978 ○ Santos Demoiselle, 2005 ○ Santos Skelton, 2009

Rotonde de Cartier
Central Chronograph, 2009

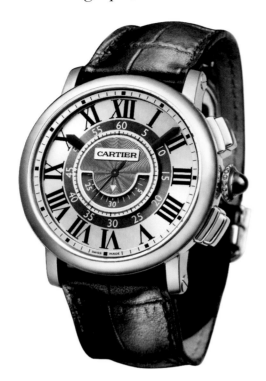

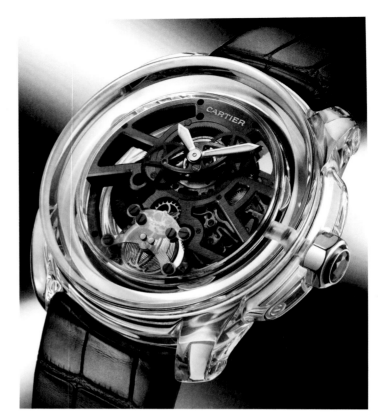

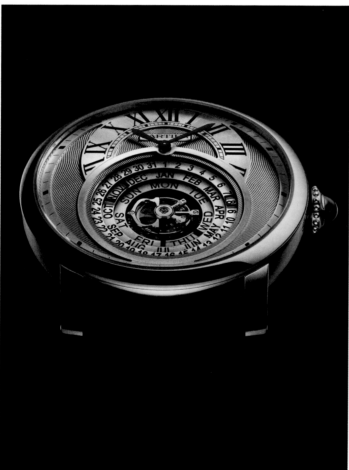

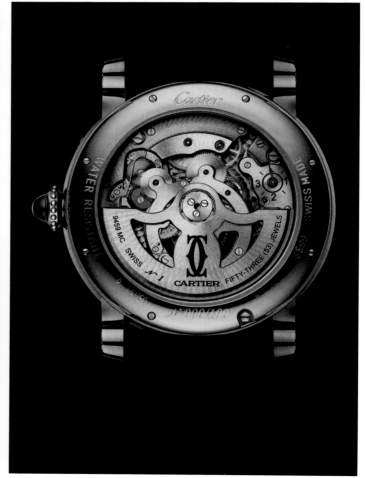

Top: *Cartier ID Two Concept Watch, 2012* ○ Bottom: *Rotonde de Cartier Astrocalendaire, 2014*

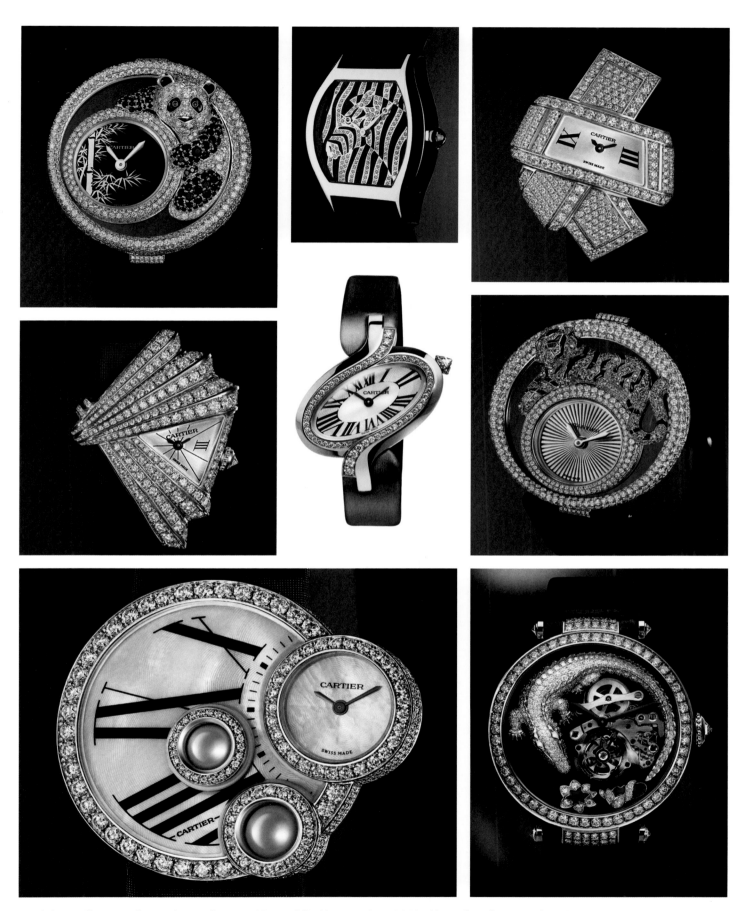

Top left: Jewellery Watch, panda motif, 2008 ◦ Top middle: Tortue Zebra, 2002 ◦ Top right: Libre Noeud, 2008 ◦
Middle left: Libre Montre Froissée, 2008 ◦ Middle: Delices de Cartiers, 2010 ◦ Middle right: Jewellery Watch, tiger motif, 2008 ◦
Bottom left: Perles de Cartier, 2008 ◦ Bottom right: Tourbillon and Crocodile, 2011

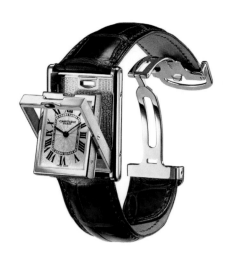 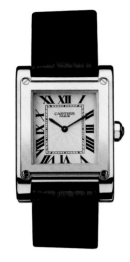 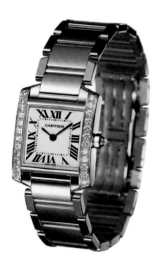 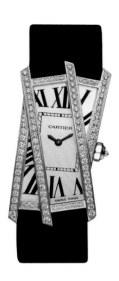

From left: Tank Basculante, 1932 ◦ Tank à Vis, 2011 ◦ Tank Française, 1996 ◦ Tank Crash, 2006

Vom Juwelier der Könige zur Uhrenmanufaktur

Der Name Cartier ist in Frankreich oder der Westschweiz keineswegs selten. Weltbekannt machte ihn der 1819 geborene Louis-François Cartier. Als der Sohn eines Pariser Pulverhornmachers das Juweliergeschäft seines Lehrmeisters Adolphe Picard übernahm, schrieb man das Jahr 1847. Nach gründlicher Ausbildung kannte der 28-jährige Geschäftsmann die Wünsche nobler Kundschaft sehr genau. Ziselierte oder gravierte Steine sowie Kameen standen hoch im Kurs. Und genau damit wagte der Newcomer seine ersten selbstständigen Schritte.

Eine zweite geschäftliche Säule initiierte sein Vater. Pierre Cartier ermutigte den Sohn, neben Kunstobjekten aus Silber, Bronze oder Elfenbein auch schmückende Uhren externer Lieferanten ins Verkaufssortiment aufzunehmen. Der Enkel Alfred Cartier, der 1874 ins aufstrebende Geschäft einstieg, veräußerte bis 1892 nicht weniger als 408 Uhren und Châtelaines, also Uhrketten. Auch Armbanduhren gehörten vermutlich ab 1888 dazu. Zehn Jahre später, 1898, schloss sich Alfred mit Louis, seinem ältesten Sohn, zu „Alfred Cartier et Fils" zusammen. 1899 erfolgte der Umzug in die weltweit erste Luxus-Einkaufsstraße, die Rue de la Paix. Hier im Herzen des mondänen Paris fanden sich die bedeutendsten Juweliere und Modehäuser. Im Zuge dessen weitete Louis Cartier auch das Uhrensortiment kräftig aus, denn es passte ideal zum erlesenen Schmuck. 1905 näherte sich der lange gehegte Traum von eigenen Cartier-Uhren endlich der Erfüllung. Durch einen Vertrag mit Edmond Jaeger, einem der wichtigsten Pariser Uhrenfabrikanten, sicherte sich Cartier eine exklusive Kollektion an Damenuhren. Louis Cartier, einer der begnadetsten Uhrendesigner, und ein vorzüglicher Uhrentechniker hatten zueinander gefunden. Aus dieser intensiven Kooperation resultierten zum einen gleichermaßen kostbare wie aufsehenerregende Pendulen. Andererseits verhalf sie der Armbanduhr zu größerer Anerkennung beim männlichen Geschlecht.

Hierzu trug allerdings auch die Freundschaft mit dem Flugpionier und Lebemann Alberto Santos-Dumont maßgeblich bei. Weil der Brasilianer zum Steuern seines Fluggeräts beide Hände brauchte, erbat er sich zum Ablesen der hoch in den Lüften besonders kostbaren Zeit eine spezielle Armbanduhr. Mit der ab 1911

frei verkäuflichen „Santos" leistete Louis Cartier den entscheidenden Beitrag zur gestalterischen Emanzipation des neuen Sterns am Uhrenhimmel. Das trifft auch auf die rechteckige „Tank" zu, die er im Jahr 1918 an General John Joseph Pershing sowie andere hohe Offiziere der US-Streitkräfte überreichte. 1919 trat auch diese Armbanduhr ins Stadium der Serienfertigung. Die ungemeine Tragweite des Entwicklungsschritts offenbarte sich allerdings erst in den 1930er Jahren, als alle Welt nach rechteckigen Exemplaren im Art-déco-Stil verlangte.

Mit dem Tod von Louis Cartier am 23. Juli 1942 endete das Goldene Zeitalter der Familie Cartier. Sämtliche Nachkommen waren nicht imstande, auch nur annähernd an das große Erbe anzuknüpfen. Eine Epoche der Orientierungslosigkeit und der Uneinigkeit zwischen den drei Cartier-Stämmen in Paris, London und New York brach an, was sich natürlich negativ auf den Erfolg und die Kreativität des Hauses auswirkte. Cartier zehrte vom Renommee der Gründergenerationen. Die High Society kaufte weiterhin des großen Namens wegen. Aber echte Innovationen blieben aus. Das gilt auch für die Uhren. Neben den hauseigenen Klassikern offerierte Cartier in dieser Phase des Übergangs auch Uhren von namhaften Schweizer Herstellern wie Audemars Piguet, Ernest Borel, Jaeger-LeCoultre, Movado, Patek Philippe, Piaget oder Vacheron & Constantin. Der Neubeginn erfolgte sukzessive ab 1962 mit dem schrittweisen Ausstieg der Familienmitglieder Claude und Pierre Cartier sowie der Lancierung des legendären Feuerzeugs im Jahr 1968.

Mit Übernahme der Cartier-Paris-Aktien durch eine Gruppe cleverer Finanziers um Joseph Kanoui brach ab 1972 an der Pariser Rue de la Paix endgültig eine neue Epoche an. Alain-Dominique Perrin, der 1969 zur Briquet Cartier S.A. gestoßen war, kümmerte sich um das Marketing. Die französische Presse jubelte: „Ein Hauch von Dynamik und Jugend weht durch die Nummer 13 an der Rue de la Paix." Ab 1973 schnellte die Accessoire-Linie „Must de Cartier" steil nach oben. Uhren dieses Namens waren ab 1976 erhältlich. Mit der Bündelung aller weltweiten Beteiligungen unter dem Dach des Konzerns Cartier Monde erfolgte 1979 ein wichtiger Schritt

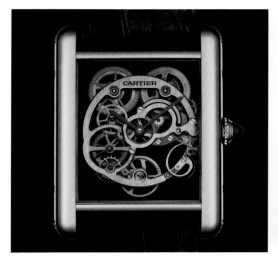
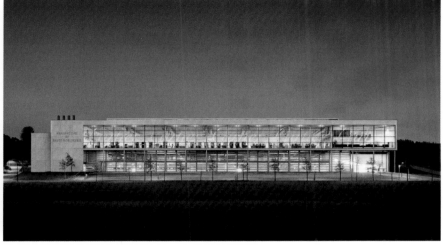

Left: Tank LC Skeleton, 2015 ◦ Right: the new factory between La Chaux-de-Fonds and Le Locle ◦ Rechts: Die neue Fabrik zwischen La Chaux-de-Fonds und Le Locle ◦ À droite : La nouvelle manufacture entre La Chaux-de-Fonds et Le Locle

in Richtung Global Player. Ein Übriges taten 1981 die Fusion der Cartier Joaillerie S.A. mit der Les Must de Cartier S.A. sowie die Ernennung von Alain-Dominique Perrin zum Cartier-Präsidenten. 1993 ging die Cartier Monde in der Vendôme Luxury Group auf, aus der 2000 die Luxus-Holding Richemont S.A. wurde.

Ein ganz neues Uhr-Zeitalter bei Cartier brach 2009 mit der Entwicklung und Fertigung eigener Uhrwerke in einem imposanten Manufakturgebäude an. Alles, was seitdem bei Cartier in La Chaux-de-Fonds geschieht, steht unter den Schlagwörtern Vision, Kreativität und Qualität. Das bereits 2000 errichtete Bauwerk auf halber Strecke zwischen La Chaux-de-Fonds und Le Locle umfasst mehr als 30000 Quadratmeter Fläche. Dort konnte sich die größte aller Richemont-Töchter trefflich vom Uhren-Etablisseur, der seine Produkte bei unterschiedlichen Zulieferern einkaufte, zu einer Manufaktur mit wachsendem Anteil eigener Kaliber entwickeln. Ausdruck eines geradezu unglaublichen Kraftakts sind die mehr als 30 exklusiven Uhrwerke innerhalb von nur sechs Jahren. Eine vorläufige Krönung brachte das Jahr 2015 mit der „Rotonde de Cartier Grande Complication Squelette". Diese Armbanduhr besitzt ewigen Kalender, Minutenrepetition und ein „fliegendes" Tourbillon. Ihr Automatikkaliber 9406 MC, zusammengefügt aus 578 Komponenten, blickt auf eine fünfjährige Entwicklungszeit zurück. Das imageträchtige Genfer Siegel für umfassende Qualität und die Präzision dieser Platinarmbanduhr sprechen für sich: Nach einer Woche darf sie nicht mehr als 60 Sekunden von der astronomischen Norm abweichen. ◦

Pasha 42MM, 2001

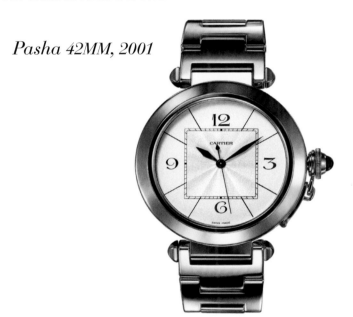

Rotonde de Cartier Grande Complication, Platinum, 2006

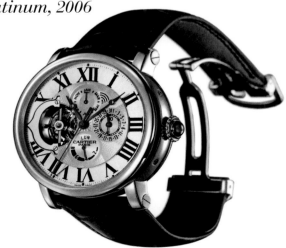

Tortue, 2006

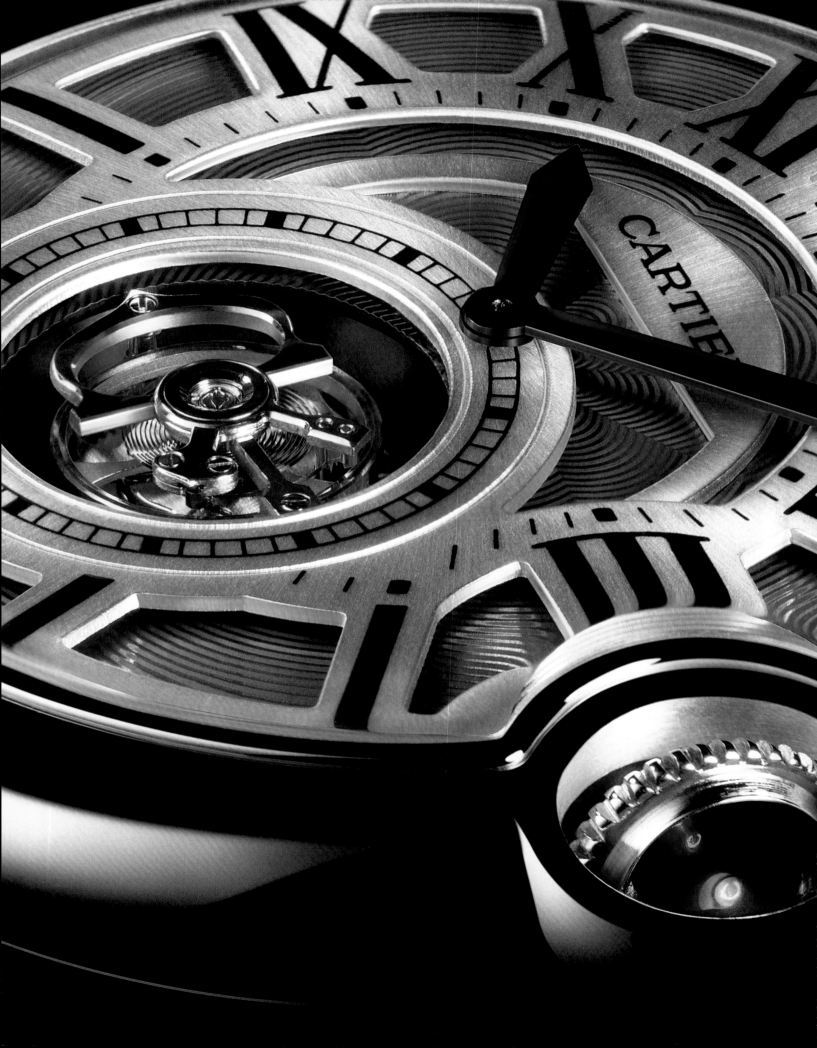

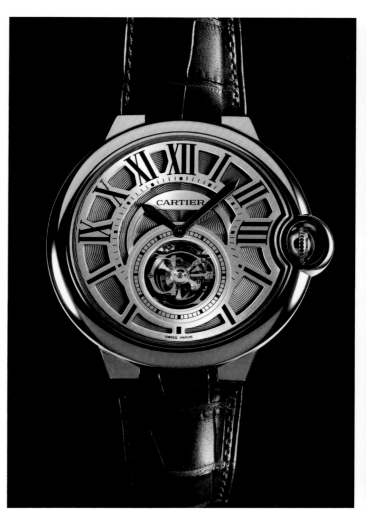
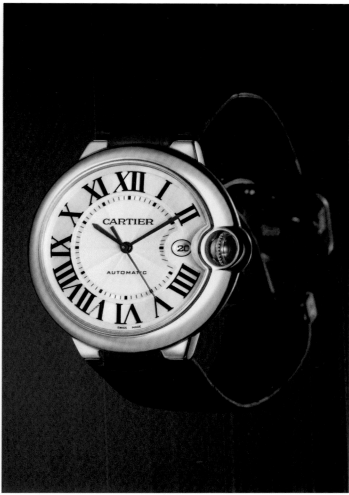
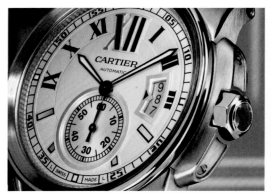
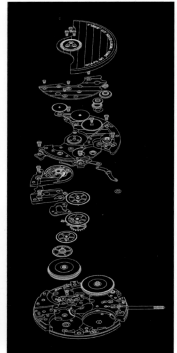
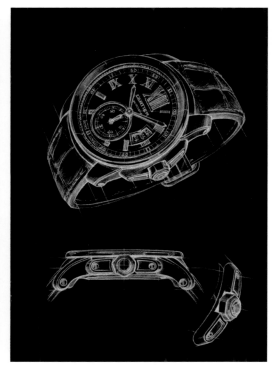
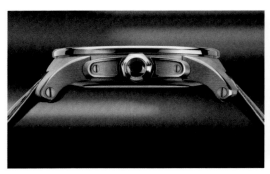

Left page and top left: Tourbillon Ballon Bleu, 2009 ◦ *Top right: Ballon Bleu, 2007* ◦
Bottom: Calibre de Cartier, 2010: exploded-view drawing and design sketch ◦ *Unten: Calibre de Cartier, 2010:*
Explosions- und Entwurfszeichnung ◦ *En bas : Calibre de Cartier, 2010 : vue éclatée et esquisse*

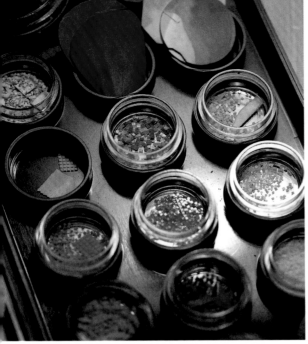

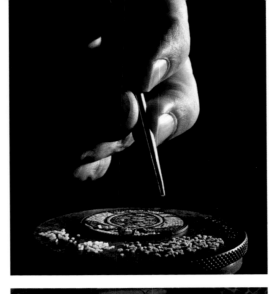

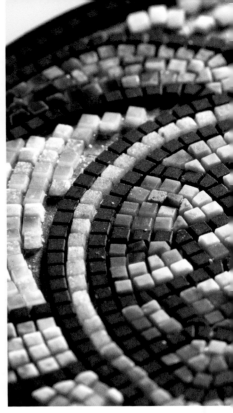

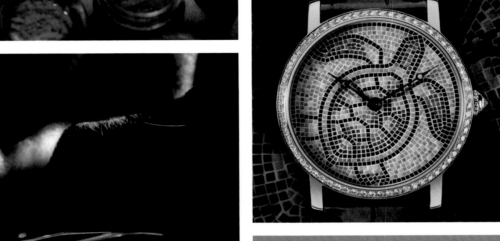

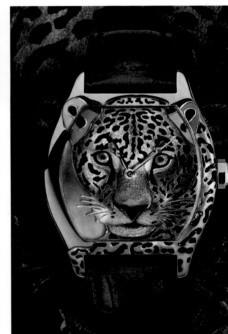

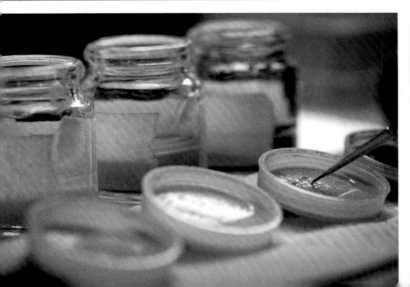

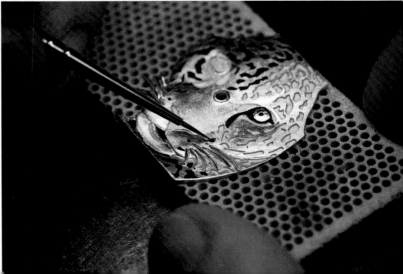

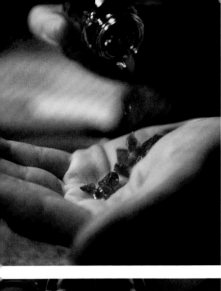
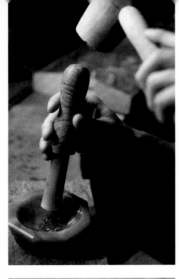
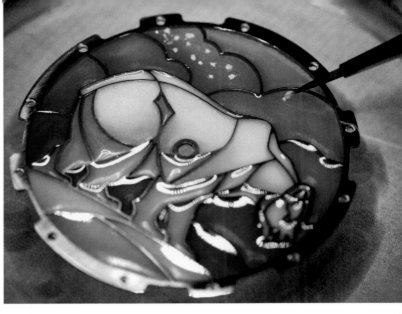
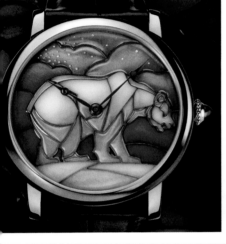
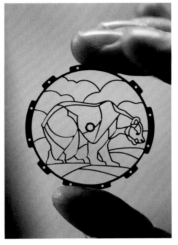
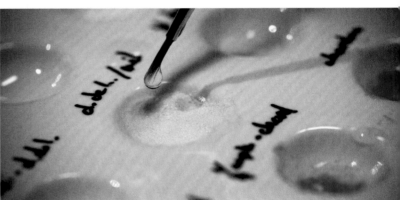
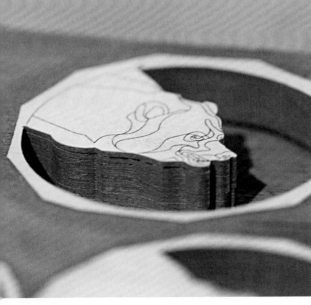
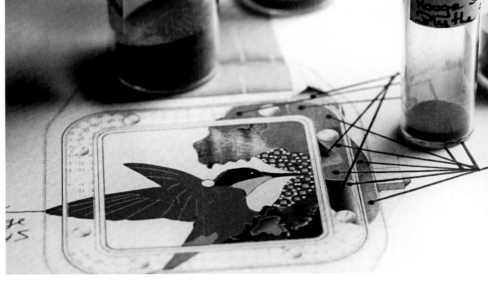
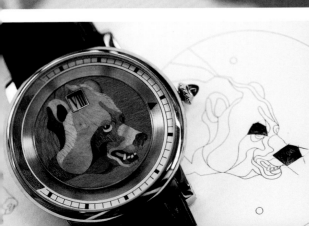
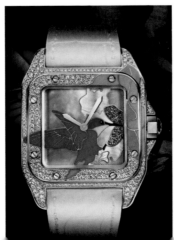
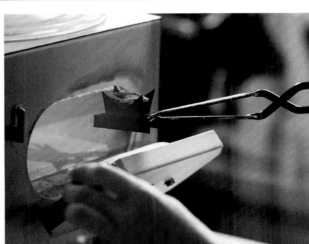

Du joailler des têtes couronnées à la manufacture de montres

En France ou en Romandie, le patronyme Cartier n'est pas rare. Né en 1819 d'un artisan spécialisé dans la fabrication de poires à poudre, Louis-François Cartier contribue à sa renommée internationale. En 1847, ce dernier reprend la joaillerie de son maître d'apprentissage Adolphe Picard. Doté d'une solide formation, ce commerçant de 28 ans connaît très précisément les désirs de sa noble clientèle. Les pierres taillées ou gravées, ainsi que les camées sont alors très en vogue. C'est précisément par la vente de ces objets que ce nouveau venu se lance en fondant la maison Cartier.

Pierre Cartier, son père, l'incite à développer un deuxième pôle commercial ; il l'encourage à proposer à la vente, outre des objets d'art en argent, bronze ou ivoire, des montres décoratives de fournisseurs externes. Le petit-fils Alfred Cartier, qui rejoint en 1874 l'affaire florissante, vendra jusqu'en 1892 pas moins de 408 montres et châtelaines, petites chaînes d'orfèvrerie supportant notamment des montres. À partir de 1888, il semble que les montres-bracelets font également partie de la gamme de produits. Dix ans plus tard, en 1898, Alfred s'associe à son fils aîné Louis pour former « Alfred Cartier et Fils ». En 1899, la boutique déménage dans la première rue du commerce de luxe dans le monde entier, la rue de la Paix. C'est ici, au cœur du Paris chic, que se trouvent les joailliers et les couturiers les plus importants. Louis Cartier élargit alors considérablement la gamme de montres, qui accompagnent à la perfection les bijoux haut de gamme. En 1905, le très vieux rêve de produire des montres Cartier est enfin près d'être exaucé. En faisant appel aux services d'Edmond Jaeger, l'un des horlogers parisiens parmi les plus connus, Cartier s'assure une collection exclusive de montres pour femmes. Louis Cartier, l'un des plus talentueux dessinateurs de montres, et Edmond Jaeger, l'un des plus grands spécialistes de technique horlogère, se sont trouvés. Leur intense collaboration engendre des pendules précieuses qui font sensation. Elle contribue également à ce que la montre-bracelet soit mieux acceptée par la gent masculine.

L'amitié avec le pionnier de l'aviation et bon vivant Alberto Santos-Dumont joue ici un rôle non négligeable. Le Brésilien ne peut consulter sa montre de poche lorsqu'il pilote et demande une montre-bracelet spéciale pour lire l'heure si précieuse dans les airs. La « Santos » est commercialisée dès 1911. Avec ce nouvel astre au firmament des montres, Louis Cartier contribue de manière décisive à l'évolution du design des garde-temps. Cela vaut aussi pour la « Tank », une montre à cadran rectangulaire, qu'il remet en 1918 au général John Joseph Pershing et à d'autres hauts gradés de l'armée américaine. En 1919, cette montre-bracelet est elle aussi fabriquée en série. Mais il faut attendre les années 1930 pour que l'incroyable portée de cette évolution soit vraiment visible, le monde entier s'arrachant alors les montres à cadran rectangulaire de style Art déco.

Le décès de Louis Cartier le 23 juillet 1942 marque la fin de l'âge d'or de la famille Cartier. Aucun de ses descendants ne sera un tant soit peu en mesure de renouer avec le grand héritage. Débute alors une époque de perte de repères et de désaccords entre les trois lignées Cartier de Paris, Londres et New York, ce qui se répercute bien sûr négativement sur la réussite et la créativité de la maison. Cartier ne vit plus que de la renommée des générations des fondateurs. La haute société continue d'acheter pour la célébrité du nom, mais il n'y a plus de véritables innovations, notamment dans le domaine des montres. Outre les classiques de la maison, Cartier propose dans cette phase de transition des montres de fabricants suisses renommés, comme Audemars Piguet, Ernest Borel, Jaeger-LeCoultre, Movado, Patek Philippe, Piaget ou Vacheron Constantin. Le renouveau se dessine peu à peu à partir de 1962 avec le retrait progressif des membres de la famille Claude et Pierre Cartier puis le lancement du légendaire briquet Cartier en 1968.

Avec le rachat de Cartier-Paris par un groupe d'habiles financiers dirigés par Joseph Kanoui, une toute nouvelle époque commence en 1972 au 13 rue de la Paix, à Paris. Entré en 1969 chez Cartier pour développer la vente de briquets, Alain-Dominique Perrin prend en charge le marketing. La presse française salue le vent de dynamisme et de jeunesse qui souffle du n° 13 de la rue de la Paix. À partir de 1973, la ligne d'accessoires « Must de Cartier » connaît une ascension fulgurante. Des montres portant cette signature arrivent sur le marché en 1976. Avec la concentration des participations du monde entier entre les mains de Cartier Monde en 1979, la maison Cartier franchit un pas important vers la position d'acteur économique mondial. Un autre est accompli en 1981 avec la fusion de Cartier Joaillerie SA et Les Must de Cartier SA, ainsi que la nomination d'Alain-Dominique Perrin à la présidence de Cartier. En 1993, Cartier Monde intègre Vendôme Luxury, groupe qui devient en 2000 la holding du luxe Richemont SA.

Une toute nouvelle ère débute pour les montres Cartier en 2009 avec le développement et la fabrication par la société de mouvements dans un imposant bâtiment. Tout ce qui se déroule dès lors chez Cartier à La Chaux-de-Fonds est marqué par les maîtres mots vision, créativité et qualité. L'édifice érigé dès l'an 2000 et situé à mi-chemin entre La Chaux-de-Fonds et Le Locle couvre plus de 30 000 mètres carrés. Dans ces lieux, la plus grande filiale de Richemont peut parfaitement se développer, et passer du statut d'établisseur, assemblant des éléments de montre achetés auprès de différents fournisseurs, à celui de manufacture produisant ses propres calibres en proportion croissante. Les 30 mouvements exclusifs produits en seulement six ans constituent un tour de force tout simplement incroyable. Sortie en 2015, la montre « Rotonde de Cartier Grande Complication Squelette » marque le couronnement provisoire de la marque. Ce garde-temps associe quantième perpétuel, répétition minutes et tourbillon volant. Composé de 578 pièces, son calibre à remontage automatique 9406 MC a nécessité cinq années de développement. Le prestigieux Poinçon de Genève, gage d'excellence, ainsi que la précision de ce garde-temps en platine parlent d'eux-mêmes : sur une semaine, l'écart de marche ne doit pas être supérieur à 60 secondes.

Monica Bellucci, Captive XL, 2010

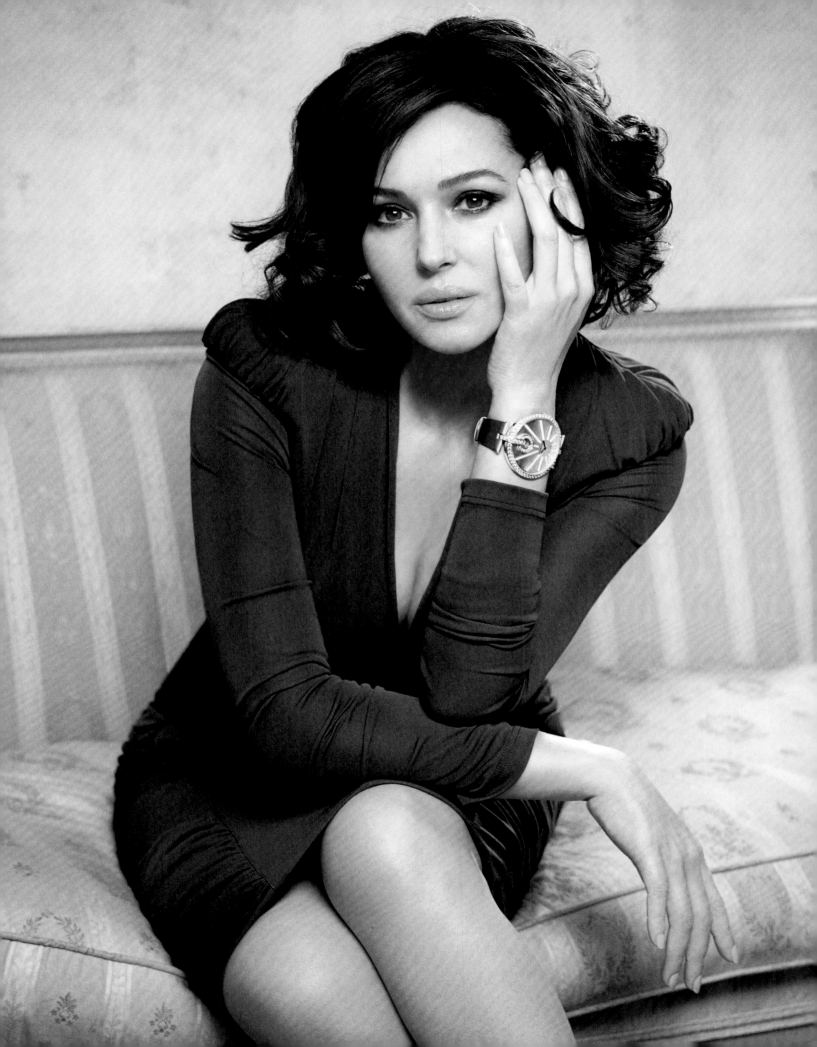

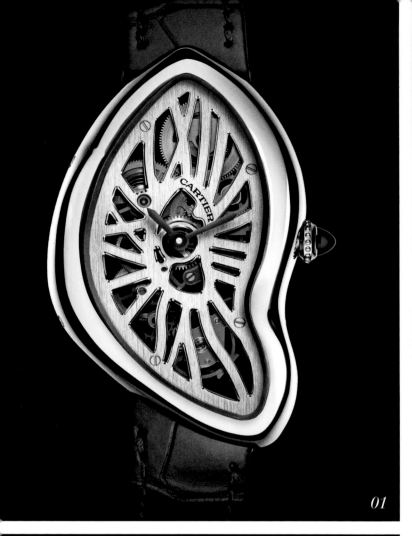

01

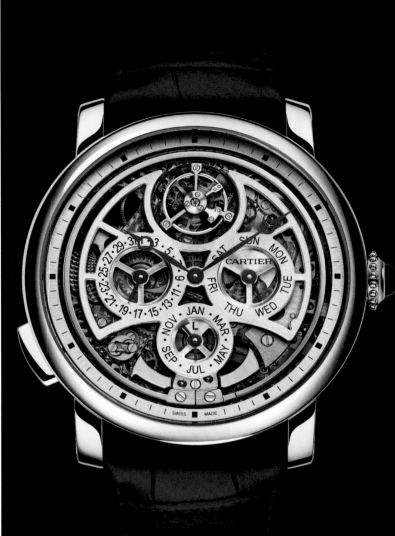

05

02

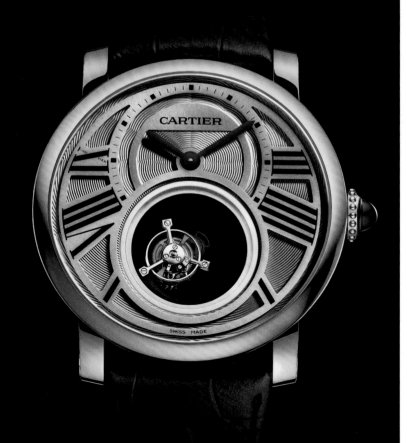

06

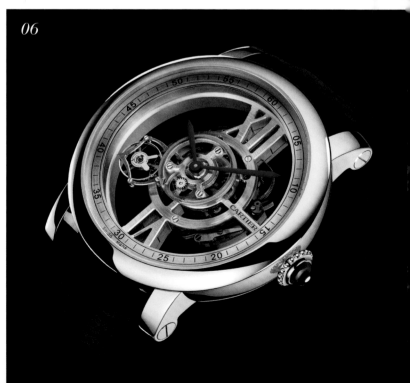

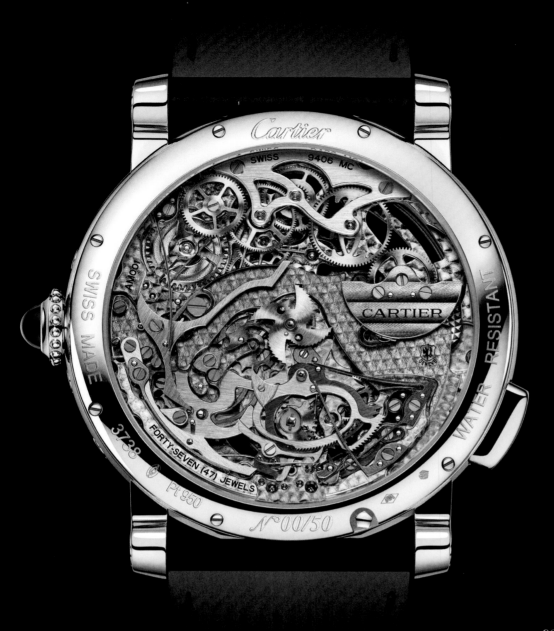

03

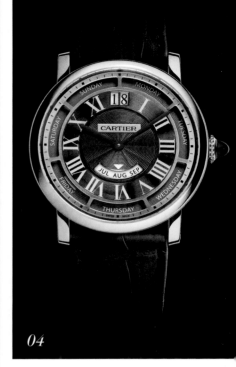

04

01. Crash Skeleton, 2015
02./03. Rotonde de Cartier
    Grande Complication
    Skeleton, 2015
04. Rotonde de Cartier
    Annual Calendar,
    40 mm, 2015
05. Rotonde de Cartier Mysterious
    Double Tourbillon, 2015
06. Rotonde de Cartier
    Astrotourbillon Skeleton, 2015
07. Clé de Cartier, 2015
08. Rotonde de Cartier
    Flying Tourbillon
    Reversed Dial, 2015

07

08

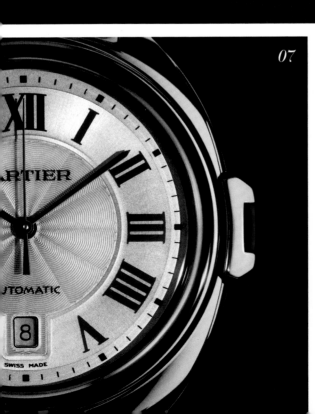

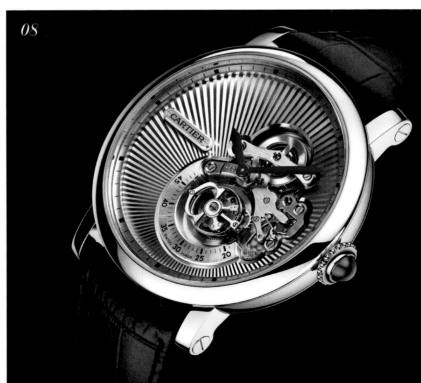

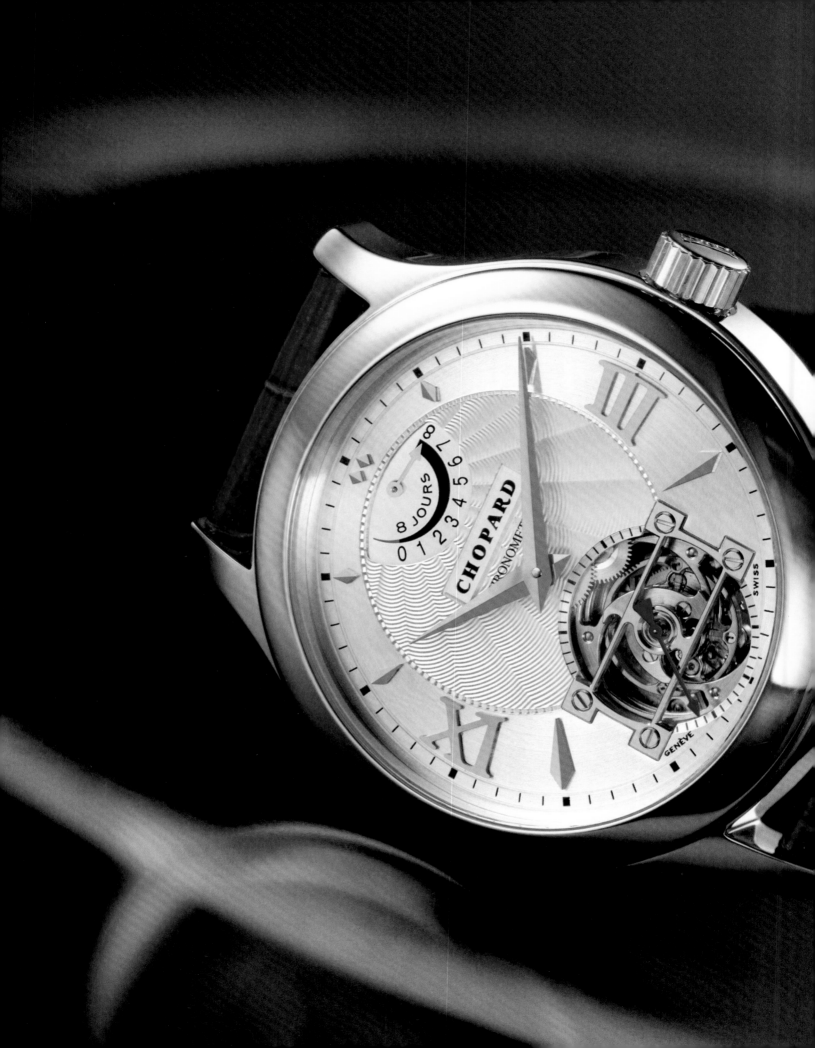

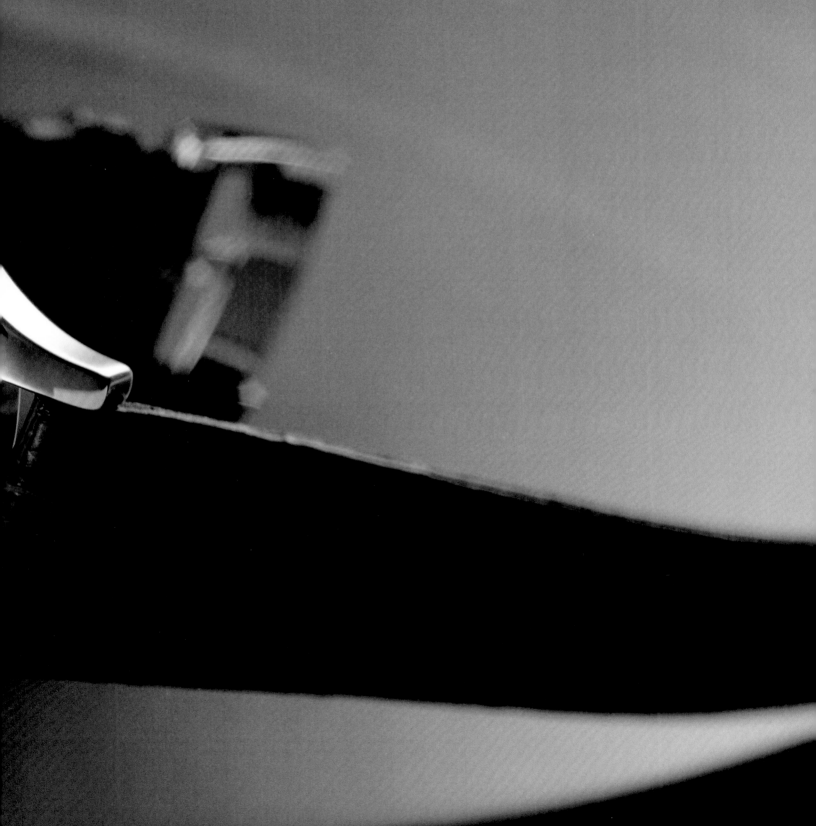

Chopard

From Family to Family

Fortunately it's purely fictive to speculate about what might have happened to Chopard without the Scheufele family in 1963, when the nearly 80-year-old Paul-André Chopard, a grandson of the firm's founder, was urgently seeking a buyer for his business. His sons showed no interest in burdening themselves with the arduous tasks of running a watch company. A German couple with ample experience in the world of jewelry and watches appeared on the scene. The chemistry between them and the elderly Monsieur Chopard was right. And a few biographical details surely contributed their fair share too.

Born in 1836, Louis-Ulysse Chopard went into business on his own at age 24 in the Jura community of Sonvilier. Although almost no papers survive to document his early horological oeuvre, there can be no doubt that he prioritized precision and quality. In 1912 he packed his most beautiful watches into two suitcases and embarked on a sales trip through Poland, Hungary, the Baltic States, and Russia. By the end of his journey, some of his products were keeping time at the court of Czar Nicolas II. This success motivated him to expand his business and led to the relocation of Chopard's headquarters to cosmopolitan Geneva in 1920. Chopard's motto—"quality through handcraftsmanship"—impressed potential customers so the firm was able to weather each successive crisis. Under the aegis of his son André Chopard, the brand primarily sold its ticking merchandise in Scandinavia and Eastern Europe. This situation remained unchanged under the direction of grandson Paul-André, who candidly told the German Karl Scheufele all the details of his success story and lamented his fate of having found no successors in his own family.

The businessman from Pforzheim likewise had plenty of tales to tell. At nearly the same time that Louis-Ulysse Chopard had packed his valises and journeyed to Russia, the young goldsmith Karl Scheufele had similarly filled his suitcases with his collection of fine jewelry watches and set out to conquer the New World. Subsequent success proved him right. The luxurious merchandise with the name ESZEHA was warmly welcomed, especially in New York. All of this had taken place nearly half a century before the grandsons of the two firms' ambitious founders met and came to appreciate one another in Geneva. Mutually aware that both parties had made the right decisions, "Le Petit-fils de L. U. Chopard" became the property of the Scheufele family. As the last watchmaker in this dynasty, Paul-André Chopard kept his old worktable, where he continued to work entirely at his own discretion until his death in 1968.

Chopard's production skyrocketed under the new leadership. With Swabian thoroughness and purposeful consistency, Karl Scheufele and his wife Karin piloted their "baby" into the upper class of sophisticated jewelry and time-keeping artistry. The company was bursting at its seams in 1968. Starting in 1974 and after a transitional solution, the firm's repeatedly enlarged building in Meyrin, a suburb of Geneva, became a new and permanent home for high-quality jewelry and luxurious watches encasing movements sourced from extramural suppliers. But this outsourcing couldn't lastingly satisfy Karl Scheufele and his son Karl-Friedrich. If Chopard wanted to offer veritably first-class watches, the label would have to evolve into a genuine manufactory. This realization catalyzed an extraordinary feat.

GPMH Chronograph, 2010

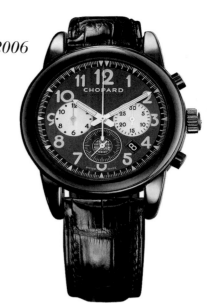

Beginning in the mid 1990s, a manufactory with ultramodern equipment and a remarkable spectrum of mechanical calibers were created in Fleurier, a little town in the isolated Val de Travers that had been badly battered by the Quartz Crisis. After a lengthy preparatory phase, Chopard unveiled a slim model named "L.U.C 1.96" with a bidirectionally winding microrotor made of 22-karat gold. There was no stopping Chopard now. Within just two decades, its design engineers and watchmakers not only mastered the entire spectrum of complications from chronographs through calendars and repeater movements to tourbillons, but also gave noteworthy impulses to the time-honored art of watchmaking. For example, they developed high-frequency balances to achieve a more accurate rate and they began using innovative materials such as silicon. As a second ébauche cornerstone, Chopard opened Fleurier Ébauches SA in the nearby neighborhood. Growing numbers of its products are destined to animate wristwatches in the "Classic Racing" and "Mille Miglia" lines. Incidentally: the beautifully renovated and energetically restored building complies with all of the rules specified by the Swiss Minergie standard. ○

Happy Sport Chrono, 2009

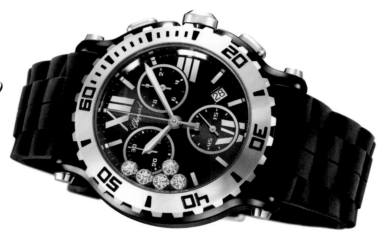

Elton John Watch, 2009

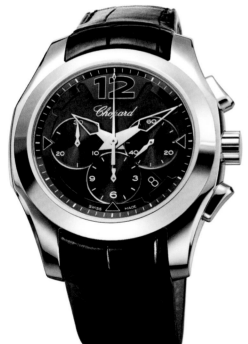

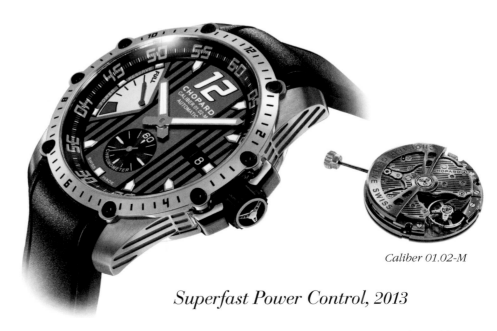

Caliber 01.02-M

Superfast Power Control, 2013

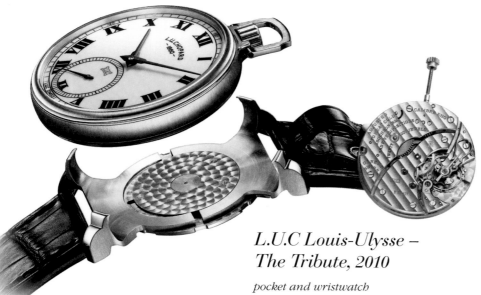

L.U.C Louis-Ulysse –
The Tribute, 2010

pocket and wristwatch

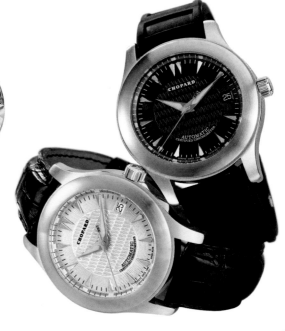

L.U.C Sport, 2000

L.U.C SW, 2008

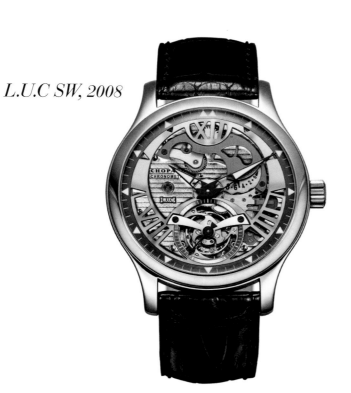

L.U.C Pro One, 2001

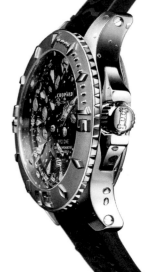

L.U.C Chrono One
Flyback, 2008

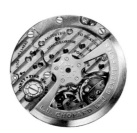

Caliber 11 C.F.

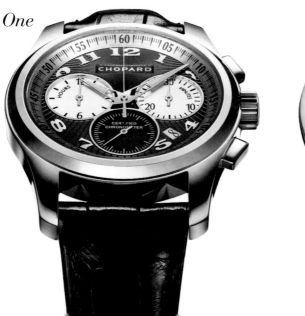

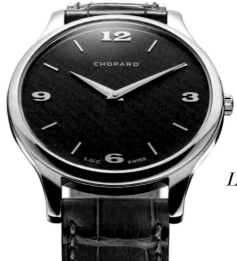

L.U.C XP, 2008

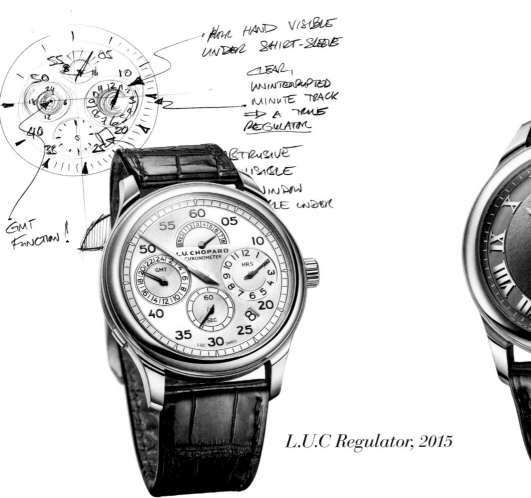

L.U.C Quattro, 2015

L.U.C Regulator, 2015

L.U.C Esprit de Fleurier, 2015

L.U.C 1963 Tourbillon, 2015

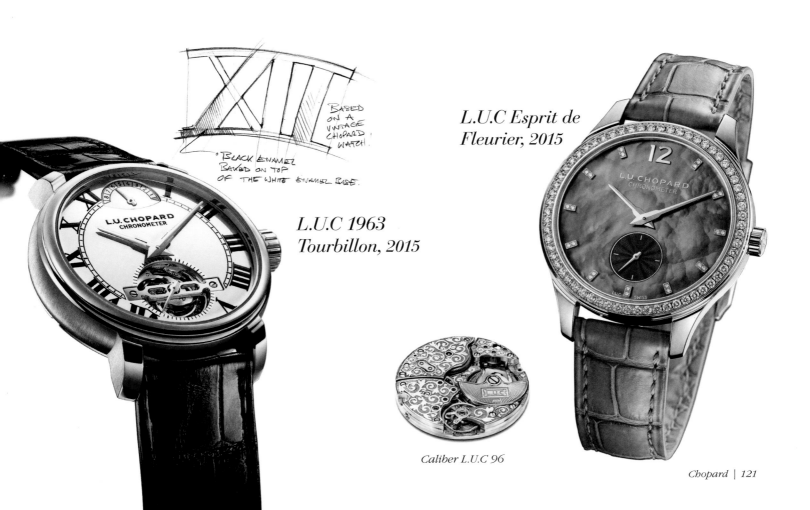

Caliber L.U.C 96

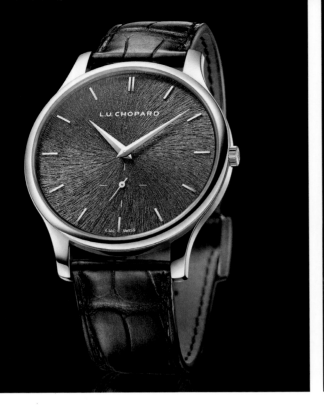

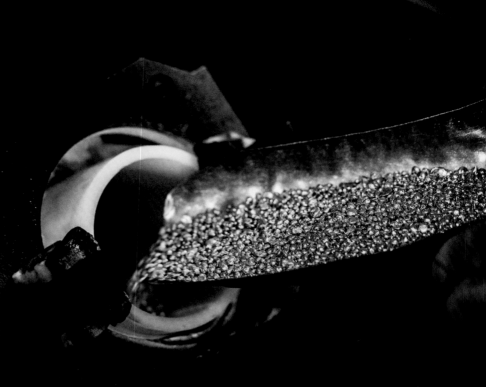

Von Familie zu Familie

Deutsch

Zum Glück ist die Frage, was 1963 mit Chopard ohne die Familie Scheufele passiert wäre, rein fiktiver Natur. In jenem Jahr musste der beinahe 80-jährige Paul-André Chopard, seines Zeichens Enkel des Firmengründers, ganz dringend die Nachfolge regeln. Seine Söhne zeigten nämlich kein Interesse am anstrengenden Uhr-Unternehmertum. Am Ende kamen die in Sachen Schmuck und Uhren erfahrenen Deutschen zum Zuge, weil neben den Konditionen auch die Chemie stimmte. Und daran dürften einige biographische Details nicht unschuldig gewesen sein.

Louis-Ulysse Chopard, Jahrgang 1836, hatte sich im Alter von 24 in der Juragemeinde Sonvilier selbstständig gemacht. Über sein frühes uhrmacherisches Œuvre existieren so gut wie keine Aufzeichnungen. Sicher ist nur, dass Präzision und Qualität einen hohen Stellenwert besaßen. 1912 packte er seine schönsten Uhren in zwei Kollektionskoffer, um mit ihnen eine Verkaufsreise durch Polen, Ungarn und die baltischen Staaten bis nach Russland zu machen. Am Ende tickten einige seiner Erzeugnisse am Hof des Zaren Nikolaus II. Das motivierte zur geschäftlichen Expansion, die 1920 mit der Verlegung des Firmensitzes ins international geprägte

Genf begann. Chopards Motto „Qualität durch Handarbeit" beeindruckte die potenzielle Kundschaft. So durchstand die Firma allerlei Krisen. Unter der Ägide von Sohn André Chopard verkauften sich die tickenden Erzeugnisse vor allem in Skandinavien und Osteuropa. Das blieb auch unter der Leitung des Enkels Paul-André so, der schließlich 1963 dem Deutschen Karl Scheufele sowohl die Erfolgsgeschichte als auch das Schicksal, keine familiären Nachfolger gefunden zu haben, in allen Details berichtete.

Der Pforzheimer Unternehmer hatte seinerseits auch einiges zu erzählen: Zu der Zeit, als sich Louis-Ulysse Chopard auf den Weg nach Russland machte, packte der junge Goldschmied Karl Scheufele ebenfalls seine Koffer. Mit seiner Kollektion feiner Schmuckuhren suchte er das Glück in der Neuen Welt. Und der Erfolg gab ihm Recht. Speziell in New York kam die luxuriöse Ware mit der Signatur ESZEHA bestens an. All das lag fast ein halbes Jahrhundert zurück, als sich die Enkel ambitionierter Firmengründer in Genf kennen und schätzen lernten. Im Bewusstsein, die für beide Seiten richtigen Entscheidungen getroffen zu haben, gelangte „Le Petit-fils de L. U. Chopard" ins Eigentum der Familie Scheufele.

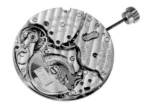 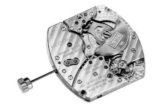 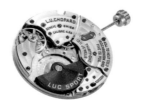 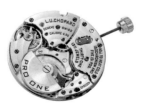

Calibers: L.U.C 96.01-L ∘ L.U.C 97.03-L ∘ L.U.C 96.04-L ∘ L.U.C 96.04-L2

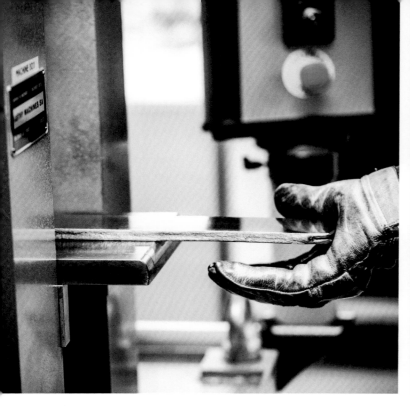
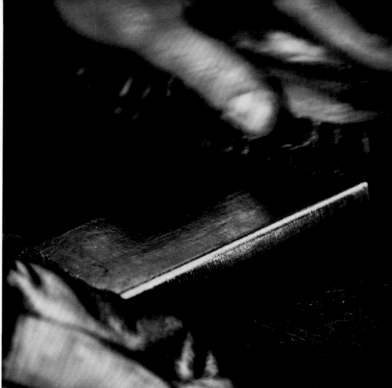

Left page: L.U.C XPS Fairmined, 2015

Paul-André Chopard behielt als letzter Uhrmacher dieser Dynastie seinen alten Werktisch. Dort arbeitete er bis zu seinem Tod im Jahr 1968 nach ganz eigenem Ermessen weiter.

Unter neuen Vorzeichen ging es bei Chopard zum einen hinsichtlich der Stückzahlen stürmisch aufwärts. Außerdem dirigierten Karl Scheufele und seine Frau Karin ihr „Baby" mit schwäbischer Gründlichkeit und zielgerichteter Konsequenz in die Upperclass anspruchsvoller Juweliers- und Zeitmesskunst. 1968 wurde der Platz zu eng. Nach einer Übergangslösung wurde das eigene, zwischenzeitlich immer wieder erweiterte Firmengebäude im Genfer Vorort Meyrin ab 1974 ein dauerhaftes Zuhause für hochkarätigen Schmuck und luxuriöse Uhren unter Verwendung zugekaufter Uhrwerke. Das stellte Karl Scheufele und seinen Sohn Karl-Friedrich auf Dauer allerdings nicht zufrieden. Uhren des obersten Segments verlangten förmlich nach eigener Manufaktur. Diese Überzeugung löste einen Kraftakt ohnegleichen aus.

Ab Mitte der 90er Jahre entstanden in Fleurier, einer durch die Quarzkrise mächtig gebeutelten Kleinstadt im abgelegen Val de Travers, nicht nur eine eigene, nach modernsten Gesichtspunkten ausgestattete Manufaktur, sondern auch ein bemerkenswertes Spektrum mechanischer Kaliber. Nach langwierigen Vorarbeiten ging 1996 das flache Modell „L.U.C 1.96" mit beidseitig aufziehendem Mikrorotor aus 22-karätigem Gold an den Start. Danach gab es kein Halten mehr. Die Konstrukteure und Uhrmacher bewältigten innerhalb von zwei Jahrzehnten nicht nur das gesamte Spektrum an Komplikationen vom Chronographen über Kalendarien und Schlagwerke bis hin zum Tourbillon, sondern verliehen der überlieferten Uhrmacherkunst auch bemerkenswerte Impulse: zum Beispiel durch die hochfrequente, der Ganggenauigkeit sehr dienliche Unruhfrequenz oder durch die Verwendung innovativer Materialien wie Silizium. Als zweites Rohwerkestandbein hat Chopard in unmittelbarer Nachbarschaft mittlerweile die Fleurier Ébauches S.A. eröffnet. Ihre Produkte werden in wachsendem Umfang Armbanduhren der Linien „Classic Racing" und „Mille Miglia" beseelen. Das dafür nach allen Regeln der Kunst renovierte und energetisch sanierte Gebäude erfüllt übrigens sämtliche Vorgaben des Schweizer Minergie-Standards. ◦

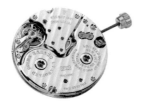
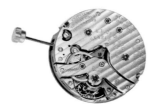
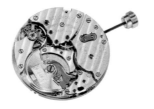

Calibers: L.U.C 98.01-L ◦ L.U.C 02.01-L ◦ L.U.C 96.06-L

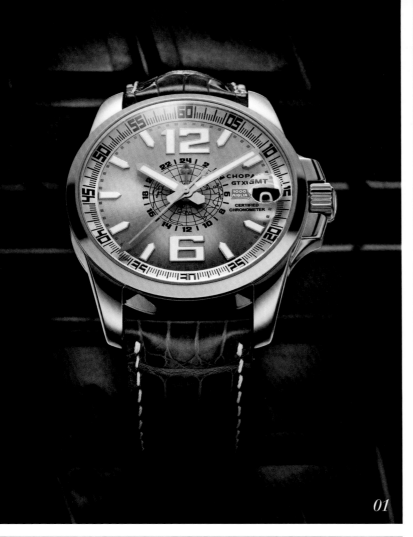

01

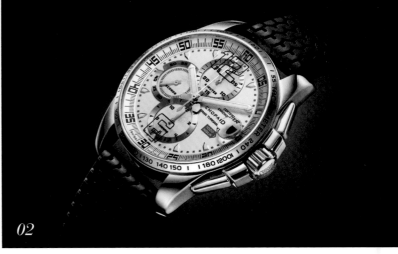

02

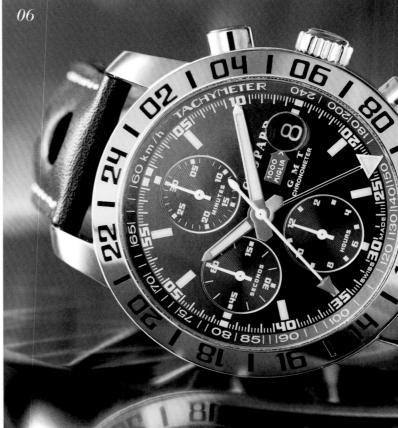

06

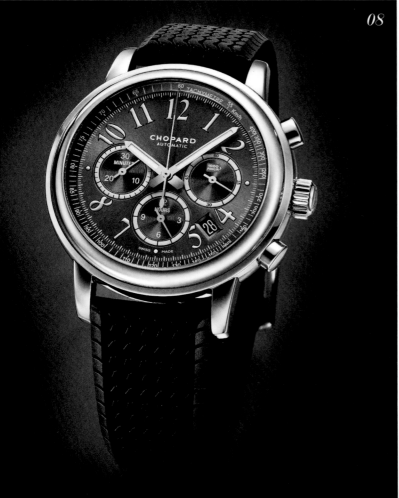

08

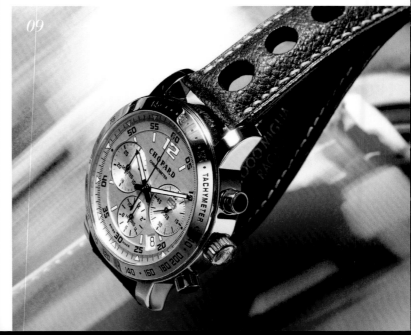

09

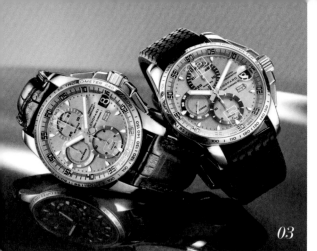

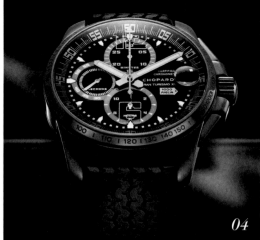

03

04

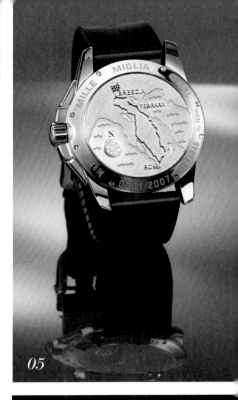

05

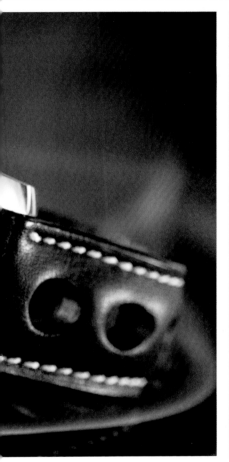

07

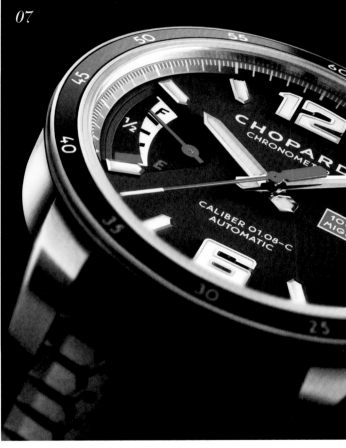

01. Mille Miglia GT XL GMT
    Chronometer, 2008
02. Mille Miglia GT XL
    Chronometer, 2008
03./05. Mille Miglia GT XL
    Chronometer, 2007
04. Mille Miglia GT XL Chronometer
    Speed Black, 2008
06. Mille Miglia GMT
    Chronometer, 2004
07. Mille Miglia GTS
    Power Control, 2015
08. Mille Miglia Chronometer, 2009
09. Mille Miglia Chronometer, 2003
10. Mille Miglia GMT
    Chronometer, 2004
11. Mille Miglia GMT
    Chronometer, 2005

10

11

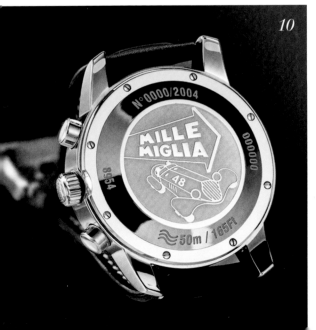

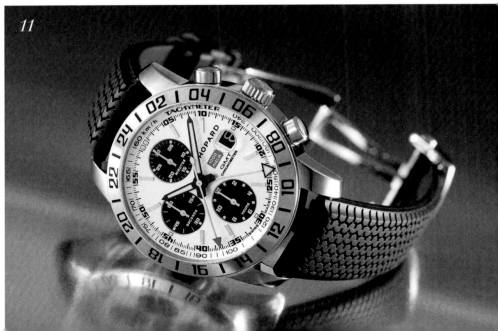

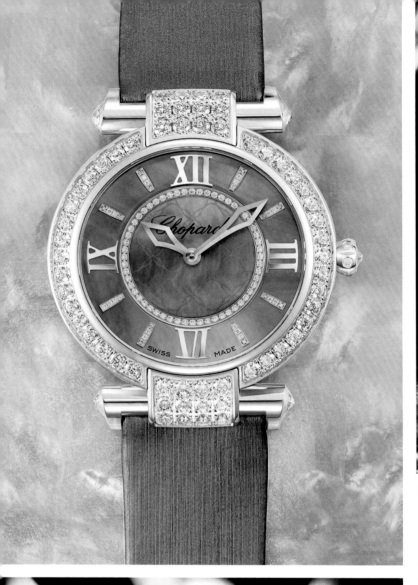

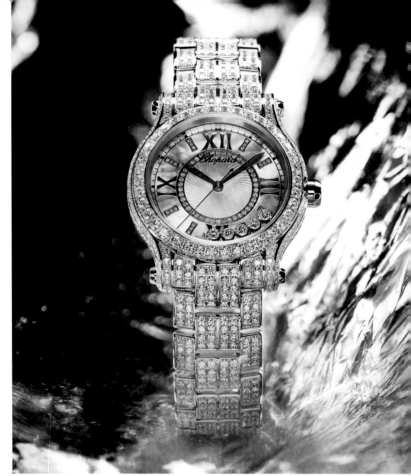

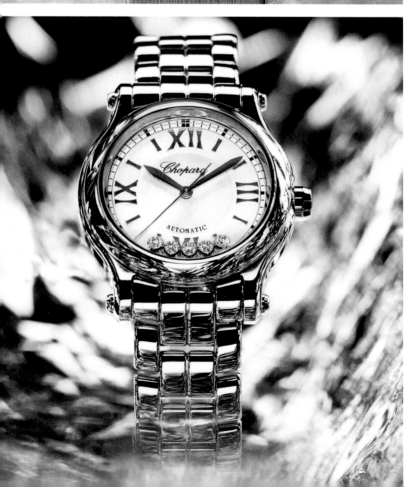

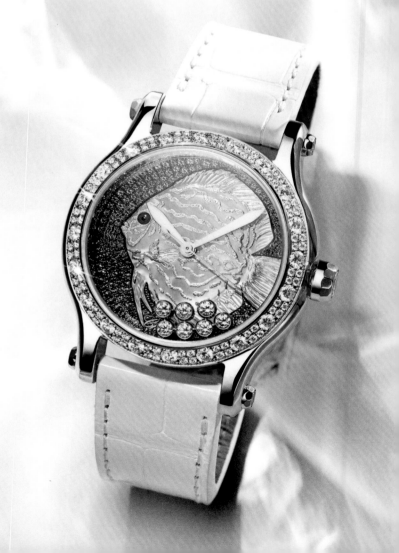

D'une famille à l'autre

Heureusement, la question de savoir ce qu'il serait advenu de la maison Chopard en 1963 sans la famille Scheufele n'est que pure rhétorique. Cette année-là, Paul-André Chopard, qui a alors presque 80 ans, lui-même petit-fils du fondateur de la marque, doit de toute urgence trouver un successeur. Ses fils ne manifestent aucun intérêt pour la gestion d'une entreprise d'horlogerie, qui n'est en aucun cas une sinécure. Finalement, c'est la dynastie allemande d'horlogers et de joailliers qui reprend l'affaire, car les conditions lui conviennent et le courant passe avec Paul-André Chopard. Certains détails biographiques y sont peut-être pour quelque chose.

Né en 1836, Louis-Ulysse Chopard fonde à 24 ans son atelier à Sonvilier, commune du Jura suisse. On ne dispose quasiment d'aucune trace documentaire sur ses premières réalisations horlogères. Une chose est sûre, la précision et la qualité en sont des piliers essentiels. En 1912, il range ses plus belles montres dans deux valises de présentation, pour les vendre lors d'un voyage qui le conduit à travers la Pologne, la Hongrie, les États baltes et jusqu'en Russie. À la fin de son périple, certains de ses produits font tic tac à la cour du tsar Nicolas II. Cela incite l'entreprise à développer ses activités commerciales, une expansion qui débute en 1920 par le transfert du siège de la société dans la Genève cosmopolite. La devise de Chopard, « Qualité par le plus de travail manuel possible », impressionne ses clients potentiels. La société surmonte ainsi toutes sortes de crises. Sous l'égide d'André Chopard, son fils, les montres se vendent en Scandinavie et en Europe de l'Est. Il en va de même sous la direction de son petit-fils Paul-André, qui finit en 1963 par raconter dans les moindres détails à l'Allemand Karl Scheufele l'histoire de sa réussite mais aussi l'infortune de ne pas avoir trouvé de successeur dans sa famille.

L'entrepreneur de Pforzheim a lui aussi vécu bien des choses : à l'époque où Louis-Ulysse Chopard se met en route pour la Russie, le jeune joailler Karl Scheufele fait lui aussi ses valises. Avec sa collection de montres joaillères raffinées, il tente sa chance dans le Nouveau Monde. Et le succès lui donne raison. À New York tout particulièrement, ses produits de luxe portant la signature ESZEHA reçoivent un excellent accueil. Tout cela se déroule près d'un demi-siècle avant que les petits-fils des ambitieux fondateurs d'entreprise ne fassent connaissance et sympathisent à Genève.

Considérant que les bonnes décisions ont été prises pour les deux parties, « Le Petit-fils de L. U. Chopard » devient la propriété de la famille Scheufele. En tant que dernier maître horloger de la dynastie Chopard, Paul-André conserve son ancien établi, sur lequel il travaillera de sa propre volonté jusqu'à sa mort en 1968.

Sous ces nouveaux auspices, la maison Chopard voit sa production augmenter massivement. Karl Scheufele et sa femme Karin dirigent par ailleurs leur « bébé » avec toute la rigueur souabe et une détermination ciblée qui le font admettre parmi l'élite des joailliers et horlogers d'exception. En 1968, l'espace vient à manquer. Après une solution temporaire, le bâtiment sans cesse agrandi de l'usine de Meyrin, dans les environs de Genève, abrite à partir de 1974 de façon définitive la fabrication de pièces de haute joaillerie et celle de montres de luxe à partir de mouvements achetés. Mais Karl Scheufele et son fils Karl-Friedrich ne peuvent s'en contenter. Ils tiennent à ce que les montres les plus complexes soient fabriquées dans leur propre manufacture. Cette conviction les conduit à un tour de force peu commun.

À partir du milieu des années 1990, Fleurier, petite ville reculée du val de Travers fortement ébranlée par la crise du quartz voit apparaître, non seulement cette manufacture ultramoderne, mais aussi une remarquable gamme de calibres mécaniques. Au terme de longs et difficiles travaux préparatoires, Chopard sort le modèle plat « L.U.C 1.96 » à remontage bidirectionnel, qui est équipé d'un microrotor en or 22 carats. Ensuite, les choses s'accélèrent. Les concepteurs et les maîtres-horlogers maîtrisent en l'espace de deux décennies non seulement toute la gamme des complications, des chronographes au tourbillon, en passant par les calendriers et les sonneries, ils donnent également des impulsions remarquables à la haute horlogerie traditionnelle : notamment la fréquence élevée du balancier, très utile à la précision de marche, ou l'utilisation de matériaux innovants, comme le silicium. Comme second site de production d'ébauches, Chopard ouvre à proximité immédiate Fleurier Ébauches SA. Les produits fabriqués par cette filiale animeront de plus en plus fréquemment des montres-bracelets des lignes « Classic Racing » et « Mille Miglia ». Rénové dans les règles de l'art et assaini sur le plan énergétique, le bâtiment affecté à cette production répond en outre à toutes les exigences du label suisse Minergie. ◦

L.U.C Tonneau, 2001

Top left: Imperiale, 2015 ◦
Top right and bottom left:
Happy Sport Automatic 30 mm, 2015 ◦
Bottom right: Happy Fish, 2015

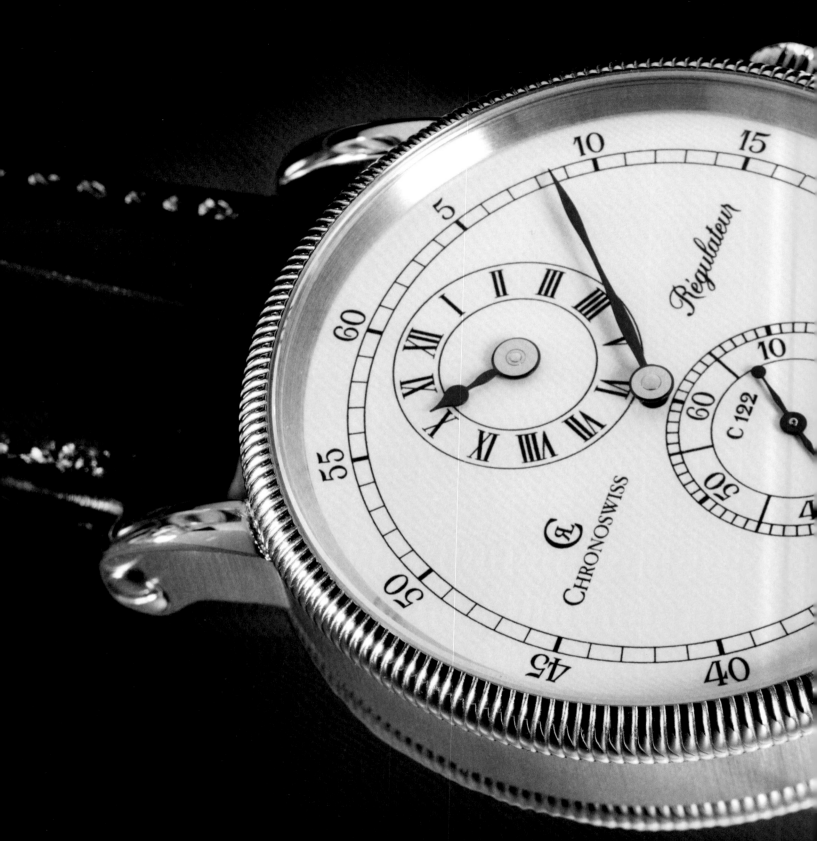

Chronoswiss

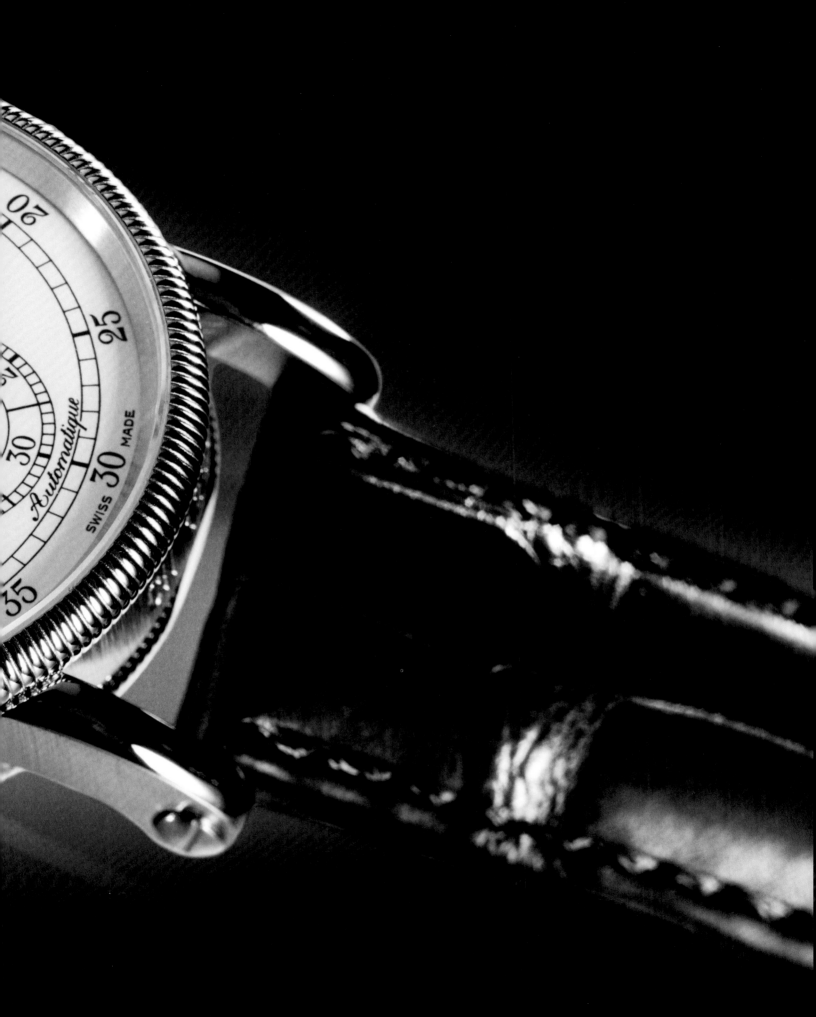

Under the Sign of the "Régulateur"

Born in 1943, Gerd-Rüdiger Lang took the daring step of starting his own business in 1981. The trained watchmaker and former director of Heuer Time (Germany) had become unemployed when that company went out of business. Necessity turned out to be the mother of invention, because Lang founded his new company in Munich with plenty of ideas, courage, decisiveness and willingness to take risks. His starting capital also included a huge stock of spare parts, including components for chronographs. The Quartz Revolution had left lasting traces in the watch scene, but this left Gerd-Rüdiger Lang singularly unimpressed and firmly convinced that good old mechanical watchmaking still held ample potential. He acted on his conviction by buying and selling conventional ticking wristwatches in retro look. His first collections of his own debuted in 1983. Although many of the wristwatches bearing Chronoswiss' name were already equipped with transparent backs, Gerd-Rüdiger Lang hadn't yet developed a characteristic style. That's why the real history of Chronoswiss began in 1987 with the premiere of the so-called "Régulateur." Not only did this wristwatch have an unconventional dial, it also featured Chronoswiss' distinctive case, which combines nineteen individual parts. The historical rationale for removing the hour-hand from the dial's center and repositioning it upward at the "12" can be traced to precise regulator clocks which watchmakers formerly relied on when finely adjusting other timepieces. For this purpose, they primarily needed to refer to the clock's second-hand. An hour-hand at the center of the clock's face made it more difficult to view the little second-hand. Clockmakers accordingly relocated the former so it could no longer obstruct their view of the latter. For the first edition of his "Régulateur," the ambitious entrepreneur used a correspondingly modified version of Unitas' hand-wound Caliber 6376 Z, which was last produced in 1984. Only a few of these calibers were available, so Chronoswiss was obliged to impose strict limits on the number of watches in this first edition. Lady Luck rewarded Gerd-Rüdiger Lang on another occasion, when the assiduous businessman's excellent connections enabled him to acquire a large stock of an automatic movement that was no longer in use. Enicar had developed the rotor-wound caliber in the 1970s, but was afterwards swallowed by the "Quartz Maelstrom." After subjecting this caliber to necessary revisions, Lang was able to launch his "Régulateur Automatique" in 1990. The "Fascination of Mechanisms," a phrase which served as the title of Chronoswiss' catalogue in 1992, was subsequently celebrated by an extensive collection of watches in which chronographs played a significant role. A revised version of Progress 6361 ticked inside the "Régulateur à Tourbillon" when it debuted in 2000. And in 2001, the aforementioned Enicar movement animated the "Chronoscope," a stopwatch with a regulator dial. Gerd-Rüdiger Lang offered collectors and fans of his brand regularly recurring opportunities to purchase strictly limited "Régulateur" editions encasing movements that were no longer manufactured. The German chapter of Chronoswiss' history ended in 2012 and the brand's future shifted to Switzerland. Due to his advanced age, Gerd-Rüdiger Lang entrusted his "baby," which had conquered an impressive eighth place in the ranking of German luxury brands, into the arms of Swiss entrepreneurs Eva Maria and Oliver Ebstein. The brand is presently headquartered in the picturesque city of Lucerne on the shores of the homonymous lake, where the House of Chronoswiss handles all of the label's activities. The brand's sporty side is represented by the distinctive "Timemaster." After the well-known case underwent a facelift, the classical line is now known as "Sirius." This eccentric icon is highlighted again in 2016, but its several versions—some with classical dials, others with three-dimensional faces—have been renamed "Regulator." ○

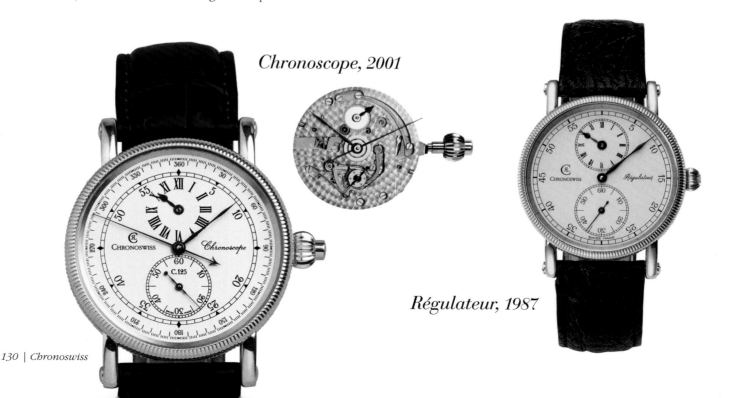

Chronoscope, 2001

Régulateur, 1987

Im Zeichen des „Régulateur"

Deutsch

Als Gerd-Rüdiger Lang, Jahrgang 1943, gezwungenermaßen den Sprung in die berufliche Selbständigkeit wagte, schrieb man das Jahr 1981. Die Liquidation der von ihm geleiteten Heuer Time (Deutschland) trieb den gelernten Uhrmacher in die Arbeitslosigkeit. Der Not gehorchend startete der Geschasste mit Ideen, Mut, Risikofreude und Entschlossenheit in München neu durch. Sein Anfangskapital: ein riesiges Lager an Ersatzteilen, darunter viele für Chronographen. Ziemlich kalt ließ ihn dabei, dass die Quarz-Revolution nachhaltige Spuren in der Uhrenszene hinterlassen hatte. Gerd-Rüdiger Lang erkannte das Zukunftspotenzial der guten alten Mechanik. Und er reagierte durch den An- und Verkauf konventionell tickender Armbanduhren im Retrolook. Ab 1983 entstanden eigene Kollektionen. Viele der Chronoswiss signierten Armbanduhren verfügten zwar schon über einen Sichtboden, zeichneten sich aber noch nicht durch eine eigenen Handschrift aus. Deshalb startete die wirkliche Chronoswiss-Geschichte im Jahr 1987 mit dem sogenannten „Régulateur". Diese Armbanduhr brachte nicht nur ein außergewöhnliches Zifferblatt, sondern auch das signifikante, aus 19 Teilen zusammengefügte Chronoswiss-Gehäuse. Natürlich hatte es seine Bewandtnis mit der Verlagerung des Stundenzeigers aus der Mitte nach oben zur „12". Diese Art der Zeitanzeige leitete sich ab von alten Präzisions-Regulatoren, welche Uhrmacher zum Regulieren ihrer Uhren nutzten. Dabei kam es in erster Linie auf die Sekunden an. Weil der mittig positionierte Stundenzeiger das Ablesen des kleinen Sekundenzeigers beeinträchtigte, musste er dorthin weichen, wo er nicht mehr stören konnte. Für die erste Edition des „Régulateur" nutzte der ambitionierte Unternehmer eine entsprechend modifizierte Version des Handaufzugskalibers Unitas 6376 Z, dessen Produktion 1984 eingestellt worden war. Wegen der nur noch begrenzten Verfügbarkeit dieses Uhrwerks musste Chronoswiss die Edition strikt limitieren. Ein weiteres Mal belohnte das Glück den Tüchtigen, als der gut vernetzte Gerd-Rüdiger Lang größere Restbestände eines nicht mehr genutzten Automatikwerks erwerben konnte. Enicar hatte dieses Rotor-Kaliber in den 1970er Jahren

entwickelt, war dann aber auch in den Quarz-Strudel geraten. Nach entsprechendem Umbau ging 1990 der „Régulateur Automatique" an den Start. In den folgenden Jahren zelebrierte Chronoswiss die 1992 zum Katalogtitel erwählte „Faszination der Mechanik" durch eine breite Uhrenkollektion, in der Chronographen eine bedeutende Rolle spielten. Auf besagtem Enicar basierte auch das 2001 vorgestellte „Chronoscope", ein Stopper mit Regulator-Zifferblatt. Ein Jahr zuvor debütierte ferner das „Régulateur à Tourbillon" unter Verwendung des umgebauten Progress 6361. Sammlern und Markenfans offerierte Gerd-Rüdiger Lang in regelmäßigen Abständen streng limitierte „Régulateur"-Editionen mit nicht mehr produzierten Uhrwerken. 2012 endete das deutsche Chronoswiss-Kapitel. Die Zukunft spielte sich in der Schweiz ab. Aus Altersgründen hatte Gerd-Rüdiger Lang sein „Baby", das im Ranking deutscher Luxusmarken einen bemerkenswerten achten Platz erobern konnte, in die Arme des Schweizer Unternehmerpaars Eva Maria und Oliver Ebstein gelegt. Mittlerweile residiert die Marke im Herzen der malerischen Stadt Luzern am Vierwaldstätter See. Alle Aktivitäten gehen im House of Chronoswiss über die Bühne. Die sportliche Seite der Marke repräsentiert der markante „Timemaster". Nach einem Facelift des bekannten Gehäuses nennt sich die klassische Linie „Sirius". 2016 steht einmal mehr im Zeichen der exzentrischen Ikone. Allerdings bekamen die verschiedenen Ausführungen – teils mit klassischen, teils mit dreidimensionalen Zifferblättern – einen neuen Namen: „Regulator".

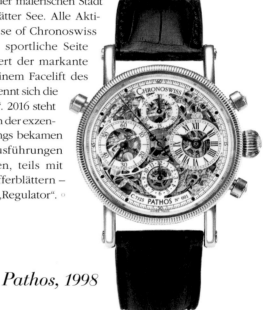

Pathos, 1998

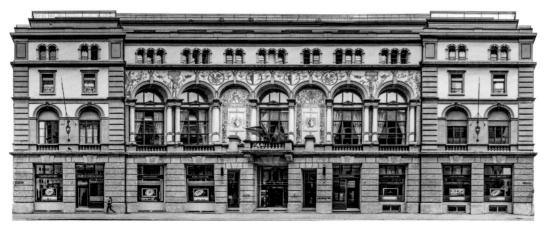

Opposite page, top:
Grand Régulateur, 1994

House of Chronoswiss, Lucerne

Sous le signe du « Régulateur »

Français

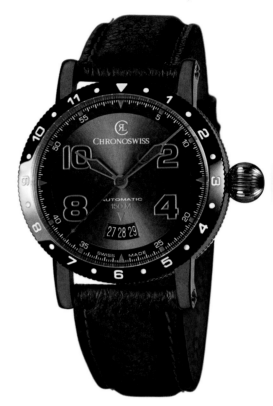

Timemaster 150, 2016

En 1981, Heuer Time Allemagne est mis en liquidation, entraînant le licenciement de son directeur, Gerd-Rüdiger Lang. Horloger de métier, il s'installe contraint et forcé à son compte, à 38 ans. Nécessité faisant loi, cet homme déchu prend un nouveau départ à Munich grâce à ses idées, son courage, son goût du risque et sa détermination. Son capital de départ : un énorme stock de pièces détachées, essentiellement pour chronographes. Assez indifférent à l'empreinte durable laissée sur le monde de l'horlogerie par la révolution du quartz, Gerd-Rüdiger Lang reconnaît le potentiel d'avenir de la bonne vieille mécanique et décide d'acheter et de vendre des montres-bracelets mécaniques au look rétro. Puis, à partir de 1983, il crée ses propres collections. Nombre des premières montres-bracelets signées Chronoswiss possèdent déjà un fond transparent, mais pas encore de style propre. L'histoire de Chronoswiss ne débute donc vraiment qu'en 1987, avec le « Régulateur ». Ce modèle marque l'apparition non seulement d'un cadran inédit, mais aussi du célèbre boîtier Chronoswiss à 19 composants. Son signe caractéristique, c'est bien sûr l'aiguille des heures, qui n'est plus au centre mais à midi. Ce type d'indication du temps est inspiré des anciennes horloges de précision que les horlogers utilisaient pour régler leurs montres. Les secondes jouaient alors un rôle prépondérant. L'aiguille des heures au centre gênant la lecture de la petite aiguille des secondes, il fallait la placer à un endroit où elle ne pouvait plus déranger. Pour la première édition de son « Régulateur », l'ambitieux entrepreneur utilise une version adaptée du calibre à remontage manuel Unitas 6376 Z, dont la production a été arrêtée en 1984. Avec les quelques exemplaires encore disponibles de ce mécanisme d'horlogerie, Chronoswiss sort une édition strictement limitée. Puis, la chance sourit une nouvelle fois aux audacieux. Gerd-Rüdiger Lang, qui dispose d'un bon réseau, fait l'acquisition des stocks inutilisés d'un mouvement mécanique à remontage automatique, un calibre à rotor mis au point dans les années 1970 par la marque Enicar, avant qu'elle aussi ne soit happée par le tourbillon du quartz. Ce mouvement, modifié comme il se doit, donne naissance en 1990 au « Régulateur Automatique ». En 1992, pour célébrer la sortie du catalogue intitulé « Fascination de la mécanique », Chronoswiss présente une vaste collection de montres, essentiellement des chronographes. En 2000, le « Régulateur à Tourbillon », fabriqué à partir d'un calibre Progress 6361 revisité, fait ses débuts. Présenté en 2001, le « Chronoscope », un chronographe à cadran de type régulateur, repose également sur le calibre Enicar évoqué plus haut. Gerd-Rüdiger Lang propose ainsi régulièrement aux collectionneurs et aux inconditionnels de la marque des éditions du « Régulateur » en séries très limitées, en s'appuyant sur des mouvements qui ne sont plus produits. Chronoswiss ayant refermé le chapitre allemand de son histoire en 2012, l'avenir de la marque se joue désormais en Suisse. En raison de son âge, Gerd-Rüdiger Lang confie son « bébé », qui a atteint une remarquable huitième place au classement des marques horlogères de luxe en Allemagne, au couple d'entrepreneurs suisses Eva Maria et Oliver Ebstein. Toutes les activités se déroulent désormais au nouveau siège de la marque (House of Chronoswiss), dans la pittoresque ville de Lucerne, sur les rives du lac des Quatre-Cantons. Si la « Timemaster » représente le volet sportif de la marque, « Sirius » incarne la ligne classique après le lifting du célèbre boîtier. L'année 2016 est elle aussi sous le signe de l'emblématique cadran à indication des heures décentrée. Ses nouvelles variantes – classiques ou à cadrans tridimensionnels – reçoivent toutefois un nouveau nom : « Regulator ». ○

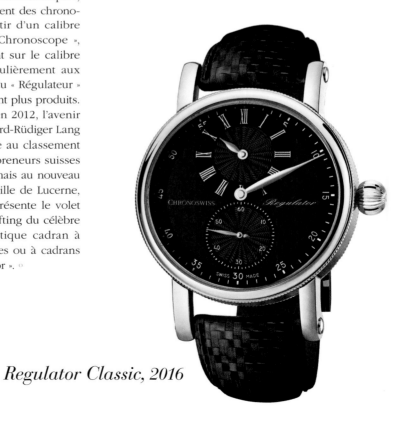

Regulator Classic, 2016

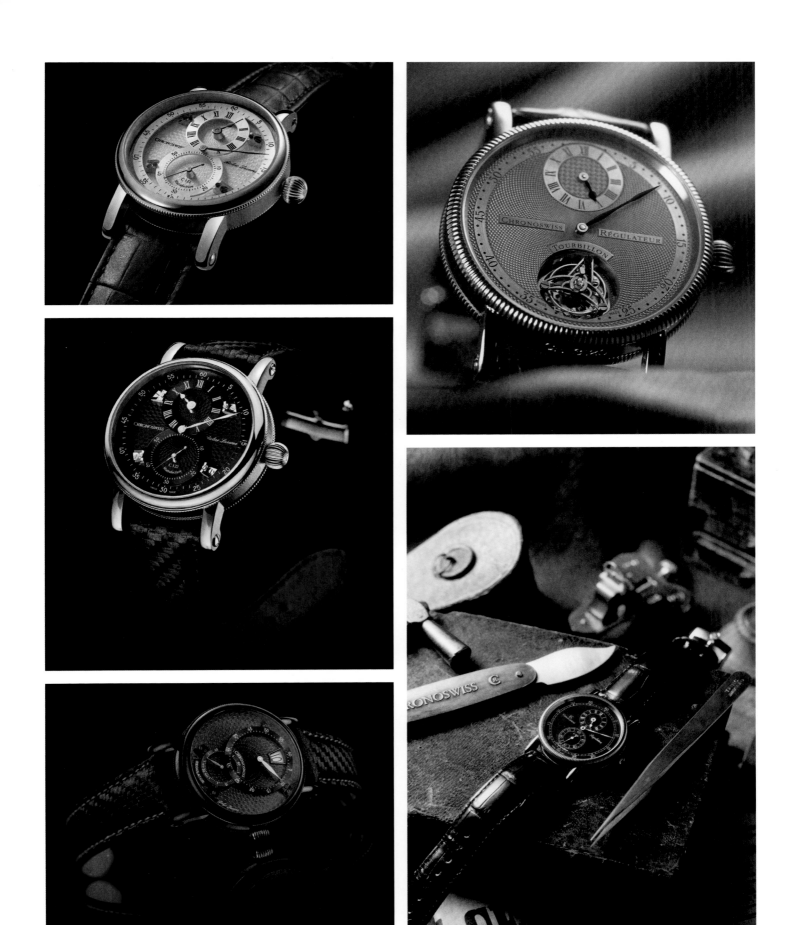

Clockwise from top left: Flying Regulator Manufacture, 2016 ◦ Régulateur à Tourbillon, 2000 ◦ Régulateur Automatique, 1990 ◦ Flying Regulator Jumping Hour, 2016 ◦ Flying Regulator Manufacture, 2016

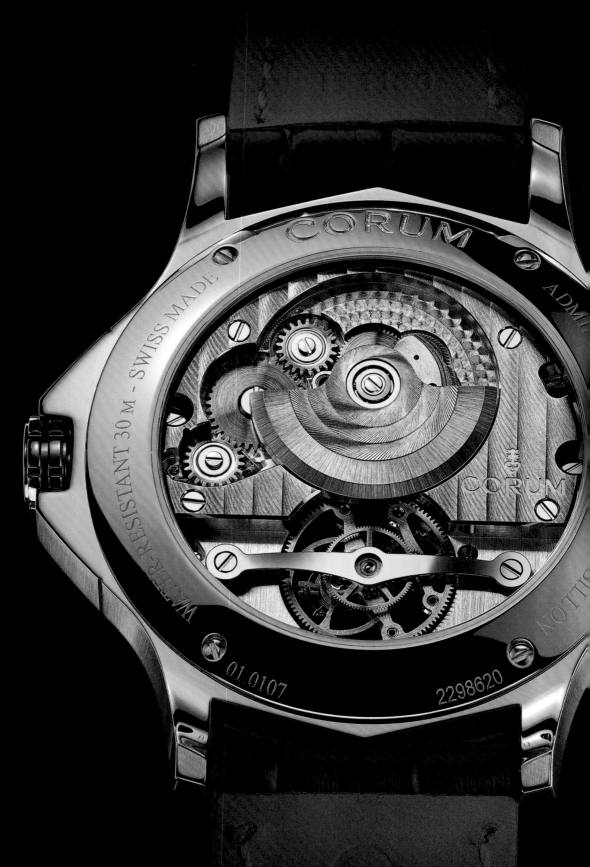

Corum

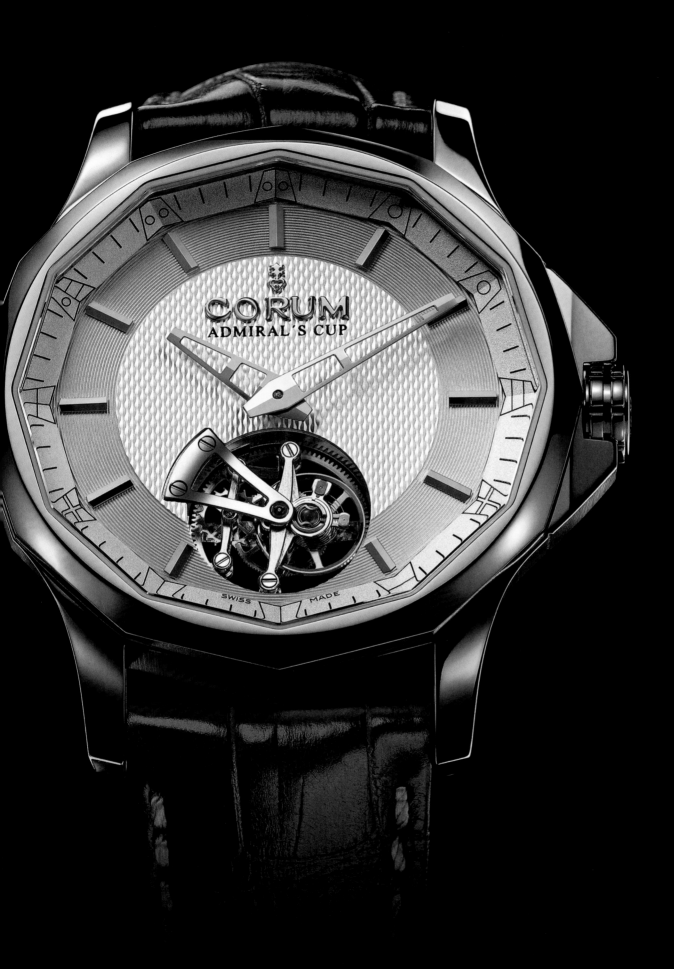

Key to Success

English

Watch design was a consuming passion for René Bannwart, a Swiss native who gained relevant experiences at Patek Philippe starting in 1933 and at Omega beginning in 1940. In 1955, he eagerly accepted an invitation from his uncle Gaston Ries and Ries's daughter Simone to join them as co-director of the watch factory that Gaston had established at La Chaux-de-Fonds in 1924. The small business delivered perfect merchandise, but because of the tradition of making watches on commission for third parties, it had neither a distinctive profile nor an attractive name. The latter, initially "Quorum," evolved from the team spirit of the executive board, but the name was difficult to pronounce in some languages and less than optimally readable on the international stage so it morphed into "Corum." The easily remembered logo (a vertical key) symbolizes "the key to beautiful time." This same key also opened the door to Corum's success, which has been based since 1956 on creating designs that set themselves appealingly apart from their competitors. This tradition continued with the 1958 debut of the "Chapeau Chinois": this model was aptly named because its broad bezel, which rises toward the middle part of the case, resembles a Chinese hat. Avidly discussed models followed with the "Sans Heures" in 1958, the "Coin Watch" in 1964 and the "Romulus" in 1966. A definitive classic debuted in 1977, when the young watchmaker Vincent Calabrese showed Bannwart the prototypes of a baton-shaped movement. After thorough evaluation, this led to the "Golden Bridge," which remained the brand's undisputed leader until 1980. Corum had to develop special machines to fabricate this watch's delicate mechanisms, but the effort earned the family business the right to describe itself as a full-fledged manufactory. The "Admiral's Cup" premiered in 1960 and has enriched the market with new interpretations since 1983. The Asian market had become increasingly important to Corum, but this dependency precipitated a crisis in the 1990s. Assistance from the Middle Eastern Alfardan Group in 1998 brought only temporary improvement. Corum was sold in 1999 to the American businessman Severin Wunderman, who had sold his Gucci Timepieces to the Amsterdam-based Gucci Group for 170 million dollars in 1997. The new owner developed a significantly more fashionable yet nonetheless idiosyncratic style that noticeably changed the face of the Corum brand. After Severin Wunderman's death in 2008, his American foundation began to entertain the idea of selling Corum, which came under the aegis of the Chinese Citychamp Group in May 2013. The company presents exclusive mechanisms in the several variants of the "Bridge" line. And the timekeeping "Bubble," which debuted in 2000, celebrated a successful comeback in 2016. ○

From top: Simone Ries and René Bannwart, 1955 ○
Chapeau Chinois, 1958 ○ Coin Watch, 1964

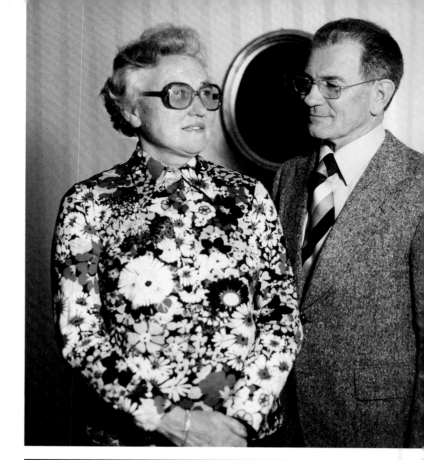

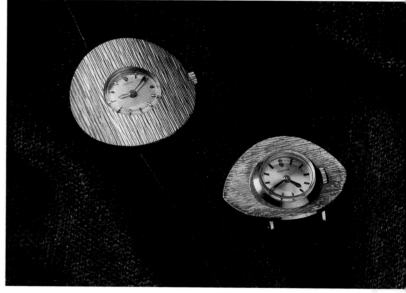

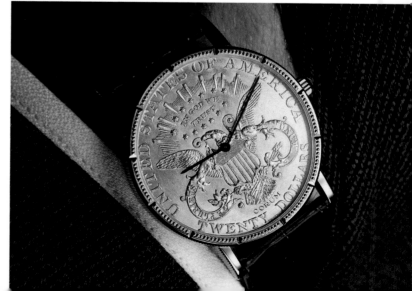

Schlüssel zum Erfolg

Uhrendesign war die große Passion von René Bannwart. Einschlägige Erfahrungen hatte der Schweizer ab 1933 bei Patek Philippe und insbesondere ab 1940 im Hause Omega gesammelt. Deshalb kam 1955 der Lockruf seines Onkels Gaston Ries und dessen Tochter Simone gerade recht, zusammen mit ihnen die Leitung der 1924 von Ries gegründeten Uhrenfabrik in La Chaux-de-Fonds zu übernehmen. Das kleine Unternehmen leistete zwar perfekte Arbeit, besaß aber wegen der Tradition, Uhren im Lohnauftrag für andere zu fertigen, weder ein ausgeprägtes Profil noch einen zugkräftigen Namen. Letzterer erwuchs aus dem Teamgeist der Führungsmannschaft und lautete zunächst Quorum. Weil das „Qu" Aussprache und Lesbarkeit auf internationaler Bühne beeinträchtigen würde, mutierte das Wort zu Corum. Das einprägsame Firmenlogo, der stehende Schlüssel, symbolisiert „The key to beautiful time". Der Schlüssel der schönen Zeitmessung ist zugleich auch jener des Erfolgs, welcher ab 1956 darin bestand, sich gestalterisch vom Üblichen abzuheben. Dem trug beispielsweise der 1958 vorgestellte „Chapeau Chinois" Rechnung. Die breite, zur Gehäusemitte ansteigende Lünette erinnert tatsächlich an einen Chinesenhut. 1958 brachte mit „Sans Heures", 1964 mit der „Coin Watch" und 1966 mit „Romulus" heiß diskutierte Uhrenmodelle. Der Klassiker schlechthin geht auf das Jahr 1977 zurück, als der junge Uhrmacher Vincent Calabrese mit dem Prototypen eines stabförmigen

Uhrwerks bei Bannwart vorsprach. Nach gründlicher Evaluation entstand daraus bis 1980 der unangefochtene Markenleader „Golden Bridge". Zur Herstellung der delikaten Mechanik musste Corum spezifische Maschinen entwickeln. Fortan durfte sich das Familienunternehmen Manufaktur nennen. In der „Admiral's Cup"-Linie kommen seit 1983 Neuinterpretationen des erstmals 1960 erschienenen Modells auf den Markt. In den 1990er Jahren geriet die immer stärker auf das Asiengeschäft ausgerichtete Firma kräftig ins Schlingern. Das Engagement der mittelöstlichen Alfardan-Gruppe brachte 1998 nur eine vorübergehende Linderung. 1999 erfolgte der Verkauf an Severin Wunderman, einen amerikanischen Geschäftsmann, welcher 1997 seine Gucci Timepieces für 170 Millionen Dollar an die Amsterdamer Gucci-Gruppe veräußert hatte. Der neue Eigentümer entwickelte einen deutlich modischeren, aber auch eigenwilligen Stil, welcher das Gesicht der Marke Corum merklich veränderte. Nach dem Tod von Severin Wunderman im Jahr 2008 trug sich dessen amerikanische Stiftung immer stärker mit Ausstiegsgedanken. Seit Mai 2013 befindet sich Corum unter dem Dach der chinesischen Citychamp. Exklusive Mechanik praktiziert das Unternehmen bei den verschiedenen Varianten der „Bridge"-Kollektion. Und 2016 feierte die 2000 lancierte Zeit-Blase namens „Bubble" ein erfolgreiches Comeback. ○

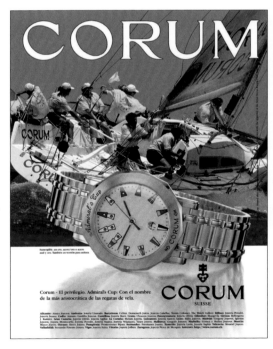

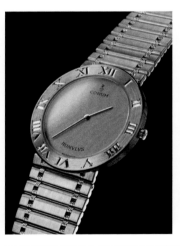

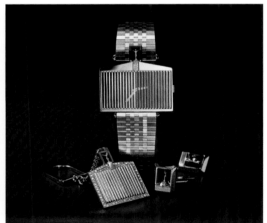

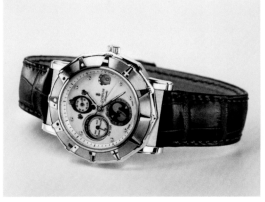

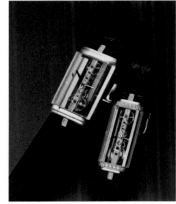

Clockwise from top left: Admiral's Cup, 1983 ○
Romulus, 1966 ○ Rolls-Royce, 1976 ○
Golden Bridge, 1980 ○ Admiral's Cup Tides, 1993

La clé du succès

Passionné de design de montres, le Suisse René Bannwart fait ses premières armes chez Patek Philippe, où il entre en 1933. Il parfait son expérience à partir de 1940 au sein de la maison Omega. Aussi, en 1955, la proposition alléchante que lui font son oncle Gaston Ries et sa fille Simone de tenir avec eux les rênes de la fabrique de montres fondée par Gaston Ries en 1924 à La Chaux-de-Fonds vient-elle à point nommé. Cette petite entreprise excelle dans son travail mais elle n'a ni une identité affirmée – elle réalise depuis toujours des montres pour le compte de tiers – ni un nom accrocheur. Elle s'appelle au départ Quorum, par référence à l'esprit d'équipe de sa direction, mais la combinaison de lettres « qu » se révélant difficile à prononcer et à lire sur la scène internationale, on adopte la graphie Corum. Son logo marquant, une clé dressée vers le ciel, symbolise « The key to beautiful time ». Cette « clé de la belle mesure du temps » est aussi celle du succès, qui tient à partir de 1956 à des designs d'avant-garde, comme le « Chapeau Chinois », présenté en 1958 : sa large lunette qui va s'évasant jusqu'à mi-hauteur du boîtier rappelle effectivement le fameux couvre-chef conique. Sortis respectivement en 1958, 1964 et 1966, les modèles « Sans Heures », « Coin Watch » et « Romulus » déchaînent les passions. Le modèle classique par excellence remonte à 1977, année où le jeune horloger Vincent Calabrese se présente chez Bannwart avec un prototype de mouvement baguette. Après une évaluation approfondie, Corum met au point sur cette base « Golden Bridge », le fer de lance incontesté de la marque, qui sort en 1980. Pour la fabrication de la délicate mécanique, Corum doit développer des machines spéciales. L'entreprise familiale peut désormais prétendre à l'appellation manufacture. Le modèle « Admiral's Cup » sorti en 1960 sera réinterprété à plusieurs reprises à partir de 1983. L'entreprise, qui mise de plus en plus sur les marchés asiatiques, entre en zone de fortes turbulences dans les années 1990. L'intervention en 1998 du groupe Alfardan, basé au Moyen-Orient, n'apporte qu'une amélioration passagère. Corum est vendue en 1999 à l'homme d'affaires américain Severin Wunderman, qui a cédé deux ans plus tôt ses garde-temps Gucci au groupe Gucci d'Amsterdam pour la somme de 170 millions de dollars. Le nouveau propriétaire développe un style nettement plus tendance, mais aussi plus original, qui change en profondeur le visage de la marque. Après le décès de Severin Wunderman en 2008, sa fondation aux États-Unis envisage de plus en plus sérieusement de céder l'affaire. Dans le giron du groupe chinois Citychamp depuis 2013, Corum équipe de mouvements mécaniques exclusifs ses différentes variantes de la collection « Bridge ». En 2016, la montre « Bubble » à verre bombé lancée en l'an 2000 fait un retour triomphal. ∘

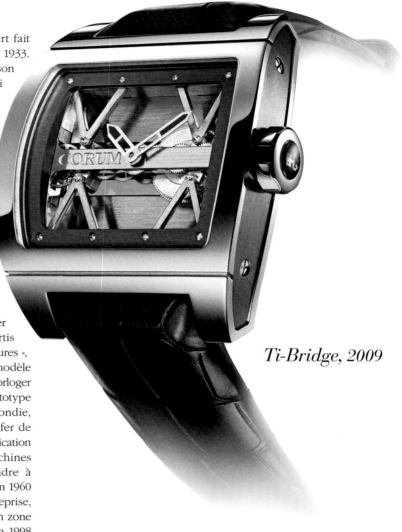

Ti-Bridge, 2009

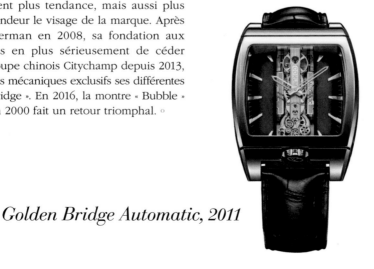

Golden Bridge Automatic, 2011

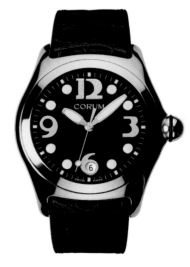

Bubble, 2000

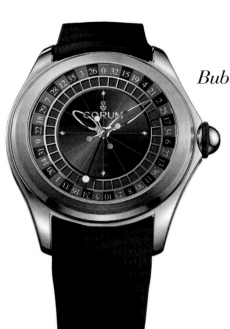

Bubble Roulette, 2016

Bubble Casino Chip, 2016

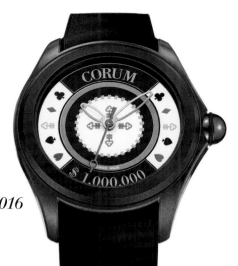

Bubble Joker, 2016

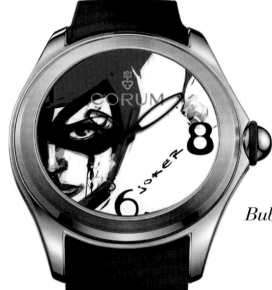

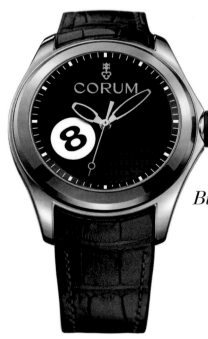

Bubble 8 Ball, 2016

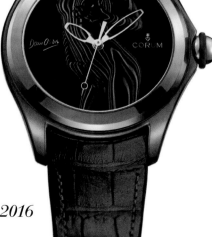

Bubble Dani Olivier, 2016

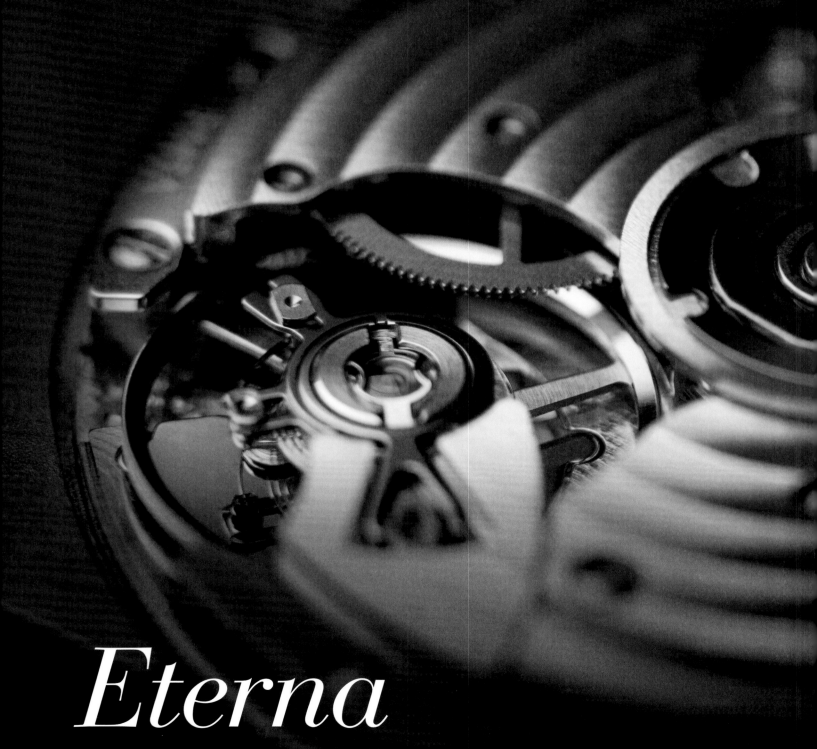

Eterna

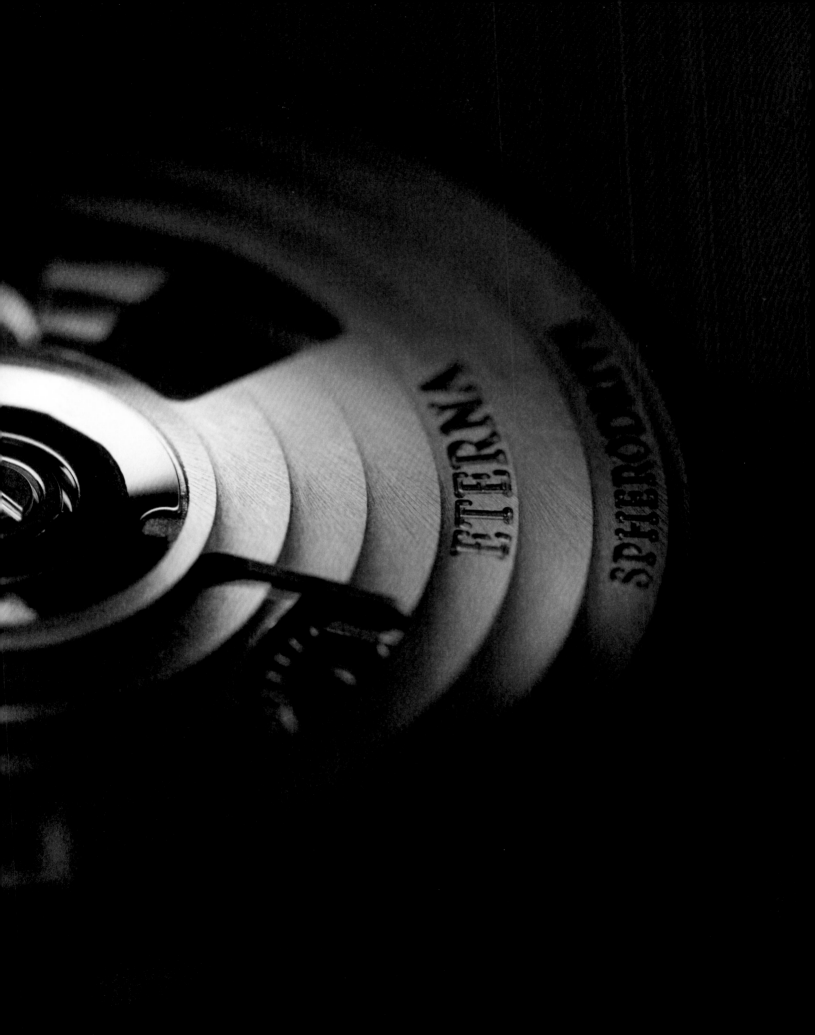

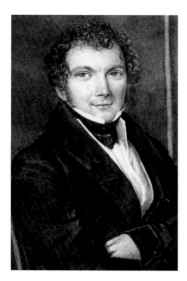
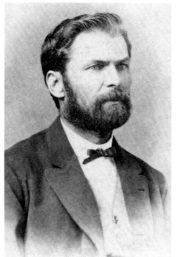
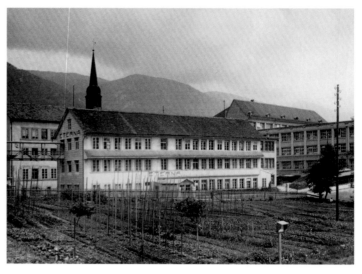

From left: Dr. Josef Girard ∘ Urs Schild ∘ Eterna's factories, 1910 ∘ Von links: Dr. Josef Girard ∘ Urs Schild ∘ Eterna-Werke, 1910 ∘ De gauche à droite : Le docteur Josef Girard ∘ Urs Schild ∘ La fabrique Eterna, 1910

Back to the Future

Eterna has faced many challenges during the past few decades. Turbulence repeatedly shaped events at this venerable Swiss watch brand. The most recent chapter began when the family of the late designer Ferdinand Alexander Porsche (who acquired Eterna in October 1995 and died in 2012) announced the brand's sale to the China Haidian corporate group, now known as Citychamp. Eterna had been the property of the industrialist Franz Wassmer, who took it over from the forerunner of the Swatch Group in September 1984. The enterprise was established by the physician Dr. Josef Girard and the schoolteacher Urs Schild, who were eager to reduce joblessness in the formerly insignificant town of Grenchen near Solothurn. On November 7, 1856, the duo established the "Dr. Girard & Schild" ébauche factory, from which the Eterna watch manufactory would later emerge. Noteworthy successes ensued in 1863 and thereafter. Girard left the company in 1866. Automated machinery was installed in 1870. And the first Eterna watches left the ateliers in 1876. When "Eterna-Werke, Gebrüder Schild & Co." was established as a limited partnership in 1906, the name that had been written on the dials was also transferred to the company per se. Numerous inventions and world premieres earned worldwide fame for Eterna in the 20th century. For example, the label debuted its first wristwatch with a built-in alarm in 1914, followed by the smallest serially manufactured wristwatch with a baguette movement in 1930. Series production of a relatively small and slim self-winding movement began in 1942. Eterna set a global standard, which remains valid today, with the 1948 premiere of a self-winding movement with a rotor borne atop reliable, low-maintenance, miniature ball bearings. Another innovation was a click-wheel polarizing mechanism to efficiently use the rotor's kinetic energy in both its directions of rotation. The "Golden Heart," which debuted in 1958, encased its epoch's smallest automatic movement with solid gold oscillating weight. Caliber 3000 followed in 1962 as the slimmest wristwatch with a centrally positioned oscillating weight and a date display. Theodor Schild agreed to divide his watch factory into two separate joint-stock companies in 1932: Eterna AG for the fabrication of precise watches and Eta AG for the production of ébauches. In recognition of its contributions to Europe's reputation throughout the world, an international jury in 1980 awarded Eterna the Grand Prix de l'Excellence Européenne, which had been established by the Noble Peace Prize laureate René Cassin.

Under the direction of its Chinese owners, Eterna now plans to pursue a two-track strategy. The entry-level price segment will be covered by watches encasing movements from third-party suppliers, e.g., calibers from Eterna's former affiliate Eta (which has belonged to the Swatch Group and its predecessor for more than 30 years) or clones from Sellita. Alongside this first track, Eterna will continue to function as a veritable manufacture. Uncommon versatility distinguishes movement family 39xx: it's fabricated by the Eterna Movement Company (EMC), which was established in 2012. Eighty-eight variants are currently available: the spectrum ranges from a simple hand-wound movement to an automatic chronograph. The latter, column-wheel Caliber 3916A, animates the "Super KonTiki." This sporty wristwatch was born in 1947, when the then 32-year-old Norwegian ethnologist Thor Heyerdahl sailed a balsawood raft across the Pacific. Eterna honored him in 1958 by launching the "KonTiki 20," which resists water pressure to 20 bar. It's not surprising to find that the "KonTiki" line ranks among the icons of the house of Eterna. ∘

Alarm wristwatch, patented in 1914

Zurück in die Zukunft

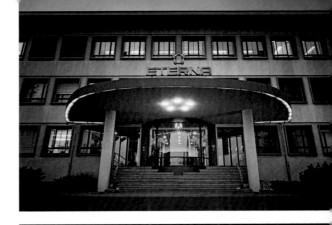

Leicht hatte es Eterna während der vergangenen Jahre wirklich nicht. Turbulenzen bestimmten das Geschehen rund um die altehrwürdige Schweizer Uhrenmarke. Das vorerst letzte Kapitel begann, als die Familie des 2012 verstorbenen Designers Ferdinand Alexander Porsche, welche Eterna im Oktober 1995 erworben hatte, den Verkauf an die Unternehmensgruppe China Haidian, heute Citychamp, verkündete. Seit September 1984 hatte sich das Unternehmen im Eigentum des Industriellen Franz Wassmer befunden, welcher es seinerseits vom Vorläufer der heutigen Swatch Group übernommen hatte. Die Gründung des Unternehmens geht zurück auf Dr. Josef Girard und Urs Schild. Dem Arzt und dem Schullehrer war es ein Anliegen, die gravierende Arbeitslosigkeit in der damals eher bedeutungslosen Ortschaft Grenchen nahe Solothurn zu mindern. Am 7. November 1856 rief das Duo die Rohwerkefabrik Dr. Girard & Schild ins Leben, aus der später die Uhrenmanufaktur Eterna hervorgehen sollte. Nennenswerte Erfolge waren ab 1863 zu berichten. 1866 zog sich der Mediziner zurück. Fertigungsautomaten hielten 1870 ihren Einzug. Und 1876 verließen erste Eterna-Uhren die Ateliers. Mit Gründung der Kommanditgesellschaft Eterna-Werke, Gebrüder Schild & Co. wurde der eingeführte Zifferblattname 1906 auch auf die Firma selbst übertragen. Im 20. Jahrhundert trugen zahlreiche Erfindungen und Weltpremieren diesen Namen in alle Welt.

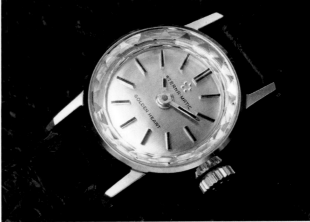

In diesem Sinne präsentierte Eterna 1914 die erste Armbanduhr mit Wecker oder 1930 die damals kleinste serienmäßig hergestellte Armbanduhr mit Baguettewerk. 1942 startete die Serienproduktion eines relativ kleinen und flachen Automatikwerks. Den bis heute gültigen Weltstandard für Uhrwerke mit Rotoraufzug lancierte Eterna 1948 durch die Verwendung eines zuverlässigen und wartungsarmen Miniatur-Kugellagers. Ebenfalls neu: ein Klinkenrad-Wechselgetriebe zur effizienten Nutzung der kinetischen Energie des Rotors in beiden Drehrichtungen. In der 1958 vorgestellten „Golden Heart" tickte das seinerzeit kleinste Automatikwerk mit massivgoldener Schwungmasse. Schließlich führte das 1962 fertiggestellte Kaliber 3000 zur flachsten Armbanduhr mit zentral angeordneter Schwungmasse und Datumsanzeige. Bereits 1932 hatte Theodor Schild nolens volens der Aufteilung seiner Uhrenfabrik in zwei getrennte Aktiengesellschaften zugestimmt: die Eterna AG zur Fertigung von Präzisionsuhren und die Eta AG, welche sich um die Fabrikation von Rohwerken kümmerte. Für ihre Verdienste um das Ansehen Europas in der ganzen Welt verlieh eine international besetzte Jury der Eterna 1980 den vom Friedensnobelpreisträger René Cassin gestifteten Grand Prix de l'Excellence Européenne.

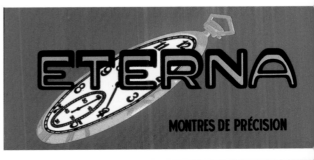

Unter chinesischer Regie wird Eterna zukünftig zweigleisig fahren. Im Einstiegs-Preissegment ticken durchweg zugekaufte Uhrwerke, also beispielsweise Kaliber der ehemaligen Schwester Eta, welche seit mehr als 30 Jahren zur heutigen Swatch Group gehört, oder Klone von Sellita. Daneben betätigt sich Eterna inzwischen auch wieder als Manufaktur. Durch besondere Vielseitigkeit zeichnet sich hier die Werkefamilie 39xx aus. Für deren Fabrikation zeichnet die 2012 gegründete Eterna Movement Company (EMC) verantwortlich. Derzeit sind 88 Varianten erhältlich, angefangen vom einfachen Handaufzugswerk bis hin zum Automatik-Chronographen. Letzterer, konkret das Schaltrad-Kaliber 3916A, beseelt die „Super KonTiki". Diese sportliche Armbanduhr erinnert ans Jahr 1947, als der damals 32-jährige norwegische Ethnologe Thor Heyerdahl mit einem Balsaholzfloß über den Pazifik segelte. Ihm zu Ehren hatte Eterna u. a. 1958 die bis 20 Bar wasserdichte „KonTiki 20" auf den Markt gebracht. Kein Wunder also, dass die „KonTiki"-Linie zu den Ikonen des Hauses Eterna gehört. ∘

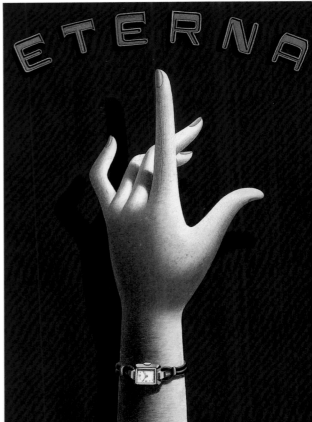

From top: Eterna's headquarters in Grenchen ∘ Golden Heart, 1958 ∘
Enamel sign, 1920 ∘ Advertisement, 1949 ∘ Von oben: Eterna-Hauptsitz
in Grenchen ∘ Golden Heart, 1958 ∘ Emailleschild, 1920 ∘ Werbung, 1949 ∘
De haut en bas : Le siège d'Eterna à Granges ∘ Golden Heart, 1958 ∘
Pancarte émaillée, 1920 ∘ Publicité, 1949

Retour vers le futur

Français

Les choses n'ont pas été faciles du tout pour Eterna au cours de ces dernières années, quelque peu tumultueuses pour cette vénérable marque horlogère suisse. Ce qui est provisoirement le dernier chapitre dans son histoire débute lorsque la famille du concepteur de voitures Ferdinand Alexander Porsche (décédé en 2012), qui a racheté Eterna en 1995, annonce sa cession au groupe chinois Haidan, connu aujourd'hui sous le nom de Citychamp. Depuis septembre 1984, l'entreprise appartient à l'industriel Franz Wassmer, qui l'a rachetée à la SMH, devenue par la suite le Swatch Group. La création de l'entreprise remonte au docteur Josef Girard et à l'instituteur Urs Schild. Soucieux de résoudre le grave problème d'emploi dans la localité plutôt insignifiante de Granges, près de Soleure, ces deux hommes fondent le 7 novembre 1856 la fabrique d'ébauches Dr. Girard & Schild, qui donnera naissance à la manufacture horlogère Eterna. L'entreprise enregistre ses premiers succès notables en 1863. Le médecin se retirera des affaires en 1866.

La production est mécanisée en 1870, et la première montre de manufacture sort des ateliers en 1876. À la fondation de la société en commandite Eterna-Werke, Gebrüder Schild & Co, en 1906, le nom Eterna apposé sur le cadran est aussi adopté comme nom de société. Au XIXᵉ siècle, de nombreuses inventions et premières mondiales feront connaître ce nom dans le monde entier. Eterna présente ainsi en 1914 la première montre-bracelet réveil ou encore en 1930 la plus petite montre-bracelet équipée d'un mouvement baguette qui soit fabriquée en série. En 1942, la société lance la production en série d'un mouvement automatique relativement petit et plat. En 1948, elle introduit ce qui est demeuré une référence mondiale en matière de mouvement à remontage automatique : un dispositif à roulement à billes miniature fiable et requérant peu de maintenance. Autre innovation, une transmission dotée d'une roue à rochet permet de tirer parti de l'énergie cinétique du rotor

dans les deux sens de rotation. La « Golden Heart » présentée en 1958 est animée par ce qui est à l'époque le plus petit mouvement mécanique doté d'une masse oscillante en or massif. Enfin, le calibre 3000 parachevé en 1962 permet de réaliser la montre-bracelet la plus plate au monde, à date et masse oscillante centrale. Dès 1932, Theodor Schild avait accepté bon gré mal gré la scission de sa fabrique horlogère en deux sociétés anonymes distinctes : Eterna SA, fabricant de montres de précision, et ETA SA, spécialisée dans la fabrication d'ébauches. En 1980, un jury international décerne à Eterna pour sa contribution au rayonnement de l'Europe dans le monde entier le Grand Prix de l'excellence européenne, récompense créée par le lauréat du prix Nobel de la paix René Cassin.

Désormais en des mains chinoises, Eterna poursuivra une stratégie reposant sur deux piliers. Les produits d'entrée de gamme sont animés par des mouvements achetés, notamment des calibres de l'ancienne entreprise sœur Eta, qui fait partie depuis plus de 30 ans de l'actuel Swatch Group, ou leurs clones livrés par Sellita. Parallèlement, Eterna a repris son activité de manufacture. La famille de calibres 39xx, dont la fabrication est entièrement assurée par Eterna Movement Company (EMC), société fondée en 2012, se distingue par sa très grande polyvalence : elle comprend à l'heure actuelle 88 variantes, du mouvement à remontage manuel simple jusqu'aux chronographes automatiques. C'est d'ailleurs le calibre 3916A équipé d'une roue à colonnes qui anime la « Super KonTiki ». Cette montre-bracelet sportive évoque le souvenir du radeau en rondins de balsa sur lequel l'ethnologue norvégien Thor Heyerdahl, alors âgé de 32 ans, a traversé le Pacifique en 1947. En son honneur, Eterna a commercialisé entre autres en 1958 la montre « KonTiki 20 » étanche jusqu'à 20 bars. Rien d'étonnant donc à ce que la ligne « KonTiki » fasse partie des icônes de la maison Eterna. ◦

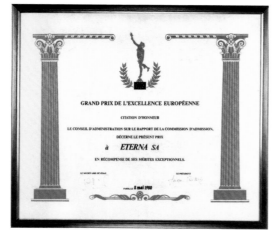
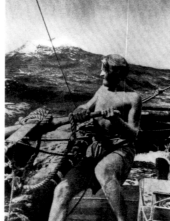
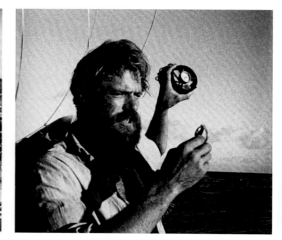

Top: Cal. 3916A ◦ From left: Grand Prix de l'Excellence Européenne, 1980 ◦ Thor Heyerdahl

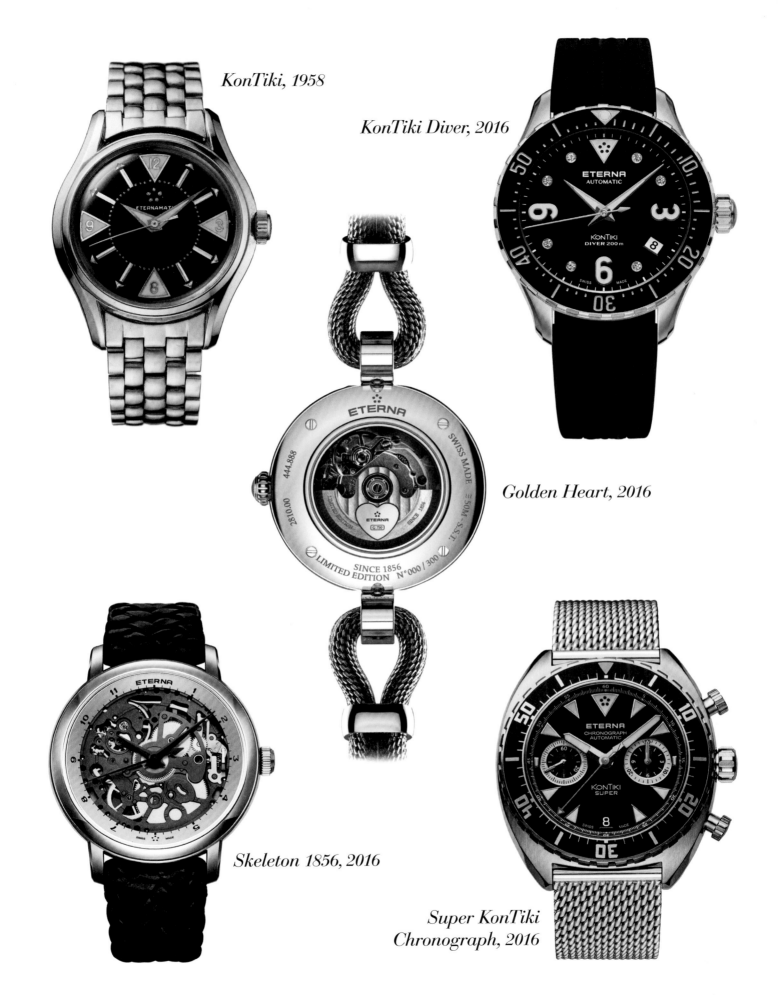

KonTiki, 1958

KonTiki Diver, 2016

Golden Heart, 2016

Skeleton 1856, 2016

*Super KonTiki
Chronograph, 2016*

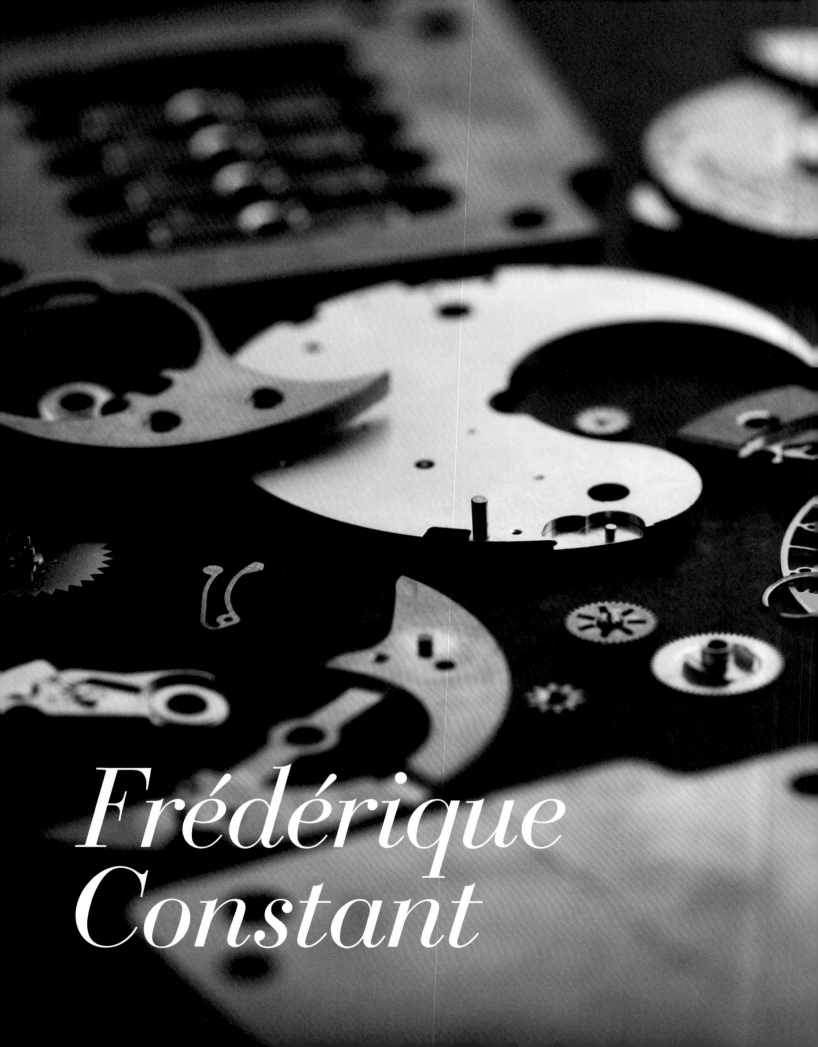

Frédérique
Constant

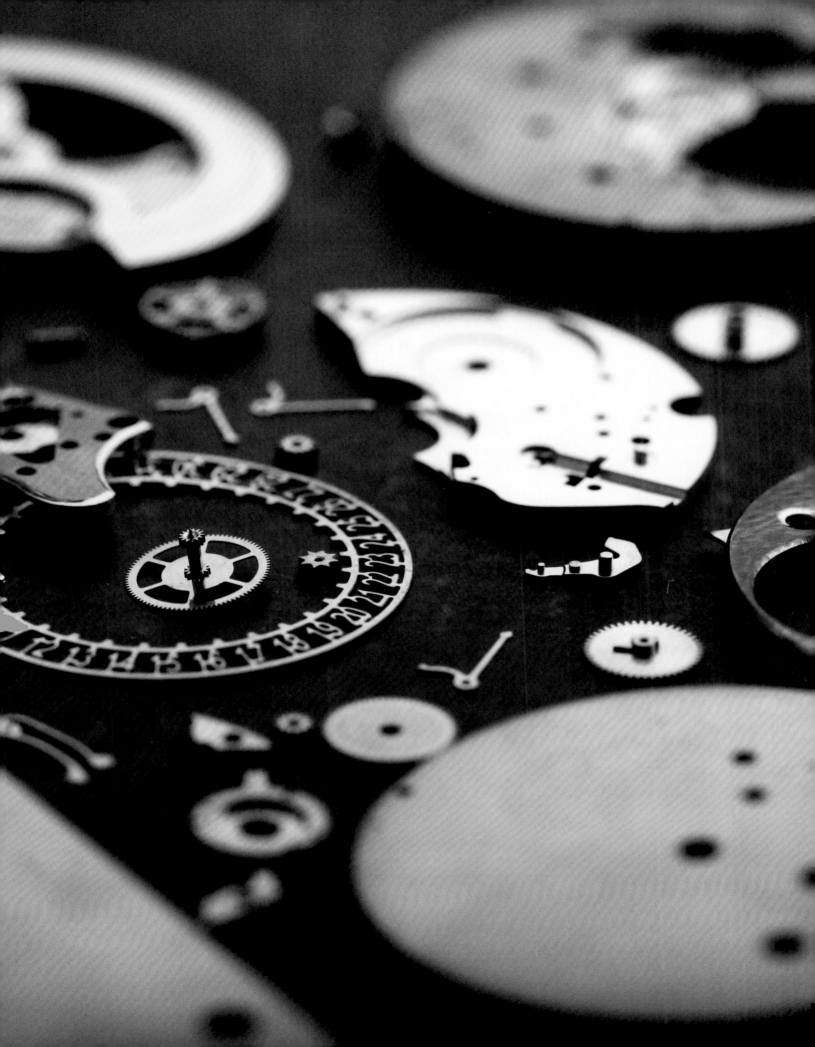

The Passionate Heartbeat of a Youthful Manufactory

A memorable line in the novel *War and Peace* declares: "It is impossible to eradicate the passions; we can only strive to pilot them toward a noble goal." Leo Tolstoy's words were evidently taken to heart by Peter C. Stas and Aletta Stas-Bax, whose entrepreneurial activities began with an insatiable passion for wristwatches, e.g., the ones this Dutch couple admired during skiing holidays in the Swiss Alps. At some point, they had nearly no other choice but to sacrifice their secure jobs at global corporations on the altar of watch-related activities. But let's take our story one step at a time.... The mid 1980s found the two young people living in Hong Kong, a very special watch Mecca. Alongside his "day job," Peter Stas started a company for design software in 1988. Among other projects, he used this software to design wristwatches because he felt that watches produced in China looked absolutely awful. Old books and catalogues served him as a basis. Aletta Stas chose the watch fair in Hong Kong as the venue for the premiere of the first prototypes, which she and Peter had screwed together, so to speak, in their kitchen. Shortly before the end of the fair, a Japanese wholesaler ordered 350 timepieces, which sold surprisingly well and led to follow-up orders for another 1,100 watches. The next year witnessed a collection of prototypes consisting of six models: each was equipped with a quartz movement, but all six were assembled in Switzerland from Swiss components. These "children" likewise thrived and soon needed a suitable name. Aletta's great-grandmother Frédérique Schreiner and Peter's great-grandfather Constant Stas served as namesakes, and the two ancestors' first names were combined to create "Frédérique Constant." Founded with starting capital of 60,000 Swiss francs, the Frédérique Constant SA and its external partners collaboratively produced more than 1,000 Swiss-made wristwatches in 1992. Brisk sales motivated the company

to attempt greater exploits. A stroke of good luck came in the form of a meeting with Miguel Garcia, who was sales director at Sellita at this time. In the course of their conversation, they realized that the dial side of Eta's relatively tall self-winding Caliber 2836 could be opened to directly reveal "the heartbeat of human culture." This formed the basis for the 1994 debut of the "Heart Beat" collection, which was not protected by patent and thus soon copied. The youthful company achieved larger quantities and more favorable cost structures through private-label production, which was later discontinued. A business plan drawn up in 1996 envisioned Geneva as the location of the firm's headquarters, so the Stases applied for permission to work in Switzerland. Their request was granted, but only under the condition that a staff of at least twelve employees would be established within five years. This goal was reached within just 2½ years—and entirely through self-financed growth. Clear visions, and the desire to make something of their own that could stand beside calibers purchased from third parties, led in 2001 to the start of developmental work for their first hand-wound manufacture caliber. Peter Stas initiated cooperation with specialists from two watchmaking schools, one in Geneva and the other in Holland, for this ambitious project. Frédérique Constant surprised the watch world in 2004 with the premiere of the Heart Beat FC-910-1. Its astonishingly large balance oscillated under an aperture cut in the dial at the "6." The moon-phase version FC-915 debuted in 2005, followed in 2006 by an exclusive self-winding model designated as FC-930. New materials debuted in 2007. A specially designed silicon escape-wheel oscillates in automatic Caliber FC-935 Silicium with moon-phase and hand-type date. A visionary developmental step is embodied in the variant with the cryptic designation "FC-935SZABS4H9" that was sold at the

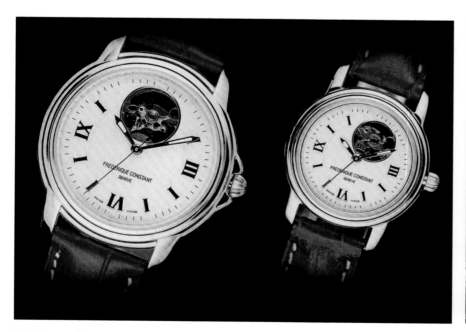
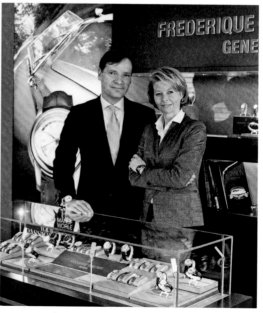

Left: Original Heart Beat models from 1994 ○ *Links: Heart Beat, Originalmodelle von 1994* ○ *À gauche : Modèles originaux Heart Beat de 1994* ○
Right: Peter C. Stas and Aletta Stas-Bax

"One of a Kind" charity auction in Monte Carlo in September 2007. Alongside its silicon escape-wheel, this unique timepiece is also equipped with a totally innovative balance made from Zerodur, a glass-ceramic composite. Another of the brand's own watches, the "Heart Beat Tourbillon," debuted in a limited edition of 188 specimens in 2008. Its special features: balance-stop function, intelligent Smart-Screw system to perfectly poise the tourbillon's cage, and an antimagnetic silicon escape-wheel. Over 80 percent of the components needed for automatic Caliber FC-980-1 were fabricated at the firm's headquarters in Plan-les-Ouates, which was inaugurated in 2006 and commands a view of several renowned traditional Genevan watch manufactories. Despite ongoing improvements and expansions in its machine park, Frédérique Constant continues to rely on external suppliers to satisfy its demand for larger quantities. However, fine processing, assembly and quality control all take place exclusively under the firm's own roof. The youthful manufactory produced 90,000 watches in 2008. Revenues doubled within just a few years' time, partly thanks to greater net value added. Of course, compared to the industry as a whole, manufactory-made products occupied and continue to occupy a small niche. The philosophy of the firm's founders, i.e., to offer affordable luxury for customers around the globe, is primarily based on elegant timepieces encasing third-party movements. If the calibers are mechanical, they're ennobled according to standards defined by the firm. Most such calibers are supplied by Selitta. Frédérique Constant celebrated the tenth anniversary of its own manufacture caliber in 2014. The product spectrum already includes fifteen calibers in two families. Calibers numbered 9xx show their escapements under the dials. Rotors automatically wind the mainsprings of FC-7xx calibers, which do not reveal their beating hearts. Basic Caliber FC-710 with central second-hand combines 137 components, has a diameter of 30 millimeters and is 6.2 millimeters tall. Its central rotor winds the mainspring in both directions of rotation. The fully wound barrel stores sufficient energy to keep the balance oscillating at a pace of four hertz for 42 hours.

A sensational manufactory highlight debuts in 2016. The "Slimline Perpetual Calendar Manufacture," which includes a calendar mechanism that won't need manual correction until 2100, encases exclusive automatic Caliber FC-775. Watchmakers need fully 78 components to create the complex switching mechanism under this model's dial. The price is unique too: circa 8,000 euros.

A wholly new chapter—namely, an electronic chapter—began in 2015 with the "Horological Smartwatch," which has analog displays for the time and the functions. Connectable with Android and iOS telephones, this Swiss-made watch resulted from collaboration with specialists in the USA. Frédérique Constant was able to bring more than 16,000 of these watches to male (75 percent) and female wrists in just one year. Thanks to the dedication of twelve engineers in Switzerland, Peter Stas would like to significantly expand this project. The numbers of units are likewise scheduled to increase. If all goes as planned, the counters will tally circa 145,000 wristwatches by the end of 2016. There's only one word for that: Success! ○

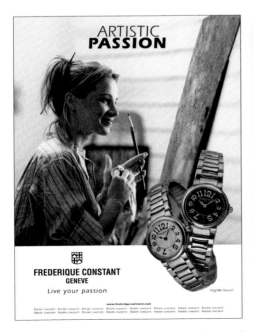

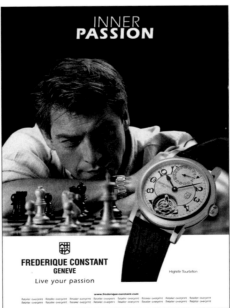

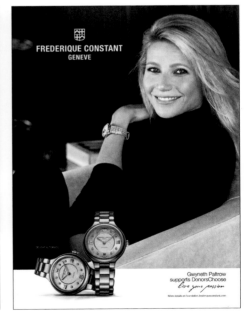

From left: Advertisement, 2004 ○ 2004 ○ 2016 ○ Von links: Werbung, 2004 ○ 2004 ○ 2016 ○ De gauche à droite : Publicité, 2004 ○ 2004 ○ 2016

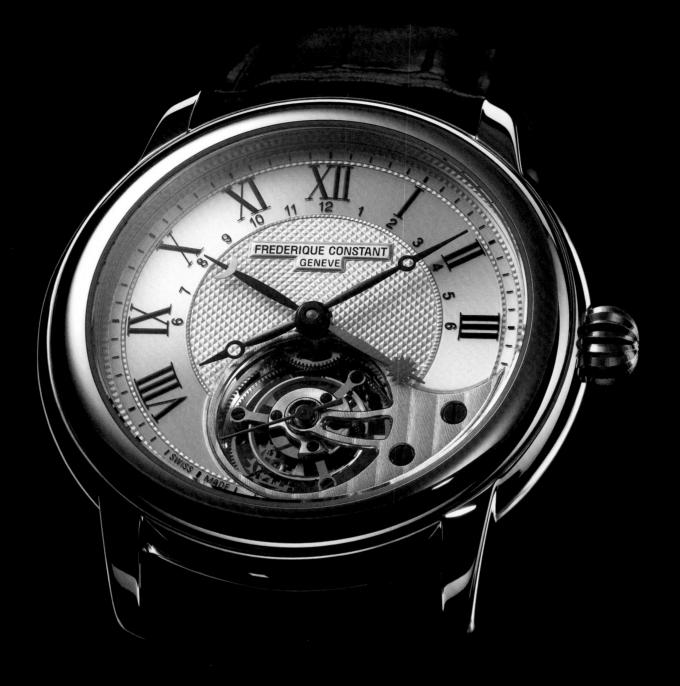

Manufacture Tourbillon,
FC-98OMC4M9, 2008

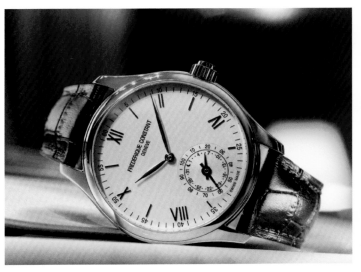

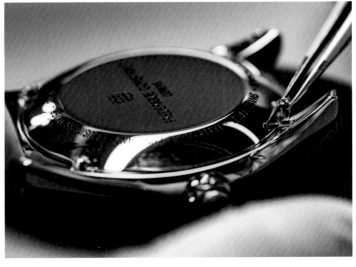

Horological Smartwatch, FC-285V5B4, 2015
When tradition meets technology
Wenn Tradition auf Technologie trifft
Le mariage entre tradition et technologie

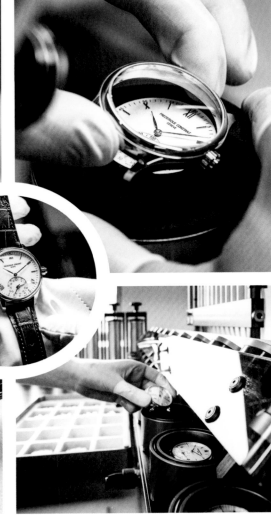

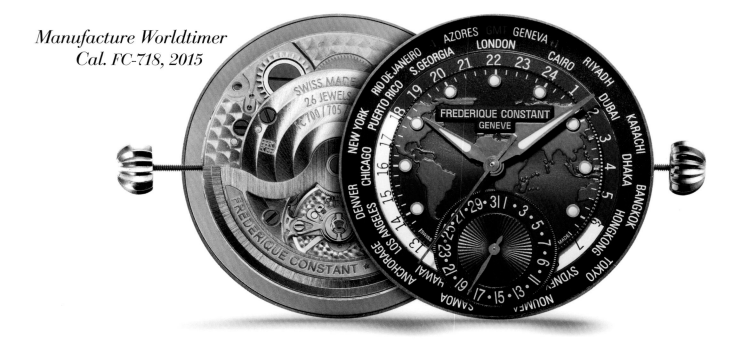

Passionierter Herzschlag einer jungen Manufaktur

Deutsch

Im Roman *Krieg und Frieden* steht zu lesen: „Es ist unmöglich, die Leidenschaften auszurotten; wir müssen nur darauf bedacht sein, sie auf ein edles Ziel zu lenken." Diese Worte von Leo N. Tolstoi haben Peter C. Stas und Aletta Stas-Bax ganz offensichtlich verinnerlicht. Ihr unternehmerisches Handeln begann nämlich mit einer unstillbaren Passion für Armbanduhren. Solche bewunderte das holländische Ehepaar während seiner Ski-Urlaube in den Schweizer Bergen. Irgendwann blieb fast keine andere Wahl, als die sicheren Jobs bei Weltkonzernen auf dem Altar eigener Uhr-Aktivitäten zu opfern. Aber schön der Reihe nach: Mitte der 1980er Jahre lebten die beiden in Hongkong, einem sehr speziellen Uhrenmekka. 1988 gründete Peter Stas nebenberuflich eine kleine Firma für preiswerte Design-Software. Letztere nutzte er auch zur Gestaltung von Armbanduhren, denn das in China Produzierte sah in seinen Augen ganz schrecklich aus. Als Grundlage dienten alte Bücher und Kataloge. Die allerersten, sozusagen in der heimischen Küche zusammengeschraubten Prototypen präsentierte Aletta Stas 1991 während der Uhrenmesse in Hongkong. Kurz vor Schluss orderte ein japanischer Grossist 350 Exemplare, die sich erstaunlich gut verkauften und Orders für weitere 1 100 Uhren nach sich zogen. Das Folgejahr stand bereits im Zeichen einer kleinen Prototypen-Kollektion mit insgesamt sechs Modellen. Alle wiederum ausgestattet mit Quarzwerken, montiert nun aber schon in der Schweiz aus eidgenössischen Komponenten. Auch diese Kinder gediehen prächtig, brauchten jedoch irgendwann einen passenden Namen. Als Paten dienten Alettas Urgroßmutter Frédérique Schreiner und Peters Urgroßvater Constant Stas. Beide Vornamen verschmolzen zu Frédérique Constant. Die mit einem Kapital von 60 000 Schweizer Franken gegründete Frédérique Constant SA produzierte 1992

zusammen mit externen Partnern schon mehr als 1 000 Swiss-Made-Armbanduhren. Zügige Abverkäufe motivierten zu größeren Taten. Als Glücksfall erwies sich die Begegnung mit Miguel Garcia, damals noch Verkaufsleiter von Sellita. Im Gespräch zeigte sich die Möglichkeit, das relativ hoch bauende Automatikkaliber Eta 2836 vorne zu öffnen und den Herzschlag der menschlichen Kultur unmittelbar vor Augen zu führen. Auf dieser Grundlage ging 1994 die nicht geschützte und deshalb rasch kopierte Kollektion namens „Heart Beat" an den Start. Größere Quantitäten und günstigere Kostenstrukturen erreichte das junge Unternehmen durch die später wieder eingestellte Private-Label-Produktion. Mit dem 1996 verfassten Businessplan für eine Genfer Firmenzentrale ging der Antrag auf Arbeitserlaubnis in der Eidgenossenschaft einher. Erteilt wurde sie unter der Bedingung, innerhalb von fünf Jahren einen Stamm von mindestens zwölf Beschäftigten aufzubauen. Bereits nach zweieinhalb Jahren war dieses Ziel ausschließlich durch selbst finanziertes Wachstum erreicht. Klare Visionen und der Drang, den zugekauften Kalibern etwas Eigenes zur Seite zu stellen, führten 2001 zum Start der Entwicklung eines ersten Manufaktur-Handaufzugskalibers. Für dieses ehrgeizige Projekt suchte Peter Stas die Kooperation mit Spezialisten zweier Uhrmacherschulen, der in Genf und einer in Holland. 2004 überraschte Frédérique Constant mit dem Heart Beat FC-910-1. In einem prominent bei der „6" positionierten Zifferblattausschnitt oszillierte eine erstaunlich große Unruh. Auf die 2005er Mondphasen-Version FC-915 folgte 2006 ein exklusives Automatikkaliber namens FC-930. 2007 stand im Zeichen neuer Materialien. Das Automatikkaliber FC-935 Silicium mit Mondphasenindikation und Zeigerdatum besitzt ein speziell gestaltetes Silizium-Ankerrad. Einen visionären Entwicklungsschritt markierte

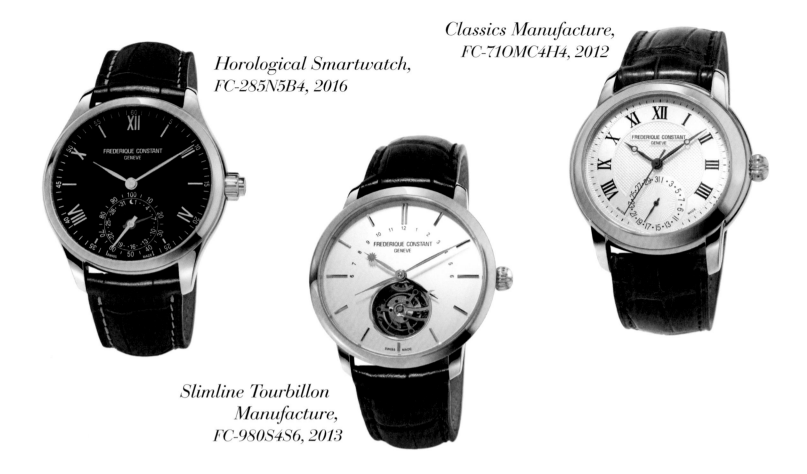

Horological Smartwatch,
FC-285N5B4, 2016

Classics Manufacture,
FC-710MC4H4, 2012

Slimline Tourbillon
Manufacture,
FC-980S4S6, 2013

die Variante mit der kryptischen Bezeichnung FC-935SZABS4H9, welche im September 2007 während der Wohltätigkeitsauktion „One of a Kind" in Monte Carlo unter den Hammer gelangte. Neben dem Silizium-Ankerrad verfügt das Unikat über eine völlig neuartige Unruh aus dem glaskeramischen Werkstoff Zerodur. Das ebenfalls hauseigene „Heart Beat Tourbillon" debütierte 2008 in limitierter Edition von 188 Stück. Seine Besonderheiten: Unruhstopp, intelligentes Smart-Screw-System zur perfekten Balance des Käfigs sowie ein amagnetisches Silizium-Ankerrad. Im 2006 eingeweihten Genfer Firmendomizil im Ortsteil Plan-les-Ouates mit Blick auf renommierte Traditionsmanufakturen entstanden für das Automatikkaliber FC-980-1 bereits gut 80 Prozent der nötigen Komponenten. Bei größeren Quantitäten muss sich Frédérique Constant trotz des kontinuierlich aufgerüsteten und ausgebauten Maschinenparks bis heute der Hilfe externer Teilezulieferer bedienen. Die Feinbearbeitung, Montage und Qualitätskontrolle geschieht jedoch ausnahmslos unter dem eigenen Dach. 2008 verbuchte die junge Manufaktur bereits 90 000 fertiggestellte Uhren. Nicht zuletzt auch dank höherer Wertschöpfung hatte sich der Umsatz innerhalb weniger Jahre verdoppelt. Im Vergleich zum Ganzen bewegten und bewegen sich die Manufakturprodukte verständlicherweise in einer Nische. Die Philosophie der Firmeninhaber, ihren Kundinnen und Kunden rund um den Globus erschwinglichen Luxus bieten zu wollen, stützt sich in erster Linie auf elegante Zeitmesser mit zugekauften und, sofern Mechanik, entsprechend selbst definierter Standards veredelten Uhrwerken. Die meisten der Kaliber stammen in diesem Fall von Selitta. 2014 feierte Frédérique Constant das zehnjährige Jubiläum der eigenen Manufakturkaliber. Mittlerweile umfasste das Produktspektrum bereits 15, in zwei Familien gebündelte Kaliber. Alle mit 9xx stellen den Gangregler vorne zur Schau. Ohne sichtbaren Herzschlag präsentieren sich die ausschließlich mit Rotor-Selbstaufzug ausgestatteten FC-7xx. Das Basiskaliber FC-710 mit Zentralsekunde besteht aus 137 Komponenten, besitzt einen Durchmesser von 30 Millimetern und misst 6,2 Millimeter in der Höhe. Der zentral angeordnete Rotor spannt die Zugfeder in beiden Drehrichtungen. Nach Vollaufzug lässt der Energiespeicher die Unruh 42 Stunden lang mit vier Hertz oszillieren.

Einen echten Manufakturhöhepunkt und eine Sensation zugleich bringt das Jahr 2016. In der „Slimline Perpetual Calendar Manufacture", deren Kalenderwerk erst 2100 einer manuellen Kalender-Korrektur bedarf, tickt das exklusive Automatikkaliber FC-775. Allein für das aufwendige Schaltwerk unter dem Zifferblatt benötigen die Uhrmacher 78 Komponenten. Einzigartig ist der Preis: rund 8000 Euro.

Auf einem ganz anderen, nämlich elektronischem Blatt Papier steht schließlich die 2015 vorgestellte „Horological Smartwatch" mit analoger Zeit- und Funktionsanzeige. Diese Armbanduhr eidgenössischer Provenienz, welche sich mit Android- und iOS-Telefonen verbinden lässt, entstammt einer Zusammenarbeit mit US-amerikanischen Spezialisten. In nur einem Jahr konnte Frédérique Constant mehr als 16 000 Exemplare an männliche (75 Prozent) und weibliche Handgelenke bringen. Durch das Engagement von zwölf Ingenieuren in der Schweiz möchte Peter Stas dieses Projekt fortan noch deutlich ausbauen. Weiter auf Wachstumskurs sind schließlich auch die Stückzahlen. Ende 2016 sollen die Zähler bei rund 145 000 Armbanduhren stehen. Das nennt man Erfolg. ○

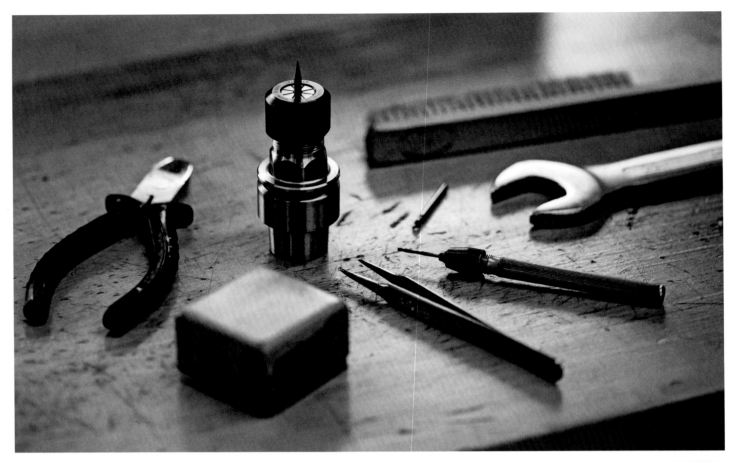

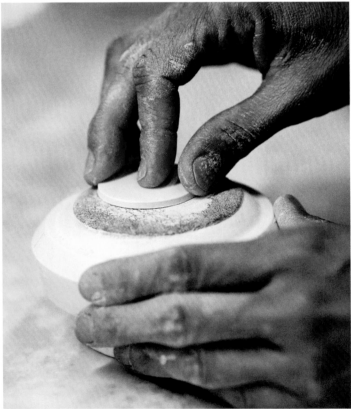

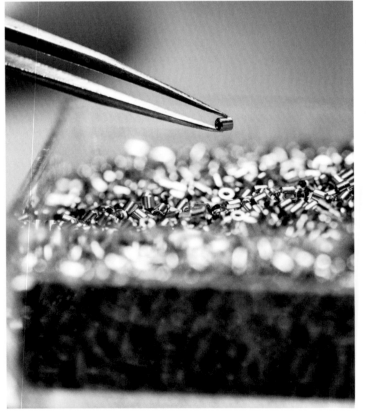

Clockwise, from top: Watchmaker's tools ◦ *Components* ◦ *Manufacturing process of a porcelain dial* ◦ *Von oben im Uhrzeigersinn:*
Uhrmacherwerkzeuge ◦ *Komponenten* ◦ *Herstellung eines Porzellanzifferblatts* ◦ *Dans le sens horaire, en partant du haut :*
Outils d'horlogers ◦ *Composants* ◦ *Processus de fabrication d'un cadran en porcelaine*

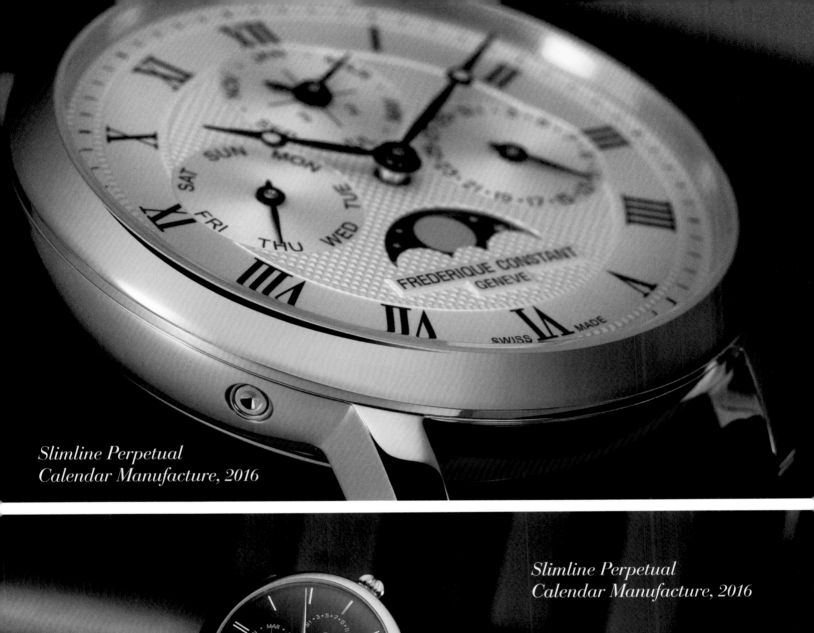

*Slimline Perpetual
Calendar Manufacture, 2016*

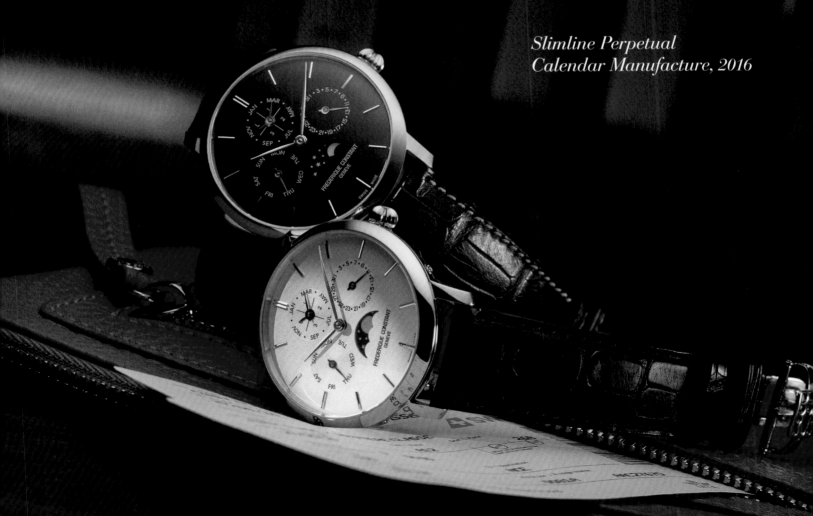

*Slimline Perpetual
Calendar Manufacture, 2016*

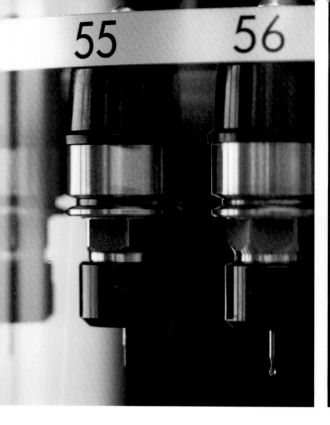

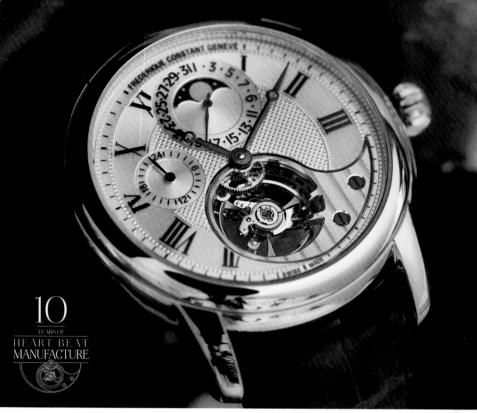

Le pouls vibrant d'une jeune manufacture

Français

Dans le roman *Guerre et Paix*, on peut lire : « Il est impossible d'éradiquer les passions ; nous devons simplement nous attacher à les mobiliser en vue d'atteindre un objectif noble. » Peter C. Stas et Aletta Stas-Bax ont fait leurs ces paroles de Léon Tolstoï. Leur activité en entreprise débute en effet par une passion insatiable pour les montres-bracelets, celles que ce couple de Hollandais admire durant des vacances au ski en Suisse. Un beau jour, un choix s'impose à eux, et ils sacrifient leurs emplois sûrs dans des entreprises internationales pour démarrer leurs propres activités horlogères. Mais prenons une chose après l'autre. Au milieu des années 1980, ils vivent tous deux à Hong Kong, haut lieu des montres plutôt singulier. En 1988, Peter Stas crée parallèlement à son activité salariée une petite entreprise de logiciels de CAO abordables. Il s'en servira plus tard pour créer des montres-bracelets, car l'apparence de celles produites en Chine le rebute vraiment. Il puise son inspiration dans de vieux livres et catalogues. Les tous premiers prototypes, pour ainsi dire assemblés sur un coin de table, sont présentés en 1991 par Aletta au Salon de l'horlogerie de Hong Kong. Peu avant la clôture de la manifestation, un grossiste japonais prend commande de 350 montres. Celles-ci se vendant étonnamment bien, il en commandera 1 100 exemplaires supplémentaires. L'année qui suit est marquée par une petite collection de prototypes qui compte au total six modèles. S'ils sont encore tous dotés de mouvements à quartz, ils sont déjà assemblés en Suisse, à partir de composants locaux. Ces nouvelles productions prospèrent bien elles aussi, mais elles ont bientôt besoin d'un nom intéressant. Frédérique Schreiner, l'aïeule d'Aletta, et Constant Stas, l'aïeul de Peter, serviront de marraine et de parrain, la réunion de leurs deux prénoms donnant Frédérique Constant. Constituée avec un capital de 60 000 francs suisses, Frédérique Constant SA produit dès 1992, avec des partenaires externes, plus de 1 000 montres-bracelets de fabrication suisse. La rapidité à laquelle elles se vendent les motive à voir plus grand. La rencontre avec Miguel Garcia, alors directeur des ventes chez Sellita, est un coup de chance. En discutant, ils découvrent qu'il est possible d'ouvrir sur le devant le calibre à remontage automatique Eta 2836 relativement épais pour dévoiler le pouls de la culture humaine. Ainsi naît en 1994 « Heart Beat », une collection non protégée et par conséquent rapidement copiée. La fabrication sous label privé (plus tard abandonnée) et en grandes quantités permet à la jeune manufacture de produire avec des structures de coûts plus avantageuses. En 1996, celle-ci dépose en Suisse un projet d'entreprise accompagné d'une demande d'autorisation d'exercer dans la Confédération en vue de transférer le siège de la société à Genève. L'autorisation leur est accordée, à la condition qu'elle constitue un socle d'au moins douze employés en cinq ans. Cet objectif est atteint en seulement deux ans et demi, et ce exclusivement par une croissance autofinancée. Des visions claires et le désir d'ajouter une touche personnelle aux calibres achetés conduisent Frédérique Constant en 2001 à développer un premier calibre de manufacture pour montres-bracelets. Pour ce projet ambitieux, Peter Stas s'adjoint la collaboration de spécialistes de deux écoles horlogères, celle de Genève et une autre en Hollande. En 2004, Frédérique Constant crée la surprise avec Heart Beat FC-910-1. Dans un large guichet à 6 heures oscille un balancier étonnamment grand. La version FC-915 Phase de lune et date de 2005 est suivie en 2006 par le FC-930, un modèle à mouvement mécanique et remontage automatique. L'année 2007 est placée sous le signe des nouveaux matériaux. Le calibre automatique FC-935 Silicium à indication de phase de lune et quantième à aiguille possède une roue d'échappement spécialement conçue

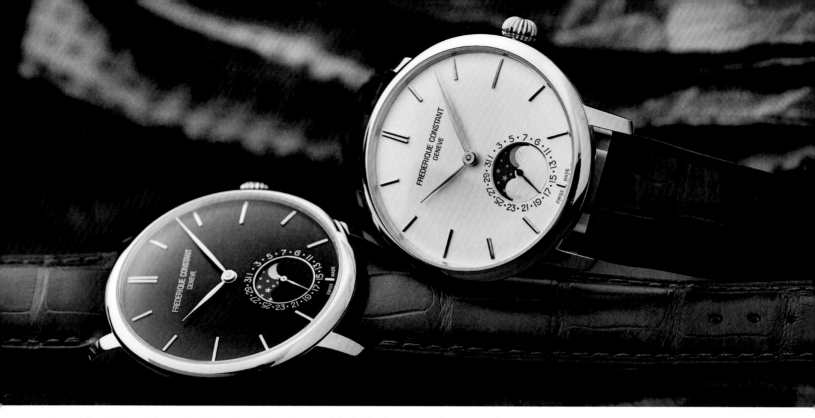

From left: CNC machine tools ◦ Heart Beat Manufacture, 2014 ◦ Slimline Moonphase Manufacture, 2014 ◦ Von links: CNC-Werkzeugmaschinen ◦ Heart Beat Manufacture, 2014 ◦ Slimline Moonphase Manufacture, 2014 ◦ De gauche à droite : Machines-outils à commande numérique ◦ Heart Beat Manufacture, 2014 ◦ Slimline Moonphase Manufacture, 2014

en silicium. La variante portant un nom aux allures de code secret FC-935SZABS4H9, adjugée lors de la vente aux enchères caritative « One of a Kind » de septembre 2007 à Monte Carlo, est caractérisée par une évolution visionnaire. Outre une roue d'échappement en silicium, cette pièce unique dispose d'un balancier d'un type entièrement nouveau constitué d'un matériau céramique appelé Zerodur. Également dotée d'un calibre de manufacture, la montre « Heart Beat Tourbillon » fait ses débuts en 2008 dans une édition limitée à 188 exemplaires. Ses particularités : un stop balancier, un système intelligent « Smart-Screw » pour un parfait équilibre de la cage et une roue d'échappement amagnétique en silicium. C'est au siège genevois de la société, inauguré en 2006 dans le quartier de Plan-les-Ouates avec ses célèbres manufactures, que sont créés plus de 80 pour cent des composants du calibre à remontage automatique FC-980-1. Même avec un parc de machines sans cesse modernisé et étoffé, la société Frédérique Constant doit encore et toujours faire appel à des fournisseurs de pièces externes pour les grands volumes. Les opérations de finissage, d'assemblage et de contrôle qualité sont toutefois toutes réalisées au sein de la manufacture genevoise. En 2008, la jeune entreprise s'enorgueillit d'une production de 90 000 montres, son chiffre d'affaires doublant en quelques années, notamment grâce à la grande valeur ajoutée de ses produits. Sur l'ensemble du marché, ils occupent et continuent bien sûr d'occuper un créneau privilégié. La philosophie des propriétaires de la société, qui consiste à offrir à leur clientèle du monde entier un luxe abordable s'appuie essentiellement sur des garde-temps élégants dont les mouvements ont été produits par des tiers et améliorés au niveau mécanique, conformément aux normes définies en interne. La plupart de ces mouvements proviennent de Sellita. Mais en 2014, Frédérique Constant célèbre la dixième année de production de calibres de manufacture.

La gamme de produits comprend désormais 15 mouvements regroupés en deux familles. Sur les calibres numérotés 9xx, le régulateur de marche est visible sur le devant. Sur les calibres numérotés FC-7xx, tous dotés d'un remontage automatique par rotor, il ne l'est pas. Le calibre de base FC-710 à seconde centrale comporte 137 composants, mesure 30 millimètres de diamètre et 6,2 millimètres de haut. Le rotor en position centrale tend les spiraux dans les deux sens de rotation. Après remontage complet, l'énergie emmagasinée permet au balancier d'osciller 42 heures à une fréquence de quatre Hertz.

L'année 2016 est marquée par une avancée sensationnelle en termes de fabrication. La « Manufacture Quantième Perpétuel » de la collection Slimline, dont le quantième perpétuel ne devrait pas exiger de correction manuelle avant 2100, est animée par le calibre de manufacture FC-775. Pour le seul mécanisme complexe sous le cadran, 78 pièces sont nécessaires aux horlogers. Le tout à un prix exceptionnel : environ 8 000 euros.

Présentée en 2015, la montre « Horological Smartwatch », qui intègre une partie électronique, donne le temps en mode analogique et propose diverses fonctions. Cette montre-bracelet, qui communique avec les smartphones Android et iPhone, est le fruit d'une collaboration avec des spécialistes américains. En seulement un an, 16 000 exemplaires de ce modèle Frédérique Constant seront portés par des poignets masculins (75 pour cent) et féminins. Peter Stas souhaite développer encore davantage ce projet en engageant douze ingénieurs en Suisse. La tendance est également à la hausse pour le volume de production. Fin 2016, les ventes devraient être de l'ordre de 145 000 montres-bracelets. Voilà ce qu'on peut appeler une réussite. ◦

Girard-Perregaux

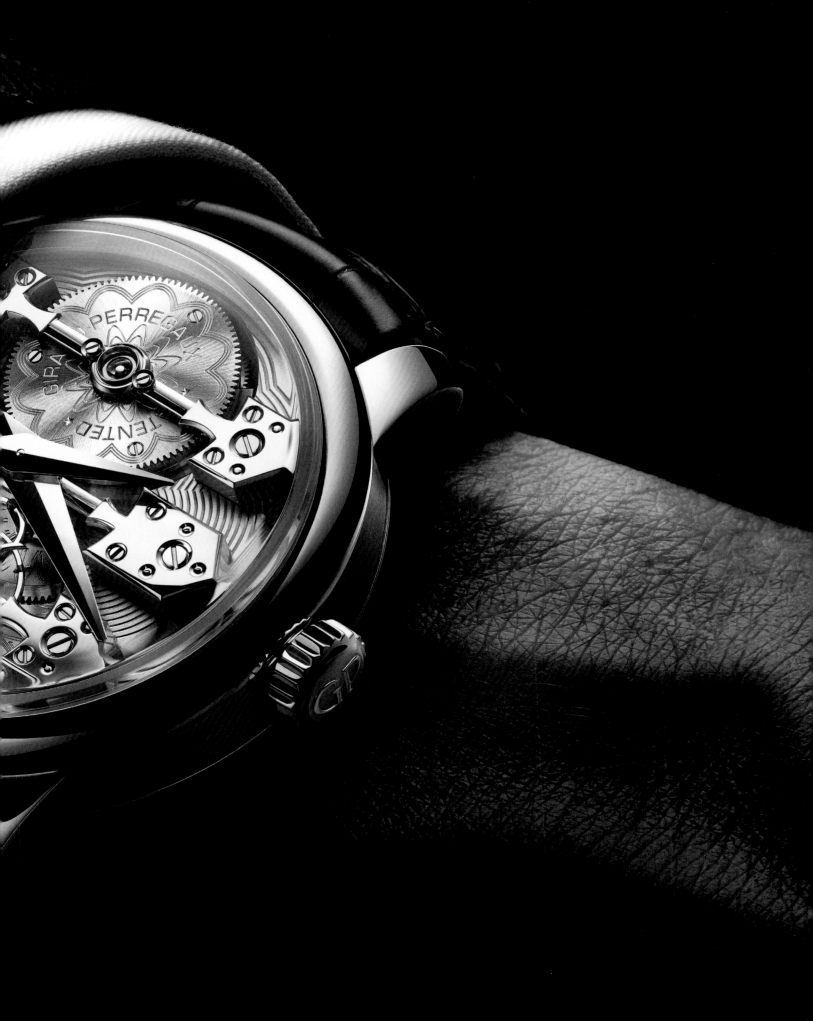

One of the Oldest Swiss Watch Manufactories

English

There is undoubtedly more than one answer to the question about the origins of the wristwatch. One of them involves the Great Industrial Exposition of Berlin, which took place on the exhibit area at Lehrter railway station in 1879. Girard-Perregaux displayed numerous watches as its contributions to this exhibition of internationally relevant technological innovations. This appearance earned an interesting commission for the Swiss manufacture: the imperial German navy needed readily legible timepieces for its officers, and Girard-Perregaux had the solution in the guise of a circular wristwatch with a chain wristband and a protective grid to shield the fragile glass crystal. The prototype impressed the German decision makers, who duly ordered 2,000 of these watches, each of which encased a 13-ligne hand-wound movement. This timepiece, which is arguably the first serially produced wristwatch, is only one of many chronometric milestones in the long and occasionally turbulent history of the house of Girard-Perregaux. The chronicle began in 1791 with the watchmaker Jean-François Bautte and the signature on his first pocket watches. Bautte's master Jacques Dauphin Moulinier was so impressed by the quality of his apprentice's work that he soon invited the assiduous young man to become a junior partner at his watch business in Geneva. Moulinier's noble clientele was enchanted by Bautte's elegant savoir vivre: for example, he perfumed the stairs of the shop and arranged to have his firewood artistically turned on lathes. After Bautte's death in 1837, the company was taken over by his son and son-in-law. The reins passed to Felipe Hecht in 1897 and later to Hecht's son Juan Hecht. The latter transferred his inheritance in 1906 to a friend and relative: Constant Girard-Gallet, who had operated a watch factory in La Chaux-de-Fonds since 1852 and who, together with his wife Marie Perregaux, had cofounded Girard-Perregaux (GP) in 1856. Wholly devoted to uncompromising precision, they repeatedly won chronometer competitions at Neuchâtel Observatory between 1851 and 1876. By 1889, GP had collected no fewer than 13 gold medals and diplomas at international exhibitions. In 1903, their eldest son Louis-Constant took the helm as the last in the Girard line. In the wake of liquidation in 1928, the brand rights and Jean-François Bautte's old ledger became the property of the Société de Banque Suisse (SBS). The Graef Family, which was of German extraction, purchased both for 30,000 Swiss francs the following year. Under their aegis, watches were produced with the Girard-Perregaux signature, as well as other timepieces labeled "MIMO." Approximately 200,000 watches were manufactured in 1951. For lack of heirs, the family business was sold to Desco von Schulthess, an international mercantile house, in 1979. A management buyout followed in 1988, but the new owners weren't very successful. Fortunately, the Italian Luigi Macaluso and his Sowind Group arrived with calmness and visionary ideas for the future in 1992. PPR, the French luxury conglomerate (now Kering) acquired a 23 percent share in the spring of 2008. Annual production at this time totaled circa 20,000 watches in the luxury price segment. Most of them encased movements that had been made at the brand's own manufacture in La Chaux-de-Fonds. The purchase contract already envisioned a further increase in the share to more than 50 percent. After Luigi Macaluso's unexpected death in October 2010, Kering availed itself of this option in the wake of a recapitalization.

The manufactory's horological highlights inarguably include the legendary tourbillon with three bridges. Girard-Perregaux registered 27 specimens of this model at Neuchâtel Observatory between 1865 and 1911. The patent for the parallel arrangement of three solid gold bridges on the plate dates from March 25, 1884. An especially precious version known as "La Esmeralda" was presented at the Fourth Paris International Exposition of 1889. Before GP repurchased this watch in the 1960s, it had belonged to Mexico's President Porfirio Díaz (1830–1915). This exceptional tourbillon finally debuted in the case of a wristwatch in 1991. Various versions with manual or automatic winding would become fixed features in Girard-Perregaux's collection in ensuing years. "La Esmeralda" found its way to the wrist in 2016, just in time to celebrate the firm's 225th anniversary. The filigreed tourbillon cage, which concatenates 80 components, weighs a mere 0.305 gram.

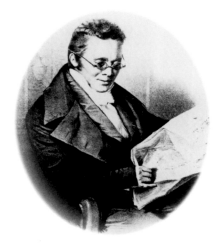 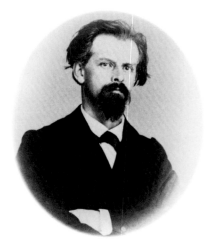 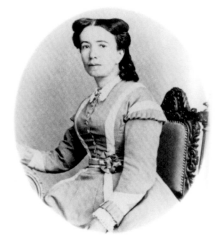

From left: Jean-François Bautte ○ Constant Girard-Gallet ○ Marie Perregaux

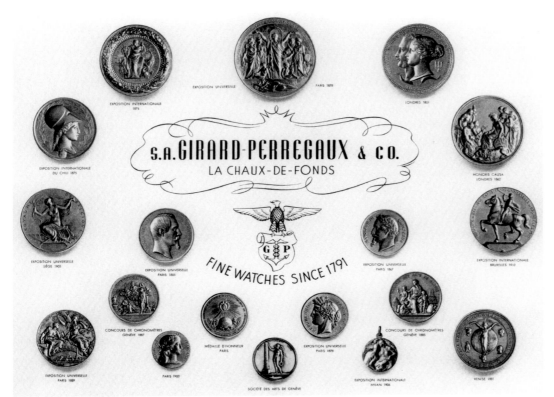

S.A. GIRARD-PERREGAUX & CO.
LA CHAUX-DE-FONDS

FINE WATCHES SINCE 1791

From left: Awards and medals ◦
La Esmeralda pocket watch ◦
Von links: Auszeichnungen
und Medaillen ◦ Taschenuhr
La Esmeralda ◦ De gauche
à droite : Diplômes et médailles ◦
Montre de poche La Esmeralda

GP launched a mechanical world premiere in 1966: automatic Calibers 31.7 and 32.7 use the innovative Clinergic escapement, which has a balance that completes 36,000 semi-oscillations per hour. The manufacture was awarded Neuchâtel Observatory's Jubilee Prize in recognition of this superlative achievement in the service of a more highly accurate rate. Unlike many other brands in the early 1970s, GP opted not to rely on its own competence for the development of an electronic quartz movement. Exclusive Caliber GP 350 anticipated the currently valid industrial standard, i.e., its quartz resonator oscillates inside a little vacuum tube at a frequency of 32,768 hertz. This caliber was encased inside the "GP Quartz," which debuted at the watch fair in Basel in 1971. Greater emphasis was again placed on the theme of mechanical timekeeping after Luigi Macaluso acquired the business. He consistently expanded the necessary ateliers, purchased new machinery and, after approximately two years of developmental work, brought caliber family 3xxx to the market in 1994. The basic caliber (GP3000) is 23.9 millimeters in diameter and a mere 2.98 millimeters tall, which made it one of the slimmest of its kind. Despite its low height, it provides sufficient space for a jumping date display with rapid switching. Caliber GP3200 is 3.2 millimeters tall and 26.2 millimeters in diameter. The third movement in the trio is the equally tall GP3300: its diameter is 25.6 millimeters. The largest of GP's own automatic movements (GP1800) spent many years "on ice." It finally debuted with a diameter of 30.6 millimeters and a height of 4.16 millimeters. Girard-Perregaux currently offers it in an "ordinary" and a skeletonized version. It's housed in the classically round cases of the longstanding and highly successful "1966" line which debuted, needless to say, in 1966. Aficionados of angular cases appreciate the nostalgic-looking "Vintage 1945." The purest retro look is also embodied by the "Heritage Anniversary 1957," which debuts in 2016. Its constant-force escapement, which premiered in 2009 and has undergone continual improvements ever since, bears witness to unflagging power of innovation. This complex ensemble would have been absolutely impossible without the use of silicon as its raw material.

Only 225 specimens exist of the anniversary model "Heritage Anniversary Place Girardet." The name recalls the square in La Chaux-de-Fonds where Girard-Perregaux maintains its headquarters. A uniquely individualized dial is the special feature of this wristwatch, which encases automatic Caliber GP1800-0005 and displays its Microvar balance on the dial side. Each dial is marked with a date between 1791 and 2016, together with an event that occurred in that year. For example, 1808 recalls Beethoven and his Fifth Symphony, and 1963 commemorates the first woman to fly in outer space. The current apex of the watch portfolio is represented by the "Minute Repeater Tourbillon with Bridges," a tourbillon watch with minute repeater and innovative bridges. Due to the timepiece's extreme complexity, Girard-Perregaux has limited production of this mellifluously named titanium wristwatch to just 30 specimens. In 2017 and thereafter, Girard-Perregaux will display many aspects of its rich past and creative present in the firm's stylish museum at number 129 Rue du Progrès in La Chaux-de-Fonds. Would-be visitors are requested to schedule their visits in advance. ◦

Vintage 1945 Small Second, 2015

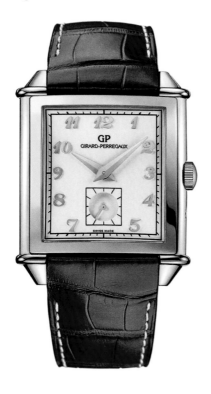

Vintage 1945 Tourbillon, 2015

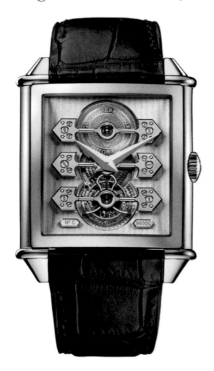

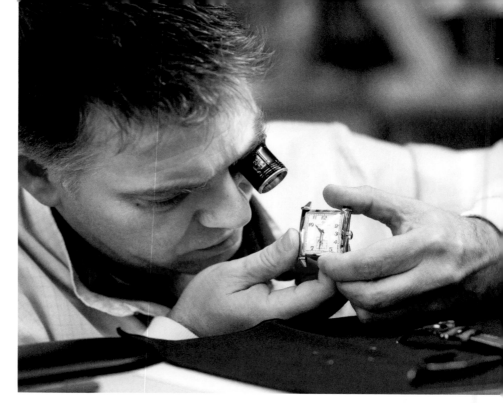

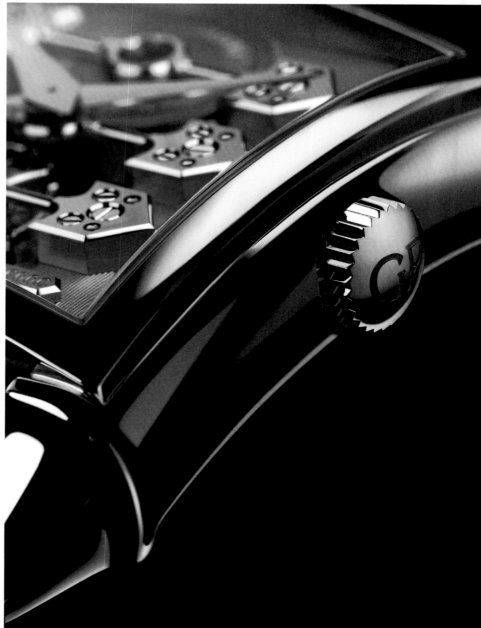

Top: Final assembly and quality control
Oben: Endmontage und Kontrolle
En haut : Assemblage final et contrôle

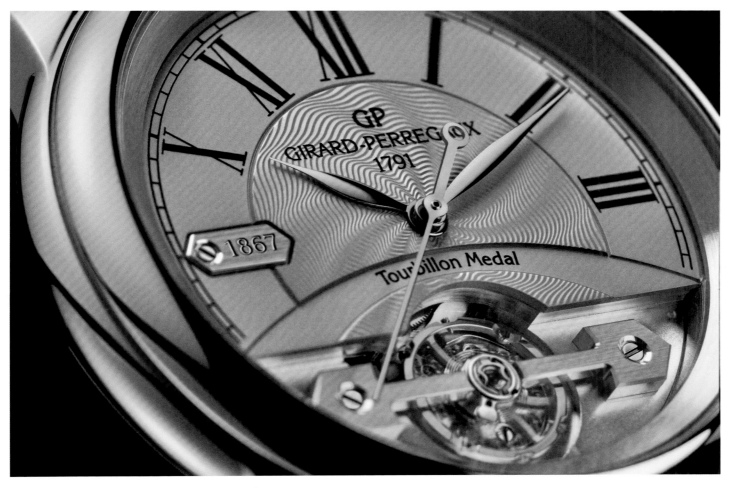

Heritage Anniversary Place Girardet 1867, 2016

*Heritage Anniversary
Place Girardet 1999, 2016*

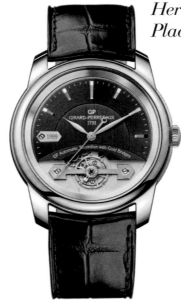

*Heritage Anniversary
Place Girardet 2016, 2016*

*Heritage Anniversary
Place Girardet 1880, 2016*

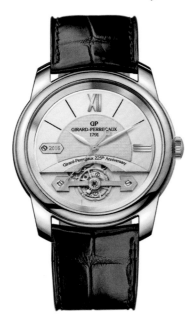

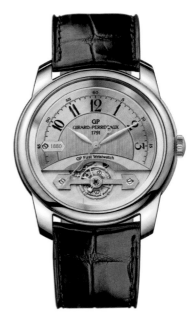

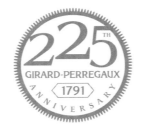

Eine der ältesten Schweizer Uhrenmanufakturen

Deutsch

Die Frage nach den Anfängen der Armbanduhr lässt zweifellos mehrere Antworten zu. Eine hängt mit der Berliner Gewerbeausstellung zusammen, welche 1879 im Ausstellungspark am Lehrter Bahnhof stattfand. Girard-Perregaux bereicherte die Leistungsschau technischer Innovationen von überregionaler Bedeutung mit etlichen Uhren. Dieser Auftritt bescherte der Schweizer Manufaktur einen interessanten Auftrag. Für ihre Offiziere benötigte die deutsche Kaiserliche Marine leicht ablesbare Zeitmesser. Und Girard-Perregaux hatte die Lösung in Gestalt einer runden Armbanduhr mit Kettenband und Schutzgitter über dem bruchempfindlichen Kristallglas. Der positiv bewertete Prototyp führte zu einer Bestellung von 2000 Exemplaren mit 13-linigem Handaufzugswerk. Die vermutlich erste Serien-Armbanduhr ist nur einer von vielen chronometrischen Meilensteinen in der langen, teilweise recht turbulenten Biographie des Hauses Girard-Perregaux. Sie beginnt 1791 mit dem Uhrmacher Jean-François Bautte und der Signatur erster Taschenuhren. Lehrmeister Jacques Dauphin Moulinier überzeugte die Qualität so sehr, dass er seinem geschäftstüchtigen Schüler schon kurz darauf Anteile an seinem Genfer Uhrengeschäft offerierte. Bautte entzückte die adelige Kundschaft durch seine elegante Lebensart. Beispielsweise parfümierte er die Treppen seines Geschäfts oder ließ das Brennholz kunstvoll drechseln. Nach seinem Tod im Jahr 1837 übernahmen sein Sohn und Schwiegersohn, 1897 Felipe Hecht und schließlich dessen Sohn Juan Hecht die Firma. Letzterer übereignete sein berufliches Erbe 1906 einem Freund und Verwandten. Constant Girard-Gallet hatte sich seit 1852 in La Chaux-de-Fonds als Uhrenfabrikant betätigt und 1856 zusammen mit Ehefrau Marie Perregaux die Firma Girard-Perregaux (GP) gegründet. Deren Maximen galten kompromissloser Präzision. Zwischen 1851 und 1876 gewann sie mehrfach die Chronometer-Wettbewerbe des Observatoriums Neuchâtel. Bis 1889 konnte GP nicht weniger als 13 Goldmedaillen und Diplome bei internationalen Ausstellungen erringen. 1903 nahm der älteste Sohn Louis-Constant als letzter Girard das Ruder in die Hand. 1928 gingen im Zuge der Liquidation die Markenrechte und das alte Hauptbuch von Jean-François Bautte an die Société de Banque Suisse (SBS). Beides kaufte die deutschstämmige Familie Graef im Folgejahr für 30 000 Schweizer Franken. Unter ihrer Ägide entstanden parallel Uhren mit den Signaturen Girard-Perregaux und MIMO. Für das Jahr 1951 zeigten die Produktionszähler circa 200 000 Uhren. Mangels familiärer Nachkommen gelangte die Traditionsmanufaktur 1979 unter das Dach des internationalen Handelshauses Desco von Schulthess. 1988 stand im Zeichen eines Management-Buyout. Die neuen Eigentümer wirtschafteten jedoch wenig erfolgreich. Ruhe und visionäre Gedanken für die Zukunft brachten 1992 der Italiener Luigi Macaluso und seine Sowind-Gruppe. Im Frühjahr 2008 beteiligte sich der französische Luxus-Multi PPR (heute Kering) mit 23 Prozent. Zu diesem Zeitpunkt lag die Jahresproduktion bei rund 20 000 im preislichen Luxussegment angesiedelten Uhren. In den meisten tickten Uhrwerke aus eigener Manufaktur in La Chaux-de-Fonds. Schon der damalige Kaufvertrag sah eine weitere Erhöhung des Anteils über 50 Prozent hinaus vor. Von dieser Option machte Kering nach dem überraschenden Ableben von Luigi Macaluso im Oktober 2010 im Zuge einer Kapitalerhöhung Gebrauch.

Zu den uhrmacherischen Highlights der Manufaktur gehört zweifellos das legendäre Tourbillon mit drei Brücken. Zwischen 1865 und 1911 ließ Girard-Perregaux 27 Exemplare beim Observatorium Neuchâtel registrieren. Das Patent für die parallele Anordnung der drei massivgoldenen Brücken auf der Platine datiert auf den 25. März 1884. Besonders kostbar präsentierte sich 1889 während der Pariser Weltausstellung „La Esmeralda". Bevor GP die Uhr in den 1960er Jahren zurückkaufte, hatte sie u. a. dem mexikanischen Präsidenten Porfirio Díaz (1830–1915) gehört. 1991 debütierte dieses Ausnahme-Tourbillon endlich auch in einem Armbanduhrgehäuse. Seitdem ist es in unterschiedlichen Ausführungen mit manuellem oder automatischem Aufzug fester Bestandteil der Girard-Perregaux-Kollektion. Zum 225. Firmenjubiläum findet „La Esmeralda" 2016 auch ans Handgelenk. In diesem Fall wiegt der filigrane, aus 80 Komponenten zusammengefügte Tourbillonkäfig nur 0,305 Gramm.

1966 lancierte GP eine mechanische Weltpremiere. In den Automatikkalibern 31.7 und 32.7 mit innovativer Clinergic-Hemmung vollzog der Gangregler stündlich 36 000 Halbschwingungen. Für diese überragende Leistung im Dienste höherer Ganggenauigkeit erhielt die Manufaktur den Jubiläumspreis des Observatoriums Neuchâtel. Anders als viele andere Marken setzte GP Anfang der 1970er bei der Entwicklung eines elektronischen Quarzwerks nicht auf eigene Kompetenz. Das exklusive Kaliber GP 350 nahm den heutigen Industriestandard vorweg, das heißt der Quarzresonator oszillierte in einer kleinen Vakuumröhre bereits mit 32 768 Hertz. Die damit ausgestattete „GP Quartz" gab 1971 während der Basler Uhrenmesse ihren Einstand. Dem Thema Mechanik-Manufaktur verlieh Luigi Macaluso nach dem Erwerb des Unternehmens beachtliche Impulse. Er baute die entsprechenden Ateliers konsequent aus, kaufte neue Maschinen und brachte 1994 nach rund zweijähriger Entwicklungsarbeit die Kaliberfamilie 3xxx auf den Markt. Das Basis-Kaliber GP3000 besitzt einen Durchmesser von 23,9 Millimetern. Mit nur 2,98 Millimetern Bauhöhe gehört es zu den flachsten seiner Art. Immerhin ist darin auch eine springende Datumsanzeige mit Schnellschaltung enthalten. Das Kaliber GP3200 misst bei 3,2 Millimetern Bauhöhe insgesamt 26,2 Millimeter. Dritter im Bunde ist das gleich hohe GP3300 mit 25,6 Millimetern Durchmesser. Jahrelang auf Eis lag das größte der eigenen Automatikwerke, GP1800 genannt. Sein Durchmesser beträgt 30,6 Millimeter, seine Höhe 4,16 Millimeter. Girard-Perregaux bietet es aktuell in „normaler" sowie skelettierter Version an. Und zwar in der klassisch runden, seit Jahren höchst erfolgreichen Linie „1966", deren Ursprünge in besagtem Jahr zu finden sind.

Full-calendar movement GP3300

Liebhaberinnen und Liebhaber kantiger Gehäuseformen kommen bei der nostalgischen „Vintage 1945" zu ihrem Recht. Retro-look in Reinkultur verstrahlt seit 2016 schließlich auch die „Heritage Anniversary 1957". Von ungebrochener Innovationskraft zeugt eine 2009 vorgestellte und seitdem kontinuierlich verbesserte Hemmung mit konstanter Kraft. Ohne Verwendung des Werkstoffs Silizium wäre das komplexe Gebilde absolut unmöglich.

Lediglich 225 Exemplare gibt es von dem Jubiläumsmodell „Heritage Anniversary Place Girardet". An diesem Platz in La Chaux-de-Fonds ist Girard-Perregaux zu Hause. Die Besonderheit dieser Armbanduhr mit dem Automatikkaliber GP1800-0005 und vorne sichtbarer Microvar-Unruh besteht in der einzigartigen Individualisierung des Zifferblatts. Auf jedem sind eine Jahreszahl zwischen 1791 und 2016 sowie eine dazu passende Begebenheit verewigt. 1808 erinnert beispielsweise an Beethoven und seine 5. Sinfonie, 1963 an die erste Frau im Weltraum. Die gegenwärtige Spitze des Uhren-Portfolios repräsentiert das „Minute Repeater Tourbillon with Bridges", ein Tourbillon mit Minutenrepetition und Neo-brücken. Wegen der hohen Komplexität hat Girard-Perregaux die Produktion dieser wohlklingenden Titan-Armbanduhr auf nur 30 Exemplare limitiert. Viele Aspekte der reichen Vergangenheit und kreativen Gegenwart wird Girard-Perregaux ab 2017 im stilvollen Firmenmuseum, Rue du Progrès 129, La Chaux-de-Fonds, zeigen. Ein Besuch ist allerdings nur nach vorheriger Anmeldung möglich. ○

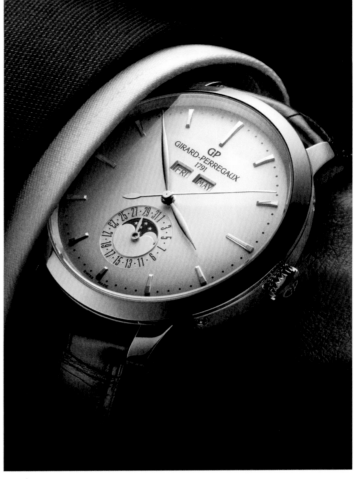

1966 Skeleton, 2016

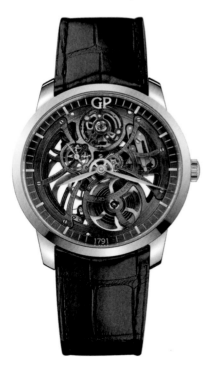

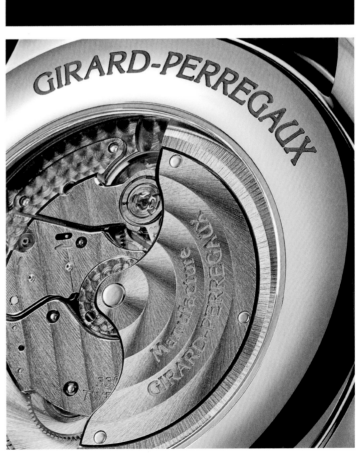

1966 Full Calendar, 2016

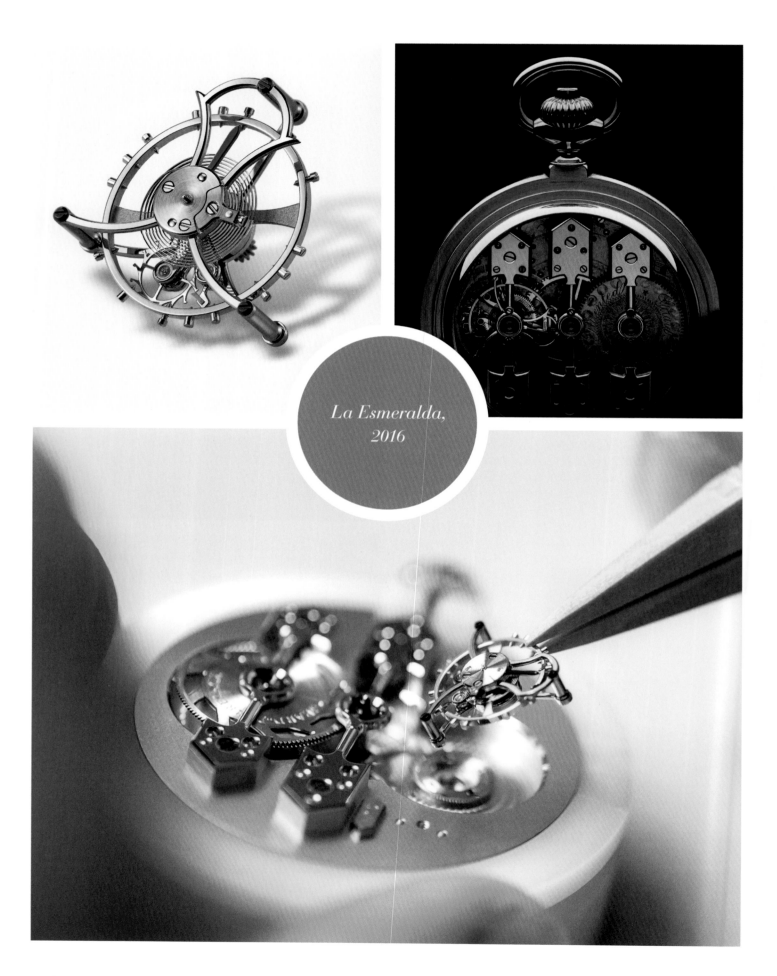

La Esmeralda,
2016

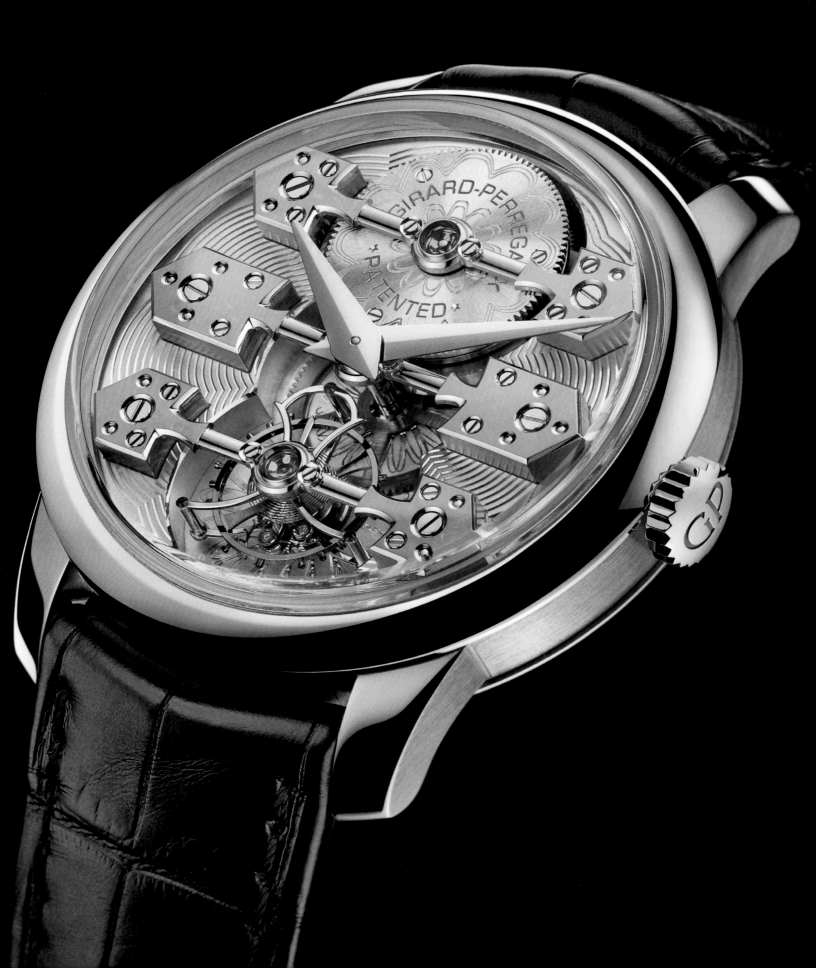

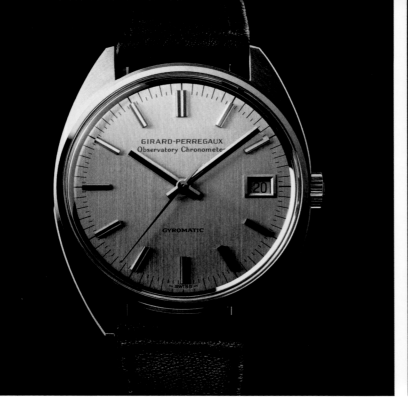
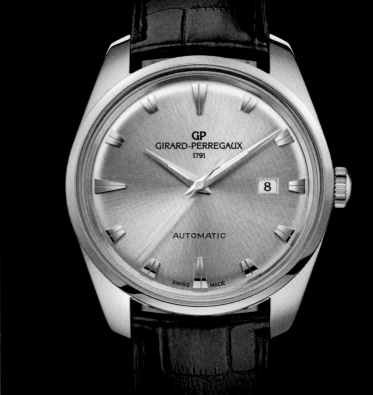

L'une des plus anciennes manufactures horlogères suisses

La question de l'origine des montres-bracelets appelle assurément diverses réponses. Celles-ci pourraient remonter à la foire-exposition organisée en 1879 à Berlin dans le parc proche de la gare de Lehrte. Girard-Perregaux contribue alors par diverses montres à cette exposition d'innovations techniques d'une portée internationale, prestation qui lui rapporte une commande intéressante. Les officiers de la marine impériale ont besoin de garde-temps faciles à consulter. Girard-Perregaux leur apporte la solution : une montre-bracelet ronde à chaîne et grille protégeant le fragile verre en cristal. Convaincue par le prototype, l'armée commande 2 000 unités équipées de mouvements 13''' à remontage manuel. Cette montre-bracelet, peut-être la première fabriquée en série, n'est qu'un jalon dans l'histoire riche et parfois tumultueuse de la maison Girard-Perregaux. Elle débute en 1791 avec l'horloger Jean-François Bautte, qui signe les premières montres de poche. Fortement séduit par la qualité proposée par son élève Bautte, également très doué pour les affaires, le maître-horloger Jacques Dauphin Moulinier lui offre peu après des parts dans sa boutique d'horlogerie à Genève. Bautte ravit la clientèle noble par son style de vie élégant, faisant par exemple parfumer les escaliers de sa boutique ou façonner artistiquement au tour son bois de chauffage. À son décès en 1837, la boutique est reprise par son fils et son gendre. Elle revient en 1897 à Felipe Hecht, qui la transmettra à son fils Juan Hecht. En 1906, ce dernier transfère la propriété de son héritage professionnel à un ami proche, Constant Girard-Gallet. Fabricant de montres depuis 1852 à La Chaux-de-Fonds, il a fondé en 1856 avec sa femme Marie Perregaux la manufacture Girard-Perregaux (GP). Leur devise : une précision sans compromis. De 1851 à 1876, la manufacture remporte plusieurs concours de chronomètres de l'observatoire de Neuchâtel. Jusqu'en 1889, GP reçoit pas moins de 13 médailles d'or et diplômes lors d'expositions internationales. En 1903, Louis-Constant, le fils aîné, prend les rênes de la manufacture, en tant que dernier des Girard. En 1928, la manufacture est mise en liquidation. Les droits de propriété des marques et l'ancien grand-livre de Jean-François Bautte échoient à la Société de Banque Suisse (SBS). La famille d'origine allemande Graef les rachète l'année suivante pour 30 000 francs suisses. Sous son égide, des montres signées Girard-Perregaux et des garde-temps MIMO sont fabriqués en parallèle. En 1951, environ 200 000 montres sont produites. Faute d'héritier, cette manufacture de tradition passe en 1979 dans le giron de la maison de commerce international Desco von Schulthess. 1988 est marquée par un rachat par les dirigeants, mais les nouveaux propriétaires n'ont pas une gestion satisfaisante. En 1992, l'Italien Luigi Macaluso et le groupe Sowind ramènent calme et stabilité, et proposent des idées visionnaires. Au printemps 2008, la multinationale du luxe PPR (actuelle Kering) entre dans le capital à hauteur de 23 pour cent. La production annuelle s'élève à environ 20 000 montres haut de gamme en termes de prix. La plupart sont animées par des mouvements fabriqués à La Chaux-de-Fonds. Le contrat de vente prévoyant à l'origine la possibilité d'augmenter la participation à plus de 50 pour cent, Kering saisit cette occasion dans le cadre d'une augmentation de capital, après le décès soudain de Luigi Macaluso en octobre 2010.

Le légendaire Tourbillon sous trois ponts d'or compte assurément parmi les produits phares de la manufacture. Entre 1865 et 1911, Girard-Perregaux en fait certifier 27 exemplaires par l'observatoire de Neuchâtel. Le brevet portant sur la disposition parallèle des trois ponts en or massif sur la platine est délivré le 25 mars 1884. En 1889, « La Esmeralda », montre de poche particulièrement précieuse, est présentée à l'Exposition universelle de Paris.

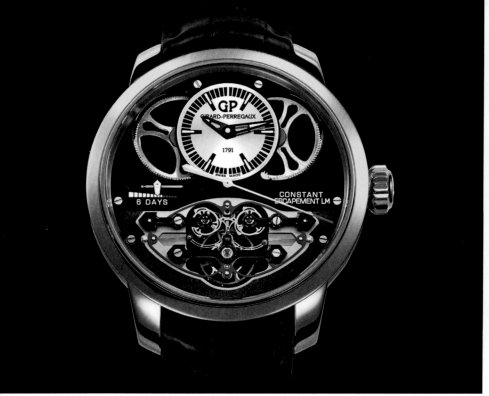
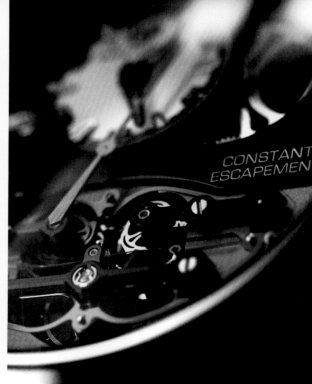

Opposite page: Gyromatic HF, 1965 ◦ Heritage Anniversary 1957, 2016 ◦ This page: Constant Escapement, 2016

Changeant plusieurs fois de mains, notamment celles du président mexicain Porfirio Díaz (1830–1915), elle est rachetée dans les années 1960 par GP. En 1991, ce tourbillon d'exception trouvera enfin le chemin d'un boîtier de montre-bracelet. Déclinée en différentes versions, avec remontage manuel ou automatique, c'est un pilier de la collection. Pour les 225 ans de GP en 2016, « La Esmeralda » est aussi proposée en montre-bracelet. Le délicat tourbillon constitué de 80 composants ne pèse que 0,305 gramme.

L'année 1966 est marquée par une première mondiale pour GP dans le domaine mécanique. Dans les calibres automatiques 31.7 et 32.7 dotés d'un échappement innovant baptisé Clinergic, le balancier réalise 36 000 alternances par heure. Cette extraordinaire performance permettant d'augmenter la précision de marche est récompensée par le prix du Centenaire de l'observatoire de Neuchâtel. Au début des années 1970, à l'inverse de bien des marques, GP ne s'appuie pas que sur ses propres compétences pour créer un mouvement à quartz. Avec son résonateur à quartz oscillant à 32 768 Hertz dans un tube à vide, le calibre exclusif GP 350 préfigure le standard universel pour les montres à quartz. Ainsi équipée, la « GP Quartz » fait ses débuts en 1971 au Salon mondial de l'horlogerie à Bâle. Luigi Macaluso reprend alors la manufacture et donne une impulsion considérable au développement de mouvements mécaniques, par un agrandissement considérable des ateliers concernés et l'achat de nouvelles machines. En 1994, après environ deux ans de développement, la famille de calibres 3xxx est sur le marché. Le diamètre du calibre de base GP3000 est de 23,9 millimètres. Avec un record de minceur de 2,98 millimètres, c'est l'un des plus plats du genre. Malgré cette finesse, il dispose de la fonction date sautante. Le calibre GP3200 mesure 26,2 millimètres de diamètre pour une épaisseur de 3,2 millimètres. Troisième de la série, le GP3300 est aussi mince que

le GP3200, mais avec un diamètre de 25,6 millimètres. Plus grand mouvement automatique de manufacture GP avec un diamètre de 30,6 millimètres et une épaisseur de 4,16 millimètres, le GP1800 restera longtemps en attente. Girard-Perregaux le propose actuellement en version « normale » ou squelette, ajouré en suivant les formes rondes traditionnelles de la ligne à grand succès « 1966 », lancée cette même année. Les amoureux de boîtiers anguleux apprécient un modèle empreint de nostalgie, le « Vintage 1945 ». Mais depuis 2016, un look rétro en concentré irradie également du modèle « Heritage Anniversary 1957 ». L'échappement à force constante présenté en 2009, depuis sans cesse amélioré, témoigne d'une inlassable capacité d'innovation. Sans le silicium, ce mécanisme complexe n'aurait en aucun cas pu être réalisé.

Pour célébrer ses 225 ans, la manufacture sort en 2016 en édition limitée à 225 exemplaires la montre « Heritage Anniversary Place Girardet », du nom de la place de La Chaux-de-Fonds où se situe le siège de GP. Cette montre-bracelet à calibre automatique GP1800-0005 et balancier Microvar visible à 6 heures se distingue par une individualisation très singulière des cadrans. Chacun d'eux porte une date de 1791 à 2016 et la mention d'un événement marquant associé. Ainsi, le cadran 1808 rappelle-t-il Beethoven et sa 5e symphonie ou 1963 la première femme dans l'Espace. Le chef-d'œuvre actuel du portefeuille horloger est assurément la « Répétition minutes Tourbillon sous ponts d'or ». Compte tenu de sa haute complexité, Girard-Perregaux a limité à seulement 30 exemplaires la production de cette montre-bracelet au boîtier en titane et au son pur. À partir de 2017, Girard-Perregaux présentera de nombreuses facettes de son riche passé et de son actualité créative dans l'élégant musée de la marque, 129 rue du Progrès, à La Chaux-de-Fonds. Une réservation préalable sera obligatoire. ◦

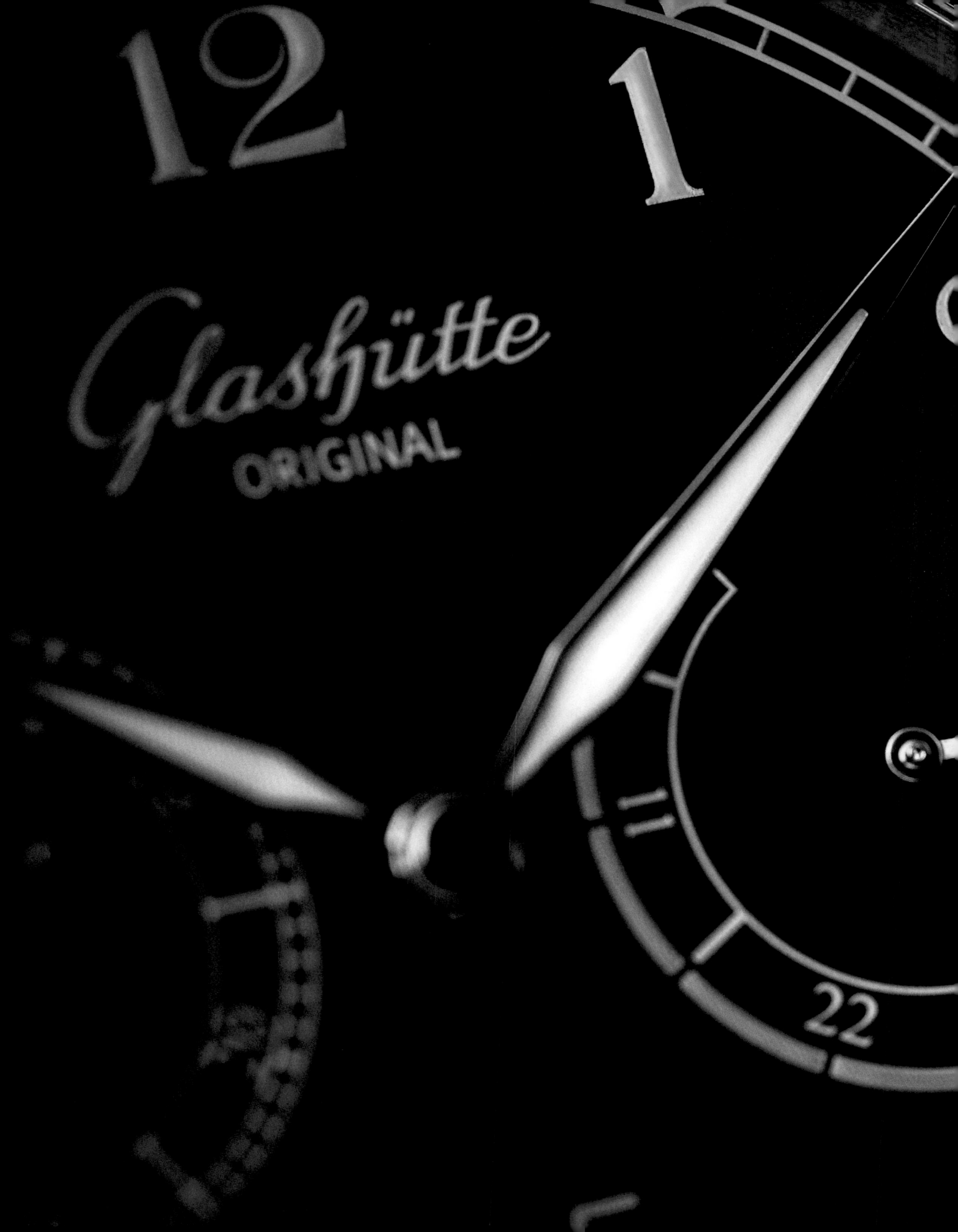
12
1
Glashütte
ORIGINAL
22

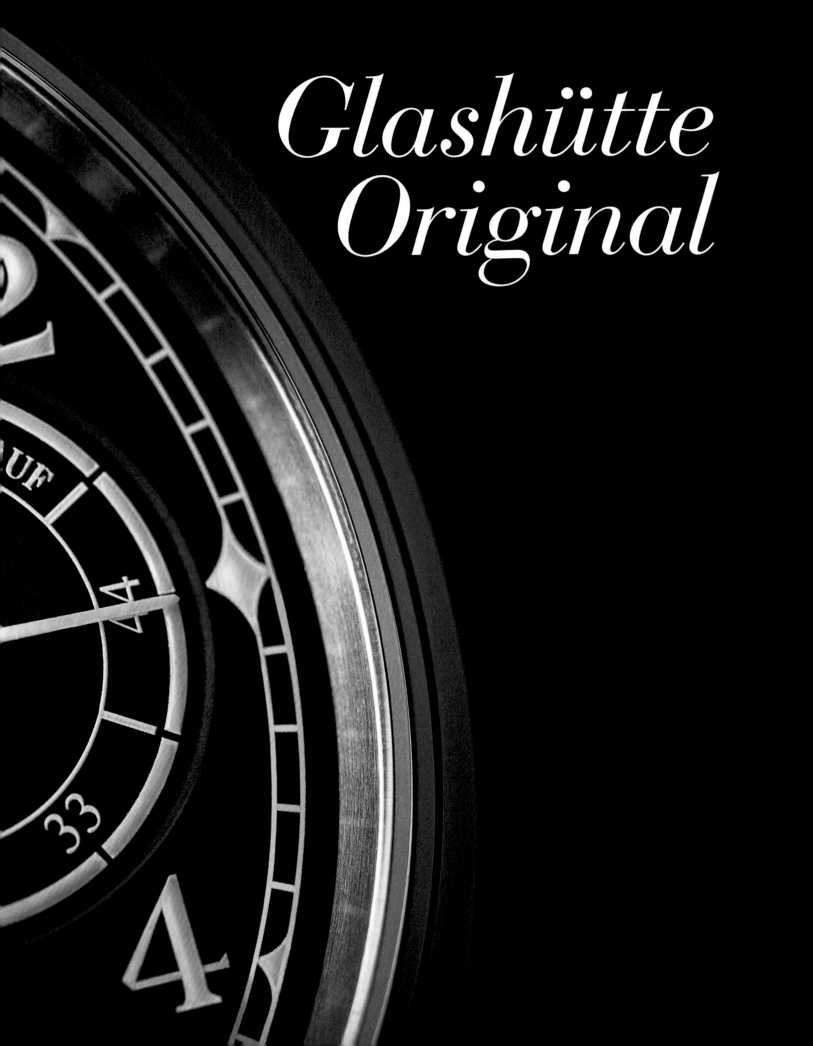

Glashütte Original

Reborn Like a Phoenix from the Ashes

May 8, 1945 was a disastrous day for the city of Glashütte. Although the German Armed Forces had unconditionally surrendered the day before, Russian bombers caused further damage to Germany's Mecca of watchmaking. An even greater calamity struck the watch industry after the war's end, when the Soviet occupying forces began dismantling all of the watch companies. But the resilient Saxons lived up to their reputation and refused to be discouraged. By 1946, they had already produced their first postwar watch movements. On July 1, 1951, the GDR's socialist government created the state-owned Glashütter Uhrenbetriebe, which brought together all of Glashütte's watchmaking activities, but without their erstwhile glamour because the new regime needed watches for the common people and high-quality marine chronometers. Luxury products were not made.

Glashütte's myth and name survived primarily in West Germany thanks to the town's glorious past. When the country was reunified in 1989, Glashütte Original dreamt of returning to its tradition of precise mechanical timepieces, but these dreams initially remained unfulfilled. After sluggish negotiations and several failed attempts at reviving the business, two Franconian businessmen appeared on the scene in October 1994. A visionary 86-page paper pointed the way for the formerly renowned Glashütter Uhrenbetriebe to regain its lost glory. The manufactory and its name, Glashütte Original, shone in a new light in the spring of 1995. Existing mechanisms were painstakingly optimized and joined by new calibers that upheld the highest standards. Topping the list of superlatives was a so-called "flying" tourbillon that debuted in November 1995 and traced its ancestry to Glashütte's master watchmaker Alfred Helwig. The highly complex "PanoRetroGraph" with mechanical countdown and alarm signal delighted connoisseurs in the spring of 2000. And the "Senator Klassik" boasted a perpetual calendar that had been developed on the brand's premises.

The Glashütte Original manufactory gradually expanded its activities. Greater depth of production soon included all components needed for watch movements, including balance-springs, mainsprings, and jewels. Glashütte Original, which had joined Switzerland's Swatch Group in 2000, proudly celebrated the grand opening of its watch factory in 2003. This modern, light-flooded building is much more than a functionally dominated factory for high-quality watches. It's conceived as a genuine experiential world into which watch lovers can immerse themselves without protracted formalities. A quick phone call suffices to reserve a place for oneself on a guided tour. Watch aficionados warmly welcomed the concept of transparency and open doors. By the end of 2004, the manufactory was annually hosting more than 7,000 visitors from around the globe. Guests can stroll among the various levels surrounding the 75-foot-tall atrium, gaze through spacious panes of glass, observe watchmakers at work, and listen to experts who competently explain all the details. For example, visitors can learn about the new movements that Glashütte Original regularly develops. These include the 100 family of calibers, which has a cleverly designed zero-return mechanism for the central elapsed-seconds hand.

Following Glashütte Original's initiative, the German Timepiece Museum opened its doors for the first time on May 22, 2008 in the completely renovated building that had formerly housed Glashütte's school of watchmaking. Steeped in history, these old walls now surround 11,000 square feet of floor space on two levels, where visitors can learn about the heydays and doldrums that the art of watchmaking experienced over the years in the secluded Müglitz Valley.

Of course, the design engineers and watchmakers were not idle while the museum building was undergoing renovation. The manufactory introduced an innovative wristwatch with a display for all 52 calendar weeks in 2006. A split-seconds chronograph encasing hand-wound caliber 99-01 followed in 2007. A wristwatch like the "Senator Diary," which debuted in 2010, was similarly unprecedented in the history of mechanical timekeeping: its month-long reminder function ably replaced the proverbial knot in the corner of one's handkerchief. The "Grande Cosmopolite Tourbillon" was another world premiere that caused a furor in 2012: one of the most complicated wristwatches ever built in Germany, its mechanical inner life combines over 500 components and powers a perpetual calendar programmed for 37 time zones. A gap in the spectrum of calibers was closed in 2014 by an entirely newly developed chronograph caliber with automatic winding, outsize panorama date and a chronograph controlled by a column-wheel. Here too, the watchmakers didn't scrimp on components: each specimen concatenates no fewer than 450 parts. The "Senator Cosmopolite," which appeared on the market in 2015, shows the aforementioned 37 time zones, but makes do without a perpetual calendar and a tourbillon. ∘

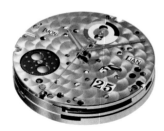

Senator Panorama Date Moon Phase, 1999

At the assembly worktable
In der Montage
Devant l'établi, lors de l'assemblage

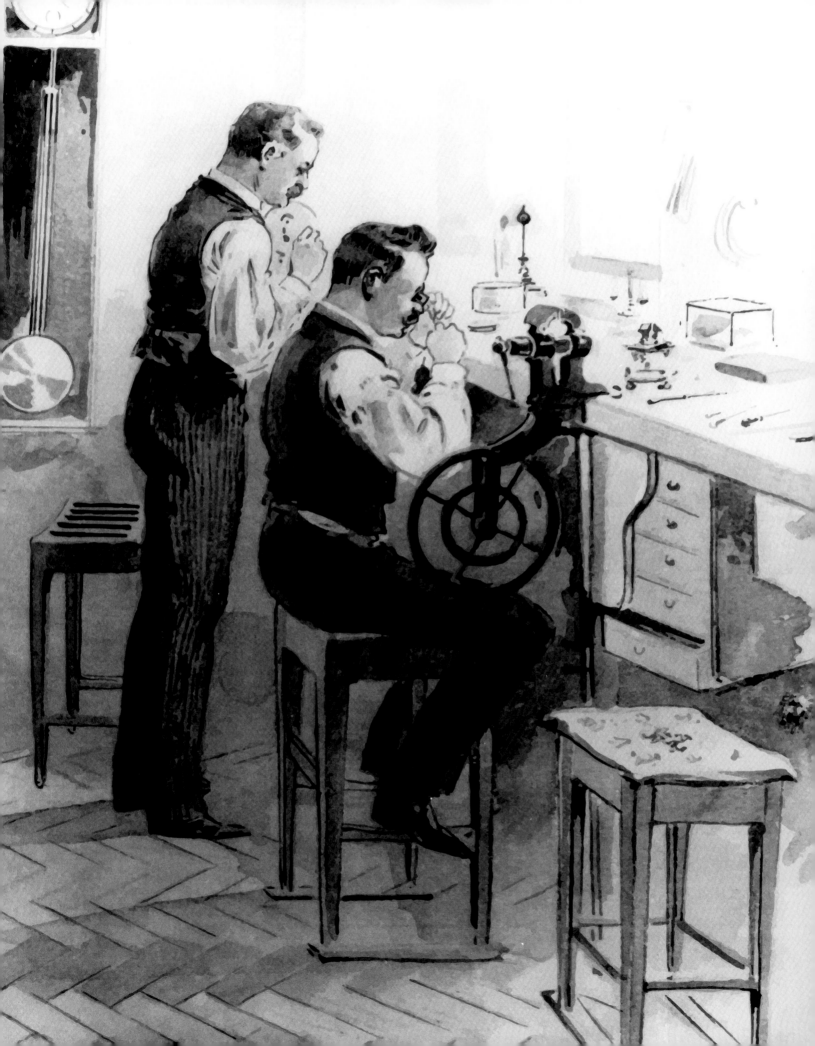

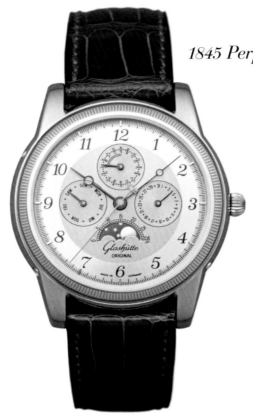

1845 Perpetual Calendar, 1996

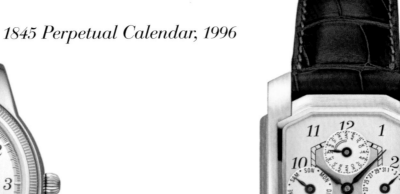

1845 Karree
Perpetual Calendar, 1998

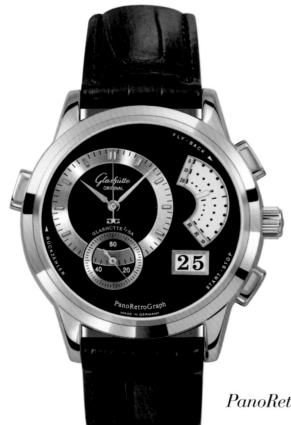

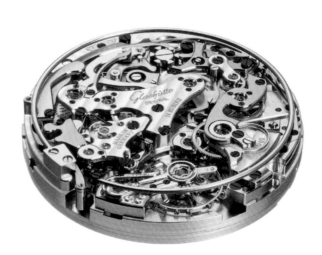

PanoRetroGraph, launch 2000

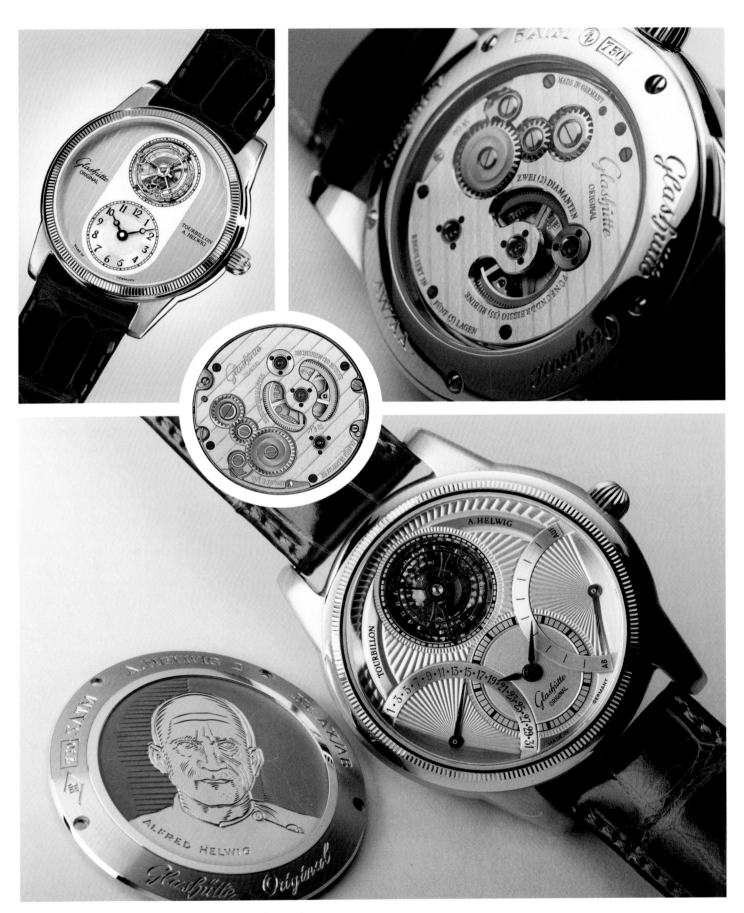

Alfred Helwig Tourbillon, 1996 (top) and 2001 (bottom)

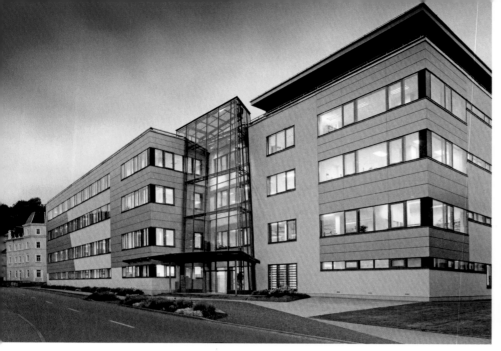
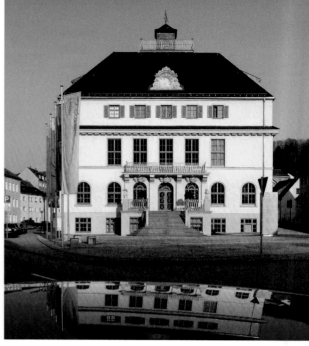

Auferstanden wie Phoenix aus der Asche

Deutsch

In die Glashütter Stadtgeschichte ist der 8. Mai 1945 als schicksalsträchtiger Tag eingegangen: Obwohl am Tag zuvor die deutsche Wehrmacht ihre bedingungslose Kapitulation unterzeichnet hatte, fügten russische Jagdflieger dem Mekka der deutschen Präzisionsuhrmacherei noch Schäden zu. Was viel schwerer für die Uhrenindustrie wog: Unmittelbar nach Kriegsende begannen die sowjetischen Besatzer mit der umfangreichen Demontierung aller Uhrenfirmen. Doch die zähen Sachsen wurden ihrem Ruf gerecht und ließen sich davon nicht entmutigen: Schon 1946 produzierten sie die ersten Nachkriegsuhrwerke. Am 1. Juli 1951 entstand der VEB Glashütter Uhrenbetriebe, in dem das sozialistische DDR-Regime alle ehemaligen Glashütter Uhr-Aktivitäten zusammenfasste. Allerdings ohne den einstigen Glanz, denn die neuen Machthaber benötigten Uhren fürs Volk sowie hochwertige Marinechronometer. Luxusprodukte waren kein Thema.

Jahrzehntelang lebten Mythos und Name Glashütte durch die großartige Vergangenheit vorwiegend im Westen Deutschlands weiter. Mit der politischen Wende 1989 erhoffte man sich bei Glashütte Original die Rückkehr zur Tradition mechanischer Präzisionsuhren – doch die vollzog sich anfangs nur in hochfliegenden Träumen. Nach mehreren gescheiterten Wiederbelebungsversuchen und überaus zähen Verhandlungen kamen im Oktober 1994 zwei fränkische Unternehmer zum Zug. Ein 86-seitiges Zukunftspapier wies einen überzeugenden Weg aus der Misere des einst renommierten Glashütter Uhrenbetriebs. Ab Frühjahr 1995 präsentierten sich die Manufaktur und der Name Glashütte Original in neuem Licht. Vorhandene Mechanik erlebte eine sorgfältige Optimierung. Hinzu gesellten sich neue Kaliber, die höchsten uhrmacherischen Ansprüchen genügten. Ganz oben rangierte ein „fliegendes" Tourbillon, das konstruktiv auf den Glashütter Meisteruhrmacher Alfred Helwig zurückging. Sein Debüt erfolgte im November 1995. Im Frühjahr 2000 machte der äußerst komplexe

„PanoRetroGraph" mit mechanischem Countdown und Alarmsignal von sich reden. Und die „Senator Klassik" gab es mit selbst entwickeltem ewigem Kalendarium.

Zug um Zug baute Glashütte Original danach die Aktivitäten der Manufaktur aus. Steigende Fertigungstiefe umfasste bald schon alle Werkskomponenten bis auf Unruhspirale, Zugfeder und natürlich die Steine. 2003 konnte Glashütte Original, seit 2000 ein Mitglied der eidgenössischen Swatch Group, nicht ohne Stolz die Eröffnung der gläsernen Uhrenfabrik zelebrieren. Das moderne, lichtdurchflutete Gebäude ist weit mehr als eine funktionsdominierte Fabrik für hochwertige Uhren. Es versteht sich als echte Erlebniswelt, in die Uhrenliebhaber ohne größere Formalitäten eintauchen können. Ein kurzer Anruf genügt, um sich für eine Manufakturführung anzumelden. Das Konzept der offenen Tür kam bei Interessierten spontan und bestens an. Ende 2004 konnte die Manufaktur schon gut 7000 Besucher aus aller Welt vermelden. Sie haben die verschiedenen Stockwerke rund um das 23 Meter hohe Atrium durchwandert, durch große Glasscheiben geblickt, den Werktätigen zugeschaut und fachkompetenten Erläuterungen gelauscht. Zum Beispiel auch zu den neuen Uhrwerken, die Glashütte Original mit schöner Regelmäßigkeit entwickelt. So besticht die 2005 lancierte Kaliberfamilie 100 durch einen ausgeklügelten Nullstellmechanismus für den zentralen Sekundenzeiger.

Am 22. Mai 2008 ging unter Federführung von Glashütte Original die Eröffnung des Deutschen Uhrenmuseums im komplett renovierten Gebäude der ehemaligen Uhrmacherschule von Glashütte über die Bühne. Auf zwei Etagen und 1000 Quadratmetern Fläche lässt sich innerhalb der geschichtsträchtigen Mauern die abwechslungsreiche Historie der Uhrmacherkunst im abgeschiedenen Müglitztal verfolgen.

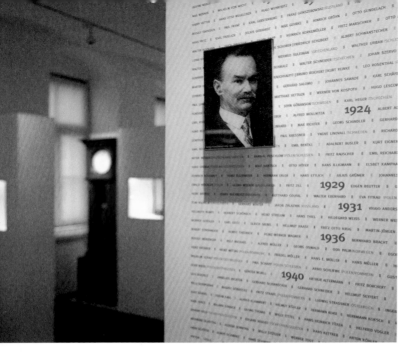

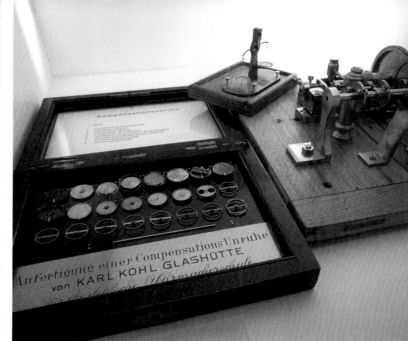

From left: production and administration 2014 ∘ School of Watchmaking/German Watch Museum ∘ Names of all students of the German School of Watchmaking on a panel in the watch museum ∘ Artifacts on display ∘ Von links: Produktion und Verwaltung 2014 ∘ Uhrmacherschule/Deutsches Uhrenmuseum ∘ Namen aller Schüler der Deutschen Uhrmacherschule auf einer Tafel im Uhrenmuseum ∘ Museumsexponate ∘ Depuis la gauche : Fabrication et administration, 2014 ∘ École d'horlogerie/Musée allemand de l'horlogerie ∘ Les noms de tous les élèves de l'école allemande d'horlogerie sont inscrits sur une plaque dans le musée ∘ Pièces d'exposition du musée

Während der umfangreichen Arbeiten am und im Museumsgebäude waren die Konstrukteure und Uhrmacher natürlich nicht untätig geblieben. Schon 2006 hatte die Manufaktur eine neuartige Armbanduhr mit Anzeige der 52 Kalenderwochen vorgestellt. 2007 brachte einen Schleppzeiger-Chronographen mit dem Handaufzugskaliber 99-01. Eine Armbanduhr wie den 2010 vorgestellten „Senator Terminkalender" hatte es in der Geschichte der mechanischen Zeitmessung ebenfalls noch nie gegeben. Die Erinnerungsfunktion über einen Monat hinweg ersetzt den sprichwörtlichen Knoten im Taschentuch. Als weitere Weltpremiere sorgte 2012 das „Grande Cosmopolite Tourbillon" für Furore. Dabei handelt es sich um eine der kompliziertesten Armbanduhren deutscher Provenienz. Ihr mechanisches Innenleben mit ewigem Kalender, programmiert für 37 Zeitzonen, besteht aus mehr als 500 Komponenten. Eine Lücke im Kaliberspektrum schloss 2014 das von der Pike auf neu entwickelte Chronographenkaliber mit Selbstaufzug, Panorama-Großdatum und schaltradgesteuertem Chronographen. Auch hier haben die Uhrmacher nicht an Bauteilen gespart: Summa summarum benötigen sie für ein Exemplar nicht weniger als 450. 2015 kam die „Senator Cosmopolite" auf den Markt, welche die besagten 37 Zeitzonen auch ohne ewigen Kalender und Tourbillon anzeigt. ∘

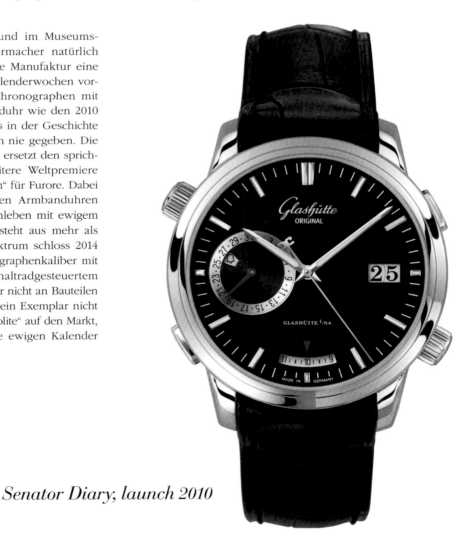

Senator Diary, launch 2010

Un phénix renaît de ses cendres

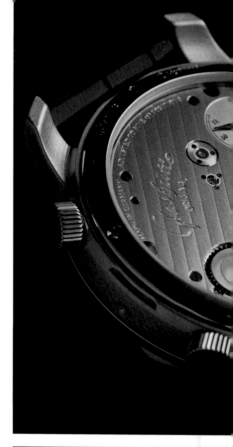

Dans l'histoire de Glashütte, le 8 mai 1945 est une date fatidique : alors que la veille, l'Allemagne a capitulé sans condition, les chasseurs russes continuent d'infliger les pires dégâts à la Mecque de la haute horlogerie allemande. Fait plus grave encore pour l'industrie horlogère : immédiatement après la fin des hostilités, les occupants soviétiques entreprennent le démontage en règle de toutes les entreprises d'horlogerie. Pourtant, fidèles à leur réputation de ténacité, les Saxons ne se découragent pas : dès 1946, ils produisent les premiers mouvements de l'après-guerre. Le 1er juillet 1951, le régime socialiste de la RDA regroupe toutes les anciennes activités horlogères de Glashütte au sein de l'entreprise d'État Glashütter Uhrenbetriebe. Mais cette dernière n'a pas l'éclat du passé, car les nouveaux dirigeants du pays n'ont besoin que de montres pour le peuple et de montres de marine de grande qualité. Les produits de luxe ne sont pas à l'ordre du jour.

Des décennies durant, le mythe et le nom Glashütte perdurent grâce à un grandiose passé, essentiellement en RFA. À la chute du Mur en 1989, Glashütte Original espère un retour à la tradition des montres mécaniques de précision. Après plusieurs tentatives avortées de relance et des négociations très tendues, deux entrepreneurs de Franconie reprennent la main en 1994. Ils proposent une méthode convaincante pour sortir l'entreprise de sa situation désastreuse. À partir du printemps 1995, la manufacture et le nom Glashütte Original se présentent sous un jour nouveau. La mécanique existante est soigneusement optimisée. À cela viennent s'ajouter de nouveaux calibres répondant aux exigences horlogères les plus élevées. Le summum est un tourbillon « volant », dont la conception doit beaucoup au maître horloger de Glashütte, Alfred Helwig, et qui sort en novembre 1995. Au printemps 2000, c'est le très complexe « PanoRetroGraph », premier chronographe mécanique doté d'un compte à rebours avec alarme acoustique qui fait parler de lui. La « Senator Klassik » s'enorgueillit pour sa part d'un calendrier perpétuel développé en interne.

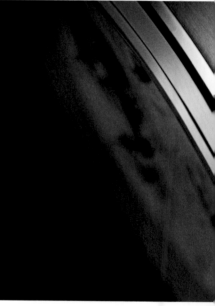

Progressivement, Glashütte Original étend les activités de la manufacture. La production en interne s'intensifie pour couvrir bientôt tous les composants des mécanismes, excepté les spiraux de balancier, les ressorts de barillet et bien sûr les pierres. En 2003, Glashütte Original, membre depuis 2000 du groupe helvétique Swatch Group, peut non sans fierté célébrer l'ouverture de sa fabrique de montres « transparente ». Ce bâtiment moderne inondé de lumière est bien plus qu'une manufacture dédiée à la fabrication de montres d'exception. Elle se veut véritable espace de découverte dans lequel les amateurs de montres peuvent plonger sans grandes formalités. Un coup de téléphone suffit pour s'inscrire à une visite guidée. Fin 2004, la manufacture peut s'enorgueillir de déjà totaliser plus de 7 000 visiteurs du monde entier, qui déambulent dans les étages donnant sur un atrium de 23 mètres de haut, scrutent par de grandes vitres et observent les employés à l'œuvre en écoutant les explications de spécialistes. Ils découvrent par exemple les nouveaux mouvements que crée Glashütte Original avec une belle régularité. Ainsi, la famille des calibres 100, lancée en 2005, les séduit-elle par son ingénieux mécanisme de remise à zéro de la trotteuse centrale.

Le 22 mai 2008, sous l'égide de Glashütte Original, le musée allemand de la Montre ouvre ses portes dans le bâtiment totalement rénové de l'ancienne école d'horlogerie de Glashütte. Sur les 1 000 mètres carrés des deux étages, on peut suivre dans ces murs prestigieux la riche histoire de l'art horloger, dans cette petite ville reculée de la vallée de la Müglitz.

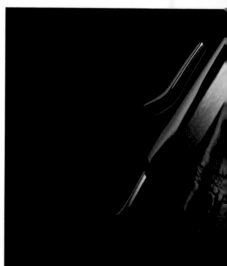

Durant les importants travaux à l'extérieur comme à l'intérieur du musée, les concepteurs et les maîtres horlogers ne restent bien sûr pas inactifs. Dès 2006, la manufacture présente une montre-bracelet d'un type nouveau avec affichage des 52 semaines du calendrier. En 2007 apparaît un chronographe à rattrapante équipé du calibre à remontage manuel 99-01. Présentée en 2010, la montre-bracelet « Senator Diary » est inédite dans l'histoire des garde-temps mécaniques. La fonction de rappel un mois à l'avance remplace le traditionnel nœud au mouchoir. Autre première mondiale en 2012, la « Grande Cosmopolite Tourbillon » fait sensation. C'est de fait l'une des montres-bracelets d'origine allemande comportant le plus de complications. Sa mécanique interne, avec son calendrier perpétuel programmé pour 37 fuseaux horaires, comporte plus de 500 composants. Une lacune dans la gamme des calibres est comblée en 2014 par un calibre de chronographe automatique de conception inédite, à affichage de la grande date panorama et roue à colonnes. Encore une fois, les horlogers n'ont pas lésiné sur le nombre de composants : au total, il faut pas moins de 450 pièces pour une de ces merveilles. En 2015, la « Senator Cosmopolite » sort sur le marché. Elle affiche les 37 fuseaux horaires précédemment évoqués, même sans quantième perpétuel ni tourbillon. ○

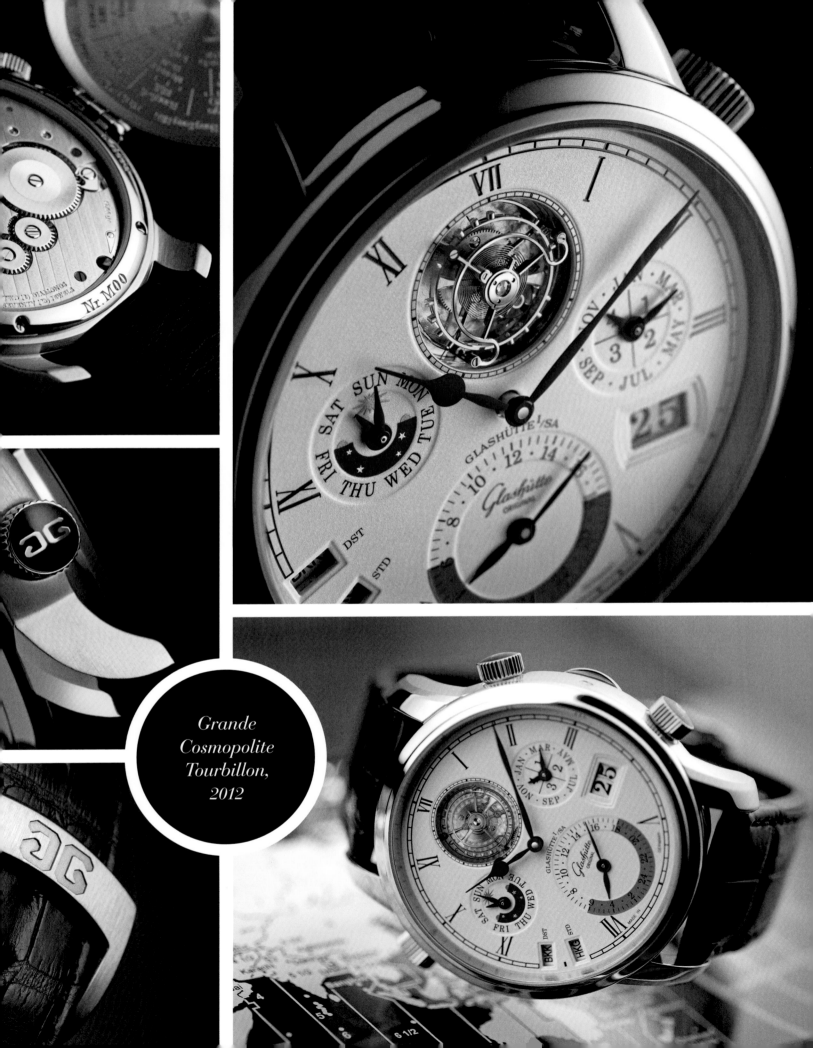

Grande Cosmopolite Tourbillon, 2012

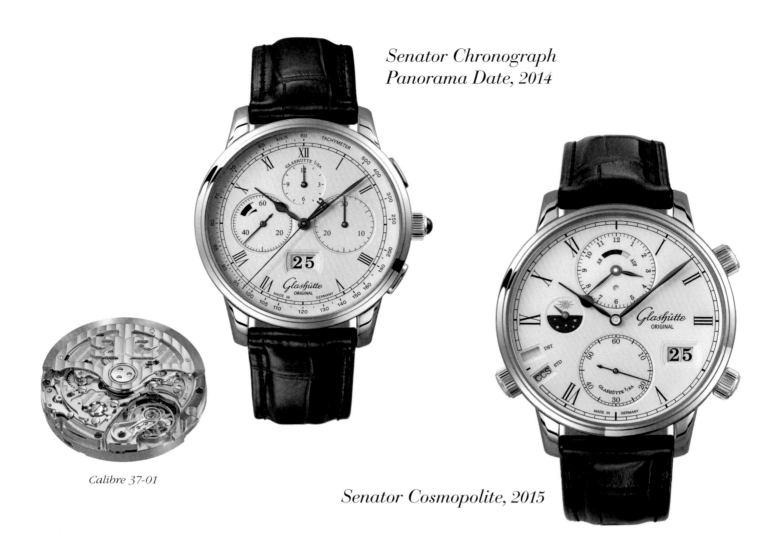

*Senator Chronograph
Panorama Date, 2014*

Calibre 37-01

Senator Cosmopolite, 2015

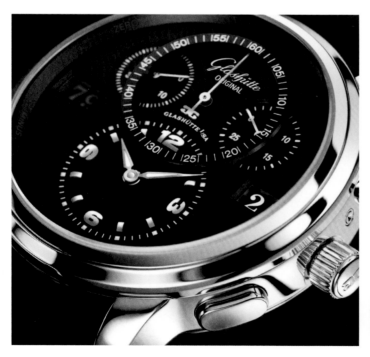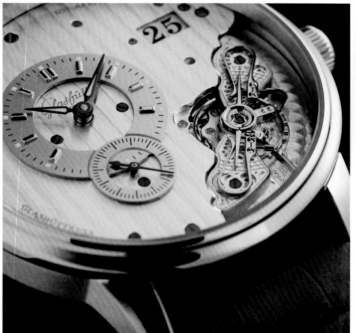

Left: PanoMaticCounter XL, 2010 ○ Right: PanoMaticInverse, 2014

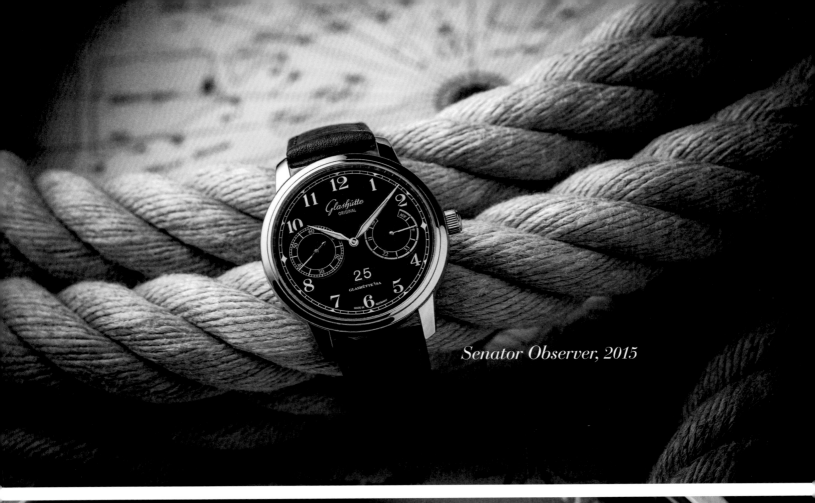

Senator Observer, 2015

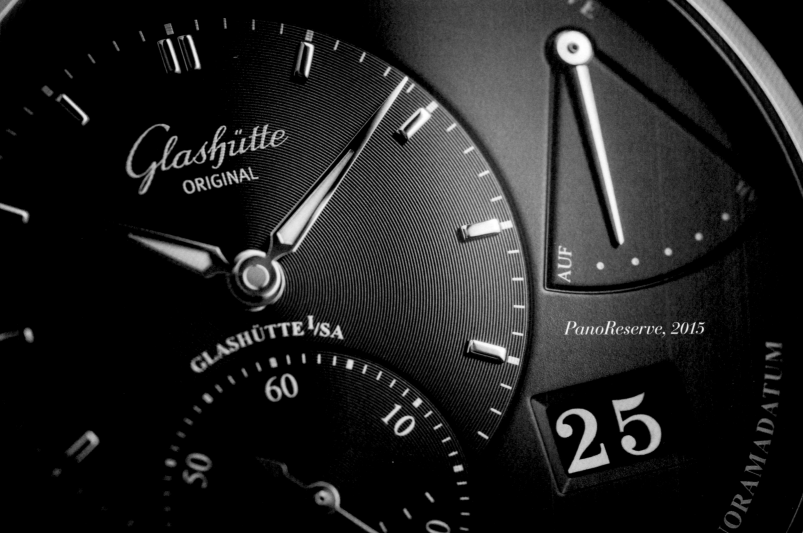

PanoReserve, 2015

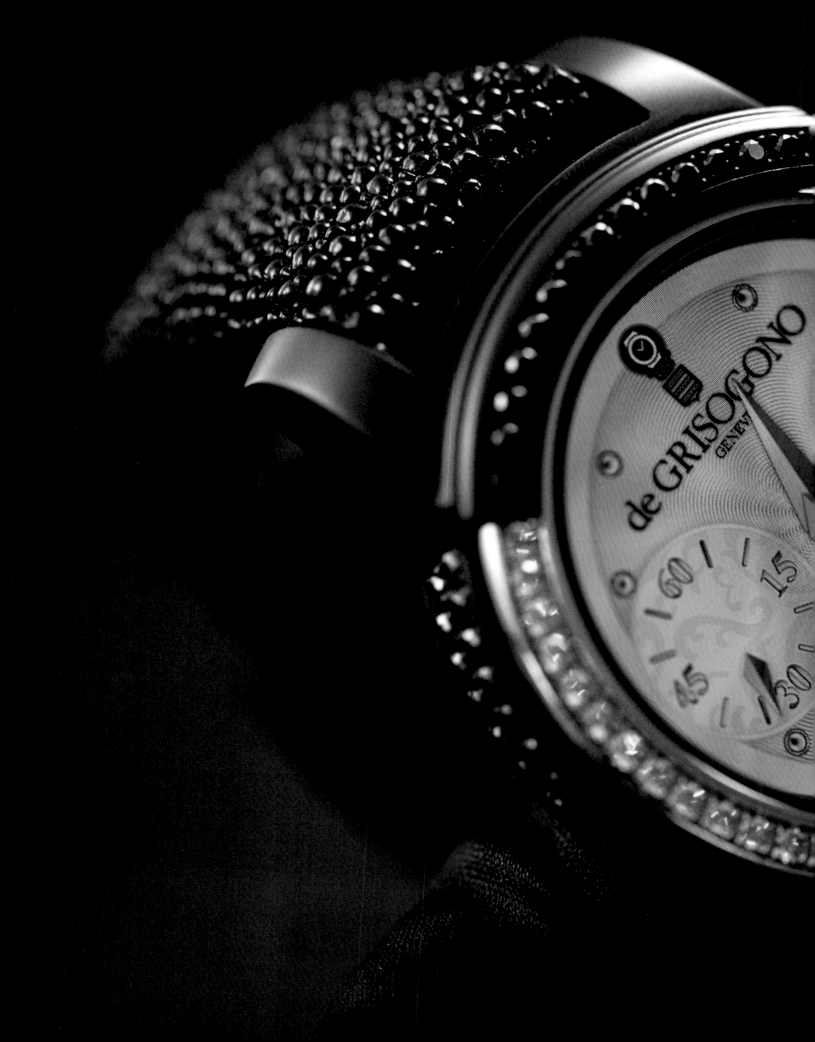

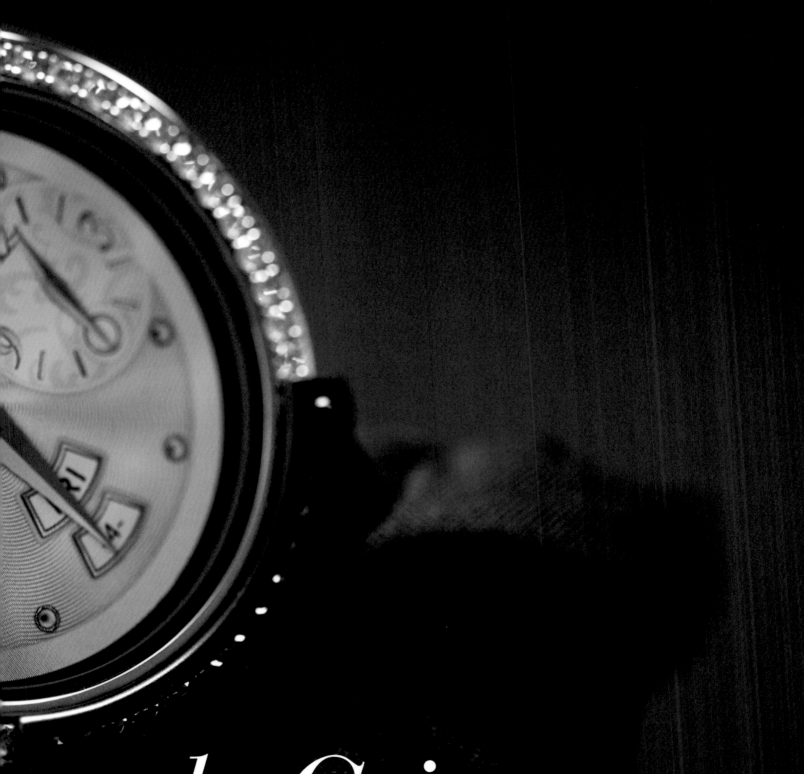

de Grisogono

Preferably Creative

English

Fawaz Gruosi, an Italian native, is an intuitive man, and the course of his career reflects this innate practicality. Florence, his hometown, helped shape his insatiable love for the fine arts, so it wasn't surprising that his first employer was a Florentine jeweler. His subsequent professional life took him to London, Saudi Arabia, New York and Geneva. Fawaz Gruosi worked for Harry Winston and later for Bulgari. When Gianni Bulgari left the family business, Gruosi likewise felt that the time had come to venture onto paths of his own. Together with his partners, he founded his own jewelry store in the center of Geneva in 1993. The dream of his own brand came closer to fruition. Of course, each partner wanted the label to bear their own name. One of the partners ultimately prevailed with the name of her mother, a marquise de Grisogono. The label's name also remained unchanged after Fawaz Gruosi separated from his erstwhile partners in 1995. He had been responsible for product design from the brand's earliest beginnings, and he continued in this capacity as sole entrepreneur. The next chapter in the firm's biography is distinguished by the "Black Orlov" because this 195-carat gem enflamed the passion for black diamonds. De Grisogono celebrated its watch debut at Baselworld in 2000. "Instrumento N° Uno," which combined a reliable self-winding Eta caliber and an exclusive additional mechanism, immediately appealed to a surprisingly large number of buyers: Fawaz Gruosi was able to fabricate and sell more than 5,000 specimens. "My strength in the watch genre is that I'm not really an expert, so I can tackle things relatively without bias. Although I'm surrounded by many competent people, I'm not easily convinced that something is impossible." The truth of this was reconfirmed by a minute-repeater wristwatch in 2005: the "Occhio" (Italian for "eye") was created in cooperation with Christophe Claret, an acknowledged specialist for horological mechanisms. When the wearer slides the trigger on the flank of the case, an aperture consisting of twelve ceramic flaps glides open and the movement, which combines 324 individual components, chimes the time on three gongs. Gruosi / de Grisogono premiered the rectangular "FG One" with automatic movement and unconventional time-zone mechanism in 2006. The following year, the brand's watchmakers assembled more than 651 parts to create a hand-wound movement encased in the limited-edition "Meccanico dG," which has an unprecedented double time display. Women as well as men were impressed by the remarkable "Otturatore": prioritizing understatement, this angular wristwatch relies on a patented aperture mask behind the hour-hand and minute-hand. The owner presses a large push-piece to select among several indicators in the background: small seconds, date window, moon phase, or power-reserve display. The creative mastermind suffered no lack of bright ideas in ensuing years. This is clearly confirmed by three current wristwatches. For gents, Fawaz Gruosi created the "New Retro," a 50- by 44-millimeter self-winding watch. Its rectangular case, which lies along rather than across the wrist, and its crown at the "12" call to mind the automobile driver's watches that were very popular in the 1930s. Ladies appreciate the lavishly gemstone-encrusted watches in the "Grappoli" line: the total weight of the gems on each of these timepiecs is approximately 73 carats. And for aficionadas with an affinity for high tech, the creative Italian collaborated with Samsung to develop a dressy smartwatch: compatible with most Samsung-brand smartphones, the "Samsung Gear S2 by de Grisogono" brings a broad spectrum of functions to its wearer's wrist. ○

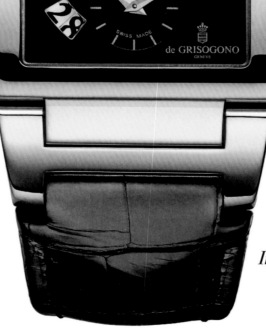

Instrumento N° Uno, 2000

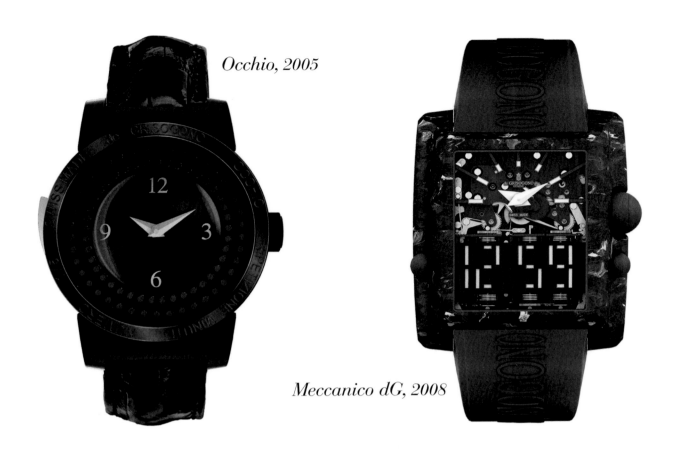

Occhio, 2005

Meccanico dG, 2008

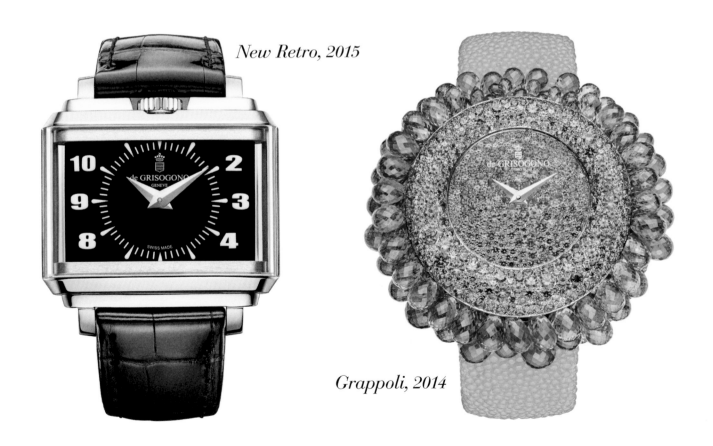

New Retro, 2015

Grappoli, 2014

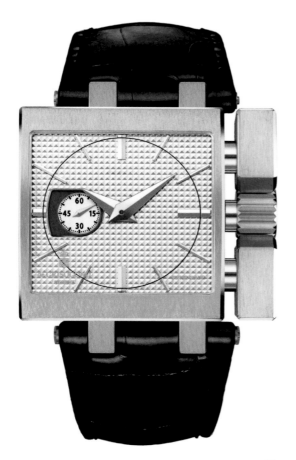

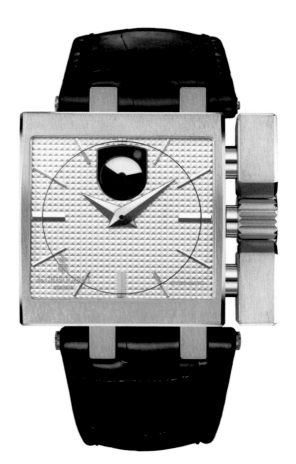

Otturatore, 2011

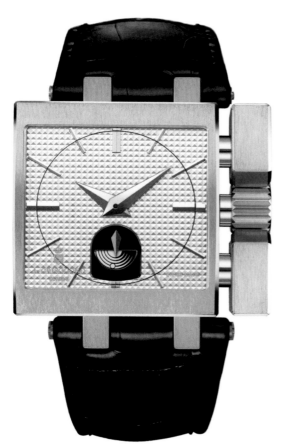

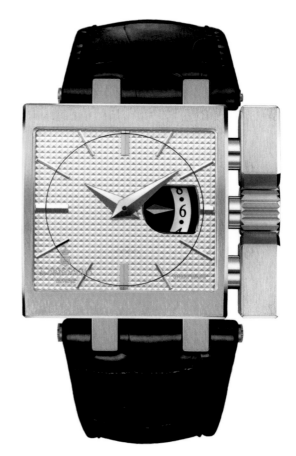

Am liebsten kreativ

Deutsch

Der gebürtige Italiener Fawaz Gruosi war und ist ein Mann der Praxis. Dementsprechend verlief auch die Karriere. Seine Heimatstadt Florenz prägte die unstillbare Liebe zu den schönen Künsten. Da lag der erste Job bei einem Florentiner Juwelier förmlich auf der Hand. Weitere Stationen des beruflichen Lebens: London, Saudi-Arabien, New York und Genf. Fawaz Gruosi arbeitete für Harry Winston und später für Bulgari. Als sich Gianni Bulgari vom Familienunternehmen trennte, waren auch für Gruosi eigene Wege angezeigt. Gemeinsam mit zwei Partnern gründete er 1993 ein eigenes Juweliergeschäft im Zentrum von Genf. Der Traum von einer eigenen Marke rückte näher. Natürlich wünschte sich damals jeder seinen eigenen Namen. Am Ende setzte sich der Name der Mutter von einem der Partner durch, einer Marquesa de Grisogono. Dabei blieb es auch 1995 nach der Trennung von den Partnern. Fawaz Gruosi hatte von Anbeginn für das Produktdesign verantwortlich gezeichnet und tat das fortan auch als Alleinunternehmer. Die weitere Firmenbiographie prägte der „Black Orlov", denn die 195-Karat-Preziose weckte die Leidenschaft für schwarze Diamanten. 2000 ging das Uhrendebüt während der Baselworld über die Bühne. „Instrumento N° Uno", ein Mix aus bewährter Eta-Automatik und exklusiver Zusatz-Mechanik, brachte es spontan auf erstaunliche Stückzahlen. Mehr als 5000 Exemplare konnte de Grisogono herstellen und verkaufen. „Meine Stärke auf dem Gebiet der Uhren ist, dass ich kein wirklicher Experte bin. Somit packe ich die Dinge relativ unvoreingenommen an. Obwohl ich viele kompetente Menschen um mich habe, lasse ich mir ungern einreden, dass irgendetwas nicht geht." Der Kern dieser Aussage bestätigte sich 2005 in einer Armbanduhr mit Minutenrepetition. „Occhio", das Auge, entstand in Kooperation mit dem anerkannten Mechanik-Spezialisten Christophe Claret.

Nach Betätigung des Auslöseschiebers im Gehäuserand öffnet sich eine Blende aus zwölf Keramikklappen. Gleichzeitig beginnt das aus 324 Teilen komponierte Œuvre die Zeit auf insgesamt drei Tonfedern zu schlagen. 2006 brachte Gruosi / de Grisogono die rechteckige „FG One" mit Automatikwerk und außergewöhnlichem Zeitzonen-Mechanismus auf den Markt. Mehr als 651 Teile assemblierten die Uhrmacher ab dem Folgejahr für ein Handaufzugswerk in der limitierten „Meccanico dG" mit nie dagewesener Doppel-Zeitanzeige. Selbst Frauen beeindruckte der markante „Otturatore", denn die kantige Armbanduhr übte sich in Zurückhaltung, hervorgerufen durch den Effekt einer patentierten Wechselblende hinter den Zeigern für Stunden und Minuten. Mit ihrer Hilfe können die Besitzer nach Belieben aus dem hintergründig vorhandenen Anzeigespektrum wählen: kleine Sekunde, Fensterdatum, Mondphasenindikation oder Gangreserveanzeige. Auch danach mangelte es dem Kreativchef nicht an Ideen. Was zu belegen ist an drei aktuellen Armbanduhren: Für die Herren der Schöpfung hat Fawaz Gruosi die „New Retro" geschaffen, eine 50 x 44 Millimeter große Automatik-Armbanduhr. Das quer am Handgelenk liegende Rechteck und die Krone bei der „12" wecken Erinnerungen an die unter anderem in den 1930er Jahren sehr beliebten Autofahrer-Armbanduhren. Frauen kommen einmal durch „Grappoli" zu ihrem Recht, eine opulent mit Schmucksteinen ausgefasste Linie. Jedes Exemplar zieren Edelsteine mit einem Wert von ungefähr 73 Karat. Für technikaffine Zeitgenossinnen entwickelte der kreative Italiener zusammen mit Samsung eine schmückende Smartwatch. Die „Samsung Gear S2 by de Grisogono" verbindet sich am liebsten mit Smartphones dieses Herstellers und bietet am Handgelenk ein breites Funktionsspektrum. ○

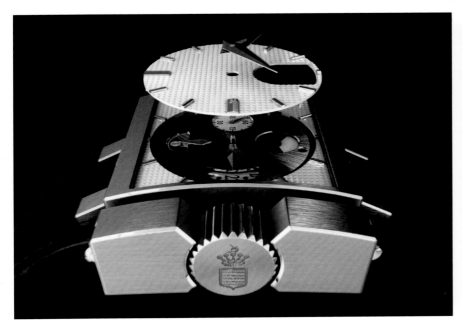

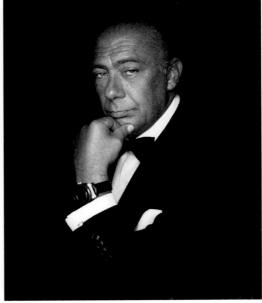

From left: Otturatore, 2011 ○ Fawaz Gruosi

Un net penchant pour la créativité

Français

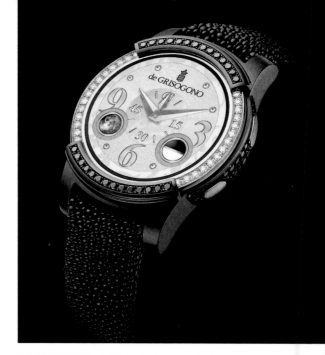

De mère italienne, Fawaz Gruosi est un homme intuitif, à l'esprit résolument pratique, comme en témoigne sa carrière. Florence, sa ville d'élection, a forgé son amour inextinguible pour les beaux-arts. Rien de surprenant donc à ce que son premier employeur soit un joaillier florentin. Sa vie professionnelle le conduit ensuite à Londres, en Arabie saoudite, à New York et à Genève. Fawaz Gruosi travaille pour Harry Winston puis pour Gianni Bulgari. Lorsque ce dernier se sépare de son entreprise familiale, c'est pour Fawaz Gruosi le signe qu'il doit suivre sa propre voie. Avec des partenaires, il fonde en 1993 une joaillerie dans le centre de Genève, et le rêve d'une marque en propre est prêt de se réaliser. Chaque associé veut bien sûr qu'elle porte son nom. Mais l'un de ses partenaires finira par imposer celui de sa mère, la marquise de Grisogono. Ce nom ne changera plus, même après la séparation des associés en 1995. Responsable dès le début de la création artistique, Fawaz Gruosi le restera une fois seul aux commandes. La suite de l'histoire de la société est marquée par « Black Orlov », une pierre précieuse de 195 carats qui déchaîne une passion pour les diamants noirs. Fawaz Gruosi fait ses débuts dans l'horlogerie lors du salon Baselworld en 2000. « Instrumento N° Uno », qui associe un mouvement automatique Eta éprouvé et des éléments mécaniques complémentaires raffinés, trouve rapidement un nombre étonnant d'acheteurs. Plus de 5 000 exemplaires sont ainsi fabriqués et vendus. « Ma force dans le domaine des montres tient au fait que je ne suis pas vraiment un spécialiste. J'aborde donc les choses plutôt sans préjugé, et les nombreuses personnes compétentes qui m'entourent ont bien du mal à me convaincre que quelque chose est impossible. » Le créateur confirme ces propos en créant en 2005 une montre-bracelet à répétition minute. « Occhio » (œil, en italien) naît de la coopération avec Christophe Claret, éminent spécialiste en mécanique horlogère. L'activation du poussoir sur la tranche du boîtier entraîne l'ouverture d'un diaphragme composé de douze volets en céramique. Le mouvement composé de 324 éléments sonne alors l'heure sur trois timbres. En 2006, Fawaz Gruosi / de Grisogono sort sur le marché la « FG One », une montre rectangulaire à mouvement automatique et guichet affichant un second fuseau horaire. L'année suivante, les horlogers de la marque assemblent plus de 650 composants pour créer le mouvement de la montre-bracelet en édition limitée « Meccanico dG », dotée d'un double affichage du temps unique en son genre. Les femmes comme les hommes sont impressionnés par la remarquable « Otturatore ». Cette montre-bracelet anguleuse joue la carte de la retenue grâce à un obturateur séquentiel breveté à l'arrière-plan des aiguilles des heures et des minutes. Au moyen de larges poussoirs, le porteur de la montre sélectionne une des complications suivantes : petites secondes, guichet de date, phase de lune ou réserve de marche. Le directeur artistique continue de faire preuve d'inventivité, ce que démontrent trois nouvelles montres-bracelets. Pour la gent masculine, Fawaz Gruosi crée la « New Retro », une montre-bracelet à remontage automatique de 50 x 44 millimètres. Le boîtier rectangulaire plus large que haut et la couronne à midi font penser aux montres-bracelets très populaires, dans les années 1930 notamment, auprès des automobilistes. Les femmes apprécient la ligne « Grappoli », richement sertie de pierres précieuses, pour un poids d'environ 73 carats par montre. Pour la femme moderne passionnée de technique, le créateur italien met au point avec Samsung une élégante montre connectée. Parfaitement compatible avec les smartphones de ce fabricant, la « Samsung Gear S2 by de Grisogono » offre un large éventail de fonctions directement au poignet de sa propriétaire. ○

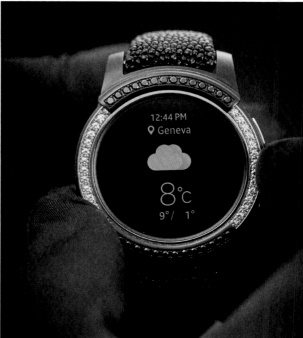

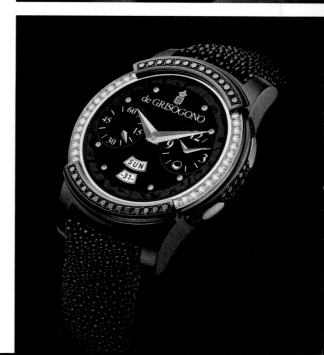

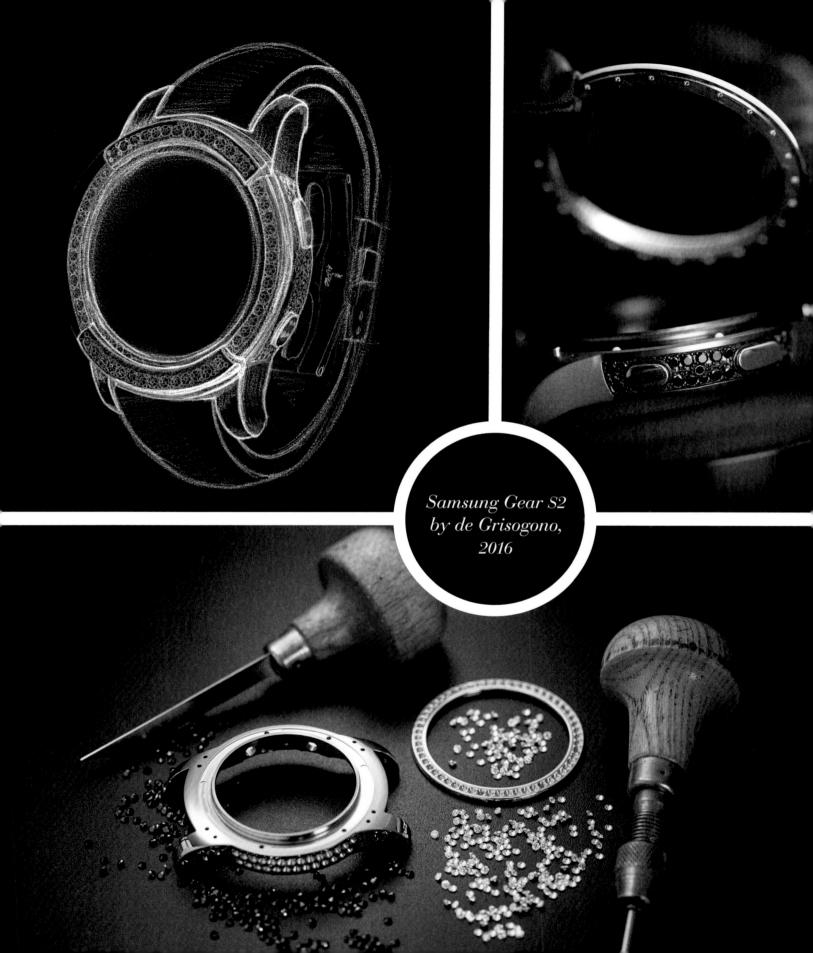

Samsung Gear S2 by de Grisogono, 2016

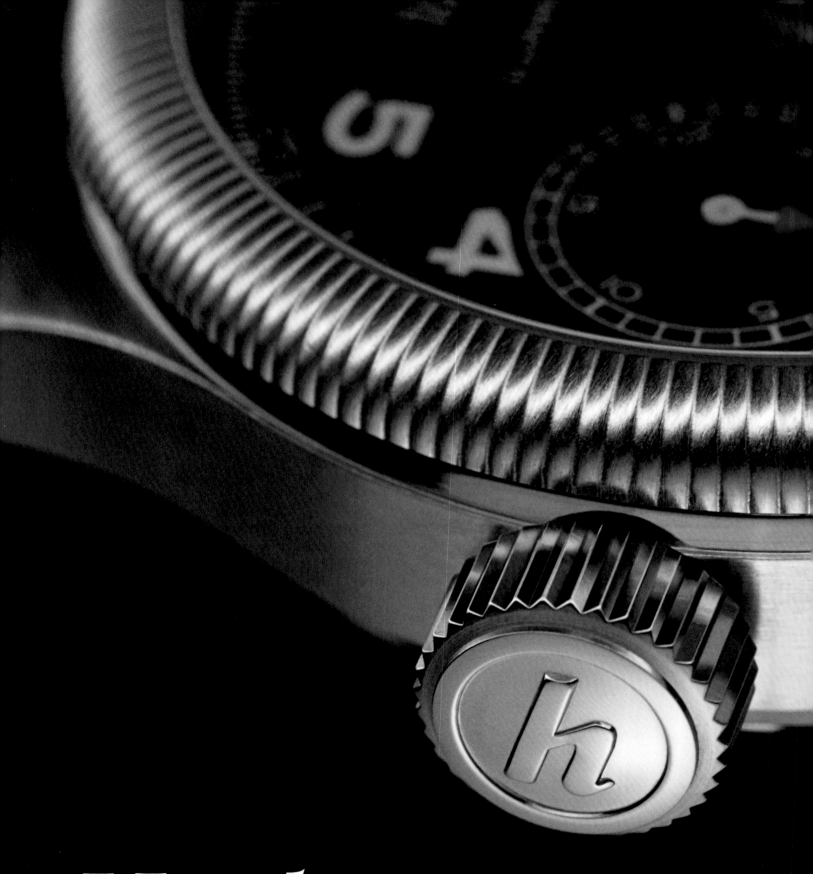

Hanhart

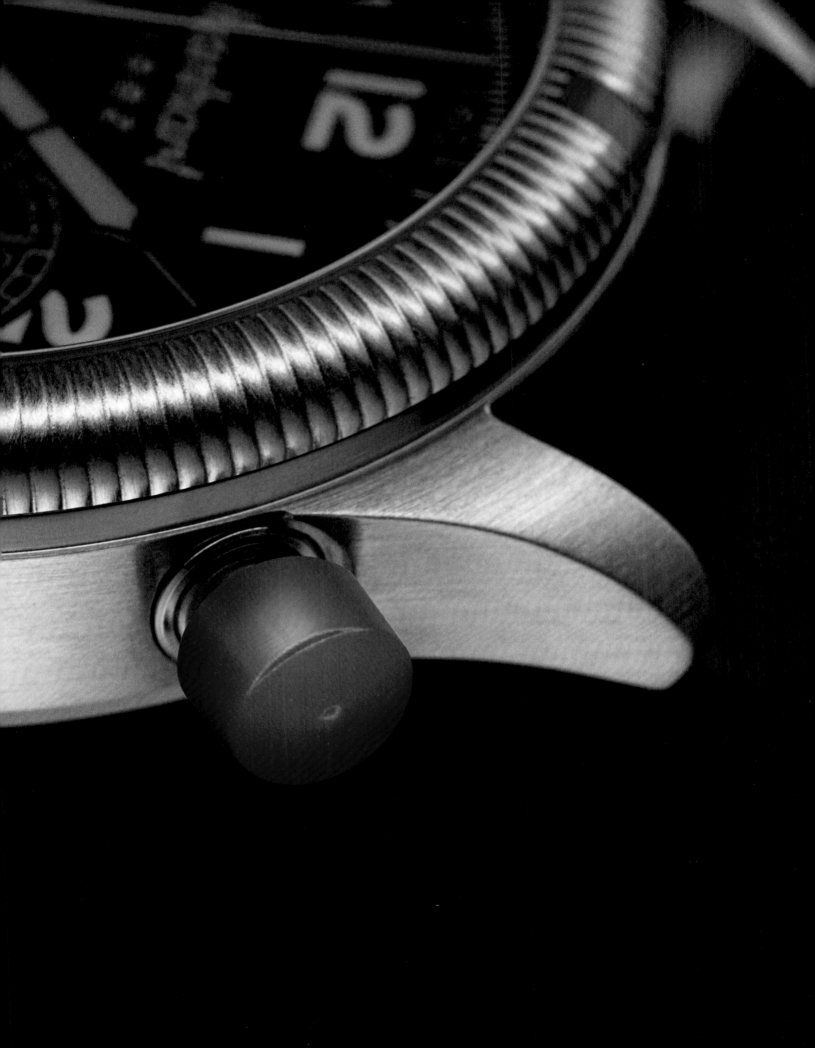

Memories of the Future

When watch aficionados hear the name "Hanhart," they usually think of a masculine chronograph with a red zero-reset button. The history of this distinctive push-piece began in the 1940s. The manufactory, which was established in 1882, already offered a very comprehensive portfolio of products prior to World War Two. Prewar catalogues listed 48 wristwatch models, six pocket watches, eight stopwatches, two sport watches, and one watch in a carrying case. The spectrum accordingly provided a clear preview of future product priorities. Hanhart began developing its first chronographs with the brand's own caliber family 4x in 1938. The diameter of these hand-wound movements was 15½ lignes, i.e., 35 millimeters. Hanhart offered: Caliber 40 with a single button to control the chronograph's start, stop and zero-return functions; Caliber 41 with two buttons for additive stopping; and top-of-the-line Caliber 42 with speedy switching, nowadays known as the "flyback" function. All three calibers included classical chronographic attributes such as column-wheel control and horizontal geared coupling. Caliber 41 can be recognized by the asymmetrical positioning of the two buttons above and below the winding and hand-setting crown. The distinctive red button at "4 o'clock" enlivened specimens destined for duty in military cockpits. To minimize the danger that the zero-return button might be inadvertently operated in the heat of battle to mistakenly trigger the flyback function, a pilot's wife simply daubed it with her red nail polish. Thus began a chronographic legend that was further distinguished by a bright red dot on the fluted rotatable bezel. This mark reminded the user that an interval had elapsed which was longer than could be tallied by the 30-minute totalizer.

Much more can be learned about this brand's colorful history by visiting its little museum at Gütenbach in the Black Forest. Hanhart, which once led the global market for stopwatches, remains the most important manufacturer of mechanical handheld stopwatches, which are fabricated in the brand's own manufactory. The cornerstone for this ticking empire was laid in Diessenhofen in 1882, when the label's founder Johann Adolf Hanhart opened a modest little shop selling watches and jewelry. The small town wasn't large enough to satisfy Hanhart's big ambitions, so he began searching for a new location with greater potential. In 1902, Hanhart relocated to the up-and-coming watch city of Schwenningen am Neckar, where his company thrived and became the region's largest artisanal business. In 1920, Hanhart's 18-year-old son Wilhelm "Willy" Julius declared that only strong familial pressure could persuade him to join the burgeoning family business. He simply wasn't interested in watchmaking or in selling time-measuring instruments. Despite his initial disinterest, Willy went on to successively lead the house of Hanhart into an exceedingly flourishing future because the athletic young man discovered a valuable niche: the shortage of affordably priced stopwatches on the German market. Hanhart capitalized on this opportunity with the 1924 launch of the first German-made "people's stopwatch." Although this model surely couldn't claim to embody haute horlogerie, its movement boasted extremely reliable mechanisms and could run for 24 hours between windings. The stopwatch business was primarily seasonal, leaving the factory with unprofitable downtime in the offseason, so Willy Hanhart also began making inexpensive pocket watches and wristwatches. His spectrum of products included, for example: 10-ligne Caliber 34 with four jewels and a pin-lever escapement; 19-ligne pocket-watch Calibers 22 and 27; and 8¾-ligne lever Caliber 36 with 15 jewels. Hanhart augmented the operation in Schwenningen by opening a small subsidiary in Gütenbach in 1934. The Black Forest community soon made him a tempting offer to relocate his firm's headquarters. Hanhart accepted and settled his business there in 1938.

Hanhart debuted the innovative "Sans Souci" alarm wristwatch toward the end of the 1940s. A product catalogue from 1950 lauded this time-honored pilot's chronograph as a "practical watch for engineers and plant managers." The American actor Steve McQueen numbered among the most famous fans of the "Hanhart 417 ES Flyback". Hanhart ceased production of wristwatches in 1962 and continued to produce only mechanical stopwatches. Electronic timekeeping became dominant in Gütenbach in the 1970s. Although Hanhart created its own quartz movement, which was manufactured in the millions, the family business couldn't lastingly compete with cheaply priced Far Eastern competitors. Jack W. Heuer proposed a merger in the early 1980s, but Hanhart refused. Stopwatches rescued its independence. Klaus Eble, a son-in-law of the Hanhart Family, joined three entrepreneurs from Munich as majority shareholders of Hanhart in 1992. The ensuing reorientation was aptly embodied by the "Replika" (1997) as a representative of the new generation. The specialists in Gütenbach revived the traditional watchmaker's art and launched this pilot's chronograph as an authentic replica of the historical model from 1939. An intermezzo began under the aegis of the Swiss investment company Gaydoul Group in 2010, but in 2014 Hanhart again became the property of the same Munich-based businessmen's group that had piloted it from 1992 to 2008. Once again headquartered in Gütenbach, Hanhart presently pursues a two-track sales strategy: cooperation with traditional specialized dealers is one track; the other is online sales, although the latter includes the former.

Considering the turbulent historical events of the past fourteen decades, it's all the more remarkable that Hanhart has existed and manufactured its products uninterruptedly since 1882. An important contributor to its success is the fact that Hanhart's products are traditionally "made in Germany." Some 12,000 mechanical stopwatches are built annually, and each encases a genuine manufacture caliber. These timepieces are highly appreciated as "classic timers" by fans of vintage automobiles and participants in rallies for classic motorcars. The distinctive collection of wristwatch chronographs is divided into three lines, each with its own specific aims. "Primus" combines traditional elements (e.g., the well-known red button) with progressive design and high-quality technologies. With boldly styled dials, the "Racemaster" line highlights close affiliations with motorsport of the 1960s and '70s. Unostentatious but effective high technology adds impressive scratch-resistance to the cases of the models in the "Racemaster GT," "Racemaster GTM" and "Racemaster GTF" lines. Hanhart uses patented, nickel-free, extremely corrosion-resistant steel known as HDS-Pro for these high-end cases. The surface of this alloy is at least three times harder than the surface of conventional stainless steel, which makes the cases approximately 100 times more resistant to scratches. Incidentally: the thermal annealing process doesn't detract from the appearance of the material, which still looks like classical steel.

Last but not least, the "Pioneer" line celebrates a contemporary retro look. Like the three aforementioned "Racemaster" models, the "TwinDicator" recalls Hanhart's grand chronographic legacy. The manufactory's hand-wound movements have become things of the past, so Hanhart cooperates with the experienced Swiss mechanical specialists at La Joux-Perret, who modify reliable Valjoux/Eta 7750 or Sellita SW 500 calibers by adding additional gears or other special features in harmony with the original look, e.g., the push-pieces are positioned at unequal distances from the crown, just as they were on the original models. The front-mounted module suggests that, as in the good old days, here too a voluminous caliber ticks inside the case. An additional counter for 12 elapsed hours is positioned concentrically with the small seconds at "9 o'clock." In keeping with tradition, one button is colored bright red. Needless to say, Hanhart again offers a version with the time-honored flyback function to delight aficionados of perfect nostalgia on the wrist. ◦

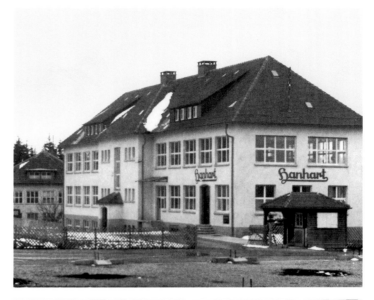

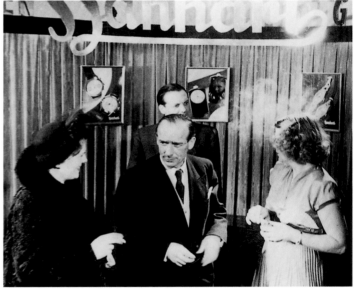

From top: The Hanhart building in Schwenningen, c. 1902 ◦
Willy Hanhart at Baselworld, 1952 ◦ A view of production at
the Hanhart watch factory in Gütenbach, c. 1930 ◦
Von oben: Hanhart-Gebäude in Schwenningen, ab 1902 ◦
Willy Hanhart bei der Baselworld, 1952 ◦ Einblick in die Produktion
der Hanhart-Uhrenmanufaktur in Gütenbach, um 1930 ◦
De haut en bas : Site Hanhart de Schwenningen, à partir
de 1902 ◦ Willy Hanhart au Salon mondial de l'horlogerie
à Bâle en 1952 ◦ Aperçu de la production dans la manufacture
de Gütenbach, vers 1930

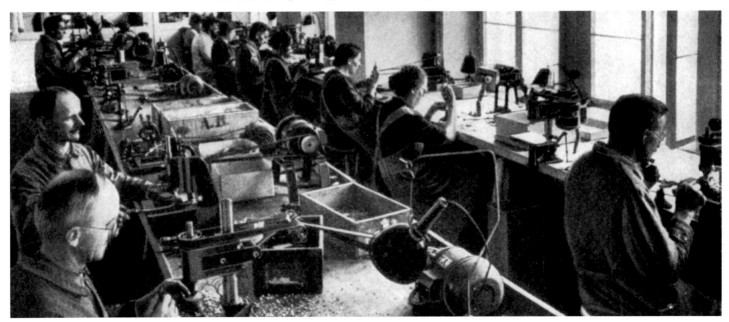

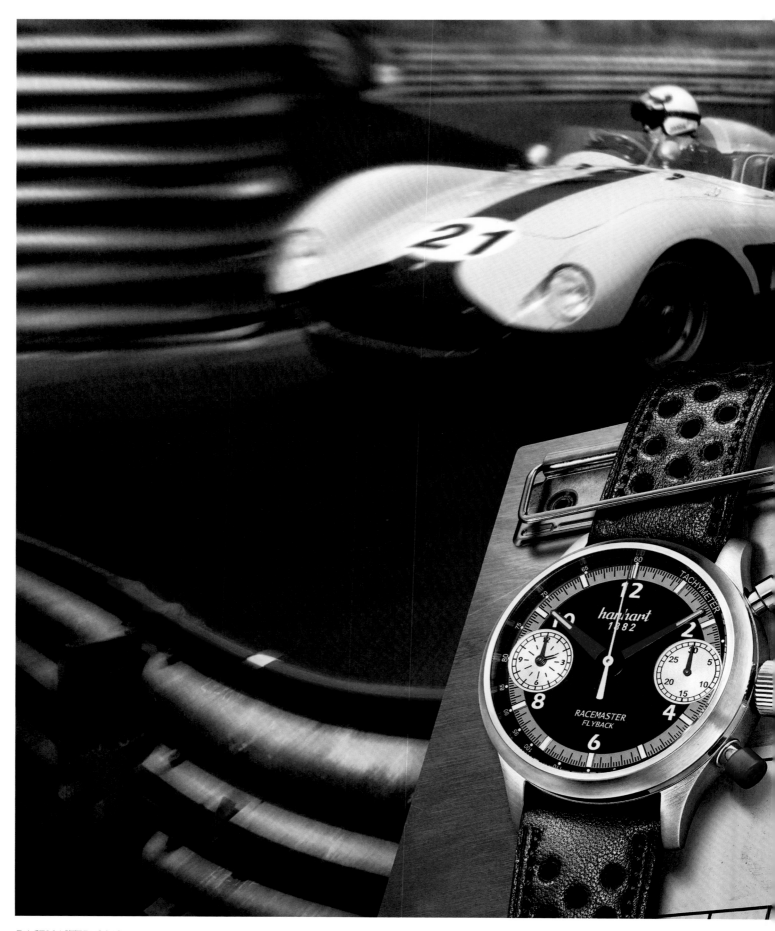

RACEMASTER, 2013

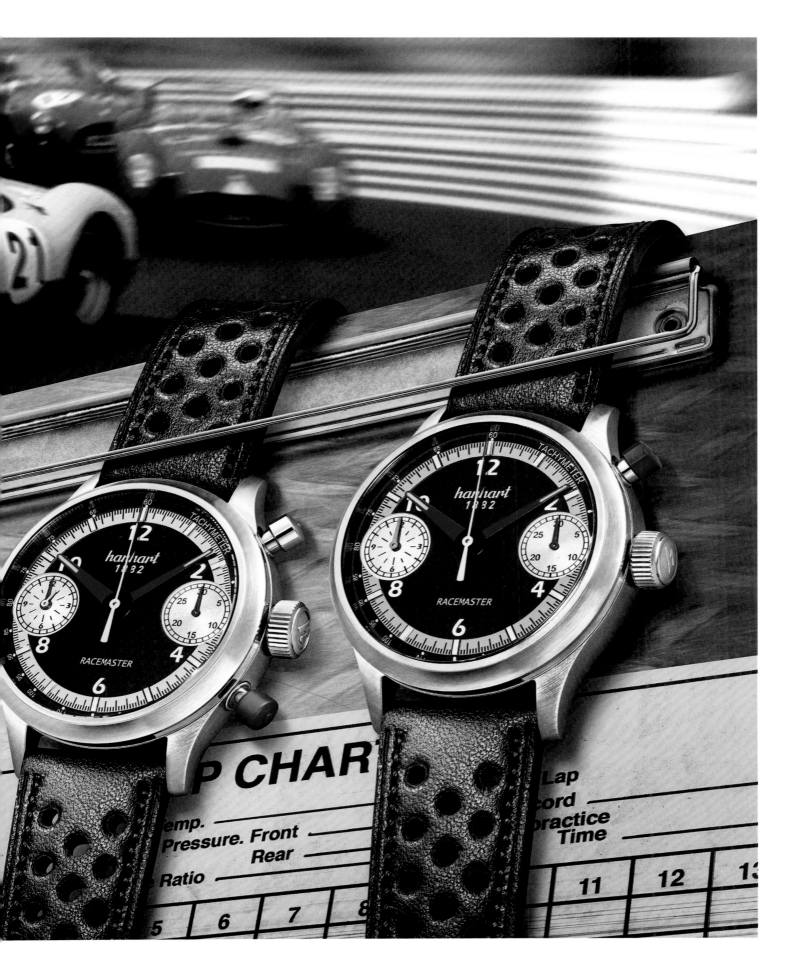

Erinnerungen an die Zukunft

Deutsch

Hanhart – mit dieser deutschen Traditionsmarke verknüpfen Uhrenliebhaber in aller Regel maskuline Chronographen mit rotem Nullstelldrücker. Selbiger hat natürlich seine spezielle Geschichte, welche in den 1940er Jahren spielt. Bereits vor Ausbruch des Zweiten Weltkriegs hatte die 1882 gegründete Manufaktur einen neuen, sehr umfassenden Verkaufskatalog präsentiert. Darin fanden sich insgesamt 45 Armbanduhr-Modelle, sechs Taschen-, acht Stopp- und zwei Sportuhren sowie eine Etuiuhr. Die künftigen Produktprioritäten waren damit ganz klar vorgezeichnet. 1938 startete die Entwicklung erster Chronographen mit der hauseigenen Kaliberfamilie 4x. Die gemäß Uhrmacherjargon 15½-linigen Handaufzugswerke maßen stattliche 35 Millimeter. Hanhart offerierte sie als Kaliber 40 mit nur einem Drücker für alle drei Funktionen Start, Stopp und Nullstellung, als 41 mit zwei Drückern für Additionsstoppungen sowie als Top-Version 42 mit Temposchaltung, was im modernen Sprachgebrauch eine sogenannte Flyback-Funktion meint. Besondere Kaliber-Kennzeichen des Trios waren klassische Chronographen-Merkmale wie Schaltradsteuerung und horizontale Räderkupplung. Das Kaliber 41 erkannte man an asymmetrisch ober- und unterhalb der Aufzugs- und Zeigerstellkrone positionierten Drückern. Bei den für militärische Cockpits bestimmten Exemplaren kommt nun der signifikant rote Drücker bei der „4" ins Spiel. Hierbei handelt es sich um den Nullstelldrücker, welchen man speziell bei der Ausführung mit Temposchaltung im Eifer des Gefechts bloß nicht versehentlich betätigen sollte. Also färbte ihn die besorgte Gattin eines Piloten kurzerhand mit ihrem roten Nagellack ein. Und damit war eine zeitschreibende Legende geboren, welche sich auch noch durch einen roten Farbtupfer auf der gerändelten Drehlünette auszeichnete. Dieser Merkpunkt erinnerte beispielsweise an den Ablauf eines längeren Zeitintervalls, das der Chronograph mit seinem 30-Minuten-Totalisator nicht erfassen konnte.

Mehr über die lange, sehr bewegte Geschichte des Unternehmens erfährt man bei einem Besuch des kleinen Museums in Gütenbach im Schwarzwald. Hanhart war einstmals Weltmarktführer im Stoppuhrenbereich und ist auch heute noch der bedeutendste Hersteller mechanischer Handstopper, die in der hauseigenen Manufaktur am Firmensitz gefertigt werden. Den Grundstein für dieses tickende Imperium hatte der Gründer Johann Adolf Hanhart 1882 in Diessenhofen durch ein schlichtes Uhren- und Schmuckgeschäft gelegt. Weil der ausgeprägte Geschäftssinn Hanharts in der relativ kleinen Ortschaft nur unzureichend zur Geltung kam, hielt der ambitionierte Unternehmer Ausschau nach adäquaten Entfaltungsmöglichkeiten. 1902 begab er sich in die aufstrebende Uhrenstadt Schwenningen am Neckar. Dort gedieh Hanhart zum größten handwerklichen Unternehmen der Region. 1920 erklärte sich der gerade einmal 18-jährige Sohn Wilhelm Julius, genannt Willy, nur unter familiärem Druck bereit, in das aufstrebende Unternehmen einzusteigen. Er hielt nicht sonderlich viel von der Uhrmacherei und dem Handel mit Zeitmessinstrumenten. Das anfängliche Desinteresse änderte nichts an der Tatsache, dass gerade er das Haus Hanhart sukzessive in eine überaus erfolgreiche Zukunft steuern sollte. Denn der sportive Juniorchef erkannte eine Lücke. Dem deutschen Markt mangelte es an preiswerten Stoppuhren. Hanhart nutzte dieses Defizit und lancierte 1924 die erste Volks-Stoppuhr deutscher Provenienz. Von hoher Uhrmacherkunst war dieses Modell weit entfernt. Dafür zeichnete sich das Uhrwerk mit einer Gangautonomie von 24 Stunden durch eine höchst zuverlässige Mechanik aus. Weil das primär saisonal ausgerichtete Stoppuhren-Geschäft den Fertigungsbetrieb nicht auslastete, setzte Willy Hanhart auf preiswerte Taschen- und Armbanduhren. Sein Fertigungsspektrum umfasste beispielsweise das 10-linige Kaliber 34 mit vier Steinen und Stiftanker-Hemmung, die 19-linigen Taschenuhr-Kaliber 22 und 27 sowie das 8¾-linige Anker-Kaliber 36 mit 15 Steinen. 1934 ergänzte Hanhart den Schwenninger Betrieb zunächst nur um eine kleine Filiale in Gütenbach. Als die abgelegene Schwarzwald-Gemeinde ein verlockendes Angebot zur Verlagerung des Firmensitzes unterbreitete, zog Hanhart im Jahr 1938 dorthin um.

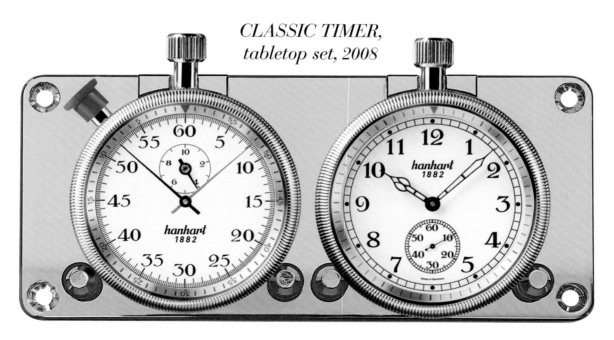

CLASSIC TIMER,
tabletop set, 2008

Ende der 1940er Jahre stellte Hanhart den innovativen Armband-wecker „Sans Souci" vor. Den bewährten Flieger-Chronographen pries der Produktkatalog 1950 als „zweckmäßige Uhr für den Ingenieur und Betriebsleiter" an. Zu den berühmtesten Fans der „Hanhart 417 ES Flyback" gehörte der amerikanische Schauspieler Steve McQueen. 1962 endete vorerst einmal die Herstellung von Armbanduhren zugunsten mechanischer Stoppuhren. Ab den 1970er Jahren dominierte in Gütenbach die Elektronik. Hanhart kreierte ein eigenes, millionenfach verkauftes Quarzwerk. Trotz-dem konnte das Familienunternehmen im permanenten Wettkampf mit fernöstlichen Billigkonkurrenten nicht dauerhaft bestehen. Anfang der 1980er Jahre lehnte es ein Fusionsangebot von Jack W. Heuer ab. Stoppuhren retteten die Unabhängigkeit. 1992 übernahmen Klaus Eble, Schwiegersohn der Familie Hanhart, sowie drei Münchner Unternehmer die Mehrheit der Hanhart-Anteile. Im Zeichen des Aufbruchs stand 1997 die „Replika" als Repräsentant einer neuen Generation. Die Gütenbacher griffen die traditionelle Uhrmacherkunst wieder auf und stellten mit dem Fliegerchronographen einen originalgetreuen Nachbau des historischen Modells von 1939 vor. Nach einem Intermezzo unter dem Dach der schweizerischen Beteiligungsgesellschaft Gaydoul Group, welches 2010 begann, befindet sich Hanhart seit 2014 erneut im Eigentum der Münchner Unternehmergruppe, welche das Unternehmen bereits von 1992 bis 2008 begleitet hatte. Aktuell fährt das wieder in Gütenbach ansässige Unternehmen eine zweigleisige Vertriebsstrategie. Zur Kooperation mit dem tradierten Fachhandel gesellt sich eine Online-Schiene, welche ersteren jedoch angemessen einbindet.

Angesichts turbulenter geschichtlicher Ereignisse gerät oftmals in Vergessenheit, dass Hanhart seit 1882 ohne jede Unterbrechung existiert und produziert hat. Hoch im Kurs steht jenes „Made in Germany", welches traditionsgemäß zu der Marke gehört. Jährlich entstehen weiterhin rund 12000 mechanische Stoppuhren mit echten Manufakturkalibern, als „Classictimer" hoch geschätzt unter anderem von Oldtimer-Fans und Vintage-Rallye-Clubs. Die markante Arm-bandchronographen-Kollektion gliedert sich in drei Linien mit ganz unterschiedlichen Ansprüchen. „Primus" kombiniert traditio-nelle Elemente wie den bekannten roten Drücker mit progressivem Design und hochwertigen Technologien. „Racemaster" weist beispiels-weise durch markante Zifferblätter auf die engen Verbindungen zum Motorsport in den 1960er und 1970er Jahren hin. Unauffällige,

aber wirkungsvolle Hochtechnologie verhilft den Gehäusen der Modelle „Racemaster GT", „Racemaster GTM" und „Racemaster GTF" zu beeindruckender Kratzfestigkeit. Für die High-End-Schalen verwendet Hanhart einen patentierten, nickelfreien und extrem korrosionsbeständigen Stahl namens HDS-Pro. Selbiger besitzt eine mindestens drei Mal härtere Oberflächenstruktur als konven-tioneller Edelstahl, was die Gehäuse rund 100-mal kratzbeständiger macht. Die Optik leidet unter der thermischen Vergütung übrigens nicht. Das Ganze sieht weiterhin aus wie klassischer Stahl.

Schließlich präsentiert sich die „Pioneer"-Linie in zeit-gemäßem Retrolook. Wie bei den drei genannten „Racemaster"-Modellen erinnert auch hier der „TwinDicator" an das große chronographische Erbe. Weil das Manufaktur-Handaufzugswerk aus mehreren Gründen der Vergangen-heit angehört, kooperiert Hanhart mit dem erfahrenen Schweizer Mechanik-Spezialisten La Joux-Perret, welcher die zuverlässigen Kaliber Valjoux/Eta 7750 oder Sellita SW 500 durch zusätzliche Zahnräder und andere Kniffe im Sinne der ursprünglichen Optik modifiziert, ergo stehen die Drücker wie bei den Originalen unterschiedlich weit von der Krone ab. Und das vorderseitig montierte Modul lässt vermuten, im Gehäuse-inneren ticke – wie einst – ein voluminöseres Kaliber. Konzentrisch zur kleinen Sekunde bei „9" rotiert ein zusätzlicher 12-Stunden-Totalisator. Getreu großer Tradition erstrahlt ein Drücker in leuchtendem Rot. Selbstverständlich offeriert Hanhart Verfechtern perfekter Nostalgie am Handgelenk auch wieder ein Exemplar mit überlieferter Flyback-Funktion. ◦

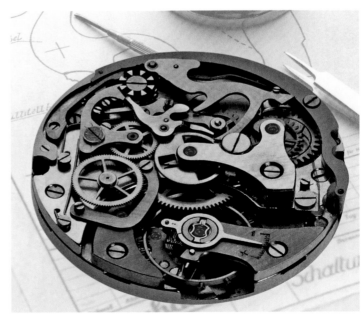

Top: Cal. 40 ◦ Right: Cal. 41

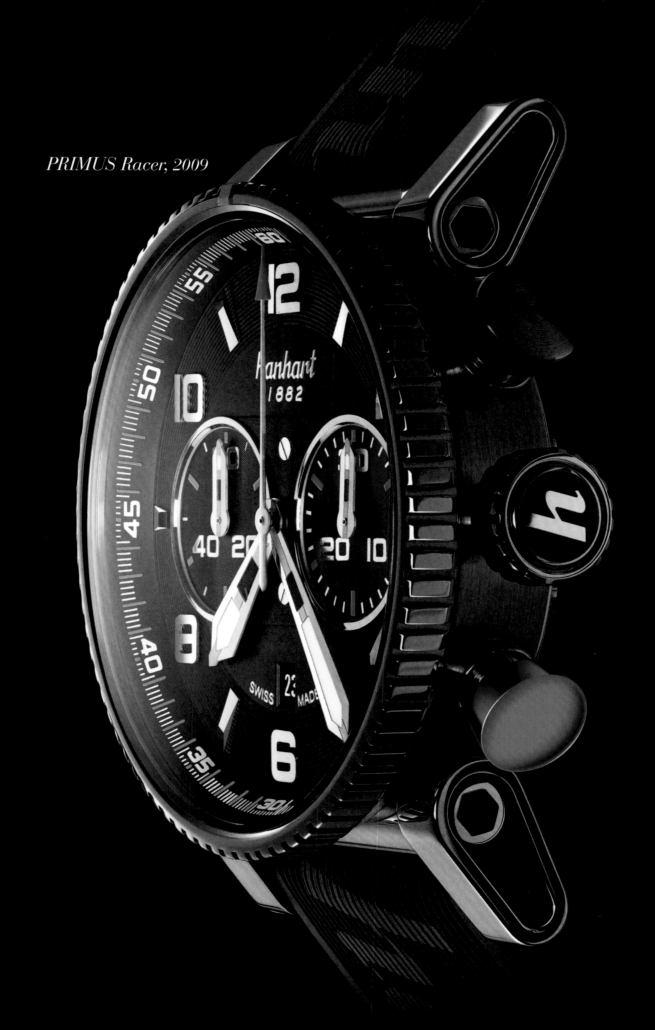

PRIMUS Racer, 2009

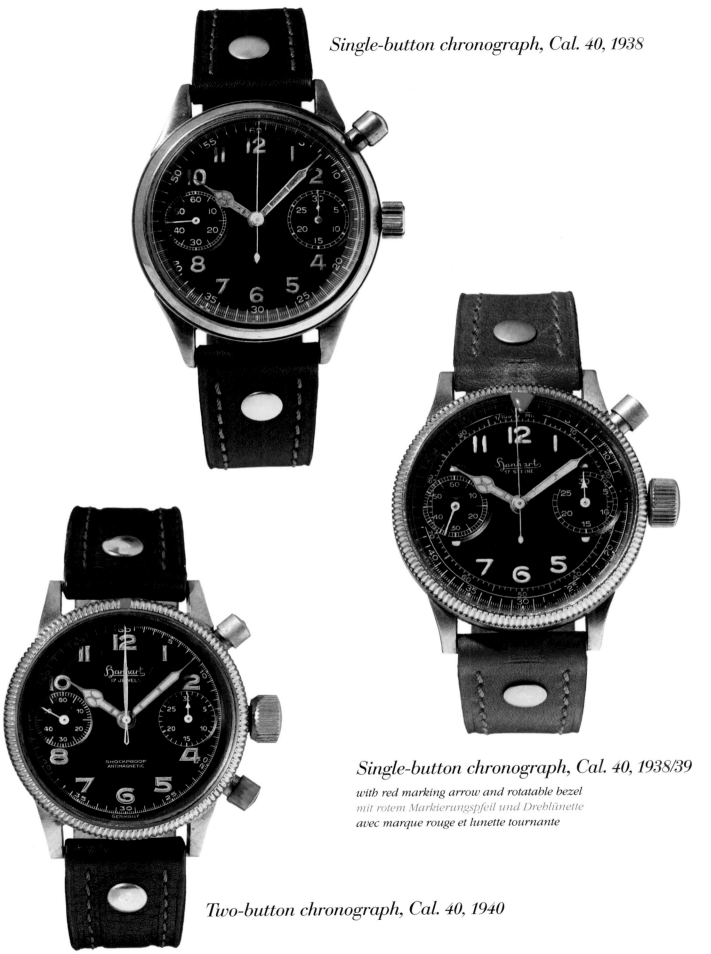

Single-button chronograph, Cal. 40, 1938

Single-button chronograph, Cal. 40, 1938/39

with red marking arrow and rotatable bezel
mit rotem Markierungspfeil und Drehlünette
avec marque rouge et lunette tournante

Two-button chronograph, Cal. 40, 1940

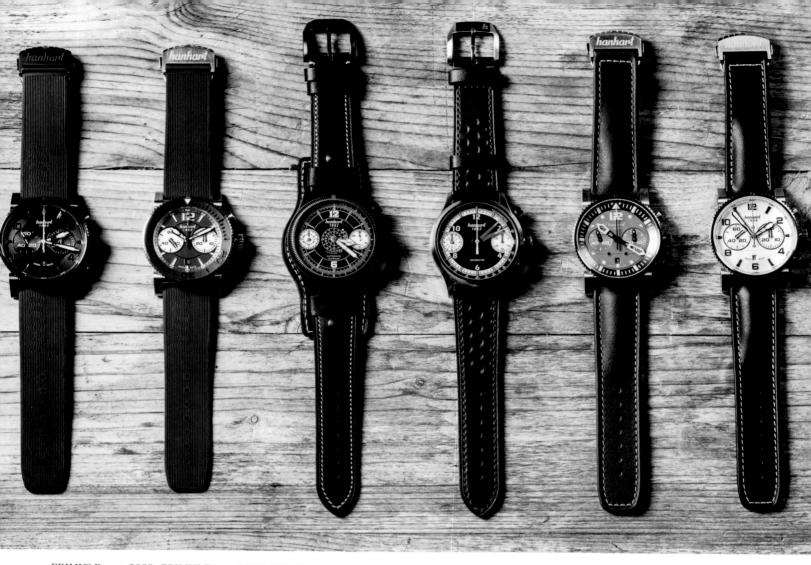

PRIMUS Racer, 2009; PRIMUS Diver, 2009; PIONEER Stealth 1882, 2012; RACEMASTER GT, 2013; PRIMUS Pilot, 2009; PRIMUS Racer, 2009

Souvenirs du futur

Français

Les passionnés de montres associent généralement à la marque allemande de tradition Hanhart des chronographes pour hommes équipés d'un poussoir de remise à zéro rouge. Pour la petite histoire, il faut remonter aux années 1940 : avant même la Seconde Guerre mondiale, cette manufacture fondée en 1882 sort un nouveau catalogue, très complet, qui comporte au total 45 montres-bracelets, six montres de poche, huit chronomètres et deux montres pour sportifs ainsi qu'une montre-étui. Les priorités futures en termes de production sont ainsi clairement esquissées. Les premiers chronographes développés par la marque à partir de 1938 sont équipés de la famille de calibres 4x produits en interne. Les mouvements à remontage manuel 15'''½, comme on les appelle dans le jargon horloger, mesurent pas moins de 35 millimètres de diamètre. Hanhart les décline en plusieurs versions : avec le calibre 40 et un poussoir unique pour les trois fonctions mise en marche, arrêt et remise à zéro ; avec le calibre 41 et deux poussoirs pour les mesures successives ; et enfin, une version hautement sophistiquée avec le calibre 42 doté ce qu'on appelle communément la fonction Flyback ou retour-en-vol. Ces trois calibres présentent des caractéristiques typiques des chronographes, à savoir la roue à colonnes et l'engrenage horizontal. Le calibre 41 se reconnaît à la disposition asymétrique des poussoirs au-dessus et au-dessous de la couronne de remontage et de mise à l'heure. C'est alors qu'entre en scène le poussoir rouge significatif situé à 4 heures, observable sur les exemplaires destinés à l'aviation militaire. Il s'agit là du poussoir de remise à zéro, qu'il ne faut en aucun cas actionner par inadvertance dans le feu du combat, en particulier sur les modèles dotés de la fonction Flyback. Aussi une femme de pilote inquiète l'a-t-elle tout simplement badigeonné de vernis à ongles rouge, donnant ainsi naissance à un chronographe de légende. Ce dernier se caractérise aussi par un point rouge sur la lunette tournante crantée, lequel signale notamment un intervalle de temps supérieur aux 30 minutes que le chronographe peut enregistrer avec son totalisateur.

Pour en savoir plus sur la longue histoire très mouvementée de l'entreprise, on ne peut que recommander une visite au petit musée municipal de Gütenbach, en Forêt-Noire. Jadis leader mondial sur le marché des chronographes, Hanhart est aujourd'hui encore le premier fabricant de chronomètres à main, qui sont produits dans la manufacture maison au siège même de l'entreprise. Tout commence en 1882 à Diessenhofen, par la création d'une simple bijouterie-horlogerie. Ne pouvant guère réaliser tout son potentiel commercial dans cette petite localité, l'ambitieux Johann Adolf Hanhart transfère son entreprise en 1902 à Schwenningen am Neckar (capitale horlogère dans le sud de l'Allemagne), où elle deviendra la plus grande manufacture de la région. Cédant à la pression parentale, Wilhelm Julius Hanhart dit Willy rejoint en 1920 l'entreprise en plein essor. Il a tout juste 18 ans et il fait peu de cas de l'horlogerie et du commerce des garde-temps. En dépit de ce désintérêt initial, il sera justement l'artisan de la réussite exceptionnelle de la maison Hanhart. En effet, ce sportif identifie un créneau dans le marché allemand, celui des chronographes abordables. Sautant dans la brèche, il lance en 1924 le premier chronographe grand public d'origine allemande. Si ce modèle est loin de répondre aux exigences de la haute horlogerie, son mouvement qui lui assure une réserve de marche de 24 heures n'en est pas moins extrêmement fiable. Avant tout saisonnière, la demande en chronographes ne permet pas de faire tourner les ateliers de fabrication à plein rendement. Willy Hanhart mise alors sur les montres de poche et les montres-bracelets abordables, parmi lesquelles le calibre 34 10''' renfermant quatre pierres et un échappement à ancre à chevilles, les calibres 22 et 27 19''' pour montres de poche, ainsi que le calibre 36 8'''¾ à ancre doté de 15 pierres. En 1934, Willy Hanhart commence modestement par adjoindre au site de production de Schwenningen une petite succursale à Gütenbach. Conquis par une offre séduisante de cette municipalité retirée de Forêt-Noire, il y transfère le siège de l'entreprise en 1938.

À la fin des années 1940, Hanhart présente la « Sans Souci », montre-bracelet innovante à réveil intégré. Le catalogue de 1950 vante le chronographe d'aviation éprouvé dans les termes suivants : « une montre qui répond aux besoins de l'ingénieur et du chef d'entreprise ». Les inconditionnels du modèle « 417 ES Flyback » comptent dans leurs rangs des célébrités, parmi lesquelles l'acteur de cinéma américain Steve McQueen. En 1962, la fabrication de montres-bracelets cède pour un temps la place à celle des chronomètres mécaniques. À partir des années 1970, Gütenbach entre dans l'ère de l'électronique, et Hanhart crée son propre mouvement à quartz, qui se vend à des millions d'exemplaires. L'entreprise familiale ne peut cependant pas contrer durablement la concurrence permanente de fabricants asiatiques pratiquant des prix très bas. Elle reçoit au début des années 1980 une proposition de fusion de la part de Jack W. Heuer, mais elle la refuse et conserve son indépendance grâce aux chronomètres. En 1992, trois entreprises munichoises prennent une participation majoritaire dans la manufacture aux côtés de Klaus Eble, gendre de la famille Hanhart. Ouvrant une ère nouvelle en 1997, les « répliques » sont les emblèmes d'une nouvelle génération. Renouant avec la haute horlogerie traditionnelle, la manufacture de Gütenbach présente un chronographe d'aviation qui est la fidèle réplique du modèle original de 1939. Après un intermède de 2010 à 2014 dans le giron de la holding suisse Gaydoul Group, la manufacture Hanhart est

de nouveau propriété du groupe d'entreprises munichoises qui l'a déjà accompagnée de 1992 à 2008. De nos jours, Hanhart, qui a de nouveau son siège à Gütenbach, s'appuie sur deux canaux de distribution, qui coopèrent étroitement : son réseau bien établi de revendeurs spécialisés et une boutique en ligne.

Les vicissitudes de l'Histoire font souvent oublier que depuis sa fondation, en 1882, Hanhart a existé et produit sans aucune interruption. Le « Made in Germany », dont la marque s'est fait une tradition, est très prisé. Hanhart continue de produire annuellement environ 12 000 chronomètres mécaniques équipés d'authentiques calibres de manufacture ; ces « Classictimer » sont très recherchés des amateurs de voitures anciennes et des clubs organisant des rallyes d'automobiles de collection. Par ailleurs, la gamme remarquable de chronographes-bracelets se décline en trois collections répondant à des attentes très différentes. « Primus » allie des éléments traditionnels comme le célèbre poussoir rouge à un design progressiste et à des technologies de pointe. « Racemaster » rend hommage aux liens étroits unissant la marque au sport automobile dans les années 1960 et 1970, notamment avec ses cadrans caractéristiques. Grâce à une technologie avancée, aussi discrète qu'efficace, les boîtiers des modèles « Racemaster GT », « Racemaster GTM » et « Racemaster GTF » présentent une résistance aux éraflures impressionnante. Ces boîtiers haut de gamme sont en HDS-PRO, un acier breveté, sans nickel et extrêmement résistant à la corrosion. Sa structure de surface étant au minimum trois fois plus dure que celle de l'acier inoxydable conventionnel, les boîtiers sont environ 100 fois plus résistants aux éraflures. Le traitement thermique n'altère en rien l'aspect du matériau, qui a encore tout d'un acier classique. La troisième collection, « Pioneer », arbore un look rétro dans l'air du temps. Comme les trois modèles « Racemaster », le « TwinDicator » de cette collection s'inscrit lui aussi dans la lignée des chronographes Hanhart. Le mouvement à remontage manuel de la manufacture étant pour différentes raisons dépassé, Hanhart coopère avec le grand spécialiste de mécanique horlogère suisse La Joux-Perret, qui ajoute aux calibres éprouvés Valjoux/ETA 7750 ou encore Sellita SW 500 des engrenages et divers détails, de façon à recréer l'apparence d'origine, avec notamment la disposition asymétrique des poussoirs de part et d'autre de la couronne. La face avant laisse supposer que le boîtier abrite, comme autrefois, un calibre relativement volumineux. Le cadran auxiliaire situé à 9 heures comprend, outre la petite seconde, un totalisateur 12 heures supplémentaire. Dans la plus grande tradition, un poussoir rouge lumineux brille de mille feux. Pour les adeptes de la nostalgie totale à leur poignet, Hanhart propose bien sûr encore un modèle doté de la fonction retour-en-vol conventionnelle. ○

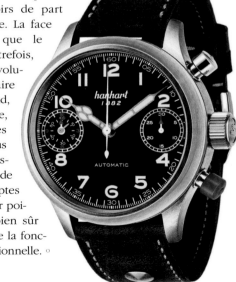

PIONEER TwinDicator, 2011

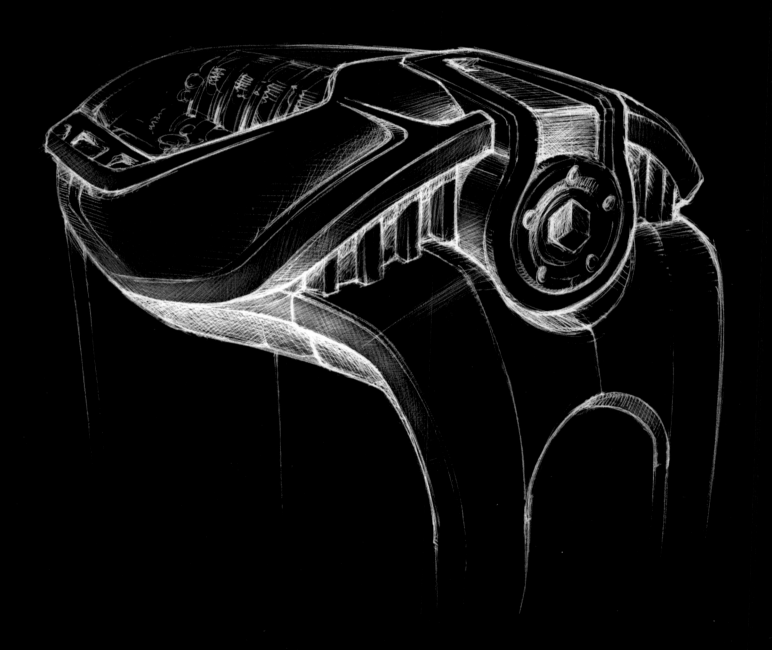

Hublot

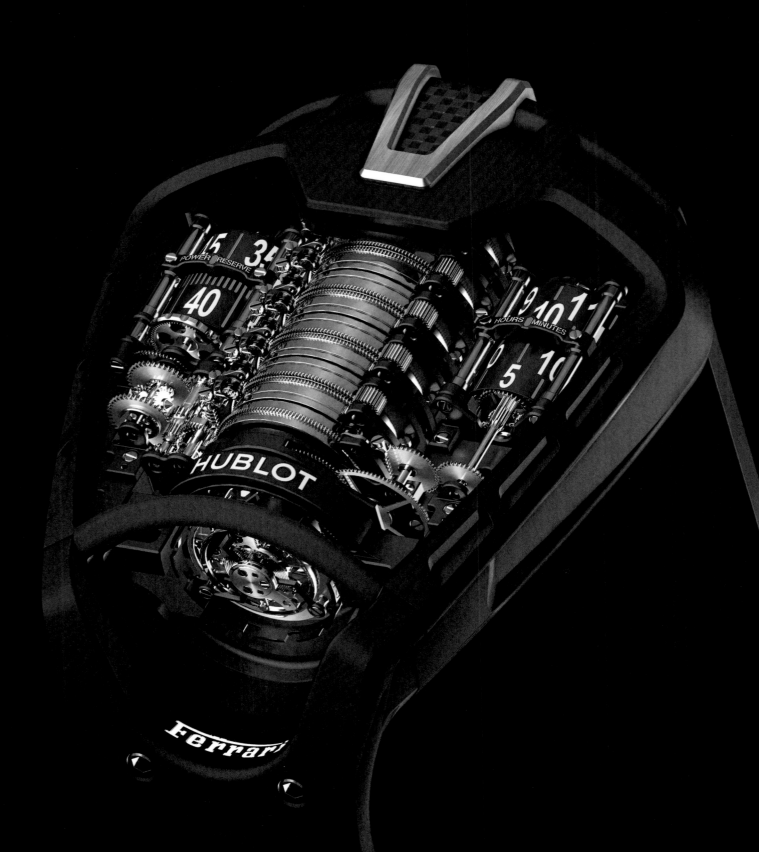

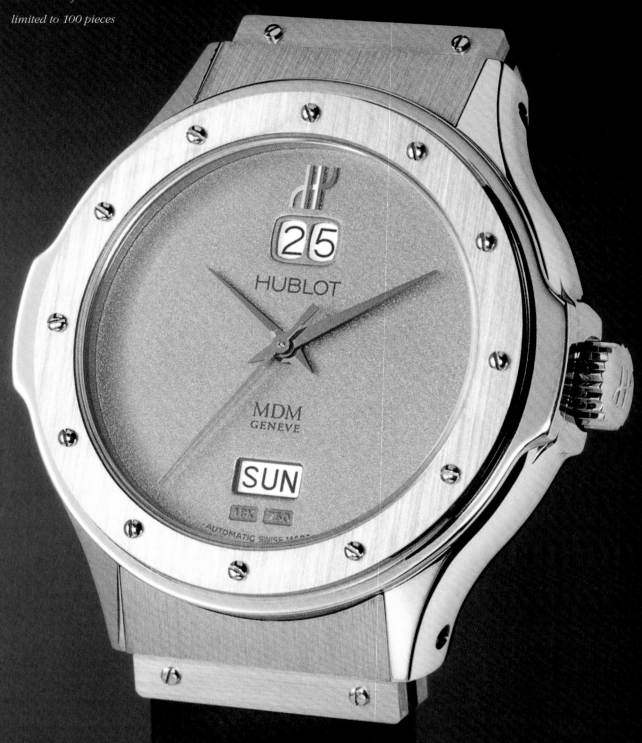

Grand Quantième
Or Noir, 2000
limited to 100 pieces

Success through Fusion

The Swiss watch industry was suffering one of its most severe crises in 1976, when Carlo Crocco, the scion of a Milanese watch dynasty, took the risky step into a freelance career. But the 32-year-old let nothing dissuade him from his daring plans and accordingly founded MDM Genève. The three letters originally stood for Marie-Danielle Montres (his wife's first name), but the acronym was later translated as Montre des Montres. The "watch of watches" was meant to be a sporty timepiece with a long-lived, resilient, rubber wristband. This was unprecedented, so the first movement didn't debut until four years later at the watch fair in Basel in 1980. Steel fibers woven into the innovative rubber strap made it very resistant to tearing and its distinctive vanilla aroma appealed to the sense of smell. The case too was inspired by Crocco's refusal to imitate his competitors. Its unusual styling explains the French name of the premiere model: "hublot" means "porthole." The unmistakable design was warmly received.

Members of several royal families, including Juan Carlos of Spain, opted for timepieces bearing this signature. Until 2003, the firm's founder clung to his original plan of offering an iconic line of watches in various visual and technical variants. But Carlo Crocco's certainty that he had created an absolutely immortal product which was lastingly resistant to fashion's changing influences ultimately led him to overlook the fact that time, especially in the watch business, never stands still. Mono products likewise have their half-lives. And the competition never sleeps: competitors vied with Hublot for possession of a niche that had seemed secure at first. Despite its legendary watch concept, Hublot found itself with its back to the proverbial wall in 2004.

Fortunately, there now appeared a man named Jean-Claude Biver, whose extroverted personality and marketing genius stood in stark contrast to Carlo Crocco's character traits. This native of Luxembourg, who had worked for Audemars Piguet, Blancpain, and Omega, began assuring Hublot's continued survival in 2004. His ingenious fusion concept first became reality with the successful "Big Bang" model in 2005. The modularly designed case makes it possible to combine diverse materials in a single wristwatch. Biver's choices of materials triggered more than one gigantic explosion. Scratch-resistant "Magic Gold," colored ceramic, carbon fibers, and an ecological composite material containing linen are only a few examples. Under Jean-Claude Biver's aegis, the acquisition of complication specialist BNB and the development of its own calibers (e.g., the "Unico" chronographs) enabled Hublot to morph from an établisseur into a dyed-in-the-wool watch manufactory that was very attractive for a possible takeover.

LVMH made an offer in 2008. The French corporate group, which specializes in luxury items, won the bid with 500 million Swiss francs: 400 million went to Carlo Crocco and the remainder to Jean-Claude Biver, who had gradually acquired a 20 percent share. But retirement was out of the question because this visionary entrepreneur still had ambitious plans for Hublot. These included simultaneously strengthening the sporty yet unmistakably more elegant "Classic Fusion" line along with the "Big Bang," which celebrated its tenth birthday in radiant good health in 2015. In the meantime, Jean-Claude Biver also brought to fulfillment a wish that he had first cherished in 2004: namely, partnership with Ferrari. Explicitly conceived for the sporty automobile brand, the first "Big Bang" initially encountered a cold shoulder at the Italian car manufacturer. Hublot? Who's that? But the cooperation succeeded in the proverbial "flick of a wrist" in 2012. There simply cannot be a better compliment for the successful renaissance of a watch brand. ○

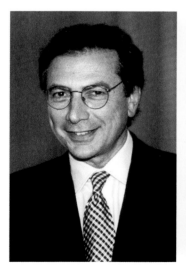
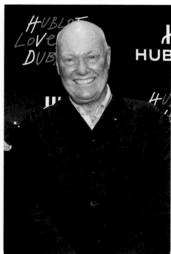
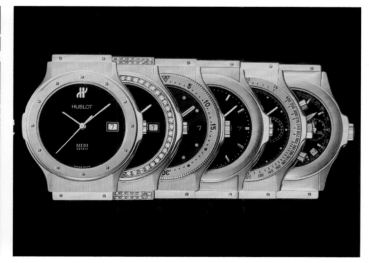

From left: Carlo Crocco ○ Jean-Claude Biver ○ Several different versions of Hublot's classic port-hole watch ○
Von links: Carlo Crocco ○ Jean-Claude Biver ○ Die klassische Bullaugenuhr von Hublot in verschiedenen Varianten ○
Depuis la gauche : Carlo Crocco ○ Jean-Claude Biver ○ Diverses versions du motif caractéristique des montres Hublot

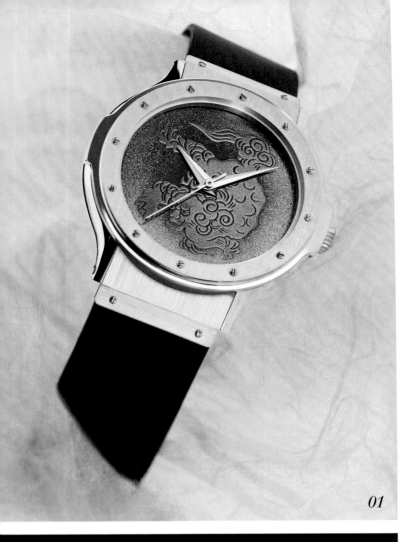

01

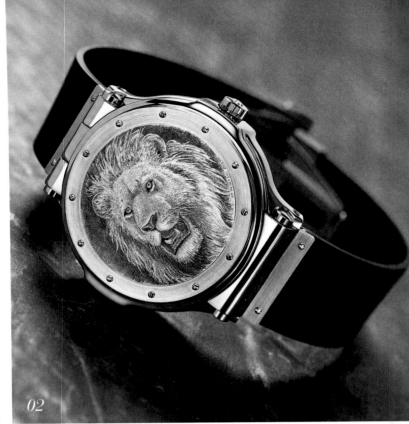

02

01. *The Art Collection, Urushi Ryu (dragon), 2003*
02. *The Art Collection, Engraving by Gianfanco Pedersoli, 2000*
03. *The Art Collection, 2000*
04. *The Art Collection, Urushi Rakucho (bird of paradise), 2003*

03

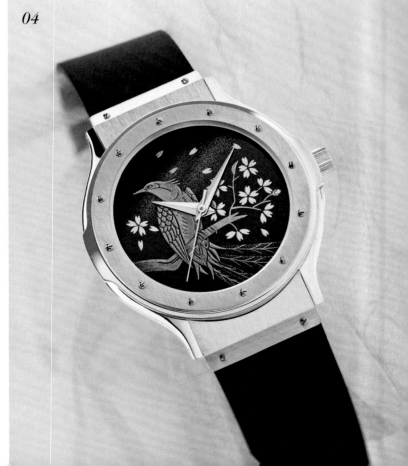

04

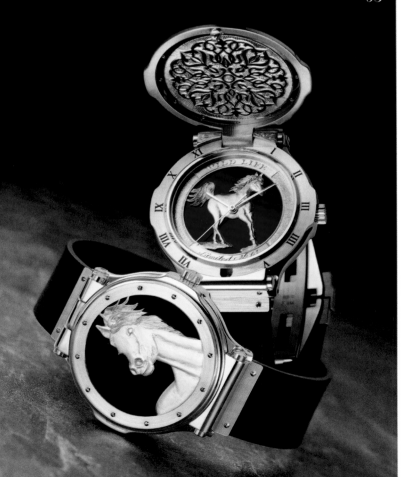

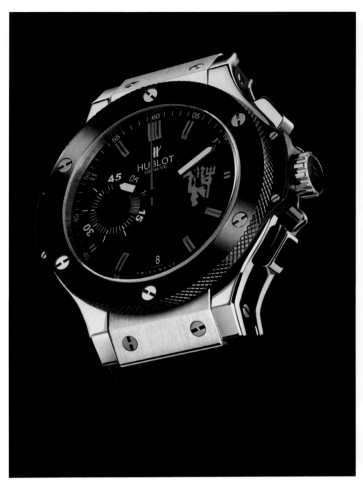

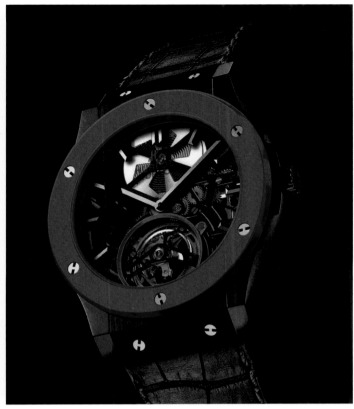

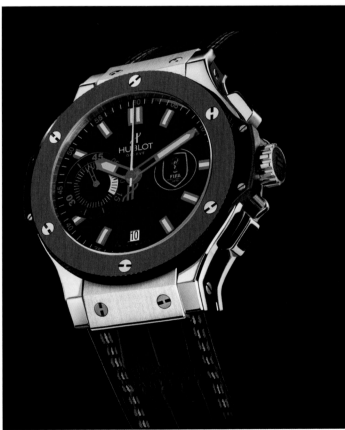

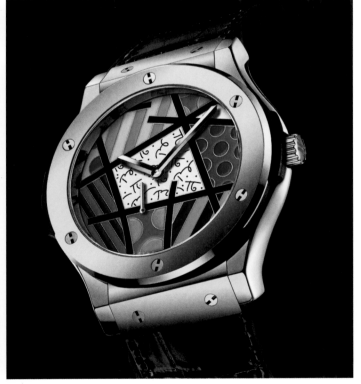

Top left: Big Bang Red Devil II, 2009 ○ Top right: Classic Fusion Skeleton Tourbillon Red Ceramic, 2013 ○ Bottom left: Big Bang FIFA World Cup Winner's Watch, 2010 ○ Bottom right: Classic Fusion Enamel by Romero Britto, 2015

Erfolg durch Fusion

Deutsch

Die Schweizer Uhrenindustrie ging durch eine ihrer größten Krisen, als Carlo Crocco, Spross einer Mailänder Uhrendynastie, 1976 den riskanten Schritt in die berufliche Selbstständigkeit wagte. Der 32-Jährige ließ sich durch nichts von seinen gewagten Plänen abbringen und gründete MDM Genève. Die drei Buchstaben standen für „Marie-Danielle Montres" (der Vorname seiner Ehefrau), später wurde dies mit „Montre des Montres" übersetzt. Die „Uhr unter den Uhren" sollte ein sportlicher Zeitmesser mit langlebigem und hoch belastbarem Kautschukarmband sein. Weil es so etwas bis dahin noch nicht gegeben hatte, dauerte die Präsentation des Erstlingswerks bis zur Basler Uhrenmesse im Jahr 1980. Eingewebte Stahlfasern sorgten beim neuartigen Kautschukband für hohe Reißfestigkeit. Duftorientierte Kunden ließen sich zudem von einem ausgeprägten Vanillegeruch betören. Auch das Gehäuse war der hartnäckigen Weigerung entsprungen, es den Mitbewerbern gleichzutun. Mit Blick aufs Design lag der französische Name des Premierenmodells nahe: „Hublot" – zu Deutsch „Bullauge". Das unverkennbare Design kam an.

Mitglieder mehrerer königlicher Familien, darunter auch Juan Carlos von Spanien, entschieden sich für einen Zeitmesser mit dieser Signatur. Am Plan, eine ikonographische Uhrenlinie in verschiedenen optischen und technischen Varianten anzubieten, hielt der Firmengründer bis 2003 eisern fest. Aber in der Überzeugung, ein absolut unsterbliches und gegen modische Einflüsse dauerhaft resistentes Produkt geschaffen zu haben, übersah Carlo Crocco letzten Endes dann doch, dass die Zeit gerade im Uhrenbusiness niemals stehen bleibt. Auch Mono-Produkte besitzen ihre Halbwertszeit. Und die Konkurrenz schläft nie. Sie besetzte und verdrängte Hublot aus seiner vermeintlich sicheren Nische. Deshalb stand Hublot 2004 trotz eines legendären Uhrenkonzepts mit dem Rücken zur Wand.

Zum Glück fand sich ein Mann namens Jean-Claude Biver, dessen extrovertierte Persönlichkeit und Marketinggenie in krassem Gegenteil stehen zu Carlo Croccos Charakteristika. Der Luxemburger mit Karriere-Etappen bei Audemars Piguet, Blancpain und Omega sicherte Hublot ab Herbst 2004 das Überleben. Sein geniales Fusionskonzept verwirklichte er 2005 im Erfolgsmodell „Big Bang".

Dessen modular gestaltetes Gehäuse lässt in der Tat die Kombination unterschiedlichster Materialien in einer Armbanduhr zu. Bei den Werkstoffen sorgte Biver für den einen oder anderen Urknall: kratzfestes „Magic Gold", farbige Keramik, Carbonfaser und ein ökologischer Verbundwerkstoff mit Leinen sind nur einige Beispiele. Durch den Erwerb des Komplikationen-Spezialisten BNB und die Entwicklung eigener Kaliber wie des „Unico"-Chronographen mauserte sich Hublot unter der Ägide von Jean-Claude Biver vom Etablisseur zur waschechten, für eine Übernahme höchst interessanten Uhrenmanufaktur.

2008 machte LVMH eine Offerte, und der französische Luxusmulti bekam den Zuschlag für 500 Millionen Schweizer Franken. 400 Millionen davon bekam Carlo Crocco, den Rest Jean-Claude Biver, der nach und nach einen 20-prozentigen Anteil erworben hatte. Von Ruhestand war dennoch keine Rede, denn der visionäre Sanierer hatte noch viel vor mit Hublot. Zu seinen Plänen gehörte auch die Stärkung der weiterhin sportlichen, aber doch unverkennbar eleganteren Uhrenlinie „Classic Fusion" parallel zur „Big Bang", die 2015 ihren zehnten Geburtstag bei bester Gesundheit zelebrieren konnte. Zwischenzeitlich ging für Jean-Claude Biver ein bereits 2004 gehegter Wunsch in Erfüllung: die Partnerschaft mit Ferrari. Mit der ersten „Big Bang", explizit konzipiert für die sportive Automarke, hatte er bei den Italienern noch auf Granit gebissen. Hublot, wer ist das? 2012 gelang die Kooperation sozusagen im Handumdrehen. Ein besseres Kompliment für die erfolgreiche Renaissance einer Uhrenmarke kann es schlichtweg nicht geben. ◦

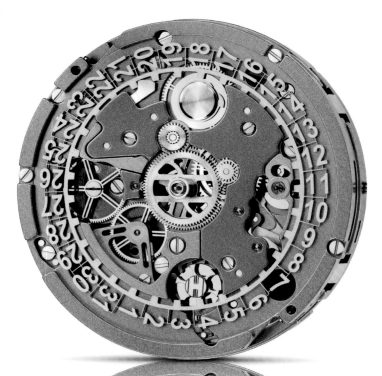

Left: Caliber HUB 1240, 2010 ◦ *Right: Self-winding column-wheel chronograph* ◦
Rechts: Automatischer Säulenrad-Chronograph ◦
À droite : Chronographe automatique avec roue à colonnes

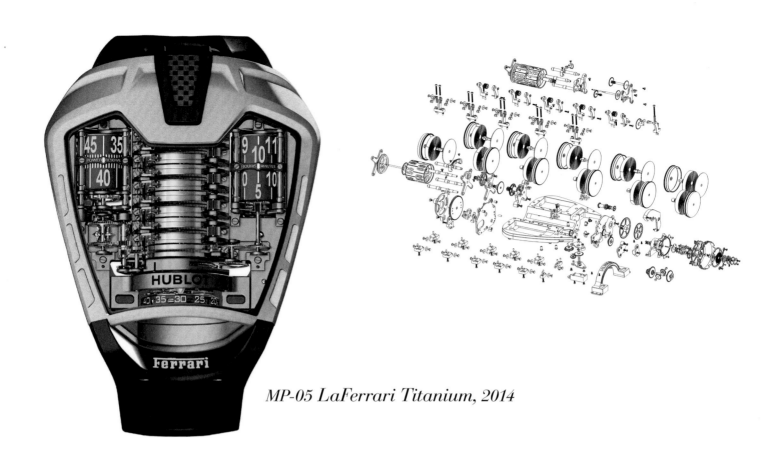

MP-05 LaFerrari Titanium, 2014

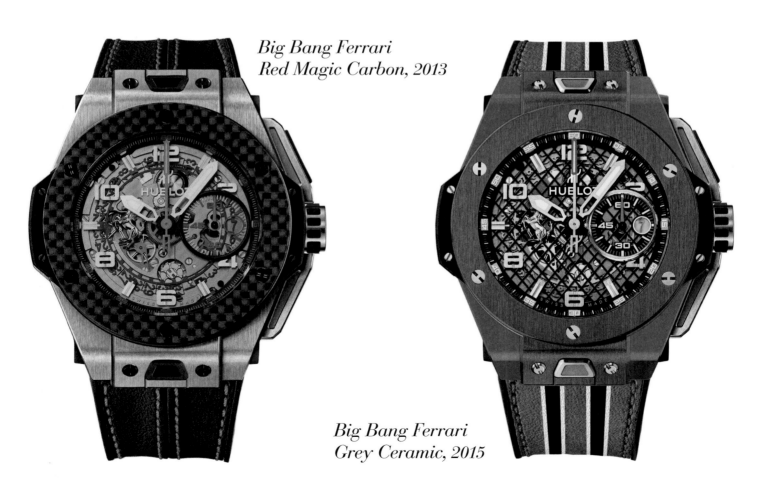

Big Bang Ferrari
Red Magic Carbon, 2013

Big Bang Ferrari
Grey Ceramic, 2015

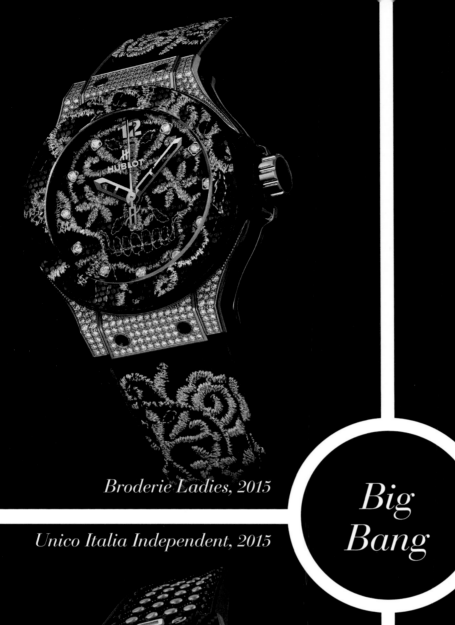

Broderie Ladies, 2015

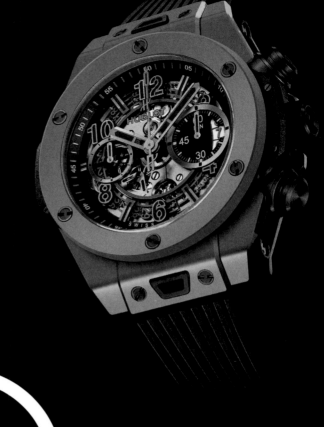

Unico Full Magic Gold, 2015

Big Bang

Unico Italia Independent, 2015

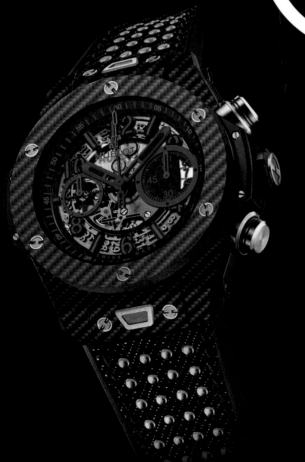

All Black Carbon, 2011

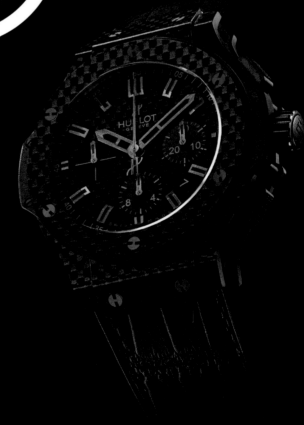

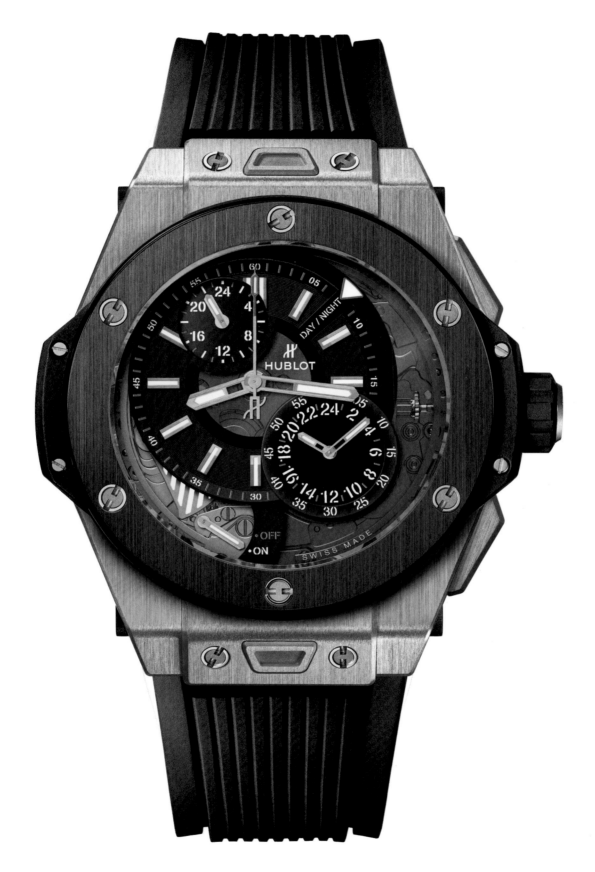

Alarm Repeater, 2015

L'art de la fusion

Alors que l'industrie horlogère suisse traverse l'une de ses plus graves crises, Carlo Crocco, descendant d'une dynastie d'horlogers milanais, tente en 1976 une aventure risquée en s'établissant à son compte. Cet homme alors âgé de 32 ans ne se laisse en rien détourner de ses plans audacieux et fonde MDM Genève. Le sigle MDM ou « Marie-Danielle Montres » (MD comme Marie-Danielle, le prénom de sa femme), sera plus tard traduit par la « Montre des Montres ». Il s'agit alors de concevoir un garde-temps sportif au bracelet en caoutchouc durable et hautement résistant. Rien de pareil n'existant encore à l'époque, il faut attendre le Salon de l'horlogerie de Bâle en 1980 pour assister à la présentation de cette première œuvre. Des fibres d'acier tissées donnent au bracelet en caoutchouc d'un nouveau type une grande résistance à l'usure. Les clients sensibles aux parfums se laissent en outre envoûter par un arôme de vanille marqué. Le boîtier naît quant à lui du refus obstiné de s'aligner sur la concurrence. Le nom de cette première création est suggéré par sa forme : le hublot. Ce style reconnaissable entre tous rencontre un accueil favorable.

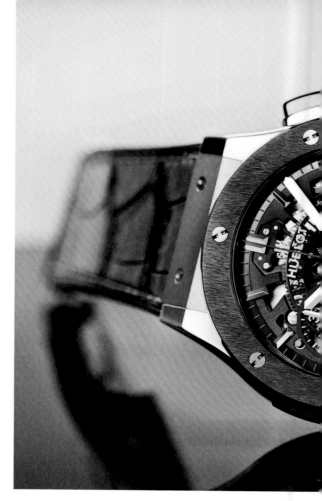

Des membres de familles royales, parmi lesquels Juan Carlos d'Espagne, se portent acquéreurs d'un garde-temps portant la signature Hublot. Le fondateur de l'entreprise persiste obstinément jusqu'en 2003 dans son projet de créer une ligne de montres emblématique, déclinée selon diverses variantes optiques et techniques. Mais Carlo Crocco, convaincu d'avoir créé un produit résolument éternel et indémodable, finit pourtant par oublier que le temps n'est jamais figé, en particulier dans le commerce des montres. Les monoproduits ont eux aussi une demi-vie. En outre, la concurrence ne dort jamais. Elle vient à occuper le créneau supposé inviolable par Hublot et à l'en chasser. Aussi, malgré son légendaire concept de montre, Hublot se retrouve-t-il en 2004 dos au mur.

Heureusement intervient alors un homme nommé Jean-Claude Biver, que sa personnalité extrovertie et son génie du marketing opposent radicalement à Carlo Crocco. Ce Luxembourgeois, qui a travaillé durant sa carrière au sein de Audemars Piguet, Blancpain et Omega assure à partir de 2004 la survie de Hublot. En 2005, il concrétise son génial concept de fusion dans le modèle à succès « Big Bang », dont le boîtier de conception modulaire permet de combiner les matériaux les plus divers dans une montre-bracelet. Les matériaux utilisés par Biver sont tous révolutionnaires : alliage résistant aux rayures « Magic Gold », céramique colorée, fibre de carbone et composite écologique à base de lin, pour n'en citer que quelques-uns. Avec l'acquisition du spécialiste des complications BNB et le développement de ses propres calibres, comme celui du chronographe « Unico », l'établisseur Hublot se mue, sous l'impulsion de Jean-Claude Biver, en une véritable manufacture horlogère, éminemment intéressante pour d'éventuels repreneurs.

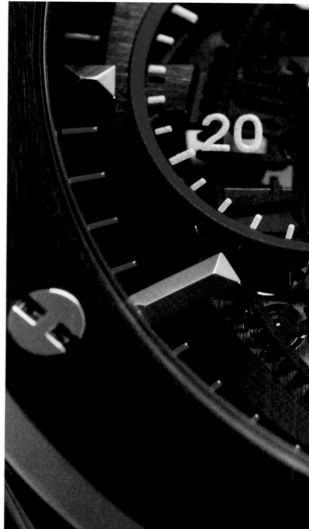

En 2008, la multinationale française du luxe LVMH fait une offre et remporte l'affaire pour 500 millions de francs suisses. 400 millions vont à Carlo Crocco, le reste à Jean-Claude Biver, qui est peu à peu parvenu à détenir 20 pour cent du capital de la société. De retraite, il n'est cependant encore pas question : ce visionnaire qui a redressé la société a encore bien des projets pour Hublot. Ainsi, il envisage par exemple de renforcer la ligne de montres « Classic Fusion », encore une fois sportive, mais sans conteste plus élégante que la ligne « Big Bang », laquelle célèbre en 2015 son dixième anniversaire au meilleur de sa forme. Dans l'intervalle, Jean-Claude Biver réalise un vœu formulé dès 2004 : devenir partenaire de Ferrari. Avec la première « Big Bang », expressément conçue pour la marque automobile sportive italienne, il s'était une première fois cassé les dents auprès des Italiens, qui l'avaient ignoré. En 2012, la collaboration a marché comme sur des roulettes. On ne peut imaginer plus beau compliment pour la renaissance de cette marque horlogère. ○

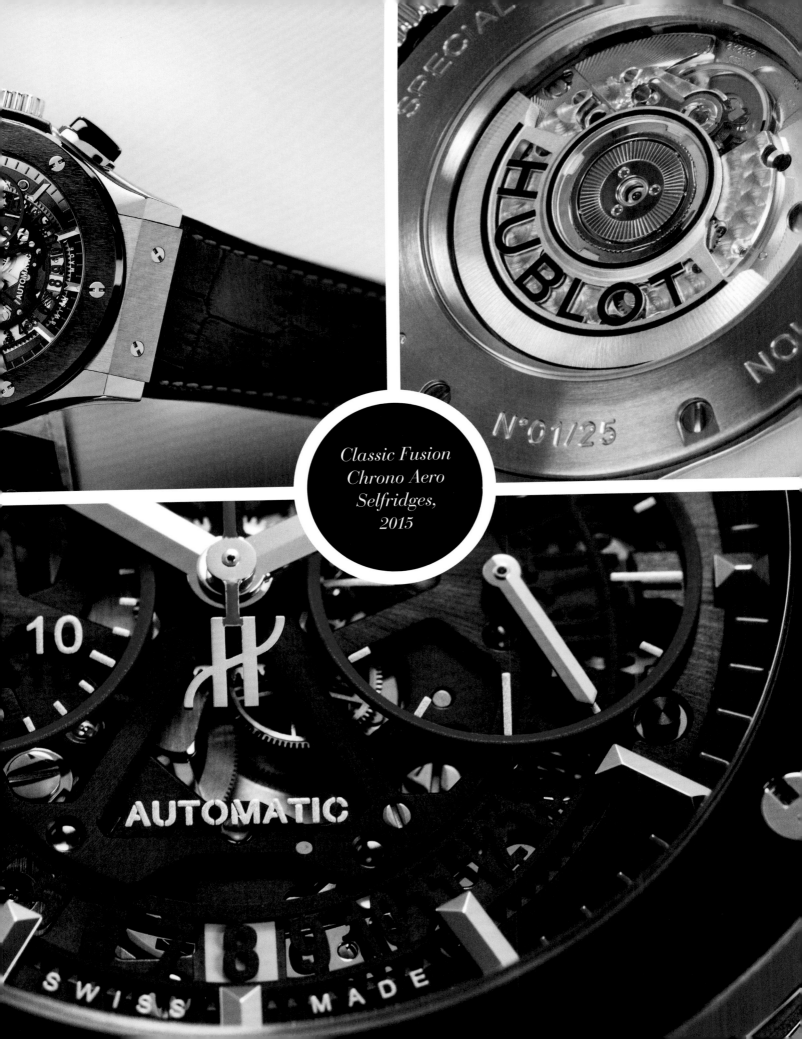

*Classic Fusion
Chrono Aero
Selfridges,
2015*

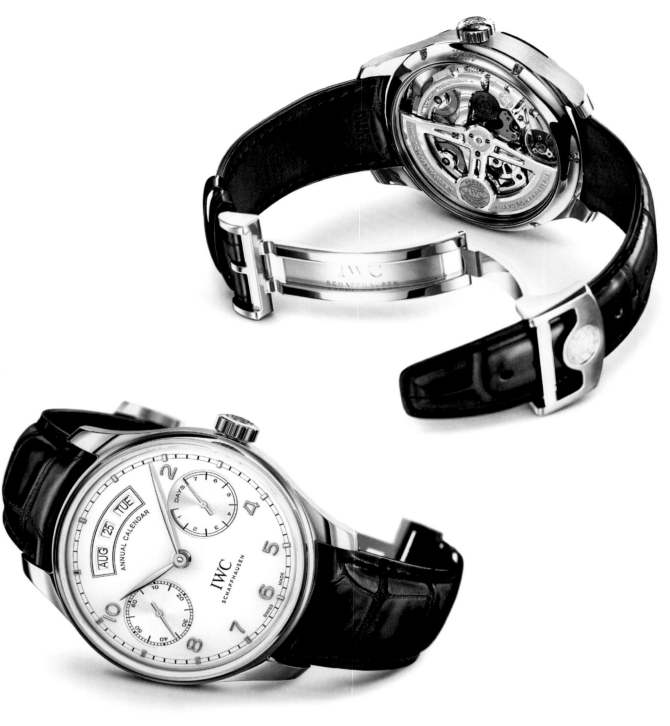

IWC
Schaffhausen

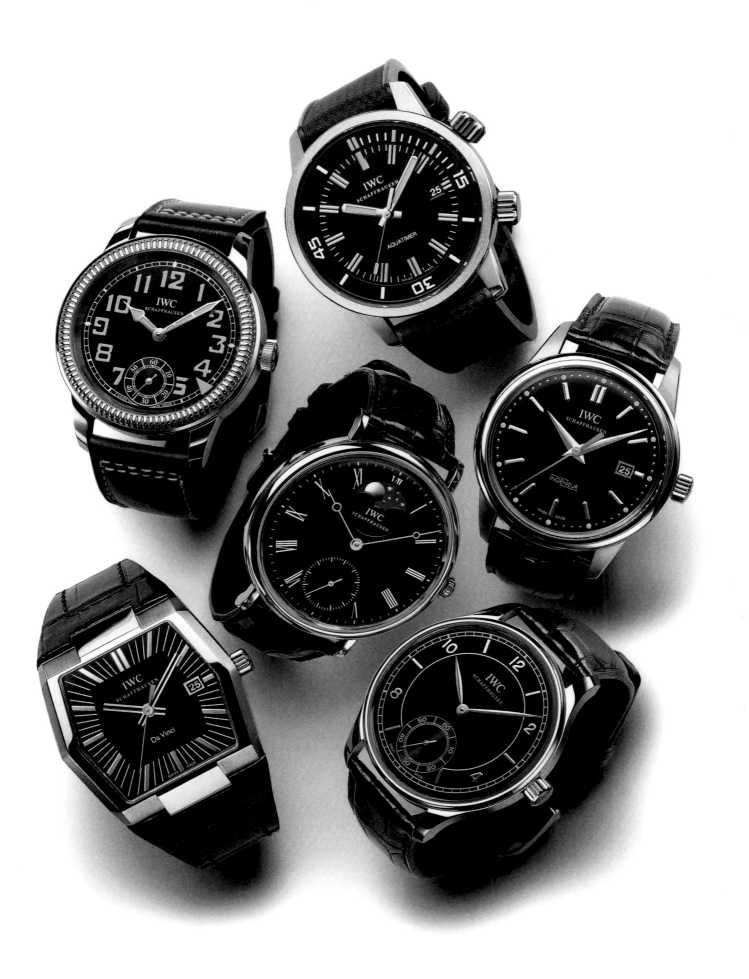

Probus Scafusia—Good Things from Schaffhausen

English

The early historical development of the International Watch Company followed a meandering path. Better known as IWC, the young watch manufactory suffered two setbacks. And the fact that its founder was a US citizen was quite unusual in those years. Likewise peculiar was the choice of Schaffhausen, a town with little or no watchmaking tradition, to be the venue for a company that Florentine Ariosto Jones established in 1868 with the goal of inexpensively manufacturing precise pocket-watches in Europe for subsequent export to the American market. The entrepreneur, who had come to Switzerland from Boston, had devoted plenty of fore-thought to his bold project. Jones had heard that for commercial purposes, the successful businessman Johann Heinrich Moser had harnessed the potential energy of the Rhine River, which plummets over a waterfall near Schaffhausen. Jones settled nearby in 1869 and two years later IWC announced the completion of its first watch movements, which were named after the brand's daring founder.

The so-called "Jones" calibers are avidly sought collector's items. It seems that they were available in seven distinct versions which differed in their dimensions, their architectures (Lépine or Savonnette) and their winding mechanisms (via key or crown). Alongside the constructive differences, the products were available in up to eight different quality grades. Jones dreamt of annually exporting at least 10,000 pocket-watches encasing these movements to the New World, but a liquidity bottle-neck brought him close to bankruptcy in 1875. To escape the unpleasant consequences, Jones absconded rather dishonorably. His compatriot Ferdinand F. Seeland, who was charged with restructuring the company, likewise met with ill luck: his shortsighted choices again plunged IWC toward ruination and like his predecessor, he too ran off. Now the man of the hour was

Johannes Rauschenbach, a Swiss investor who transformed the faltering company into a horological pearl that remained largely in family ownership until 1978, when IWC was suffering a severe crisis precipitated by the Quartz Revolution. The German tachometer manufacturer VDO came onto the scene at this time. Mannesmann acquired VDO in 1991 and the Richemont Group invested large sums of money in IWC, Jaeger-LeCoultre, and A. Lange & Söhne in July 2000.

Despite recurrent turbulences, IWC repeatedly contributed important impulses to precise mechanical watchmaking during these years. For example, the brand's history is associated with a large number of impressive pilot's watches, which IWC first marketed in 1935. A few of the highlights include the "Große Fliegeruhr" ("Big Pilot's Watch") which was unveiled with a 55-mm-diameter case in 1940 and the legendary "Mark XI," which debuted in 1948.

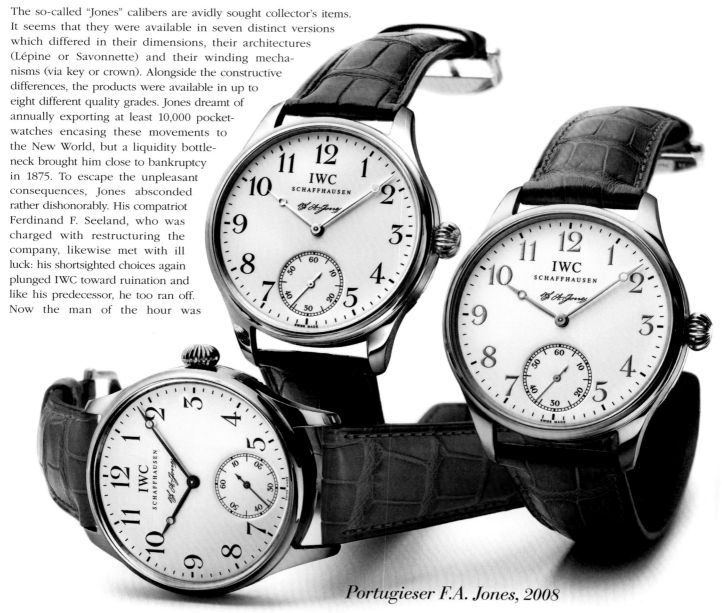

Portugieser F.A. Jones, 2008

Specimens of this model destined for the Royal Air Force bore on their dials the so-called "Broad Arrow," a historical British symbol. Before they left the manufactory, all timepieces of this type were subjected to a 44-day test program for "Navigator Wrist Watches." The examiners tested the accuracy of their timekeeping in five positions and at temperatures between –5° and +46° Celsius. The "Da Vinci," which was unveiled in 1985, was the world's first self-winding wristwatch with a chronograph and a perpetual calendar. The highly complicated "Warhorse from Schaffhausen," better known as "Il Destriero Scafusia," debuted in 1993. All this and more can be admired at IWC's museum. The exhibition is housed on the ground floor of the firm's headquarters, which is a protected architectural landmark. Visitors can simultaneously experience IWC here: the museum's west wing is devoted to the company's first 100 years, while its eastern counterpart represents the successful watches of IWC's recent history, in which manufactory craftsmanship and the brand's own movements again enjoy high priority.

Each "Portugieser Sidérale Scafusia" is built only by special request. Amassing a 96-hour power reserve, this watch has indicators on both sides of its case. A large tourbillon dominates the front and carries an escapement that works with constant force. Mean solar time is shown by a second and smaller pair of hands at "12 o'clock." The back of the case boasts an artistically rendered starry firmament that rotates in synchrony with sidereal time. Each emulated sky differs depending on the geographical location of the owner's home, so an expert individually calculates each star map. A pair of polarizing filters gives a grayish hue to the daytime sky and a blue color to the nighttime heavens. A yellow zone demarcates the horizon of visibility and a red line indicates the ecliptic. The times of sunrise and sunset naturally depend on the location of the future owner's home. The back of the watch also shows the average solar time and daylight saving time. To facilitate correct setting, the timepiece includes a perpetual calendar that shows the number of days that have elapsed since the beginning of each year and also includes the cycle of leap years in its calculations. Circa ten years were needed to develop the movement, which combines more than 520 components.

Energy provided by a waterfall in the Rhine River first gave an essential basis for ecologically impeccable watch manufacturing in Schaffhausen in 1868. Heat pumps and the use of groundwater have contributed to a positive energy footprint since 2005. Thanks to innovative technologies, the manufactory can operate almost entirely without the consumption of fossil fuels. ○

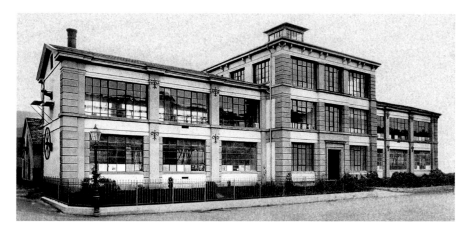

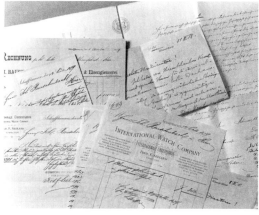

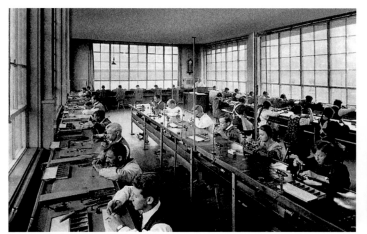

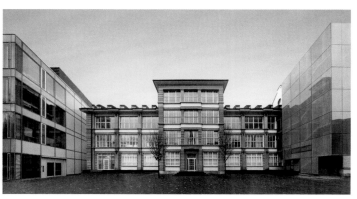

IWC production site—yesterday and today ○ *Die Produktionsstätte von IWC – damals und heute* ○ **Manufacture d'IWC – hier et aujourd'hui**

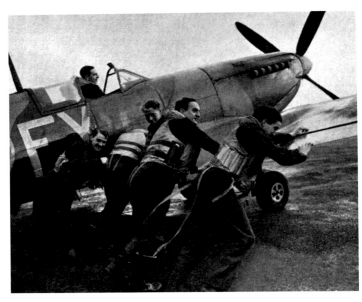

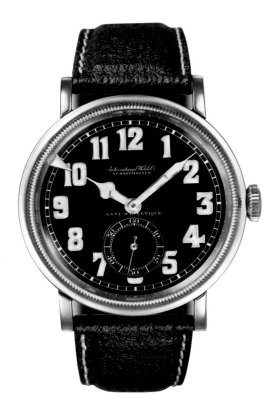

Top from left: at the takeoff of a Spitfire ◦ Special watch for pilots, 1939 ◦
Oben von links: Beim Start einer Spitfire ◦ Spezialuhr für Flieger, 1939 ◦
En haut depuis la gauche : Lors du décollage d'un Spitfire ◦ Montre
spéciale pour aviateur, 1939

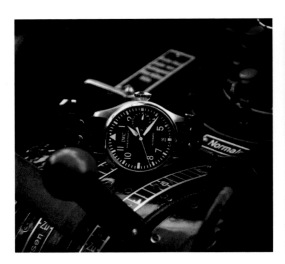

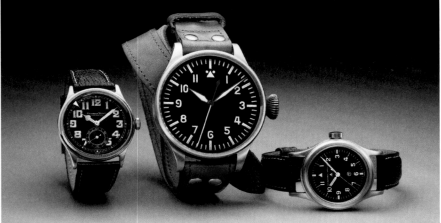

Top: Big Pilot, 2006 ◦ Top Right: Pilot's Watch,
1939; Big Pilot's Watch, 1940; Mark XI, 1948 ◦
Bottom: Pilot's Watch Chronograph Edition
Antoine de Saint Exupéry, 2006

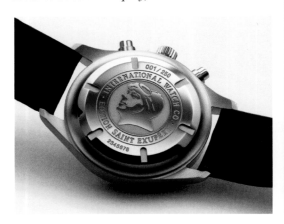

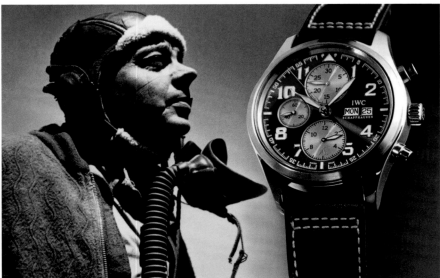

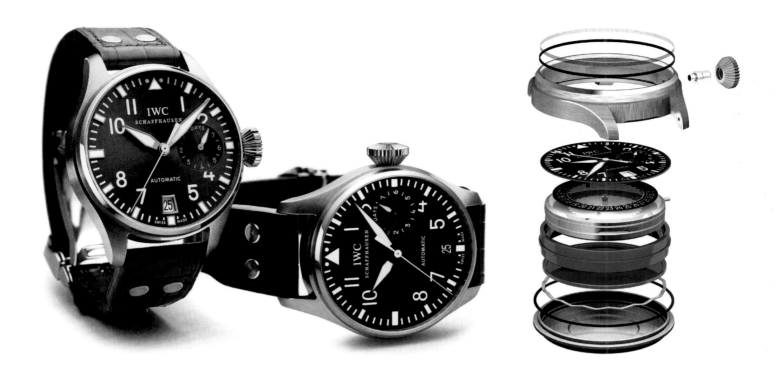

Top: Big Pilot, white gold and steel, 2006 ∘ Bottom: Mark XVI Pilot Spitfire, 2006

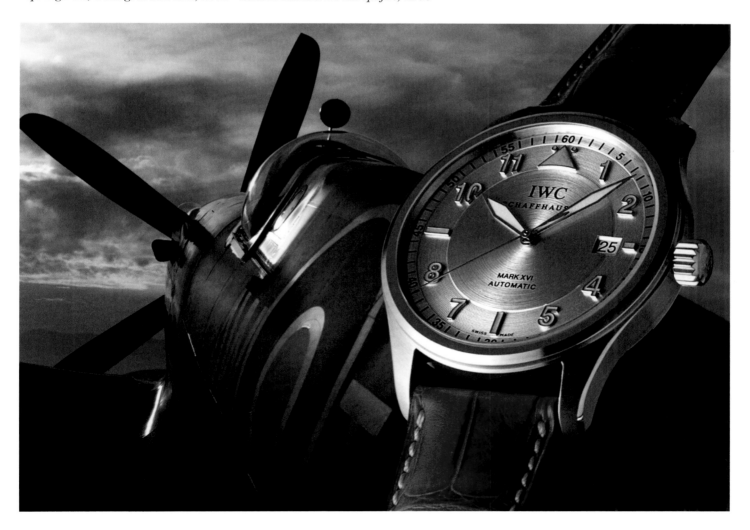

Probus Scafusia – Gutes aus Schaffhausen

Deutsch

Geradlinig ist genau das Gegenteil dessen, wie die frühe geschichtliche Entwicklung der International Watch Company verlief. Den Anfang der besser als IWC bekannten Uhrenmanufaktur markieren nämlich zwei Pleiten. Auch die US-amerikanische Nationalität des Firmengründers mutet für damalige Verhältnisse eher ungewöhnlich an. Bleibt die Wahl des Firmensitzes. 1868, als ein gewisser Florentine Ariosto Jones den Entschluss fasste, Präzisionstaschenuhren für den amerikanischen Markt vergleichsweise kostengünstig in Europa produzieren zu lassen, war Schaffhausen in chronometrischen Dingen ein völlig unbedeutender Ort. Gleichwohl handelte der aus Boston stammende Unternehmer keinesfalls unüberlegt. Ihm war zu Ohren gekommen, dass der erfolgreiche Geschäftsmann Johann Heinrich Moser das Energiepotenzial des Rheinfalls, der bei Schaffhausen in die Tiefe stürzt, für gewerbliche Zwecke nutzbar gemacht hatte. 1869 zog Jones bei ihm ein, und schon zwei Jahre später verkündete IWC die Fertigstellung erster Uhrwerke, benannt nach der schillernden Gründerpersönlichkeit.

Die sogenannten Jones-Kaliber stehen bei Sammlern hoch im Kurs. Insgesamt muss es wohl sieben verschiedene Ausführungen gegeben haben. Diese unterscheiden sich durch ihre Größe, Lépine- oder Savonnette-Bauweise und schließlich die Art des Aufzugs per Schlüssel oder Krone. Zu den konstruktiven Unterschieden gesellte sich schließlich eine Graduierung in bis zu acht Qualitätsstufen. Von hehrem Wunschdenken geprägt war freilich der Plan, jährlich mindestens 10 000 Taschenuhren mit diesen Werken in die Neue Welt zu liefern. Mangels Liquidität stand 1875 ein erster Konkurs vor der Tür, dessen unliebsamen Konsequenzen sich Jones durch wenig ehrenhafte Flucht entzog. Auch der als Sanierer eingesetzte Ferdinand F. Seeland, abermals amerikanischer Provenienz, besaß keine Fortune. Durch sein wenig zukunftsweisendes Handeln stürzte er die IWC abermals ins Verderben. Danach machte auch er sich aus dem Staub. Nun schlug die Stunde des Schweizers Johannes Rauschenbach. Er kaufte die marode Firma und machte daraus eine uhrmacherische Perle, die bis 1978 weitgehend in Familienbesitz blieb. In besagtem Jahr – IWC befand sich in einer tiefen, durch die Quarzuhren-Revolution ausgelösten Krise – trat der deutsche Tachometerhersteller VDO auf den Plan. 1991 hieß die VDO-Herrin Mannesmann und im Juli 2000 machte der Richemont-Konzern sehr viel Geld locker für IWC, Jaeger-LeCoultre sowie A. Lange & Söhne.

Ungeachtet der teilweise sehr turbulenten Jahre haben die Schaffhauser der mechanischen Präzisionsuhrmacherei viele positive Impulse erteilt. Beispielsweise verknüpft sich mit ihrer Geschichte eine Vielzahl beeindruckender Fliegerarmbanduhren. Spezifische Konstruktionen sind seit 1935 am Markt vertreten.

Zu den Highlights gehören die „Große Fliegeruhr" mit 55 Millimetern Gehäusedurchmesser, vorgestellt 1940, oder die legendäre „Mark XI", deren Geburtsstunde im Jahr 1948 schlug. Für die Royal Air Force bestimmte Exemplare trugen auf dem Zifferblatt den sogenannten „Broad Arrow", ein historisches britisches Pfeilsymbol. Ausnahmslos alle Zeitmesser dieses Typs mussten vor der Lieferung ein 44-tägiges Testprogramm für „Navigator Wrist Watches" durchlaufen. Dabei wurden sie in fünf Positionen sowie bei Temperaturen zwischen −5 und +46 °C auf ihre Ganggenauigkeit hin überprüft. „Da Vinci" hieß 1985 der weltweit erste Armbandchronograph mit automatischem Aufzug und ewigem Kalender. Besonders kompliziert präsentierte sich 1993 das „Schlachtross aus Schaffhausen", besser bekannt als „Il Destriero Scafusia". Das alles und noch mehr gibt es im firmeneigenen Museum zu bestaunen. Die Ausstellung findet sich im Erdgeschoss des denkmalgeschützten Stammhauses. Wer sie besucht, erlebt gleichzeitig auch die IWC. Den ersten hundert Jahren der Geschichte widmet sich der westliche Museumsflügel. Das östliche Pendant repräsentiert die erfolgreiche Uhren-Neuzeit, in der Manufakturarbeit mit eigenen Uhrwerken wieder einen hohen Stellenwert genießt.

Nur auf persönliche Bestellung entsteht die „Portugieser Sidérale Scafusia" mit 96 Stunden Gangautonomie sowie Anzeigen auf Vorder- und Rückseite. Vorne dominiert ein großes Tourbillon. Die darin verbaute Hemmung arbeitet mit konstanter Kraft. Ein zweites, kleines Zeigerpaar bei der 12 bildet mittlere Sternzeit ab. Die Rückseite besticht mit einem kunstvoll gestalteten Sternenhimmel, der sich im Takt der Sternzeit dreht. Weil der jeweils sichtbare Himmel vom Wohnort abhängt, berechnet ein Gelehrter das Abgebildete ganz individuell. Dank zwei Polarisationsfiltern zeigt sich der Taghimmel in Grau, der nächtliche in Blau. Eine gelbe Zone markiert den Sichthorizont, eine rote Linie zeigt die Ekliptik. Die Zeiten des Sonnenauf- und -untergangs beziehen sich natürlich auf den jeweiligen Wohnort des künftigen Besitzers. Schließlich bildet die Rückseite nochmals die mittlere Sonnenzeit und die Sommerzeit ab. Zum korrekten Einstellen gibt es ein ewiges Kalendarium, das die Zahl der Tage seit Jahresbeginn anzeigt und den Schaltjahreszyklus einberechnet. Die Entwicklung des aus mehr als 520 Teilen zusammengefügten Uhrwerks beanspruchte übrigens rund zehn Jahre.

1868 bildete das Energiepotenzial des Rheinfalls eine wesentliche Grundlage für eine ökologisch einwandfreie Uhrenfertigung in Schaffhausen. Seit 2005 tragen Wärmepumpen und Grundwassernutzung stark zu einer positiven Energiebilanz bei, denn durch den Einsatz innovativer Technologien kann die Manufaktur weitgehend auf fossile Brennstoffe verzichten.

Wenn diese Uhr stehen bleibt, rufen Sie bitte einen Arzt. Denn dann haben Sie sich sieben Tage nicht bewegt.

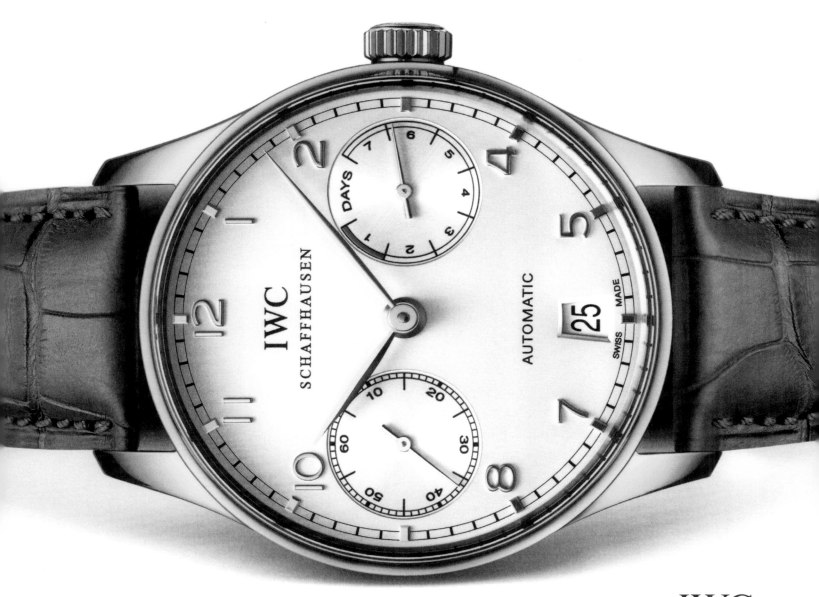

IWC
SCHAFFHAUSEN
SINCE 1868

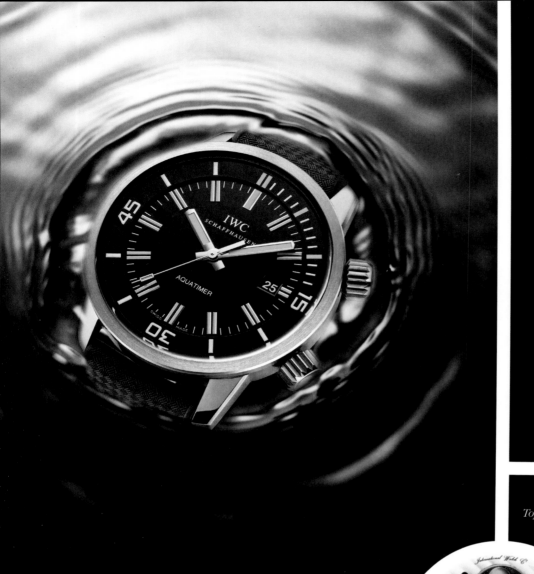

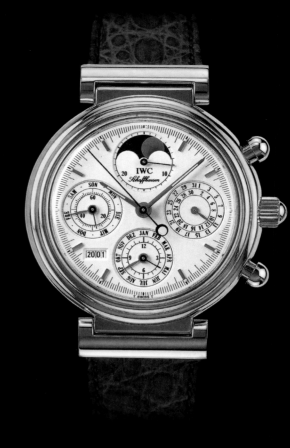

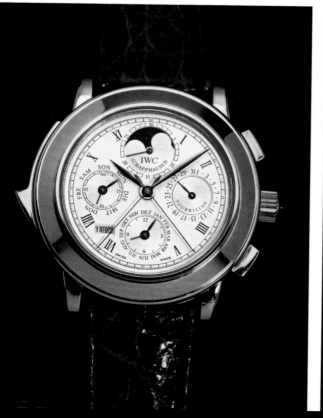

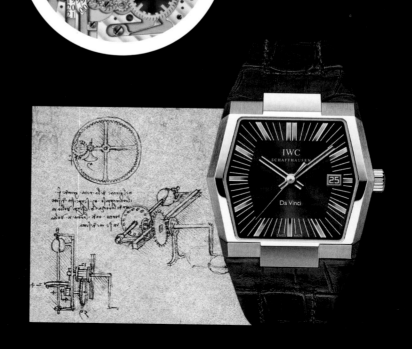

Top left: Aquatimer, 2008
Top right: Da Vinci Perpetual Calendar, launch 1985
Middle: dial with calendar
Bottom left: Il Destriero Scafusia, 1992
Bottom right: Da Vinci
Automatic, 2008

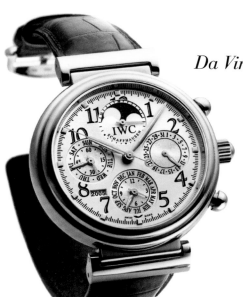

Da Vinci Perpetual Calendar

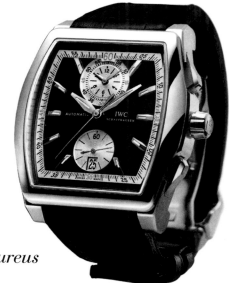

Da Vinci Chrono Laureus

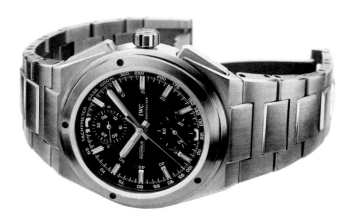

Ingenieur Chronograph, 2005

Portugieser Perpetual Calendar
Double Moon, 2015

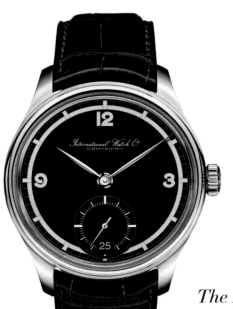

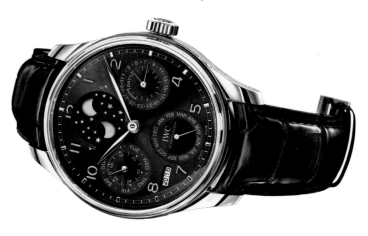

The Portugieser Hand-Wound Eight Days,
75ᵗʰ Anniversary Edition, 2015

Probus Scafusia – produit éprouvé de Schaffhouse

Français

Dans les premiers temps, le parcours d'International Watch Company est tout sauf linéaire. Les débuts de la manufacture horlogère mieux connue sous le sigle IWC sont de fait marqués par deux faillites. La nationalité américaine du fondateur semble par ailleurs assez inhabituelle pour l'époque. Reste le choix du siège. En 1868, lorsqu'un dénommé Florentine Ariosto Jones décide de faire fabriquer en Europe des montres de poche de précision à des prix relativement avantageux pour le marché américain, Schaffhouse – comme les Suisses romands appellent Schaffhausen – est un site totalement insignifiant sur le plan horloger. Néanmoins, cet entrepreneur originaire de Boston n'agit nullement de manière irréfléchie. Il a entendu dire que le riche homme d'affaires Johann Heinrich Moser exploite le potentiel énergétique des chutes du Rhin à Schaffhouse à des fins industrielles. En 1869, Jones installe son usine dans un bâtiment industriel appartenant à Moser. Tout juste deux ans plus tard, IWC annonce la fabrication de ses premiers mouvements, qui reçoivent leur nom de leur fondateur aux multiples facettes.

Très prisés des collectionneurs, les calibres Jones seront produits en pas moins de sept versions différentes. Elles se différencient par la taille, le mouvement, Lépine ou Savonnette, et enfin le type de remontoir, à clé ou à couronne. À ces différences dans le mode de construction viendront s'ajouter un classement comptant jusqu'à huit niveaux de qualité. Le projet visant à livrer chaque année au Nouveau Monde au moins 10 000 montres de poche équipées de ces mouvements est assurément empreint d'un optimisme irréaliste. Un défaut de liquidités entraîne en 1875 la première faillite, avec ses désagréables conséquences, auxquelles Jones se soustrait par une fuite peu glorieuse. L'homme chargé de redresser la situation, Ferdinand F. Seeland, est lui aussi américain. Mais il n'est pas fortuné et son action sans vision pour l'avenir fait à nouveau plonger IWC dans le désastre. Lui aussi s'évanouit dans la nature. C'est alors qu'entre en scène le Suisse Johannes Rauschenbach. Il rachète la société en faillite pour en faire un fleuron de haute horlogerie, qui restera jusqu'en 1978 dans une large mesure dans sa famille. Cette même année – IWC traverse alors une grave crise liée à la révolution des montres à quartz – VDO, fabricant allemand de tachymètres, rachète la manufacture. En 1991, VDO est repris par Mannesmann. En juillet 2000, le groupe Richemont débourse de grosses sommes pour IWC, Jaeger-LeCoultre et A. Lange & Söhne.

En dépit de ces années en partie très tumultueuses, les représentants schaffhousois de l'horlogerie mécanique de précision seront à l'origine de bon nombre d'impulsions positives. Leur histoire est par exemple liée à un nombre impressionnant de montres-bracelets d'aviateur. Des constructions spécifiques sont présentes sur le marché dès 1935. Au rang des produits phares figurent « La Grande Montre d'Aviateur », habillée d'un boîtier de 55 millimètres de diamètre et présentée en 1940, ainsi que la légendaire « Mark XI », créée en 1948. Des exemplaires destinés à la Royal Air Force arborent sur leur cadran le « Broad Arrow » ou phéon, symbole héraldique britannique en forme de flèche. Les garde-temps de ce type sont tous soumis sans exception avant livraison à un programme d'essais de 44 jours pour « Navigator Wrist Watches » ou montres-bracelets de navigation. Leur précision de marche est alors vérifiée dans cinq positions différentes et à des températures comprises entre −5 et +46 °C. En 1985 paraît « Da Vinci », le premier chronographe à bracelet, à remontage automatique et à calendrier perpétuel. Créé en 1993, « Le destrier de Schaffhausen », plus connu sous le nom « Il Destriero Scafusia », affiche une extrême complexité. On peut admirer ces modèles et bien d'autres au musée de la manufacture. L'exposition se tient au rez-de-chaussée du bâtiment principal, classé monument historique. En le visitant, vous vivrez l'histoire d'IWC. L'aile ouest couvre les cent premières années. L'aile est

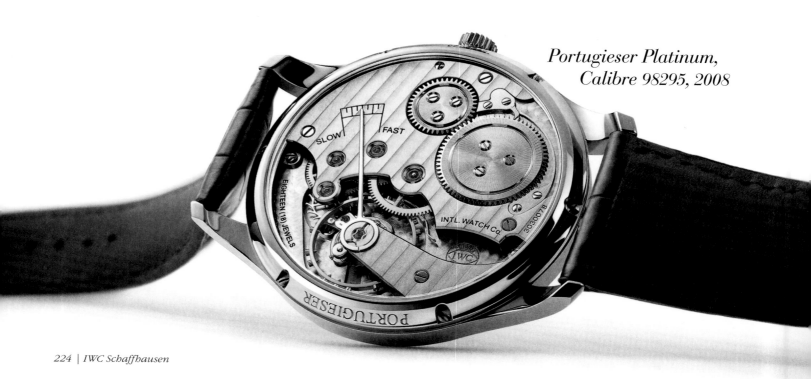

Portugieser Platinum,
Calibre 98295, 2008

retrace la nouvelle ère du succès pour les montres, période au cours de laquelle le travail avec des mouvements fabriqués dans la manufacture revêt à nouveau une grande importance.

Disposant de 96 heures de réserve de marche et d'affichages sur la face comme au revers, la « Portugieser Sidérale Scafusia » est fabriquée uniquement sur commande et selon les desiderata du client. Le cadran est dominé par un grand tourbillon, dans lequel est monté un échappement à force constante. Une deuxième petite paire d'aiguilles à 12 heures indique le temps sidéral moyen. Le revers est fascinant, avec son disque céleste artistiquement ouvragé et tournant au rythme du temps sidéral. Comme ce que le client voit du ciel dépend de son lieu de résidence, la carte du ciel est calculée individuellement. Grâce à deux filtres polarisants, le ciel diurne apparaît en gris, tandis que le ciel nocturne s'affiche en bleu. Une zone jaune marque l'horizon apparent, une ligne rouge indique l'écliptique. Les heures du lever et du coucher du soleil se réfèrent bien sûr au lieu de résidence du futur propriétaire de la montre. Enfin, le revers indique encore une fois le temps solaire moyen et l'heure d'été. Pour garantir un réglage correct, un calendrier perpétuel indique le nombre de jours écoulés depuis le début de l'année et tient compte des années bissextiles. Une dizaine d'années a été nécessaire au développement de ce mouvement composé de 520 pièces.

En 1868, le potentiel énergétique des chutes du Rhin a beaucoup compté dans la fabrication de montres respectueuses de l'environnement à Schaffhouse. Depuis 2005, les pompes à chaleur et l'utilisation des eaux souterraines contribuent fortement à un bilan énergétique positif, car le recours à ces technologies innovantes permet à la manufacture de se passer dans une large mesure de combustibles fossiles. ○

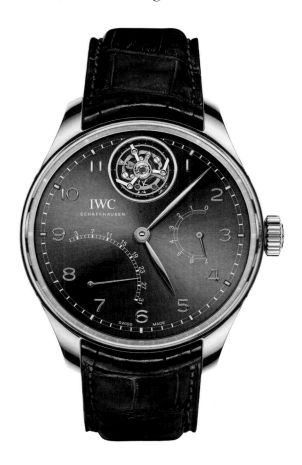

Portugieser
Perpetual Calendar, 2002

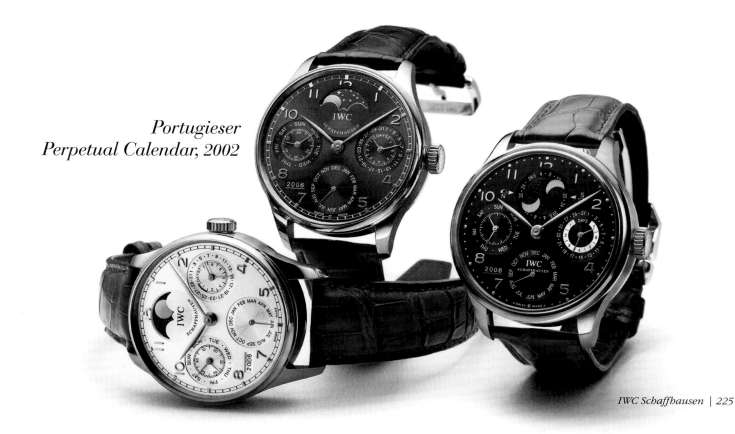

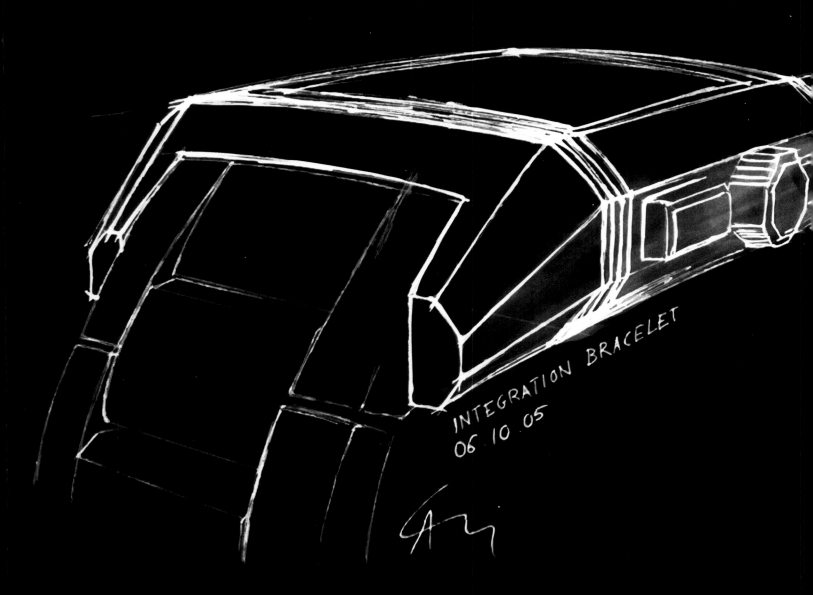

INTEGRATION BRACELET
06 10 05

Jaeger-LeCoultre

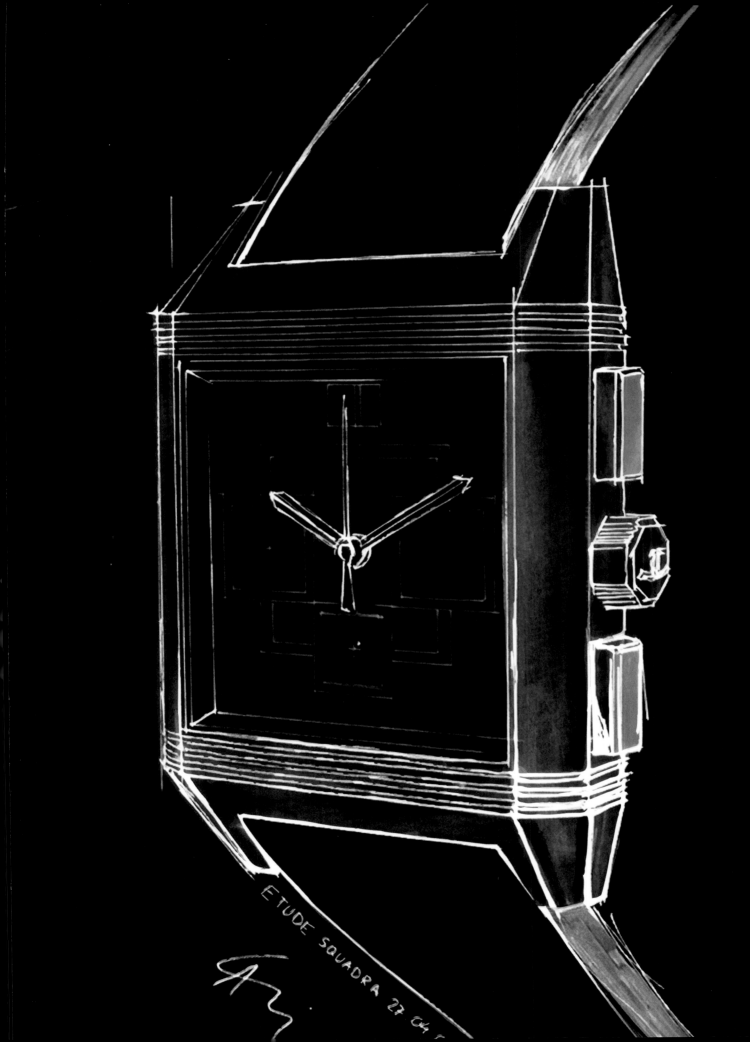

ETUDE SQUADRA 27 04

Manufactory Work in the Vallée de Joux

Four words succinctly describe the watches that Jaeger-LeCoultre makes nowadays: "everything from the manufactory," which has stood at the same location ever since the brand was founded in 1833. Namely, in Le Sentier, 3,300 feet above sea level in the Vallée de Joux, where an inventive 30-year-old mechanician and craftsman named Charles-Antoine LeCoultre began fabricating gears. Inspired by an insatiable passion for precise tools and mechanisms, he soon presented his first gear-making machines. A decisive contribution toward greater precision was made by the so-called "millionometer," which enabled watchmakers to measure components to the nearest 1,000th of a millimeter. At the Great Exhibition in London in 1851, Charles-Antoine LeCoultre displayed a remarkable variety of merchandise, including tools, pinions, gears, ébauches, and pocket-watch chronometers. The leading Swiss watch brands bought his creative products, so LeCoultre soon became the largest employer in the economically underdeveloped region. Chronographs, alarms, calendars, repeater striking trains, and tourbillons numbered among the complicated movements that the sons of the firm's founder began designing and fabricating in 1877. According to in-house statistics, some 60,000 movements equipped with diverse complications left the factory between 1860 and 1925. Unconventional items regularly appeared on the market afterwards, e.g., the double-level "Duoplan" movement that debuted in 1925. An unprecedentedly small mechanical movement was ready for serial manufacturing in 1929: Caliber 101 combines 74 components and measures a mere 14 by 4.8 by 3.4 millimeters; together with its dial and hands, this 228.48 cubic millimeter microcosm weighs just 0.9 gram. Meanwhile, British officers in faraway India complained to a watch importer that the protective crystals above their watches' dials repeatedly shattered when they played polo. Their criticisms bore fruit and led in 1931 to the creation of the "Reverso," a wristwatch with a swiveling case. In the heat of equestrian battle, the wearer could simply turn this watch's fragile face toward his wrist so the case's sturdy steel back would be exposed to whatever impacts came its way.

Jaeger in Paris and LeCoultre in Le Sentier put a seal on their long-term collaboration in 1937, when they officially introduced the Jaeger-LeCoultre brand name for finished watches. Jaeger-LeCoultre surprised aficionados after the Second World War with the 1953 debut of the "Futurematic," a wristwatch with a power-reserve display and an automatic winding mechanism which was so efficient that the crown could be hidden on the back of the case and the wearer could entirely dispense with manual winding. The trailblazing "Movado," a nearly inaudible wristwatch with an alarm mechanism, was available with manual winding in 1951 and with automatic winding in 1959. The Quartz Revolution in the early 1970s brought Jaeger-LeCoultre to a low point in its long history. The saviors came from Germany, which the Swiss jocularly describe as "the large canton." The German VDO took over 60 percent of the share capital in 1978; the remainder went to Audemars Piguet, a longstanding buyer of Jaeger-LeCoultre's movements. The visionary Günter Blümlein came aboard to restructure the business in 1986. In 1991, the tachometer manufacturer VDO came under the roof of the Mannesmann Group, which combined its watch-related activities (IWC, Jaeger-LeCoultre, and A. Lange & Söhne) in the LMH-Holding (Les Manufactures Horlogères) in 1996. As a member of this group, Jaeger-LeCoultre was taken over by the Richemont Group in July 2000. Prior to this gigantic deal, Jaeger-LeCoultre had informed its ébauche customers that they could no longer expect deliveries in the foreseeable future. On the other hand, Jaeger-LeCoultre consistently expanded its own manufactory activities and thus also the development of highly complicated movements.

The latter would have been unthinkable without specialized competence in diverse and highly dissimilar fields. The complexity increases almost exponentially, especially when horological complications are involved. The roofs of the various parts of Jaeger-LeCoultre's building in Le Sentier accordingly shelter floor space totaling 270,000 square feet on which artisans with some 180 different professional qualifications are employed. Among other specialists, the workforce includes engineers, technicians, watchmakers, creative designers, artists, artisans, and craftspeople. It goes without saying that Jaeger-LeCoultre makes its own sophisticated levers and delicate balance-springs for its escapement system. Equally skilful are the men and women who decorate the movements, dials, and cases. On premises where each artisan pursues his or her own special métier, environmental protection takes high priority. The facility uses solar energy to heat its water. The machines are cooled via a closed water cycle and by free cooling: the latter is a system for recycling fresh air. These and similar measures have earned Jaeger-LeCoultre the Minergie certificate and the Hydro Locale label, which confirms that renewable water power provides the necessary energy.

In accord with the motto that "trust is good, but verification is better," each and every wristwatch is tested for a total of 41 days and 16 hours (i.e., 1,000 hours) before it's allowed to leave the manufactory. Unlike the official Swiss chronometer tests, these examinations scrutinize the completely finished timepiece, including its dial, hands, and case. The tests are performed in six positions, at various temperatures, and in conformity with explicitly defined parameters. Strong shocks are also part of the ordeal. This guarantees that the ticking candidates will be subjected to rougher treatment in this first epoch of their lives than they will be likely to encounter later on their wearers' wrists around the world. ◦

Dial, viewed from below, with wreath of numerals for the patented outsize date display
Zifferblattansicht von unten mit dem patentierten Großdatums-Zahlenkranz
Vue de dessous du cadran avec couronne de chiffres pour l'affichage grande date

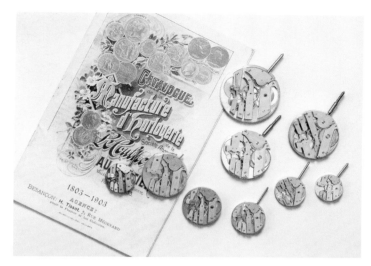

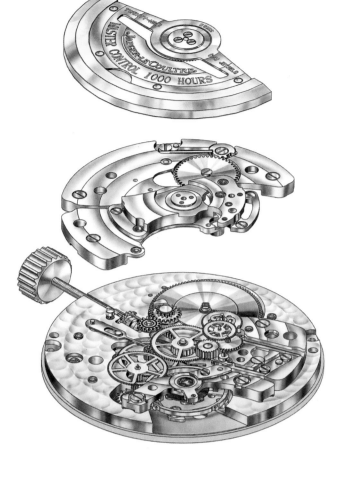

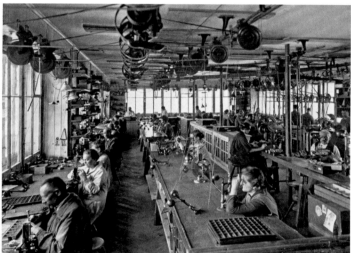

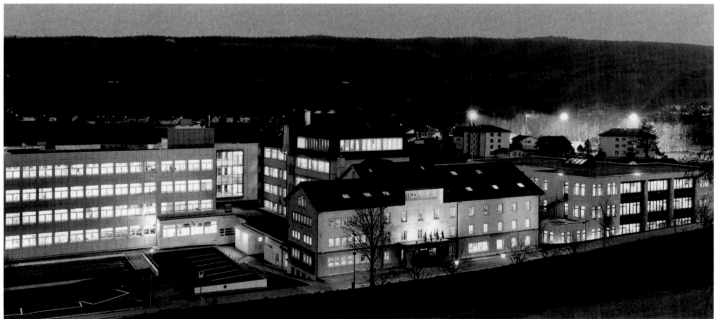

Clockwise from top left: catalogue from 1903 with nine ébauches ◦ Automatic-winding subassembly of the caliber 975 ◦ Factory building in 2010 ◦ Assembly hall in the early 20th century ◦ Von oben links im Uhrzeigersinn: Katalog von 1903 mit neun Rohwerken ◦ Automatikgruppe des Kalibers 975 ◦ Fabrikgebäude 2010 ◦ Montagehalle Anfang 20. Jahrhundert ◦ Sens horaire depuis la gauche : Catalogue de 1903 avec de nouvelles ébauches ◦ Module de remontage automatique du calibre 975 ◦ Manufacture en 2010 ◦ Atelier d'assemblage au début du XXe siècle

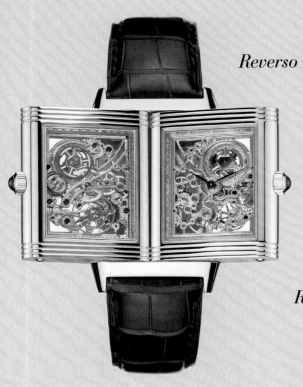

Reverso Platinum Number One, 2001

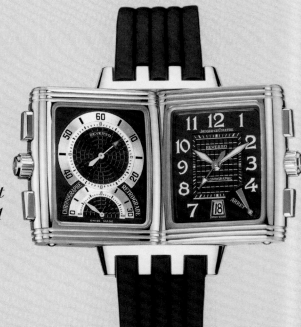

Reverso Gran'Sport
Automatic, 2001

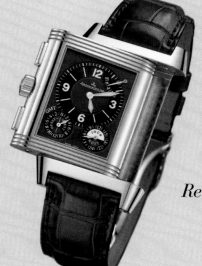

Reverso Grande GMT, 2004

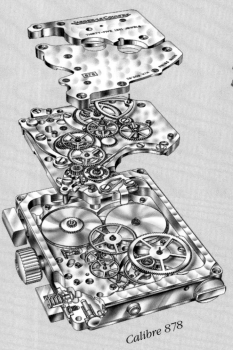

Calibre 878

Reverso à Éclipses,
Scène d'Amour, 2006

single piece

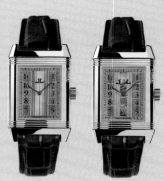

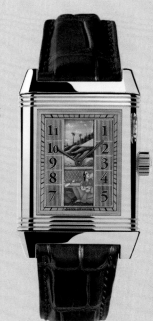

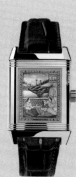

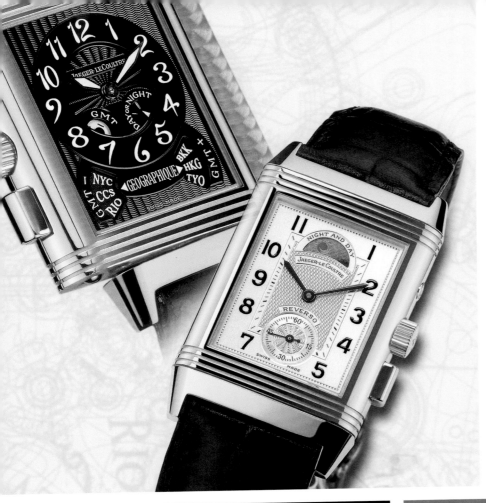

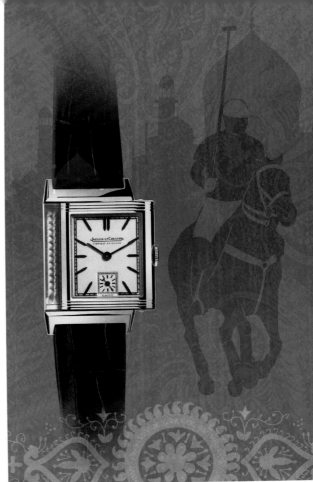

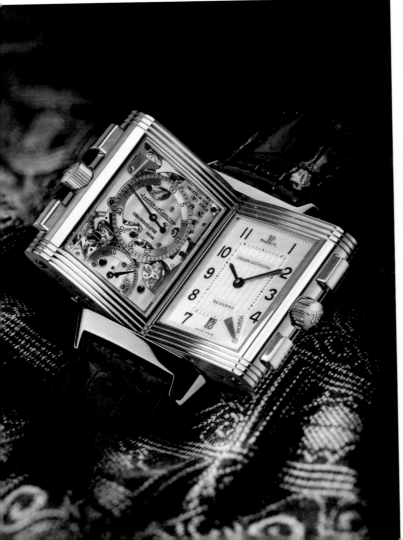

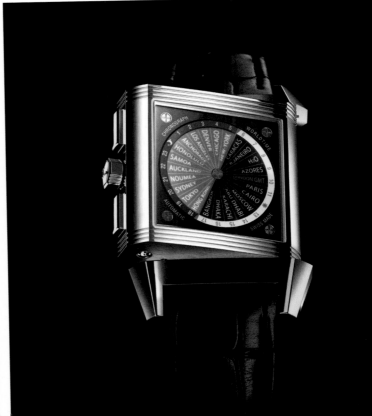

Top left: Reverso Geographique, 1998
Top right: Grande Reverso Ultra Thin, 1948, 2001
Bottom left: Reverso Chronograph, 1996
Bottom right: Reverso Squadra World Chrono Ti,
limited to 1,500 pieces, 2008

Manufakturarbeit im Vallée de Joux

Deutsch

Die heutigen Uhren von Jaeger-LeCoultre lassen sich mit nur vier Wörtern beschreiben: Alles aus eigener Manufaktur. Und die befindet sich seit der Gründung im Jahr 1833 stets am gleichen Fleckchen Erde. Gemeint ist Le Sentier im 1000 Meter hoch gelegenen Vallée de Joux. Dort begann der einfallsreiche Mechaniker und Handwerker Charles-Antoine LeCoultre im Alter von 30 Jahren mit der Fertigung von Zahnrädern. Beseelt von unstillbarer Passion für präzise Werkzeuge und Mechanismen präsentierte er schon bald erste Maschinen zur Zahnradfabrikation. Einen entscheidenden Beitrag zur Steigerung uhrmacherischer Präzision leistete das „Millionometer", mit dem die Branche erstmals auf den Tausendstelmillimeter genau messen konnte. Im Rahmen der ersten Weltausstellung 1851 in London zeigte Charles-Antoine LeCoultre bereits eine bemerkenswerte Vielfalt von Produkten: Werkzeuge, Zahntriebe, Zahnräder, Uhrenrohwerke und Taschenchronometer. Durch seine kreativen Erzeugnisse, welche die bedeutendsten Schweizer Uhrenmarken bezogen, entwickelte sich LeCoultre zum größten Arbeitgeber in der strukturschwachen Region. 1877 starteten die Söhne des Firmengründers mit der Konstruktion und Fabrikation komplizierter Uhrwerke, also Chronographen, Wecker, Kalendarien, Repetitionsschlagwerke und auch Tourbillons. Laut einer internen Statistik verließen zwischen 1860 und 1925 rund 60000 Uhrwerke mit verschiedensten Komplikationen die Fabrik. Mit schöner Regelmäßigkeit gelangten auch danach außergewöhnliche Entwicklungen auf den Markt, wie 1925 das zweistöckige Uhrwerk „Duoplan". 1929 war das bis heute kleinste mechanische Uhrwerk zur Serienreife gediehen. Das spektakuläre „Kaliber 101" besteht aus 74 Teilen und misst 14 x 4,8 x 3,4 Millimeter. Samt Zifferblatt und Zeiger wiegt der 228,48 Kubikmillimeter große Mikrokosmos gerade einmal 0,9 Gramm. Weil die schützenden Kristallgläser über dem Zifferblatt beim Polospielen immer wieder brachen, beschwerten sich britische Offiziere im fernen Indien bei einem Uhrenimporteur. Die Kritik fruchtete und führte 1931 zur Kreation der „Reverso", einer Armbanduhr mit Wendegehäuse. Wenn es hart herging, konnte das Glas mit einem Handgriff nach unten und die stählerne Rückseite nach oben gekehrt werden.

1937 besiegelten Jaeger in Paris und LeCoultre in Le Sentier ihre langjährige Kooperation ganz offiziell durch die Einführung des Markennamens Jaeger-LeCoultre für fertige Uhren. Nach dem Zweiten Weltkrieg überraschte Jaeger-LeCoultre 1953 zunächst mit der „Futurematic", einer Armbanduhr mit Gangreserveanzeige und hocheffizienter Aufzugsautomatik, bei der auf ein Handaufzugsystem gänzlich verzichtet und die Krone am Boden versteckt werden konnte. Die wegweisende „Memovox", eine beinahe unüberhörbare Armbanduhr mit Wecker, gab es ab 1951 mit manuellem und 1959 auch mit automatischem Aufzug. Bedingt durch die Quarz-Revolution erlebte Jaeger-LeCoultre in den frühen 1970er Jahren einen Tiefpunkt in der langen Geschichte. Die Retter kamen aus dem „großen Kanton". 1978 übernahm die deutsche VDO 60 Prozent des Aktienkapitals, Audemars Piguet, ein langjähriger Werkekunde, den Rest. 1986 kam der visionäre Günter Blümlein als Sanierer ins Haus. Der Tachometerhersteller VDO gelangte 1991 unter das Dach des Mannesmann-Konzerns, welcher seine Uhr-Aktivitäten (IWC, Jaeger-LeCoultre und A. Lange & Söhne) 1996 in der LMH-Holding (Les Manufactures Horlogères) vereinigte. Als Mitglied dieser Gruppierung ging Jaeger-LeCoultre im Juli 2000 an den Richemont-Konzern. Bereits vor diesem Mega-Deal hatte Jaeger-LeCoultre seine Rohwerkekunden wissen lassen, dass mit einer Belieferung in absehbarer Zeit nicht mehr gerechnet werden könne. Konsequent ausgebaut wurden hingegen die eigenen Manufakturaktivitäten und damit auch die Entwicklung hoch komplizierter Uhrwerke.

Master Grand Tourbillon
Continents Asia, 2009

limited to 20 pieces

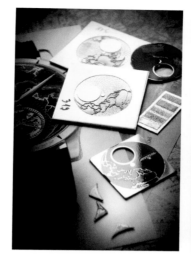

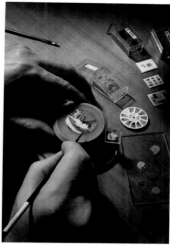

Letzteres wäre ohne fachliche Kompetenz auf vielen sehr unterschiedlichen Gebieten undenkbar. Speziell bei uhrmacherischen Komplikationen steigen die Anforderungen beinahe exponentiell. Aus diesen Gründen beschäftigt Jaeger-LeCoultre unter den Dächern der verschiedenen Gebäudeteile in Le Sentier auf 25 000 Quadratmetern Fläche ausgesuchte Fachleute, die insgesamt rund 180 verschiedene berufliche Qualifikationen vorweisen können. Grob zusammengefasst handelt es sich um Ingenieure, Techniker, Uhrmacher, Kreative, kunstgewerblich Schaffende und Handwerker. Jaeger-LeCoultre betrachtet es nämlich als Selbstverständlichkeit, beispielsweise auch die besonders anspruchsvollen Anker des Hemmungssystems selbst herzustellen. Oder die filigrane Unruhspirale. Dahinter stehen jene Mitarbeiterinnen und Mitarbeiter, welche das Dekorieren von Werken, Zifferblättern und Gehäusen beherrschen, nicht zurück. Dort, wo alle ganz unterschiedlichen Tätigkeiten nachgehen, wird Umweltschutz besonders großgeschrieben. Heißes Wasser gewinnt das Unternehmen durch Sonnenenergie. Die Maschinenkühlung erfolgt durch einen geschlossenen Wasserkreislauf und Free Cooling, ein System zur Frischluftrückgewinnung. Für diese und andere derartige Maßnahmen gab es das Minergie-Zertifikat sowie das Hydro-Locale-Label, welches die Energieversorgung durch erneuerbare Wasserkraft belegt.

Nach dem Motto, dass Vertrauen gut, Kontrolle aber deutlich besser ist, muss jede einzelne Armbanduhr vor dem Verlassen der Manufaktur insgesamt 41 Tage und 16 Stunden auf den Prüfstand. Zusammen sind das 1 000 Stunden. Anders als bei der offiziellen Schweizer Chronometerprüfung bezieht sich der Check auf den komplett fertiggestellten Zeitmesser inklusive Zifferblatt, Zeiger und Gehäuse. Die Tests vollziehen sich nach klar definierten Maßgaben in sechs Positionen und bei verschiedenen Temperaturen. Auch harte Stöße gehören zum Programm. Auf diese Weise müssen die tickenden Kandidaten in der ersten Epoche ihres Lebens vermutlich härteren Anforderungen standhalten, als während ihrer späteren Verwendung an einem Handgelenk irgendwo auf dieser Welt. ○

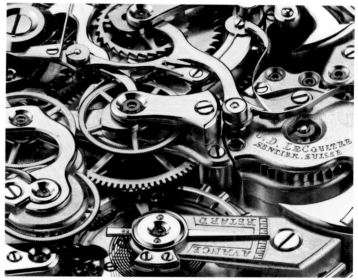

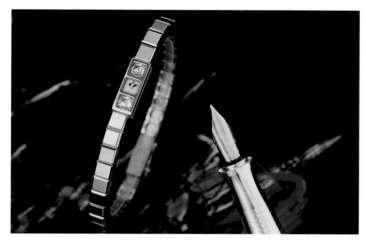

Top: an enamel painter at work ○ Middle: close-up view of a complicated movement with chronograph and repeater ○ Bottom: the world's smallest mechanical movement, caliber 101, 1929 ○ Oben: Bei einer Email-Malerin ○ Mitte: Blick in ein kompliziertes Uhrwerk mit Chronograph und Repetition ○ Unten: Kleinstes mechanisches Uhrwerk der Welt, Kaliber 101, 1929 ○ En haut : Peintre émailleuse à l'œuvre ○ Au milieu : Vue de détail d'un mouvement avec chronographe et répétition ○ En bas : Plus petit mouvement mécanique du monde, calibre 101, 1929

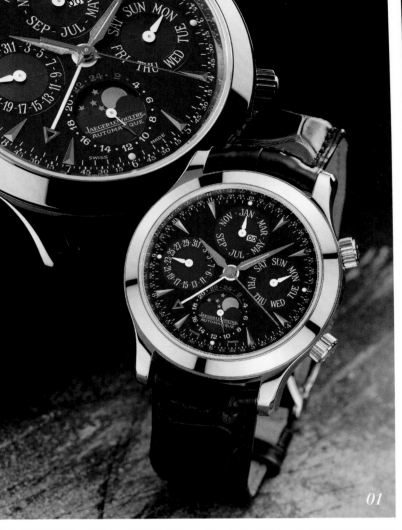

01

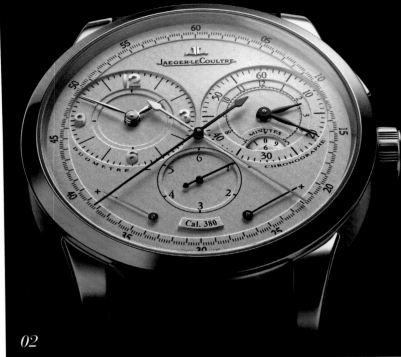

02

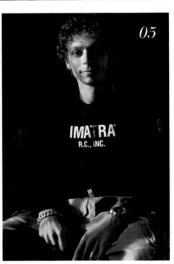

04

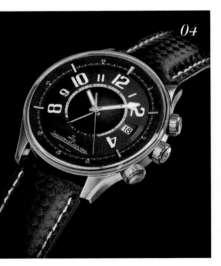

05

IMATRA
R.C., INC.

06

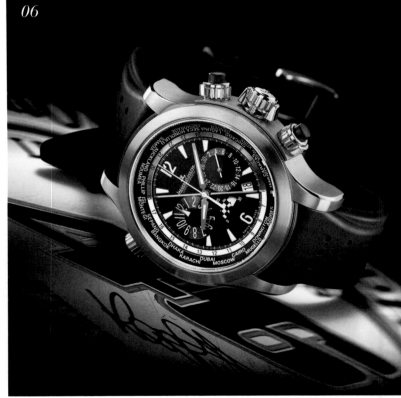

08

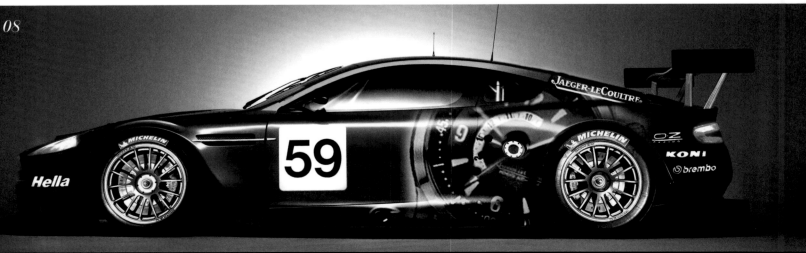

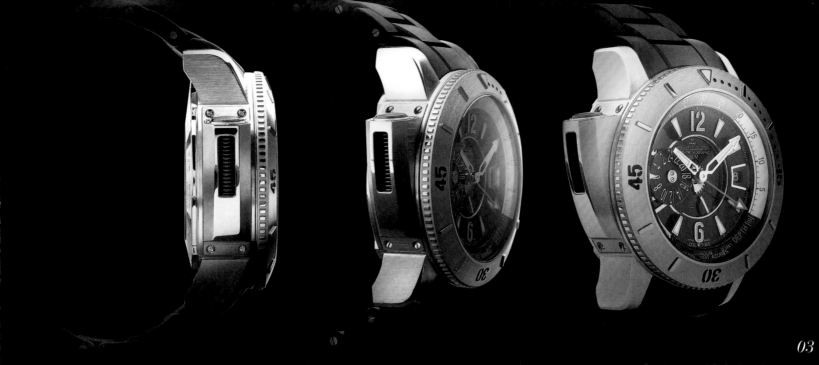

03

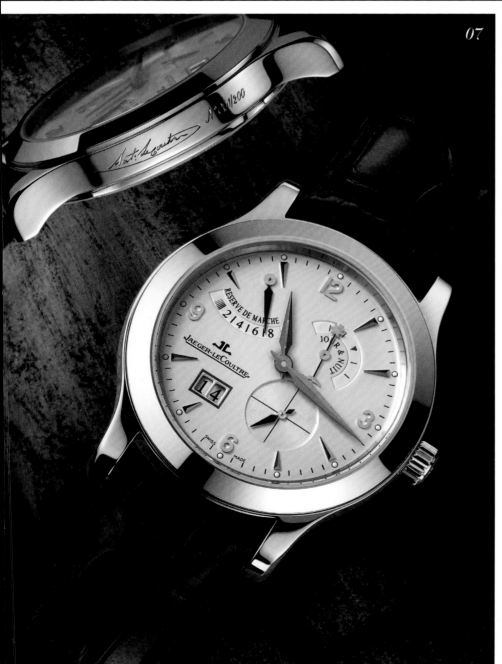

07

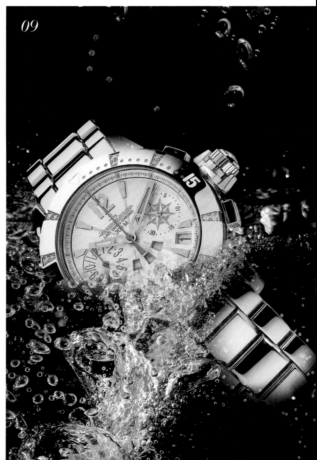

09

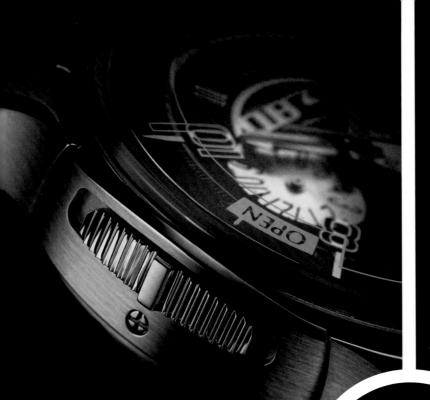

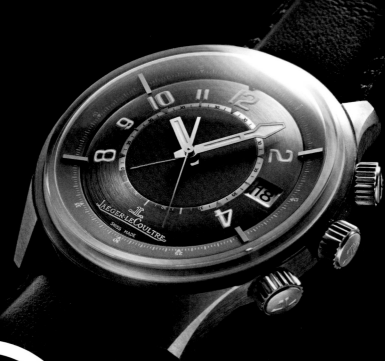

AMVOX2 DBS Transponder, 2000

AMVOX Alarm Titanium, 2005

limited to 1,000 pieces

AMVOX

AMVOX5 World
Chronograph Ceramic, 2010

limited to 200 pieces

AMVOX2 Racing Paris
Titanium, 2009

limited to 24 pieces

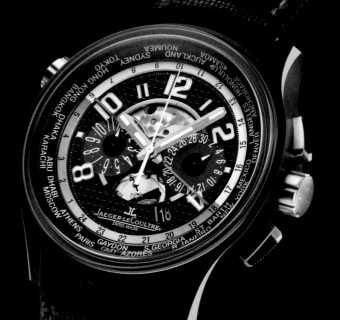

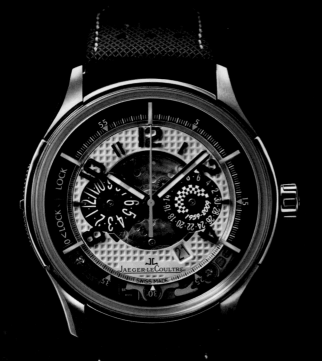

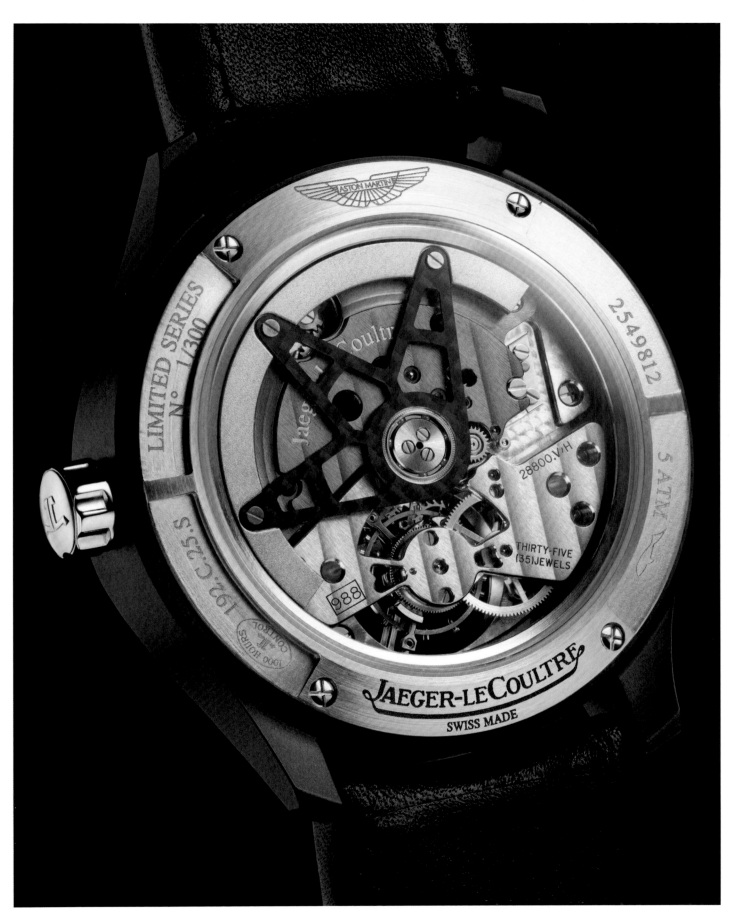

AMVOX3 Tourbillon GMT, ceramic and rose gold, limited to 300 pieces, 2008

Production manufacturière dans la vallée de Joux

Cinq mots suffisent à décrire les montres Jaeger-LeCoultre actuelles : des produits de notre manufacture. Celle-ci se trouve au même endroit depuis sa création, en 1833 : dans un village appelé Le Sentier, à 1 000 mètres d'altitude dans la vallée de Joux. C'est là que l'ingénieux mécanicien et artisan Charles-Antoine LeCoultre débute à l'âge de 30 ans dans la fabrication de pignons. Animé par une insatiable passion pour les outils et les mécanismes de précision, il invente assez vite des machines pour fabriquer ces mêmes pignons. On lui doit aussi un instrument décisif dans l'amélioration de la précision horlogère : le « millionomètre » permet pour la première fois de réaliser des mesures au micron près. Lors de la première Exposition universelle, à Londres en 1851, Charles-Antoine LeCoultre présente un remarquable assortiment de produits : outils, engrenages, pignons, ébauches de mouvement d'horlogerie et chronomètres de poche. Grâce à ses innovations, que lui achètent les plus grandes marques de montres suisses, LeCoultre devient le plus grand employeur dans cette région défavorisée. En 1877, les fils du fondateur de la marque se lancent dans la conception et la fabrication de mouvements d'horlogerie complexes, notamment des chronographes, des réveils, des calendriers, des répétitions et des tourbillons. D'après des statistiques internes, environ 60 000 mouvements équipés des complications les plus diverses sortent de la manufacture entre 1860 et 1925. Avec une belle régularité, d'autres produits représentant une évolution extraordinaire apparaissent sur le marché, notamment en 1925 la montre « Duoplan », dont les composants sont disposés sur deux étages. En 1929, le mouvement mécanique jusqu'ici le plus petit est désormais fabriqué en série. Le spectaculaire Calibre 101, qui mesure 14 x 4,8 x 3,4 millimètres, comporte 74 pièces. Cadran et aiguilles compris, ce microcosme de 228,48 millimètres cubes pèse à peine 0,9 gramme. Comme les verres en cristal protégeant le cadran n'arrêtent pas de se briser lorsque les officiers britanniques jouent au polo dans l'Inde lointaine, ces derniers se plaignent auprès de l'importateur de montres. La critique porte ses fruits et conduit en 1931 à la création de la « Reverso », une montre-bracelet à système de retournement. Dans le feu de l'action, un simple mouvement suffit pour retourner le boîtier.

En 1937, Jaeger à Paris et LeCoultre au Sentier scellent officiellement leur collaboration de longue date en introduisant la signature Jaeger-LeCoultre pour les montres de leur fabrication. Au lendemain de la Seconde Guerre mondiale, Jaeger-LeCoultre crée la surprise, d'abord en 1953, avec la « Futurematic », une montre-bracelet avec indicateur de réserve de marche et 100 pour cent automatique, dépourvue de couronne de remontoir apparente, celle-ci étant sur le revers. L'innovante « Memovox », montre-bracelet à réveil qu'il est quasiment impossible de ne pas entendre, est proposée en 1951 avec remontage manuel, puis en 1959 avec remontage automatique. Au début des années 1970, la révolution du quartz touche de plein fouet Jaeger-LeCoultre. Les sauveurs viennent du « grand canton », comme les Suisses se plaisent à appeler l'Allemagne voisine. En 1978,

la société allemande VDO Adolf Schindling AG acquiert 60 pour cent du capital en actions, Audemars Piguet, qui se fournit depuis longtemps en mouvements auprès de la société, les 40 pour cent restants. En 1986, le visionnaire Günter Blümlein rejoint la société pour la redresser. Le fabricant de tachymètres VDO est repris en 1991 par le groupe Mannesmann, lequel regroupe en 1996 ses activités horlogères (IWC, Jaeger-LeCoultre et A. Lange & Söhne) au sein d'une holding, Les Manufactures Horlogères. En juillet 2000, Jaeger-LeCoultre devient, en tant que membre de ce regroupement, la propriété du groupe Richemont. Avant même cette colossale opération financière, Jaeger-LeCoultre a informé les clients qui lui commandent des ébauches qu'ils ne doivent plus escompter être livrés dans des délais raisonnables. En revanche, les activités propres de la manufacture sont résolument étendues et par conséquent, le développement de mouvements d'une grande complexité.

Cette dernière activité serait impensable sans expertise dans de très nombreux domaines. Pour les complications en particulier, les exigences augmentent de manière quasi exponentielle. Aussi Jaeger-LeCoultre emploie-t-il dans les différentes parties du bâtiment du Sentier, qui s'étend sur 25 000 mètres carrés, des spécialistes triés sur le volet et qui peuvent se réclamer d'environ 180 métiers différents, dans les grandes catégories suivantes : ingénieurs, techniciens, maîtres horlogers, créatifs, artisans et ouvriers. Jaeger-LeCoultre considère comme une évidence de fabriquer en interne par exemple les ancres particulièrement complexes de l'échappement, ou encore les délicats spiraux. Les spécialistes de l'ornementation des mouvements, cadrans et boîtiers n'ont rien à leur envier. Dans cet endroit où tous s'adonnent à des activités totalement différentes, on attache la plus grande importance à la protection de l'environnement. L'entreprise se fournit en eau chaude grâce à l'énergie solaire. Le refroidissement des machines s'effectue à l'aide d'un circuit d'eau fermé et par refroidissement naturel, via un système qui régénère l'air. Pour ces mesures et d'autres encore, le bâtiment est certifié Minergie mais aussi Hydro locale, un label qui atteste d'une production d'électricité par l'énergie hydraulique, qui est renouvelable.

Suivant l'aphorisme, la confiance c'est bien, le contrôle c'est mieux, chaque montre-bracelet subit avant de quitter l'usine des tests durant 41 jours et 16 heures, soit 1 000 heures au total. À la différence du contrôle officiel suisse des chronomètres (COSC), cette batterie de tests porte sur le garde-temps terminé, avec son cadran, ses aiguilles et son boîtier. Les essais sont effectués suivant des prescriptions bien précises, dans six positions différentes et à différentes températures. Des tests de résistance à des chocs importants sont aussi au programme. Les candidates à la carrière de montre sont ainsi soumises durant la première période de leur existence à des contraintes plus sévères qu'elles ne le seront plus tard au poignet de leur propriétaire quelque part dans le monde. ○

Hybris Mechanica
à Grande Sonnerie, 2009

limited to 30 pieces

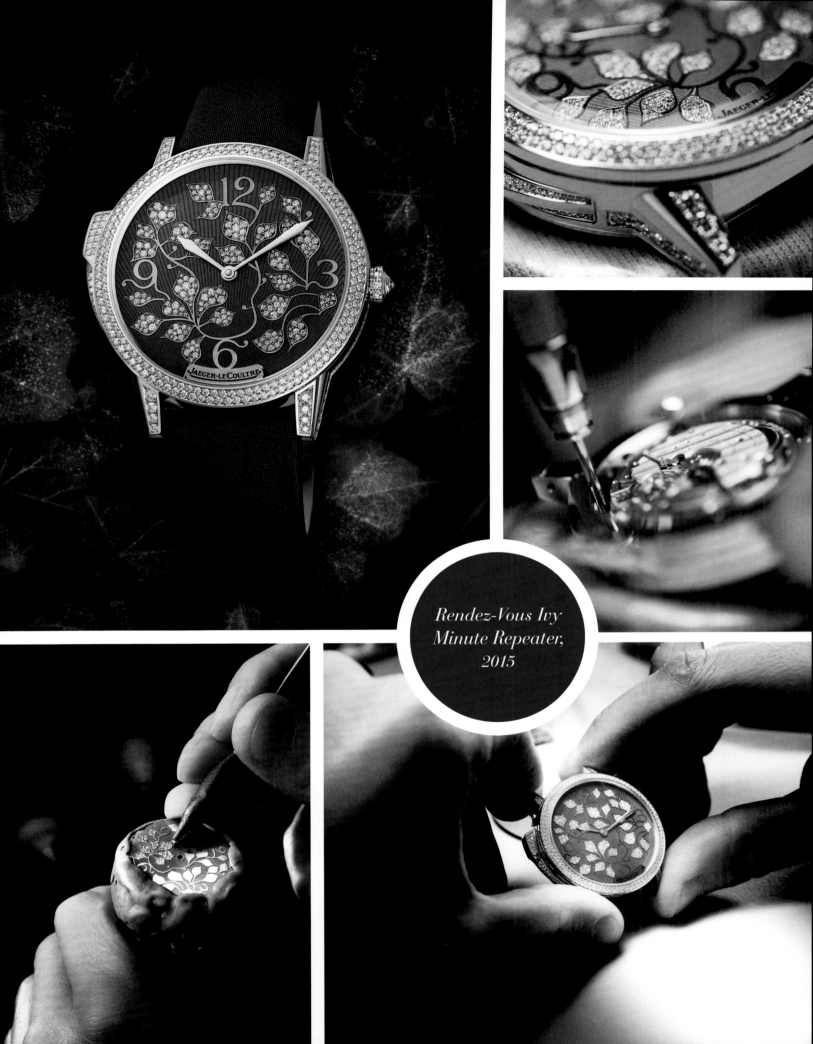

*Rendez-Vous Ivy
Minute Repeater,
2015*

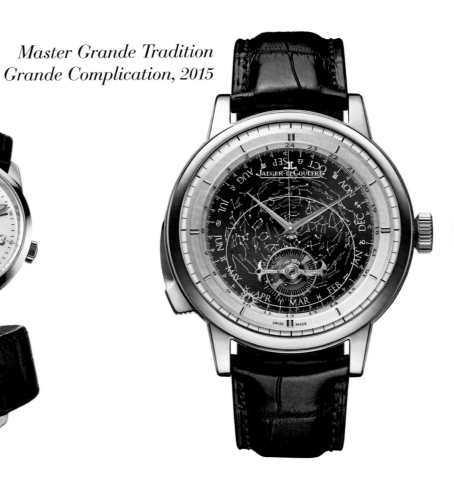

Master Grande Tradition Grande Complication, 2015

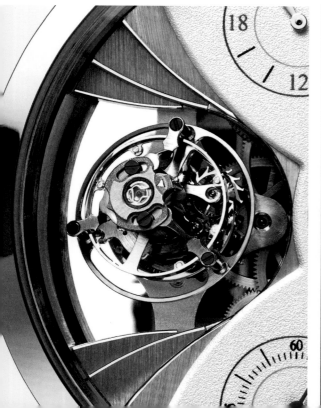

Geophysic Tribute to 1958, 2014

Left: Sphérotourbillon ∘ Right: Geophysic, 1958

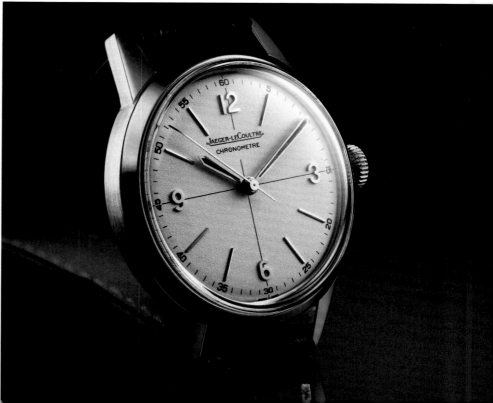

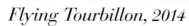

Master Ultra Thin Minute Repeater
Flying Tourbillon, 2014

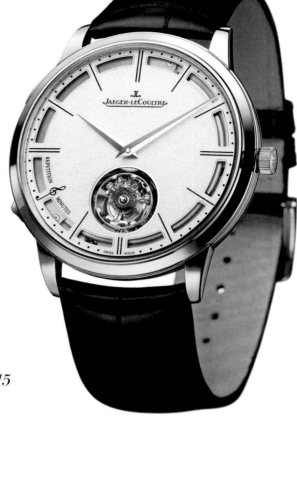

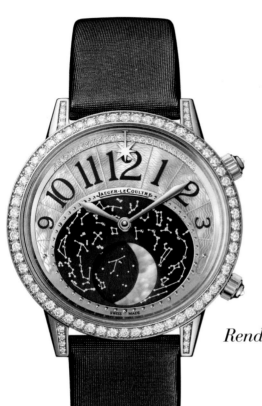

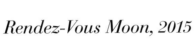

Rendez-Vous Moon, 2015

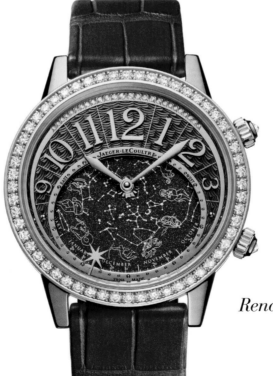

Rendez-Vous Celestial, 2015

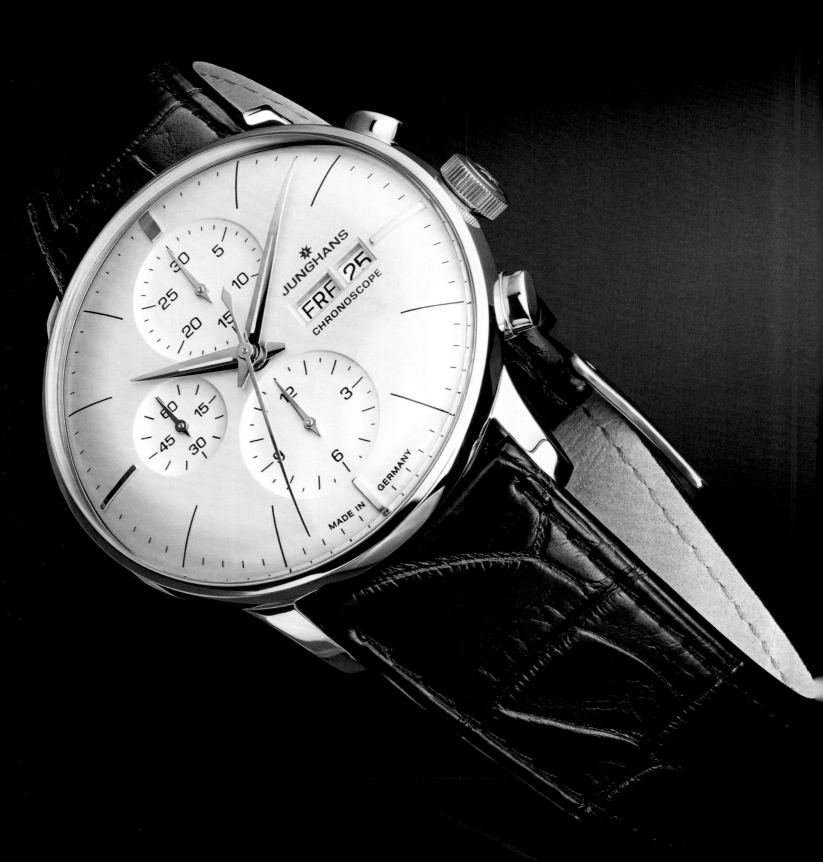

Junghans

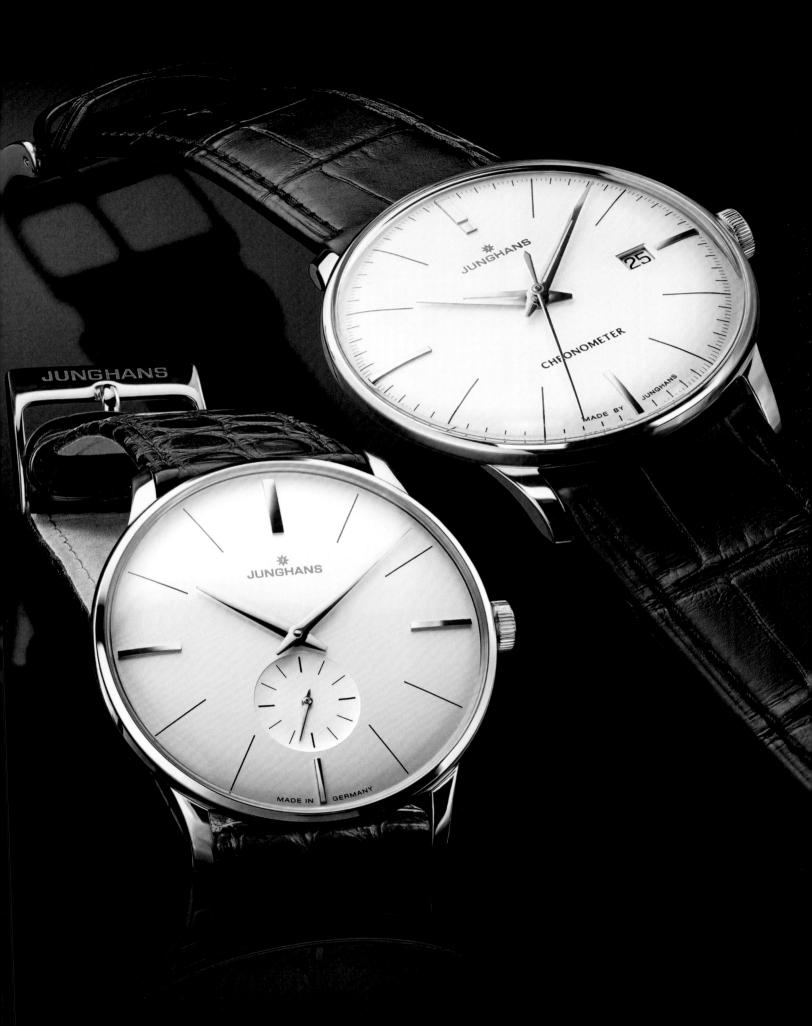

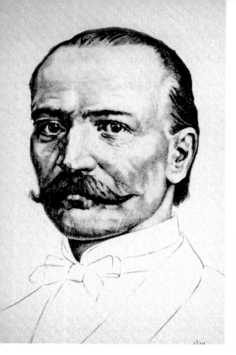
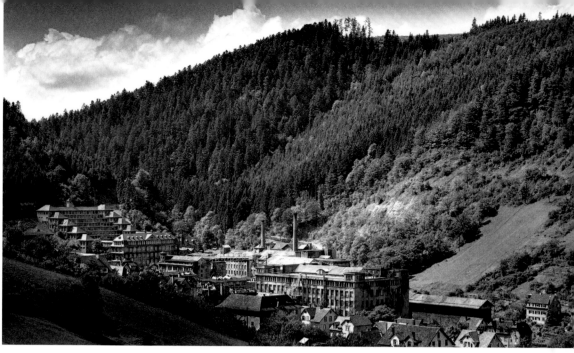

A Traditional German Brand
with Innovative Dynamism

Erhard Junghans was equally uninterested in traditional horological values and in the fabrication of pocket watches. The factory that he established in Schramberg in 1861 initially produced affordably priced components for popular Black Forest clocks. American mass-production methods were introduced as early as 1862. The next step was to build complete clocks using efficient American division of labor and precise machinery. Larger quantities went hand in hand with a broader spectrum of products. With over 3,000 employees producing an incredible 9,000 timepieces per day, Junghans was the world's largest manufacturer of timepieces in 1903. German watch aficionados welcomed the first pocket watches with luminous dials in 1912, but no one in Schramberg had begun thinking about wristwatches at this early date because the decision makers at Junghans were initially dubious about this variety of timepiece. This skepticism was finally overcome under the aegis of Helmut and Siegfried Junghans. On November 22, 1930, the advertising department ran an advertisement with the headline: "The Junghans wristwatch is here!" From this moment on, watches for the wrist would be inseparably linked with the brand's further evolution. The management announced the establishment of a research center for timekeeping in 1944. Caliber J88, the manufactory's first postwar movement, debuted in 1949: equipped with a classical column wheel to control its chronograph and a Breguet hairspring, this caliber first animated the official pilot's chronograph for the newly founded German military in 1955. No fewer than 852 patent applications between 1949 and 1956 bear witness to Junghans' extraordinary creativity. At least one of these patents related to the "Minivox" alarm wristwatch, which rang loudly on its wearer's wrist for the first time in 1951, the same year that Junghans premiered its first self-winding caliber. A long-term change in standards of precision for one part of the collection

began with the launch of the legendary Junghans chronometer in 1951. Prior to sale, each of these watches underwent an official test to verify the accuracy of its timekeeping. The success of this initiative proved Junghans right: by the end of 1954, the manufactory's ledgers recorded no fewer than 7,000 Junghans chronometers. The manufactory had quietly evolved into Germany's largest manufacturer of officially certified wristwatch chronometers with hand-wound and automatic movements. Junghans occupied an honorable third place behind Rolex and Omega in the international ranking of producers of officially certified timepieces. This fact and many others obviously fascinated a Nuremberg businessman named Karl Diehl, whose company acquired the majority of shares in Junghans on December 15, 1956. The Quartz Era dawned at Junghans in the 1970s. With the debut of the "Astro-Quartz" in 1970, the German flagship company began vying with powerful Japanese and Swiss competitors. As it had at the 1936 Olympics, Junghans again served as official timekeeper at the Summer Olympics in Munich in 1972. Needless to say, quartz-controlled technology for athletic timekeeping had progressed in Schramberg. In accord with the spirit of the times, Junghans discontinued production of mechanical calibers for wristwatches in 1976. Over 150 exclusive mechanical movements had been manufactured during the preceding 46 years—a feat which made Junghans an extraordinarily versatile manufactory of mechanical timepieces. The company first caused an appreciative furor with ultra-precise radio-controlled quartz clocks in 1985. These high-performance receivers were miniaturized to wristwatch format five years later in 1990: the wristband of the "Mega 1" provided space for its indispensable antenna and all data were presented digitally on a liquid crystal display. Ecologically advanced solar technology was integrated into radio-controlled

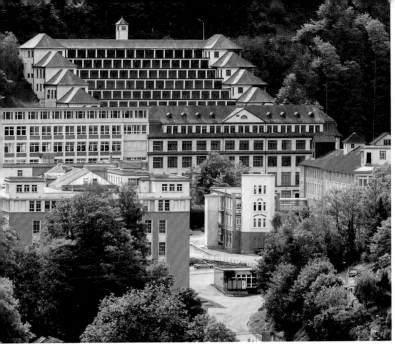

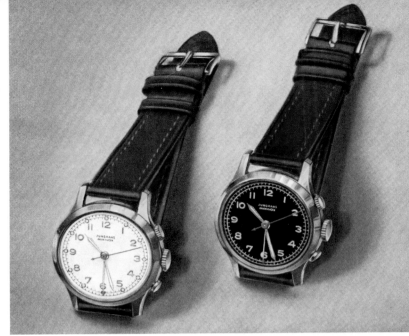

From left: Erhard Junghans, 1861 ∘ The company's campus, 1950s ∘ Minivox alarm wristwatch, 1951 ∘ Von links: Erhard Junghans, 1861 ∘ Firmengelände, 1950er ∘ Firmengelände, 2014 ∘ Armbandwecker Minivox, 1951 ∘ De gauche à droite : Erhard Junghans, 1861 ∘ La manufacture dans les années 1950 ∘ La manufacture en 2014 ∘ Montre-bracelet réveil Minivox, 1951 ∘ Opposite page, bottom: Cal. J325

wristwatches in 1993. Another quantum leap followed in 1995, when Junghans premiered the "Mega Solar Ceramic" with analog time display, solar technology, six-month power reserve, time-zone adjustment and scratch-resistant ceramic case. Under the aegis of Egana Goldpfeil, Junghans occupied itself with the most widely diverse state-of-the-art technologies at the beginning of the 21st century. But the specialists in the Black Forest simultaneously realized that good old mechanical timekeeping cannot be killed. Thanks to the rich trove in its archive, Junghans readily participated in the mechanical renaissance. An especially important legacy was bequeathed to Junghans by Max Bill (1908–1994), a Swiss architect, sculptor, painter and product designer who ranked among the best-known protagonists and theorists of Concrete art. He expressed his philosophy of creating "useful objects that are humble in a beautiful way" through a series of four dials which he designed in the early 1960s. Wristwatches with these handsome dials are coveted collectors' items nowadays, so it's not surprising that they inspired Junghans' designers to create the classical and bestselling wristwatches in the "Max Bill by Junghans" line. The instantaneous success of this line of watches proved that genuine classics never go out of style.

Paying due homage to the firm's founder, the "Erhard Junghans" embodied the top of the line in the mechanical collection in 2006. Its distinctively designed case houses a Seiko movement that's reserved for Junghans. The entry level is defined by Caliber J830 with a rotor for the self-winding mechanism and a date window. The undisputed pinnacle in this line remains chronograph Caliber J890, a meticulously finished chronograph movement with a column wheel. Junghans is by no means content merely to purchase finished calibers: instead, the brand's specialists fabricate and finely finish a large number of components (e.g., bridges, ratchets, gold *châtons*, and fine adjustment mechanisms) for Calibers J325 and J330. The latter is also equipped with a blue

Breguet hairspring made from Nivarox by the Carl Haas firm in Schramberg. These movements tick inside the "Erhard Junghans 1" and "Erhard Junghans 2," each of which is available in a limited edition of twelve models.

The insolvency of its mother company Egana Goldpfeil in 2008 also had some inarguably positive aspects for Junghans. The reins were passed early in 2009 to the Steims, a family of business-people in Schramberg. The new owners recognized the existing and partly dormant values of the venerable watch brand. The Steims' mechanical timekeeping philosophy cultivates tradition and offers affordable horological luxury of German provenance. Alongside the "Erhard Junghans" and the "Max Bill by Junghans," the "Junghans Meister" evokes many memories of this brand's mechanical tradition. Clear, frill-free design had distinguished the watches in this collection in the past. Today too, there isn't the slightest reason to depart from this winning formula. In this sense, the "Meister Driver" models that debuted at Baselworld in 2016 celebrate nostalgia in its purest form, either with a classical hand-wound movement or as the "Chronoscope," in which Junghans encases automatic Caliber J.880.3, a combination of time-honored Eta 2892-A2 and chronograph module 2030 from Dubois Dépraz. Equally reliable basic Caliber Eta 2824-2 ticks inside the "Meister Pilot," which harkens back to the year 1955 and the official chrono-graph of the German military. All this expresses loyalty to the motto which asserts that those who can look back upon strong past needn't be concerned about the future. ∘

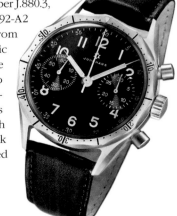

Pilot's watch, Cal. J88, 1955

Deutsche Traditionsmarke mit Innovationskraft

Mit tradierten uhrmacherischen Werten hatte Erhard Junghans 1861 ebenso wenig am Hut wie mit der Herstellung von Taschenuhren. Die von ihm in Schramberg gegründeten Uhrenfabriken Gebrüder Junghans AG lieferten zunächst preiswerte Komponenten für die beliebten Schwarzwälder Uhren. Bereits 1862 hielten amerikanische, d.h. großserielle Fertigungsmethoden Einzug. In einem nächsten Schritt folgte die Produktion kompletter Zeitmesser nach amerikanischen, sprich arbeitsteiligen Vorgehensweisen unter Verwendung präziser Maschinen. Mit steigenden Quantitäten weitete sich auch die Produktpalette aus. 1903 durfte sich Junghans mit über 3000 Mitarbeitern und einer unglaublichen Tagesproduktion von 9000 Zeitmessern weltweit größter Uhrenhersteller nennen. Gleichzeitig herrschte an Innovationen niemals Mangel. Das Jahr 1912 bescherte deutschen Uhrenliebhabern die ersten Taschenuhren mit Leuchtzifferblättern. Von Armbanduhren war in Schramberg allerdings noch keine Rede. Bei Junghans glaubte man zuerst nicht an diesen Typus Zeitmesser. Die Skepsis legte sich erst unter der Ägide von Helmut und Siegfried Junghans. Am 22. November 1930 schaltete die Werbeabteilung eine Anzeige mit der Überschrift: „Die Junghans Armband-Uhr ist da!" Ab da verknüpfte sich dieser Uhrentyp unlösbar mit der weiteren Entwicklung des Hauses. 1944 verkündete das Management die Einrichtung eines Forschungszentrums für Zeitmessung. Und 1949 präsentierte die Manufaktur das erste Nachkriegs-Kaliber J88. Dieses Uhrwerk mit klassischer Schaltradsteuerung und Breguetspirale beseelte ab 1955 auch die offiziellen Fliegerchronographen der damals neu gegründeten deutschen Bundeswehr. Von ungemeiner Kreativität zeugen nicht weniger als 852 Patentanmeldungen zwischen 1949 und 1956. Mindestens eine davon bezog sich auf den Armbandwecker „Minivox", der ab 1951 lautstark an den Handgelenken bimmelte. Im gleichen Jahr debütierte auch das erste Junghans-Kaliber mit automatischem Aufzug. Mit der Lancierung der legendären Junghans-Chronometer änderte sich ab 1951 der Präzisionsanspruch für einen Teil der Kollektion. Vor dem Verkauf musste jede dieser Uhren eine amtliche Genauigkeitsprüfung bestehen. Der Erfolg dieser Initiative war durchschlagend. Ende 1954 standen bereits 7000 Junghans-Chronometer in den Büchern. Still und leise hatte sich die Manufaktur zum größten deutschen Hersteller offizieller Armbandchronometer mit Handaufzugs- oder Automatikwerken entwickelt. Im internationalen Ranking der Produzenten amtlich zertifizierter Zeitmesser belegte Junghans einen ehrenvollen dritten Platz hinter Rolex und Omega. Das und etliches mehr faszinierte ganz offensichtlich den Nürnberger Unternehmer Karl Diehl. Am 15. Dezember 1956 übernahm seine Firma die Aktienmehrheit bei Junghans. In den 1970er Jahren hielt das Quarz-Zeitalter seinen Einzug. Mit der „Astro-Quartz" trat das deutsche Vorzeigeunternehmen ab 1970 gegen die mächtige japanische und schweizerische Konkurrenz an. Bei den Olympischen Sommerspielen 1972 in München tat sich Junghans zum zweiten Mal nach 1936 als offizieller Zeitnehmer hervor. Die Entwicklung der durchweg quarzgesteuerten Technologie zur Sportzeitmessung war selbstverständlich in Schramberg über die Bühne gegangen. Den Zeichen der Zeit folgend, stellte Junghans 1976 die Fertigung mechanischer Armbanduhrkaliber unwiderruflich ein. Im Laufe von 46 Jahren waren mehr als 150 exklusive Werke entstanden, was den Titel einer äußerst vielseitigen Mechanik-Manufaktur rechtfertigt. Ab 1985 sorgte Junghans mit ultrapräzisen, weil funkgesteuerten Quarz-Zeitmessern für Furore. Nur fünf Jahre später, also 1990, passte der Hochleistungs-Empfänger erstmals in ein Armbanduhr-Gehäuse. Das Armband der bahnbrechenden „Mega 1" bot Platz für die unverzichtbare Antenne. Alle Informationen erschienen digital auf einem Flüssigkristall-Display. Die ökologisch fortschrittliche Solartechnologie hielt 1993 ihren Einzug in die Funkarmbanduhren. Einen weiteren Quantensprung verkörperte 1995 die Junghans

From left: Astro-Quartz, 1970 ◦ Official timekeeper at the Olympic Games in Munich, 1972 ◦ Mega 1, the world's first radio wristwatch, 1990 ◦ Von links: Astro-Quartz, 1970 ◦ Offizieller Zeitnehmer bei den Olympischen Spielen in München, 1972 ◦ Weltweit erste Funkarmbanduhr Mega 1, 1990 ◦ De gauche à droite : Montre Astro-Quartz, 1970 ◦ Chronométreur officiel des Jeux olympiques à Munich, 1972 ◦ Mega 1, la première montre-bracelet radiopilotée au monde, 1990

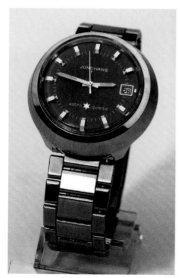 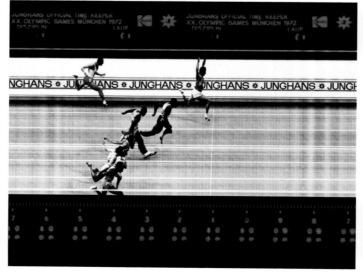 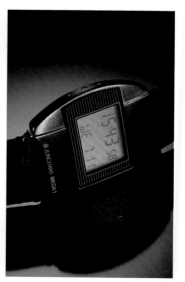

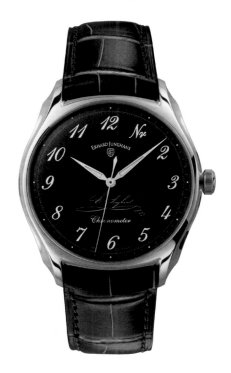

Cal. J890

„Mega Solar Ceramic" mit analoger Zeit-anzeige, Solartechnologie, sechsmonatiger Gangautonomie, Zonenzeitverstellung sowie kratzfestem Keramikgehäuse. Unter dem Dach der Holding Egana Goldpfeil beschäftigte sich Junghans zu Beginn des 21. Jahrhunderts einerseits mit unterschied-lichsten Zukunftstechnologien. Zum anderen zeigte sich auch im Schwarzwald, dass die gute alte Mechanik nicht umzubringen ist. Angesichts prall gefüllter Archive fiel die Renaissance nicht sonderlich schwer. Als besonders bedeutsam erwies sich das Erbe von Max Bill (1908–1994). Der Schweizer Architekt, Bildhauer, Maler und Produktgestalter gehörte bekanntlich zu den wichtigsten Vertretern und Theoretikern der konkreten Kunst. Seine Philosophie, „das Nützliche, das auf schöne Art Bescheidene" zu schaffen, hatte er Anfang der 1960er Jahre in einer Serie von vier Zifferblättern zum Ausdruck gebracht. Die damit ausgestatteten Armbanduhren galten und gelten als gesuchte Sammlerobjekte. Kein Wunder, dass sich die Produktverantwortlichen ihrer besannen und „Max Bill by Junghans" kreierten. Diese spontan erfolgreiche Uhrenlinie trat den schlagkräftigen Beweis an, dass echte Klassiker nie aus der Mode kommen.

Als Hommage an den Firmengründer repräsentierte ab 2006 „Erhard Junghans" die Spitze der Mechanik-Kollektion. Im Inneren der markanten Gehäuse tickten für Junghans reservierte Kaliber aus dem Hause Seiko. Den Einstieg markierte das J830 mit Rotor-Selbstaufzug und Fensterdatum. Ganz oben rangiert bis heute der sorgfältig feinbearbeitete Schaltrad-Chronograph J890. Dabei begnügt sich Junghans keineswegs mit dem Einkauf fertiger Werke, sondern fertigt und finissiert für die Kaliber J325 und J330 eine

Vielzahl der Komponenten, unter anderem Brücken, Gesperre, Goldchatons und Fein-regulierungen. Das Kaliber J330 ist zudem mit einer blauen, aus Nivarox gefertigten Breguetspirale der Schramberger Firma Carl Haas ausgestattet. Diese Werke finden sich in den auf jeweils zwölf Exemplare limitierten Modellen „Erhard Junghans 1" und „Erhard Junghans 2".

Die Insolvenz des Mutterkonzerns Egana Goldpfeil im Jahr 2008 hatte für Junghans auch ihre unbestreitbar positiven Seiten. Anfang 2009 übernahm die Schramberger Unternehmerfamilie Steim das Ruder. Die neuen Eigentümer erkannten die existenten und teilweise brach liegenden Werte der altehrwürdigen Uhrenmarke. Ihre Mechanik-Philosophie zielt auf Pflege der Tradition und bezahlbaren uhrmacherischen Luxus aus deutschen Landen. Neben „Erhard Junghans" und „Max Bill by Junghans" weckt „Junghans Meister" jede Menge Erinnerungen an die Mechanik-Tradition des Hauses. Klares, schnörkelloses Design zeichnete die Uhren dieser Kollektion in früheren Jahren aus. Und es gibt heutzutage nicht den geringsten Grund, daran irgendwie zu rütteln. In diesem Sinn verstrahlen die 2016 während der Baselworld vorgestellten „Meister Driver"-Modelle Nostalgie pur – mit klas-sischem Handaufzugswerk oder als „Chronoscope". Hierin verbaut Junghans das Automatikkaliber J.880.3, einen Mix aus dem bewährten Eta 2892-A2 und dem Chronographenmodul 2030 von Dubois Dépraz. Das nicht minder zuverlässige Basiskaliber Eta 2824-2 findet sich im „Meister Pilot". Hier leben das Jahr 1955 und der Dienst-Chronograph der deutschen Bundeswehr wieder auf. Getreu dem Motto, dass sich um die Zukunft keine Sorgen machen muss, wer eine derart starke Vergangenheit besitzt. ○

Left: Aerious Chronoscope, 2007 ○ *Right: Erhard Junghans 2, 2011, limited to* **twelve timepieces** ○ *Erhard Junghans 2, 2011, limitiert auf zwölf Exemplare* ○ *Erhard Junghans 2, 2011, limitée à douze exemplaires*

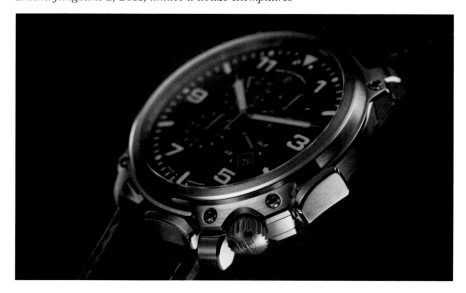

Tradition et force d'innovation à l'allemande

Français

En 1861, les valeurs horlogères traditionnelles sont tout aussi étrangères à Erhard Junghans que la fabrication de montres de poche. Gebrüder Junghans AG, la fabrique de montres qu'il fonde à Schramberg, commence par fournir des composants bon marché pour les célèbres coucous de Forêt-Noire. Dès 1862, il adopte les méthodes de production américaines : il introduit la fabrication en grandes séries, avant de passer à la production complète d'instruments de mesure du temps selon le principe de division du travail et le recours à des machines de précision. L'augmentation du volume de production va de pair avec l'élargissement de la palette de produits. En 1903, avec plus de 3 000 salariés et une impressionnante fabrication journalière de 9 000 produits horlogers, Junghans peut s'arroger le titre de plus grand fabricant de montres au monde. Parallèlement, les innovations vont bon train. L'année 1912 voit la sortie des premières montres de poche à cadrans luminescents, à la grande joie des amateurs de montres allemandes. À Schramberg, les montres-bracelets sont encore une chimère. Dans un premier temps, Junghans ne croit pas à ces produits horlogers. C'est seulement sous l'égide de Helmut et Siegfried Junghans que le scepticisme se dissipe. Le 22 novembre 1930, le département publicité fait passer une annonce intitulée : « La montre-bracelet Junghans est arrivée ! » Ce type de montre deviendra indissociable de l'évolution future de la maison. En 1944, la direction annonce la création d'un centre de recherche sur la mesure du temps. En 1949, la manufacture présente son premier calibre de l'après-guerre, le J88. Ce mouvement à roue à colonnes et spiral Breguet animera à partir de 1955 également les chronographes d'aviateurs officiels de la Bundeswehr créée cette année-là. Le chiffre impressionnant de 852 dépôts de brevet effectués entre 1949 et 1956 témoigne d'une créativité inouïe. L'un d'entre eux concerne la montre-bracelet réveil « Minivox » qui, à partir de 1951, retentit au poignet de ceux qui la possèdent. La même année voit la sortie du premier calibre Junghans à remontage automatique. Avec le lancement de ses légendaires chronomètres, la manufacture revoit alors à la hausse ses exigences en matière de précision pour une partie de la collection. Avant sa commercialisation, chaque exemplaire des modèles concernés est testé par un organisme officiel. Cette initiative est on ne peut plus heureuse : avec déjà 7 000 chronomètres vendus fin 1954, Junghans est devenu sans tambour ni trompette le fabricant allemand numéro un de chronomètres-bracelets à remontage manuel ou automatique certifiés chronomètres. Au classement international des fabricants de garde-temps officiellement certifiés, Junghans occupe alors une honorable troisième place, derrière Rolex et Omega. Visiblement fasciné par ce score et bien d'autres prouesses, Karl Diehl, entrepreneur de Nuremberg, acquiert une participation majoritaire dans la manufacture le 15 décembre 1956. À l'ère du quartz, qui débute dans les années 1970, l'entreprise allemande modèle se mesure, notamment avec l'« Astro-Quartz », à des concurrents de taille, japonais et suisses. À l'occasion des Jeux olympiques de 1972 à Munich, Junghans s'illustre comme chronométreur officiel de la manifestation, pour la deuxième fois depuis 1936. Schramberg est bien sûr passé au développement du tout-quartz pour le chronométrage sportif. Sacrifiant à l'air du temps, Junghans cesse définitivement en 1976 de produire des calibres de montres-bracelets mécaniques. Avec plus de 150 mouvements exclusifs développés en 46 ans, Junghans peut prétendre au titre de manufacture mécanique extrêmement diversifiée. À partir de 1985, Junghans fait sensation avec des horloges à quartz radiopilotées et de ce fait ultraprécises. Seulement cinq ans plus tard, en 1990, le récepteur de haute performance est miniaturisé pour pouvoir loger dans un boîtier de montre-bracelet. L'indispensable antenne est intégrée dans le bracelet de l'innovante « Mega 1 », à affichage numérique sur écran à cristaux liquides. Le solaire, qui représente un progrès sur le plan écologique, s'impose dans les montres-bracelets radio-guidées en 1993. Incarnant une formidable avancée technologique, la « Mega Solar Ceramic », sortie en 1995, s'appuie sur la technologie solaire et dispose d'un affichage de l'heure analogique, d'une réserve de marche de six mois, d'une fonction GMT ainsi que d'un boîtier céramique inrayable. Dans le giron de la holding Egana Goldpfeil, Junghans s'intéresse en ce début de XXIe siècle aux techniques d'avenir les plus diverses. Parallèlement, il s'avère aussi en Forêt-Noire que la bonne vieille mécanique n'a pas dit son dernier mot. Grâce à des archives très riches, la renaissance n'est pas trop difficile. L'héritage de Max Bill (1908–1994) se révèle particulièrement intéressant. Cet architecte, sculpteur, peintre et designer de produits suisse compte parmi les représentants et théoriciens majeurs de l'Art concret. Il avait exprimé sa philosophie consistant à créer des « objets utiles, modestes mais beaux » dans une série de quatre cadrans. Les montres-bracelets qui en disposaient étaient et sont encore considérées comme des objets de collection recherchés. Aussi n'est-il pas étonnant que les responsables produits se soient souvenus de ces cadrans pour créer « Max Bill by Junghans ». Le succès inattendu de cette ligne de montres apporte la preuve éclatante que les vrais classiques ne se démodent pas. Hommage au fondateur de l'entreprise, « Erhard Junghans » représente à partir de 2006 le summum de la collection mécanique. Les boîtiers remarquables abritent des calibres fournis par la maison Seiko spécifiquement pour Junghans. L'entrée de gamme est représentée par le J830 à remontage automatique et guichet dateur, tandis que le chronographe à roue à colonnes J890 ayant bénéficié d'une finition très soignée figure, aujourd'hui encore, en tête de gamme. Or, Junghans ne se contente aucunement d'acheter des mouvements finis, mais assure en interne la fabrication et le finissage de nombreux composants pour les calibres J325 et J330, notamment des ponts, des arrêtages, des chatons or et des dispositifs de réglage de précision. Le calibre J330 est en outre pourvu d'un spiral Breguet en Nivarox bleu fabriqué par Carl Haas, spécialiste de la mécanique de précision également établi à Schramberg. Ce sont ces mouvements qui animent les modèles « Erhard Junghans 1 » et « Erhard Junghans 2 », limités chacun à douze exemplaires.

Meister Kalender, 2012

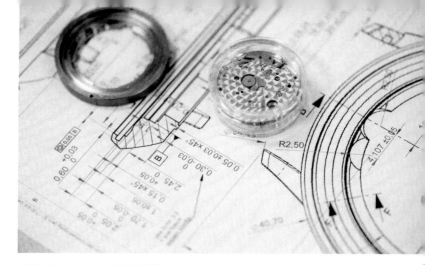

La mise en liquidation de la maison-mère Egana Goldpfeil en 2008 présente sans conteste également des avantages pour Junghans, qui est reprise au début de l'année 2009 par une entreprise familiale de Schramberg, Steim. Les nouveaux propriétaires reconnaissent les atouts existants, en partie inexploités, de la vénérable marque horlogère. La philosophie des Steim en matière de mécanique repose sur la promotion de la tradition et les produits horlogers haut de gamme made in Germany. Aux côtés de « Erhard Junghans » et de « Max Bill by Junghans », « Junghans Meister » s'inscrit parfaitement dans la tradition de garde-temps mécaniques de Junghans. Les toutes premières montres de cette collection se caractérisaient déjà par un design épuré, sans fioritures, et il n'y a de nos jours pas la moindre raison d'y changer quoi que ce soit. En ce sens, les modèles « Meister Driver » – avec remontage manuel classique ou en version « Chronoscope » – présentés en 2016 lors du Salon mondial de l'horlogerie à Bâle (Baselworld) font renaître toute une époque. La version « Chronoscope » est équipée du calibre automatique J.880.3, synthèse du mouvement Eta 2892-A2 éprouvé et du module de chronographe 2030 de Dubois Dépraz. C'est le calibre de base non moins fiable Eta 2824-2 qui anime le « Meister Pilot », réminiscence de l'année 1955 et du chronographe de service de la Bundeswehr. Qui a un passé aussi solide n'a pas à se soucier de l'avenir. ○

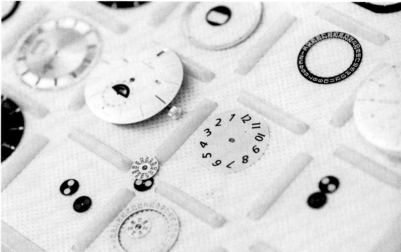

*Meister
Telemeter,
2013*

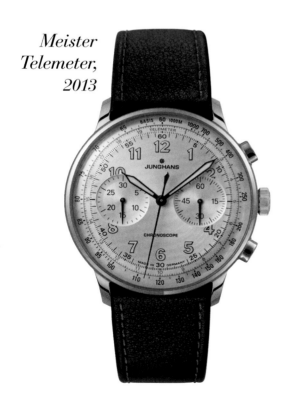

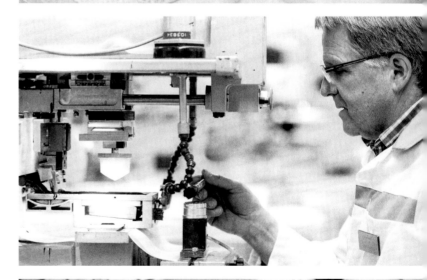

*Engineered, components fabricated, and designed in
Schramberg ○ Konstruktion, hauseigene Teilefertigung und
Design in Schramberg ○ Conception, fabrication des pièces
et design en interne à Schramberg*

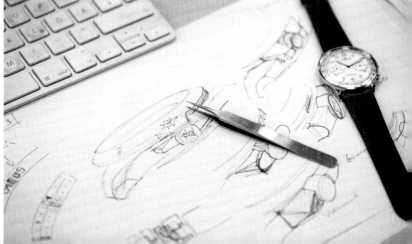

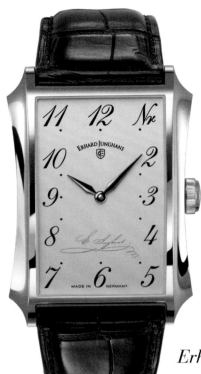

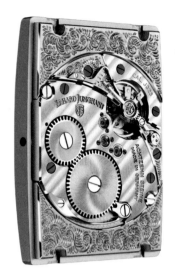

Cal. J325

Erhard Junghans 1, 2008

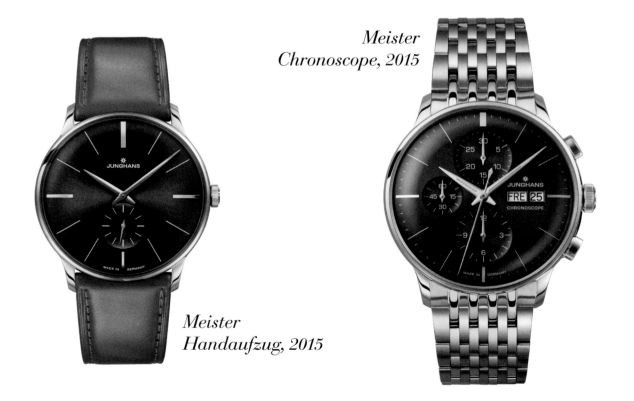

Meister Chronoscope, 2015

Meister Handaufzug, 2015

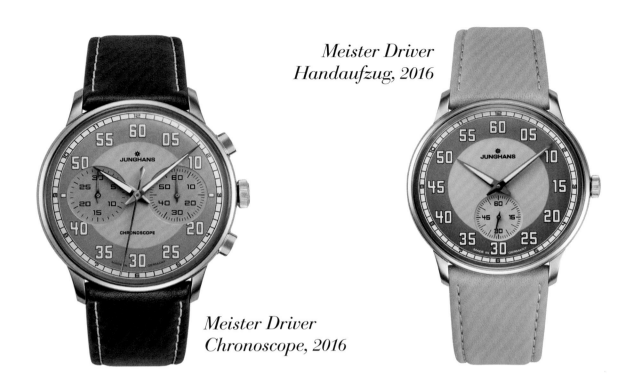

*Meister Driver
Handaufzug, 2016*

*Meister Driver
Chronoscope, 2016*

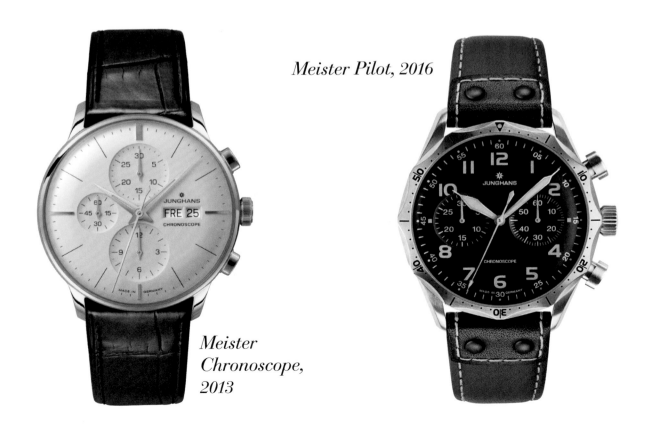

Meister Pilot, 2016

*Meister
Chronoscope,
2013*

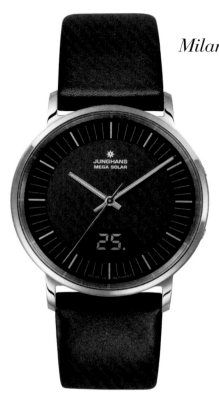

Milano Mega Solar, 2012

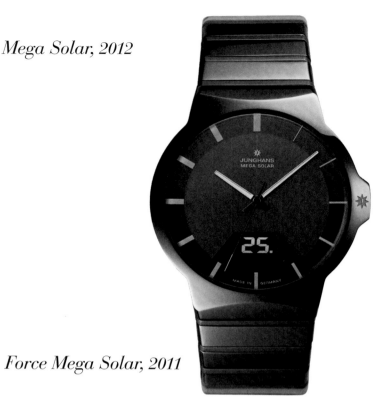

Force Mega Solar, 2011

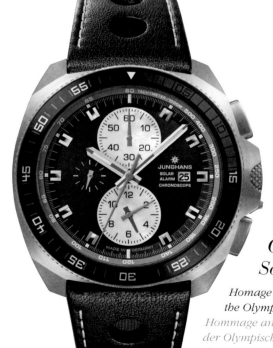

*Spektrum
Mega Solar, 2014*

*1972
Chronoscope
Solar, 2012*

**Homage to timekeeping at
the Olympic Games in 1972**
*Hommage an die Zeitmessung
der Olympischen Spiele 1972*
*L'hommage au chronométrage
des Jeux olympiques de 1972*

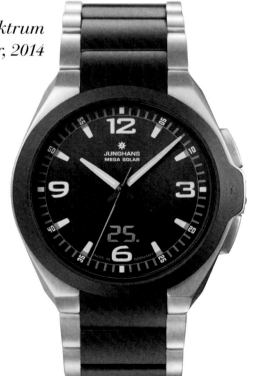

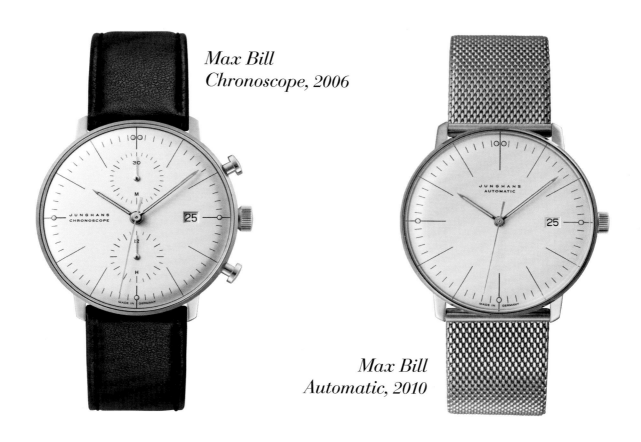

Max Bill
Chronoscope, 2006

Max Bill
Automatic, 2010

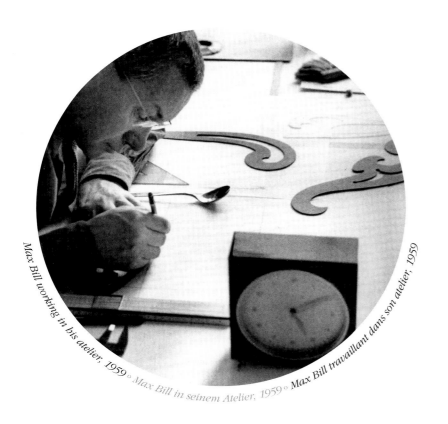

Max Bill working in his atelier, 1959 ∘ Max Bill in seinem Atelier, 1959 ∘ Max Bill travaillant dans son atelier, 1959

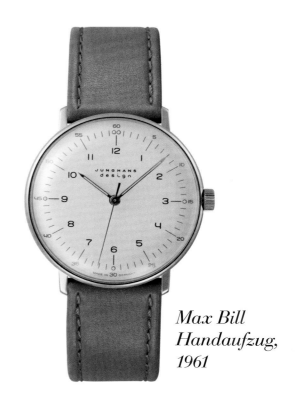

Max Bill
Handaufzug,
1961

A. Lange & Söhne

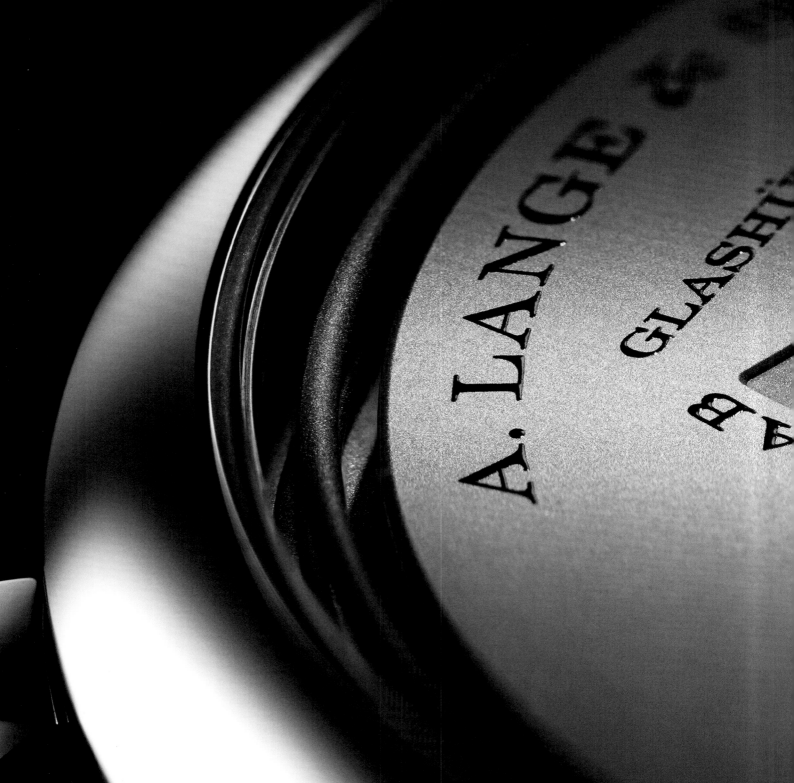

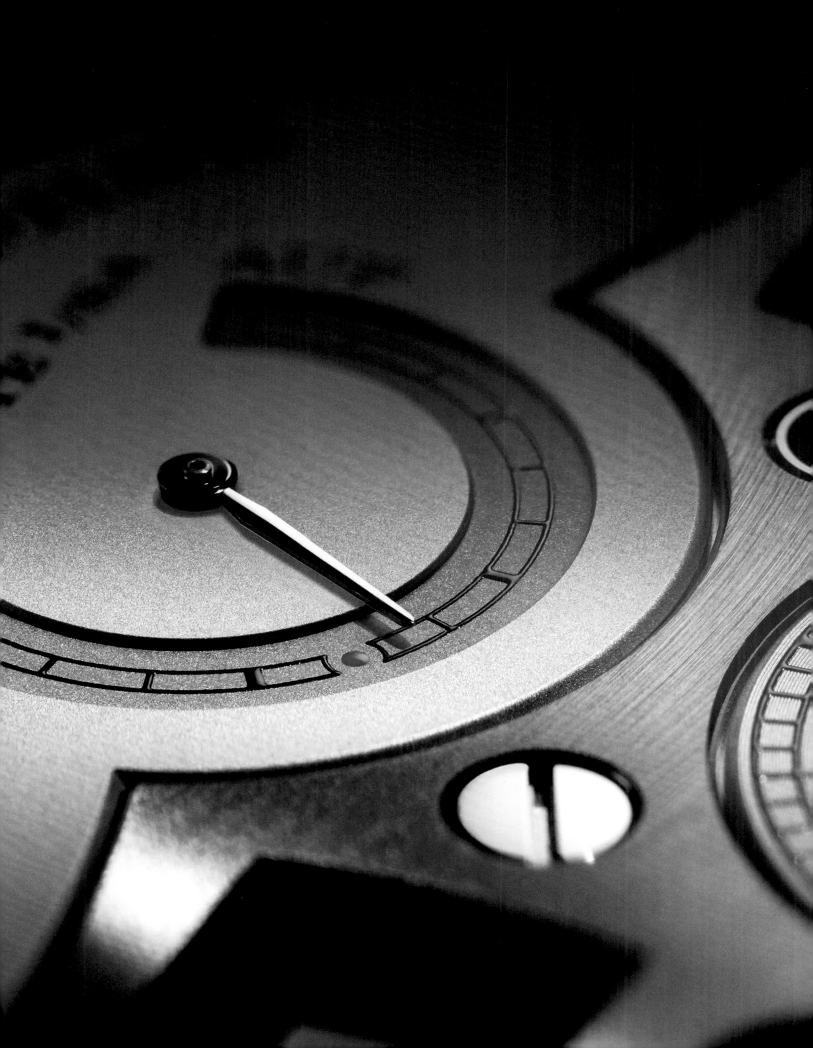

Risen from Ruins

English

Ferdinand Adolph Lange, born in Dresden on February 18, 1815, had no interest in luxury watches at first. After technical school and an apprenticeship under court watchmaker Johann Christian Friedrich Gutkaes, the young watchmaker embarked on journeyman years in France, England, and Switzerland. When he returned to Saxony, he was eager to bring his skilled craft to the economically underdeveloped Erzgebirge region, where poverty and unemployment were rampant. To improve the situation, the philanthropist penned numerous letters and entreaties to the government, respectfully requesting official support for his project of establishing watch manufacturing in Glashütte.

The mills of bureaucracy turn quite slowly, but Lange persevered. A contract was signed with the royal Saxon interior ministry on May 31, 1845. This document stipulated that Lange would offer three years of training to 15 young men from Glashütte, who would hopefully graduate as full-fledged watchmakers. In return, the state agreed to loan him 6,700 thalers, plus 1,120 thalers to purchase the necessary tools. After they had completed their training, the men were expected to work in Lange's production facility for an additional five years, during which time they would gradually repay their student loans in weekly installments of 24 neugroschen. The first list of trainees sounds more like a farm crew than a watchmaking school: a painter's assistant, twelve straw weavers, four messenger boys, a farmhand, a quarry worker, and a vintner's laborer. Lange unfortunately had no choice but to dismiss several

of these country lads as unqualified for the intended apprenticeship. A few others left voluntarily because they didn't enjoy the work at his school. The remainder completed their training and became the core of Lange's first team, which soon grew to 30 young professionals.

Despite Adolph Lange's unflagging dedication, his business strayed perilously close to failure several times. On the one hand, there was a severe shortage of qualified personnel. On the other hand, money too was in short supply. In the course of time, Lange invested his own personal savings, his wife's estate, and all other available funds in the teetering business.

When Adolph Lange died in 1875, he left his sons and grandsons a flourishing company with some 100 employees, as well as a large number of the highest international awards. Lange served for 18 years as the mayor of Glashütte, which owes him credit for transforming the town into a Mecca of fine German watchmaking. Countless small businesses and special workshops for watches and watch components were established under his aegis; other manufactories for watches and clocks were likewise founded.

But none of these businesses achieved international renown to rival that of the house of A. Lange & Söhne. The list of complicated Lange timepieces is lengthy and illustrious. These were affordable, of course, only for the most prosperous contemporaries. The

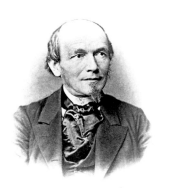
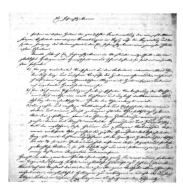
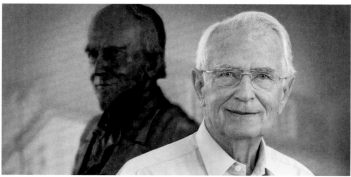
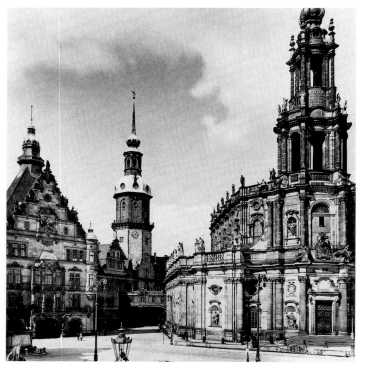

Clockwise from top left: Ferdinand Adolph Lange (1815–1875) ◦ Application for a loan from the government of Saxony ◦ Tower of the palace of the royal residence in Dresden, 1930s ◦ Walter Lange, * 1924 ◦ Von oben links im Uhrzeigersinn: Ferdinand Adolph Lange (1815–1875) ◦ Darlehensantrag an die sächsische Regierung ◦ Schlossturm der Stadtresidenz Dresden, 1930er Jahre ◦ Walter Lange, * 1924 ◦ Dans le sens horaire depuis la gauche : Ferdinand Adolph Lange (1815–1875) ◦ Demande de prêt au gouvernement de Saxe ◦ Tour du château de la Résidence de Dresde, années 1930 ◦ Walter Lange, * 1924

normal models, which encased a movement with a three-quarters plate that was designed in 1864 and used until 1948, were likewise too costly for many potential purchasers. Alongside fine and highly precise pocket-watches and marine chronometers, equally memorable are Lange's big wristwatches for professional pilots: these models encased a movement that was originally intended for pocket-watches. Lange also made a large series of wristwatches for civilians. These models encased either Glashütte ébauches or Swiss movements, for example, from Altus. For 100 years, A. Lange & Söhne offered everything that raised the pulse rates of watch aficionados.

In 1945, a dark chapter in Germany's history put an end to the activities of this important German watch manufactory. Fortunately, this end was only temporary. A. Lange & Söhne existed only through its horological legacy until 1990. The memorable events of 1989 brought the fall of the Berlin Wall and the arrival of ambitious visionaries: Walter Lange, who was the great-grandson of the firm's founder, joined forces with Günter Blümlein, who headed IWC and Jaeger-LeCoultre, and with Albert Keck, a trained watchmaker and the head of the advisory board of the VDO tachometer factory, whose roof sheltered the two Swiss traditional manufacturing capabilities. A contract signed on November 29, 1990 established the Lange Uhren GmbH, which would again make Glashütte into what it had been before the years of the German Democratic Republic: a recognized Mecca of the finest precision watchmaking.

The first collection of watches from the manufactory's new era debuted in 1994. These included icons such as the "Lange 1," which is still produced today, and the unique "Richard Lange Tourbillon Pour le Mérite," which is an avidly sought and costly collector's item. With these and other models, A. Lange & Söhne enabled the town of Glashütte to revive its industrial culture of precision timekeeping. In 1996, Mannesmann—which now owned IWC, Jaeger-LeCoultre, and A. Lange & Söhne—combined its three watch brands into a holding company, Les Manufactures Horlogères, which came under the roof of the Richemont Group in the wake of a spectacular purchase in July 2000.

Assistance from Swiss experts was indispensable during the first years after Germany's reunification because much know-how about the topic of luxury watches had been lost in Saxony during the socialist-influenced era. But thoroughly German manufacturing survived in the firm's long tradition. In the isolated Müglitz Valley, which had earned the unflattering nickname "Valley of Know-Nothings" under the GDR's communist regime because Western radio and television broadcasts couldn't be received there, A. Lange & Söhne now produces simple or complex timepieces, all of which rely on sophisticated mechanical movements. ◦

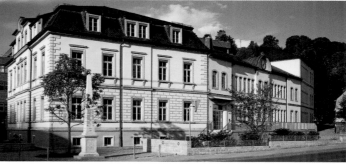

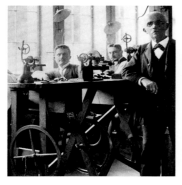
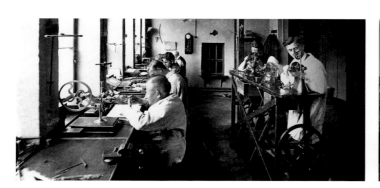
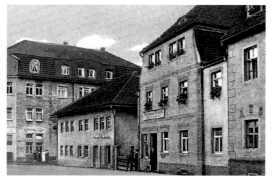
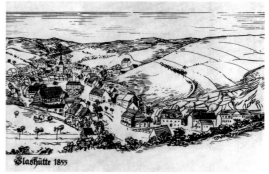

Clockwise from top left: assembling chronographs in 1907 ◦ The original company building in 2013 ◦ A view of Glashütte in 1855 ◦ German Watchmaking School ◦ Testing the rate and final quality control in 1907 ◦ Von oben links im Uhrzeigersinn: Bei der Chronometermontage 1907 ◦ Stammhaus 2013 ◦ Stadtansicht Glashütte 1855 ◦ Deutsche Uhrmacherschule ◦ Gangprüfer und Endkontrolle 1907 ◦ Dans le sens horaire depuis la gauche : Assemblage de chronographes, en 1907 ◦ Maison mère, en 2013 ◦ Vue de la ville de Glashütte, en 1855 ◦ École d'horlogerie allemande ◦ Contrôle de la marche et vérification finale, en 1907

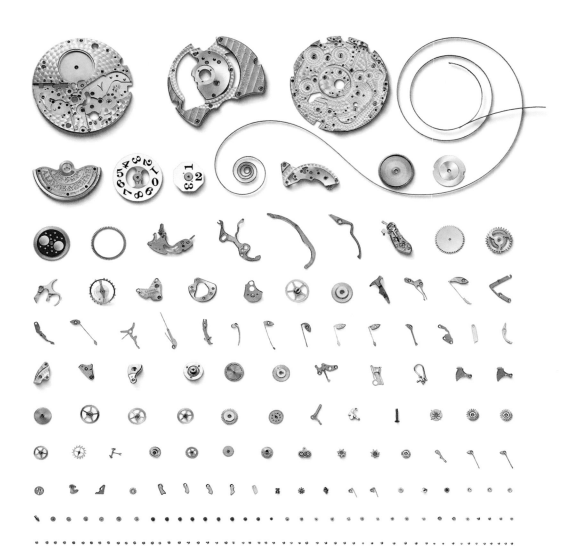

Left: the components of the
Langematik Perpetual
Links: Stückliste der
Langematik Perpetual
À gauche : Nomenclature
de la Langematik Perpetual

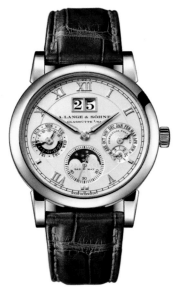

Top: Perpetual Calendar, 2002
Bottom: 1815 Tourbillon, 2014

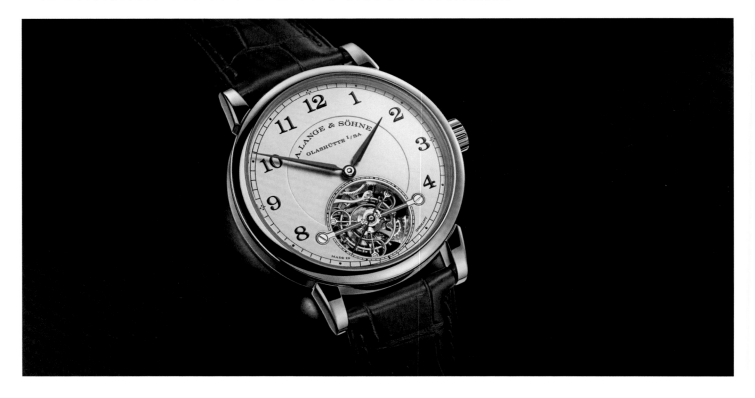

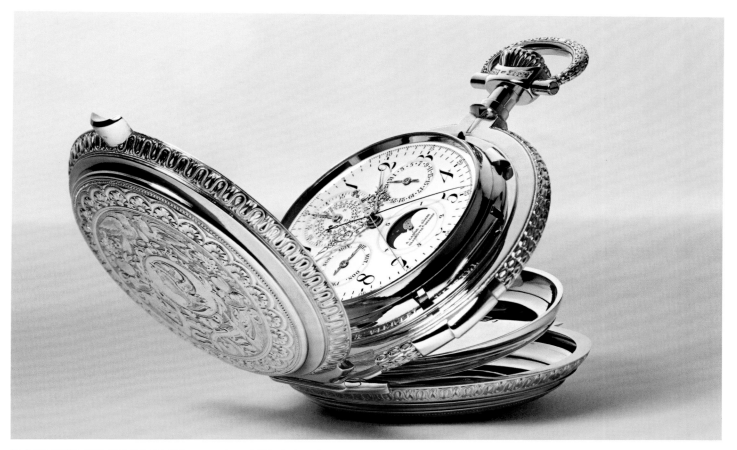

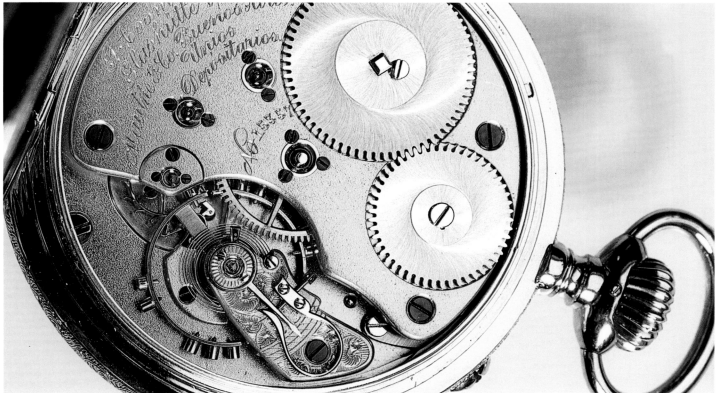

Top: Grande Complication, no. 42500, the most complicated pocket-watch ever built by A. Lange & Söhne, 1902 ◦ Bottom: the movement for a pocket-watch built for export to South America, ca. 1880/85 ◦ Oben: Grande Complication, Nr. 42500, die komplizierteste von A. Lange & Söhne je gebaute Taschenuhr, 1902 ◦ Unten: Werk einer für Südamerika gefertigten Taschenuhr, ca. 1880/85 ◦ En haut : Grande Complication, n° 42500, la montre de poche la plus complexe jamais fabriquée par A. Lange & Söhne, en 1902 ◦ En bas : Mouvement d'une montre fabriquée pour l'Amérique du Sud, vers 1880/1885

Auferstanden aus Ruinen

Mit Luxusuhren hatte er anfänglich nicht das Geringste im Sinn, der am 18. Februar 1815 in Dresden geborene Ferdinand Adolph Lange. Auf den Besuch der Technischen Bildungsanstalt zusammen mit einer gründlichen Ausbildung beim Hofuhrmacher Johann Christian Friedrich Gutkaes folgten uhrmacherische Wanderjahre nach Frankreich, England und in die Schweiz. Zurück in Sachsen war es ihm ein Anliegen, der strukturschwachen Erzgebirgsregion sein Handwerk näherzubringen. Dort war allerdings Schmalhans Küchenmeister. Um die große Not und Arbeitslosigkeit ein wenig lindern zu können, wandte sich der Philanthrop in mehreren Briefen und Eingaben an seine Regierung. Er bat darum, ihn doch gnädigst bei der Errichtung einer Uhrenproduktion in Glashütte zu unterstützen.

Bekanntlich mahlen Behördenmühlen langsam. Aber Lange war zäh und hartnäckig. Am 31. Mai 1845 kam es zwischen ihm und dem Königlich Sächsischen Ministerium des Innern zum Abschluss eines Vertrags. Dieser legte fest, dass Lange 15 Glashütter Jugendliche innerhalb von drei Jahren zu Uhrmachern ausbilden würde. Im Gegenzug stellte der Staat einen rückzahlbaren Vorschuss in Höhe von 6700 Talern sowie 1 120 Taler zur Anschaffung von Werkzeugen bereit. Nach Abschluss der Ausbildung sollten die Jugendlichen weitere fünf Jahre in Langes Betrieb arbeiten. In Wochenraten zu je 24 Neugroschen hatten sie währenddessen die Kosten ihrer Ausbildung abzustottern. Die erste Liste der Auszubildenden klang eher nach einem Landwirtschaftsbetrieb, als nach einer Uhrmacherei: ein Malgehilfe, zwölf Strohflechter, vier Dienstburschen, ein Landwirtschaftsgehilfe, ein Steinbruch- und ein Winzerarbeiter. Einigen der Naturburschen musste Lange zu seinem Leidwesen bereits nach kurzer Zeit die berufliche Nichteignung bescheinigen. Andere gingen von alleine, weil ihnen die Arbeit nicht behagte. Der Rest hielt durch und bildete den Stamm der ersten Lange-Mannschaft, die schon bald auf 30 Berufsanfänger wuchs.

Trotz des übermäßigen Engagements von Adolph Lange drohte das Unternehmen mehrfach zu scheitern. Zum einen stand kaum qualifiziertes Personal zur Verfügung. Zum anderen fehlte es immer wieder an Geld. Im Laufe der Zeit investierte Lange sein ganzes Vermögen, das seiner Frau sowie sonstige verfügbare Mittel in die holprig anlaufende Firma.

Als Adolph Lange 1875 starb, hinterließ er seinen Söhnen und Enkeln nicht nur einen florierenden Betrieb mit rund 100 Mitarbeitern, sondern auch eine Vielzahl höchster internationaler Auszeichnungen. Die Ortschaft Glashütte verdankt Lange, der hier 18 Jahre lang auch Bürgermeister war, die Entwicklung zum Mekka der deutschen Feinuhrmacherei. Unzählige Kleinbetriebe und Spezialwerkstätten für Uhren und Uhrenbestandteile entstanden unter seiner Ägide. Dazu kamen weitere Manufakturen für Klein- und Großuhren.

Keines dieser Unternehmen erreichte allerdings das internationale Renommee des Hauses A. Lange & Söhne. Immerhin füllt die Palette komplizierter Lange-Zeitmesser Bände. Diese konnten sich freilich nur ausgesprochen betuchte Zeitgenossen leisten. Doch auch die normalen Modelle mit dem 1864 konstruierten und bis 1948 verwendeten Dreiviertelplatinenwerk waren rein preislich nicht jedermanns Kragenweite. Neben den feinen Präzisionstaschenuhren und Marinechronometern dürfen die großen professionellen Fliegerarmbanduhren mit Taschenuhrwerk nicht vergessen werden. Daneben gab es bei Lange auch eine ganze Reihe ziviler Armbanduhren, in denen jedoch neben Glashütter Rohwerken häufig auch Schweizer Fabrikate tickten, z. B. von Altus. Somit offerierte A. Lange & Söhne 100 Jahre lang nahezu alles, was das Herz ausgewiesener Uhr-Aficionados höherschlagen lässt.

Im Jahr 1945 beendete ein dunkles Kapitel deutscher Geschichte die Aktivitäten der bedeutenden deutschen Uhrenmanufaktur. Fürs Erste jedenfalls. Bis 1990 existierte A. Lange & Söhne nämlich allein durch das uhrmacherische Vermächtnis. Mit den bewegenden Ereignissen des Jahres 1989 und dem darauf folgenden Mauerfall schlug abermals die Stunde ambitionierter Visionäre: Walter Lange, Urenkel des Firmengründers, Günter Blümlein, Chef von IWC und Jaeger-LeCoultre, sowie Albert Keck, gelernter Uhrmacher und Aufsichtsratchef des Tachometerfabrikanten VDO, unter dessen Dach sich besagte Schweizer Traditionsmanufakturen befanden. Am 29. November 1990 kam es zur Unterzeichnung eines Vertrags zur Gründung der Lange Uhren GmbH, welche aus Glashütte wieder das machen würde, was es zu Zeiten vor der Deutschen Demokratischen Republik gewesen war: ein anerkanntes Mekka feinster Präzisionsuhrmacherei.

1994 debütierte die erste Uhrenkollektion der Neuzeit. Zu ihr gehörten Ikonen wie die bis heute produzierte „Lange 1" oder die einzigartige, von Sammlern teuer bezahlte „Richard Lange Tourbillon Pour le Mérite". Mit diesen und anderen Modellen verhalf A. Lange & Söhne der Ortschaft Glashütte erneut zu einer industriellen Kultur rund um die Präzisionszeitmessung. 1996 vereinigte Mannesmann – wozu IWC, Jaeger-LeCoultre und A. Lange & Söhne mittlerweile gehörten – seine drei Uhrenmarken in der Holding Les Manufactures Horlogères, welche im Juli 2000 im Zuge eines spektakulären Verkaufs unter das Dach des Richemont-Konzerns gelangte.

Ohne tatkräftige Schweizer Hilfe ging es damals in der Zeit nach der „Wende" nicht, denn während der sozialistisch geprägten Ära war in Sachsen viel Know-how zum Thema Luxusuhren verlorengegangen. Was in der langen Firmentradition erhalten blieb, ist dagegen die durchweg deutsche Manufaktur. Im einstigen „Tal der Ahnungslosen", wie das abgeschiedene Müglitztal zu Zeiten des DDR-Regimes hieß, weil westliche Radio- und Fernsehsender dort nicht zu empfangen waren, entwickelt und fertigt A. Lange & Söhne auch heute seine einfachen oder komplizierten, in jedem Fall aber anspruchsvollen mechanischen Uhrwerke. ∘

Assembling a perpetual calendar and (bottom right) engraving the balance-bridge ∘ Montage eines ewigen Kalenders und (unten rechts) Gravur der Unruhbrücke ∘ Assemblage d'un quantième perpétuel et (en bas à droite) gravure du pont de balancier

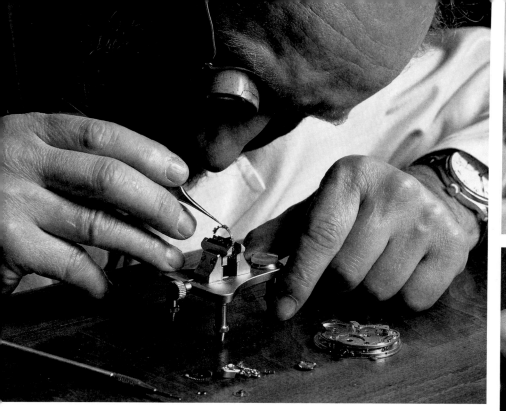
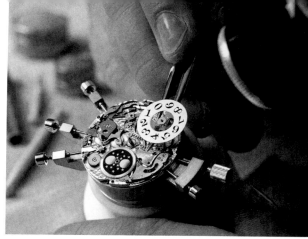
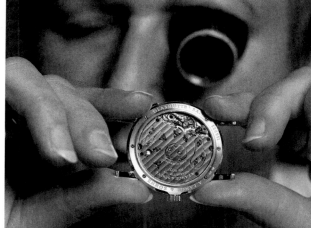
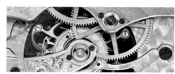
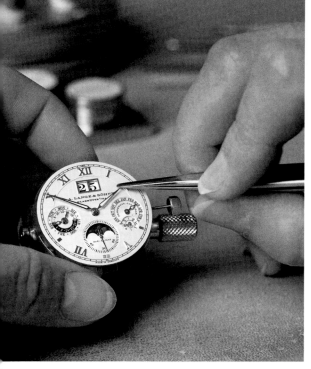
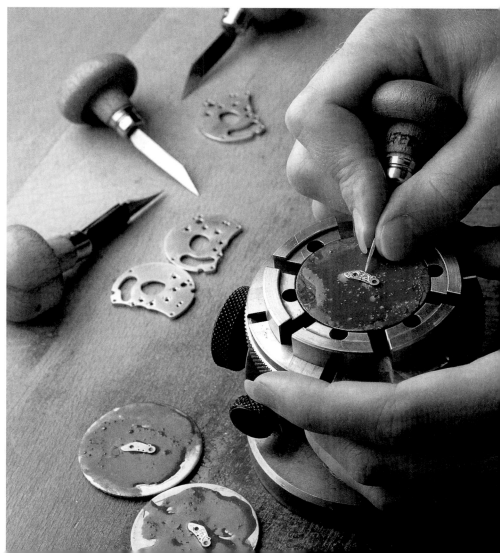
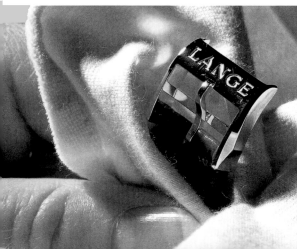

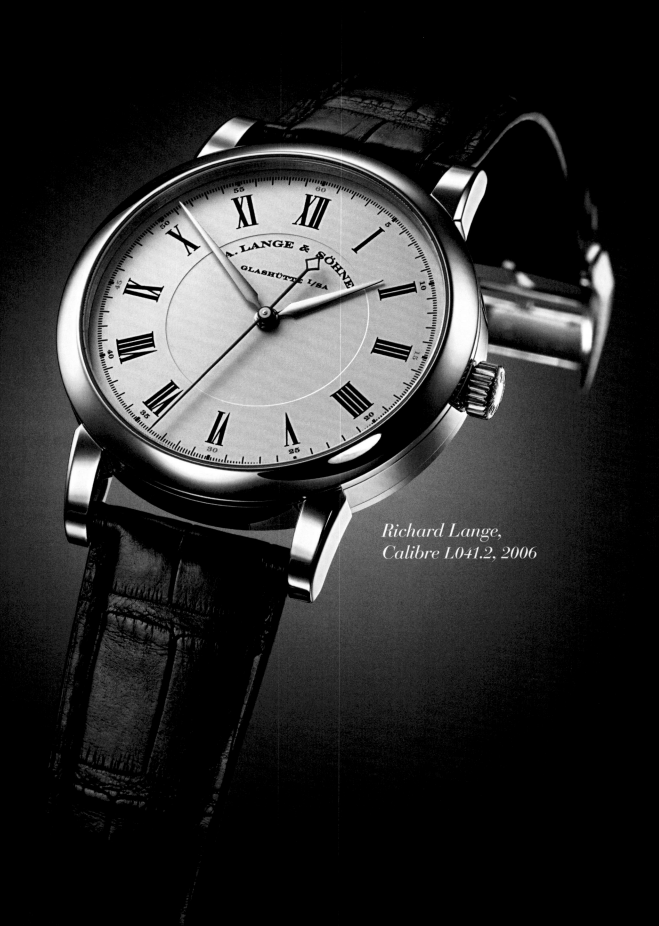

Richard Lange,
Calibre L041.2, 2006

Ressuscité des ruines

Français

Les montres de luxe sont, à l'origine, à mille lieues des préoccupations de Ferdinand Adolph Lange, né le 18 février 1815 à Dresde. Ayant reçu une solide formation auprès de Johann Christian Friedrich Gutkaes, horloger à la cour de Saxe, puis à l'Institut technique de Dresde, il parfait ses connaissances horlogères pendant son voyage de compagnonnage, qui le mène en France, en Angleterre et en Suisse. Il revient en Saxe animé par la volonté de diffuser son savoir-faire dans les monts Métallifères. Les habitants de cette région en crise sont alors des crève-la-faim. Pour atténuer la disette et le chômage, le philanthrope adresse lettres et requêtes au gouvernement de Saxe, sollicitant de sa haute bienveillance une aide pour créer une manufacture de montres à Glashütte.

Comme chacun sait, il faut affronter les lenteurs de l'administration. Mais à force de ténacité et d'obstination, Lange signe le 31 mai 1845 un contrat avec le ministère de l'Intérieur du royaume de Saxe. Ce document stipule que Lange s'engage à former en trois ans 15 jeunes gens au métier d'horloger. En contrepartie, l'État lui accorde un crédit de 6 700 thalers ainsi que 1 120 thalers pour l'achat d'outils. Les horlogers ainsi formés doivent rester au service de Lange pendant cinq ans, le temps de rembourser les frais de leur formation, à raison de 24 nouveaux groschen par semaine. La toute première liste d'apprentis évoque davantage une exploitation agricole qu'une manufacture horlogère : un apprenti peintre, douze pailleurs, quatre commis, un aide agricole, un carrier et un vigneron. À son grand regret, Lange se voit rapidement contraint de remercier quelques-uns de ces gars de la campagne, inaptes à exercer ce métier. Certains, n'ayant guère de goût pour ce travail, partent d'eux-mêmes. Quant aux autres, ils tiennent le coup et constitueront le gros de la première équipe Lange, qui ne tardera pas à compter dans ses rangs 30 novices dans le métier.

Malgré l'investissement personnel hors pair de Ferdinand Adolph Lange, l'entreprise frôle la faillite à plusieurs reprises : il y a pénurie de personnel qualifié, et les fonds viennent régulièrement à manquer. Au fil des années, Lange investit toute sa fortune, à savoir celle de sa femme ainsi que diverses liquidités, dans cette entreprise qui marche cahin-caha.

Adolph Lange disparaît en 1875, laissant à ses fils et ses petits-enfants non seulement une manufacture florissante qui compte une centaine de salariés, mais s'enorgueillit aussi d'une kyrielle de distinctions internationales éminentes. Grâce à Lange, maire de Glashütte pendant 18 ans, cette localité devient la Mecque de la haute horlogerie allemande. Une infinité de petites entreprises et d'ateliers spécialisés dans les montres et leurs composants y voient le jour sous son égide. Et d'autres manufactures horlogères s'y installent également.

Cependant, aucune de ces entreprises n'atteindra la renommée internationale de la maison A. Lange & Söhne. Il faut dire que sa palette de garde-temps à complications est très étendue. Mais seuls des contemporains très fortunés en ont les moyens, les modèles standard équipés jusqu'en 1948 de la platine trois quarts conçue en 1864 n'étant pas non plus à la portée de toutes les bourses. Outre les montres de poche de précision raffinées et les chronomètres de marine, il ne faut pas non plus oublier les grandes montres-bracelets équipées d'un mouvement de montre de poche. Lange propose en outre toute une série de montres-bracelets à usage civil dans lesquelles des produits suisses, par exemple d'Altus, côtoient souvent des ébauches réalisées à Glashütte. Ainsi A. Lange & Söhne offrira pendant un siècle presque tout ce qui fait vibrer les amateurs avérés de montres.

En 1945, un chapitre sombre de l'histoire allemande met fin aux activités de la plus grande manufacture horlogère du pays, ne serait-ce que dans un premier temps. Jusqu'en 1990, A. Lange & Söhne n'existe en effet plus qu'à travers son héritage horloger. Il faut attendre les émouvants événements de 1989, qui se sont accompagnés de la chute du Mur et de l'entrée en scène, une nouvelle fois, de grands visionnaires, parmi lesquels Walter Lange, arrière-petit-fils du fondateur de l'entreprise, Günter Blümlein, à la tête d'IWC et de Jaeger-LeCoultre, mais aussi Albert Keck, horloger de formation et président du conseil d'administration du fabricant de tachymètres VDO, désormais propriétaire de ces deux manufactures suisses de tradition. Le 29 novembre 1990 voit la signature d'un contrat portant création de Lange Uhren GmbH, l'objectif étant de refaire de Glashütte ce qu'il était avant l'ère de la République démocratique allemande : le centre par excellence de l'horlogerie de précision allemande.

L'année 1994 salue la sortie de la première collection de montres de l'ère nouvelle. Celle-ci comprend des icônes, telles que la « Lange 1 », encore produite de nos jours, ou une montre sans pareille que les collectionneurs achètent à prix d'or, la « Richard Lange Tourbillon Pour le Mérite ». Grâce à ces modèles et à d'autres encore, A. Lange & Söhne aide Glashütte à se reconstituer une culture industrielle fondée sur l'horlogerie de précision. En 1996, Mannesmann – qui a entre-temps repris IWC, Jaeger-LeCoultre et A. Lange & Söhne – regroupe ces trois marques dans une holding, Les Manufactures Horlogères, dont le groupe Richemont prendra le contrôle en juillet 2000, à la suite d'un rachat spectaculaire.

Sans une aide énergique venue de Suisse, une renaissance n'est pas possible après la réunification de l'Allemagne, car sous le régime socialiste, la Saxe a perdu une grande partie de son savoir-faire en matière de montres de luxe. Et s'il reste quelque chose de la longue tradition familiale, c'est la manufacture spécifiquement allemande. Dans l'ancienne « vallée des ignorants », comme on surnommait la vallée de la Müglitz à l'époque de la RDA parce qu'on n'y captait pas les émissions de radio et de télévision de RFA, A. Lange & Söhne développe et produit aujourd'hui encore des mouvements mécaniques, simples ou compliqués, mais qui répondent toujours aux plus hautes exigences. ∘

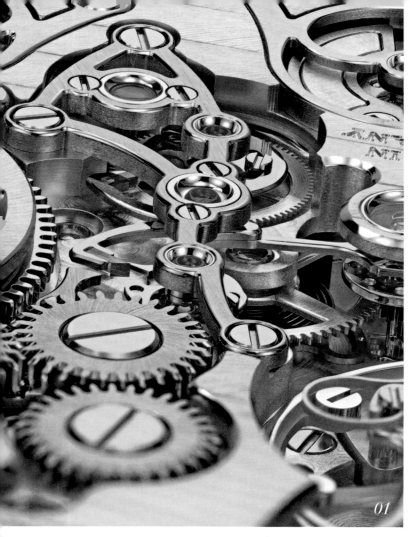

01

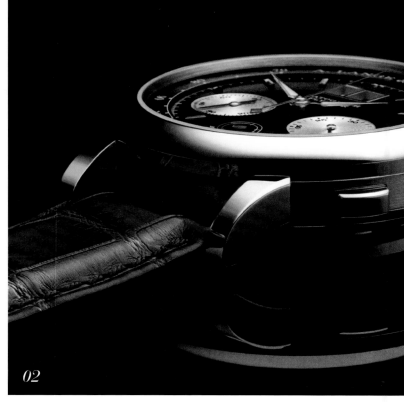

02

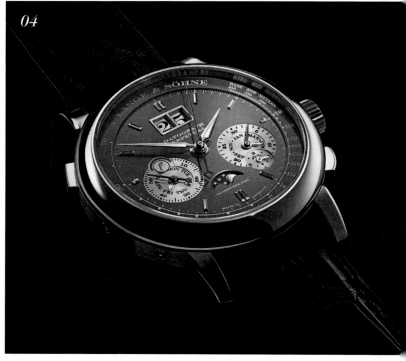

04

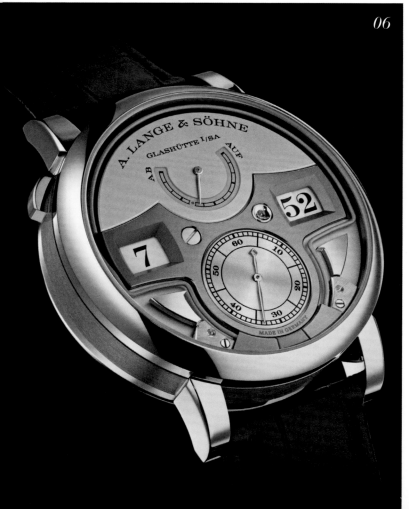

06

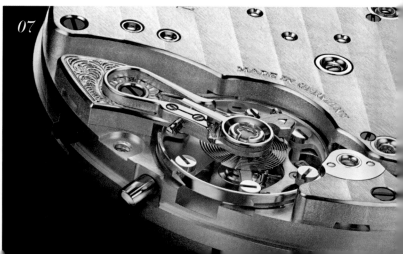

07

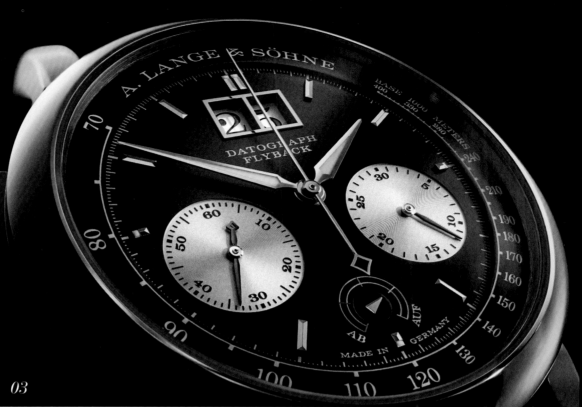

03

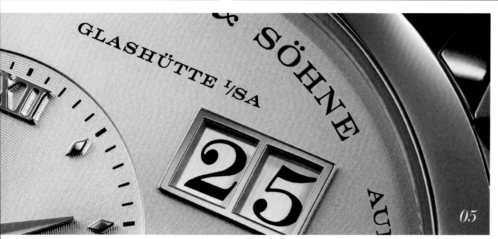

05

01. Calibre L043.1
02./03. Datograph Fly, 2006
04. Datograph Perpetual/Moon, 2015
05. Outsize Date, Grand Lange 1, 2015
06. Zeitwerk Minute Repeater, 2009
07. Lange 1, caliber detail, 2015
08. Saxonia Dual Time, Saxonia,
   Saxonia Automatic, 2015

08

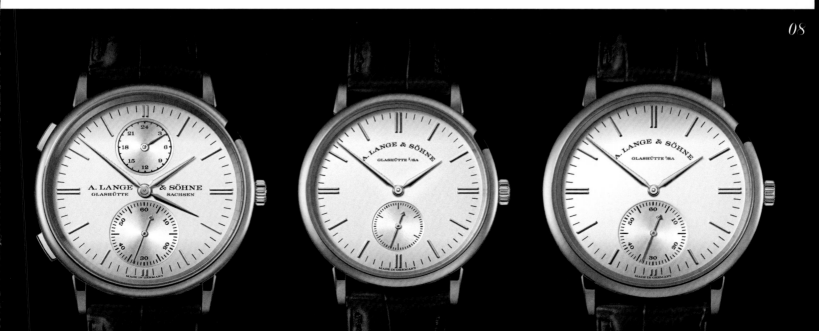

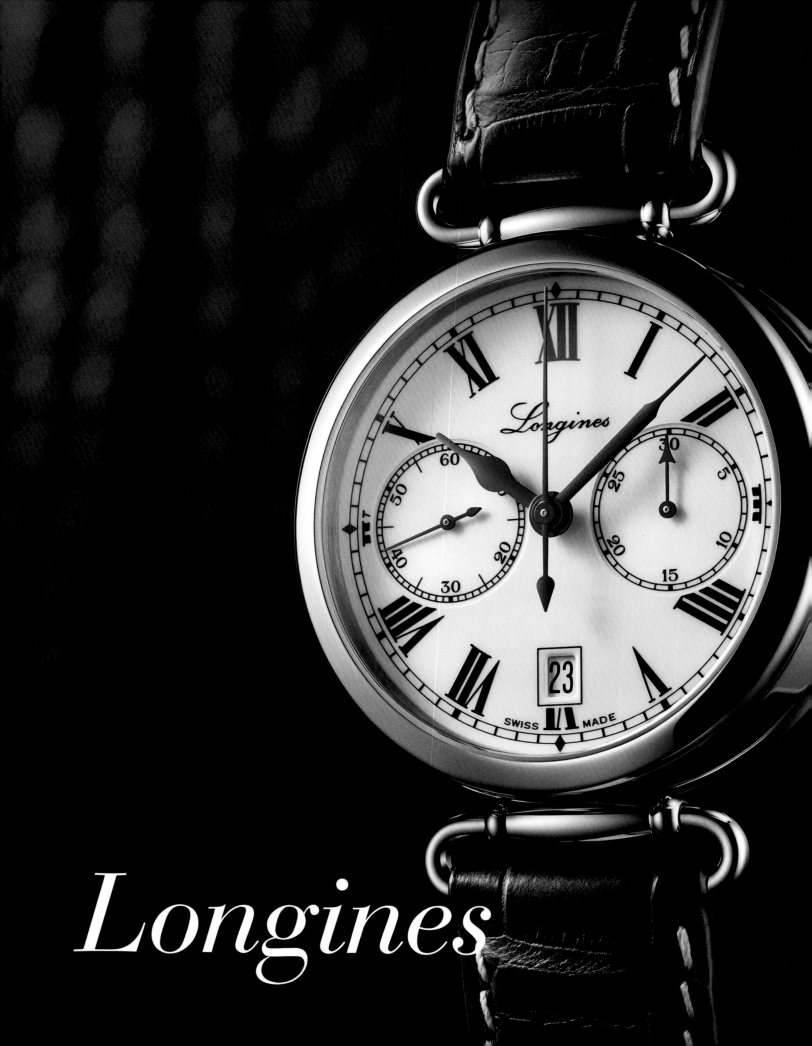

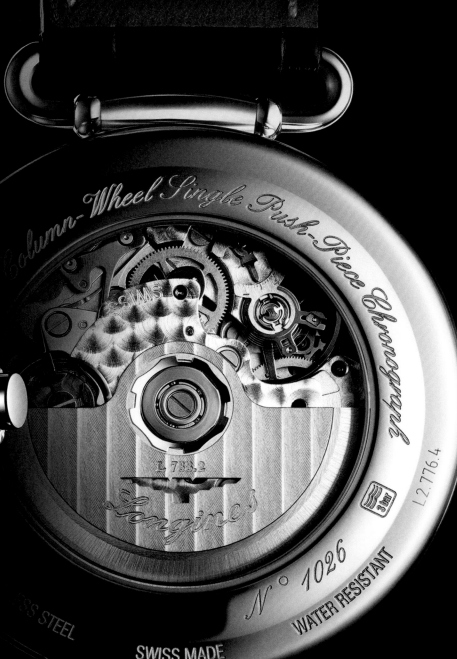

Column-Wheel Single Push-Piece Chronograph

L.788.2

Longines

N° 1026

WATER RESISTANT

L2.776.4

SWISS STEEL

SWISS MADE

In Saint-Imier Since 1832

English

Sometimes a statistic is proof of success: for example, the fact that Longines manufactured approximately one million timepieces in 2015. The manufacturing spectrum primarily includes wristwatches characterized by a remarkably good value for the money. A look into the history of Longines, which was founded on August 14, 1832, reveals that this company has alternately undergone fat and lean years. More or less resounding successes regularly recurred. The greatest crisis, which was sparked by the Quartz Revolution in the 1970s, brought Longines in 1983 under the aegis of what is now the Swatch Group. The traditional Longines brand, which annually earns an estimated 1.5 billion Swiss francs, currently ranks second behind first-place Omega as the Swatch Group's strongest source of revenue. Longines' history began with Auguste Agassiz, a 23-year-old merchant who went into business for himself by opening a flourishing watch *comptoir* in the village of Saint-Imier in Switzerland's Jura region in 1832. Health issues prompted him take Ernest Francillon on board in 1852. Two years later, this 20-year-old newcomer was already serving as executive director. Despite thriving business deals, Francillon had to admit that *comptoir* watchmaking had become obsolete. One way or another, each watch arrived in its owner's hands as a handmade one-of-a-kind artifact. Series manufacturing, problem-free interchangeability of components for repairs, and speedy customer service were accordingly inconceivable. With the goal of improving this situation and ultimately establishing large-scale industrialized watchmaking, Ernest Francillon erected his first factory building on a plot of land known as "Les Longines" (i.e., "The Long Meadows") on the shores of the Suze River in 1866. He displayed the first watches bearing the signature "E. Francillon, Longines, Suisse" at the Exposition Universelle in Paris one year later. To protect his products against imitators, Francillon began using the winged sandglass as his logo in 1874. Official protection was granted to Longines' signature in combination with the eye-catching logo in 1889. The Swiss Federal Institute of Intellectual Property cites this as the oldest registration, which has undergone only marginal alternations over the years. The trademark is also registered with the World Intellectual Property Organization (WIPO). Longines unveiled its first chronographs, which now comprise a specialty for Longines, in 1878. In subsequent decades, the manufactory developed diverse chronograph calibers, all of which are avidly coveted by contemporary collectors. The legendary 13.33Z for wristwatches debuted in 1913. Timepieces for the wrist have been part of Longines' collection since 1905. Longines revolutionized athletic timekeeping in 1912, when it developed the broken-thread system to automatically measure time at races. Longines first served as official timekeeper for the Olympic Games in Helsinki forty years later. Collaboration with Charles A. Lindbergh, who piloted the first solo flight across the Atlantic on May 21, 1927, led to the development in 1931 of the famous "Hour Angle Watch," which helped pilots with the difficult task of airborne navigation. Longines' first self-winding movement (Caliber 22A) debuted in 1945, followed by the world's first portable quartz timepiece in 1954, the slimmest wristwatch caliber with electromagnetic power in 1960, and the world premiere of a liquid crystal display for quartz wristwatches in 1972. The manufacture put a crowning touch on its portfolio of automatic movements with the 1977 debut of ultra-slim Caliber L990. Longines announced the sale of its fifteen millionth watch in 1967. As a member of the Swatch Group, Longines no longer fabricates its own movements. This brand's cases now contain either mechanical or electronic movements supplied by Longines' affiliate Eta. These also include exclusive developments such as a chronograph caliber with a classical column-wheel and rocking-pinion coupling. Caliber L688 and its derivatives tick inside modernly styled wristwatches and in numerous retro models: the latter recall Longines' long history in service of time measurement. ◦

Self-winding movement Cal. 22A, 1945

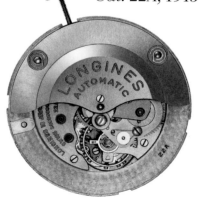

Chronograph Cal. 13.33Z, 1913

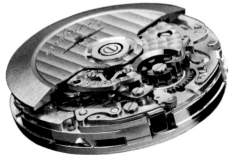

Cal. L688, 2012

Seit 1832 in Saint-Imier

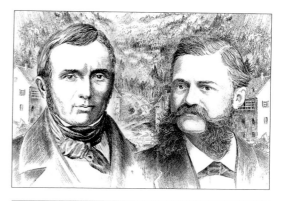

So etwas nennt man Erfolg: Rund eine Million Zeitmesser stellte Longines 2015 her. Das Fertigungsspektrum umfasst größtenteils Armbanduhren, welche sich durch ein bemerkenswert gutes Preis-Leistungs-Verhältnis auszeichnen. Beim Blick zurück in die Geschichte zeigt sich, dass Longines seit der Gründung am 14. August 1832 sehr wechselvolle Perioden erlebt hat. Mehr oder minder große Erfolge wechselten sich mit schöner Regelmäßigkeit ab. Die größte Krise, ausgelöst durch die Quarz-Revolution in den 1970er Jahren, führte Longines 1983 unter das Dach der heutigen Swatch Group. Dort findet sich das Traditionsunternehmen mit geschätzten 1,5 Milliarden Schweizer Franken Umsatz heute nach Omega auf Platz zwei. Die biographischen Anfänge sind mit Auguste Agassiz verknüpft, einem 23-jährigen Kaufmann, der sich in besagtem Jahr 1832 im Juradorf Saint-Imier mit einem florierenden Uhren-Comptoir selbständig machte. Gesundheitliche Probleme brachten 1852 Ernest Francillon an Bord. Zwei Jahre später fungierte der gerade einmal 20-Jährige bereits als verantwortlicher Direktor. Trotz blühender Geschäfte musste er eingestehen, dass sich die Comptoir-Uhrmacherei überlebt hatte. Irgendwie gelangte jede Uhr als handgearbeitetes Unikat zum Kunden. An Serienfertigung, die problemlose Austauschbarkeit der Komponenten im Servicefall und einen zügigen Kundendienst war so nicht zu denken. In diesem Sinn ließ Ernest Francillon, dem eine groß angelegte Uhrenfertigung vorschwebte, 1866 am unter dem Namen Les Longines („die länglichen Wiesen") bekannten Gelände am Ufer des Flüsschens Suze sein erstes Fabrikgebäude errichten. Schon ein Jahr später nahm er mit den ersten Uhren mit der Signatur E. Francillon, Longines, Suisse an der Weltausstellung in Paris teil. Zur Abwehr von Nachahmern nutze Francillon ab 1874 die geflügelte Sanduhr als Logo. Der offizielle Schutz des Schriftzugs Longines in Verbindung mit dem signifikanten Logo erfolgte 1889. Das Eidgenössische Institut für Geistiges Eigentum spricht von der ältesten, im Laufe der Zeit nur marginal veränderten Registrierung, welche übrigens auch bei der Weltorganisation für geistiges Eigentum (World Intellectual Property Organization, WIPO) hinterlegt ist. Den ersten Chronographen, eine anerkannte Domäne des Unternehmens, präsentierte Longines im Jahr 1878. In den anschließenden Jahrzehnten entwickelte die Manufaktur ganz unterschiedliche, bei Sammlern äußerst begehrte Kaliber dieser Art. 1913 debütierte das legendäre 13.33Z für Armbanduhren. Zeitmesser fürs Handgelenk finden sich seit 1905 in der Kollektion. 1912 revolutionierte das von Longines entwickelte Fadenriss-System zur automatischen Zeiterfassung die Sportzeitmessung. Vierzig Jahre später trat Longines in Helsinki erstmals als offizieller Zeitnehmer bei Olympischen Spielen auf. Eine Kooperation mit Charles A. Lindbergh, der am 21. Mai 1927 im Alleinflug den Atlantik überquert hatte, führte zur berühmten Stundenwinkel-Armbanduhr. Ab 1931 erleichterte sie vielen Piloten die schwierige Navigation in den Lüften. 1945 brachte das erste Automatikwerk von Longines, Kaliber 22A genannt, 1954 die weltweit erste transportable Quarzuhr, 1960 das damals flachste Armbanduhr-Kaliber mit elektromagnetischem Antrieb und 1972 als Weltpremiere die Flüssigkristallanzeige in Quarz-Armbanduhren. 1977 krönte die Manufaktur ihre Palette an Automatik-Uhrwerken mit dem ultraflachen Kaliber L990. Den Verkauf seiner 15-millionsten Uhr hatte Longines im Jahr 1967 verkündet.

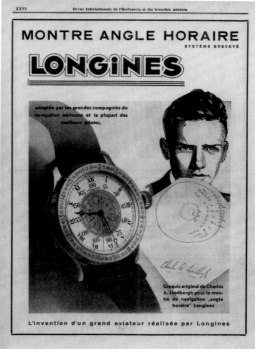

Als Mitglied der Swatch Group fertigt Longines keine eigenen Uhrwerke mehr. In definitiv allen Gehäusen finden sich heutzutage mechanische oder elektronische Uhrwerke der Schwester Eta. Dazu gehören auch exklusive Entwicklungen wie zum Beispiel ein Chronographenkaliber mit klassischer Schaltradsteuerung und Schwingtrieb-Kupplung. Das L688 und seine Derivate ticken in Armbanduhren mit modernem Outfit ebenso wie in zahlreichen Retromodellen, welche an die lange Geschichte im Dienste der Zeitmessung erinnern. ◦

First chronograph, 1878

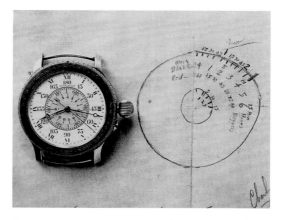

From top: Auguste Agassiz and Ernest Francillon ◦ Advertisement featuring Charles A. Lindbergh, 1931 ◦ Longines Lindbergh, 1931 ◦ Von oben: Auguste Agassiz und Ernest Francillon ◦ Werbemotiv mit Charles A. Lindbergh, 1931 ◦ Longines Lindbergh, 1931 ◦ De haut en bas : Auguste Agassiz et Ernest Francillon ◦ Publicité avec Charles A. Lindbergh, 1931 ◦ Longines Lindbergh, 1931

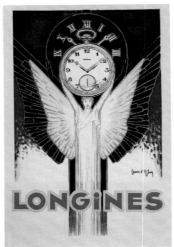

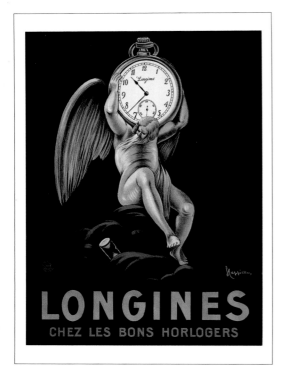

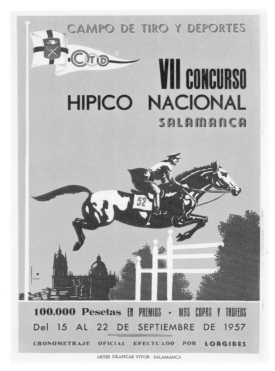

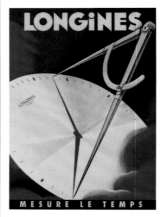

1911

1957

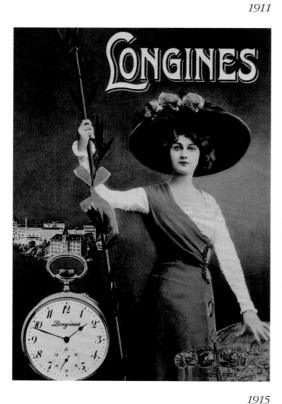

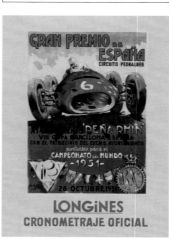

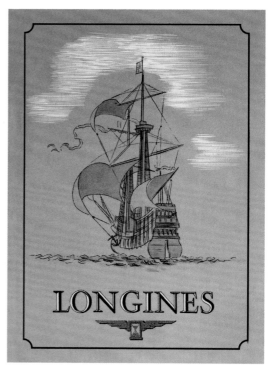

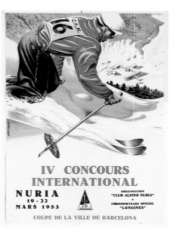

1915

1960

From top: 1929, 1945, 1951, 1953

À Saint-Imier depuis 1832

Français

Environ un million de montres en 2015, c'est indéniablement un succès. Dans la gamme de produits, on trouve en majorité des montres-bracelets au rapport qualité-prix remarquable. Un retour sur l'histoire de la marque permet de constater qu'elle a connu des hauts et des bas depuis sa création, le 14 août 1832. Des succès plus ou moins importants se sont succédé avec une belle régularité. La plus grande crise, déclenchée par la révolution du quartz dans les années 1970, a conduit Longines en 1983 dans le giron de Swatch Group. Entreprise de tradition au chiffre d'affaires estimé à 1,5 milliard de francs suisses, Longines est aujourd'hui deuxième au sein de la multinationale, derrière Omega. Les débuts remontent à Auguste Agassiz, un commerçant de 23 ans qui s'établit à son compte en cette année 1832 en ouvrant un comptoir horloger prospère dans le village jurassien de Saint-Imier. Des problèmes de santé l'amènent à engager Ernest Francillon en 1852. À peine deux ans plus tard, à tout juste 20 ans, celui-ci dirige le comptoir. Malgré des affaires florissantes, il doit reconnaître que l'horlogerie de comptoir appartient au passé. Chaque montre livrée étant un spécimen unique monté à la main, on ne peut envisager ni de fabrication en série, ni de remplacement facile de composants en cas de réparation, ni de service rapide. Aussi, Ernest Francillon, qui rêve de fabriquer des montres à grande échelle, fait édifier en 1866 une première usine sur le terrain des Longines (« les prairies allongées ») qui borde le petit cours d'eau de la Suze. Un an plus tard seulement, il présente à l'Exposition universelle de Paris les premières montres signées « E. Francillon, Longines, Suisse ». Pour se démarquer, il utilise dès 1874 comme logotype le sablier ailé. Cet emblème caractéristique et le nom Longines seront officiellement déposés en 1889. L'Office fédéral de la propriété intellectuelle (OFPI) indique que Longines est la plus ancienne marque enregistrée sans interruption ni modification majeure au cours du temps et qu'elle est par ailleurs déposée auprès de l'Organisation mondiale de la propriété intellectuelle (OMPI). En 1878, Longines présente sa première création dans un de ses domaines forts, les chronographes. Durant les décennies suivantes,

la manufacture produit une large gamme de calibres de ce type, très prisés des collectionneurs. En 1913 sort le légendaire 13.33Z pour montres-bracelets, qui figurent d'ailleurs dès 1905 dans la collection. En 1912, Longines révolutionne le chronométrage sportif par le système électromécanique selon le système du fil coupé. Quarante ans plus tard, Longines est chronométreur officiel des Jeux olympiques d'Helsinki. De la coopération avec Charles A. Lindbergh, qui le 21 mai 1927 survole l'Atlantique en solitaire et sans escale, naît la célèbre montre Lindbergh à angle horaire. Dès 1931, elle facilite la navigation pour de nombreux aviateurs. En 1945 paraît le premier mouvement à remontage automatique de Longines, référencé 22A, en 1954 la première horloge à quartz portative au monde, en 1960 le plus mince calibre de montre-bracelet à entraînement électromagnétique jamais conçu, et en 1972, première mondiale, l'affichage numérique sur écran à cristaux liquides pour montres-bracelets à quartz. En 1977, le calibre ultraplat L990 couronne la palette de composants de mouvements automatiques. La 15 millionième montre de la marque est sortie en 1967. Mais Longines, membre de la multinationale Swatch Group, ne fabrique plus elle-même ses mouvements. Tous les boîtiers sans exception sont désormais équipés de calibres mécaniques ou électroniques de la société sœur Eta SA. Parmi ces mouvements figurent des réalisations exclusives, notamment un calibre de chronographe à traditionnels roue à colonnes et pignon oscillant. Le calibre L688 et ses dérivés animent aussi bien les montres-bracelets à l'habillage moderne que de nombreux modèles rétro, témoins d'une longue histoire au service de la mesure du temps. ◦

Conquest Classic Moonphase, 2015

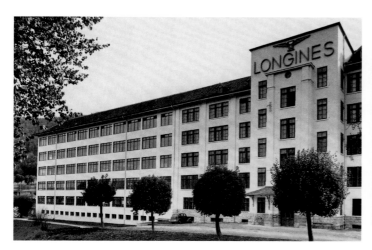
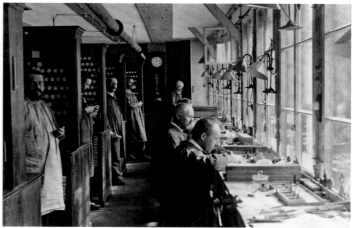

Left: Longines factory, 1950 ◦ Longines workshop, 1900 ◦ Links: Longines-Fabrik, 1950 ◦ Blick in die Werkstatt, 1900 ◦
De gauche à droite : Fabrique Longines, 1950 ◦ Atelier Longines, 1900

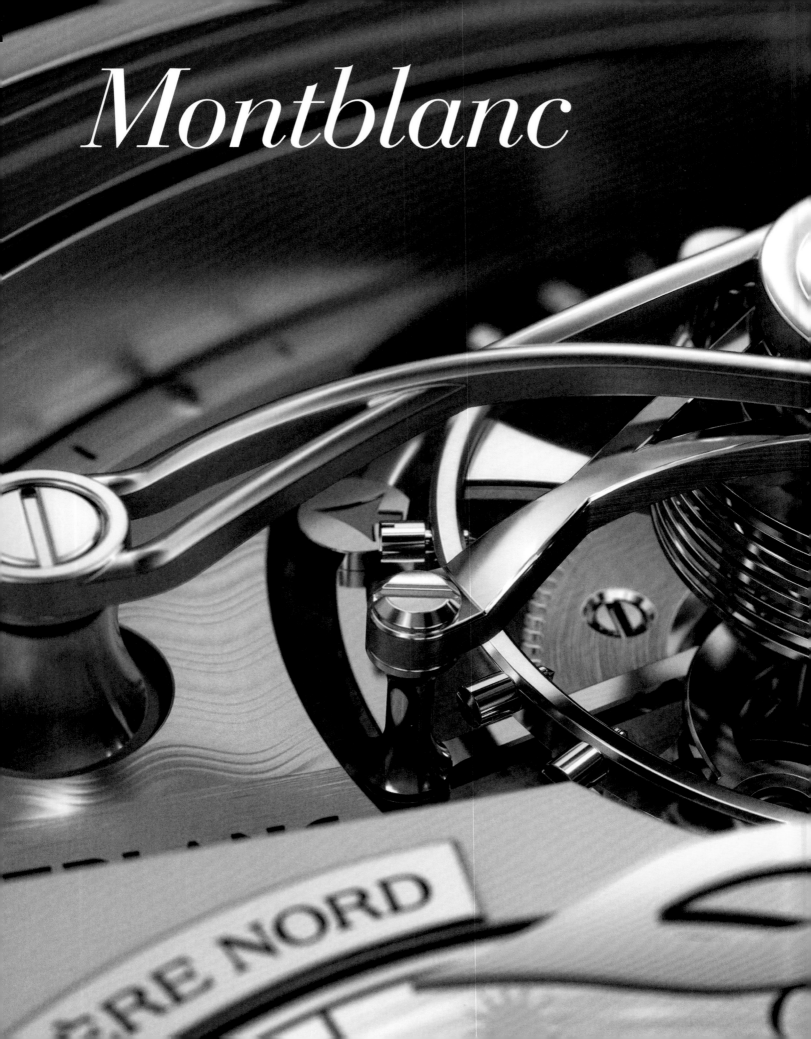

Montblanc

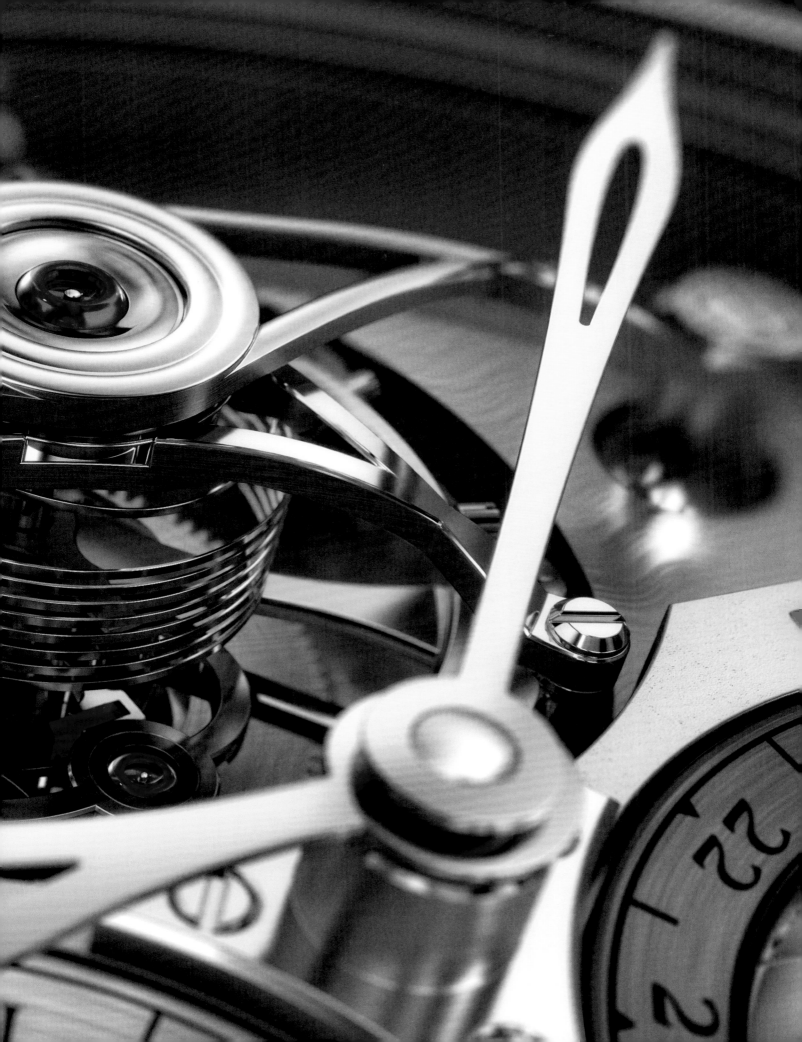

From Fountain Pens to Wristwatches

English

The name "Montblanc" almost necessarily calls to mind fountain pens, especially the legendary black "Meisterstück" that penned new chapters in the history of writing. In the 1990s, a new "Meisterstück" (the German word for "masterpiece") provided the impulse for the creation of Montblanc's own watch collection, which debuted at the SIHH watch salon in Geneva in 1997. The sarcastic question "Where does one pour in the ink?" has since been answered. From a modest start as an établisseur that assembles watches from outsourced parts, Montblanc has evolved into a watch manufacture with a rich assortment of its own calibers.

No one at the brand's headquarters in Hamburg ever thought of putting the name Montblanc on the dial of a run-of-the-mill watch. The plan had always been to develop modern manufacturing capacities and to strive toward full-fledged manufactory status. Shortly before the launch of their first watches, the Germans bought and enlarged a villa in the traditional Swiss watchmaking city of Le Locle. As fate would have it, the date carved under the house's tall gable was "1906"—the year when Montblanc was founded. Another decisive date in the firm's history was 2007. One year before, the Richemont Group (to which Montblanc had belonged sensu lato since 1974, when Alfred Dunhill Ltd. joined the group) purchased Minerva, a chronograph manufactory founded in Villeret in 1858. Ruminations about which of the group's brands would be best suited to welcome this new jewel prompted Richemont to assign Minerva to Montblanc, which not only brought Montblanc competence in classical chronographs, but also provided an entrée into the elite world of fine watchmaking. Aficionados of haute horlogerie can now find at Montblanc everything that raises their heart rates, for example: the "ExoTourbillon" with a patented rotating escapement; the "Metamorphosis II," a wristwatch with uncommonly versatile abilities; the universal "Tourbillon Cylindrique Geosphères Vasco da Gama;" and a diverse array of chronographs.

Apropos chronographs: with wise forethought, Montblanc had protected the name Nicolas Matthieu Rieussec, the French watchmaker who unveiled and patented the world's first "time-writer" (i.e., chronograph) at a horserace on the Champ de Mars in Paris in 1821. Droplets of ink deposited on this device's rotating dial could capture intervals of time. Fascinated by this invention and cognizant of Montblanc's original métier as a maker of writing instruments, the brand's executives chose the name "Nicolas Rieussec" in 2008 for the first chronographs that Montblanc developed and fabricated. Rotating discs for the continually running seconds and the totalizer on its dial unmistakably allude to the French inventor.

Montblanc divides its chronometric activities into several echelons. Watches with reliable movements sourced from external suppliers and subsequently ennobled by Montblanc appeal to entry-level customers. One step above, aficionados can choose timepieces encasing outsourced movements with more-or-less complex and highly functional modules. These feature calendars, including perpetual calendars that won't need manual correction until 2100. Newly released watches show the time in various zones and are ideal for jetsetters who want to know precisely when—and for whom—the bell tolls while they travel around the globe. The next level offers movements from Montblanc's industrialized manufactory, e.g., the "Nicolas Rieussec." The top of the pyramid is reserved for the traditionally crafted and strictly limited timepieces that Montblanc assigns to its "Collection Villeret 1858." As an expression of special craftsmanship, the delicate balance-springs that oscillate inside these watches are fabricated on the manufactory's own premises. ○

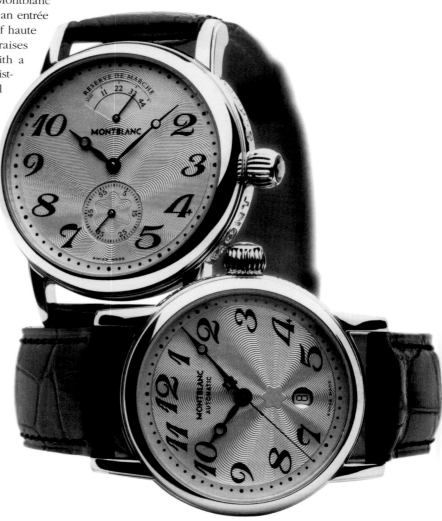

Meisterstück,
Limited Anniversary
Edition 1924, 1999

Reserve de Marche for men,
Automatic for ladies

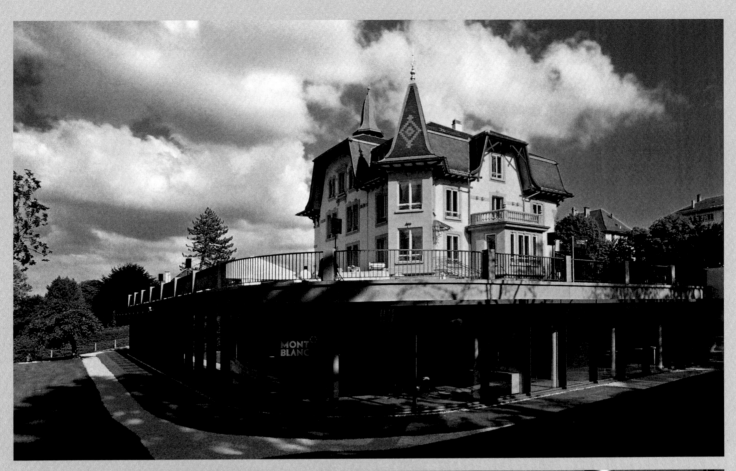

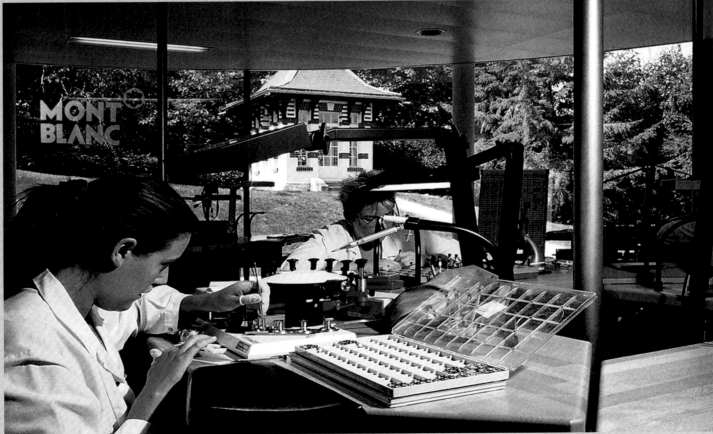

Montblanc's production site in an Art Nouveau villa dating from 1906 ∘ Die Produktionsstätte von Montblanc in einer Jugendstilvilla von 1906 ∘ Atelier de production Montblanc dans une villa de style Art nouveau de 1906

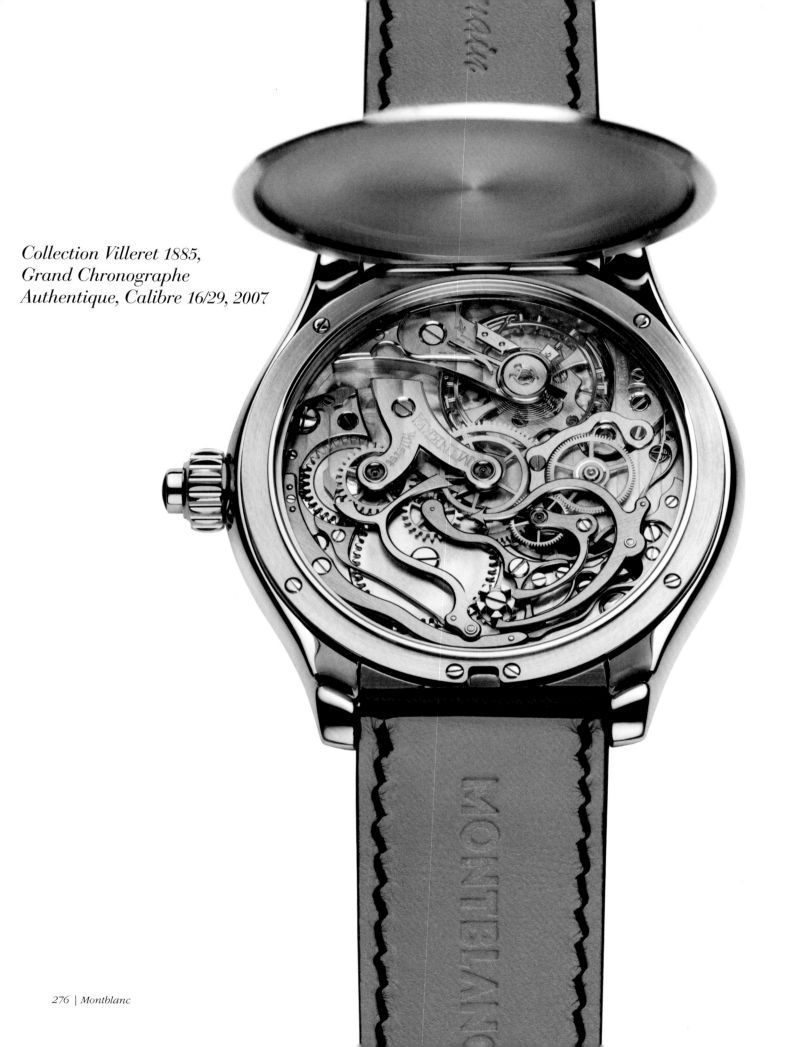

Collection Villeret 1885,
Grand Chronographe
Authentique, Calibre 16/29, 2007

Vom Füller zur Uhr

Wer den Namen Montblanc hört, denkt beinahe zwangsläufig an Füller, insbesondere an das legendäre schwarze „Meisterstück". Damit hat die Marke zweifellos Geschichte geschrieben. Eben jenes „Meisterstück" lieferte Mitte der 1990er Jahre den Impuls zur Kreation einer eigenen Uhrenkollektion, die 1997 während des Genfer Uhrensalons SIHH debütierte. Die einstige Frage, wo man denn die Tinte einfülle, hat sich inzwischen erübrigt. Aus eher bescheidenen Anfängen als sogenannter Etablisseur, also Hersteller von Uhren aus zugekauften Komponenten, ist längst eine etablierte Uhrenmanufaktur mit einer reichhaltigen Kaliberpalette geworden.

Den wohlklingenden Namen Montblanc allein zum Signieren irgendwelcher Zeitmesser zu nutzen, daran hatte am Stammsitz in Hamburg übrigens nie jemand gedacht. Von Anfang an stand die Errichtung moderner Fertigungskapazitäten ebenso fest wie der künftige Weg in Richtung echter Manufakturarbeit. Kurz vor der Lancierung der ersten Uhren konnten die Deutschen in der traditionsreichen Schweizer Uhrmacherstadt Le Locle eine prachtvolle Villa erwerben und für diesen Zweck erweitern. Wie es der Zufall so will, war unter dem hohen Giebel „1906" zu lesen – das Gründungsjahr von Montblanc. Ein weiteres einschneidendes Datum in der Firmengeschichte war das Jahr 2007. Im Vorjahr hatte der Richemont-Konzern, zu dem Montblanc im weitesten Sinne seit 1974 durch den Einstieg der englischen Alfred Dunhill Ltd. gehörte, die 1858 in Villeret gegründete Chronographenmanufaktur Minerva gekauft. Überlegungen, zu welcher der Konzernmarken das Juwel am besten passen würde, führte Minerva zu Montblanc. Diese Aktion brachte nicht nur Kompetenz bei klassischen Zeitschreibern, sondern auch den Eintritt in die Welt der höchsten Uhrmacherkunst. Heute findet sich bei Montblanc alles, was das Herz anspruchsvoller Liebhaber der Haute Horlogerie erfreut. Zum Beispiel das „ExoTourbillon", ein patentierter Drehgang, die „Metamorphosis II" als wandlungsfähige Armbanduhr, das universelle „Tourbillon Cylindrique Geosphères Vasco da Gama" oder Chronographen unterschiedlichster Prägung.

Apropos Chronographen: In weiser Voraussicht hatte sich Montblanc den Namen Nicolas Matthieu Rieussec schützen lassen. Jener französische Uhrmacher hatte 1821 anlässlich eines Pferderennens am Pariser Marsfeld den weltweit ersten und seinerzeit deshalb auch patentierten Zeitschreiber präsentiert. Mit diesem Gerät war es möglich, Zeitintervalle per Tinte auf drehende Scheiben zu bannen. Das sowie die Verbindung zum ureigenen Metier, dem Schreiben, faszinierte Montblanc so sehr, dass 2008 die ersten selbst entwickelten und gefertigten Chronographen „Nicolas Rieussec" hießen. Reminiszenz an den Erfinder: rotierende Scheiben für die Permanentsekunde und den Totalisator.

Aktuell gliedern sich die chronometrischen Aktivitäten von Montblanc in mehrere Ebenen. Einsteiger bekommen zuverlässige, aber extern zugelieferte und von Montblanc veredelte Uhrwerke. Eine Stufe darüber rangiert die Kombination aus fremden Uhrwerken und mehr oder minder komplexen, stets aber sehr funktionalen Modulen. In diese Kategorie fallen Kalendarien bis hin zum ewigen Kalender, welcher erst 2100 einer manuellen Korrektur bedarf. Brandneu sind selbst entwickelte Zeitzonen-Dispositive für Kosmopoliten, die bei ihren Reisen rund um den Globus stets genau wissen möchten, was es da und dort gerade schlägt. Nochmals höher sind die Uhrwerke aus eigener, jedoch industrialisierter Manufaktur wie beispielsweise „Nicolas Rieussec" angesiedelt. Die Spitze der Pyramide repräsentiert weitgehend in alter Manier Hergestelltes und deshalb streng Limitiertes, das Montblanc seiner „Collection Villeret 1858" zuordnet. Dort, und das ist ein Ausweis besonderer Kunstfertigkeit, oszillieren sogar zierliche Unruhspiralen aus eigener Fertigung. ○

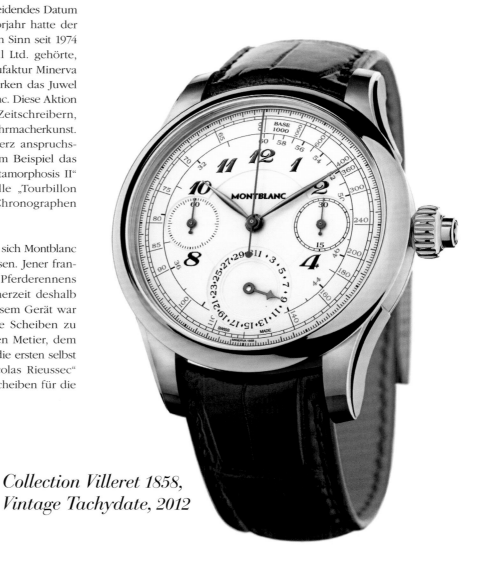

Collection Villeret 1858,
Vintage Tachydate, 2012

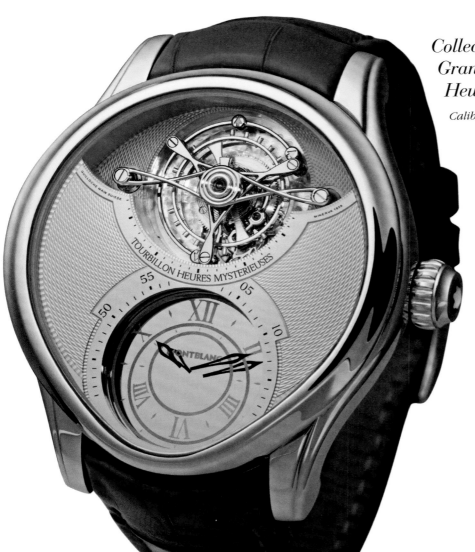

Collection Villeret 1858,
Grand Tourbillon
Heures Mystérieuses, 2009

Calibre M 65.00, limited to 8 pieces

Collection Villeret 1858,
Grande Seconde au Centre
Retoure à Zéro, 2008

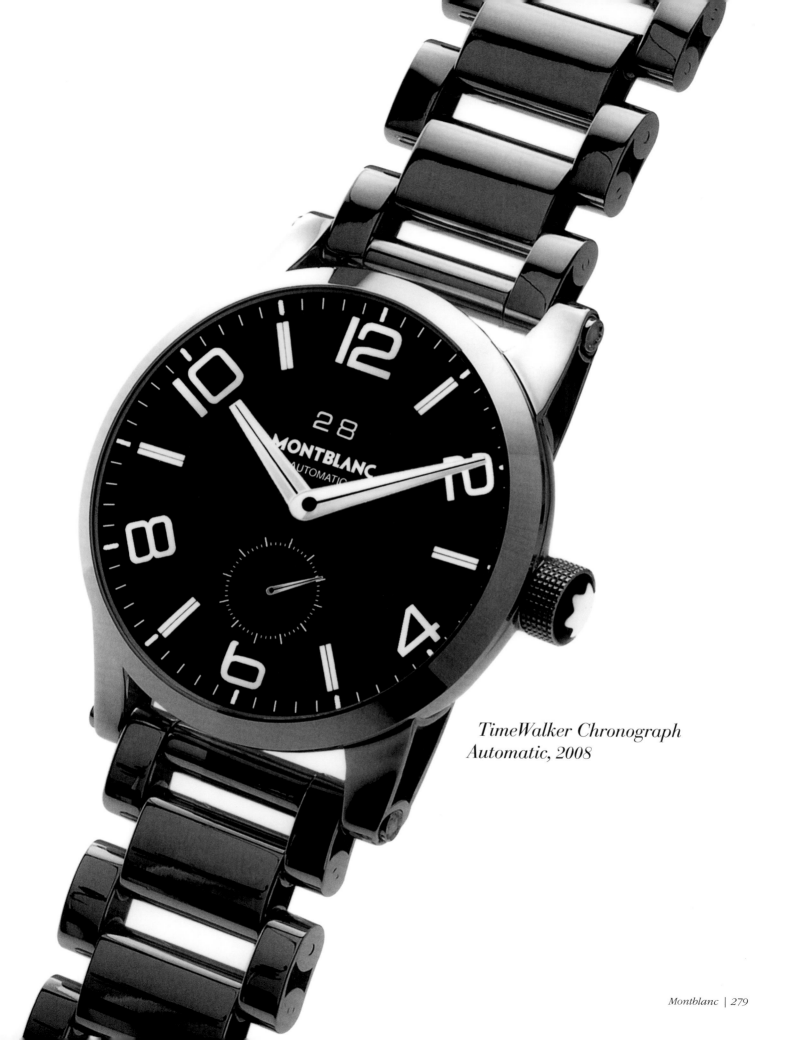

TimeWalker Chronograph Automatic, 2008

Du stylo-plume à la montre

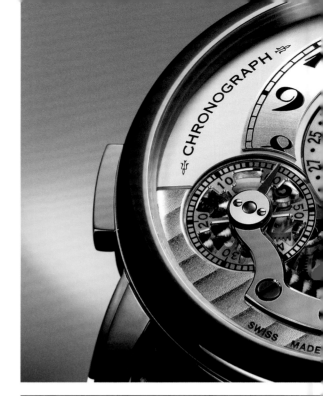

Le nom Montblanc évoque presque immanquablement des stylos-plumes, en particulier le légendaire « Meisterstück » noir, avec lequel la marque a sans nul doute écrit une page d'histoire. C'est justement ledit stylo-plume qui, au milieu des années 1990, amène la marque à créer sa propre collection de montres, lancée en 1997 à l'occasion du Salon international de la haute horlogerie de Genève (SIHH). Les mauvaises langues qui, à l'époque, demandent comment les recharger en encre se sont tues depuis : le modeste établisseur, qui assemblait des montres à partir de composants achetés auprès de fournisseurs, est très vite devenu une manufacture horlogère offrant une riche palette de calibres.

Au siège hambourgeois, on refuse de se contenter d'apposer le si doux nom de Montblanc sur des garde-temps quelconques. En effet, dès le début, les objectifs déclarés sont la mise en place de capacités de production modernes, tout comme un véritable travail de manufacture. Peu avant de lancer leurs premières montres, les Allemands acquièrent et agrandissent dans ce but une somptueuse villa au Locle, ville horlogère suisse. Et le hasard veut que sous le haut pignon figure l'inscription « 1906 », l'année de fondation de Montblanc. Une autre date clé dans l'histoire de l'entreprise est l'année 2007. Un an auparavant, le groupe Richemont, auquel Montblanc appartient indirectement depuis 1974 du fait de la prise de participation de l'entreprise britannique Alfred Dunhill Ltd., a racheté Minerva, manufacture de chronographes fondée en 1858 à Villeret. Et c'est à Montblanc qu'échoit ce joyau. Cette décision accroît ses compétences en matière de chronographes classiques et lui ouvre les portes de la haute horlogerie. De nos jours, Montblanc propose tout ce qui réjouit le cœur des amateurs les plus délicats : le mécanisme à construction brevetée « ExoTourbillon », la montre-bracelet « Metamorphosis II », qui commute du module affichage horaire au chronographe, le « Tourbillon Cylindrique Geosphères Vasco da Gama » universel et des chronographes les plus divers, pour ne citer que quelques exemples.

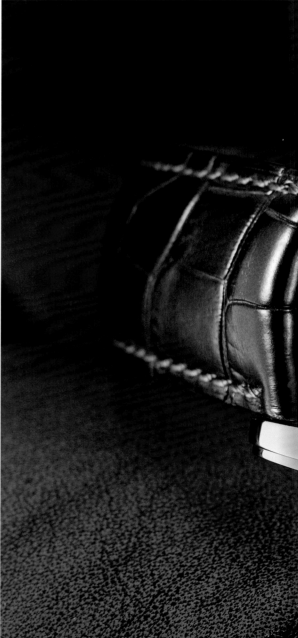

Précisons à ce sujet que Montblanc a pris la sage précaution de déposer le nom Nicolas Matthieu Rieussec. Cet horloger français a présenté en 1821, à l'occasion d'une course de chevaux sur le Champ-de-Mars à Paris, le premier chronographe au monde et par conséquent aussi le premier à être breveté. Cet instrument permet d'inscrire à l'encre des intervalles de temps sur des disques rotatifs. Ce dispositif ainsi que le rapport avec le cœur de métier de Montblanc, à savoir les instruments d'écriture, fascinent cette société à tel point que les premiers chronographes conçus et réalisés en interne en 2008 s'appellent « Nicolas Rieussec ». Les disques rotatifs pour la trotteuse permanente et le totaliseur sont par ailleurs un héritage de l'inventeur.

De nos jours, le pôle horloger de Montblanc s'articule sur plusieurs niveaux. Les modèles de base, déjà fiables en soi, sont assemblés à partir de mouvements fabriqués en externe mais ennoblis par Montblanc. Un cran au-dessus, des mouvements achetés côtoient des modules plus ou moins complexes, mais toujours très fonctionnels. On trouve dans cette catégorie des calendriers, et ce jusqu'au quantième perpétuel qui ne nécessitera pas de réglage manuel avant 2100. Citons aussi une innovation récente de Montblanc : une montre à temps universel pour les globe-trotteurs, qui veulent savoir en tout point du monde l'heure qu'il est partout ailleurs à l'instant même. Plus haut dans la hiérarchie, les mouvements, comme les « Nicolas Rieussec », sont réalisés par Montblanc dans sa manufacture selon des procédés industriels. Le sommet de la pyramide correspond à des produits fabriqués « à l'ancienne », et par conséquent en série strictement limitée, regroupés sous l'appellation « Collection Villeret 1858 ». Et gage d'une maîtrise extrême, ces garde-temps sont même animés par de délicats spiraux signés Montblanc. ○

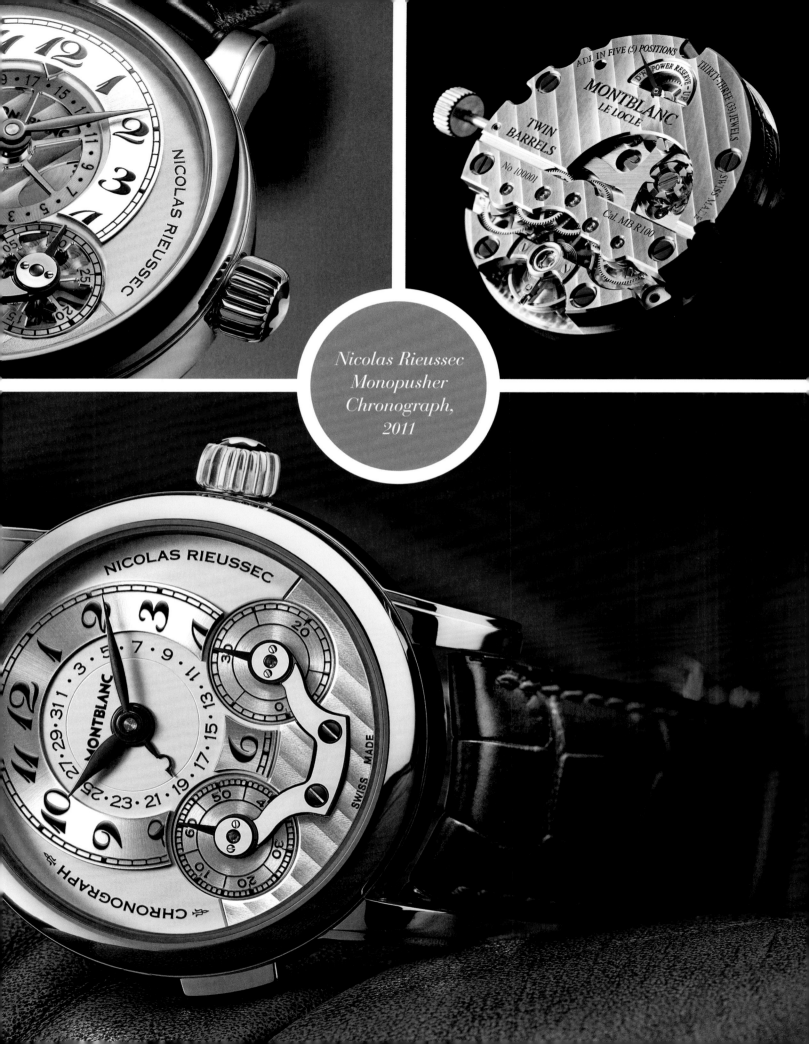

Nicolas Rieussec
Monopusher
Chronograph,
2011

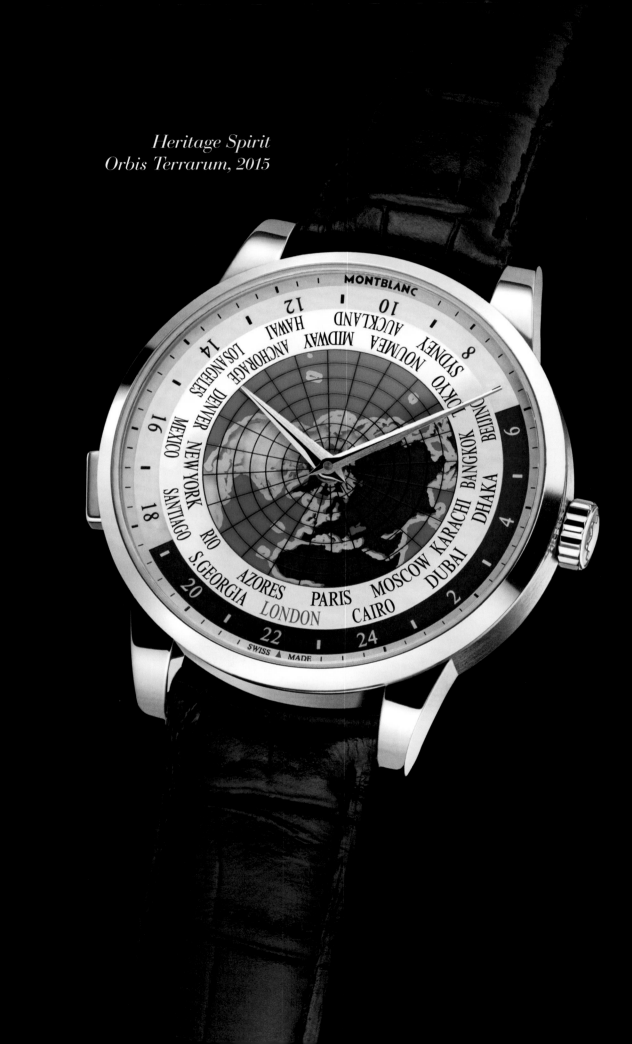

Heritage Spirit
Orbis Terrarum, 2015

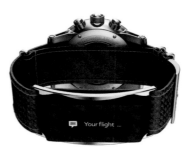

*TimeWalker Urban Speed
Chronograph e-Strap, 2015*

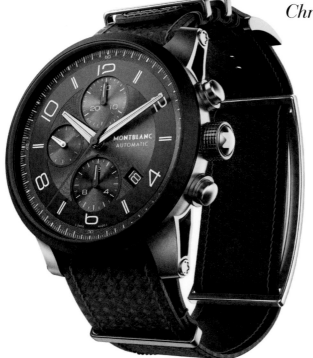

*Heritage Chronométrie
ExoTourbillon, 2015*

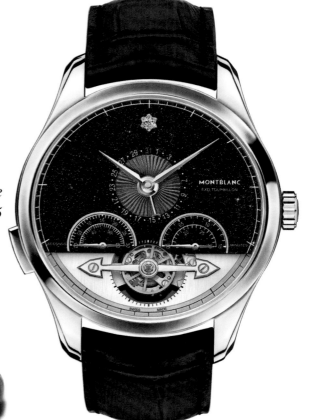

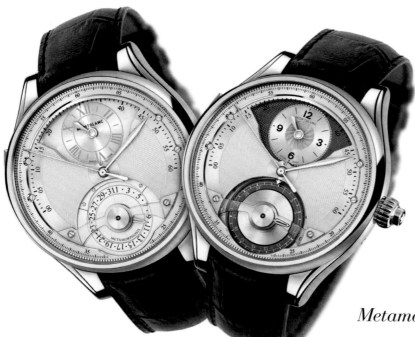

Metamorphosis II, 2015

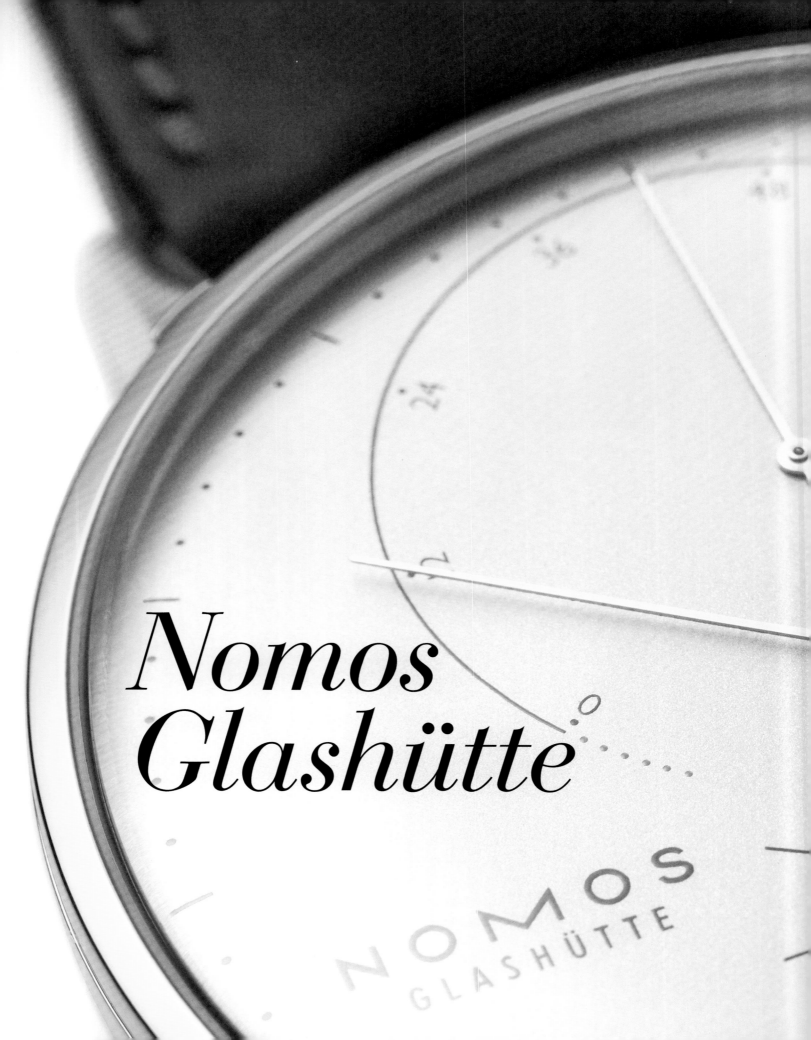

Nomos Glashütte

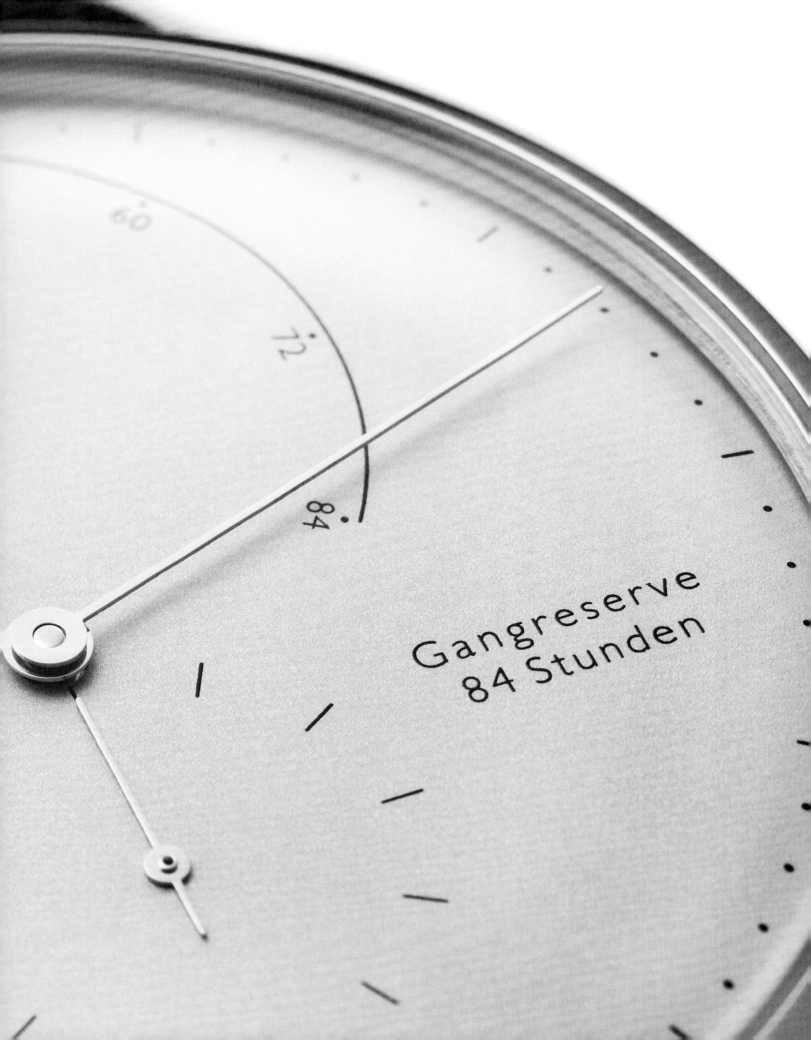

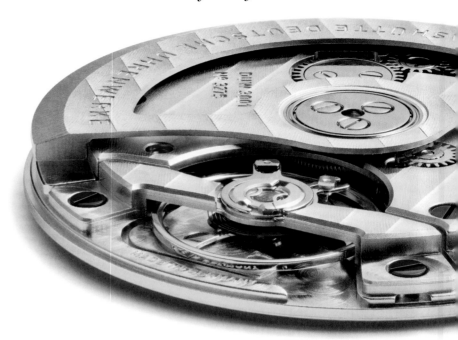

Hand-wound Cal. Alpha

Automatic manufactory Cal. DUW 3001

Affordable Luxury in Watches from Germany

English

Following A. Lange & Söhne and Glashütte Original, Nomos Glashütte inarguably holds third place among Glashütte's watch brands. The family-owned business traditionally doesn't disclose exact statistics, but it seems likely that approximately 40,000 to 50,000 watches were shipped from the brand's factories in Saxony's Müglitz Valley in 2015. The payroll currently includes more than 250 employees, and additional colleagues regularly join the growing workforce. Ever since this label's renaissance in 1990, quartz watches have never been in its portfolio. Suffice it to say the following about the brand and its history: calendars around the world read "1906" when Clemens Guido Müller and his brother-in-law Karl Nierbauer put their business idea into practice. Their concept was to import Swiss watches and to sell them with the prestigious additional signature "Glashütte" on their dials. But A. Lange & Söhne, as the firmly established top dog in town, was not pleased and filed a law suit against the upstarts. The verdict, handed down in 1910, proposed a compromise: Nomos Glashütte would be permitted to sell its remaining stock; afterwards, however, no further reference to the German watchmaking Mecca near Dresden would be permitted on its products. Bankruptcy was unavoidable. No one so much as mentioned the name of Nomos Glashütte in ensuing decades. After the fall of the Berlin Wall, Roland Schwertner discovered the forgotten watch brand. His attempt to imitate his predecessors—i.e., to put the finishing touches on Swiss movements in Glashütte, to encase the calibers, and then to sell them with the signature "Nomos Glashütte" on their dials—similarly sparked strong opposition from his big neighbors. But the West German businessman was ultimately able to prove that more than 50 percent of the value had been added to his products in Glashütte. Discussions of this sort are no longer on the agenda because Nomos Glashütte has made all of its movements at its own manufacture for more than a decade. The entry into the world of Nomos Glashütte is made possible by hand-wound Caliber α (Alpha), which was originally derived from the time-honored Peseux 7001 and which naturally has a Glashütte three-quarter plate. In the meantime, even the gears for this movement are manufactured on the brand's premises. The other end of the hand-wound scale is occupied by Caliber DUW 1001, which is reserved for the gold watches in the "Lambda" line. Incidentally: DUW is the acronym for "Deutsche Uhrenwerke." All calibers with this designation include the so-called "Nomos Swing System." The brand's own escapement subassembly—which consists of the balance, hairspring, pallet lever and escape wheel—gave Nomos Glashütte its long-sought independence from Swiss suppliers. The indispensable blanks for the hairsprings are produced in a cooperative venture between Nomos Glashütte and Haas, a recognized specialist. The newest member of the family of Nomos Glashütte's own calibers is the ultra-slim DUW 3001 with ball-borne central rotor. This movement is 28.8 millimeters in diameter and a mere 3.2 millimeters tall. A total of 157 components are assembled to create each one. The sleekly simple "Tangente," which was launched in 1992, remains a longstanding bestseller in the product portfolio. Its purist design can be traced back to the 1930s. Other watch lines are named "Ahoi," "Club," "Ludwig," "Lux," "Metro," "Tetra", and "Zürich." Finally, Nomos Glashütte also offers a horological complication: thanks to a well-conceived time-zone display, cosmopolites can take an uncomplicated journey through time with either the "Tangomat GMT" or the "Zürich Weltzeit." Over 130 awards and commendations from around the world are the well-earned rewards for more than two decades of dedication to the art of Glashütte-style watchmaking. □

Bezahlbarer Uhr-Luxus aus deutschen Landen

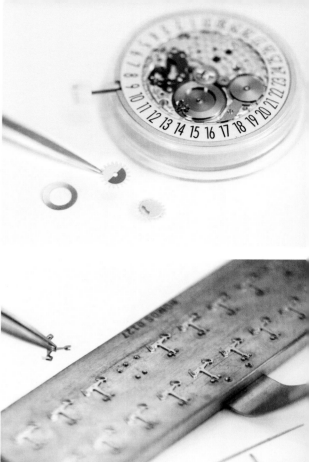

Unter den Glashütter Uhrenmarken ist Nomos Glashütte nach A. Lange & Söhne sowie Glashütte Original die unangefochtene Nummer drei. 2015 dürften schätzungsweise 40 000 bis 50 000 Uhren die Fabrikationsstätten im sächsischen Müglitztal verlassen haben. Genaue Zahlen gibt das Familienunternehmen traditionsgemäß nicht bekannt. Auf der Gehaltsliste stehen gut 250 Mitarbeiterinnen und Mitarbeiter, Tendenz weiterhin steigend. Quarzuhren sind für die Sachsen seit der Renaissance des Labels im Jahr 1990 kein Thema. Zur Marke und ihrem geschichtlichen Hintergrund nur so viel: Man schrieb das Jahr 1906, als Clemens Guido Müller und sein Schwager Karl Nierbauer ihre Geschäftsidee realisierten. Diese basierte auf dem Import Schweizer Uhren samt deren Vertrieb mit der imageträchtigen Zusatzsignatur Glashütte. Genau das ließ sich der etablierte Platzhirsch A. Lange & Söhne nicht gefallen. Seine Klage endete 1910 mit einen Vergleich. Ihm zufolge durfte Nomos Glashütte seinen Lagerbestand noch verkaufen. Danach hatte jeglicher Hinweis auf das deutsche Uhrenmekka nahe Dresden zu unterbleiben. Ein Konkurs der Firma war unausweichlich. In den folgenden Jahrzehnten sprach niemand mehr von Nomos Glashütte. Nach dem Mauerfall entdeckte Roland Schwertner die vergessene Uhrenmarke. Sein Versuch, es den Vorgängern gleichzutun, also Schweizer Werke in Glashütte zu finissieren, einzuschalen und unter der Signatur Nomos Glashütte zu verkaufen, zog einmal mehr heftiges Abwehrfeuer der großen Nachbarn nach sich. Letzten Endes konnte der westdeutsche Unternehmer jedoch nachweisen, dass mehr als 50 Prozent der Wertschöpfung vor Ort erfolgt. Diskussionen dieser Art sind längst kein Thema mehr, denn seit mehr als zehn Jahren fertigt Nomos Glashütte ausnahmslos alle Uhrwerke in eigener Manufaktur. Den Einstieg in die Welt von Nomos Glashütte ermöglicht das ursprünglich einmal vom bewährten Peseux 7001 abgeleitete Handaufzugskaliber α (Alpha), das natürlich eine Glashütter Dreiviertelplatine besitzt. Bei diesem Uhrwerk stammen mittlerweile sogar die Zahnräder aus eigener Fertigung. Am anderen Ende der Handaufzugsskala findet sich das DUW 1001, welches der goldenen Uhrenlinie „Lambda" vorbehalten bleibt. Apropos: DUW ist das Kürzel für Deutsche Uhrenwerke. Alle Kaliber mit dieser Bezeichnung besitzen als Besonderheit das sogenannte Nomos-Swing-System. Das hauseigene Assortiment, bestehend aus Unruh, Unruhspirale, Anker und Ankerrad, beschert die intendierte Unabhängigkeit von eidgenössischen Zulieferungen. Die unverzichtbaren Rohspiralen sind das Produkt einer Kooperation zwischen Nomos Glashütte und dem einschlägig erfahrenen Spezialisten Haas. Neuestes Mitglied der Familie eigener Kaliber ist das ultraflache DUW 3001 mit kugelgelagertem Zentralrotor. Bei 28,8 Millimetern Durchmesser baut dieses Automatikwerk lediglich 3,2 Millimeter hoch. Für ein Exemplar braucht es 157 Komponenten. Zu den Bestsellern im Produktportfolio gehört weiterhin die schlichte, bereits 1992 lancierte „Tangente". Ihr puristisches Design geht auf die 1930er Jahre zurück. Weitere Uhrenlinien heißen „Ahoi", „Club", „Ludwig", „Lux", „Metro", „Tetra" oder „Zürich". Eine uhrmacherische Komplikation bietet Nomos Glashütte schließlich auch. Dank eines durchdachten Zeitzonen-Dispositivs können sich Kosmopoliten entweder mit der „Tangomat GMT" oder der „Zürich Weltzeit" auf unkomplizierte Zeit-Reise begeben. Verdienter Lohn für über zwei Jahrzehnte Engagement rund um die Glashütter Uhrmacherkunst sind mehr als 130 Auszeichnungen aus aller Herren Länder. ○

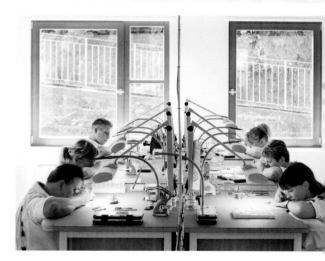

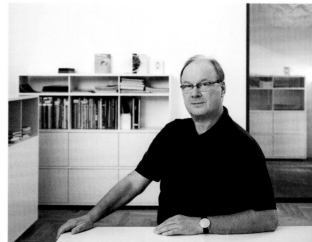

From top: Date and power-reserve displays ○ Installing the pallet lever ○
Watchmakers at work ○ Nomos' founder Roland Schwertner ○
Von oben: Datums- und Gangreserveanzeige ○ Montage der Anker ○
Uhrmacherei ○ Nomos-Gründer Roland Schwertner ○
De haut en bas : indication de la date et de la réserve de marche ○
Montage de l'ancre ○ Atelier d'horlogerie ○ Roland Schwertner, fondateur de Nomos

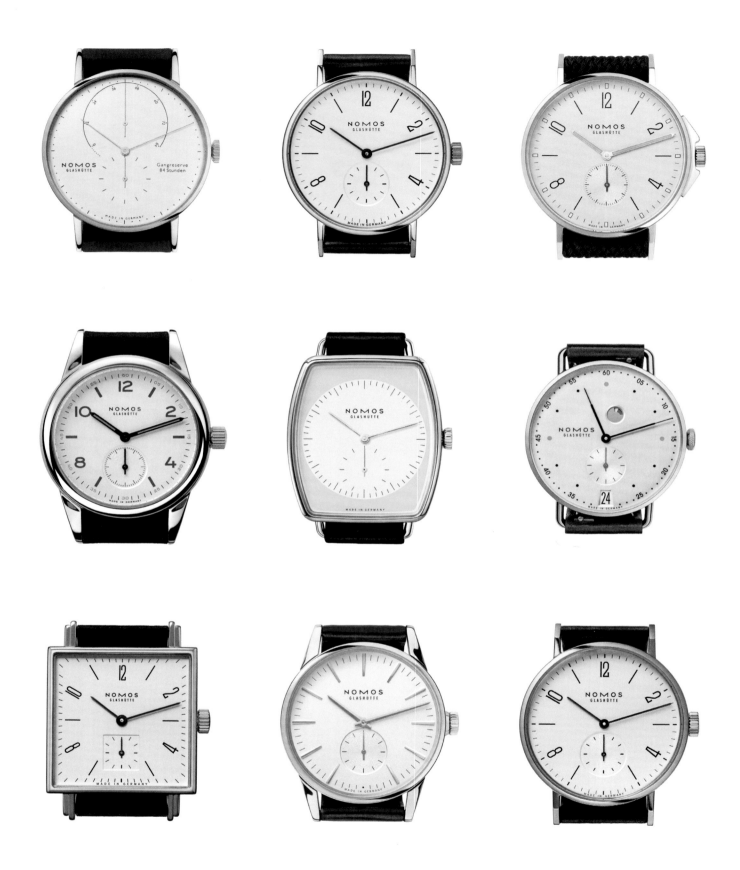

This page: A selection of watches from Nomos' collection ○ Diese Seite: Ausschnitt aus der Nomos-Kollektion ○
Cette page : Extrait de la collection Nomos ○ Opposite page, top: Cal. DUW 1001

Luxe horloger abordable made in Germany

Parmi les marques horlogères de Glashütte, Nomos Glashütte figure sans conteste à la troisième place, derrière A. Lange & Söhne et Glashütte Original. Le nombre de montres fabriquées dans ces ateliers de la localité de Müglitztal (Saxe) est estimé entre 40 000 et 50 000. Cette entreprise familiale a en effet coutume de ne jamais donner de chiffres précis. L'entreprise compte plus de 250 salariés, et la tendance est à la hausse. Depuis la renaissance du label en 1990, il n'est plus question de quartz. Mais reprenons l'histoire de la marque depuis le début. En 1906, Clemens Guido Müller et son beau-frère Karl Nierbauer concrétisent leur idée commerciale, à savoir importer et distribuer des montres suisses, qu'ils revêtent de la prestigieuse signature Glashütte. Mais c'est précisément ce que le numéro un de la place, A. Lange & Söhne, n'est pas disposé à accepter. La plainte qu'il dépose aboutira en 1910 à un arrangement. Nomos Glashütte sera autorisée à écouler ses stocks, mais toute référence au haut lieu de l'horlogerie allemande proche de Dresde lui sera dorénavant interdite. La faillite de la société est inévitable et, dans les années qui suivent, on n'entend plus parler de Nomos Glashütte. Après la chute du Mur, Roland Schwertner redécouvre la marque horlogère oubliée. Sa tentative d'imiter ses prédécesseurs, c'est-à-dire de procéder au finissage et à la mise en boîtiers de mouvements suisses puis de les vendre avec la signature Nomos Glashütte entraîne de nouveau une levée de boucliers chez ses illustres voisins. Finalement, l'entrepreneur ouest-allemand réussit à démontrer que la valeur est créée sur place à plus de 50 pour cent. Mais les débats de ce type n'ont plus cours depuis longtemps. Depuis plus de dix ans en effet, tous les mouvements Nomos Glashütte sans exception sont fabriqués en interne. En entrée de gamme,

Nomos Glashütte propose le mouvement à remontage manuel α (Alpha), lequel s'appuie sur un mouvement éprouvé, le Peseux 7001, et possède bien sûr une platine trois-quarts de Glashütte. Sur le mouvement, même les pignons sont désormais « maison ». À l'autre extrémité de la gamme de mouvements à remontage manuel, on trouve le DUW 1001, dédié à la collection de montres en or « Lambda ».

Tous les mouvements DUW (abréviation de Deutsche Uhrenwerke, « mouvements horlogers allemands ») se caractérisent par un système d'échappement propre baptisé Nomos-Swing-System. L'assortiment de pièces fabriqué en interne (balanciers, spiraux, ancres et roues d'ancre) fournit l'indépendance recherchée par rapport aux fournisseurs suisses. Les indispensables spiraux bruts sont le produit d'une collaboration entre Nomos Glashütte et Haas, spécialiste dans le domaine. Le dernier-né des calibres de manufacture est le DUW 3001 ultraplat à rotor central amorti par roulements à billes. Nécessitant 157 composants pour sa fabrication, il mesure 28,8 millimètres de diamètre et seulement 3,2 millimètres d'épaisseur. Parmi les best-sellers de la gamme de produits figure également « Tangente », collection lancée en 1992 dont les lignes épurées remontent aux années 1930. Les autres collections ont pour noms « Ahoi », « Club », « Ludwig », « Lux », « Metro », « Tetra » ou « Zürich ». La société Nomos Glashütte propose elle aussi une complication horlogère : un astucieux dispositif d'affichage de fuseaux horaires permet aux globe-trotters d'effectuer aisément un voyage dans le temps avec la « Tangomat GMT » ou la « Zürich Weltzeit ». Plus de 130 distinctions et prix du monde entier sont venus récompenser à juste titre deux bonnes décennies d'engagement en faveur de l'art horloger de Glashütte. ○

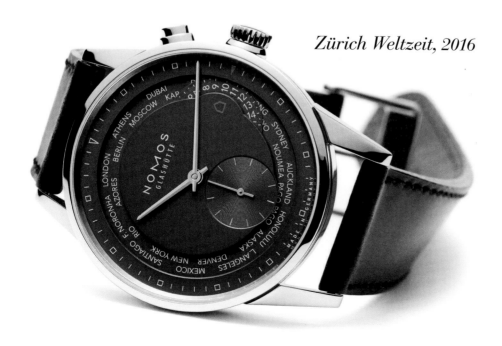

Zürich Weltzeit, 2016

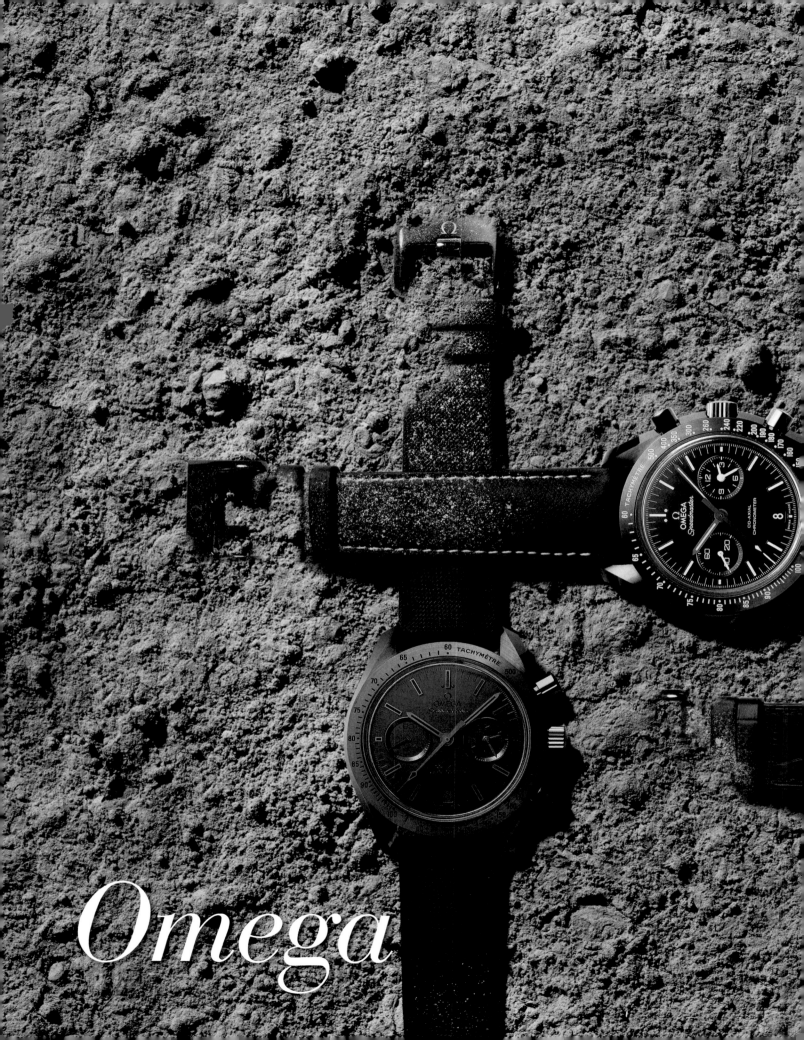

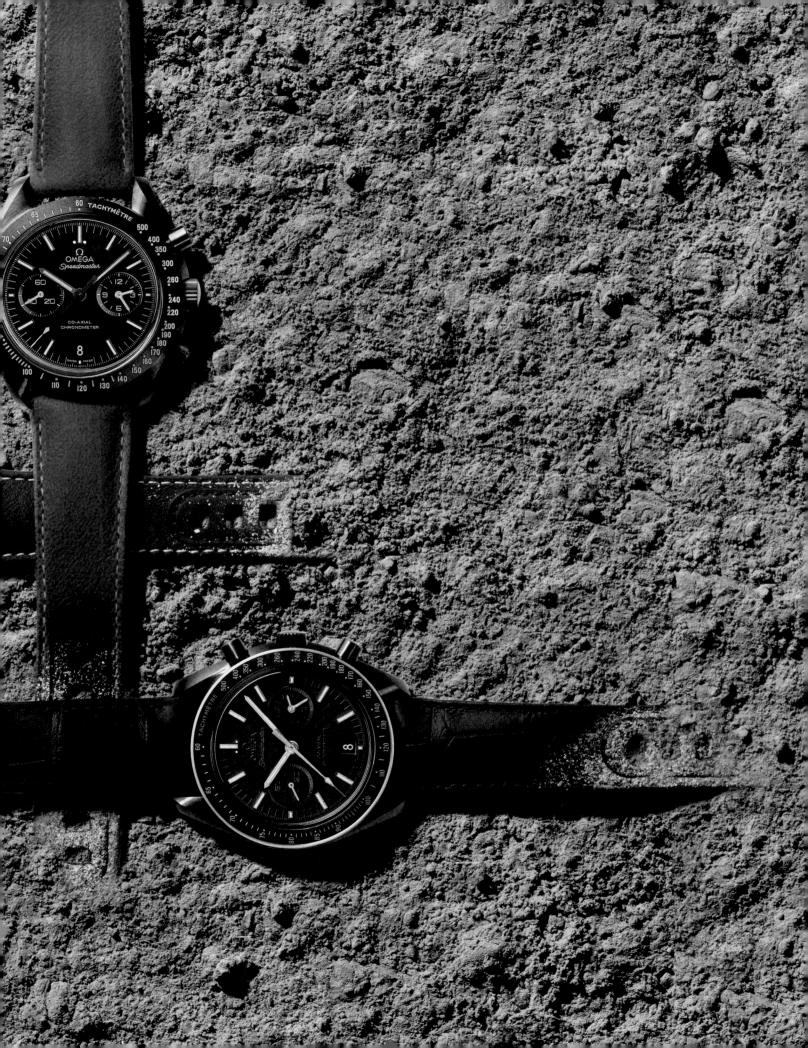

From Alpha to Omega

When its historical roots reach back to 1848, the tree that grows from those roots can weather nearly every storm. The letter Ω, the last letter in the Greek alphabet, is carved into this old tree's trunk, where it is impossible to overlook. But this mighty tree would never have sprouted without an α, which is represented here by a gentleman named Louis Brandt. To meet him, we need to turn back the clocks to 1848, when the great Gold Rush was luring hordes of adventurous young men to California. But one 23-year-old fellow was not among those hardy adventurers: Louis Brandt remained faithful to his Swiss homeland—and was richly rewarded for his loyalty. In the watchmaking town of La Chaux-de-Fonds, he discovered gold of a different sort—namely, the golden legacy of the watchmaker's craft. He founded a "comptoir d'établissage," i.e., a point of sale for watches that had been assembled by watchmakers commissioned on a piecework basis. Most of these timepieces were silver pocket-watches, which were eagerly awaited on the English market. Brisk sales contributed to the steady growth of Brandt's business. He died in 1879 and his sons Louis-Paul and César Brandt relocated the company one year later. In Bienne, which lies in a valley below the Jura Range, the brothers rented a floor in a suitable building and proceeded to transform it into a modern production site for watches. Within just four months'

time, their business was booming so strongly that they were able to take over the entire building. With an annual production of 100,000 watches in 1889, Louis Brandt & Fils had become the largest company in the Swiss watchmaking industry.

Five years later, in 1894, the α was finally joined by the Ω. Inventive technicians at the continually growing manufactory had developed a new caliber that could be serially manufactured for pocket-watches. The ticking microcosm relied on a remarkably simple but thoroughly reliable and precise construction. Its developers had also ensured that its components were interchangeable, a capability which had not yet become commonplace in those years and which made it much easier to repair a watch if and when it needed a watchmaker's attention. This was indeed the cat's meow on the contemporary watch market, but the ticking kitten needed an appropriate name. The Brandt brothers had already registered the names "Jura," "Patria," "Helvetia," "Celtic," and "Gurzelen," as well as an astonishingly accurate caliber named "Labrador" in 1885. But none of these names seemed appropriate for the newcomer. A good idea was needed—and was soon forthcoming from the mind of the company's banker Henri Rieckel, who was so interested in his client's financial well-being that he

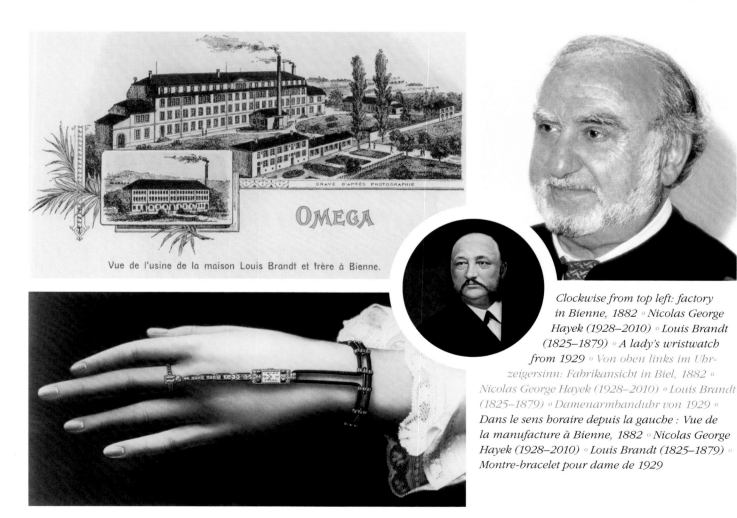

Vue de l'usine de la maison Louis Brandt et frère à Bienne.

Clockwise from top left: factory in Bienne, 1882 ∘ Nicolas George Hayek (1928–2010) ∘ Louis Brandt (1825–1879) ∘ A lady's wristwatch from 1929 ∘ Von oben links im Uhrzeigersinn: Fabrikansicht in Biel, 1882 ∘ Nicolas George Hayek (1928–2010) ∘ Louis Brandt (1825–1879) ∘ Damenarmbanduhr von 1929 ∘ Dans le sens horaire depuis la gauche : Vue de la manufacture à Bienne, 1882 ∘ Nicolas George Hayek (1928–2010) ∘ Louis Brandt (1825–1879) ∘ Montre-bracelet pour dame de 1929

refused to accept a fee for having suggested the name "Omega." The word fit perfectly because, as the last letter in an alphabet, it symbolized the utmost quality among pocket-watches. Rieckel's idea, which was accepted without delay, proved so successful that all other brand names were eliminated in 1903.

The Boer War raged in South Africa from 1899 to 1902. Thanks to Omega, British officers no longer had to clumsily pull their watches from the pockets of their uniforms—a quick glance at their wrists was all they needed to see the time. Subsequent reports confirmed that these watches performed with praiseworthy reliability. Omega was first chosen to serve as official timekeeper at the Olympic Games in Los Angeles in 1932: the IOC entrusted this brand with the same honor and responsibility more than twenty times in ensuing years.

Omega also had plenty to offer in the discipline of precision timekeeping. The brand's chronometers first won the accuracy competition at Neuchâtel Observatory in 1919. Record-breaking accuracy was also attested in 1933, 1936, 1946, and other years. After extensive testing, in 1965 NASA chose the "Speedmaster" chronograph to serve as official timepiece on its space missions. When Neil Armstrong set foot on the lunar surface at 2:56:20 (UTC) on July 21, 1969 and uttered the words "That's one small step for a man, one giant leap for mankind," he was wearing a wristwatch that would later be christened the "Speedmaster Professional."

The subsequent Quartz Revolution compelled Omega to traverse a deep vale of tears. The SSIH (Société suisse pour l'industrie horlogère SA), which also owned Tissot, suffered losses totaling 17 million Swiss francs in 1982 alone. A bankruptcy with drastic damage to the image of the Swiss watchmaking industry loomed menacingly. No other option remained in 1983 but to merge with the equally hard-pressed ASUAG (Allgemeine Schweizer Uhrenindustrie AG). Afterwards Nicolas G. Hayek relied on a combination of circumspection, farsightedness, and ironhandedness to restructure the newly created SMH (Société suisse de microélectronique et d'horlogerie SA) conglomerate, which has been known as the Swatch Group since 1998. The traditional Omega brand had regained radiant good health in 1999, when it presented a horological milestone: the coaxial escapement. This innovative rate-regulating organ is now installed in all of Omega's manufactory calibers, which have been produced in impressively large numbers. ○

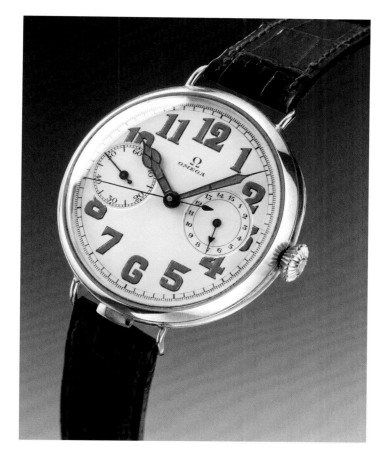

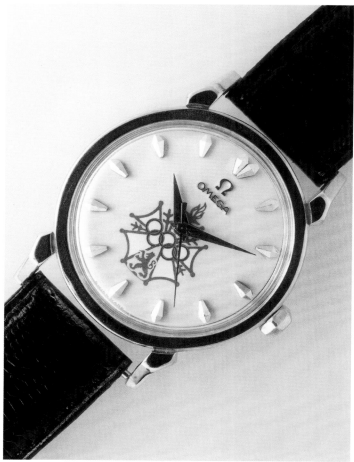

Top: chronograph with button in crown and 15-minute counter
Oben: Chronograph mit Kronendrücker und 15-Minuten-Zähler
En haut : Chronographe avec couronne à poussoir et compteur 15 minutes

Bottom: Seamaster XVI, Special Edition for the Olympic Games in Melbourne, 1956

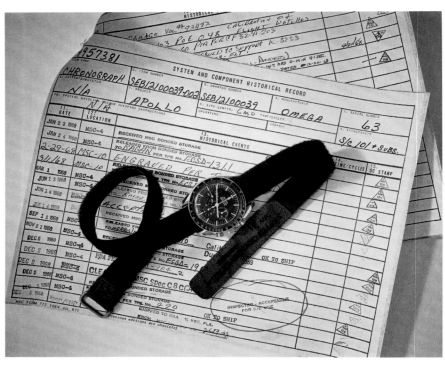

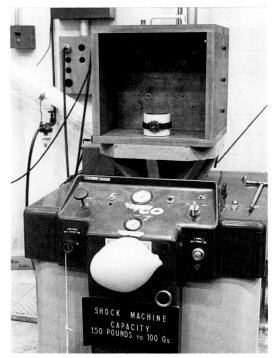

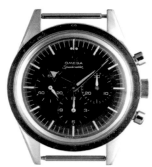

Clockwise from top left: NASA's choice: Speed-master, the moon watch, 1965 ◦ Shock machine for testing the Speedmaster, 1964 ◦ Speedmaster Racing, 2012 ◦ Speedmaster, 2000 ◦ Speedmaster "First Omega in Space," 1962 ◦ Speedmaster, Limited Edition, Caliber 321, 2009 ◦ Speedmaster Broad Arrow, 1957

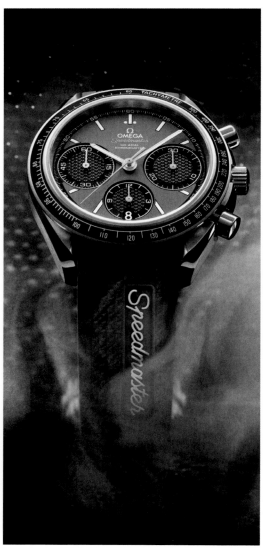

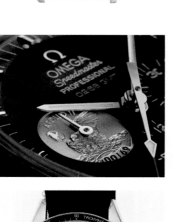

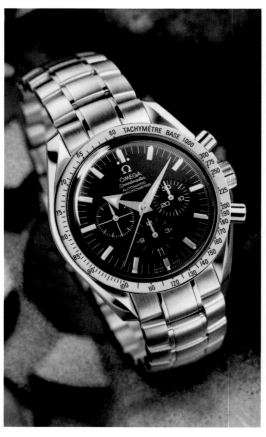

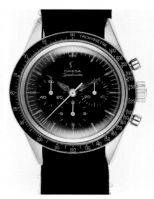

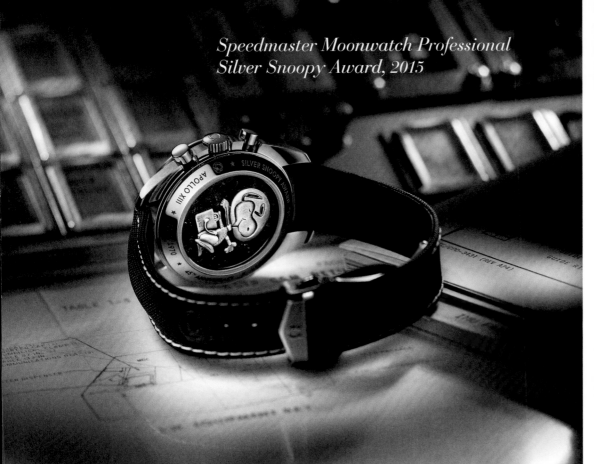

*Speedmaster Moonwatch Professional
Silver Snoopy Award, 2015*

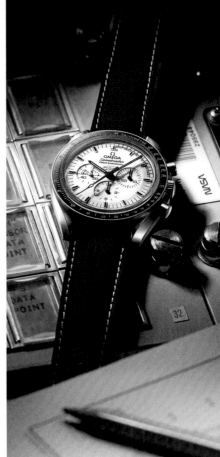

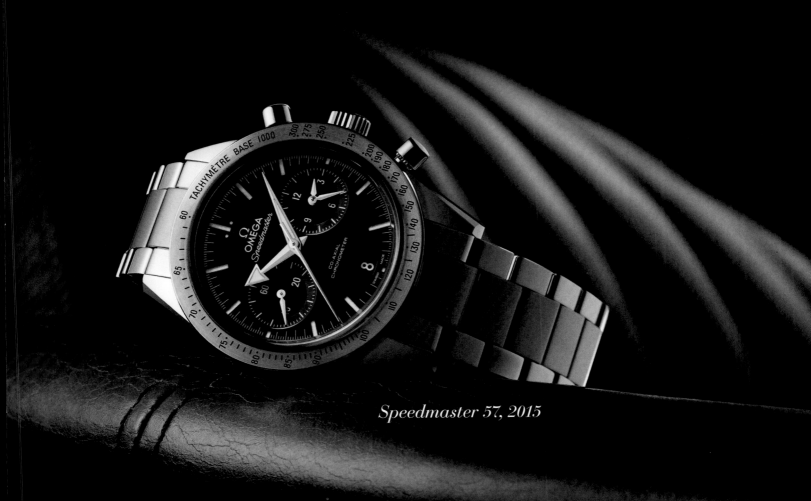

Speedmaster 57, 2015

Von Alpha zu Omega

Deutsch

Wenn geschichtliche Wurzeln zurückreichen bis 1848, kann den Baum, der daraus erwächst, so schnell kein Sturm umwerfen. An dessen Stamm steht unübersehbar ein Ω, der letzte Buchstabe des griechischen Alphabets. Ohne das α, repräsentiert durch einen Herrn Louis Brandt, wäre es dazu natürlich nie gekommen. Die Begegnung mit diesem Mann verlangt eine Rückblende in besagtes Jahr 1848. Damals lockte der berühmte Goldrausch abenteuerlustige Raubeine scharenweise nach Kalifornien. Nicht jedoch den damals 23-jährigen Louis Brandt. Der hielt seiner eidgenössischen Heimat die Stange und wurde reich belohnt. In der Uhrenmetropole La Chaux-de-Fonds entdeckte er ebenfalls Gold, und zwar den goldenen Boden des Uhrmacherhandwerks. Er rief ein „Comptoir d'établissage" ins Leben, also einen Verkaufspunkt für im Lohnauftrag zusammengebaute Zeitmesser. Meist handelte es sich um silberne Taschenuhren, auf die der aufstrebende englische Markt begehrlich wartete. Gute Umsätze ließen die Brandt'sche Unternehmung beständig wachsen. 1880, ein Jahr nach dem Tod des Gründers, zog es die Söhne Louis-Paul und César Brandt hinab ins Tal. In Biel am Fuße des Jura fanden sie eine anmietbare Etage zum Aufbau einer modernen Fabrikationsstätte für Uhren. Schon nach vier Monaten brummte das Geschäft derart stark, dass man die Aktivitäten auf das gesamte Bauwerk ausdehnen musste. Mit einer Jahresproduktion von 100 000 Exemplaren konnte sich Louis Brandt & Fils im Jahr 1889 rühmen, die größte Unternehmung der eidgenössischen Uhrenindustrie zu sein.

Fünf Jahre später, also 1894, gesellte sich zum α endlich das Ω. Und das kam so: Findige Techniker der stetig wachsenden Manufaktur hatten ein neues Kaliber für Taschenuhren zur Serienreife entwickelt. Der tickende Mikrokosmos bestach zum einen durch die bemerkenswert simple, aber in jeder Hinsicht zuverlässige und präzise Konstruktion. Andererseits hatten die Produktentwickler an die seinerzeit keineswegs selbstverständliche Austauschbarkeit der Bauteile gedacht, was Reparaturen im Fall des Falles deutlich erleichterte. Deswegen handelte es sich in der Tat um den letzten Schrei auf dem Uhrenmarkt. Natürlich brauchte dieses geniale Kind auch einen passenden Namen. An Auswahl mangelte es den Brüdern grundsätzlich nicht. Sie hatten sich bereits zahlreiche schützen lassen: „Jura", „Patria", „Helvetia", „Celtic" und „Gurzelen". Und dann war da auch noch das erstaunlich genau gehende Kaliber „Labrador" von 1885. Für den Newcomer wollte jedoch keiner der Namen so richtig passen. Guter Rat war also eigentlich teuer. Er kam erstaunlich schnell vom Hausbankier. Henri Rieckel lag das finanzielle Wohlergehen seiner Kundschaft so sehr am Herzen, dass er den Vorschlag „Omega" völlig kostenlos lieferte. Das Wort passte wie die Faust aufs Auge, denn es symbolisierte den letzten Schrei bei Taschenuhren ausgesprochen trefflich. Der überwältigende Erfolg dieser unverzüglich akzeptierten Idee führte schon 1903 zur Eliminierung aller anderen Markennamen.

Von 1899 bis 1902 wütete der Burenkrieg in Südafrika. Dank Omega mussten britische Offiziere ihre Taschenuhren nicht mehr umständlich aus dem Waffenrock zerren – zum Ablesen der Zeit genügte ein Blick aufs Handgelenk. Und die dort befestigten Armbanduhren bewährten sich späteren Berichten zufolge bestens. 1932 trat Omega erstmals als offizieller Zeitnehmer bei Olympischen Spielen auf, und zwar in Los Angeles. In den anschließenden Jahren betraute das IOC die Marke mehr als zwanzig weitere Male mit dieser verantwortungsvollen Aufgabe.

Auch auf dem Sektor Präzision hat Omega einiges zu bieten: 1919 konnten Chronometer dieser Marke erstmals den Genauigkeitswettbewerb des Observatoriums Neuenburg für sich entscheiden; Genauigkeitsrekorde waren unter anderem auch 1933, 1936 und 1946 zu verzeichnen. 1965 erkor die NASA nach ausgiebigem Prüf-Prozedere den „Speedmaster"-Chronographen zum offiziellen Zeitmesser für ihre Weltraummissionen. Als Neil Armstrong am 21. Juli 1969 um 02:56:20 Uhr (UTC) seinen Fuß mit den bedeutungsvollen Worten „Das ist ein kleiner Schritt für einen Menschen, aber ein großer Sprung für die Menschheit" auf den Mond setzte, trug er die Armbanduhr, die danach „Speedmaster Professional" getauft wurde, bei sich.

Im Zuge der Quarzuhren-Revolution musste dann auch Omega ein tiefes Tal der Tränen durchwandern. Allein 1982 verbuchte die Dachgesellschaft SSIH (Société suisse pour l'industrie horlogère S.A.), zu der auch Tissot gehörte, Verluste in Höhe von insgesamt 17 Millionen Schweizer Franken. Ein Konkurs mit beträchtlichem Imageschaden für die Schweizer Uhrenindustrie stand ins Haus. So kam es 1983 nolens volens zu einer Verschmelzung mit der nicht minder defizitären ASUAG (Allgemeine Schweizer Uhrenindustrie AG). Anschließend sanierte Nicolas G. Hayek mit Umsicht und Weitblick, aber auch eiserner Hand das neu entstandene Konglomerat SMH (Société suisse de microélectronique et d'horlogerie S.A.), welches seit 1998 Swatch Group heißt. 1999 wartete die durch und durch gesundete Traditionsmarke mit einem uhrmacherischen Meilenstein auf: der Co-Axial Hemmung. Diese findet sich heute in allen Manufakturkalibern von Omega. Und davon gibt es mittlerweile eine ganze Menge. ○

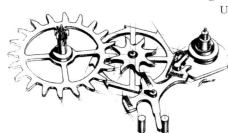

Skeleton Tourbillon Co-Axial Platinum, limited to 18 pieces, 2010

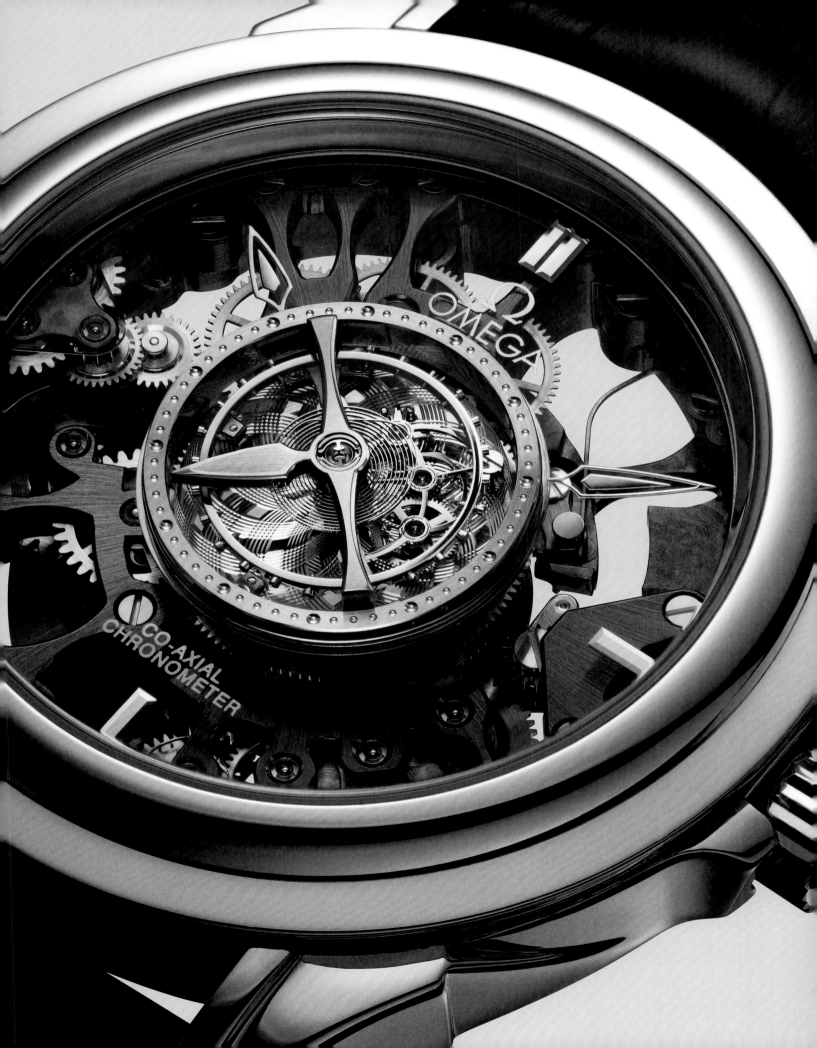

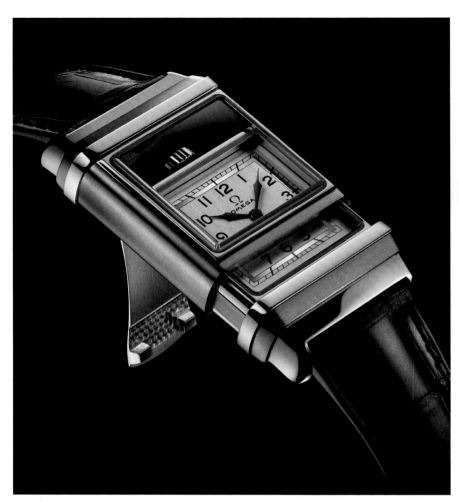

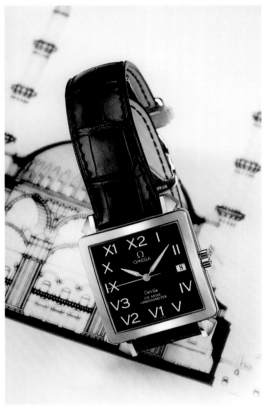

Top left: Marine 1932, Limited Edition, 2007
Top right: De Ville Byzantium, 2005
Bottom: Seamaster Ploprof 1200 M
Co-Axial, 2009

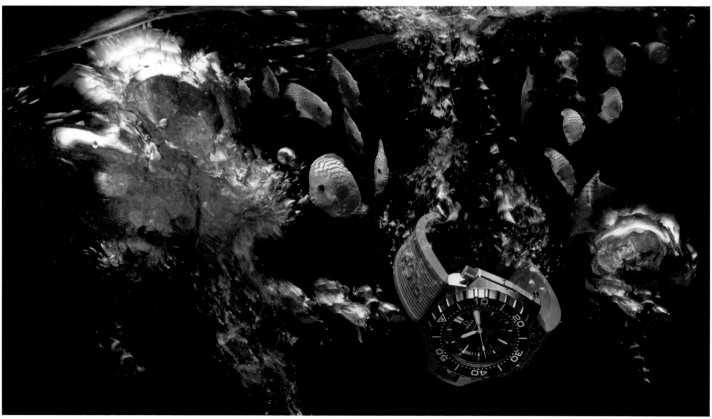

Seamaster Aqua Terra
James Bond Limited Edition, 2015

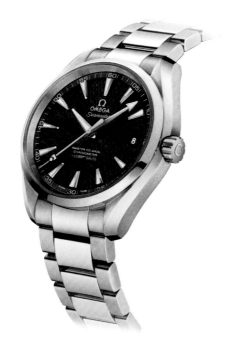

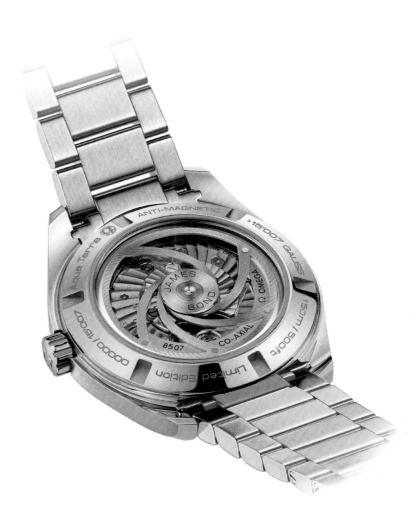

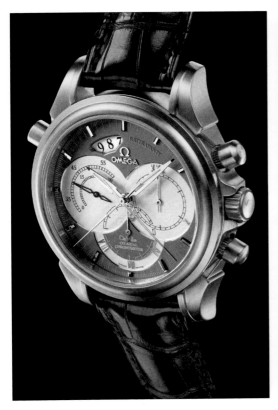

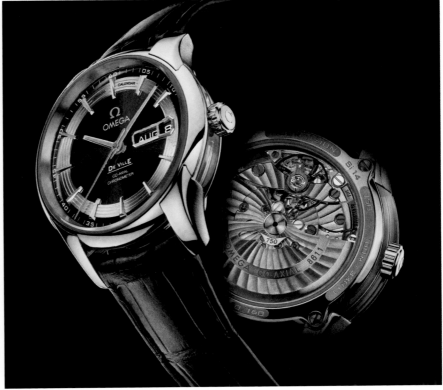

Left: De Ville Co-Axial Rattrapante, 2005 ◦ Right: De Ville Co-Axial, 2007

De l'alpha à OMEGA

L'arbre dont les racines remontent aussi loin qu'à l'année 1848 n'est pas près de succomber à une tempête. Son tronc arbore un Ω, la dernière lettre de l'alphabet grec. Mais sans l'α, en la personne d'un certain Louis Brandt, il ne serait bien sûr jamais sorti de terre. Pour rencontrer cet homme, il nous faut faire un retour en arrière, à ladite année 1848. La fièvre de l'or attire alors des flots d'aventuriers en Californie. Louis Brandt, 23 ans à l'époque, n'est pas du lot : il prend fait et cause pour sa patrie, la Suisse, et s'en trouve largement récompensé. Dans la métropole horlogère de La Chaux-de-Fonds, il découvre également un filon, sous la forme d'un métier artisanal en or, celui d'horloger. Il crée un comptoir d'établissage, c'est-à-dire un commerce de garde-temps assemblés à façon. Il s'agit pour l'essentiel de montres de poche en argent dont est avide le marché anglais en pleine expansion. Réalisant de bons chiffres d'affaires, l'entreprise de Brandt se développe régulièrement. En 1880, un an après le décès du fondateur, ses fils Louis-Paul et César Brandt décident de s'installer dans la vallée. À Bienne, au pied du Jura suisse, ils louent un étage d'un bâtiment pour y monter un atelier de fabrication de montres moderne. Au bout de quatre mois seulement, les affaires marchent si bien que les activités sont redéployées sur l'ensemble de l'édifice. Avec une production annuelle de 100 000 unités, Louis Brandt & Fils peut se glorifier en 1889 d'être leader de l'industrie horlogère suisse.

Cinq ans plus tard, en 1894, l'α est enfin rejoint par l'Ω : d'ingénieux techniciens de cette manufacture en constante expansion mettent au point un nouveau calibre de montres de poche qu'il est possible de produire en série. Ce microcosme au doux tic-tac séduit à double titre : sa construction remarquablement simple est cependant fiable et précise en tous points. De plus, chose loin d'être évidente à l'époque, les développeurs ont pensé aux pièces de rechange, ce qui facilite les réparations en cas de nécessité. Aussi est-ce la montre dernier cri sur le marché. Et cet enfant prodige a bien sûr besoin d'un nom qui lui aille à la perfection. Les frères Brandt ont en principe l'embarras du choix car ils ont déjà déposé une kyrielle d'appellations : « Jura », « Patria », « Helvetia », « Celtic » et « Gurzelen ». Il faut aussi tenir compte du « Labrador », un calibre d'une impressionnante précision de marche mis au point en 1885. Cependant, aucun nom ne convenant vraiment au nouveau-venu, l'entreprise prend conseil auprès de spécialistes. C'est son banquier, Henri Rieckel, qui fait une proposition avec une rapidité étonnante : la santé financière de son client lui tient tant à cœur qu'il suggère

![Advertisement by Paul Helleu, 1904]

Advertisement by Paul Helleu, 1904
Werbeanzeige von Paul Helleu, 1904
Publicité par Paul Helleu, 1904

« Omega » sans attendre aucune contrepartie financière. Ce nom convient à la perfection car il symbolise à merveille le dernier cri des montres de poche. L'idée est aussitôt acceptée, et face à son succès foudroyant, on abandonne dès 1903 tous les autres noms de marques.

De 1899 à 1922, la guerre des Boers fait rage en Afrique du Sud. Grâce à Omega, les officiers britanniques ne doivent plus se livrer à des contorsions pour extirper leurs montres de poche de leurs tuniques : pour lire l'heure, ils jettent un simple coup d'œil à leur poignet. Les montres-bracelets qui y sont fixées font pleinement leurs preuves, comme il sera rapporté plus tard. En 1932, Omega est le premier chronométreur officiel des Jeux olympiques, qui se tiennent alors à Los Angeles. Par la suite, le CIO confiera cette lourde responsabilité à cette société à plus de vingt reprises.

Dans le domaine des mesures de précision, Omega n'est pas non plus en reste : en 1919, ses chronomètres remportent le concours de précision de l'observatoire de Neuchâtel ; ils battront d'autres records de précision, notamment en 1933, 1936 et 1946. En 1965, au terme d'essais dans des conditions extrêmes, la NASA désigne la « Speedmaster » comme chronographe officiel pour ses missions spatiales. Quand, le 21 juillet 1969, à 2 heures 56 minutes et 20 secondes (UTC), Neil Armstrong pose le pied sur la Lune en prononçant ces paroles mémorables « C'est un petit pas pour un homme, mais un grand bond pour l'humanité », il porte la montre-bracelet plus tard baptisée « Speedmaster Professional ».

La révolution du quartz signifie une traversée du désert également pour Omega. Pour le seul exercice 1982, le regroupement d'entreprises SSIH (Société suisse pour l'industrie horlogère SA), dont Tissot fait alors aussi partie, enregistre des pertes s'élevant au total à 17 millions de francs suisses. Pour éviter une faillite fortement préjudiciable à l'image de l'industrie horlogère suisse, Omega fusionne bon gré, mal gré en 1983 avec l'ASUAG (Allgemeine Schweizer Uhrenindustrie AG), pas moins déficitaire. Avec circonspection et clairvoyance, mais d'une main de fer, Nicolas G. Hayek redresse le nouveau conglomérat SMH (Société suisse de microélectronique et d'horlogerie SA), qui prendra le nom de Swatch Group en 1998. En 1999, parfaitement rétablie, la marque de tradition lance un dispositif horloger d'avant-garde : l'échappement coaxial, qui équipe désormais tous les calibres de manufacture signés Omega. Et ces derniers sont maintenant légion. ◌

De Ville Prestige Dewdrop, 2015

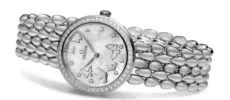

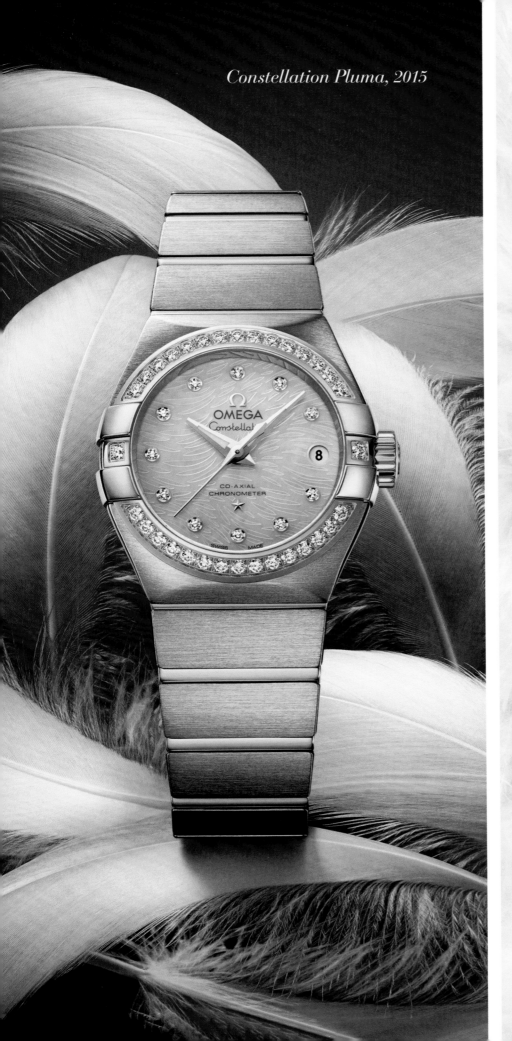

Constellation Pluma, 2015

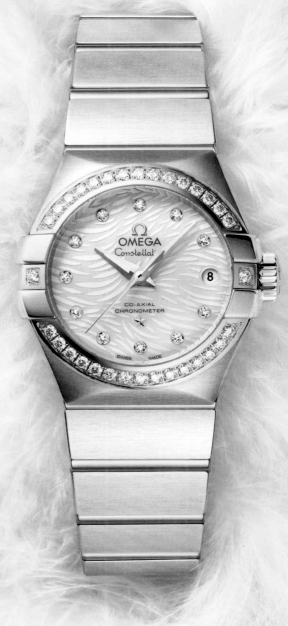

Constellation Pluma
Light Coral, 2015

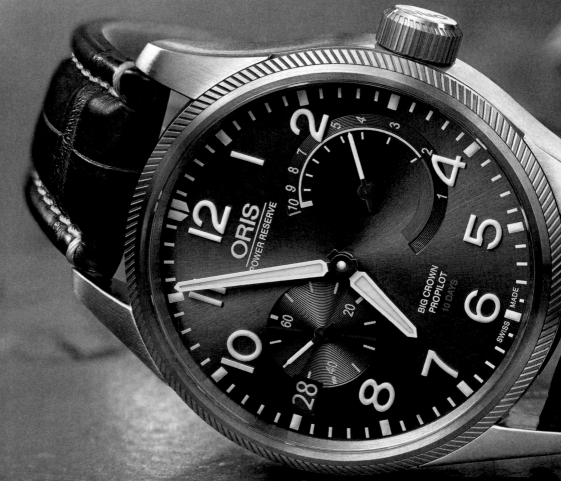

Oris

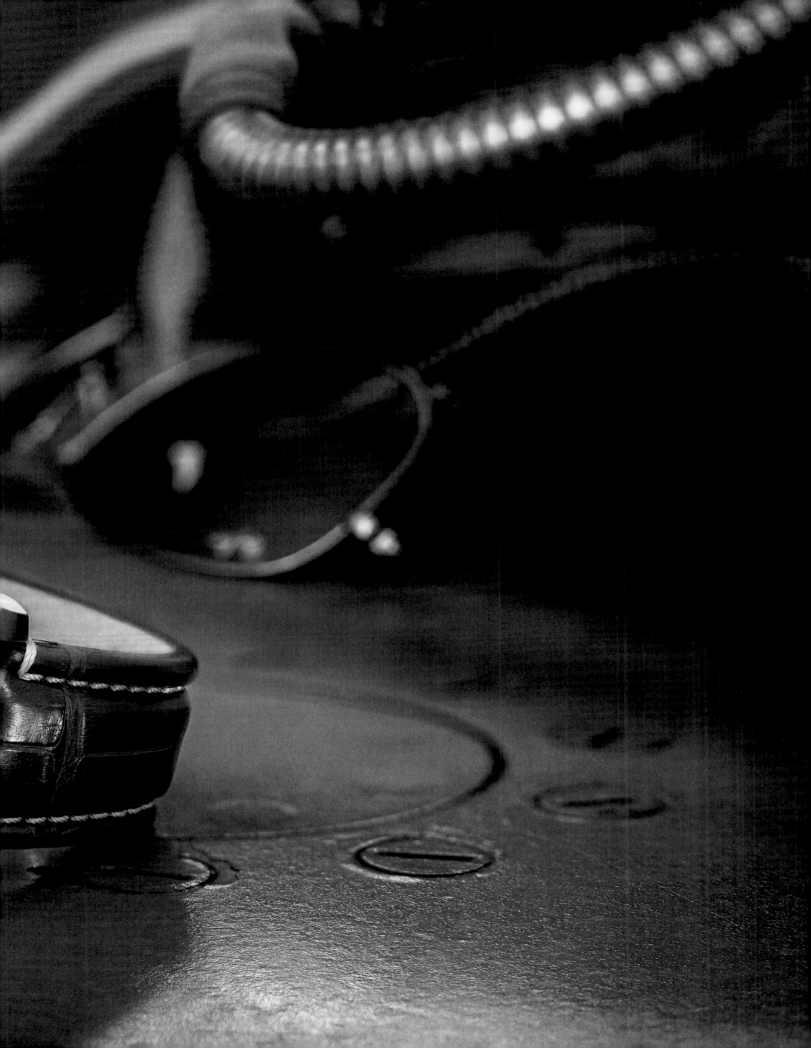

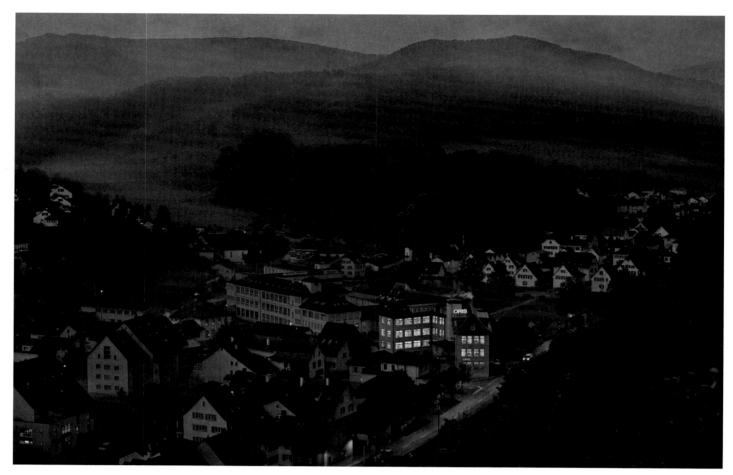

Oris factory at dawn, 2015 ◦ Oris-Fabrik in der Morgendämmerung, 2015 ◦ La manufacture Oris à l'aube, 2015

Under the Sign of "High Mech"

English

The town of Hölstein, which is located southeast of Basel, was a sleepy little hamlet in the early years of the 20ᵗʰ century. A first attempt to bolster the weak infrastructure in this region was undertaken with the opening of a watchmaking factory, but the venture failed a mere two years later. All that survived was the name "Oris," which was borrowed from a nearby stream and registered as a trademark for timepieces in 1903. Paul Cattin and Georges Christian hoped for better luck in 1904. Their venture succeeded so well that by 1910 three-hundred families were fed by breadwinners who worked for the Manufacture d'Horlogerie de Hölstein, Cattin & Christian, which made affordably priced pin-pallet watches "for the people." After its founders' deaths, the company was sold in the late 1920s to a group of investors which included Georges Christian's widow and Jacques-David LeCoultre from the Vallée de Joux, who was a friend of the Christian Family. Watches bearing the Oris insignia repeatedly proved that pin-pallet calibers, if they're painstakingly designed and meticulously assembled, can be just as reliable and precise as their counterparts with traditional Swiss lever escapements. The reward for ongoing optimization was forthcoming in 1968, when Caliber Oris 652 became the first pin-pallet caliber to earn a rate certificate from Neuchâtel Observatory.

This movement strategy had shown its Achilles' heel in 1934, when a Swiss law, designed to rescue the confederation's ailing watch industry from demise, specified that each firm must restrict its product spectrums to their current status. But these restrictions could stop neither Oris's creativity nor its success. Beginning in 1936, the company also fabricated its pin-pallet models in Holderbank, Como, Courgenay, Ziefen, Herbetswil, and Bienne. The payroll included the names of circa 1,000 employees, many of whom lived in homes owned by Oris. Others rode the company's buses each morning and evening for the 25-kilometer commute between Basel and Hölstein.

Caliber 373 "Pointer" was first marketed in 1938. Afterwards, every Oris collection included at least one model with a centrally axial date-hand. Oris's first automatic watch with a bidirectionally active rotor raised appreciative eyebrows in 1952. A power-reserve display at the "12" highlighted the efficiency of the self-winding mechanism. When the restrictive watch statute finally ended in 1966, Oris quickly took action and brought Caliber 645, its first automatic movement with a classical Swiss lever escapement, to maturity for series production. Caliber 645 and other outstanding

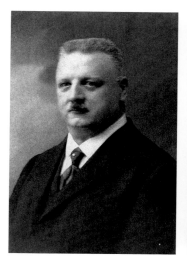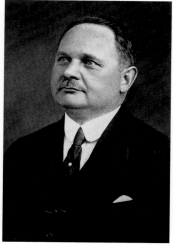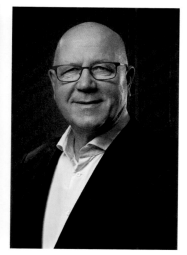

From left: Paul Cattin ◦ Georges Christian ◦ Ulrich W. Herzog

achievements elevated Oris into the elite circle of the great names in the international watch industry. Oris shipped no fewer than 1.2 million mechanical watches and alarm timepieces from its various factories in 1970.

The following year, when the long shadow of the Quartz Revolution was beginning to darken the skies over Switzerland, Oris came under the aegis of the Allgemeine Schweizer Uhren AG (ASUAG), a super-holding that had been founded in 1931 and that merged in 1985 with the equally ailing SSIH to become the SMH (Schweizerische Gesellschaft für Mikroelektronik und Uhrenindustrie AG).

To make a long story short: newfangled electronic timekeeping drove nearly 900 Swiss watchmaking companies into bankruptcy in the late 1970s. Approximately two-thirds of the employees lost their jobs. Oris was among the victims. When ASUAG wanted to shut down its factories in 1982, its CEO Dr. Rolf Portmann and its marketing director Ulrich W. Herzog assured its survival with a management buyout. The erstwhile Oris Watch Co. SA became the smaller Oris SA.

Finally, a few statistics from Oris's illustrious history: from 1919 to 1987, the firm's production sites in Hölstein and elsewhere fabricated no fewer than 96,850,000 watches. Top ratings were achieved during the glory days from 1919 to 1928 (27.5 million timepieces) and from 1929 to 1948 (39 million timepieces). The company developed a total of 279 calibers or modules for existing movements between 1904 and 1981.

Naysayers smirked patronizingly in 1985 when Oris took the almost visionary decision to manufacture only mechanical watches. Relying on time-honored Eta calibers and partly also on modules that Oris had engineered, the brand developed a wide spectrum of affordably priced wristwatches. These included collectible retro models, some with built-in alarm mechanisms, and the exclusive "Worldtimer," which debuted in 1997 with a useful display for other time zones. A totally updated look for the brand and the catchy slogan "It's High Mech" underscored the new focus on mechanical timekeeping.

Punctually in time for its 110th birthday in 2014, Oris revived its manufactory activities, which some people thought the brand had forgotten. Hand-wound Caliber 110, which was developed and fabricated in collaboration with the Technicum in Le Locle and other experienced partners, embodied Oris's first own creation in 35 years. A gaze through the transparent back of this 34-millimeter-diameter opus reveals a positively gigantic energy reservoir. The 1.80-meter-long mainspring inside the big barrel provides power for ten days of uninterrupted running. The patented power-reserve display on the dial is particularly creative. The hand on this indicator moves relatively slowly at first, but its tempo increases as the mainspring slackens. The first limited edition of the wristwatch that was equipped with this powerhouse sold out in a proverbial blink of an eye. Oris followed this debutante with the premiere of an unlimited successor in 2015: Caliber 111, which was produced entirely with the brand's own tools. Along with the previous movement's features, this newcomer also offered a large date window at the "9." The latest phase in the evolutionary process is Caliber 112, which is encased in "Artelier Calibre 112," which debuted in 2016. To the delight of frequent flyers, this version has a user-friendly time-zone function.

Of course, wristwatches with this exclusive caliber represent only the crowning mound of whipped cream atop this broadly diverse collection. In addition to these, Oris also offers reasonably priced watches to satisfy nearly every requirement. Divers, for example, can opt for the nostalgic-looking "Divers Sixty-Five." But the 2,000 watches that debuted in 2016 in the "Carl Brashear Limited Edition" are already sold out. The "Aquis Depth Gauge" is equipped with a practical bathometer. The counterpart for pilots is the "Big Crown ProPilot Altimeter," which naturally includes an altimeter. Automobile fans will appreciate Audi or Williams models. And jazz lovers will enjoy the latest "Thelonious Monk Limited Edition." Oris offers just about everything—except watches with quartz movements. If you're looking for one of those, look elsewhere. ◦

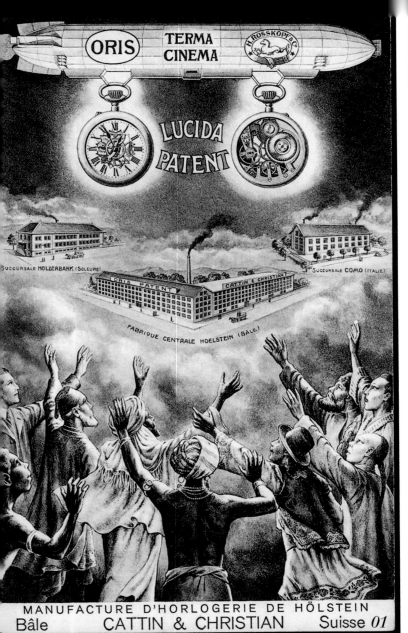

ORIS
TERMA
CINEMA
H. ROSSKOPF & Cᵒ

LUCIDA
PATENT

SUCCURSALE HOLDERBANK (Soleure)
SUCCURSALE COMO (ITALIE)
ORIS PATENT
CATTIN & CHRISTIAN
FABRIQUE CENTRALE HOELSTEIN (BÂLE)

MANUFACTURE D'HORLOGERIE DE HÖLSTEIN
Bâle CATTIN & CHRISTIAN Suisse 01

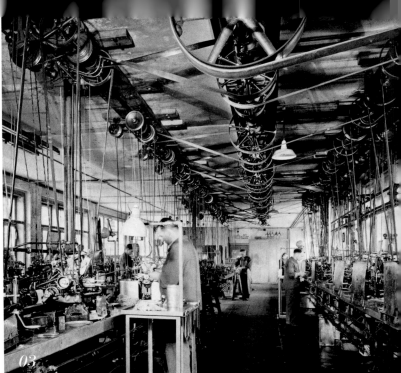

03

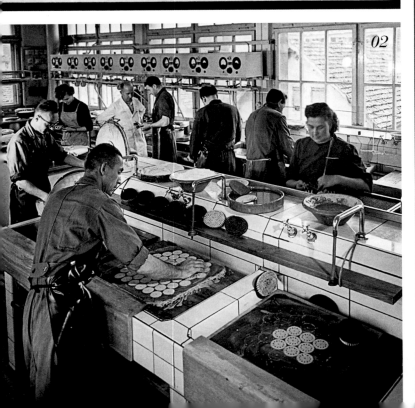

02

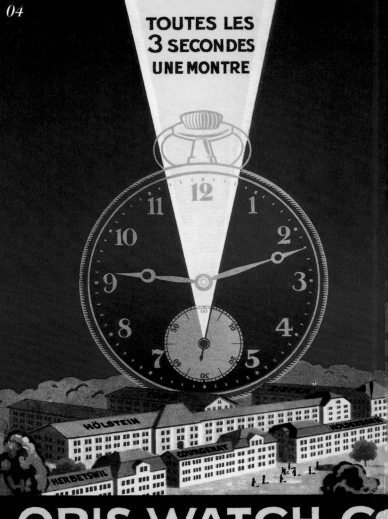

04

TOUTES LES
3 SECONDES
UNE MONTRE

HÖLSTEIN
COURGENAY
HERBETSWIL
HOLDERBANK

ORIS WATCH Cᵒ
HÖLSTEIN (SUISSE)
MONTRES ROSKOPF & CYLINDRE
MARQUES: ORIS, TERMA, BREVO, FIDES, FIDO, FIXO
COLOMBUS ELSINE, RIO, LUCIDA, BABEL, VIRTUS, VALDORI
VOIR AU VERS

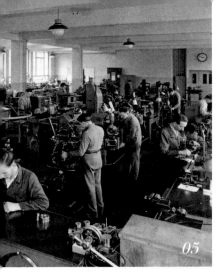

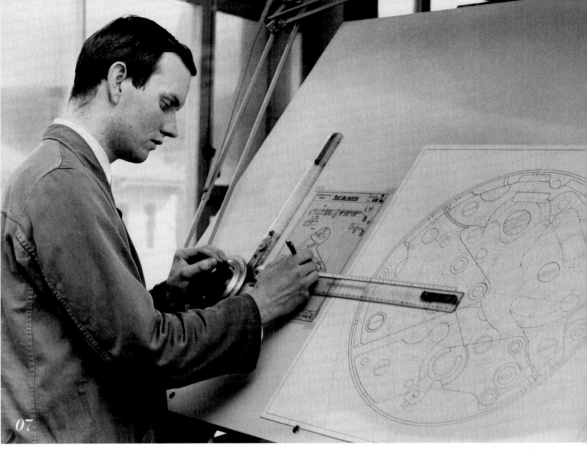

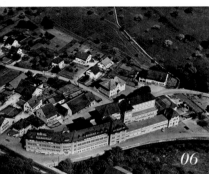

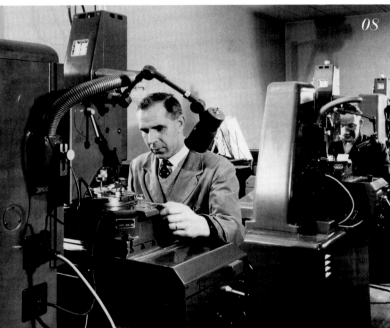

01: Postcard, 1910 ○ *Postkarte, 1910* ○ **Carte postale, 1910** ○
02: Oris electroplating factory at Herbetswil, 1925 ○ *Oris-Galvanisier-fabrik in Herbetswil, 1925* ○ **L'usine de galvanoplastie Oris à Herbetswil, 1925** ○ 03: Production line, 1930s ○ *Fertigungslinie, 1930er* ○ **La chaîne de production, années 1930** ○ 04: Advertising, 1926 ○ *Werbung, 1926* ○ **Publicité, 1926** ○ 05: Mechanical production, 1938 ○ *Mechanische Fertigung, 1938* ○ **Fabrication mécanique, 1938** ○
06: Headquarters in Hölstein, 1953 ○ *Hauptsitz in Hölstein, 1953* ○ **Le siège social à Hölstein, 1953** ○ 07: Oskar Mohler working on a technical drawing, 1968 ○ *Oskar Mohler beim Erstellen einer tech-nischen Zeichnung, 1968* ○ **Oskar Mohler planchant sur un dessin technique, 1968** ○ 08: Oris factory, 1969 ○ *Oris-Fabrik, 1969* ○ **La manufacture Oris, 1969** ○ 09: Painting of Oris factories, 1929 ○ *Gemälde der Oris-Fabriken, 1929* ○ **Peinture représentant la manufacture Oris, 1929** ○ 10: Pointer Calendar, 1938

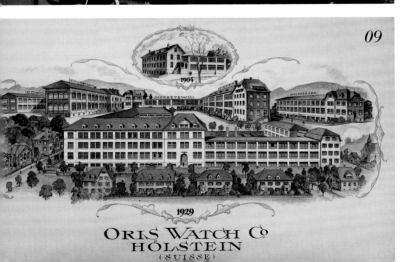

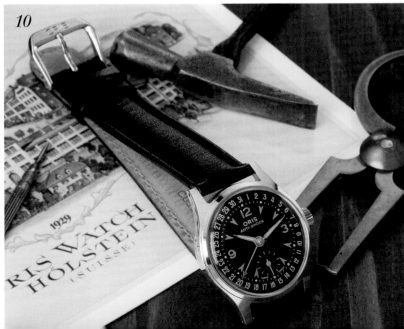

Im Zeichen von „High Mech"

Deutsch

Salopp könnte man sagen, dass sich zu Beginn des 20. Jahrhunderts in der südöstlich von Basel gelegenen Gemeinde Hölstein die Füchse und die Hasen Gute Nacht sagten. Ein erster Versuch, die missliche Situation in der strukturschwachen Gegend durch den Aufbau einer Uhrenfertigung zu bessern, scheiterte schon nach zwei Jahren. Was blieb, war der 1903 für Zeitmesser geschützte, von einem Bach in der Nähe entliehene Name Oris. 1904 wollten es die Herren Paul Cattin und Georges Christian besser machen. Und das Unterfangen gelang. 1910 ernährten die Manufacture d'Horlogerie de Hölstein, Cattin & Christian und ihre preiswerten Stiftanker-Uhren „fürs Volk" bereits 300 Familien. In den späten 1920er Jahren, nach dem Tod der Gründer, gelangte das Unternehmen ins Eigentum einer Investorengruppe um die Witwe von Georges Christian mit Jacques-David LeCoultre aus dem Vallée de Joux, einem Freund der Familie. Regelmäßig stellten die Oris signierten Uhren unter Beweis, dass Stiftanker-Kaliber bei sorgfältiger Konstruktion und Ausführung genauso zuverlässig und präzise sind wie solche mit traditioneller Schweizer Ankerhemmung. Den Lohn für kontinuierliche Optimierung gab es 1968. Als erstes Stiftanker-Kaliber überhaupt erhielt das Oris 652 ein Gangzeugnis des Observatoriums in Neuchâtel.

1934 hatte sich aber auch die Achillesferse dieser Werkestrategie gezeigt. Damals zementierte ein Schweizer Bundesgesetz, welches die kriselnde Uhrenindustrie vor dem Untergang bewahren sollte, das Produktspektrum der Firmen auf den aktuellen Status. Bei Oris tat Beschränkung weder der Kreativität noch dem Erfolg einen Abbruch. Ab 1936 produzierte das Unternehmen seine Stiftanker-Modelle auch in Holderbank, Como, Courgenay, Ziefen, Herbetswil und Biel. Auf der Gehaltsliste standen rund 1 000 Mitarbeiterinnen und Mitarbeiter. Etliche lebten in Oris-Häusern. Viele reisten jedoch auch täglich mit eigenen Buslinien aus dem 25 Kilometer entfernten Basel nach Hölstein.

1938 gelangte das Kaliber 373 „Pointer" auf den Markt. Anschließend beinhaltete jede Oris-Kollektion mindestens ein Modell mit zentralem Zeigerdatum. Die erste Oris-Automatikuhr mit beidseitig wirkendem Rotoraufzug machte 1952 von sich reden. Eine Gangreserveanzeige bei der „12" führte die Effizienz des Selbstaufzugs vor Augen. 1966, als das lähmende Uhrenstatut endlich endete, schritt Oris unverzüglich zur Tat. Noch im gleichen Jahr erlangte das erste Automatikwerk mit klassischer Schweizer Ankerhemmung seine Serienreife. Mit diesem Kaliber 645 und anderen Spitzenleistungen gelang Oris der Anschluss an die Größten des internationalen Uhrenbusiness. 1970 verließen 1,2 Millionen mechanische Uhren und Wecker die verschiedenen Fabriken.

Im Folgejahr, die Quarz-Revolution warf ihre Schatten voraus, gelangte Oris unter das Dach der 1931 gegründeten Superholding Allgemeine Schweizer Uhren AG (ASUAG), welche 1985 mit der nicht minder kränkelnden SSIH zur SMH (Schweizerische Gesellschaft für Mikroelektronik und Uhrenindustrie AG) fusionierte.

Dazu so viel in aller Kürze: In den späten 1970er Jahren hatte die zeitbewahrende Elektronik knapp 900 Schweizer Uhr-Unternehmen in den Konkurs getrieben. Etwa zwei Drittel der Belegschaft verlor ihren Job. Betroffen war auch Oris. Als die ASUAG den Betrieb 1982 schließen wollte, sicherten Geschäftsführer Dr. Rolf Portmann und der damalige Marketingleiter Ulrich W. Herzog das Überleben durch einen Management-Buyout. Aus der ehemaligen Oris Watch Co. SA wurde die natürlich kleinere Oris SA.

Zum Schluss noch einige Zahlen zur illustren Oris-Geschichte: Von 1919 bis 1987 entstanden in Hölstein und den zugehörigen Produktionsstätten nicht weniger als 96 850 000 Uhren. Spitzenränge nahmen die glanzvollen Epochen von 1919 bis 1928 (27,5 Millionen Exemplare) und 1929 bis 1948 (39 Millionen Exemplare) ein. Außerdem entwickelte das Unternehmen zwischen 1904 und 1981 insgesamt 279 Kaliber oder Module zu existenten Uhrwerken.

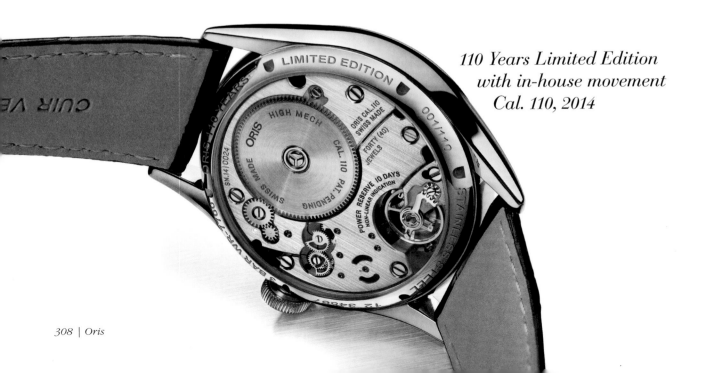

110 Years Limited Edition
with in-house movement
Cal. 110, 2014

Movement Cal. 690
Worldtimer, 1997

Als Oris im Jahre 1985 fast schon visionär entschied, nur noch mechanische Uhren herzustellen, wurde das Unternehmen von manchen Seiten belächelt. Auf der Basis bewährter Eta-Kaliber und zum Teil selbstentwickelter Module entstand ein breites Spektrum preiswerter Armbanduhren. Dazu gehörten sammelnswerte Retro-Modelle beispielsweise mit Wecker oder 1997 der exklusive „Worldtimer" mit hilfreichem Zeitzonen-Dispositiv. Ein völlig neuer Markenauftritt unter dem Slogan „It's High Mech" unterstrich die auf Mechanik fokussierte Neuausrichtung.

Pünktlich zum 110. Geburtstag im Jahr 2014 belebte Oris seine vergessen geglaubten Manufaktur-Aktivitäten. Das Handaufzugskaliber 110, entwickelt und gefertigt gemeinsam mit dem Technikum in Le Locle sowie einschlägig erfahrenen Partnern, repräsentierte die erste Eigenkreation seit 35 Jahren. Beim Blick auf die Rückseite des 34 Millimeter großen Œuvres sticht der geradezu riesige Energiespeicher ins Auge. Die darin aufgewundene 1,80 Meter lange Zugfeder liefert Kraft für zehn Tage ununterbrochenen Lauf. Besonders kreativ: die patentierte Gangreserveanzeige auf dem Zifferblatt. Ihr Zeiger bewegt sich anfangs relativ langsam und steigert sein Tempo mit nachlassender

Federkraft. Die erste limitierte Edition der damit ausgestatteten Armbanduhr war im Handumdrehen ausverkauft. 2015 präsentierte Oris das fortan mit eigenen Werkzeugen unlimitiert produzierte 111. Hier gesellt sich zum Bestehenden ein angenehm großes Fensterdatum bei der „9". Die letzte Evolutionsstufe heißt 112, zu finden in der 2016 vorgestellten „Artelier Calibre 112". Zur Freude vielfliegender Kosmopoliten besitzt diese Version eine leicht handhabbare Zeitzonen-Funktion.

Armbanduhren mit diesem exklusiven Uhrwerk repräsentieren freilich nur die Sahnehaube der breit gefächerten Kollektion. Darüber hinaus bedient Oris zu vernünftig kalkulierten Preisen nahezu jeden Anspruch. Taucher können sich an der sehr nostalgisch anmutenden „Divers Sixty-Five" erfreuen. Die 2000 Exemplare der „Carl Brashear Limited Edition" von 2016 sind jedoch längst vergriffen. Über einen praktischen Tiefenmesser verfügt die „Aquis Depth Gauge". Das Piloten-Pendant mit Höhenmesser heißt „Big Crown ProPilot Altimeter". Autofans kommen durch Audi- oder Williams-Modelle zu ihrem Recht, Jazzliebhaber zum Beispiel durch die aktuelle „Thelonious Monk Limited Edition". Alleine bei Uhren mit Quarzwerk muss Oris schlichtweg passen. ○

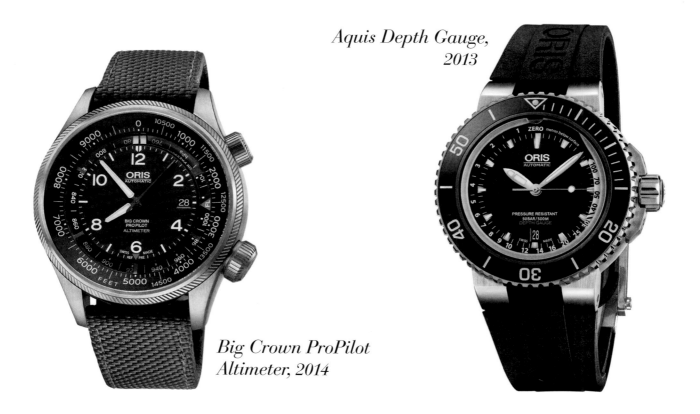

Aquis Depth Gauge,
2013

Big Crown ProPilot
Altimeter, 2014

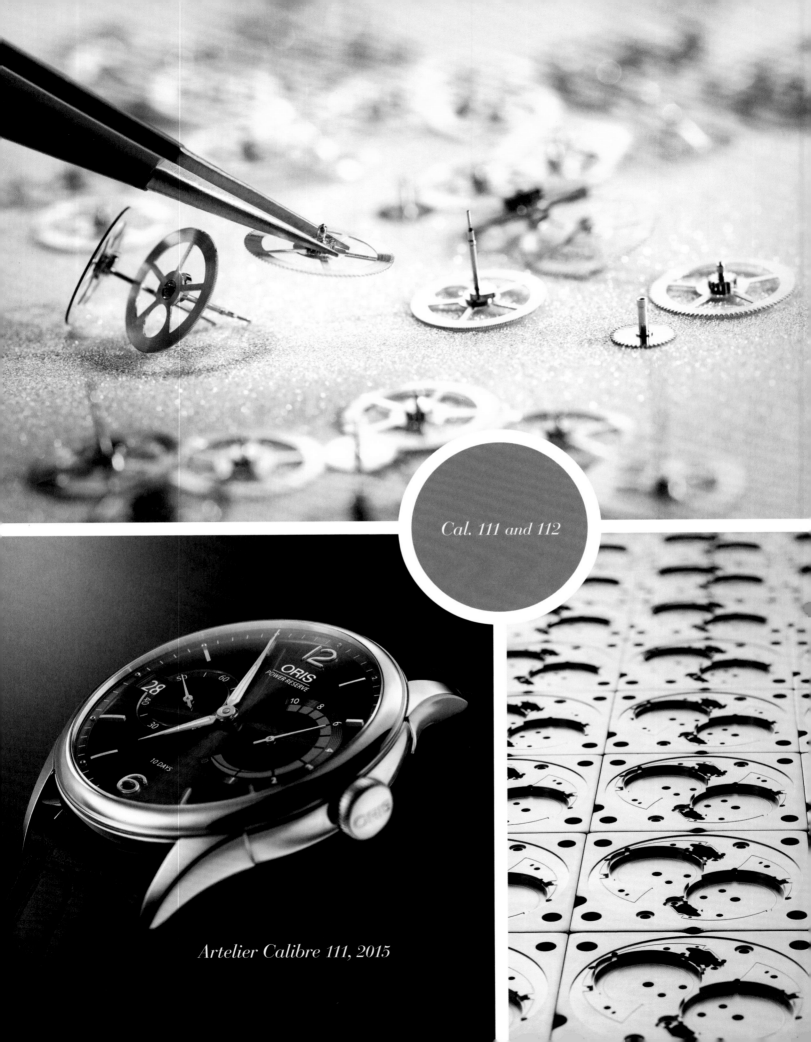

Cal. 111 and 112

Artelier Calibre 111, 2015

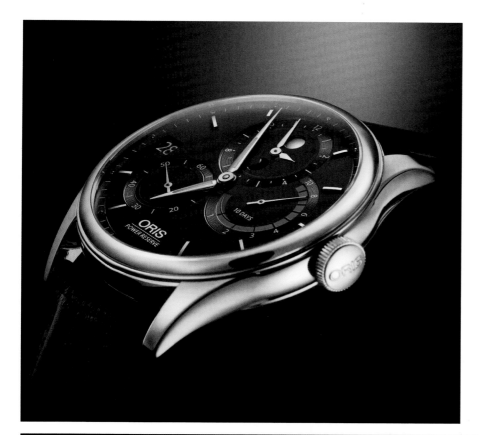

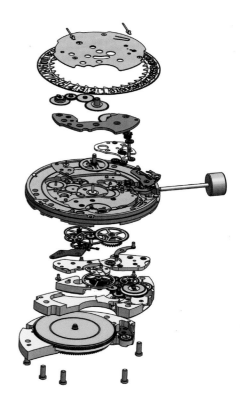

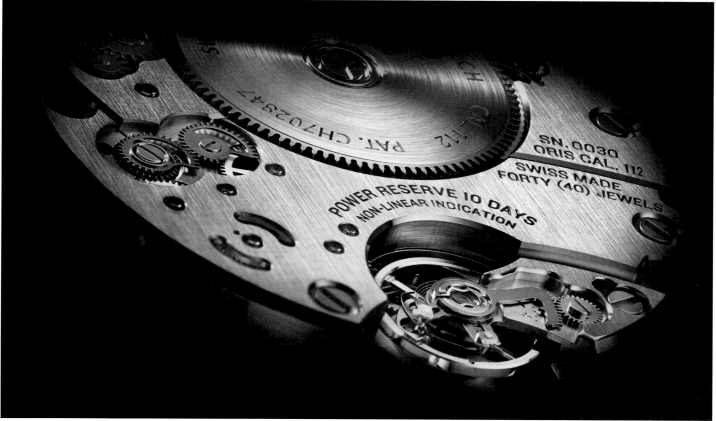

Clockwise from top left: Artelier Calibre 112, 2016 ◦ Exploded view of Caliber 111, 2016 ◦ In-house movement 112 with worm gear on the left for non-linear power-reserve indication (Oris patent) ◦ *Von oben links im Uhrzeigersinn: Artelier Calibre 112, 2016 ◦ Explosionsdarstellung von Kaliber 111, 2016 ◦ Hausintern entwickeltes Kaliber 112 mit Schneckengetriebe auf der linken Seite für nicht-lineare Gangreserveanzeige (Oris-Patent)* ◦ *Dans le sens horaire, en partant du haut, à gauche : Artelier Calibre 112, 2016 ◦ Vue en éclaté d'un calibre 111, 2016 ◦ Mouvement de manufacture 112 avec vis sans fin sur la gauche pour l'indicateur de réserve de marche à affichage non-linéaire (brevet Oris)*

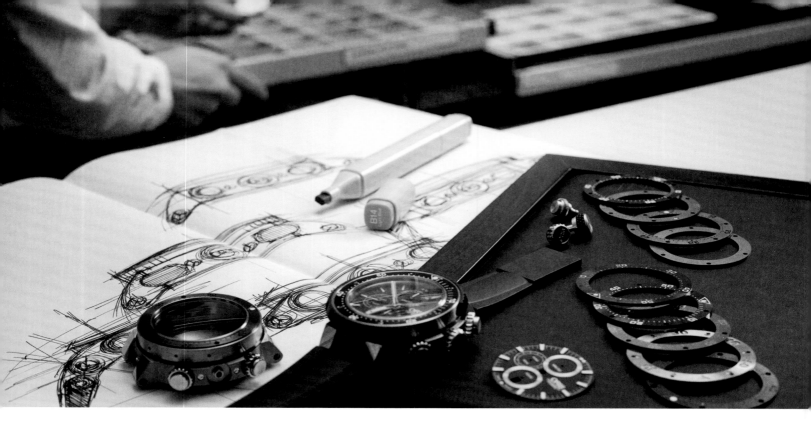

Sous le signe de la « High Mech »

Français

Au début du XXᵉ siècle, Hölstein, commune située au sud-est de Bâle, est pour ainsi dire encore en pleine cambrousse. Une première tentative pour remédier à cette fâcheuse situation en établissant une manufacture horlogère dans cette région défavorisée s'avère être un échec après seulement deux ans. Tout ce qu'il en reste, c'est Oris, le nom d'un ruisseau du voisinage déposé à titre de marque en 1903 pour des garde-temps. En 1904, Paul Cattin et Georges Christian espèrent plus de succès, et leur tentative réussit. Dès 1910, la Manufacture d'Horlogerie de Hölstein Cattin & Christian et ses montres « populaires » abordables dotées d'échappements à ancre à chevilles font vivre 300 familles. À la fin des années 1920, après le décès des fondateurs, l'entreprise devient la propriété d'un groupe d'investisseurs réunis autour de la veuve de Georges Christian et d'un ami de la famille, Jacques-David LeCoultre de la vallée de Joux. Régulièrement, les montres signées Oris démontrent que les calibres à ancre à chevilles sont tout aussi fiables et précis que ceux équipés d'un échappement à ancre suisse classique, à condition d'avoir été correctement conçus et réalisés. Des optimisations permanentes sont récompensées en 1968 : le calibre Oris 652 est le premier mouvement à échappement à chevilles à obtenir un certificat de chronomètre de l'observatoire astronomique et chronométrique de Neuchâtel.

En 1934, toutefois, cette stratégie en matière de mouvements montre ses limites. À l'époque, une loi fédérale suisse censée sauver l'industrie horlogère en crise, gèle les gammes de produits des manufactures. Mais cette limitation ne portera atteinte ni à la créativité ni au succès d'Oris. À compter de 1936, l'entreprise produit des modèles à échappement à chevilles également à Holderbank, Côme, Courgenay, Ziefen, Herbetswil et Bienne. Elle compte 1 000 employés, dont bon nombre vivent dans des logements construits par Oris. Mais nombreux sont aussi ceux qui empruntent quotidiennement la ligne privée de bus de l'entreprise pour parcourir les 25 kilomètres séparant Bâle d'Hölstein.

En 1938, le calibre 373 « Pointer » arrive sur le marché. Chaque collection de la marque comportera dès lors au moins un modèle avec calendrier à aiguille central. La première montre automatique Oris à remontage par rotor bidirectionnel sort en 1952. L'indication de réserve de marche à 12 heures dévoile l'efficacité du remontage automatique. En 1966, lorsque le carcan du Statut horloger se desserre, Oris passe aussitôt à l'action. La même année, le premier mouvement automatique à échappement à ancre suisse classique est fabriqué en série. Grâce à ce calibre 645 et à d'autres prouesses, Oris rejoint le club des grands du marché international des montres. En 1970, 1,2 million de montres et de réveils mécaniques quittent les différentes usines de l'entreprise.

L'année suivante, alors que se profile l'ombre de la révolution du quartz, Oris passe dans le giron de l'ASUAG (Allgemeine Schweizer Uhren AG), superholding de l'industrie horlogère fondée en 1931, qui fusionnera en 1985 avec la non moins chancelante SSIH (Société suisse pour l'industrie horlogère) pour former la SMH (Société suisse de microélectronique et d'horlogerie).

Mais revenons brièvement sur cette situation : à la fin des années 1970, les garde-temps électroniques conduisent près de 900 entreprises horlogères suisses à la faillite. Les deux tiers environ du personnel horloger perdent leur emploi. Et cette crise touche également Oris. Lorsque l'ASUAG veut fermer l'entreprise en 1982, le président

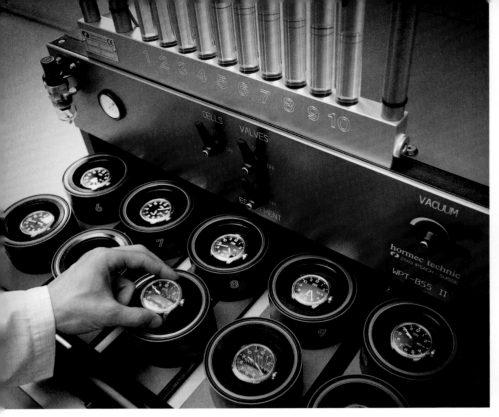
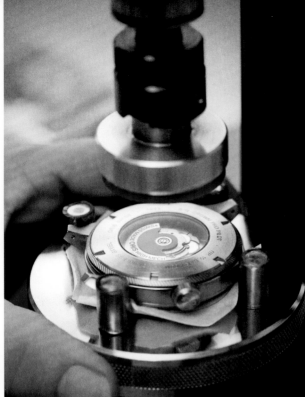

From left: Redesign work for ProDiver chronograph ○ Each watch passes a water-resistance check before leaving the factory ○
Closing the screwed case back ○ Von links: Neugestaltung des Modells ProDiver Chronograph ○ Jede Uhr wird auf Wasserdichtigkeit getestet,
bevor sie die Fabrik verlässt ○ Schließvorgang bei verschraubter Gehäuserückseite ○ De gauche à droite : Travail de refonte du chronographe
ProDiver ○ Chaque montre subit un contrôle d'étanchéité avant de quitter l'usine ○ Opération de fermeture du fond de boîtier vissé

Rolf Portmann et le directeur général Ulrich W. Herzog assurent sa survie en la rachetant. Oris Watch Co. SA devient Oris SA, avec bien sûr des effectifs moins nombreux.

Pour terminer, quelques chiffres encore sur la célèbre histoire d'Oris : de 1919 à 1987, pas moins de 96 850 000 montres sortent de la manufacture de Hölstein et des autres sites de production. Dans les époques fastes de 1919 à 1928 (27,5 millions d'exemplaires) et de 1929 à 1948 (39 millions d'exemplaires), la production est à son maximum. Entre 1904 et 1981, l'entreprise met en outre au point 279 calibres et modules pour les mouvements.

En 1985, la décision quasi-visionnaire d'Oris de ne plus fabriquer que des montres mécaniques fait sourire certains. Mais la manufacture parvient, à l'aide de calibres Eta éprouvés et de modules en partie fabriqués en interne, à se constituer une gamme respectable de montres-bracelets abordables. Parmi ces dernières figurent des modèles rétro pour collectionneurs, équipés notamment d'un réveil, ou encore le « Worldtimer » exclusif avec son dispositif bien utile d'affichage des fuseaux horaires. Un tout nouveau positionnement de la marque avec pour devise « It's High Mech » vient souligner cette nouvelle orientation résolument axée sur la mécanique.

En 2014, date précise de son 110e anniversaire, Oris fait revivre ses activités de manufacture que l'on croyait disparues. Le calibre 110 à remontage manuel, conçu et fabriqué en collaboration avec le technicum du Locle et une équipe de partenaires dûment expérimentés, constitue la première création de la manufacture depuis 35 ans. Au revers de ce chef-d'œuvre de 34 millimètres de diamètre, le dispositif assez conséquent chargé d'emmagasiner la force motrice saute aux yeux. Le ressort de 1,80 mètre enroulé fournit assez d'énergie pour dix jours d'autonomie de marche. Caractéristique particulièrement innovante : l'indication de réserve de marche brevetée, sur le cadran. L'aiguille se déplace au début plutôt lentement puis accélère au fur et à mesure que la tension du ressort diminue. La première série limitée de montres-bracelets ainsi équipées se vend en un rien de temps. En 2015, Oris présente le calibre 111, produit en série illimitée par l'entreprise avec son propre équipement. Un guichet dateur d'une taille appréciable à 9 heures vient compléter les caractéristiques habituelles. Au dernier stade de l'évolution, le calibre 112 anime la montre « Artelier Calibre 112 », présentée en 2016. À la grande joie des personnes qui enchaînent les trajets en avion, cette version dispose d'une fonction GMT simple d'utilisation.

Les montres-bracelets équipées de ce mouvement exclusif ne représentent toutefois que la crème de la crème de la gamme très étendue de la marque. Oris répond en outre quasiment à tous les désirs à des prix raisonnables. La « Divers Sixty-Five », qui incite fortement à la nostalgie, fait la joie des plongeurs. Les 2000 exemplaires de la « Carl Brashear Limited Edition », sortie en 2016, sont pour leur part depuis longtemps épuisés. Le modèle « Aquis Depth Gauge » dispose d'un profondimètre bien utile. Baptisé « Big Crown ProPilot Altimeter », son homologue pour pilotes est équipé d'un altimètre. Si les passionnés d'automobile trouvent leur compte avec les modèles Audi ou Williams, les amoureux de jazz sont comblés par l'intemporelle « Thelonious Monk Édition limitée ». Il n'y a que sur les montres à quartz qu'Oris doive tout bonnement faire l'impasse. ○

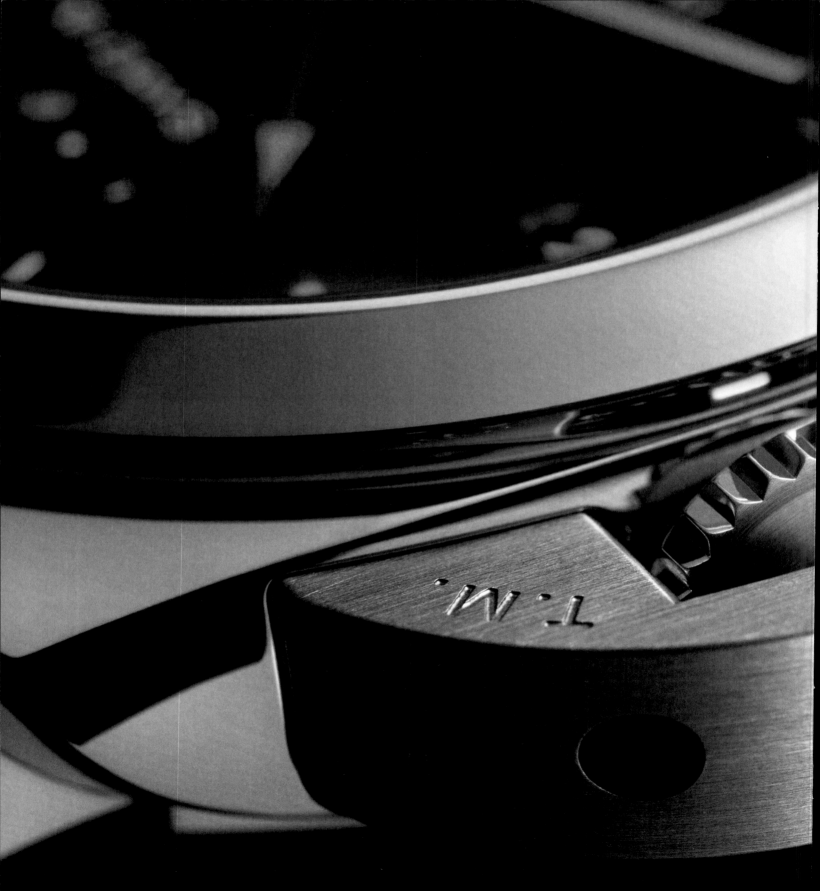

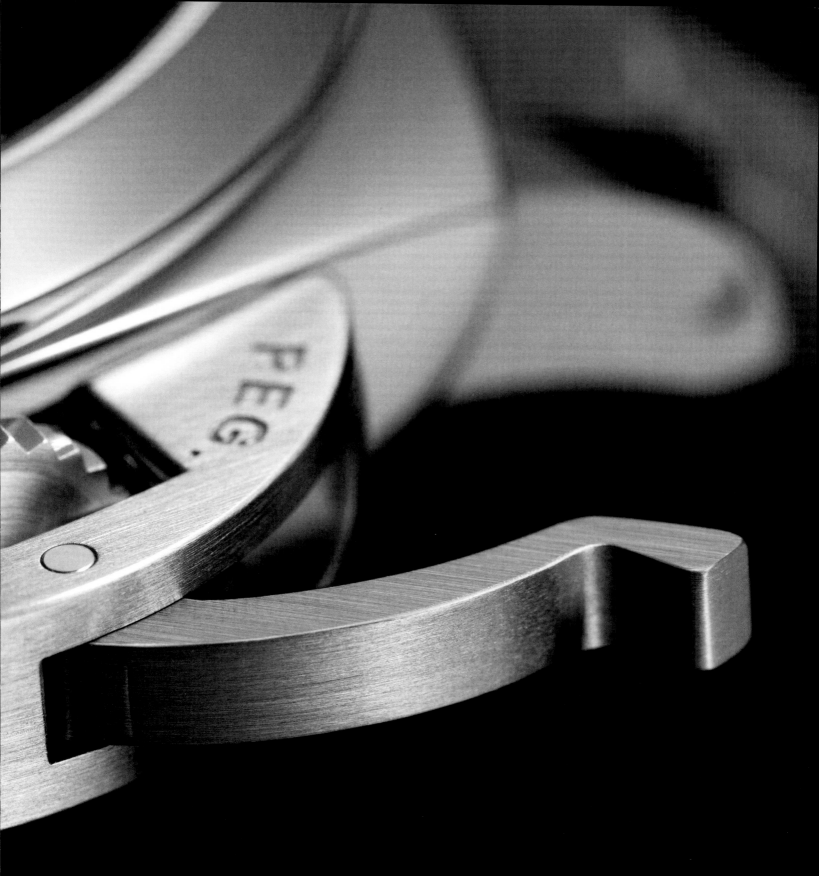

Panerai

A Swiss Manufactory with Italian Roots

The story of Officine Panerai began in a small workshop for mechanisms in Florence, where Giovanni Panerai first made precise instruments and mechanical equipment for nautical applications in the 1860s. Later, his great-grandchildren Giuseppe and Maria opened the "Orologeria Svizzera" watch shop in the shadow of Florence's majestic cathedral in the early 1930s.

Confidence in Panerai had grown among naval authorities over the decades. This prompted the naval decision-makers to commission the development of a prototype of a special wristwatch for frogmen in 1936. Panerai didn't yet have its own movements at this time, so the ambitious Italians turned to Montres Rolex SA, which had the relevant expertise in watchmaking and watertightness. The "Radiomir" satisfied all expectations of the official commission thanks to its screwed case and crown, which was combined with an innovative luminous dial that remained legible even in adverse viewing conditions. This pillow-shaped watch, followed in 1950 by the "Luminor" with a unique lever mechanism to insulate the winding and hand-setting crown, distinguished Panerai as an acknowledged specialist. The little Florentine family business always made its professional wristwatches in very small numbers, so today's collectors regularly pay high prices for the rare originals.

Beginning in the 1960s, Panerai underwent a long-lasting decline due to sickness and to consequences stemming from the end of the Cold War. Under new directors, the label continued to manufacture instruments for military purposes, but sustainable success was elusive. To broaden its clientele, Panerai began offering its big, time-honored, military watches to civilians with a penchant for the unconventional. This prompted Sylvester "Sly" Stallone to enter Panerai's shop in 1994. The Italian-American movie star bought a "Luminor"—and kept it strapped to his wrist throughout the filming of *Daylight*. A watch from a brand that had been mostly unknown outside Italy was suddenly seen on silver screens worldwide. After the filming was complete, the actor ordered an individually numbered series of watches engraved with his signature to give as presents to his friends. Soon afterwards, Arnold Schwarzenegger wore one of these rare "Slytech Submersible" models on his wrist in the film *Eraser*. Whether by chance or intention, Johann Rupert, the main shareholder of Switzerland's Richemont luxury group, was so enthusiastic that he added Panerai to his watch empire in 1997.

Panerai's trajectory began a steep ascent in 1998. An eye-catching retro look and lavishly large dimensions were exactly what the world's real men—and astonishingly many of its real women—were looking for to express their chronometric hedonism. A team of technicians and watchmakers was commissioned to develop an exclusive caliber family in 2002. With three barrels and a marathon eight-day power reserve, the first 250 examples of hand-wound P.2002 debuted in 2005, thus proving that Panerai had masterfully bridged the gap between past and present. Events now followed one another in quick succession: the manufactory's assortment grew to include approximately twenty different specimens. The spectrum includes hand-winding and automatic winding, chronograph, tourbillon, and time-zone indicator, as well as astronomical displays such as the times of sunrise and sunset. All movements tick inside emphatically nostalgic-looking cases inspired by this brand's colorful history. Panerai's watches are currently manufactured in an ultramodern building that was completed high above Neuchâtel in 2014. ◦

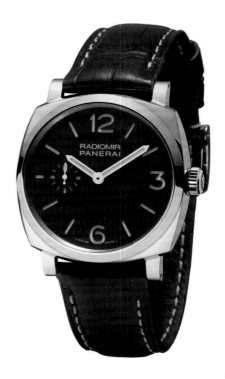

*Radiomir 1940,
PAM 00513, 2014*

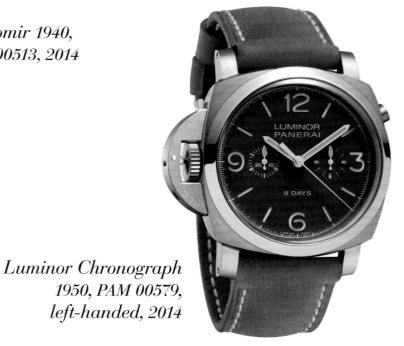

*Luminor Chronograph
1950, PAM 00579,
left-handed, 2014*

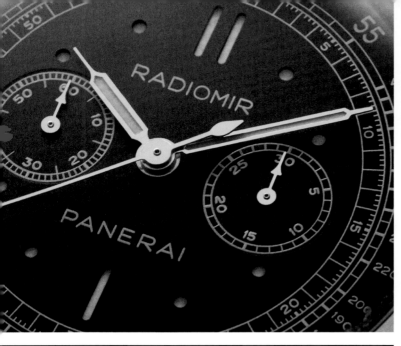

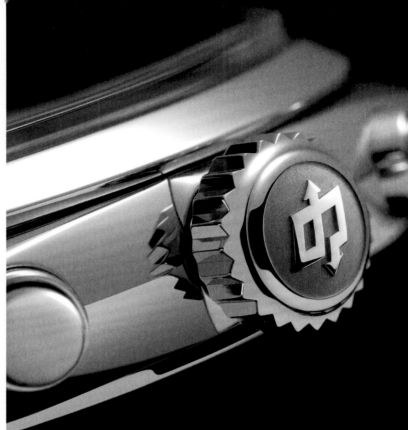

Top left: Radiomir Chronograph 1940, PAM 00520, 2014
Top right: Radiomir Chronograph 1940, PAM 00518, 2014
Bottom: Radiomir Chronograph 1940, PAM 00519, 2014

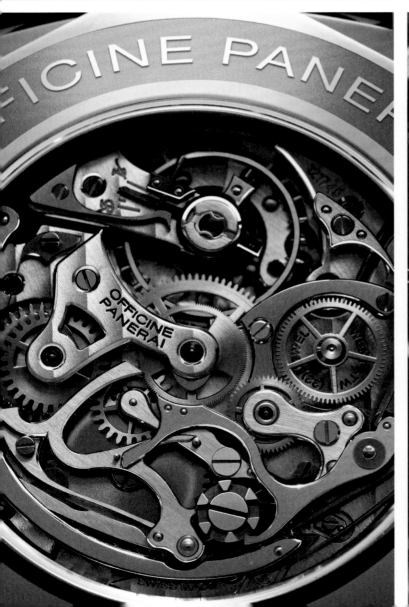

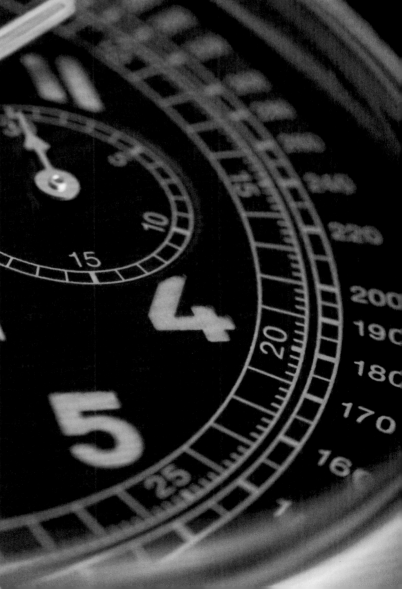

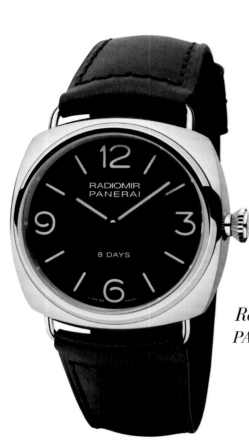

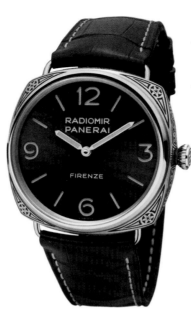

*Radiomir Firenze
3 Days Acciaio,
PAM 00604, 2015*

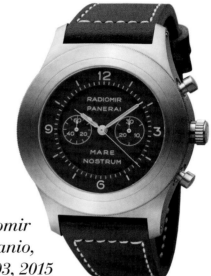

*Radiomir 8 Days Acciaio,
PAM 00610, 2015*

*Radiomir
Mare Nostrum Titanio,
PAM 00603, 2015*

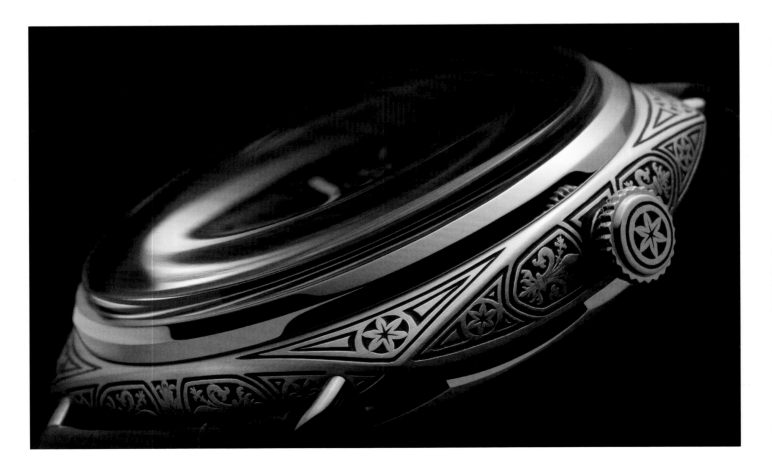

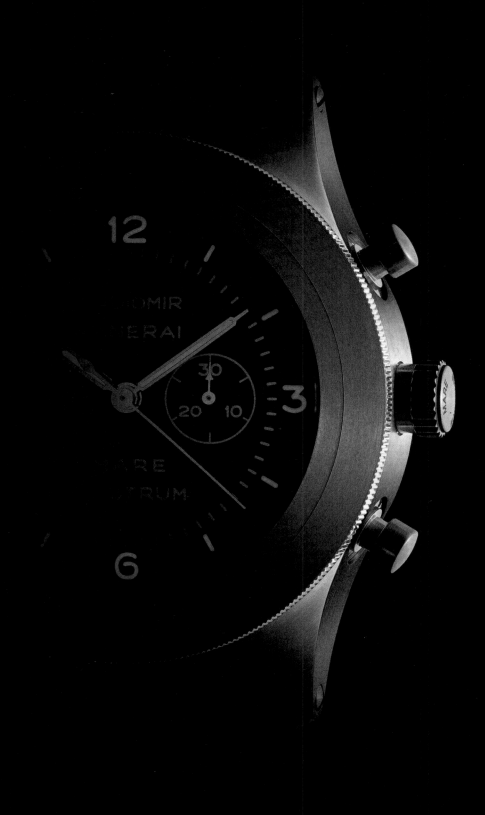

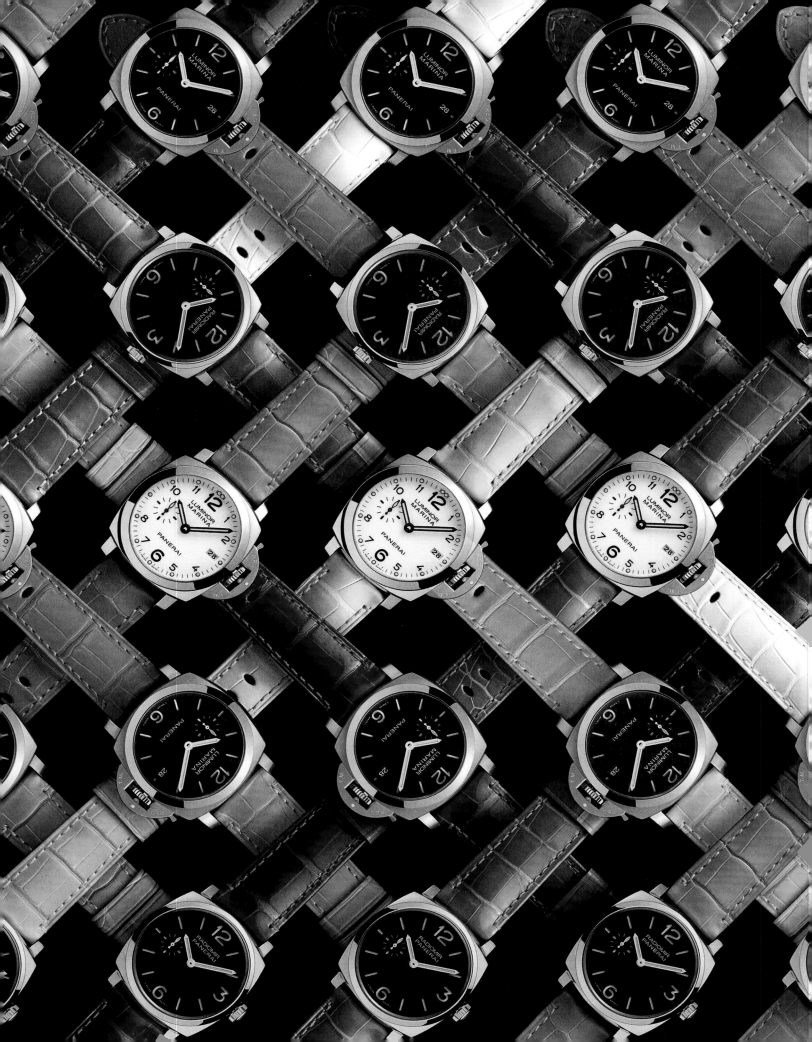

Schweizer Manufaktur
mit italienischen Wurzeln

Am Anfang dessen, was als „Officine Panerai" Uhrengeschichte schrieb und schreibt, stand eine kleine mechanische Werkstatt in Florenz. Giovanni Panerai fertigte dort ab 1860 unter anderem Präzisionsinstrumente und mechanische Gerätschaft für die Nautik. Seine Urenkel Giuseppe und Maria eröffneten zu Beginn der 1930er Jahre im Schatten des majestätischen Doms das Uhrenfachgeschäft „Orologeria Svizzera".

Das über Jahrzehnte hinweg erworbene Vertrauen der Marinebehörden führte 1936 zur Entwicklung des Prototypen einer speziellen Armbanduhr für Kampftaucher. Mangels eigener Werke kooperierten die ambitionierten Italiener mit der in Sachen Uhrmacherei und Wasserdichtigkeit äußerst erfahrenen Montres Rolex S.A. Kein Wunder, dass die „Radiomir" mit Schraubgehäuse und -krone alle Erwartungen der amtlich eingesetzten Kommission erfüllte, zumal sich ihr innovatives Leuchtzifferblatt selbst bei schlechten Sichtverhältnissen unmissverständlich ablesen ließ. Durch diese kissenförmige Armbanduhr und das 1950 lancierte Nachfolgemodell „Luminor" mit einzigartigem Hebelmechanismus zur Abdichtung der Aufzugs- und Zeigerstellkrone avancierte Panerai zu einem anerkannten Spezialisten. Weil das kleine Florentiner Familienunternehmen seine professionellen Armbanduhren immer nur in sehr begrenzten Stückzahlen fertigte und lieferte, zahlen Sammler für die Originale fast schon aberwitzige Höchstpreise.

Krankheit und später das Ende des Kalten Kriegs bescherten Panerai ab den 60er Jahren eine längere Durststrecke. Unter neuer Leitung entstanden zwar weiterhin Instrumente für militärische Zwecke; von nachhaltigen Erfolgen war jedoch keine Rede mehr. Deshalb bot das Unternehmen die bewährten Militär-Boliden ab 1993 auch Zivilisten mit einem Faible für Außergewöhnliches an. Das führte

1994 Sylvester Stallone, genannt Sly, ins winzige Panerai-Geschäft. Der italo-amerikanische Filmstar kaufte sich eine „Luminor" und legte sie auch während der Dreharbeiten zum Film *Daylight* nicht mehr ab. So kam die außerhalb Italiens kaum bekannte Uhrenmarke auf den Leinwänden rund um den Globus ganz groß raus. Nach Fertigstellung des Streifens orderte der Schauspieler für seine Freunde eine individuell nummerierte und mit seinem Namenszug gravierte Serie. Mit einer der wenigen „Slytech Submersible" am Handgelenk agierte Arnold Schwarzenegger wenig später im Film *Eraser*. Zufall oder nicht: Johann Rupert, Hauptaktionär des Schweizer Luxuskonzerns Richemont, begeisterte sich darüber so sehr, dass er Panerai 1997 seinem Uhrenimperium einverleibte.

Ab 1998 ging es steil bergauf. Markanter Retrolook und opulente Dimensionen waren genau das, was sich echte Männer – und erstaunlich viele Frauen – zur Demonstration ihres chronometrischen Hedonismus wünschten. 2002 erhielt ein Team bestehend aus Technikern und Uhrmachern den Auftrag zur Entwicklung einer exklusiven Kaliberfamilie. Bereits die ersten 250 Exemplare des sogenannten P.2002 mit Handaufzug, drei Federhäusern und stattlichen acht Tagen Gangautonomie bewiesen 2005 das Gelingen des keinesfalls leichten Spagats zwischen Vergangenheit und Gegenwart. Danach ging es Schlag auf Schlag. Mittlerweile beinhaltet die Manufakturpalette rund zwanzig verschiedene Exemplare. Zum Spektrum gehören Hand- und Selbstaufzug, Chronographen, Drehgang, Zeitzonen-Dispositive und auch astronomische Indikationen wie die Zeiten des Sonnenauf- und -untergangs. Alle Uhrwerke ticken in nostalgiebetonten, aus der bewegten Geschichte abgeleiteten Gehäusen. Die Uhrenproduktion erfolgt heute in einem hochmodernen, 2014 fertiggestellten Gebäude hoch über Neuchâtel. ∘

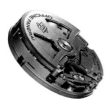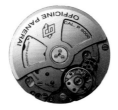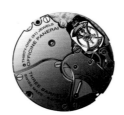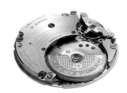

Automatic movements: 2007 P.2003 ∘ 2009 P.9000 ∘ 2013 P.9100 ∘ 2014 P.4000

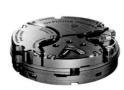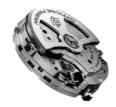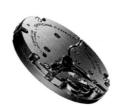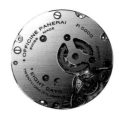

Hand-wound movements: 2005 P.2002E ∘ 2007 P.2005 ∘ 2011 P.3000 ∘ 2014 P.5000

Left page: Luminor Marina and Radiomir 1940, colored straps, 2014

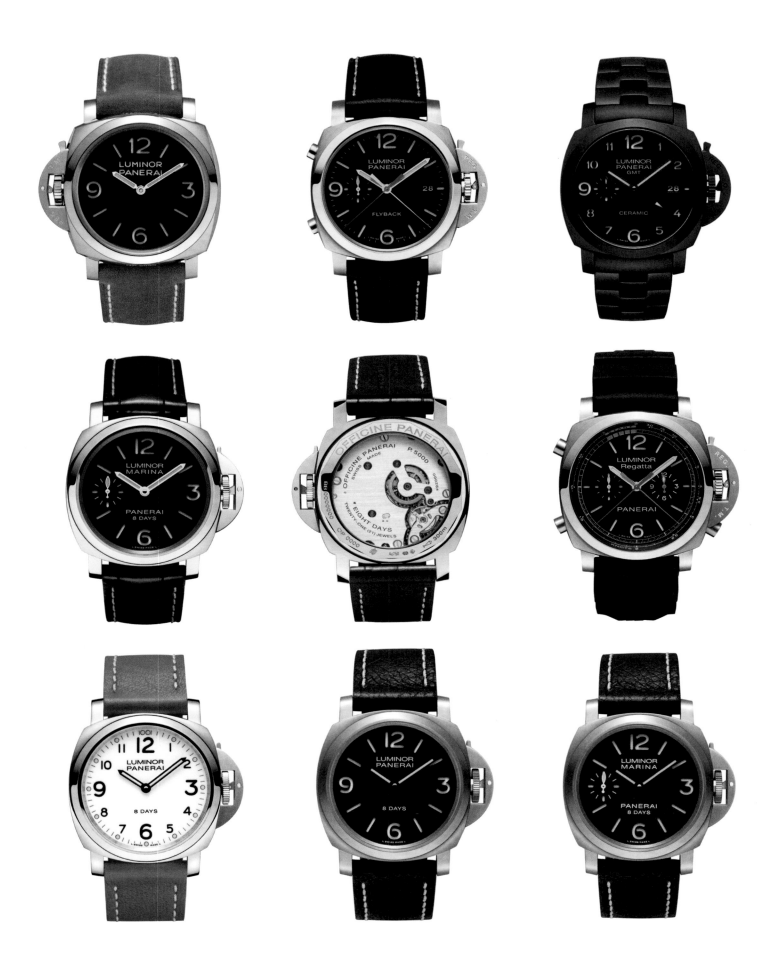

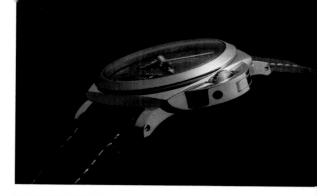

Une manufacture suisse aux racines italiennes

La boutique « Officine Panerai », qui a marqué et marque encore l'histoire de l'horlogerie, est née d'un petit atelier de mécanique, à Florence. C'est là que, dès 1860, Giovanni Panerai fabrique des instruments de précision et des équipements mécaniques pour la navigation. Ses arrière-petits-enfants Giuseppe et Maria ouvriront au début des années 1930 à l'ombre de la majestueuse cathédrale le magasin spécialisé dans l'horlogerie « Orologeria Svizzera ».

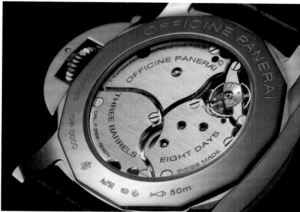

Ayant gagné la confiance des autorités navales au fil de plusieurs décennies, ils se voient confier en 1936 la mise au point d'un prototype de montre-bracelet spéciale pour plongeurs de combat. Mais ces ambitieux Italiens, ne fabriquant pas de mouvements, coopèrent avec Montres Rolex SA, qui a une très grande expérience en matière d'horlogerie et d'étanchéité. Rien d'étonnant donc à ce que la « Radiomir », à boîtier et couronne vissés, réponde aux attentes de la commission officiellement constituée, d'autant que son cadran luminescent permet une lecture sans ambiguïté même dans de mauvaises conditions de visibilité. Grâce à ce modèle au boîtier en forme de coussin, puis à la « Luminor », qui lui succède en 1950 avec un mécanisme à levier novateur pour l'étanchéité des couronnes de remontoir et de mise à l'heure, les Panerai sont promus au rang d'experts reconnus. La petite entreprise familiale florentine vend les montres-bracelets professionnelles qu'elle fabrique en série très limitée. Aussi les collectionneurs actuels déboursent-ils pour ces originaux des sommes quasiment insensées.

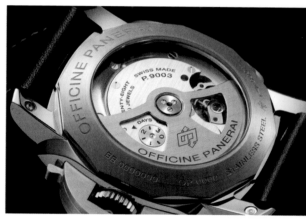

La maladie, puis la fin de la Guerre froide conduisent, à partir des années 1960, à une longue traversée du désert pour Panerai. Des instruments à usage militaire continuent de voir le jour sous la nouvelle direction, mais plus question de succès durable. La société propose alors dès 1993 au grand public friand de produits singuliers les montres luminescentes qui avaient fait leurs preuves au sein de l'armée. C'est ainsi qu'en 1994, Sylvester Stallone, dit Sly, est conduit à entrer dans la minuscule boutique des Panerai. L'acteur italo-américain acquiert une « Luminor », qu'il ne quittera pas une seconde durant le tournage de *Daylight*. Cette marque de montres guère connue hors d'Italie fait alors une apparition triomphale sur tous les écrans de cinéma du monde. Une fois le film terminé, l'acteur commande pour ses amis une série de montres numérotées et portant sa signature gravée. Peu de temps après, Arnold Schwarzenegger joue dans *L'effaceur* avec au poignet l'une des rares « Slytech Submersible ». Coïncidence ou pas, Johann Rupert, le principal actionnaire du groupe suisse du luxe Richemont, est tellement enthousiasmé à la vue de cette montre qu'il absorbe en 1997 Panerai dans son empire horloger.

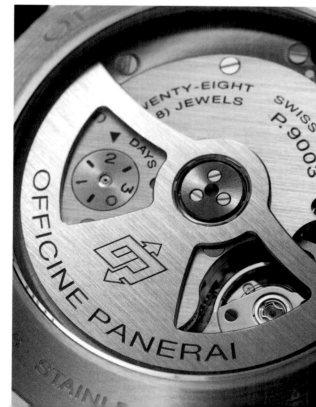

À partir de 1998, les affaires décollent. Un look résolument rétro et des dimensions très généreuses, c'est exactement ce que souhaitent les vrais hommes – et étonnamment beaucoup de femmes – pour démontrer combien ils ont plaisir à pouvoir mesurer le temps avec précision. En 2002, une équipe composée de techniciens et d'horlogers est invitée à développer une famille de calibres raffinés. En 2005, les 250 premiers exemplaires du P.2002 à remontage manuel, trois barillets et réserve de marche confortable de huit jours démontrent que l'on peut concilier passé et présent, bien que cela soit une prouesse a priori pas évidente du tout. Ensuite, les choses s'accélèrent. Aujourd'hui, la palette des créations comprend une vingtaine de modèles différents. Ces montres sont équipées du remontage manuel ou automatique, de chronographes, de tourbillons, de dispositifs d'affichage des fuseaux horaires et d'indications astronomiques, telles les heures du lever et du coucher du soleil. Les mouvements battent tous dans des boîtiers évoquant avec une certaine nostalgie une époque agitée. Les montres sont aujourd'hui fabriquées dans un bâtiment ultramoderne construit en 2014 sur les hauteurs de Neuchâtel. ◦

Left page: various Luminor models
Top: eight days and hand-wound calibre, P.2002, 2005
Bottom: eight days and automatic calibre, P.9003, 2014

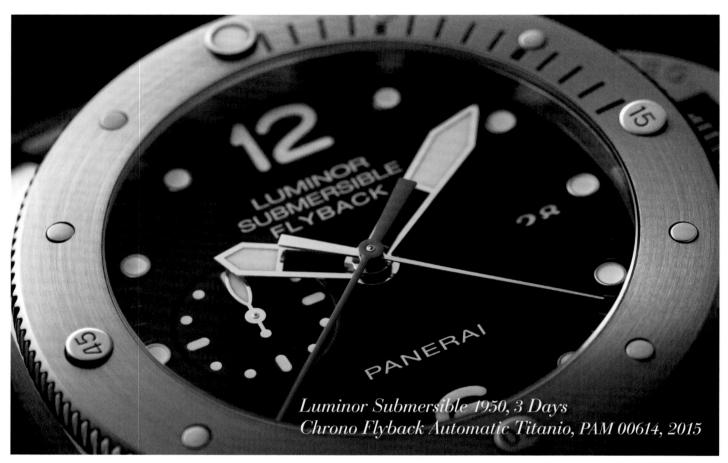

Luminor Submersible 1950, 3 Days
Chrono Flyback Automatic Titanio, PAM 00614, 2015

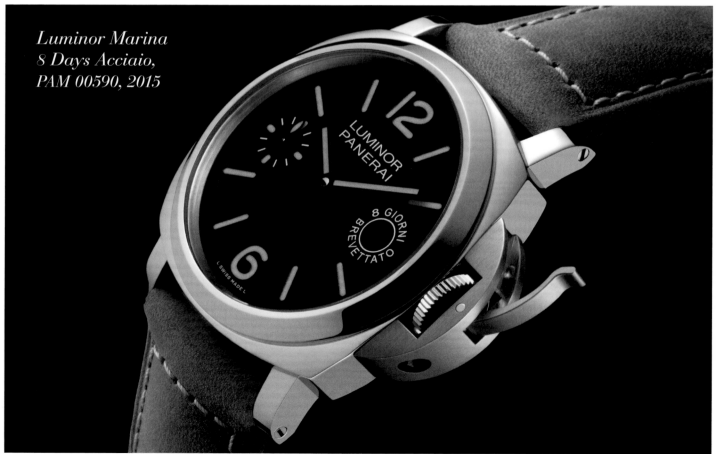

Luminor Marina
8 Days Acciaio,
PAM 00590, 2015

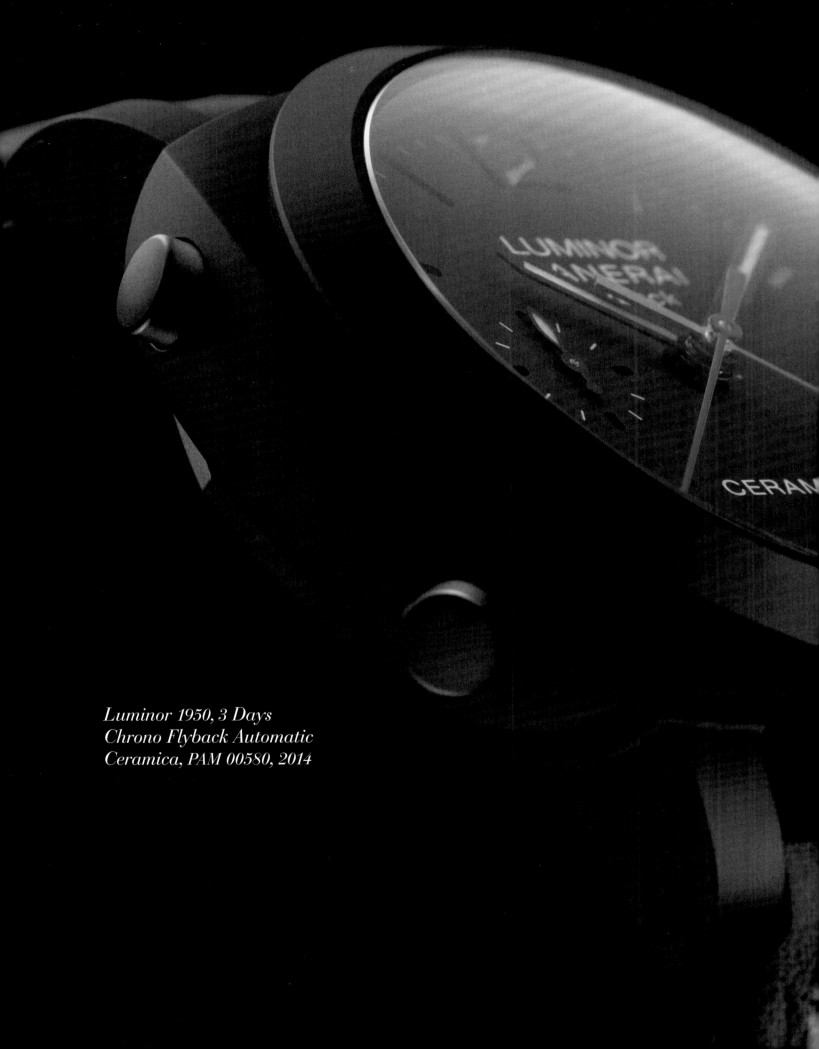

Luminor 1950, 3 Days
Chrono Flyback Automatic
Ceramica, PAM 00580, 2014

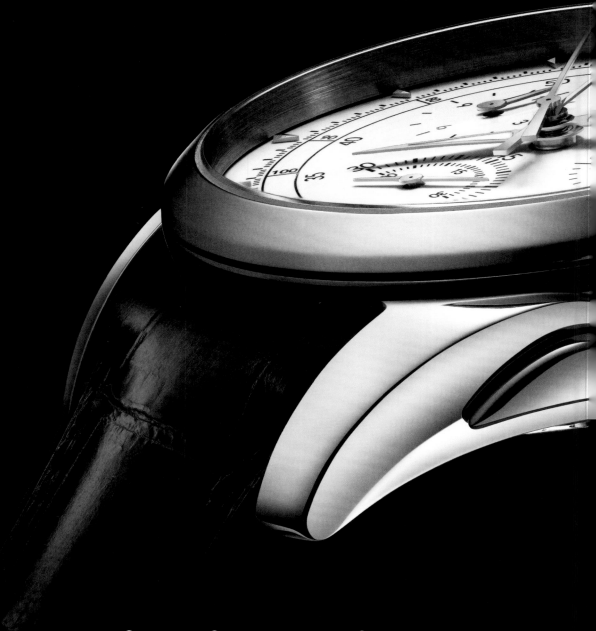

Parmigiani
Fleurier

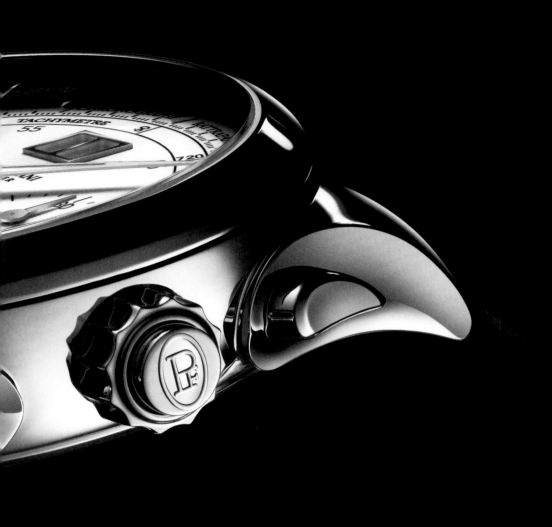

Craftsmanship in Service of Time

The first few decades of his life were quite challenging for Michel Parmigiani, who was born in 1950. He began his career at the school of watchmaking in Fleurier, the same town that's now the home of the watch brand which bears his name. The son of Italian parents, Parmigiani continued his education at the technical institute in the watch metropolis of La Chaux-de-Fonds and at the school of engineering in Le Locle. This master watchmaker was employed by a small watch factory, where he was a reluctant eye-witness to the decline of classical mechanical watchmaking in the early 1970s. But this frustrating period from 1973 to 1975 in the life of a characteristically demure horological artist was not without its positive aspects. Michel Parmigiani had no choice but to ask himself why the Quartz Revolution was shaking the foundations of traditional values. More or less overnight, functionality took precedence over aesthetics, commercial considerations eclipsed quality, and mental doggedness was valued above creativity. The talented craftsman was unwilling to participate in these supposedly innovative trends, so he shunned the mainstream and began a career as a freelance watchmaker. Now that traditional mechanisms were becoming increasingly rare in new watches, he dedicated himself to loveable artifacts from bygone epochs. A wide variety of precious timepieces around the world were in need of competent and meticulous restoration. Here was a field of endeavor in which Parmigiani, who admittedly has a perfectionist streak, could uninhibitedly pursue his calling. Fascinating commissions soon arrived. Atelier Parmigiani, which was an exclusively one-man show until 1980, accepted every request, no matter how intricate. The horological soloist achieved widespread public acclaim with his successful repair of a valuable "Pendule Sympathique," which had been crafted by Abraham-Louis Breguet and which most experts had dismissed as irreparable. His restoration of a tourbillon of the same provenance for the Museo Poldi Pezzoli in Milan not only earned cult status for the talented craftsman, but also won him the full confidence of the Sandoz Family, which not only makes pharmaceuticals, but also cultivates a penchant for the fine arts. The ticking treasures that were serviced or restored by Parmigiani and his growing team of dedicated employees formatively influenced the next phase in his career. This naturalized Swiss citizen was increasingly often asked to provide developmental assistance for the theoretical and/or practical realization of interesting time-related projects. In 1994, the watch collector Pierre Landolt urged Michel Parmigiani to devote more time and effort to his own watches. Acting in accord with the expressed will of the foundation's founder Edouard-Marcel Sandoz, the chairman of the Sandoz Foundation decided to participate in the new Michel Parmigiani watch manufacture in 1996. Needless to say, the mellifluous name of the firm's founder remained unaffected by the foundation's participation. Though it was surely not a simple task, Parmigiani's creation of his own watch collection was a stimulating and fascinating exercise. The successful premiere took place in Lausanne on May 29, 1996 in front of an audience of circa 450 connoisseurs. The spectators immediately understood the clear and impressive philosophy: small numbers of timepieces, artistic handcraftsmanship, the finest quality, the utmost fidelity to detail, and luxury in its purest form. But even more lies behind the name "Parmigiani Fleurier." Parmigini designs and engineers, fabricates, embellishes and assembles his own movements, thus fully justifying his use of the prestigious term "manufacture." The experts in his ateliers daily reconfirm the truth of the adage which states that "time respects nothing which doesn't require plenty of time in its creation." But time hasn't stood still for Parmigiani. Under Sandoz's aegis, the 21st century has penetrated into his ateliers in the form of modern computer-guided machines, which assist the artisans who craft a wide variety of diverse components and ébauches. These specialists, however, don't work only in Parmigiani's ateliers, but also in the buildings of the affiliated companies Vaucher, Atokalpa and Elwin. This team of acknowledged experts achieves a vertical depth of fabrication of more than 90 percent. The palette of products ranges from extremely elaborate plates to seemingly unimpressive but highly sophisticated *assortiments*, i.e., kits for escapements. Even the tiniest screws and pinions are made on the premises. Two years of development were required for the hairsprings, which the company now also delivers to renowned competitors. All this notwithstanding, Michel Parmigiani places great importance on the fact that an ébauche contributes less than ten percent to a completed watch. The lion's share of its value is added by the time-consuming artisanal fabrication and fine processing of the components, their assembly and fine tuning, and ultimately the meticulous adjustment of the finished movement to assure that it keeps time with an optimally accurate rate.

The diverse spectrum of Parmigiani's own calibers began in 1998 with the debut of *tonneau*-shaped Caliber PF110, a manually wound movement that can run for nearly 200 hours between windings. Mastery of automatic winding in an ultra-slim construction is proven by the 2.6-millimeter-tall microrotor Calibers PF701 and 705 (skeletonized). A centrally positioned oscillating weight automatically winds the mainspring of Caliber PF331, which is nonetheless just 3.5 millimeters tall. When an exclusive perpetual calendar is added to this movement, its name changes to Caliber PF333. Version PF377 is equipped with a practical time-zone display, which includes a day/night indicator. Caliber PF517 set a world record which will surely remain difficult to break: although it's equipped with a self-winding mechanism and a one-minute tourbillon, this superlative caliber is a mere 3.4 millimeters tall.

Apropos tourbillons: the never-ending striving toward horological perfection is beautifully expressed by hand-wound Calibers PF500 and PF511. Their tourbillon requires only 30 seconds to complete each rotation. Doubling the rotational speed goes hand in hand with markedly improved precision. A patented power-reserve display is positioned at the "12." And this extraordinary movement also boasts a centrally positioned second-hand. Incidentally: the tourbillon remains visible when one slips this watch off the wrist, turns it over, and peers through the pane of sapphire crystal in the back of its case.

Inspired by a predecessor from the late 18th century, the "Ovale Pantographe" is absolutely unique. Its slender telescoping hands automatically extend and slowly retract so they continually conform to the ovoid shape of its case. These complex hands very likely also gave a few grey hairs to the watchmakers who were involved in their development and fabrication! The lengthening and shortening of the hands is powered by time-honored hand-wound shaped Caliber PF111, which provides an eight-day power reserve.

Complex hand-wound Caliber PF361 was ready for series manufacture punctually in time for the firm's 20th anniversary in Fleurier. Its bearing parts (i.e., its plates, bridges and cocks) are made of solid gold. Alongside the chronograph with classical column-wheel control and innovative friction coupling, this horological opus includes a balance paced at five hertz and a split-seconds mechanism. The latter can be used, for example, to measure intermediate times at athletic events. Each of the 25 timepieces in the limited series of "Tonda Chronor Anniversaire" watches features a large and eye-catching date display. Simultaneously with the presentation of the ultra-sporty Bugatti Chiron at the automobile salon in Geneva in 2016, this small and very fine manufacture also unveiled a matching concept wristwatch. Caliber PF390, which is encased crosswise to serve as the "engine," has two barrels and a "flying" (i.e., cantilevered) one-minute tourbillon. A central planetary gear powers the dial train. The console-like dial enables the wearer to check time's progress without taking his hands off his car's steering wheel.

Parmigiani Fleurier has unveiled many delightful surprises during the past two decades. And we have every reason to believe that this will continue in the future. ○

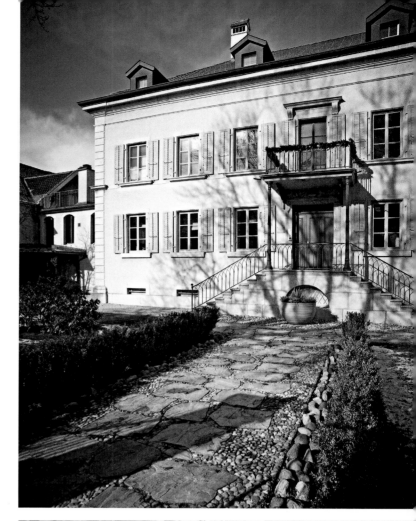

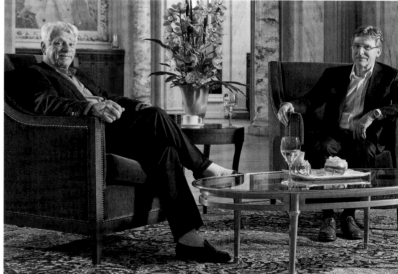

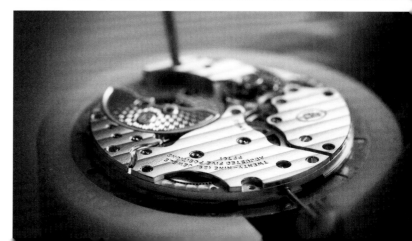

*Top: **The manufactory in Fleurier** ○ Manufaktur in Fleurier ○*
*La manufacture de Fleurier ○ **Middle and bottom: Pierre Landolt***
and Michel Parmigiani ○ Cal. PF701

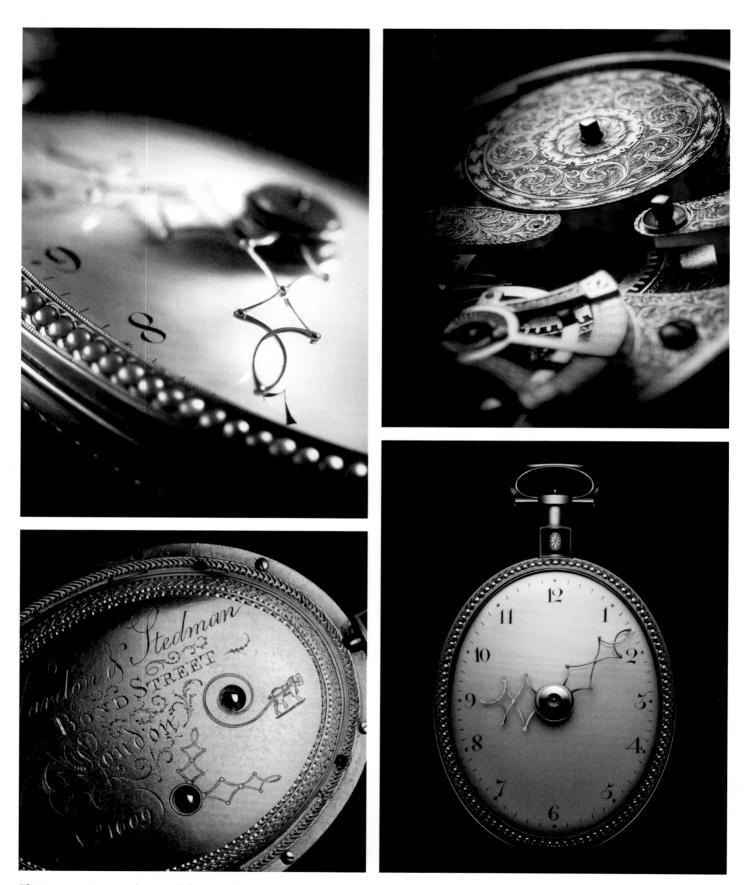

This page: Antique pocket watch from Vardon & Stedman, inspiration for the Ovale Pantographe ○ Opposite page: Making of Tonda 1950 Tourbillon, PF517 ○ Diese Seite: Antike Taschenuhr von Vardon & Stedman, Vorbild für die Ovale Pantographe ○ Gegenüberliegende Seite: Fertigung des Tourbillons Tonda 1950, Kaliber PF517 ○ *Ci-dessus : Montre de poche ancienne signée Vardon & Stedman, qui a inspiré le modèle Ovale Pantographe ○ Page ci-contre : Fabrication de la Tonda 1950 Tourbillon, PF517*

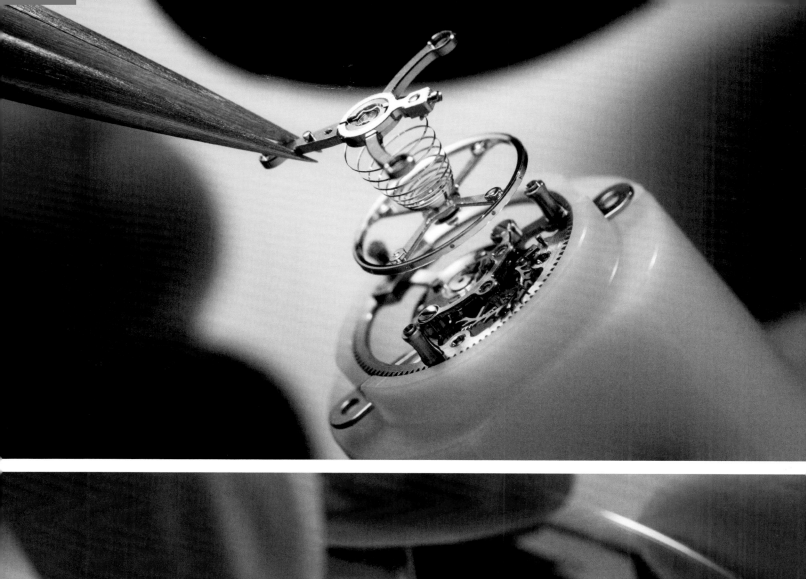

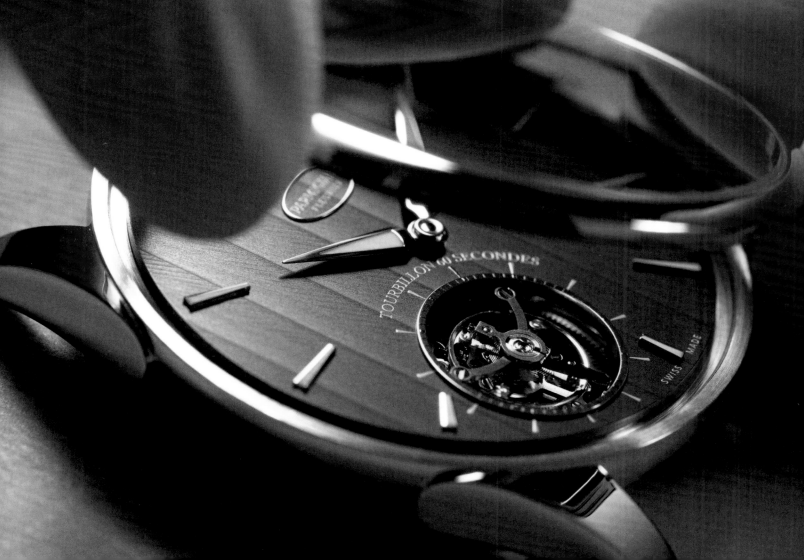

Handwerkskunst rund um die Zeit

Leicht hatte er es nicht, der 1950 geborene Michel Parmigiani. Seine Karriere startete an der Uhrmacherschule jener Ortschaft, in der man heute die Uhrenmarke seines Namens findet: Fleurier. Seine Ausbildung komplettierte der Sohn italienischer Eltern am Technikum der Uhrenmetropole La Chaux-de-Fonds sowie in der Ingenieurschule Le Locle. Anfang der 1970er Jahre musste der Meister-Uhrmacher den Niedergang der klassischen mechanischen Uhrmacherei als Mitarbeiter einer kleinen Uhrenfabrik hautnah miterleben. Diese frustrierende, von 1973 bis 1975 während Periode im Leben des stets zurückhaltend auftretenden Uhren-Künstlers hatte aber auch ihre positiven Seiten. Michel Parmigiani musste sich damit auseinandersetzen, warum die Quarz-Revolution an den Grundfesten überlieferter Werte rüttelte. Quasi über Nacht rangierte Funktionalität vor Ästhetik, Kommerzdenken vor Qualität und mentale Beharrlichkeit vor Kreativität. Diesen vermeintlich innovativen Tendenzen konnte und wollte der begnadete Handwerker nicht folgen. Also suchte er sein Heil als selbständiger Uhrmacher. Nachdem die traditionelle Mechanik in neuen Uhren nur noch ein Schattendasein fristete, widmete er sich liebenswerten Objekten aus vergangenen Epochen. Weltweit harrte ein breites Spektrum kostbarer Zeitmesser gleichermaßen kompetenter wie liebevoller Restaurierung. Auf diesem Gebiet konnte sich der ausgeprägte Hang zur Perfektion endlich ungehemmt entfalten. Faszinierende Aufträge ließen nicht lange auf sich warten. Das Atelier Parmigiani, in dem bis 1980 ausschließlich der Meister selbst tätig war, schreckte vor keinem einzigen zurück. Öffentliche Wahrnehmung erzielte die One-Man-Show durch die Wiederherstellung einer kostbaren, von Experten als irreparabel titulierten „Pendule Sympathique" von Abraham-Louis Breguet. Durch die Restaurierung eines Tourbillons gleicher Provenienz für das Museo Poldi Pezzoli in Mailand erlangte Michel Parmigiani Kultstatus und das uneingeschränkte Vertrauen der kunstsinnigen Pharma-Familie Sandoz. Deren tickende Kostbarkeiten, welche der Chef und eine kontinuierlich steigende Zahl engagierter Angestellter fachkundig betreuen, beeinflussten den weiteren Lebensweg. Immer öfter leistete der naturalisierte Schweizer Entwicklungshilfe bei der theoretischen und/oder praktischen Umsetzung interessanter Zeit-Projekte. 1994 wirkte der Uhrensammler Pierre Landolt auf Michel Parmigiani ein, künftig verstärkt an eigenen Uhren zu arbeiten. Durch eine Beteiligung an der neuen Uhrenmanufaktur Michel Parmigiani erfüllte der Vorsitzende der Sandoz-Stiftung im Jahr 1996 den erklärten Willen des Stiftungsgebers Edouard-Marcel Sandoz. Selbstverständlich blieb der klangvolle Name des Firmengründers unangetastet erhalten. Die Kreation einer eigenen Uhrenkollektion war keine leichte, dafür aber ausgesprochen spannende und faszinierende Übung. Das erfolgsgekrönte Debüt ging am 29. Mai 1996 in Lausanne vor rund 450 Kennern über die Bühne. Das Auditorium hatte die klare, einprägsame Philosophie auf Anhieb verstanden: geringe Stückzahlen, kunstfertige Handarbeit, feinste Qualität, höchste Detailtreue und Luxus pur. Hinter dem Namen Parmigiani Fleurier verbirgt sich freilich noch weit mehr. Die Konstruktion, Fertigung, Finissage und Terminage eigener Uhrwerke rechtfertigt den imageträchtigen Titel Manufaktur mit Fug und Recht. In den Ateliers gilt die bewährte Spruchweisheit, dass die Zeit nichts respektiert, was ohne sie geschaffen wurde. Gleichwohl ist die Zeit auch bei Parmigiani nicht stehen geblieben. Unter der Ägide von Sandoz hielt das 21. Jahrhundert seinen Einzug in Form moderner computergesteuerter Maschinen, welche die Produktion einer großen Vielfalt verschiedener Komponenten und Rohwerke unterstützen. Allerdings agieren sie nicht bei Parmigiani selbst, sondern in den Gebäuden der Schwesterfirmen Vaucher, Atokalpa und Elwin. Der Verbund ausgewiesener Spezialisten bringt es auf eine Fertigungstiefe von mehr als 90 Prozent. Die Produktpalette reicht von den teilweise extrem aufwendigen Platinen bis hin zu dem unscheinbaren, aber höchst anspruchsvollen Assortiment. Selbst kleinste Schrauben und Zahntriebe entstehen in eigener Regie. Hinter den Unruhspiralen, die das Unternehmen auch an renommierte Mitbewerber liefert, stecken zwei Jahre Entwicklungszeit. Ungeachtet dessen legt Michel Parmigiani größten Wert auf die Feststellung, dass der Anteil eines Rohwerks an der fertigen Uhr gerade einmal zehn Prozent beträgt. Den weitaus überwiegenden Teil der Wertschöpfungskette beanspruchen die zeitaufwendige, weil in traditioneller Weise von Hand ausgeführte Feinbearbeitung der Komponenten, die Assemblage und Feinstellung des Ganzen sowie die Regulierung des fertigen Uhrwerks auf optimale Ganggenauigkeit.

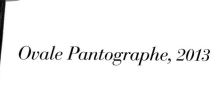

Ovale Pantographe, 2013

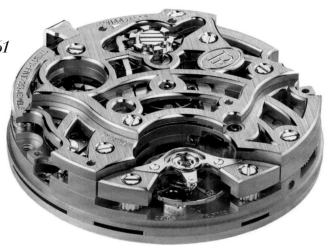

Cal. PF361

Den Anfang der mittlerweile breit gefächerten Palette eigener Kaliber markierte 1998 das tonneauförmige PF110 mit manuellem Aufzug und knapp 200 Stunden Gangautonomie. Die Beherrschung des automatischen Aufzugs in ultraflacher Bauweise demonstrieren die nur 2,6 Millimeter hoch bauenden Kaliber PF701 und 705 (skelettiert) mit Mikrorotor. Demgegenüber verfügt das PF331, Bauhöhe 3,5 Millimeter, über eine zentral positionierte Schwungmasse. Ausgestattet mit einem exklusiven ewigen Kalendarium heißt dieses Uhrwerk PF333. Die Version PF377 verfügt über ein hilfreiches Zeitzonen-Dispositiv inklusive Tag-Nacht-Indikation. Einen sicher nur schwer zu brechenden Weltrekord repräsentiert das PF517. Ausgestattet mit Selbstaufzug und Minutentourbillon misst der Superlativ lediglich 3,4 Millimeter in der Höhe.

Apropos Tourbillon: Ausdruck des fortwährenden Strebens nach uhrmacherischer Vollendung sind die Handaufzugskaliber PF500 und PF511. Ihr Drehgang benötigt für eine komplette Umdrehung nur 30 Sekunden. Die doppelte Umlaufgeschwindigkeit bringt eine spürbar bessere Präzision mit sich. Patentiert ist die Gangreserve-indikation bei der „12". Und noch etwas zeichnet dieses Uhrwerk aus: der zentral angeordnete Sekundenzeiger. Das Tourbillon bleibt übrigens auch dann sichtbar, wenn man die damit ausgestattete Armbanduhr vom Handgelenk nimmt und durch den Saphirglasboden blickt.

Absolut einzigartig ist die „Ovale Pantographe", deren Vorbild aus dem späten 18. Jahrhundert stammt. Entsprechend der Form des Gehäuses dehnen sich ihre filigranen Teleskopzeiger, deren Herstellung den involvierten Uhrmachern regelmäßig graue Haare wachsen lässt, entweder aus, oder sie ziehen sich wie von Geisterhand bewegt ganz langsam zusammen. Den Antrieb liefert das bewährte Handaufzugs-Formkaliber PF111 mit acht Tagen Gangautonomie.

Pünktlich zum 20. Firmenjubiläum ist in Fleurier das komplexe Handaufzugskaliber PF361 zur Serienreife gediehen. Seine tragenden Teile, also die Platinen, Brücken und Kloben, bestehen aus massivem Gold. Neben dem Chronographen mit klassischer Schaltradsteuerung und innovativer Friktionskupplung besitzt das mit fünf Hertz tickende Œuvre auch noch einen Schleppzeiger-Mechanismus beispielsweise zum Erfassen von Zwischenzeiten bei sportlichen Wettkämpfen. Markant schließlich auch das Großdatum der auf zwei Mal 25 Exemplare limitierten „Tonda Chronor Anniversaire". Parallel zur Vorstellung des ultrasportiven Bugatti Chiron auf dem Genfer Autosalon 2016 präsentierte die kleine, aber feine Manufaktur eine passende Konzept-Armbanduhr. PF390 heißt der quer angeordnete „Motor" mit zwei Federhäusern und „fliegendem" Minutentourbillon. Einem zentralen Planetengetriebe obliegt der Antrieb des Zeigerwerks. Das pultförmig angeordnete Zifferblatt gestattet die Wahrnehmung der fortschreitenden Zeit mit beiden Händen am Lenkrad.

In den zurückliegenden zwei Jahrzehnten war Parmigiani Fleurier für viele Überraschungen gut. Und daran wird sich auch in Zukunft nichts ändern. ∘

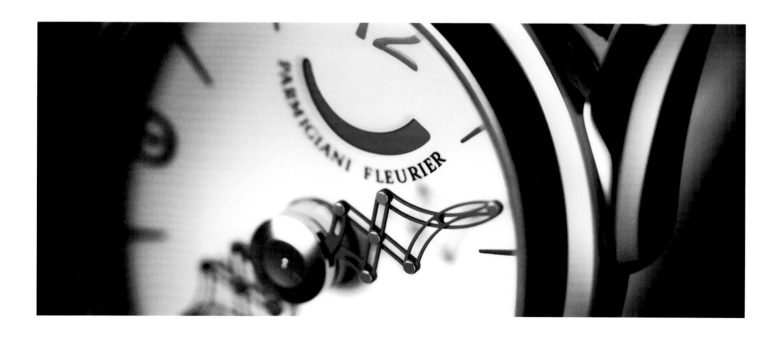

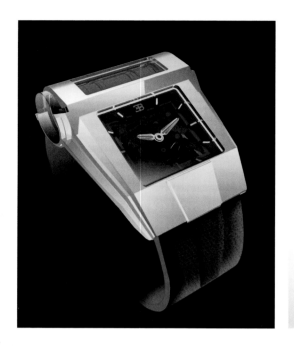

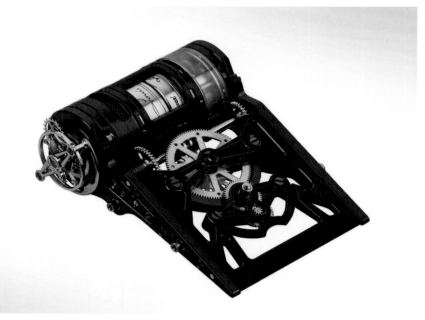

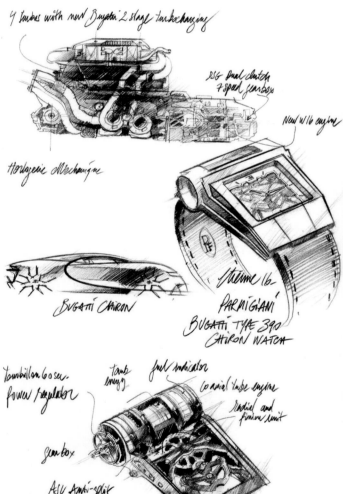

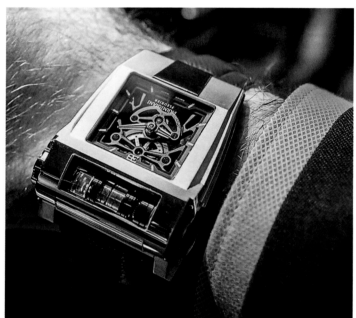

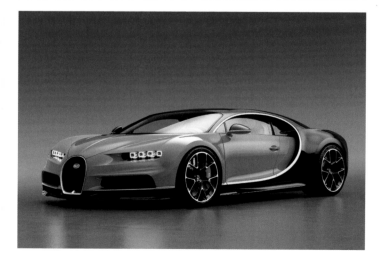

Bugatti PF390, 2016

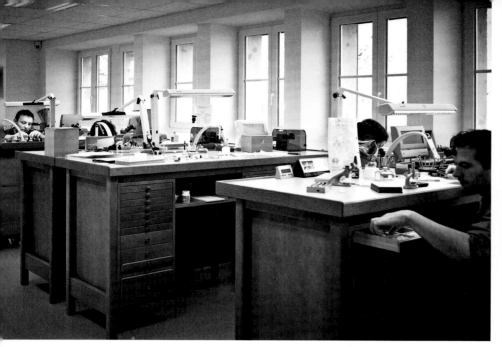

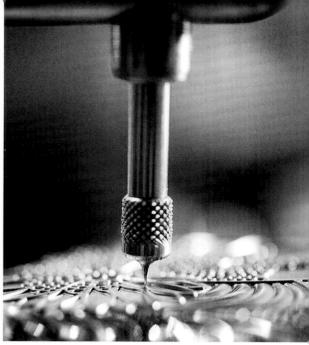

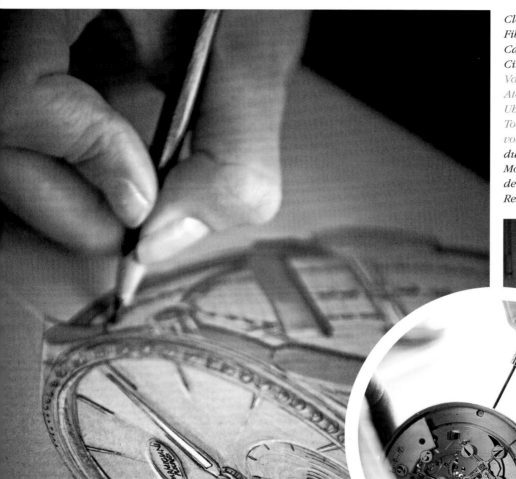

Clockwise from top left: Haute horlogerie atelier ○
Fibonacci watch ○ Haute horlogerie watchmaker ○
Case production ○ The design of a Tonda 1950 ○
Circle: Perrin Frères pocket watch restoration ○
*Von oben links im Uhrzeigersinn: Haute-Horlogerie-
Atelier ○ Taschenuhr Fibonacci ○ Haute-Horlogerie-
Uhrmacher ○ Gehäuseproduktion ○ Design einer
Tonda 1950 ○ Kreis: Restaurierung einer Taschenuhr
von Perrin Frères* **Dans le sens horaire, en partant
du haut, à gauche : Atelier de haute horlogerie ○
Montre Fibonacci ○ Maître-horloger ○ Fabrication
de boîtiers ○ Dessin d'une Tonda 1950 ○ Au centre :
Restauration d'une montre de poche Perrin Frères**

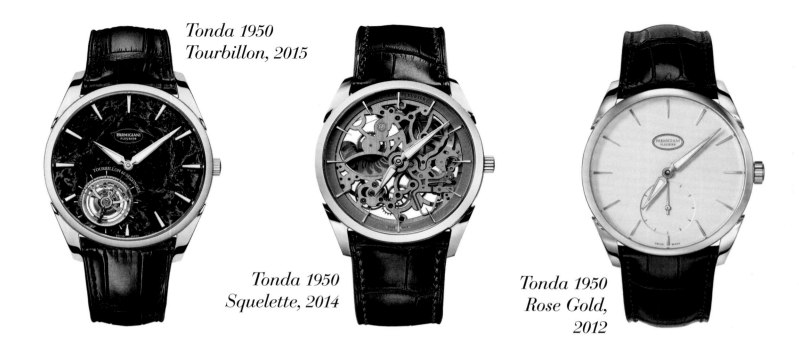

*Tonda 1950
Tourbillon, 2015*

*Tonda 1950
Squelette, 2014*

*Tonda 1950
Rose Gold,
2012*

Toute une époque d'artisanat d'art

Pour Michel Parmigiani, rien n'est facile. Né en 1950 de parents italiens, il s'initie au métier à l'école horlogère de Fleurier, la localité où se trouve aujourd'hui la manufacture qui porte son nom. Il complète sa formation dans les métropoles horlogères suisses, au technicum de La Chaux-de-Fonds puis à l'école d'ingénieurs du Locle. Au début des années 1970, ce maître-horloger alors employé d'une petite horlogerie subit de plein fouet le déclin des traditionnelles montres mécaniques. Cette période, qui s'étend de 1973 à 1975, est frustrante pour cet artisan horloger d'un naturel réservé, mais elle aura aussi des côtés positifs. Michel Parmigiani est en effet contraint d'essayer de comprendre pourquoi la révolution du quartz ébranle les bases des valeurs traditionnelles. Pour ainsi dire du jour au lendemain, l'esthétique, la qualité et la créativité ont cèdent le pas à la fonctionnalité, au mercantilisme et à l'étroitesse de vues. Cet artisan exceptionnellement doué ne peut ni ne veut suivre ces supposées tendances novatrices. C'est pourquoi il cherche le salut en tant qu'horloger indépendant. La mécanique traditionnelle ne jouant plus qu'un rôle mineur dans les nouvelles montres, il se tourne vers d'admirables objets d'époques passées. Dans le monde entier, nombre de garde-temps attendent d'être restaurés avec compétence et amour. Son inclination marquée pour la perfection peut enfin s'exprimer à loisir. Des commandes passionnantes ne tardent pas à se présenter. Aucune d'elles n'effraie l'atelier Parmigiani, dans lequel il officiera seul jusqu'en 1980. Cet homme-orchestre gagne la reconnaissance publique en restaurant une pièce considérée irréparable par les experts, la « pendule sympathique » d'Abraham-Louis Breguet. La restauration d'un tourbillon du même créateur pour le compte du musée Poldi Pezzoli de Milan lui vaut d'obtenir la consécration, ainsi que la confiance sans limite de la Fondation de famille Sandoz, mécène des arts et de la culture. Leurs précieux garde-temps, dont il s'occupe avec professionnalisme, secondé par un nombre sans cesse croissant d'employés investis, vont influencer son avenir.

Michel Parmigiani, qui a pris la nationalité suisse, contribue de plus en plus souvent à la mise en œuvre théorique et/ou pratique de projets horlogers intéressants. En 1994, le collectionneur de montres Pierre Landolt le convainc de se concentrer sur ses propres créations. En prenant en 1996 une participation dans la nouvelle manufacture de montres de Michel Parmigiani, le président de la Fondation réalise ainsi le vœu de son créateur, Édouard-Marcel Sandoz. Naturellement, le nom du fondateur de la marque est conservé tel quel, car il sonne bien. Si créer sa propre collection de montres n'est pas chose facile, c'est une tâche absolument passionnante et fascinante. Ses débuts, couronnés de succès, ont lieu le 29 mai 1996 à Lausanne, devant 450 connaisseurs. Le public comprend immédiatement la philosophie claire et simple de la nouvelle marque : séries limitées, travail artisanal minutieux, qualité exceptionnelle, souci extrême du détail et luxe à l'état pur. Mais derrière le nom Parmigiani Fleurier se cachent assurément bien d'autres valeurs encore. La conception, la fabrication, le finissage et l'achevage de ses propres mouvements justifient pleinement le titre prestigieux de manufacture. Le travail dans les ateliers est régi par la proverbiale formule selon laquelle le temps ne respecte pas ce qui se fait sans lui. Néanmoins, le temps ne s'est pas arrêté pour la manufacture Parmigiani. Sous l'égide de Sandoz, le XXIᵉ siècle fait son entrée sous la forme de machines modernes assistées par ordinateur, qui facilitent la fabrication d'un large éventail de composants et mouvements très divers. Ces machines ne tournent toutefois pas dans les locaux de Parmigiani, mais dans les bâtiments des manufactures affiliées Vaucher, Atokalpa et Elwin. Grâce à la mise en commun de compétences spécialisées, plus de 90 pour cent des pièces requises pour fabriquer les montres sont réalisées en interne. La gamme des produits concernés s'étend des platines parfois très complexes aux échappements, peu spectaculaires mais hautement sophistiqués. Même les plus petits engrenages et vis sont produits en interne. Les spiraux de balancier, que l'entreprise fournit également à des

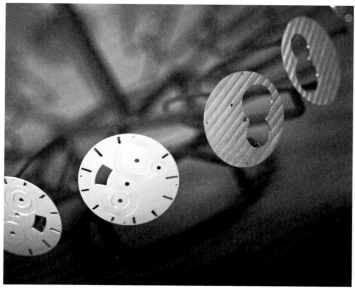
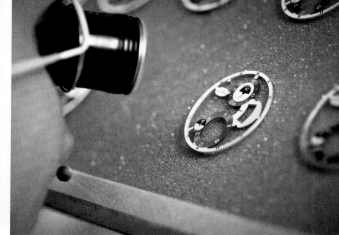

Dial production ○ *Zifferblattproduktion* ○ **Fabrication de cadrans**

concurrents renommés, exigent deux ans de développement. Mais Michel Parmigiani est parfaitement conscient du fait que l'ébauche représente tout juste dix pour cent de la montre terminée. La partie de loin la plus considérable de la chaîne de création de valeur est liée au finissage des composants, à l'assemblage et à l'ajustage de l'ensemble, opérations chronophages car réalisées à la main, mais aussi au réglage du mouvement terminé pour une fiabilité optimale.

Premier de la gamme aujourd'hui très étendue de calibres de manufacture, le PF110 à remontage manuel et environ 200 heures de réserve de marche a la forme d'un tonneau. Avec les PF701 et PF705 (squelette) à microrotor, l'entreprise démontre sa maîtrise des calibres à remontage automatique dans un boîtier ultraplat (2,6 millimètres). Le PF331, de 3,5 millimètres d'épaisseur, présente une masse oscillante en position centrale. Équipé d'un quantième perpétuel exclusif, il donne le PF333. La version PF377 est équipée d'un utile dispositif d'affichage des fuseaux horaires avec indication jour-nuit. Le PF517 enfin représente un record certainement très difficile à battre. Équipé d'un remontage automatique et d'un tourbillon minute, ce concentré de technique atteint seulement 3,4 millimètres d'épaisseur.

En parlant de tourbillon, les calibres à remontage manuel PF500 et PF511 sont l'expression de la recherche permanente de la perfection horlogère. Leur tourbillon n'a besoin que de 30 secondes pour effectuer une révolution complète. Cette vitesse de rotation doublée procure une précision bien meilleure. L'indicateur de réserve de marche, situé à 12 heures, est breveté. Une chose encore caractérise ce calibre : la position centrale de l'aiguille des secondes. Le tourbillon reste visible lorsque l'on retire de son poignet la montre-bracelet qui en est équipée pour regarder à travers le fond saphir.

Le modèle « Ovale Pantographe », qui s'inspire d'un garde-temps de la fin du XVIIIᵉ siècle, est absolument unique. Les aiguilles télescopiques, qui donnent régulièrement des cheveux blancs aux horlogers chargés de les fabriquer, s'allongent et se rétractent en suivant les contours elliptiques du boîtier, comme par enchantement. L'entraînement est assuré par un calibre éprouvé à remontage manuel disposant de huit jours d'autonomie de marche, le PF111.

Le complexe à remontage manuel PF361 a atteint le stade de la fabrication en série précisément pour le 20ᵉ anniversaire de l'entreprise sise à Fleurier. Les éléments de soutien de l'ensemble des composants, autrement dit la platine et les différents ponts, sont en or massif. Outre le chronographe doté d'une classique roue à colonnes et d'un embrayage à friction innovant, ce chef-d'œuvre dont le cœur bat à cinq Hertz possède un mécanisme à ratrappante pour mesurer les temps intermédiaires, notamment dans les compétitions sportives. La grande date du modèle « Tonda Chronor Anniversaire », sorti en édition limitée à deux fois 25 exemplaires, est remarquable. La petite mais non moins remarquable manufacture propose cette concept watch en parallèle à la présentation de l'ultra-sportive Bugatti Chiron au salon de l'automobile de Genève de 2016. Le « bloc moteur » associé au cadran positionné perpendiculairement avec deux barillets et un tourbillon minute « volant » est baptisé PF390. La transmission se fait à l'aide d'un train de rouage planétaire avec rouage central. Le cadran disposé comme un pupitre permet de voir l'heure tout en gardant les mains sur le volant.

Ces vingt dernières années, Parmigiani Fleurier nous a réservé de nombreuses surprises. Et il devrait continuer sur sa lancée. ○

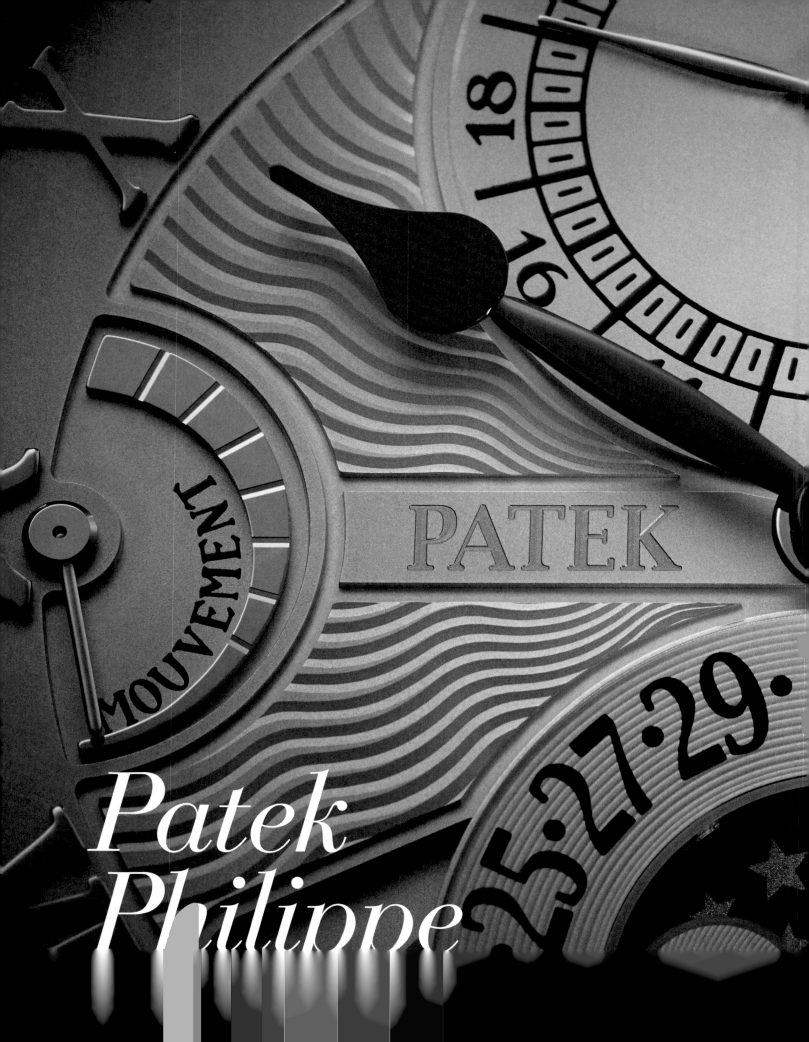

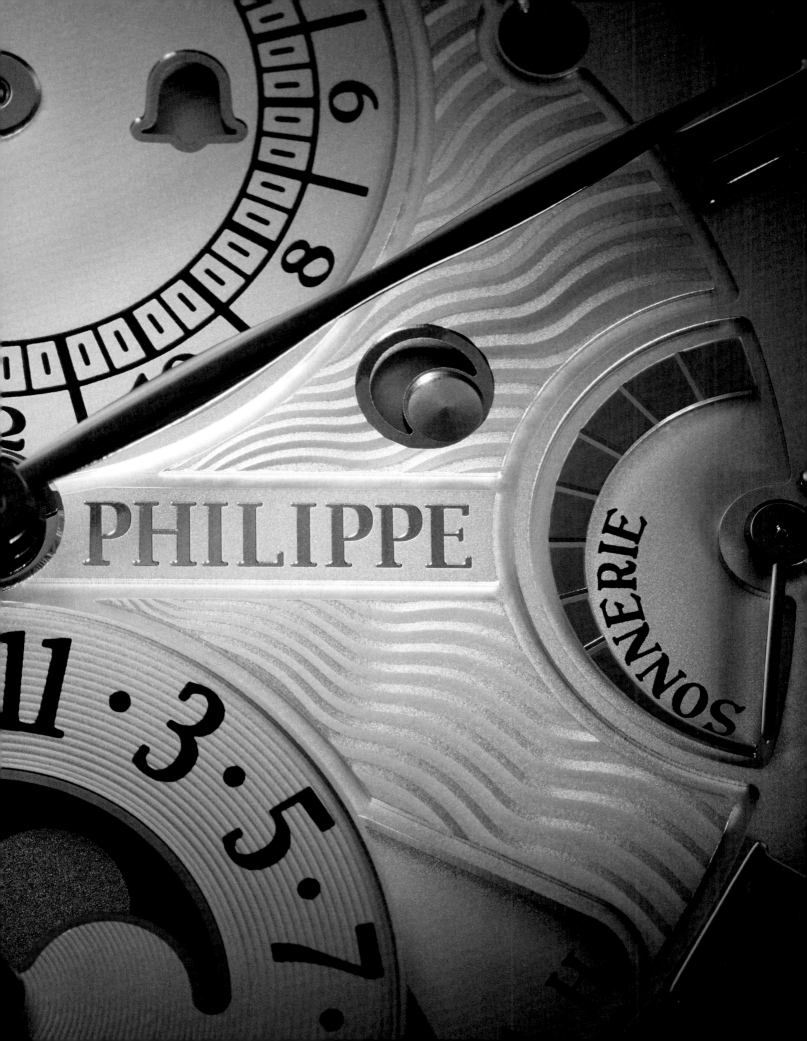

At Home in Geneva since 1839

English

Patek Philippe's archives contain three important items of information about the brand's ownership. The first dates from 1839, when Antoni Norbert Patek de Prawdzic, born in 1812, fled his native Poland and joined many other Polish emigrants who settled in Switzerland. He was in Versoix in 1835, where he married a businessman's daughter named Marie-Louise Adélaïde Elisabeth Thomasine Dénizard in 1835. One of the marriage's witnesses was Franciszek Czapek, a Bohemian with close ties to Polish culture. Czapek too was an émigré and a watchmaker who pursued his métier in collaboration with an uncle of Patek's wife. This generous uncle contributed 8,000 francs and was present as a silent partner at the signing ceremony establishing Patek, Czapek & Co. on May 1, 1839. Patek invested an equal sum as an active partner. Czapek brought his horological competence to the endeavor. The division of labor initially functioned as planned: Czapek made the watches; Patek sold the annual production of circa 200 timepieces, most of which were bought by uprooted Poles who felt well looked after in the shop at Quai des Bergues no. 29 in Geneva. But Czapek's extravagant lifestyle and his many journeys to his former homeland soon brought the young business to the verge of ruin. Liquid capital shrank to just one franc and 86 centimes. The remaining assets were several finished watches, various accounts receivable, and diverse tools. Loans were sought in vain.

This situation prompted Patek to travel by coach to Paris in December 1844 with a valise containing several watches with crown winding based on a system developed by Louis Audemars in 1833. Patek wanted to show these timepieces at an industrial exhibition on the Champs-Élysées, but was dismayed to discover in Paris that his products were far inferior to slim and keyless pocket-watches made by Jean-Adrien Philippe. Soon after this creative French watchmaker won a medal at the show, Patek made him the enticing offer of a shared future. Jean-Adrien Philippe agreed, relocated to Geneva in 1845 and enthusiastically began developing and perfecting his patented invention. After the six-year contract with Czapek expired, Philippe was promoted to technical director of Patek & Co. He remained in this position until 1851—the second important date in the firm's history—when Patek, Philippe & Co. was founded.

The productive partnership between Patek and Philippe led to the creation of outstanding pocket-watches that represented the full spectrum of precise time measurement on the highest technical and craftsmanly level. The extensive palette ranged from simple, slim specimens to watches with various complications. Incidentally: Patek, Philippe & Co. did not fabricate its own ébauches; instead, the company bought its movement-blanks from Victorin Piguet and LeCoultre in the Vallée de Joux. These kits were subsequently decorated, assembled, finely adjusted, encased, and, of course, sold to an international clientele in Geneva. Customers in the New World gained increasing importance too. In the second half of the 19th century, many watches embarked on the long journey to New York and from there to other cities in the USA, as well as to Brazil—and especially to the jeweler Carlos Gondolo. This strong focus would cause severe headaches for the Genevans after the stock market crashed in 1929.

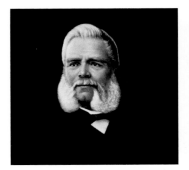

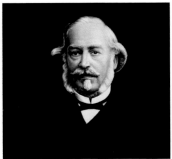

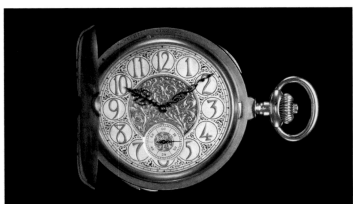

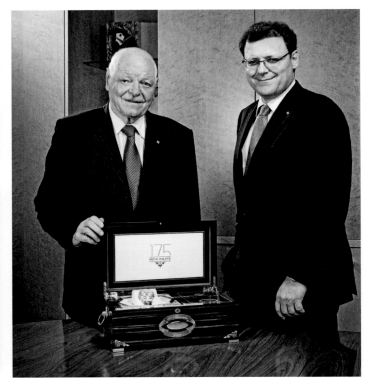

Clockwise from top left: Antoni Norbert de Patek ○ Jean-Adrien Philippe ○ Philippe Stern and his son Thierry ○ Duc de Regla, Savonnette, no. 138.285, 1910

But let's not get ahead of our story. Major upheavals in the company's structure ensued after the death of Count Antoine Norbert de Patek in 1877. His son Léon Mecyslas Vincent, who was a law student, and his daughter Marie Edwige weren't interested in joining the family business. They were initially satisfied with their roles as silent partners. When Patek, Philippe & Co. was reorganized in 1901 as Ancienne Manufacture d'Horlogerie Patek Philippe & Co. SA with capital totaling 1.6 million Swiss francs, Léon and Marie sold the rights to their family name in exchange for life annuity to the Philippe family, which had long since become dominant at the company. A contract signed with LeCoultre on August 13, 1904 guaranteed annual deliveries of 4,000 ébauches, including numerous specially designed kits to be assembled into the famous "Chronometro Gondolo." With this arrangement in place, Patek Philippe was able to survive several crisis-beset years in the early 20th century. The situation became extremely turbulent beginning in 1929, when sales to the USA and Brazil plummeted to zero more or less overnight. To make matters worse, the important customer and shareholder Gondolo & Labouriau in Rio de Janeiro was unable to pay its bills. The advisory council, to which Jacques-David LeCoultre had belonged since 1931, convened one emergency session after another. When the financial cataclysm had grown to gargantuan dimensions, the advisory committee offered to sell a majority of the shares in Patek Philippe to Monsieur LeCoultre and his holding company (SAPIC). Fortunately, the deal failed due to unsatisfactory demands and unacceptable asset valuations.

In this extraordinarily unpleasant situation, two white knights came to the rescue in 1932, which is the third important date in the company's history. Charles and Jean Stern, owners of Fabrique de Cadrans Stern Frères and long-term business associates, successfully acquired not only the freely traded stocks, but also shares in Stein (the importer in New York), the insolvent firm of Gondolo & Labouriau, and the movement supplier LeCoultre. The Stern brothers brought the competent Jean Pfister aboard as CEO, who in 1933 made the farsighted decision to begin constructing the brand's own ébauches. Through four generations of the Stern family—Charles and Jean, Henri, Philippe, and currently Thierry Stern—Patek Philippe gradually developed into a fullfledged manufactory that develops, fabricates, and encases all its movements, without exceptions, on the firm's modern factory premises in Geneva. All chronograph calibers have likewise been provided by the brand's own production since 2009. Patek Philippe has been granted over 100 patents in the course of its history, which spans more than 175 years. There's no horological complication that hasn't been mastered by this Geneva-based family business. Continual innovative dynamism also expresses itself in the use of silicon for the oscillating and escapement system, which forms the ticking heart and soul of mechanical watches. ∘

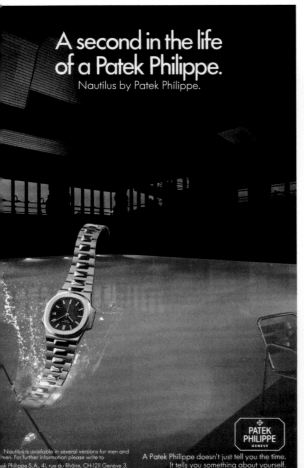

From left: Nautilus advertisement from the 1980s ∘ the new factory in Plan-les-Ouates near Geneva ∘ Von links: Reklame der 1980er Jahre für die Nautilus ∘ Die neue Fabrik in Plan-les-Ouates bei Genf ∘ Depuis la gauche : Réclame des années 1980 pour la Nautilus ∘ La nouvelle manufacture de Plan-les-Ouates près de Genève

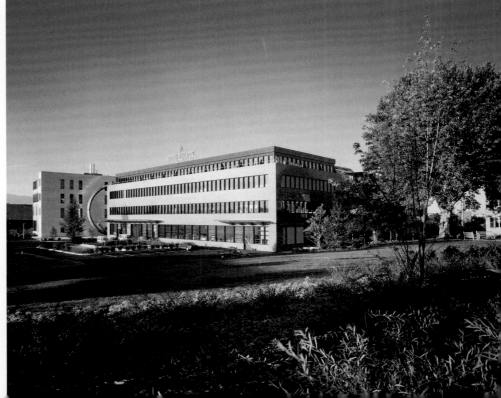

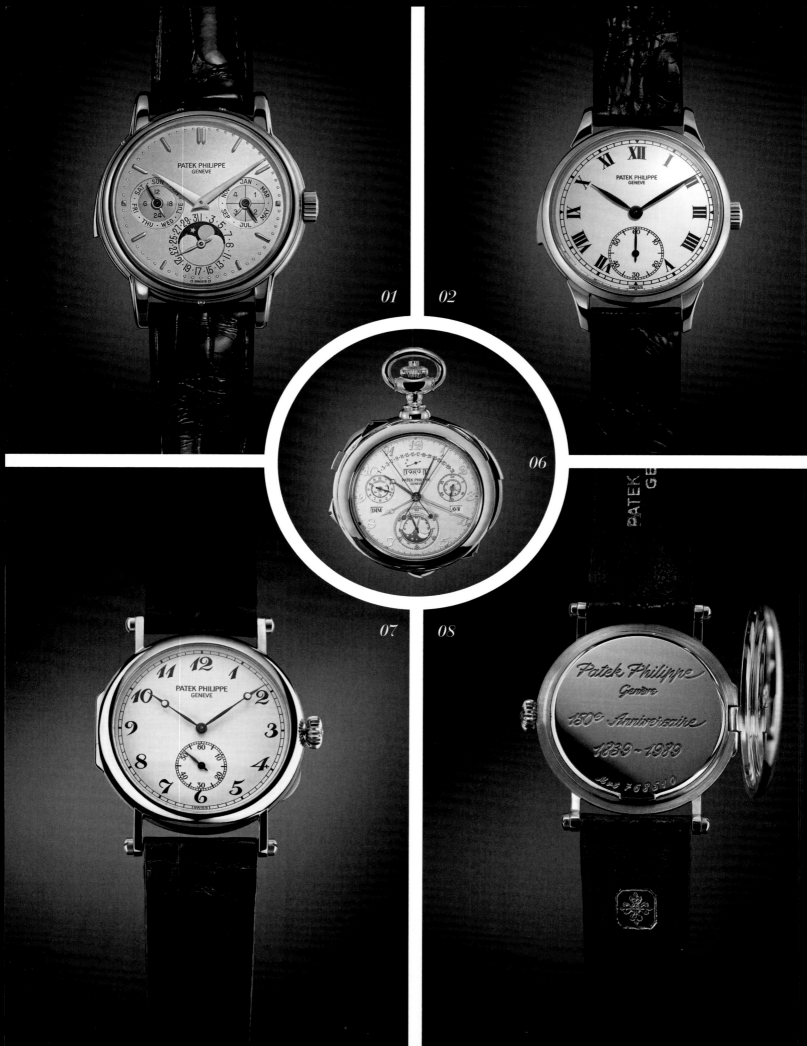

01

02

06

07

08

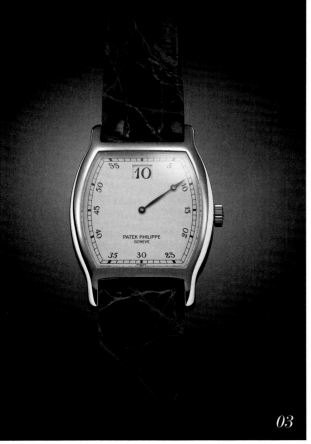

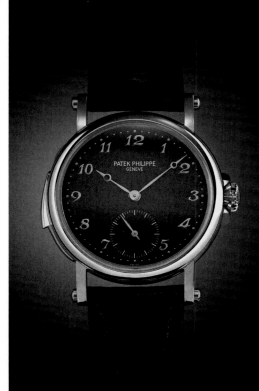

03

04

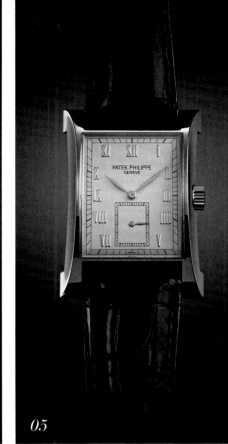

05

09

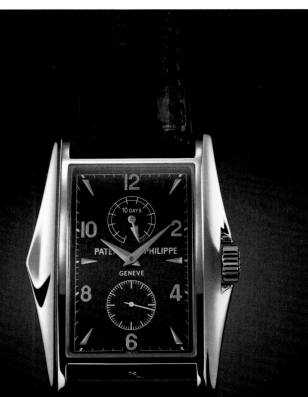

01. Minute Repeater and Perpetual Calendar, Ref. 3974, 1989
02. Minute Repeater, Ref. 3979, 1989
03. Jump Hour, Ref. 3969, 1989
04. Minute Repeater, Ref. 5029, 1997
05. Pagoda, Ref. 5500, 1997
06. Caliber 89, 1989
07./08. Officer's watch, Ref. 3960, 1989
09. 10-Day, Ref. 5100, 2000
10. Star Caliber 2000, 2000

10

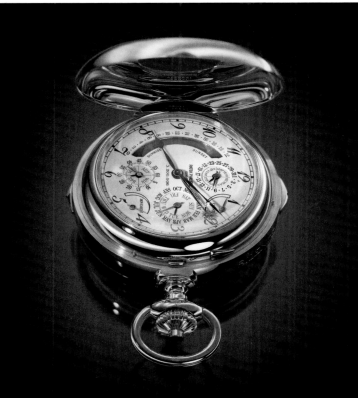

Seit 1839 in Genf zu Hause

Deutsch

Die Archive von Patek Philippe offenbaren bezüglich der Eigentumsverhältnisse drei wichtige Daten. Das erste liegt im Jahr 1839. Nach der Flucht aus seiner Heimat hatte sich der 1812 geborene Antoni Norbert Patek de Prawdzic wie viele andere polnische Emigranten in der Schweiz niedergelassen. 1835 fand man ihn in Versoix, wo er am 10. Juli 1839 die Kaufmannstochter Marie-Louise Adélaïde Elisabeth Thomasine Dénizard heiratete. Einer der Trauzeugen war der polonisierte Böhme Franciszek Czapek. Der 1831 ebenfalls geflohene Uhrmacher hatte sich in chronometrischen Dingen mit dem Onkel der Ehefrau von Patek zusammengetan. Selbiger war bei der Vertragsunterzeichnung zur Gründung von Patek, Czapek & Co. mit Wirkung vom 1. Mai 1839 durch seine Einlage von 8 000 Franken als stiller Teilhaber dabei. Patek investierte als aktiver Partner den gleichen Betrag. Czapek schließlich brachte seine uhrmacherische Kompetenz ein. Anfänglich funktionierte die Arbeitsteilung wie intendiert: Czapek verantwortete die Uhrenfertigung, Patek kümmerte sich um den Verkauf der jährlich rund 200 Zeitmesser – meist an entwurzelte Polen, die sich im Geschäft am Genfer Quai des Bergues Nr. 29 gut betreut fühlten. Bald aber trieben Czapeks verschwenderischer Lebensstil sowie seine vielen Reisen in die alte Heimat die junge Firma an den Rand des Ruins. Das flüssige Kapital reduzierte sich auf einen Franken und 86 Rappen. Die übrigen Aktivposten bestanden in einigen fertigen Uhren, verschiedenen Außenständen und Werkzeugen. Um Kredite bemühte man sich vergebens.

Notgedrungen reiste Patek im Dezember 1844 per Kutsche nach Paris. Im Gepäck: einige Uhren mit Kronenaufzug nach dem 1833 vorgestellten System von Louis Audemars, die er im Rahmen einer Industrieausstellung am Champs-Élysées zeigen wollte. An der Seine musste er allerdings erfahren, dass flache, schlüssellose Taschenuhren des Uhrmachers Jean-Adrien Philippe seine eigenen Produkte deutlich in den Schatten stellten. Nach der Verleihung einer Medaille an den kreativen Franzosen bot Patek ihm eine gemeinsame Zukunft an. Der Lockruf fruchtete. In Genf stürzte sich Jean-Adrien Philippe ab 1845 mit Verve auf die konsequente Weiterentwicklung und Vervollkommnung seiner bereits patentierten Erfindung. Nach dem Auslaufen des sechsjährigen Vertrags mit Czapek avancierte Philippe erst zum technischen Direktor von Patek & Co., bis es 1851 – und das ist die zweite wichtige Jahreszahl in der Firmengeschichte – zur Gründung von Patek, Philippe & Co. kam.

Aus der fruchtbaren Partnerschaft zwischen Patek und Philippe erwuchsen großartige Taschenuhren, welche das gesamte Spektrum der Präzisionszeitmessung auf höchstem technischem und handwerklichem Niveau repräsentierten. Die Palette reichte von schlichten flachen Exemplaren bis hin zu verschiedenen großen Komplikationen. Eigene Rohwerke fertigte Patek, Philippe & Co. übrigens nicht. Man kaufte sie im abgeschiedenen Vallée de Joux bei Victorin Piguet und LeCoultre. In Genf erfolgten danach die Finissage, Assemblage, Regulierung, Einschalung und natürlich der internationale Verkauf. Dabei gewannen Kunden in der Neuen Welt zunehmend an Bedeutung. In der zweiten Hälfte des 19. Jahrhunderts nahmen viele Uhren ihren Weg nicht nur nach New York und von dort aus in die ganzen USA, sondern auch nach Brasilien und dort speziell zum Juwelier Carlos Gondolo. Eine starke Fokussierung, die den Genfern nach dem Börsencrash von 1929 noch heftige Bauchschmerzen bereiten sollte.

Doch alles der Reihe nach: 1877 bedingte der Tod des Grafen Antoine Norbert de Patek große Umwälzungen in der Unternehmensstruktur. Sohn Léon Mecyslas Vincent, Student der Rechtswissenschaften, und Tochter Marie Edwige konnten dem Uhrengeschäft nichts abgewinnen. Zunächst begnügten sie sich mit der Rolle stiller Teilhaber. Als Patek, Philippe & Co. 1901 zur Kollektivgesellschaft Ancienne Manufacture d'Horlogerie Patek Philippe & Co. S.A. mit einem Kapital von 1,6 Millionen Schweizer Franken mutierte, veräußerten sie ihre Namensrechte gegen Leibrente an die längst von der Familie Philippe dominierte Firma. Der 13. August 1904 brachte ein Abkommen mit LeCoultre über die Lieferung von jährlich 4000 Rohwerken, darunter zahlreiche nach den spezifischen konstruktiven Vorgaben zur Herstellung der berühmten „Chronometro Gondolo". Unter diesen Vorzeichen überstand Patek Philippe einige Krisenjahre im frühen

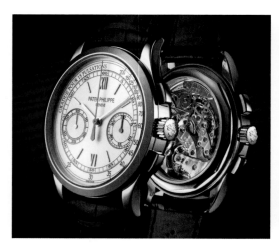
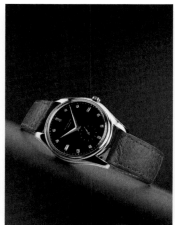
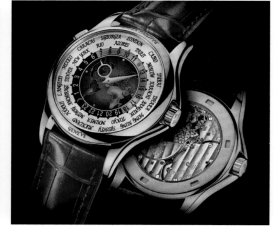

Left: Complications, Ref. 5170, 2010 ∘ Middle: Calatrava, Ref. 2526, 1960 ∘ Right: World Time, Ref. 5131, 2008

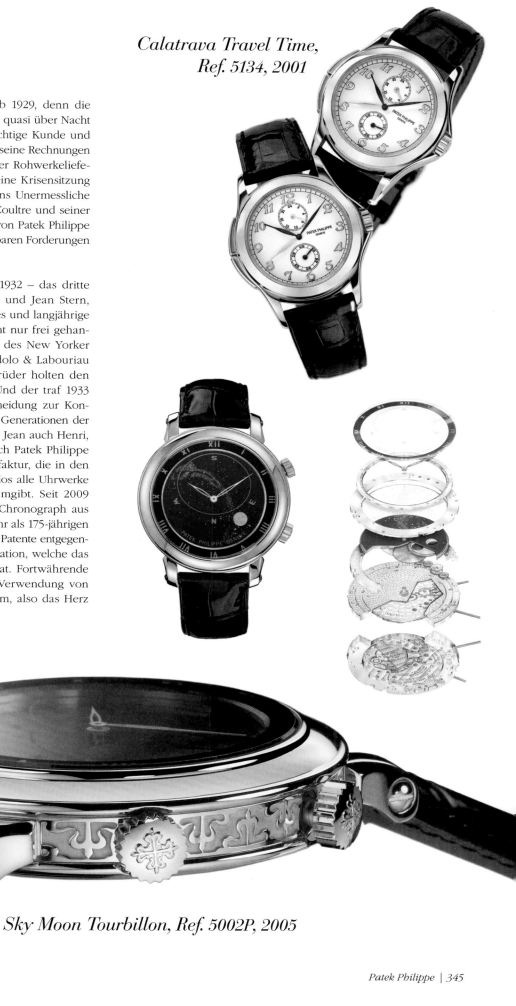

Calatrava Travel Time,
Ref. 5134, 2001

20. Jahrhundert. Richtig turbulent wurde es ab 1929, denn die Verkäufe in Richtung USA und Brasilien sanken quasi über Nacht gegen null. Zu allem Überfluss beglich der wichtige Kunde und Aktionär Gondolo & Labouriau in Rio de Janeiro seine Rechnungen nicht. Im Verwaltungsrat, dem ab 1931 auch der Rohwerkelieferant Jacques-David LeCoultre angehörte, jagte eine Krisensitzung die andere. Als die finanzielle Not fast schon ins Unermessliche wuchs, bot das Aufsichtsgremium Monsieur LeCoultre und seiner Holding SAPIC den Erwerb der Aktienmehrheit von Patek Philippe an. Zum Glück scheiterte der Deal an unannehmbaren Forderungen und inakzeptablen Vermögensbewertungen.

In dieser äußerst unangenehmen Zeit halfen 1932 – das dritte wichtige Datum – zwei Weiße Ritter. Charles und Jean Stern, Eigentümer der Fabrique de Cadrans Stern Frères und langjährige Geschäftspartner, übernahmen sukzessive nicht nur frei gehandelte Anteilsscheine, sondern auch die Aktien des New Yorker Importeurs Stein, der insolventen Firma Gondolo & Labouriau sowie des Werkelieferanten LeCoultre. Die Brüder holten den kompetenten Jean Pfister als CEO ins Boot. Und der traf 1933 die zukunftsweisende unternehmerische Entscheidung zur Konstruktion eigener Uhrenrohwerke. Während vier Generationen der Familie Stern – gemeint sind neben Charles und Jean auch Henri, Philippe sowie aktuell Thierry – entwickelte sich Patek Philippe schrittweise zu einer hundertprozentigen Manufaktur, die in den modernen Genfer Fabrikationsstätten ausnahmslos alle Uhrwerke selbst entwickelt, fertigt und mit Gehäusen umgibt. Seit 2009 stammen nämlich auch sämtliche Kaliber mit Chronograph aus eigener Produktion. Im Laufe der inzwischen mehr als 175-jährigen Firmengeschichte konnte Patek Philippe über 100 Patente entgegennehmen. Es gibt keine uhrmacherische Komplikation, welche das Genfer Familienunternehmen nicht bewältigt hat. Fortwährende Innovationskraft zeigt sich schließlich in der Verwendung von Silizium für das Schwing- und Hemmungssystem, also das Herz und die Seele mechanischer Uhren. ○

Sky Moon Tourbillon, Ref. 5002P, 2005

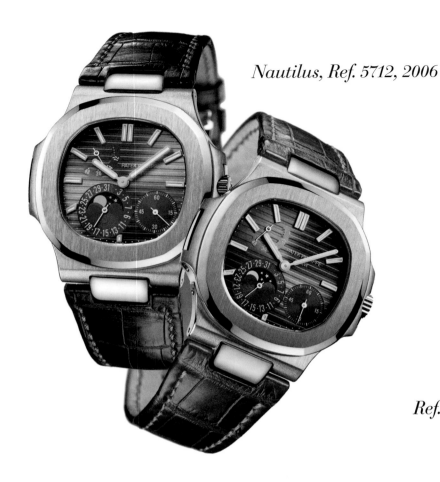

Nautilus, Ref. 5712, 2006

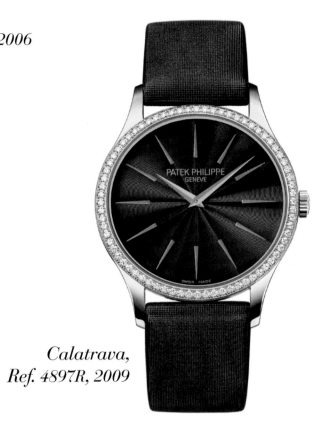

Calatrava,
Ref. 4897R, 2009

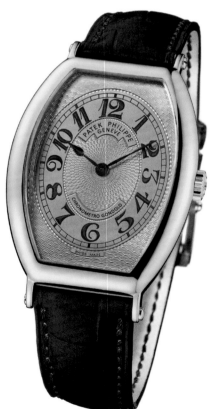

Chronometro Gondolo, Ref. 5098R, 2009

Minute Repeater
Perpetual Calendar
Tourbillon,
Ref. 5207, 2008

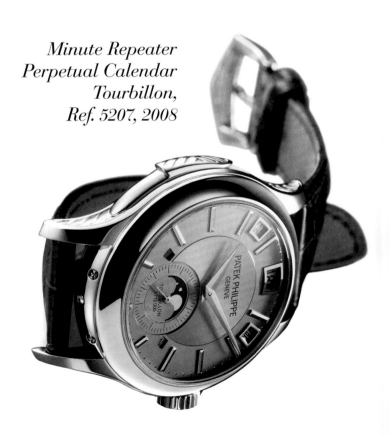

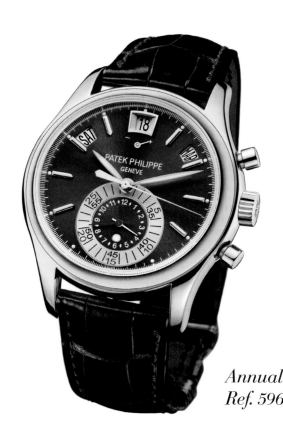

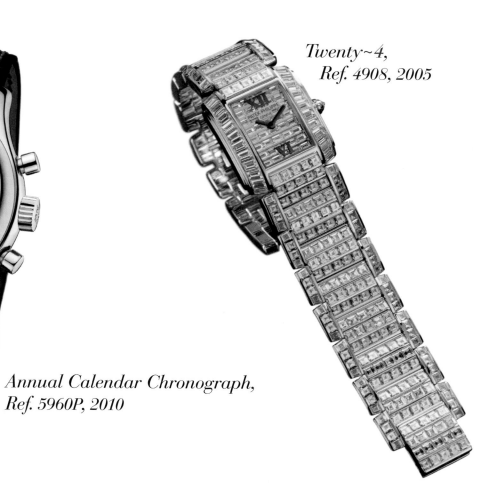

Twenty~4,
Ref. 4908, 2005

Annual Calendar Chronograph,
Ref. 5960P, 2010

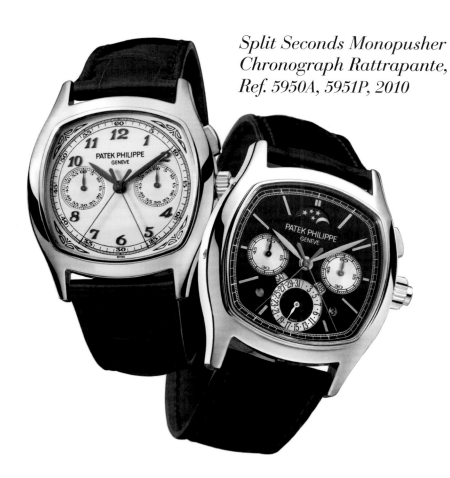

Split Seconds Monopusher
Chronograph Rattrapante,
Ref. 5950A, 5951P, 2010

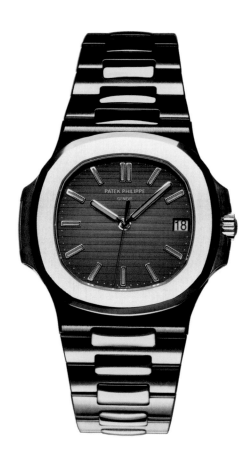

Nautilus, Ref. 370, 2006

Établi à Genève depuis 1839

Les archives de Patek Philippe livrent trois dates clés concernant ses propriétaires. La première est l'année 1839. Né en 1812, Antoni Norbert Patek de Prawdzic fuit sa patrie comme de nombreux émigrants polonais pour s'installer en Suisse. En 1835, on le trouve à Versoix, où il épouse le 10 juillet 1839 Marie-Louise Adélaïde Elisabeth Thomasine Dénizard, fille d'un commerçant. L'un des témoins est Franciszek Czapek, un Polonais originaire de Bohême. Cet horloger qui a lui aussi fui, en 1831, s'associe alors avec l'oncle de l'épouse de Patek pour ouvrir un négoce de montres. Aux termes du contrat portant création de Patek, Czapek & Co., qui prend effet au 1er mai 1839, ce dernier est actionnaire non actif, avec une mise de fonds de 8 000 francs suisses. Patek engage le même montant en qualité de partenaire actif. C'est en fait Czapek qui apporte les compétences en horlogerie. Au début, la répartition des tâches fonctionne comme prévu : Czapek assure la fabrication des montres et Patek s'occupe de vendre tous les ans environ 200 garde-temps – le plus souvent à des Polonais déracinés, qui se sentent bien accueillis dans la boutique genevoise au n° 29 du quai des Bergues. Mais bientôt, le train de vie dispendieux de Czapek ainsi que ses nombreux voyages dans son ancienne patrie conduisent la jeune société au bord de la faillite. Les liquidités se limitent à un franc et 86 centimes. Les autres postes de l'actif se résument à quelques montres terminées, diverses créances et des outils. Les demandes de crédit sont toutes rejetées.

Contraint et forcé, Patek se rend en 1844 en calèche à Paris. Dans ses bagages : quelques montres dotées d'un remontoir à couronne conçu en 1833 par Louis Audemars, qu'il veut proposer à l'Exposition nationale des produits de l'industrie qui se tient sur les Champs-Élysées. Mais arrivé à Paris, il apprend que les montres de poche plates avec mise à l'heure sans clef de l'horloger Jean-Adrien Philippe surclassent ses propres produits. Une médaille est décernée à l'horloger de génie français. Lorsque Patek l'invite à s'associer avec lui, il répond favorablement à son appel. À Genève, Jean-Adrien Philippe entreprend dès 1845 avec entrain et détermination de continuer de développer et perfectionner son invention déjà brevetée. Le contrat de six ans avec Czapek étant arrivé à son terme, Philippe est tout d'abord promu directeur technique de Patek & Co., société qui en 1851 – deuxième date clé dans l'histoire de la marque – devient Patek, Philippe & Co.

Le partenariat fécond entre Patek et Philippe donne naissance à de magnifiques montres de poche, qui couvrent tout le spectre des mesures précises du temps à un niveau technique et artisanal d'exception. La palette s'étend des montres plates simples aux montres à grandes complications les plus diverses. Patek, Philippe & Co. ne fabrique toutefois pas ses propres ébauches. La société se les procure dans la paisible vallée de Joux, chez Victorin Piguet et LeCoultre. C'est ensuite à Genève qu'ont lieu le finissage, l'assemblage, le réglage, l'emboîtage et bien sûr la vente aux clients du monde entier, toujours plus nombreux dans le Nouveau Monde. Dans la seconde moitié du XIXe siècle, un grand nombre de montres partent non seulement pour New York où elles sont distribuées dans tous les États-Unis, mais aussi vers le Brésil, plus particulièrement

à destination du joaillier Carlos Gondolo. Après le krach boursier de 1929, cette forte concentration des ventes mettra les entrepreneurs genevois dans une situation délicate.

Mais reprenons les choses dans l'ordre : en 1877, le décès du comte Antoine Norbert de Patek entraîne de grands bouleversements dans la structure de la société. Son fils Léon Mecyslas Vincent, étudiant en droit, et sa fille Marie Edwige ne retirent rien du magasin d'horlogerie. Dans un premier temps, ils se contentent du rôle d'actionnaires non actifs. En 1901, lorsque Patek, Philippe & Co. devient l'Ancienne Manufacture d'Horlogerie Patek Philippe & Co. SA, société en nom collectif au capital de 1,6 million de francs suisses, ils cèdent contre une rente viagère leurs droits sur le nom à la société, depuis longtemps dominée par la famille Philippe. Le 13 août 1904 est signé avec LeCoultre un accord portant sur la fourniture annuelle de 4000 ébauches, dont un grand nombre doivent satisfaire les exigences constructives spécifiques à la fabrication du célèbre « Chronometro Gondolo ». Patek Philippe surmonte ainsi quelques années de crise au début du XXe siècle. La situation devient franchement chaotique en 1929, lorsque les ventes à destination des États-Unis et du Brésil s'effondrent quasiment du jour au lendemain. Pour couronner le tout, Gondolo & Labouriau, actionnaire et client majeur basé à Rio de Janeiro, est en défaut de paiement. Au sein du conseil d'administration, rejoint en 1931 par le fournisseur d'ébauches Jacques-David LeCoultre, une réunion de crise chasse l'autre. Lorsque les difficultés financières finissent par atteindre un niveau qui dépasse l'imagination, le conseil de surveillance propose à monsieur LeCoultre et à sa holding SAPIC d'acquérir la majorité des actions de Patek Philippe. Heureusement, des exigences inadmissibles et une évaluation des actifs inacceptable font capoter la transaction.

Cette période éminemment inconfortable est marquée en 1932 – troisième date clé – par l'arrivée de deux chevaliers blancs. Charles et Jean Stern, détenteurs de la Fabrique de Cadrans Stern Frères et partenaires commerciaux de longue date, reprennent tour à tour non seulement les titres librement négociables, mais aussi les actions de l'importateur new-yorkais Stein, de la société insolvable Gondolo & Labouriau et du fournisseur de mouvements LeCoultre. Les frères Stern persuadent Jean Pfister, un homme compétent, de les suivre, et le nomment directeur général. En 1933, celui-ci prend une décision capitale pour l'avenir de l'entreprise, à savoir construire ses propres ébauches. Sous quatre générations de Stern – soit Charles et Jean, puis Henri, Philippe et aujourd'hui Thierry –, Patek Philippe se développe peu à peu pour devenir une manufacture à part entière, qui met au point, fabrique et habille de boîtiers la totalité de ses mouvements dans ses ateliers genevois modernes. Depuis 2009, tous les calibres de chronographes sont fabriqués en interne. Au cours d'une histoire qui s'étend sur plus de 175 ans, Patek Philippe obtient plus de 100 brevets. Il n'est pas une complication horlogère que l'entreprise familiale genevoise n'ait réussi à maîtriser. La capacité permanente d'innovation se manifeste enfin dans l'utilisation de silicium pour la masse oscillante et l'échappement, c'est-à-dire le cœur et l'âme des montres mécaniques. ○

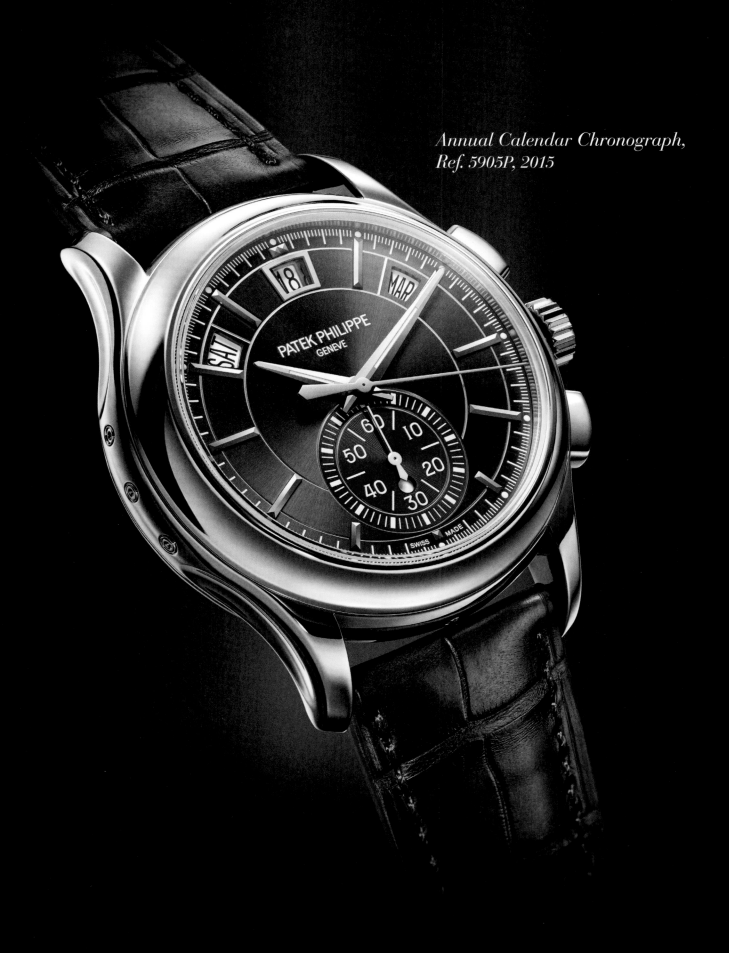

Annual Calendar Chronograph, Ref. 5905P, 2015

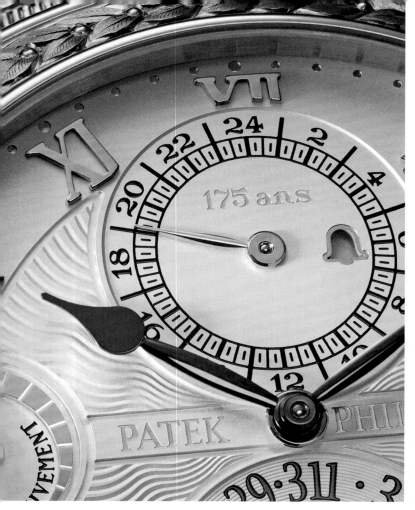
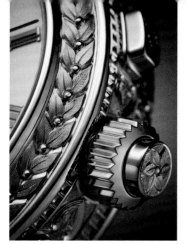
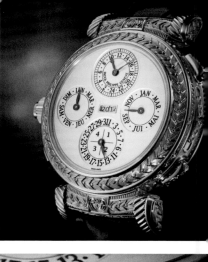
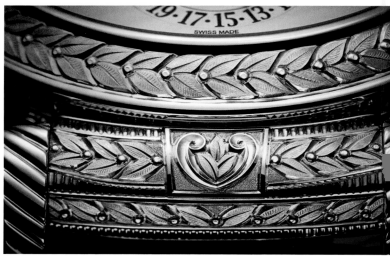
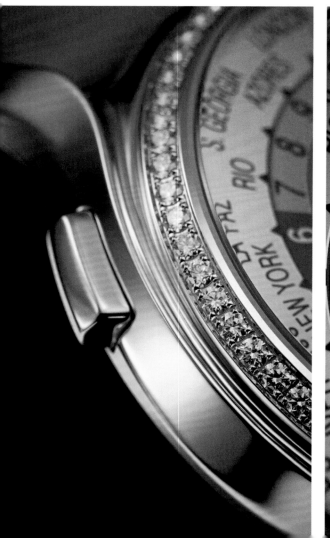
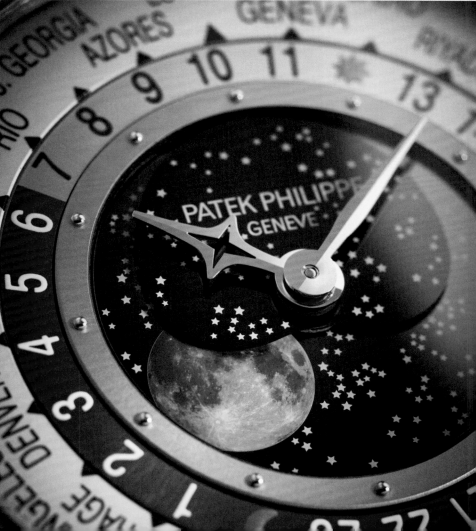

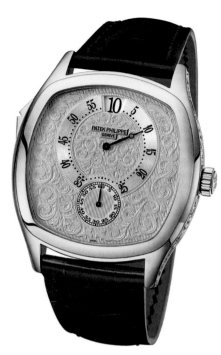

Chiming Jump Hour,
175th Collection,
Ref. 5275, 2014

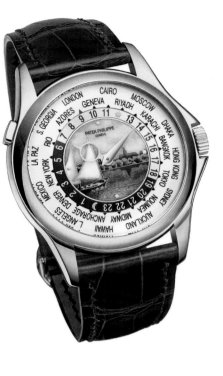

World Time
Geneva Harbor,
Ref. 5131 175G, 2014

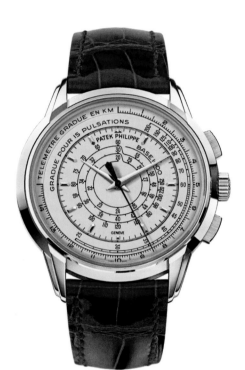

Multi-Scale Chronograph,
175th Collection, Ref. 5975, 2014

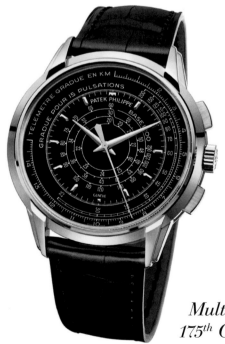

Multi-Scale Chronograph,
175th Collection, Ref. 5975, 2014

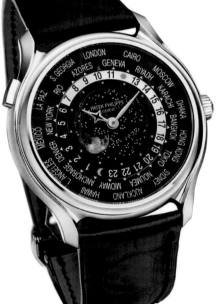

World Time Moon,
175th Collection,
Ref. 5575, 2014

Left page:
Top: Grandmaster Chime, 175th Collection, limited to 7 pieces, Ref. 5175R, 2015
Bottom: World Time Moon, 175th Collection, Ref. 7175R, 2014

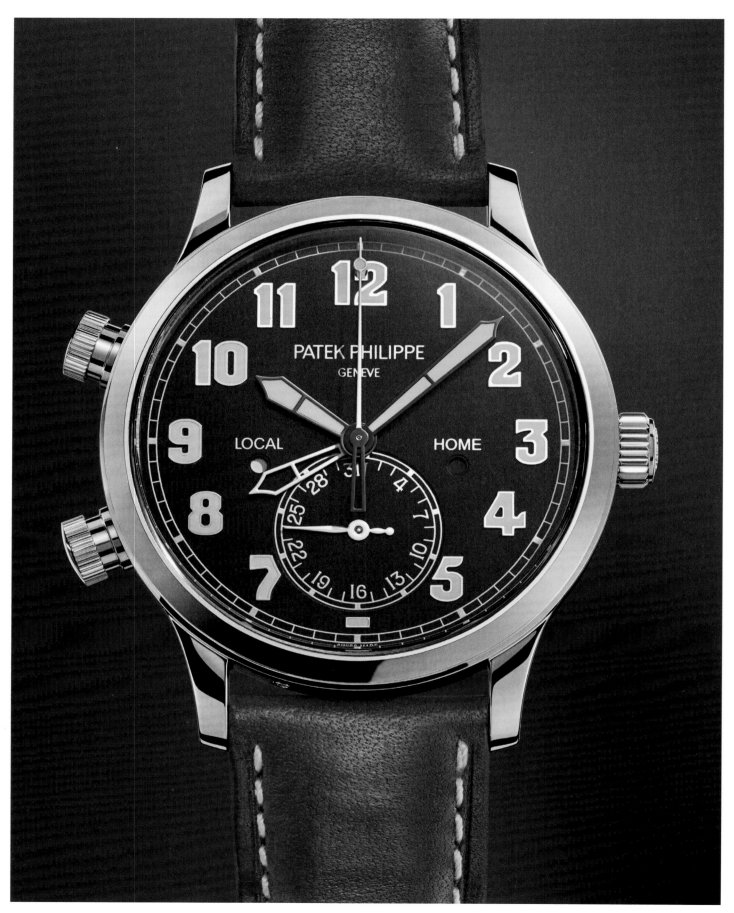

Calatrava Pilot Travel Time, Ref. 5524, 2015

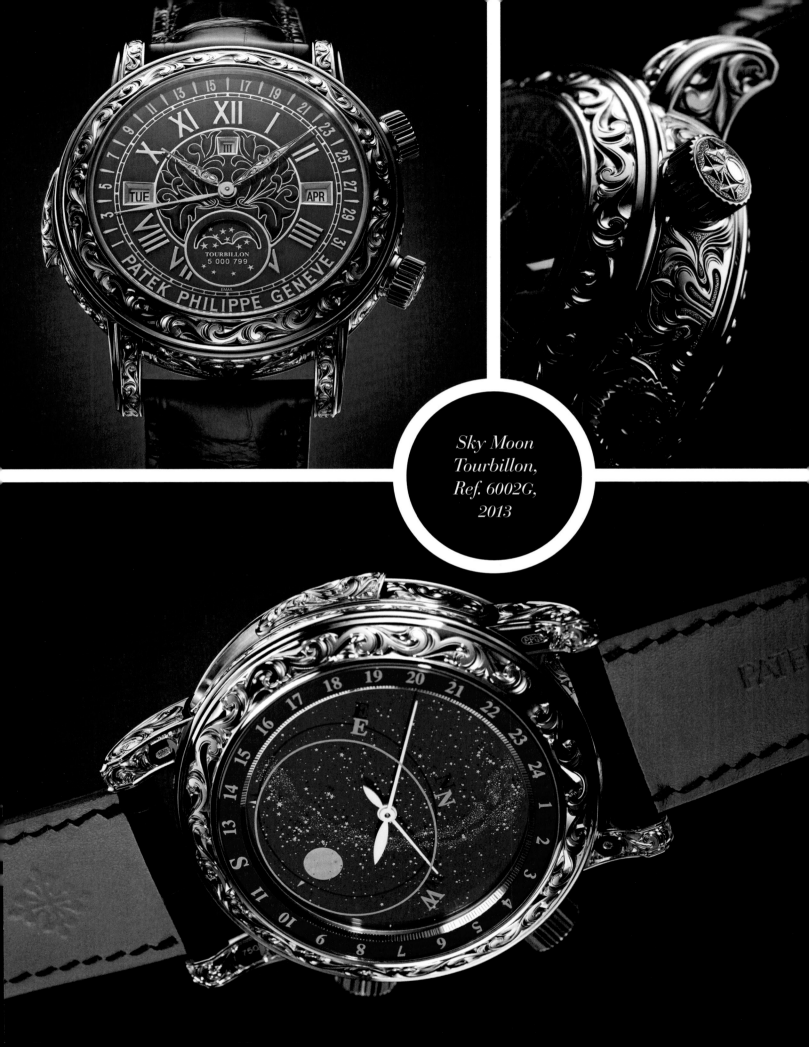

Sky Moon Tourbillon, Ref. 6002G, 2013

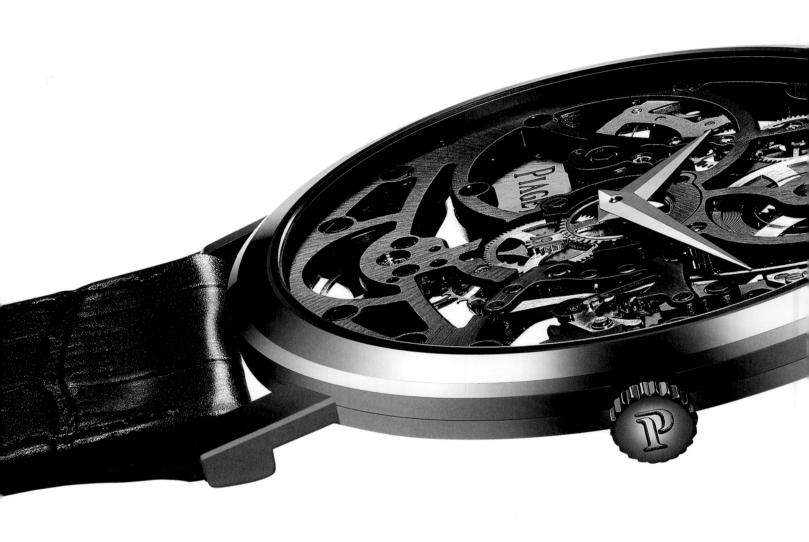

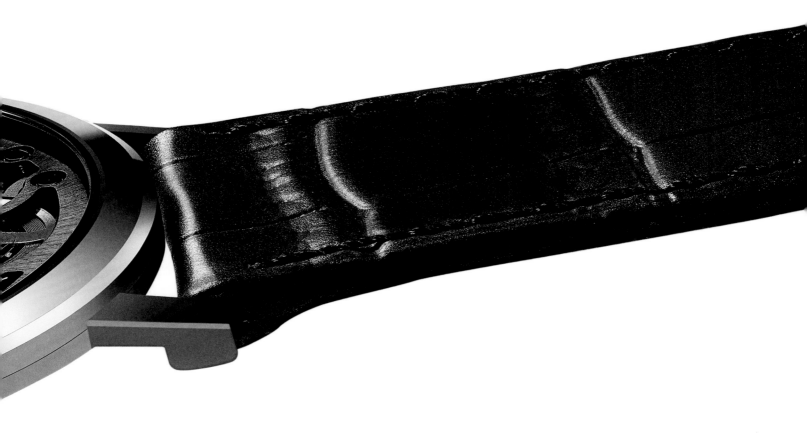

Piaget

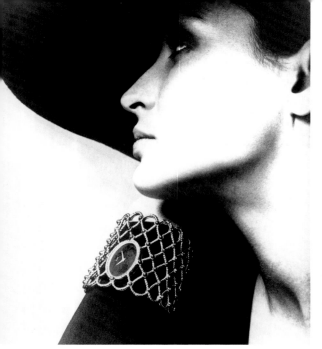

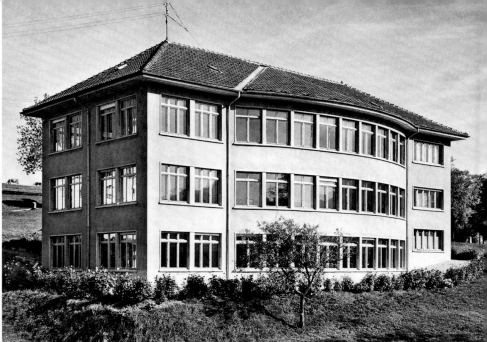

World Champion of Slim Wristwatches

English

If there were an award for minimalism in wristwatches, Piaget would surely win it. Founded by Georges-Édouard Piaget in 1874 in the Jura village of La Côte-aux-Fées at an altitude of 3,300 feet, Piaget first broke slimness records in 1957. Prior to that, the family business's watches had encased nothing but calibers sourced from third parties. Piaget's debut as a manufacture ensued in 1957 with the launch of caliber 9P. The "9" in this movement's designation refers to its diameter: nine lignes (20.5 millimeters). The patented hand-wound caliber was just two millimeters tall: it wasn't the world's slimmest movement, but it was remarkably slender. Thanks to its small dimensions, caliber 9P could fit neatly inside the cases of ladies' and gents' watches.

Piaget responded to its clientele's preference for slim, elegant, self-winding watches with the debut of caliber 12P in 1960. With an ultra-slim and bidirectionally effective oscillating weight made of solid 24-karat gold, this movement was a mere 2.3 millimeters tall, which earned it a place in the Guinness Book of Records. It retained its world record until 1978, when it was unseated by the Lassale 2000, a caliber with a central rotor and an overall height of just 2.08 millimeters. Piaget's caliber 12PC is a slightly taller variant that shows the date in a window. After the bankruptcy of Bouchet-Lassale SA, which had developed and produced the slimmest hand-wound movement of all times (just 1.2 millimeters tall), the rights to both calibers were acquired in the early 1980s by Piaget and the ébauche manufacturer Lémania, in which Piaget participated as a shareholder. Revised hand-wound caliber 20P debuted in 1982 and self-winding caliber 25P followed in 1983, but neither of these achieved more than a modest presence on the market because both were beset by constructive problems related to their micro ball bearings and flying (i.e., cantilevered) wheels.

Fortunately, Piaget still had its time-honored calibers 9P and 12P, which were produced until the early 21st century, when the tools began showing signs of wear and the watch world was demanding bigger and bolder opulence. The 50th birthday of caliber 12P brought a totally newly developed comeback in the guise of caliber 1208P. No other automatic movement at this time was slimmer than its height of 2.35 millimeters. One of its tricks: some wheels are a mere 0.12 millimeters thick. With a diameter of 30 millimeters, this newcomer amassed a power reserve that was lengthy enough to support the watch's practical use. Its comparatively small balance continued to oscillate for 40 hours without a fresh transfusion of energy.

Piaget celebrated its 140th anniversary in 2014. This provided a welcome occasion to grant retrospective honors to good old caliber 9P. But hand-wound caliber 900P, which was developed in the course of a three-year cooperation among watchmakers and casemakers, had little or nothing to do with old-fashioned technologies. To achieve an unprecedentedly slim height of just 3.65 millimeters for the movement and case, both parts had to merge. In other words, the back of the 38-millimeter gold case serves as the base plate. And the caliber that's integrated into this plate impressively pushes the horological envelope. Incidentally: Piaget also began marketing the world's slimmest hand-wound chronographs in 2015. Together with its gold case, which houses 4.65-millimeter-slim caliber 883P, this watch is only 8.24 millimeters tall.

We'll conclude this chapter with a few facts about the manufactory's history. Piaget acquired a majority share of the Baume & Mercier watch brand in 1964. Despite brisk business, or rather because of it, the family sold 60 percent of its shares in 1988. The buyer was Cartier Monde SA in Paris, which is now Richemont SA. Both brands were acquired by the Richemont Group in 1993. ○

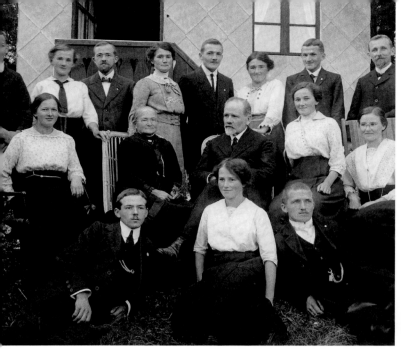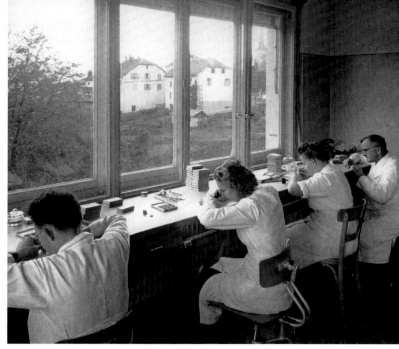

From left: advertising photo for a cuff watch from the 1970s ∘ The factory in La Côte-aux-Fées ∘ The Piaget family ∘ Watchmakers assembling watches in the 1960s ∘ Von links: Werbefoto aus den 1970ern für eine Manschetten-Uhr ∘ Die Fabrik in La Côte-aux-Fées ∘ Familie Piaget ∘ Uhrmacher bei der Montage, 1960er Jahre ∘ Depuis la gauche : Photo publicitaire des années 1970 pour une montre manchette ∘ Manufacture de La Côte-aux-Fées ∘ La famille Piaget ∘ Horlogers lors de l'assemblage, années 1960

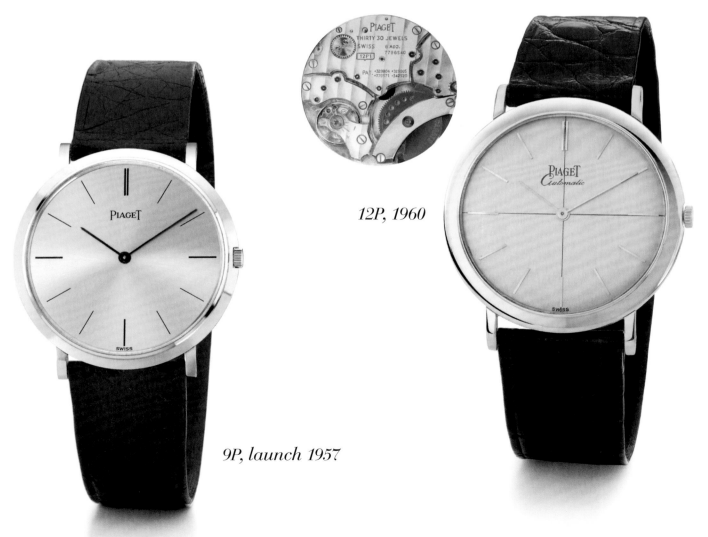

12P, 1960

9P, launch 1957

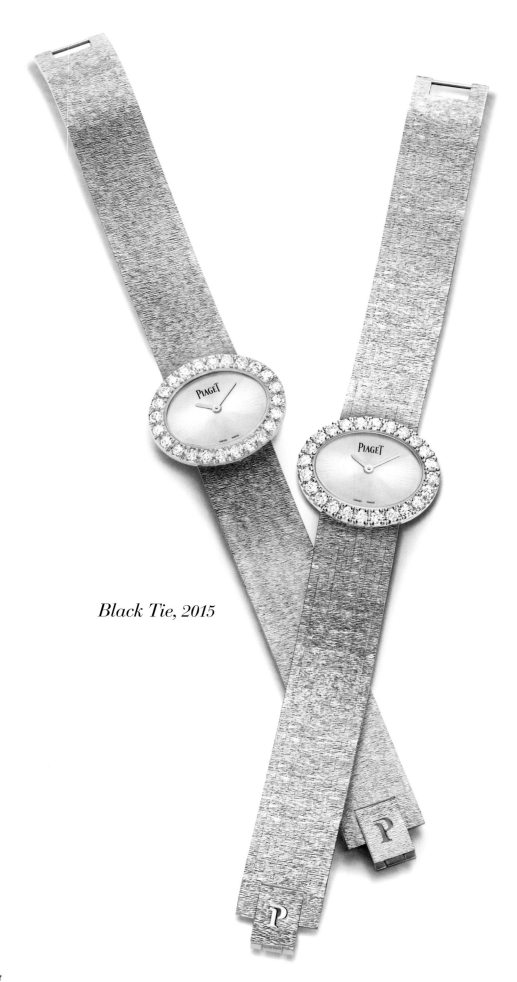

Black Tie, 2015

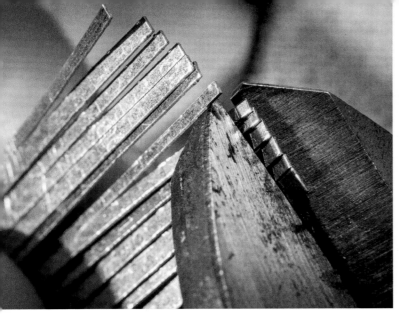
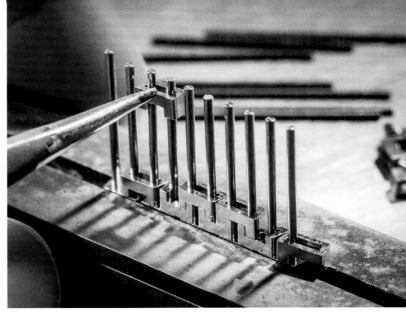
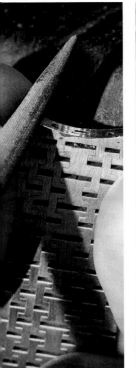
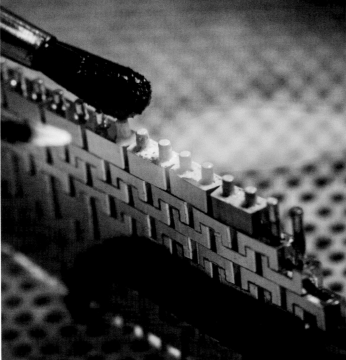

The highly elaborate process of manufacturing metal bracelets at Piaget ◦ Die sehr aufwendige Herstellung der Metallbänder bei Piaget ◦ La fabrication longue et complexe des bracelets métalliques chez Piaget

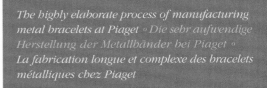
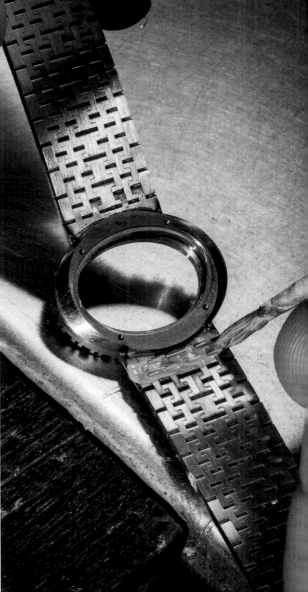
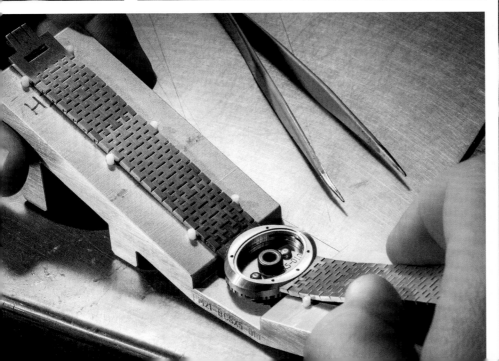

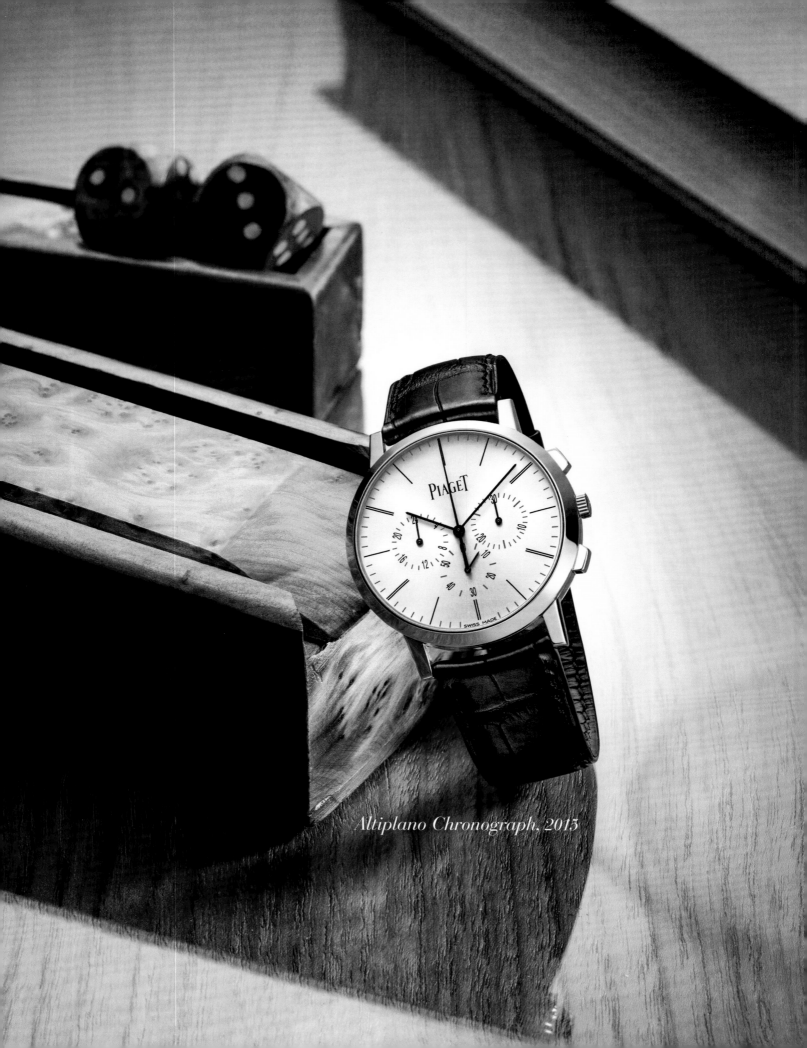

Altiplano Chronograph, 2015

Weltmeisterlich flache Armbanduhren

Wenn es einen Minimalismus-Weltmeister in Sachen Armband-uhren gibt, dann heißt er zweifelsohne Piaget. Obwohl schon 1874 im 1000 Meter hoch gelegenen Juradörfchen La Côte-aux-Fées von Georges-Édouard Piaget gegründet, begann die Jagd nach Rekorden erst 1957. Bis dahin verbaute das Familienunternehmen in den fürs Handgelenk bestimmten Uhren ausschließlich zugekaufte Uhr-werke. Dann, in jenem Jahr 1957, sorgte das Manufaktur-Debüt 9P auf Anhieb für einiges Aufsehen. Die Kaliberbezeichnung resul-tierte aus dem Durchmesser von 9 Linien oder 20,5 Millimeter. In der Höhe baute das patentierte Handaufzugswerk gerade einmal zwei Millimeter. Dieser Wert repräsentierte zwar keinen Superlativ, verdient aber dennoch das Attribut bemerkenswert. Wegen seiner moderaten Dimensionen passte das 9P in Damen- wie Herrenuhren.

Auf das zunehmende Faible für flache und elegante Automatik-armbanduhren reagierte Piaget 1960 mit dem 12P. Ausgestattet mit ultraflacher, in beiden Drehrichtungen aufziehender Schwung-masse aus 24-karätigem Massivgold baute dieses Uhrwerk lediglich 2,3 Millimeter hoch, was ihm den Eintrag ins Guinness-Buch der Rekorde verschaffte. Und es verkörperte bis zur Fertigstellung des extrem delikaten, nur noch 2,08 Millimeter hohen Zentralrotor-kalibers Lassale 2000 im Jahr 1978 einen unangefochtenen Welt-rekord. Beim Piaget 12PC handelte es sich um eine leicht höhere Werksvariante mit Fensterdatum. Nach dem Konkurs der Bouchet-Lassale S.A., die mit 1,2 Millimetern Höhe auch das flachste Hand-aufzugswerk aller Zeiten entwickelt und produziert hatte, erwarben Piaget und der Rohwerkehersteller Lémania, an dem Piaget seiner-zeit eine Beteiligung hielt, zu Beginn der 80er Jahre die Rechte an beiden Kalibern. Wegen der problembehafteten Konstruktion mit Mikrokugellagern und fliegenden Rädern war weder dem über-arbeiteten und 1982 eingeführten Handaufzugs-20P noch dem Automatik-25P von 1983 eine nachhaltige Marktpräsenz beschieden.

Aber Piaget besaß ja zum Glück noch die bewährten 9P und 12P, deren Geschichte erst im frühen 21. Jahrhundert endete, als die Werkzeuge sichtliche Abnutzungserscheinungen zeigten und die Uhrenwelt nach markanter Opulenz verlangte. Der 50. Geburts-tag des 12P brachte ein völlig neu entwickeltes Comeback in Gestalt des 1208P. Seine Mini-Bauhöhe von 2,35 Millimetern

unterschreitet derzeit keine andere Automatik. Einer der ange-wandten Tricks: Manche Zahnräder sind gerade einmal 0,12 Milli-meter dick. An alltagstauglicher Gangautonomie mangelt es dem knapp 30 Millimeter großen Newcomer trotzdem nicht. Erst nach rund 40 Stunden ohne Energienachschub stellt die vergleichsweise kleine Unruh ihre Oszillationen ein.

2014 zelebriert Piaget sein 140-jähriges Firmenjubiläum. Anlass genug, auch dem altehrwürdigen 9P retrospektive Ehre ange-deihen zu lassen. Mit dem Zurückliegenden hat das Handaufzugs-kaliber 900P, das über einen Zeitraum von drei Jahren in enger Kooperation zwischen Uhr- und Gehäusemachern entwickelt wurde, herzlich wenig zu tun. Um eine bislang unerreichte Gesamthöhe von nur 3,65 Millimetern für Werk und Gehäuse realisieren zu können, musste beides zusammenwachsen. Mit anderen Worten: Der Boden des 38-Millimeter-Goldgehäuses dient als Grundplatine. Und das darin Integrierte verlangte nach dem Ausloten aller erdenk-lichen uhrmacherischen Grenzen. 2015 brachte Piaget übrigens auch den flachsten Handaufzugschronographen auf den Markt. Samt Goldgehäuse trägt das 4,65 Millimeter flache Kaliber 883P am Handgelenk mit nur 8,24 Millimetern auf.

Zur Geschichte der Manufaktur nur noch so viel: 1964 erwarb Piaget die Mehrheit an der Uhrenmarke Baume & Mercier. Trotz oder gerade wegen guter Geschäfte trennte sich die Familie 1988 von 60 Prozent ihrer Anteile. Käufer war die Cartier Monde S.A. in Paris, heute Richemont S.A. Schließlich gingen 1993 beide Marken vollständig ins Eigentum des Richemont-Imperiums über. ○

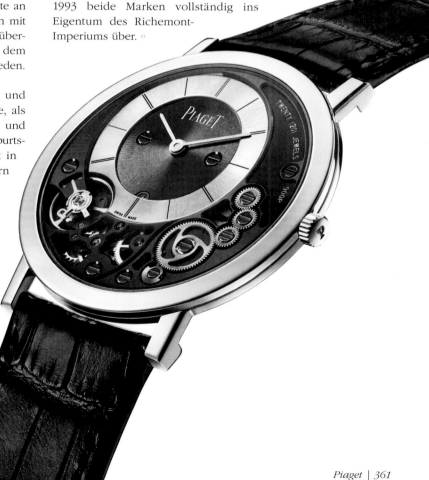

Altiplano 900P, 2013

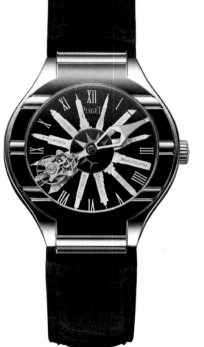

Polo Tourbillon Relatif, 2015

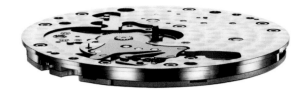

Caliber 1208P, 2009

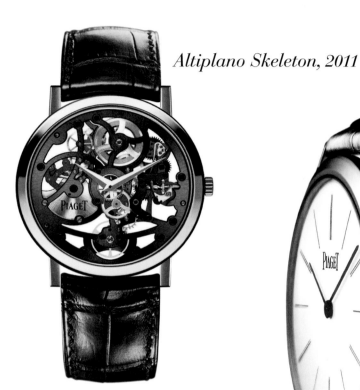

Altiplano Skeleton, 2011

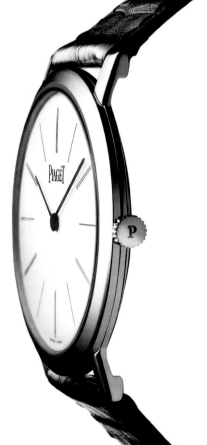

*Altiplano
Ultra-thin, 2002*

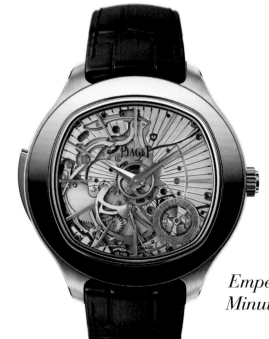

Limelight Gala, 2012

*Emperador Coussin XL
Minute Repeater, 2012*

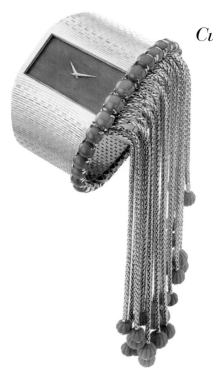

Cuff Watch, 1974

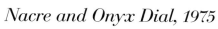

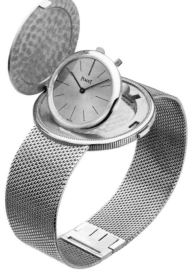

Nacre and Onyx Dial, 1975

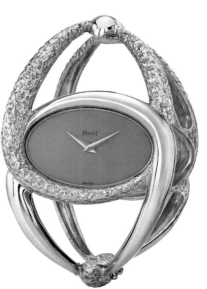

Twenty Dollars, 1979

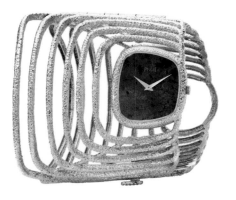

Cuff Watch, 1973

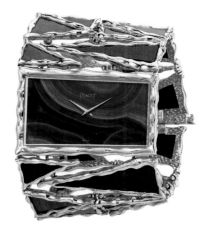

Extremely Cuff Watch, 1970

Haute Joaillerie, 1972

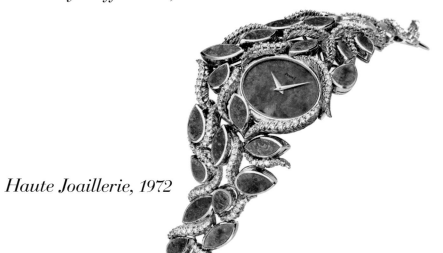

Esclave, 1971

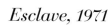

Maître mondial des montres-bracelets extra-plates

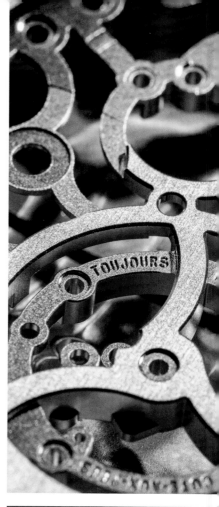

S'il est un maître mondial incontesté du minimalisme en matière de montres-bracelets, c'est bien Piaget. Georges-Édouard Piaget crée dès 1874 un atelier à La Côte-aux-Fées, un petit village à 1000 mètres d'altitude dans le Jura suisse, mais la course aux records ne débutera pas avant 1957. Jusqu'à cette date, l'entreprise familiale n'équipe ses montres-bracelets que des mouvements qu'elle achète. Puis, ladite année 1957 le mouvement 9P, le premier conçu dans l'atelier, fait d'emblée sensation. Le calibre est ainsi désigné car son diamètre d'encageage est de 9 lignes, soit 20,5 millimètres. Le remontoir manuel atteint tout juste deux millimètres d'épaisseur. Si cette valeur n'a alors certes rien d'exceptionnel, on peut toutefois la qualifier de remarquable. Grâce à ses dimensions modestes, le mouvement 9P s'intègre aussi bien dans les montres pour hommes que pour femmes.

En 1960, Piaget répond à l'intérêt croissant porté aux élégantes montres-bracelets automatiques plates par la mise au point du calibre 12P. Doté d'une masse oscillante ultraplate en or massif 24 carats à rotation bidirectionnelle, ce mouvement, avec ses 2,3 millimètres d'épaisseur, fait son entrée dans le Livre Guinness des records. Il conservera son record de minceur jusqu'à la sortie en 1978 du calibre à rotor central Lassale 2000. D'une extrême délicatesse, celui-ci ne mesure que 2,08 millimètres d'épaisseur. Légèrement plus épais, le Piaget 12PC est une variante à guichet pour la date. Après la faillite de Bouchet-Lassale SA, concepteur fabricant du plus petit mouvement à remontage manuel de tous les temps avec seulement 1,2 millimètre d'épaisseur, Piaget et Lémania, un fabricant d'ébauches dont Piaget détient des parts, acquièrent au début des années 1980 les droits sur ces deux calibres. Ni le 20P à remontage automatique plusieurs fois remanié et introduit en 1982, ni le 25P, mouvement à remontage automatique de 1983, ne resteront longtemps sur le marché : leur construction à roulements à billes microscopiques et mobiles portés par la platine se révèle défaillante.

Heureusement, Piaget dispose encore des 9P et 12P, mouvements éprouvés dont la carrière ne se termine qu'au XXIe siècle, lorsqu'ils montrent des signes visibles d'usure et que l'on réclame sur la scène horlogère des montres aux dimensions plus impressionnantes. À l'occasion de son 50e anniversaire, le 12P fait un retour sous la forme d'un mouvement totalement repensé, le 1208P. Avec 2,35 millimètres, son extrême finesse n'a encore été battue par aucun mouvement automatique. Une des astuces utilisées : certaines roues dentées font tout juste 0,12 millimètre d'épaisseur. Ce nouveau mouvement d'à peine 30 millimètres d'épaisseur dispose pourtant d'une réserve de marche largement suffisante pour un usage quotidien. C'est seulement au bout d'environ 40 heures sans apport d'énergie que le balancier, relativement petit, cesse d'osciller.

En 2014, Piaget célèbre ses 140 ans d'existence, une occasion suffisante pour rendre rétrospectivement hommage au vénérable 9P. Mais le calibre à remontage manuel 900P, mis au point au terme de trois ans d'une étroite collaboration entre fabricants de montres et de boîtiers n'a franchement pas grand-chose à voir avec son lointain ancêtre. Pour atteindre une minceur encore jamais atteinte de 3,65 millimètres pour le mouvement et le boîtier, l'un et l'autre doivent fusionner. Autrement dit : le fond du boîtier en or de 38 millimètres sert de platine. La réalisation des éléments intégrés à l'intérieur a exigé de tutoyer les confins du possible en termes d'art horloger. En 2015, Piaget présente d'ailleurs le chronographe à remontage manuel le plus plat du marché. Abritant le calibre extra-plat 883P de 4,65 millimètres d'épaisseur, le boîtier en or n'affiche que 8,24 millimètres de hauteur.

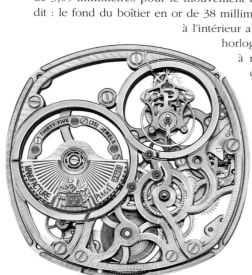

Un mot encore sur l'histoire de la manufacture : en 1964, Piaget devient actionnaire majoritaire de Baume & Mercier. Malgré ou justement à cause des affaires florissantes, la famille Piaget se défait en 1988 de 60 pour cent de ses parts. Elles sont rachetées par Cartier Monde SA Paris, aujourd'hui Richemont SA. Finalement, en 1993, les deux marques rejoignent l'empire Richemont. ○

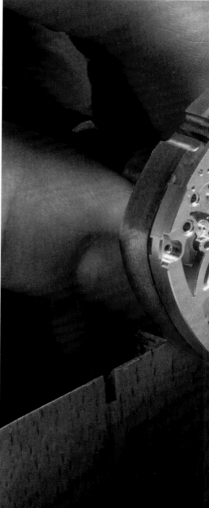

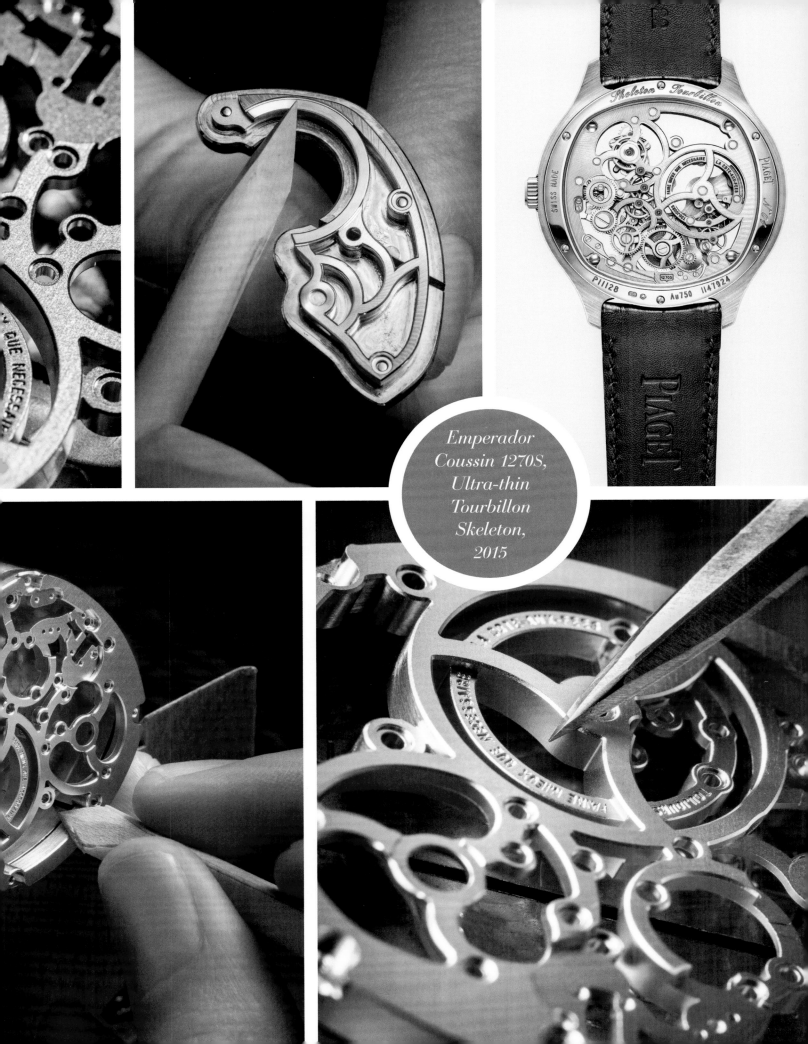

Emperador Coussin 1270S, Ultra-thin Tourbillon Skeleton, 2015

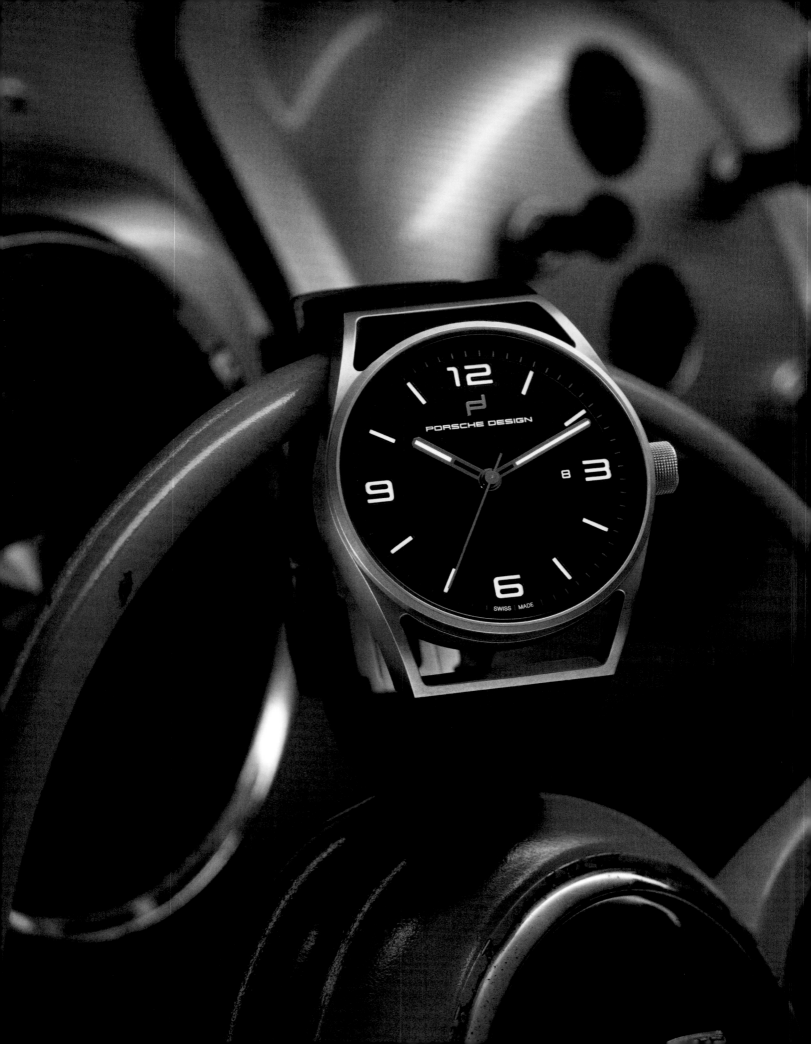

Porsche Design

Creative Power That Makes Itself Useful

English

To stand still is to take a step backward. The automobile industry is well aware of this. And the decision makers at Porsche Design are likewise keenly conscious of it too. Time never stands still when watches are involved because humankind's most precious resource has always been subject to change. Remarkable steps in a new stylistic direction were taken at Porsche Design in 2016, but the innovators never lost sight of their inheritance from the legendary Professor Ferdinand Alexander Porsche, who believed that form ought to follow function and that titanium, thanks to its excellent physical properties, is the ideal material for watch cases. These convictions are expressed in the new "1919 Collection." The date in the name was a logical choice because the famous designer had profoundly internalized the tenets of the Bauhaus, which was founded in 1919. The explicit goal of renowned Bauhaus protagonists such as Walter Gropius, Paul Klee and Mies van der Rohe was to overcome the separation between artisans and artists, a division which was often perceived as discriminating. This applied especially to the design and fabrication of objects for use in daily life. The innovative fusion led to the creation of chairs, tables, lamps, vases and many other items that "spoke" an autonomous language of forms and colors. These artifacts were functionally designed and reduced to the bare essentials, exactly as "F. A." did when he created his first wristwatch and as he continued to do on subsequent "time projects" in ensuing years. The circular shape, which he cultivated throughout his life, is archetypically appropriate for watches because the circle represents a departure from the origin and a return to it. Based on the doctrines articulated by Paul Klee and Vasily Kandinsky, color took its rightful place alongside form. Here too, Prof. F. A. Porsche had very clear notions. His ideas revolved around a puristic, instrumental look derived from the cockpit of the classic Porsche 911 automobile, which he had created in 1963.

It's no wonder that the newcomer which debuted in the "1919 Collection" at the watch fair in Basel in 2016 pays homage to the philosophy of a puristic appearance that can justifiably be described as minimalistic. The French pilot and author Antoine de Saint-Exupéry believed that "perfection doesn't happen when nothing more can be added, but when nothing more can be taken away." This is the viewpoint from which one can best admire, for example, the "1919 Datetimer Eternity." Notwithstanding all its sleek simplicity and modernity, the design of this wristwatch also reveals a kinship with the Porsche 356 and the stylistic elegance of the 1950s. Authentic and high-quality materials, including browned titanium for the lightweight case, underscore the brand's traditions and values. Self-winding Caliber Sellita SW200 is entrusted with the tasks of keeping the time and knowing the date. Like Prof. F. A. Porsche's first creation, here too black is the preferred color for the dial of the "1919 Datetimer Eternity Black Edition," which encases the abovementioned caliber. Aficionados of chronographs with understated styling will appreciate the "1919 Chronotimer." This synthesis of architectural design, sportiness and horological functionality is available in pure titanium or in PVD-coated titanium and with either a titanium bracelet or a rubber strap. A pane of sapphire crystal in the back of the case offers a clear view of time-honored self-winding Caliber Sellita SW500, a clone of the venerable Valjoux 7750, with which the history of wristwatches began at Porsche Design in 1972. This was the year when the shareholders of the Porsche AG decided to entrust the handling of the business to external managers. "This decision," Prof. F. A. Porsche said, "was congruent with my personal desire for stylistic freedom." Two years after the founding of Porsche Design, he accordingly relocated from Stuttgart-Zuffenhausen to picturesque Zell am See in Austria, where the Porsche Family has lived for many generations.

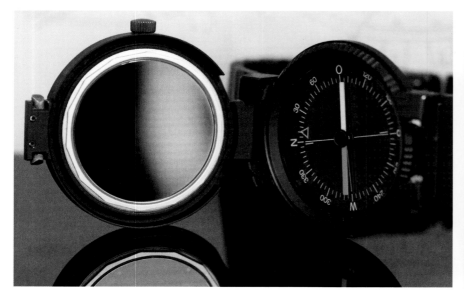
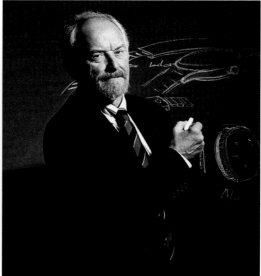

Left: *Compass watch, 1978* ◦ *Links: Kompassuhr, 1978* ◦ *A gauche : Montre-boussole, 1978* ◦
Right: *Professor Ferdinand Alexander Porsche*

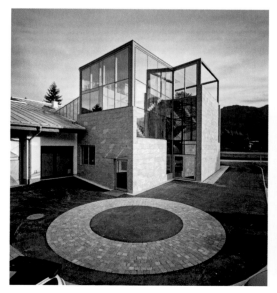

Cal. Sellita SW200

For his first chronographic opus, he asked himself how a chronograph could be made differently. Considering the Quartz Revolution, a new approach was urgently needed. "I imagined a wristwatch to match the automobile. The face would be black, like the speedometer and tachometer aboard the 911, because a black dial doesn't cause blinding glare when a driver glances at it." Prof. F. A. Porsche's certainty that black isn't a color, but a state of being, detracted not one iota from the success of this ticking timepiece with a tachymeter scale for convenient calibration of average speeds. More than 50,000 of these watches quickly found their way to the wrists of design-conscious sportcar fans. Clay Regazzoni and other Formula One drivers likewise appreciated this ticking star in the design firmament, which celebrated the product designer's profound contempt for frills and gags. "A formally consistent product needs neither embellishment nor exaggeration; it should be elevated by pure form alone. The form should live through minimalism, should present itself understandably, and should not distract from the product and its function." This same mode of thought was expressed in subsequent milestones that positively revolutionized the watch market. These included the Compass Watch, which was conceived in 1976 and ultimately realized in cooperation with IWC of Schaffhausen in 1978, as well as the titanium chronograph, which debuted in 1980 with large, planar and accordingly ergonomically shaped push-pieces. This chronograph wristwatch inaugurated a new era in watch history because it impressively demonstrated what's possible when the right partners find one another. The talented watch designer become a watch industrialist when he acquired the traditional Swiss Eterna brand in 1995. This era, which came to its end in 2013, is also recalled by a chronographic monument known as the "Indicator." A quick glance at its dial suffices to show what it has timed. It can measure intervals up to nine hours and 59 minutes in length, which it displays in gigantic numerals. Three engines (i.e., three barrels) provide the power. Three "cruise controls" (i.e., centrifugal governors) coordinate the speed with which the mainsprings release their tension. And the "gas gauge" is a power-reserve display controlled by an elaborate differential gear-train. It's not surprising that this automatic movement requires more than 800 components.

A new era began with the founding of Porsche Design Timepieces AG in Solothurn, Switzerland in 2014. This subsidiary of the Porsche Design Group is responsible for the technical development and production of understatedly designed yet nonetheless distinctive wristwatches with an extremely high recognition value. Along with this new beginning, steel as a material for cases was summarily abandoned. The present and the future belong to titanium—either purist, coated, or otherwise processed. One pillar of the product portfolio is the "Chronotimer Collection," which pays homage to the stylistic legacy of the brand's founder, which is continually updated or further refined at Studio F. A. Porsche in Zell am See. The second pillar, which closes the circle, is written in large numerals: "1919 Collection."

Without exception, all wristwatches in this collection result from German, Austrian and Swiss activities. Planned in Germany, designed in Austria and "made in Switzerland," they unite the best from three neighboring nations. ○

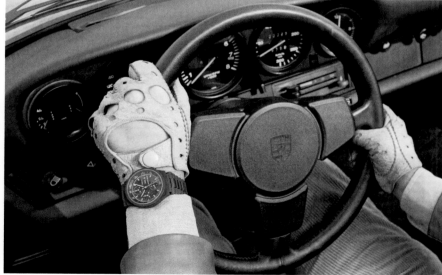

From left: Studio F. A. Porsche ○ Chronograph I (1972) in a 911 Turbo 3.0, 1975 ○ Von links: Studio F. A. Porsche ○ Chronograph I (1972) in einem 911 Turbo 3.0, 1975 ○ De gauche à droite : Studio F. A. Porsche ○ Chronograph I (1972) dans une 911 Turbo 3.0, 1975

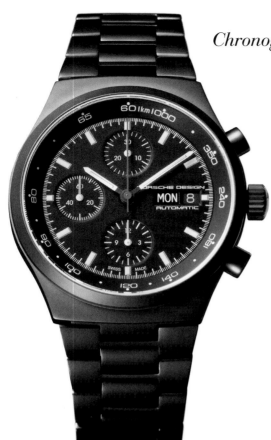

Chronograph I, 1972

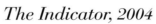

The Indicator, 2004

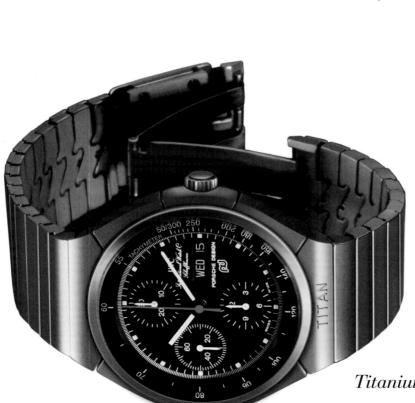

Titanium Chronograph, 1980

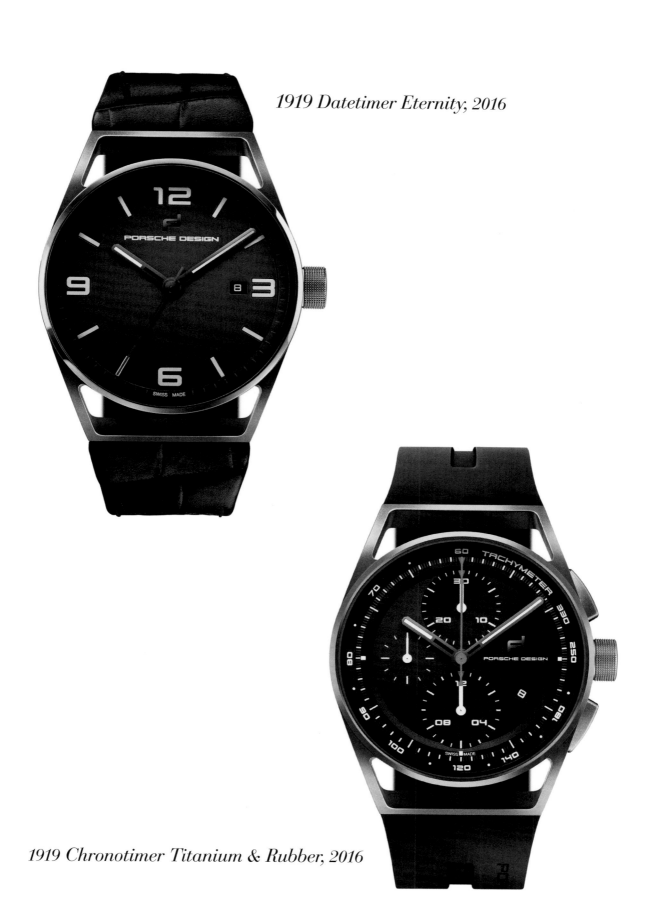

1919 Datetimer Eternity, 2016

1919 Chronotimer Titanium & Rubber, 2016

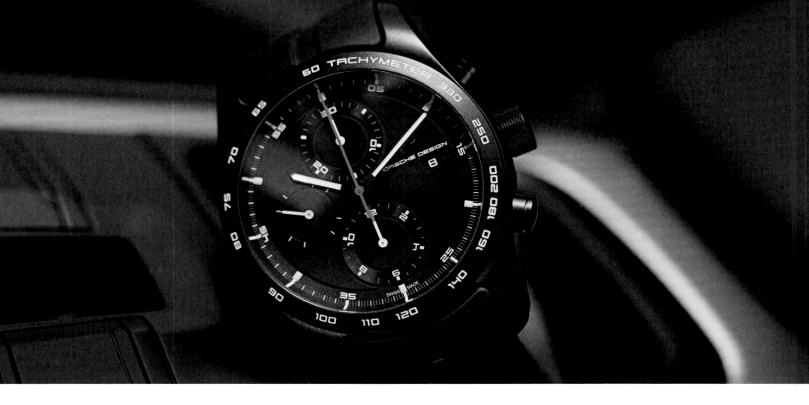

Gestaltungskraft, die sich nützlich macht

Stillstand bedeutet Rückschritt. Das weiß die Autoindustrie nur zu gut. Und die Verantwortlichen von Porsche Design sind sich dieser Tatsache ebenfalls bewusst. Wenn es um das Thema Uhren geht, bleibt die Zeit bekanntlich niemals stehen, denn zum kostbarsten Gut der Menschheit gehört seit eh und je der Wandel. Nicht zuletzt deshalb brachte das Jahr 2016 bei Porsche Design bemerkenswerte Schritte in eine neue gestalterische Richtung, ohne dass dabei das Vermächtnis des legendären Professors Ferdinand Alexander Porsche aus den Augen verloren wurde. Dessen nach wie vor gültiges Credo, dass die Form der Funktion zu folgen hat und Titan wegen seiner vorzüglichen Materialeigenschaften der ideale Werkstoff für Uhrengehäuse ist, beseelt die neue Linie „1919 Collection". Die Jahreszahl kommt nicht von ungefähr: 1919 wurde das Bauhaus gegründet, dessen Schule der Designpapst zutiefst verinnerlicht hatte. Das erklärte Ziel bekannter Bauhaus-Protagonisten wie Walter Gropius, Paul Klee und Mies van der Rohe bestand in der Überwindung der immer wieder als diskriminierend empfundenen Trennung von Handwerkern und Künstlern. Dies galt insbesondere beim Entwurf und bei der Fertigung von Gegenständen des täglichen Lebens. So entstanden in einem neuartigen Miteinander Stühle, Tische, Lampen, Vasen und vieles mehr mit eigenständiger Formen- und Farbensprache. Alles funktional gestaltet und auf das Wesentliche reduziert. Exakt so, wie es „Eff-A" bei der Kreation seiner ersten Armbanduhr und in den folgenden Jahren bei vielen anderen Zeit-Projekten tat. Seine lebenslang gepflegte Form, das Rund, passt idealtypisch zu Uhren, denn sie repräsentiert das Ausgehen von sowie die Rückkehr zu den Ursprüngen. Basierend auf den Lehren von Paul Klee und Wassily Kandinsky gesellte sich zur Form die Farbe. Auch hier hatte Prof. F. A. Porsche klare Vorstellungen. Seine Ideen kreisten um einen puristischen Instrumentenlook, abgeleitet vom Cockpit des automobilen Klassikers Porsche 911, welchen er 1963 geschaffen hatte.

Kein Wunder also, dass die während der Basler Uhrenmesse 2016 vorgestellten Newcomer in der Linie „1919 Collection" der Philosophie eines puristischen, man könnte fast sagen minimalistischen Auftritts huldigen. „Vollkommenheit entsteht nicht dann, wenn sich nichts mehr hinzufügen lässt", pflegte der französische Pilot und Schriftsteller Antoine de Saint-Exupéry zu sagen, „sondern dann, wenn man nichts mehr wegnehmen kann." Aus diesem Blickwinkel muss man beispielsweise den „1919 Datetimer Eternity" betrachten. Trotz aller Schlichtheit und Modernität lässt das Design dieser Armbanduhr auch eine gewisse Nähe zum Porsche 356 und der stilistischen Eleganz der 1950er Jahre erkennen. Authentische und vor allem hochwertige Materialien – dazu gehört gebräuntes Titan für das leichte Gehäuse – unterstreichen die Traditionen und Werte des Unternehmens. Für die Bewahrung der Uhrzeit und des Datums ist ein selbstverständlich individualisiertes Automatikwerk vom Kaliber Sellita SW200 zuständig. Schwarz wie das Erstlingswerk von Prof. F. A. Porsche präsentiert sich das Zifferblatt der „1919 Datetimer Eternity Black Edition" mit dem gleichen Uhrwerk. Liebhaber zurückhaltend gestalteter Zeitschreiber kommen beim „1919 Chronotimer" zu ihrem Recht. Diese Synthese aus architektonischem Design, Sportlichkeit und uhrmacherischer Funktionalität gibt es in purem oder PVD-beschichtetem Titan mit Titan- oder Kautschukband. Hinter dem Saphirglas-Sichtboden tickt das bewährte Automatikkaliber Sellita SW500, ein Klon des altbewährten Valjoux 7750, mit dem die Geschichte der Armbanduhren von Porsche Design 1972 begann. Damals hatten sich die Gesellschafter der Porsche AG dafür entschieden, die Geschäftsführung externen Managern anzuvertrauen. „Diese Entscheidung", so Prof. F. A. Porsche, „deckte sich mit meinem persönlichen Verlangen nach gestalterischer Freiheit." Konsequenterweise zog er bereits zwei Jahre nach der Gründung von Porsche Design in Stuttgart-Zuffenhausen ins malerische Zell am

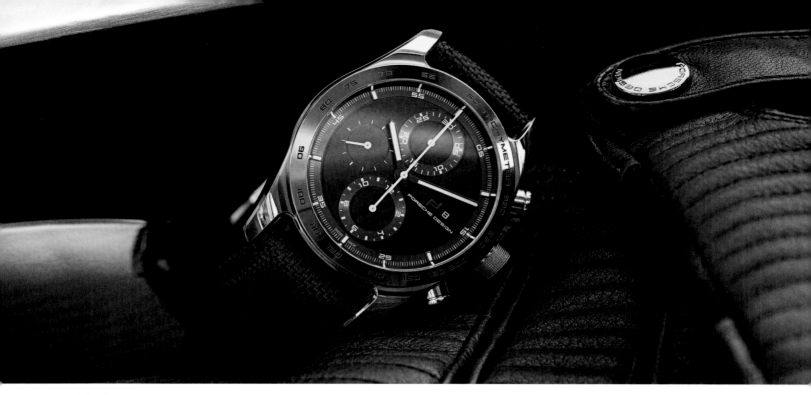

From left: Chronotimer Series 1 Sportive Black, 2015 ○ Chronotimer Series 1 Deep Blue, 2015

See in Österreich, wo die Familie seit Generationen domizilierte. Bei seinem chronographischen Erstlingswerk hatte er sich die simple Frage gestellt, wie man ein solches schlichtweg anders machen könne. Im Angesicht der Quarz-Revolution war dies auch bitter nötig. „Mir schwebte eine Uhr zum Auto vor. Schwarz wie die Tachometer und Drehzahlmesser des 911er, weil das beim Ablesen nicht blendet." Seine innere Überzeugung, dass Schwarz keine Farbe, sondern ein Zustand ist, tat dem Erfolg des tickenden Boliden mit Tachymeterskala zum unkomplizierten Erfassen von Durchschnittsgeschwindigkeiten über einen Kilometer hinweg nicht den geringsten Abbruch. Erstaunlich schnell fanden mehr als 50 000 Exemplare an die Handgelenke designbewusster Sportwagenfans. Auch Formel-1-Piloten wie Clay Regazzoni schätzten den Design-Star am Chronographenhimmel, welcher die tiefe Abneigung des Produktgestalters gegen Schnörkel und Gags zelebrierte. „Ein formal stimmiges Produkt braucht keine Verzierung, keine Erhöhung, es soll durch die reine Form erhöht werden. Die Form sollte durch das Minimum leben, sich verständlich präsentieren, nicht ablenken vom Produkt und dessen Funktion." Dieser Denkweise folgten auch die weiteren Meilensteine, welche den Uhrenmarkt förmlich revolutionierten. Dazu gehörte die bereits 1976 angedachte und in Kooperation mit der Schaffhauser IWC 1978 endlich realisierte Kompassuhr. Oder der 1980 vorgestellte Titan-Chronograph mit großflächigen und deshalb besonders ergonomischen Bedientasten. Dieser Stopper läutete eine neue Ära in der Uhrengeschichte ein, denn er demonstrierte auf eindrucksvolle Weise, was möglich ist, wenn die richtigen Partner zusammenfinden. 1995 schlüpfte der begnadete Uhrendesigner durch den Erwerb der Schweizer Traditionsmarke Eterna ins Gewand eines Uhr-Unternehmers. An diese Ära, welche 2013 zu Ende ging, erinnert unter anderem ein chronographisches Monument namens „Indicator". Hier weiß Mann mit einem Blick, was er gestoppt hat.

Die Messwerte bis zu neun Stunden und 59 Minuten stechen durch die riesigen Ziffern ins Auge. Drei Motoren, sprich Federhäuser, liefern die Kraft, drei „Tempomaten" (Fliehkraftregler) steuern die Ablaufgeschwindigkeit. Als Tankuhr dient die von einem aufwendigen Differenzialgetriebe angesteuerte Gangreserveanzeige. Kein Wunder, dass jedes Automatikwerk nach mehr als 800 Komponenten verlangt.

Die Neuzeit startete 2014 mit Gründung der Porsche Design Timepieces AG im eidgenössischen Solothurn. Diese Tochter der Porsche Design Group zeichnet verantwortlich für die technische Entwicklung und Produktion der puristisch gestalteten und trotzdem markanten Armbanduhren mit extrem hohem Wiedererkennungswert. Mit diesem Neubeginn hat Stahl als Gehäusematerial definitiv ausgedient. Die Gegenwart und Zukunft gehören Titan. Und zwar entweder puristisch, beschichtet oder wie auch immer verarbeitet. Eine Säule des Produktspektrums, „Chronotimer Collection" genannt, huldigt dem gestalterischen Nachlass der Gründerpersönlichkeit, aktualisiert oder verfeinert bei Studio F. A. Porsche in Zell am See. Auf der zweiten Säule, und damit schließt sich der Kreis, steht in großen Ziffern „1919 Collection".

Ausnahmslos alle Armbanduhren dieser Kollektion sind ebenfalls das Ergebnis deutscher, österreichischer und eidgenössischer Aktivitäten. Geplant in Deutschland, designt in Österreich und „made in Switzerland", vereinen sie so das Beste aus drei benachbarten Ländern. ○

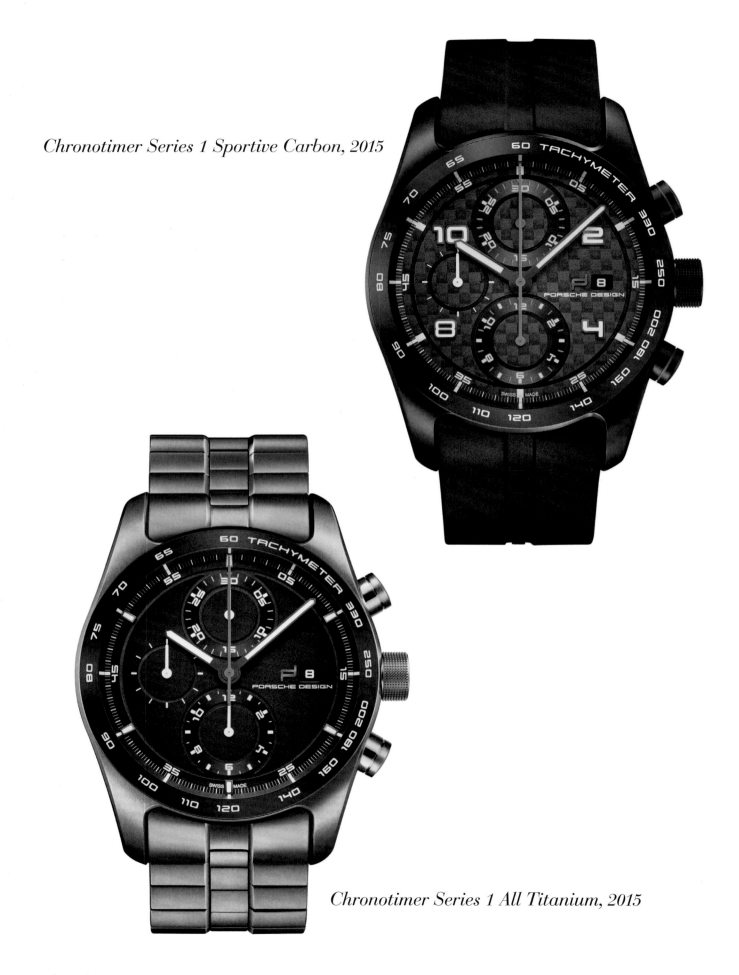

Chronotimer Series 1 Sportive Carbon, 2015

Chronotimer Series 1 All Titanium, 2015

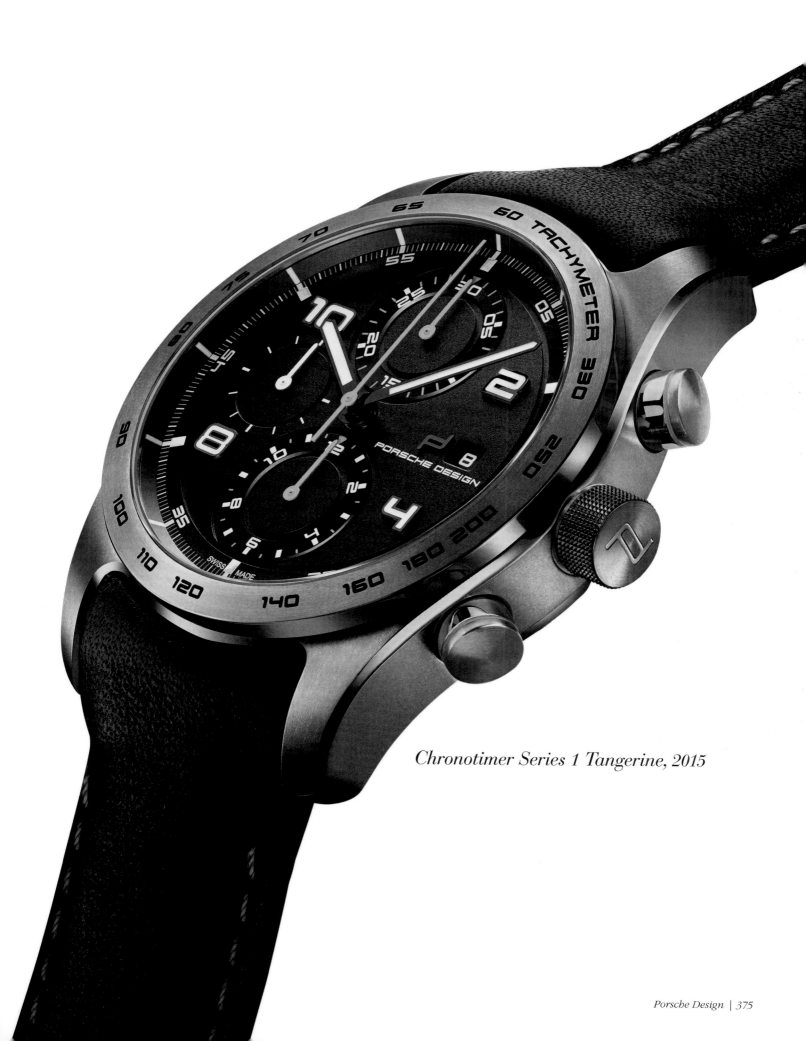

Chronotimer Series 1 Tangerine, 2015

Une capacité de conception axée sur l'utilité

Français

Stagner équivaut à régresser. Cette vérité, que l'industrie automobile connaît trop bien, n'est pas non plus inconnue des responsables de Porsche Design. Lorsqu'il s'agit de montres, on le sait, le temps ne s'arrête jamais. L'évolution est en effet depuis toujours le bien le plus précieux de l'humanité. C'est en grande partie pour cette raison que Porsche Design s'est engagé en 2016 dans une nouvelle voie créatrice, sans toutefois perdre de vue l'héritage du légendaire professeur Ferdinand Alexander Porsche. La fidélité à la maxime intemporelle selon laquelle la forme suit la fonction, et la conviction que le titane est le matériau idéal pour les boîtiers de montres, inspirent la nouvelle ligne « 1919 Collection ». Le millésime 1919 n'est pas le fruit du hasard : il correspond à l'année de la fondation du Bauhaus, mouvement dont les principes ont fortement marqué ce maître du design. L'objectif déclaré des grandes figures du Bauhaus, dont Walter Gropius, Paul Klee et Mies van der Rohe, est de surmonter la distinction partout ressentie comme discriminatoire entre artisans et artistes. Cela s'applique en particulier à la conception et à la fabrication d'objets de la vie courante. Naissent alors, grâce à une coopération d'un nouveau type, des chaises, des tables, des lampes, des vases et bien d'autres créations avec un langage propre en termes de formes et de couleurs. Toutes ces créations sont fonctionnelles et réduites à l'essentiel. C'est précisément l'esprit dans lequel « Eff-A » crée sa première montre-bracelet et réalise de nombreux projets horlogers au cours des années suivantes. La forme circulaire, qu'il affectionne depuis toujours, est idéale pour les montres, car elle représente aussi bien le départ des origines que le retour à ces dernières. S'appuyant sur les enseignements de Paul Klee et de Vassily Kandinsky, il associe couleur et forme. Une fois encore, le professeur F. A. Porsche sait précisément ce qu'il veut. Il imagine un look épuré, inspiré du tableau de bord du grand classique automobile, la Porsche 911, qu'il a conçue en 1963.

*Chronotimer Series 1
Matte Black, 2015*

Rien d'étonnant donc à ce que les nouveaux modèles de la ligne « 1919 Collection » présentés lors du Salon mondial de l'horlogerie à Bâle (Baselworld) 2016 s'inscrivent dans cette philosophie des lignes épurées, voire du minimalisme. « Il semble que la perfection soit atteinte, non quand il n'y a plus rien à ajouter, mais quand il n'y a plus rien à retrancher », aimait à dire l'aviateur et écrivain Antoine de Saint-Exupéry. C'est par ce prisme que l'on doit par exemple admirer la montre-bracelet « 1919 Datetimer Eternity ». Malgré toute sa simplicité et sa modernité, elle est proche par son design de la Porsche 356 et de l'élégance des années 1950. Des matériaux authentiques et surtout de grande qualité – parmi lesquels le titane bruni du boîtier léger – soulignent les traditions et les valeurs de l'entreprise. Pour conserver la date et l'heure, on choisit bien sûr un mouvement automatique personnalisé, le calibre Sellita SW200. Comme sur la première création horlogère du professeur F. A. Porsche, le cadran de la « 1919 Datetimer Eternity Black Edition » équipée du même mouvement est noir. Le « 1919 Chronotimer » séduit quant à lui les amateurs de chronographes au design épuré. Modèle conciliant design architectural, caractère sportif et performances horlogères, il est proposé en titane non traité ou noirci, avec un bracelet titane ou caoutchouc. Derrière le fond saphir transparent oscille le calibre automatique éprouvé Sellita SW500, clone d'un mouvement des premières montres-bracelets Porsche Design en 1972, le célèbre Valjoux 7750. À cette époque, les associés de Porsche AG ont confié la direction de la société à des responsables externes. « Cette décision, indique le professeur F.A., correspond à ma volonté de pouvoir créer en toute liberté. » Dans le droit fil de cette décision, deux ans seulement après la création de Porsche Design à Stuttgart-Zuffenhausen, il part s'installer dans la pittoresque ville autrichienne de Zell am See, où sa famille est établie depuis des générations.

En créant sa première montre, le professeur F. A. Porsche se pose une question simple, à savoir comment réaliser quelque chose de carrément nouveau. Face à la révolution du quartz, c'est d'ailleurs plus que nécessaire. « Je rêvais d'une montre qui aille avec une auto, noire comme le tachymètre et le compte-tours de la 911, que l'on peut consulter sans être ébloui. » Son intime conviction que le noir n'est pas une couleur mais un état ne porte pas le moindre préjudice au succès de ce garde-temps d'exception dont l'échelle tachymétrique permet d'obtenir aisément la vitesse moyenne au kilomètre. Avec une étonnante rapidité, plus de 50 000 exemplaires rejoindront les poignets d'amateurs de sport automobile épris de design. Des pilotes de Formule-1 comme Clay Regazzoni apprécieront eux aussi cet astre du design au firmament des chronographes, qui célèbre le profond rejet par son créateur des fioritures et gadgets. « Un produit cohérent dans sa forme n'a pas besoin d'être orné ou rehaussé, il doit se distinguer par sa seule forme. Celle-ci doit exister à minima, se présenter de manière compréhensible, ne pas détourner l'attention du produit et de sa fonction. » Cette façon de penser sera suivie d'autres jalons, qui révolutionneront littéralement le marché des montres. C'est le cas de la montre-boussole, imaginée dès 1976 et finalement réalisée en 1978 en collaboration avec IWC Schaffhausen. C'est aussi, en 1980, le cas du chronographe en titane, que ses larges boutons poussoirs rendent particulièrement ergonomique. Ce dernier annonce une nouvelle ère dans l'histoire des montres, car il démontre de manière impressionnante que tout est affaire de bon partenariat. En 1995, ce concepteur au talent exceptionnel endosse le costume d'entrepreneur en horlogerie en se portant acquéreur de la marque suisse de tradition Eterna. Cette période, qui s'achève en 2013, est entre autres marquée par un monument parmi les chronographes, le modèle « Indicator ». Avec celui-ci, on voit d'un coup d'œil le résultat du chronométrage. Les valeurs de mesure jusqu'à neuf heures et 59 minutes sautent aux yeux avec leurs chiffres énormes. Trois moteurs, autrement dit barillets, fournissent l'énergie, et trois « tempomat » (régulateurs) pilotent la vitesse de défilement. Le rôle de la jauge de carburant est joué par une indication de réserve de marche commandée par un système différentiel complexe. Rien d'étonnant à ce que chaque mouvement automatique nécessite plus de 800 composants.

Une nouvelle ère encore débute en 2014 avec la création de Porsche Design Timepieces AG à Soleure, en Suisse. Cette filiale de Porsche Design est chargée de développer et fabriquer des montres-bracelets épurées et malgré tout marquantes, reconnaissables entre toutes. Avec ce nouveau départ, l'acier disparaît définitivement des boîtiers. Le présent et l'avenir appartiennent au titane, pur, plaqué ou traité d'une toute autre manière. Un des piliers de la gamme, le « Chronotimer Collection », rend hommage à l'héritage conceptuel du fondateur, mis au goût du jour ou perfectionné à Zell am See par Studio F. A. Porsche. Bouclant la boucle, le second pilier arbore en gros caractères « 1919 Collection ».

Toutes les montres-bracelets de la collection sans exception sont le produit du travail d'intervenants allemands, autrichiens et suisses. Programmées en Allemagne, conçues en Autriche et « made in Switzerland », elles réunissent le meilleur de ces trois pays voisins. ○

Chronotimer Series 1 Matte Black, 2015

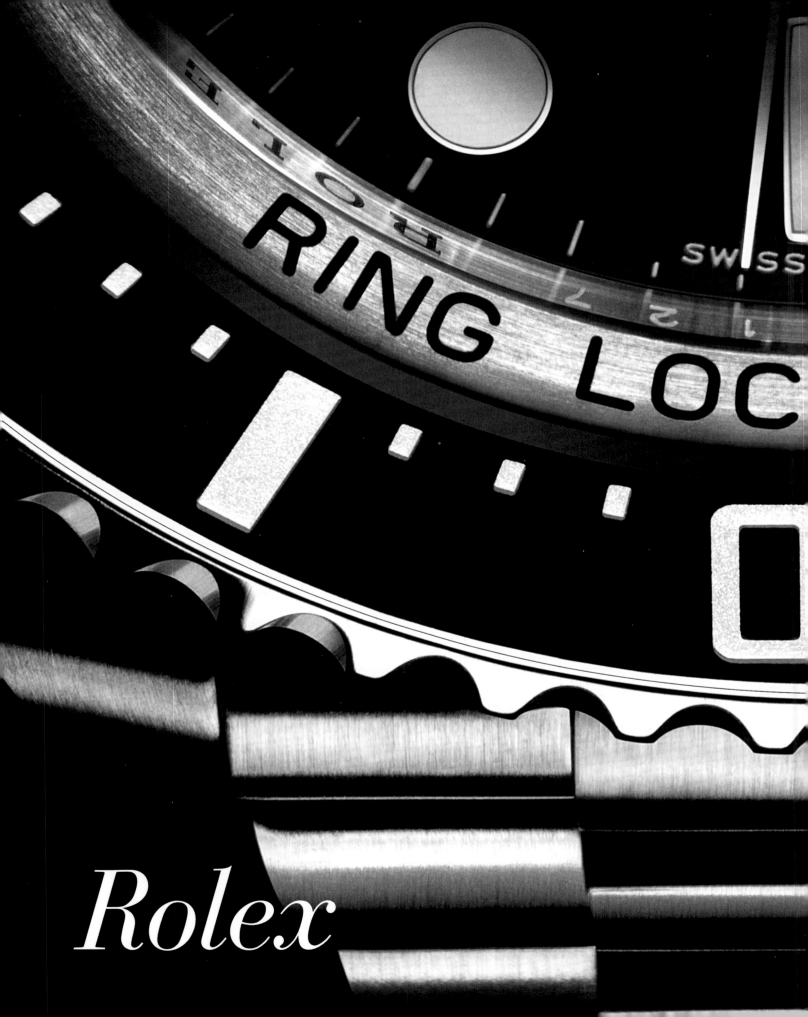

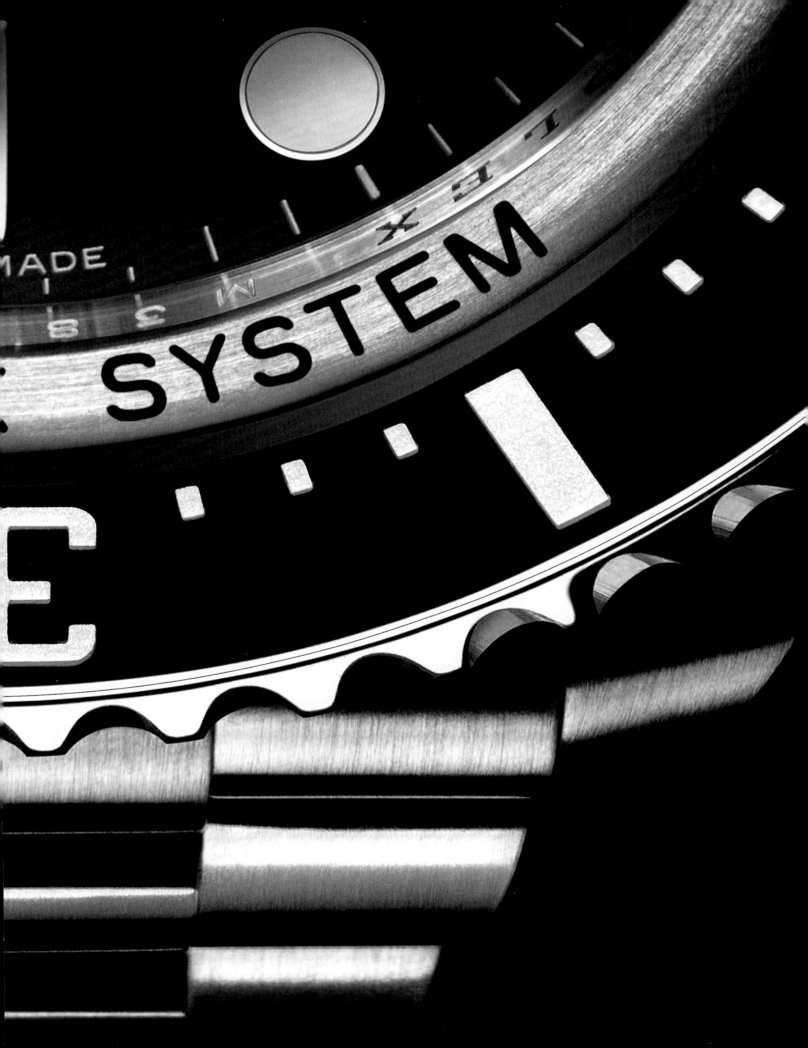

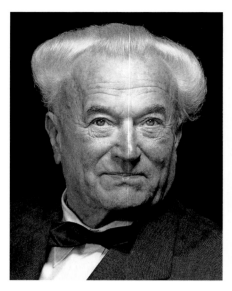
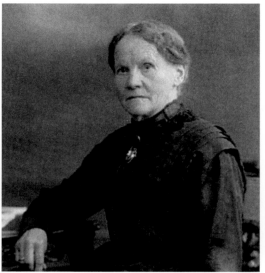
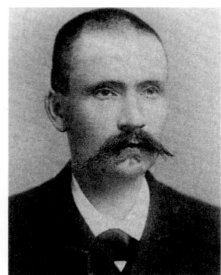

From left: Hans Wilsdorf ○ Anna Maria Aegler ○ Jean Aegler

With the "Oyster" to Global Renown

English

It's surprising, but true: Rolex, an archetypically Swiss watch brand, began with a man who initially had nothing at all to do with the Alpine confederation. Rolex traces its ancestry to Hans Wilsdorf, who was born at Kulmbach in northern Bavaria. He left his homeland at age 19 and went to Switzerland, where he first became familiar with watch exporting. He packed his bags again in 1903 and settled in London, where he worked for a watch importer.

Peeved because dealers neglected the quality of the watches they sold, he joined forces with Alfred James Davis—who later turned out to be an unscrupulous scoundrel—and founded the watch wholesaler Wilsdorf & Davis Ltd. Competition was fierce, so the company needed attractive and unconventional products. The new genre of wristwatches was exactly what Hans Wilsdorf had sought. To assure a steady supply, he activated his existing contacts in Switzerland. He was particularly impressed by Aegler SA, which had been founded in 1878. The decision-makers in Bienne were enthusiastic about his offer of exporting accurate wristwatches to England. Success soon followed and the imported timepieces sold very well.

To simplify its expanding business relationships, the young English company set up an office in Switzerland in 1907. Hans Wilsdorf, who had firmly established himself in England and become a British citizen, importantly contributed to the acceptance of the wristwatch as a genre of timepiece. Now he needed a catchy name, which he soon found in "Rolex," which is readily pronounceable by speakers of all major languages. According to his widow, the neologism was born by compressing the phrase "rolling export," which Wilsdorf's Swiss partner Aegler steadily practiced because of the booming business. The upheavals caused by World War I prompted Wilsdorf to relocate his company's headquarters first to Bienne and later to Geneva, where Montres Rolex SA was entered in the commercial register on January 17, 1920.

Wilsdorf and Rolex decisively influenced four aspects of the modern wristwatch. First, the history of the officially certified wristwatch chronometer: as early as 1914, London's Kew Observatory awarded a coveted "A-grade" chronometer certificate to one of Hans Wilsdorf's wristwatches. Today the COSC, i.e., Switzerland's official chronometer-tested authority, issues the majority of its rate certificates to this brand's watches. Over ten million Rolex movements have earned accreditation as official chronometers.

Second, Rolex deserves credit for developing the first truly watertight wristwatch. Wilsdorf was granted a patent for a multiply screwed case in 1926. When the English stenographer Mercedes Gleitze swam across the English Channel in 15 hours and 15 minutes on October 7, 1927, the new "Oyster" was strapped to her wrist. Needless to say, not a droplet of water penetrated its case. Rolex has continually optimized this ingenious system over the years, but has never made any decisive changes in it.

Rolex laid the third milestone with rotor winding for automatic wristwatches. Hans Wilsdorf unveiled the "Perpetual" in 1932: developed by Aegler, it was world's first wristwatch that used an unlimitedly turning oscillating weight to automatically wind its mainspring. Rolex evolved this revolutionary self-winding system in subsequent decades. The manufactory always gave greater priority to reliability, longevity, and precision than to minimal dimensions. Rolex is currently regarded as Switzerland's largest manufacturer of mechanical movements. All of Rolex's watches are equipped with the brand's own movements.

Finally, Hans Wilsdorf and Rolex celebrated the brand's 40th anniversary in 1945 with the development of the "Datejust," a watertight wristwatch with a self-winding movement and a date window. Dials with date windows had existed prior to this, but the combination of a watertight case, automatic winding via a rotor, and a date

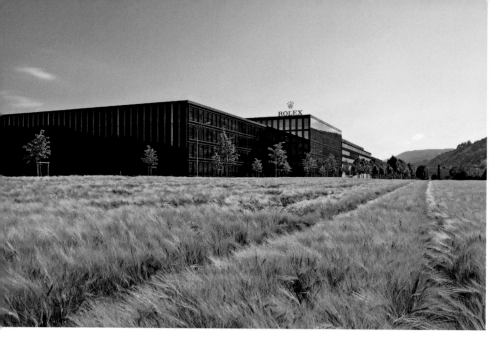

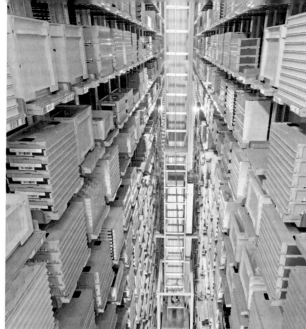

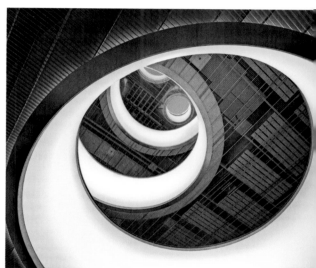

Top left: the new factory building in Bienne ◦ Right: high-shelf storage ◦ Bottom: the new factory in Plan-les-Ouates ◦ Links oben: Das neue Fabrikgebäude in Biel ◦ Rechts: Hochraumlager ◦ Darunter: Die neue Fabrik in Plan-les-Ouates ◦ En haut à gauche : Le nouveau bâtiment de la manufacture à Bienne ◦ À droite : Entrepôt à hauts rayonnages ◦ En dessous : La nouvelle manufacture à Plan-les-Ouates

window was a world premiere. The "Datejust" was named the "Wristwatch of the 20th Century" in the USA. A helpful addition was the magnifying lens above the date. Wilsdorf's widow, who had poor eyesight, explained that a fortuitous coincidence inspired her husband to invent this characteristic magnifier: in the bathroom one morning, a droplet of water chanced to land on his watch's crystal precisely above the date window. The magnifying effect prompted Wilsdorf to exclaim "Eureka!"

Other milestones of the firm's childless founder, who left his large estate to the Hans Wilsdorf Foundation, bear the names "Explorer," "GMT-Master," "Submariner," and "Day-Date." Each is unique in its own way.

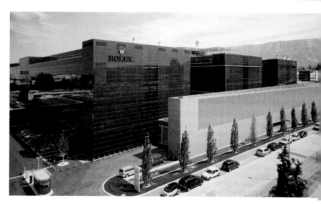

Switzerland offered Swiss citizenship on the proverbial silver platter to the cosmopolitan Bavarian and naturalized Englishman, but Wilsdorf steadfastly rejected the offer until his death on July 6, 1960. Two legally and financially separate Rolex businesses existed until 2004: nothing but watch movements were made in Bienne by the Aegler family and by the Manufacture des Montres Rolex SA, which is run by the family's descendants. Montres Rolex SA, which Hans Wilsdorf had founded in 1920, was responsible for manufacturing finished watches. Protracted negotiations ultimately led to unification under the roof of the Genevan part of the business in 2004.

The purchase cost millions for the Rolex watch manufactory, which has always been obsessed with perfection and precision, and which equips all of its calibers with tiny "Parachrom" balance-springs that it fabricates itself. Rolex later invested up to 500 million francs to bring movement production on the outskirts of Bienne up to the top international standard. The last section of construction work—for the time being—was completed on October 16, 2012. Alongside the impressive complex in Geneva, the new buildings at Bözingenfeld in Bienne house Rolex's administration, research, and development departments. The architects prioritized functionality, energy efficiency, and the well-being of the circa 2,500 employees. ◦

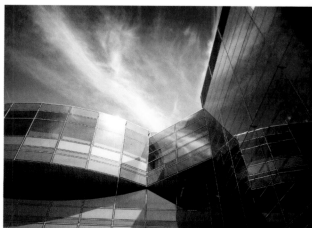

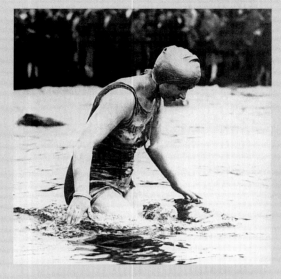

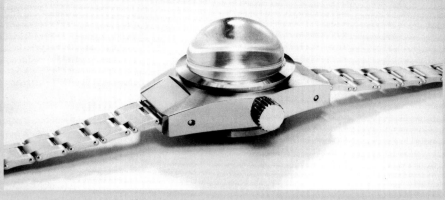

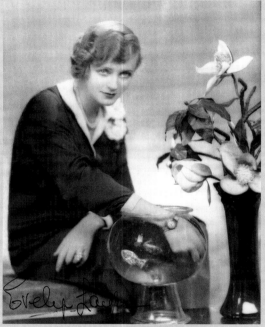

First Oyster, 1926

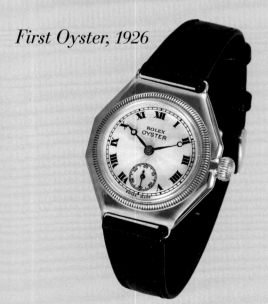

Clockwise from top left: Mercedes Gleitze at the start of her swim across the English Channel, 1927 ○ Deepsea, 1962 ○ Famous advertisement in the Daily Mail, 1927 ○ Advertising motif with the actress Evelyn Laye, 1927 ○ Von oben links im Uhrzeigersinn: Mercedes Gleitze beim Start zur Kanaldurchquerung, 1927 ○ Deepsea, 1962 ○ Berühmte Anzeige in der Daily Mail, 1927 ○ Werbemotiv mit der Schauspielerin Evelyn Laye, 1927 ○ Dans le sens horaire depuis la gauche : Mercedes Gleitze au départ de sa traversée de la Manche, en 1927 ○ Deepsea, 1962 ○ Publicité célèbre dans le Daily Mail, 1927 ○ Motif de campagne publicitaire avec l'actrice Evelyn Laye, 1927

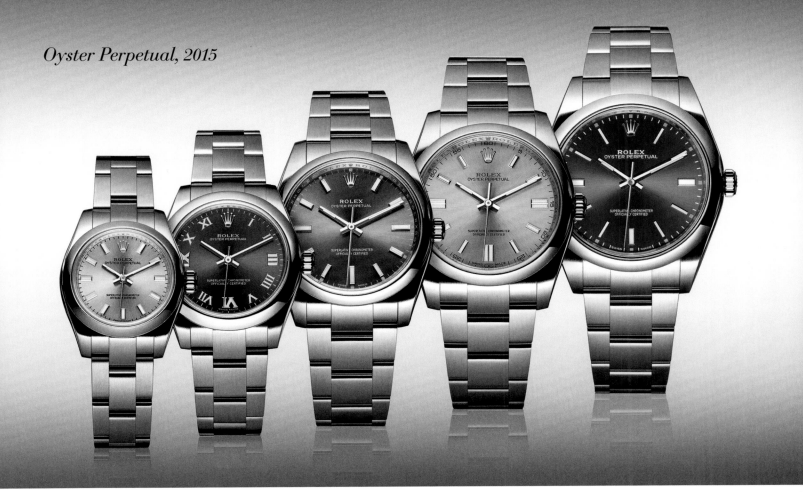

Oyster Perpetual, 2015

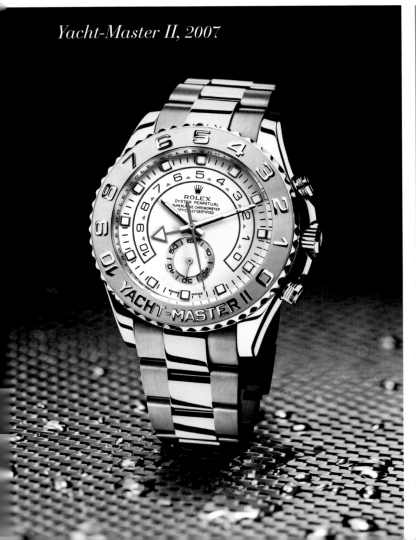

Yacht-Master II, 2007

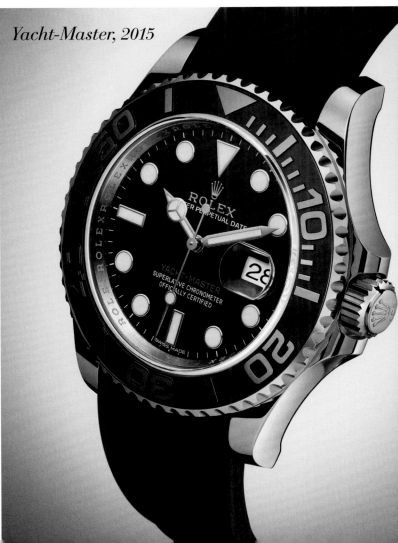

Yacht-Master, 2015

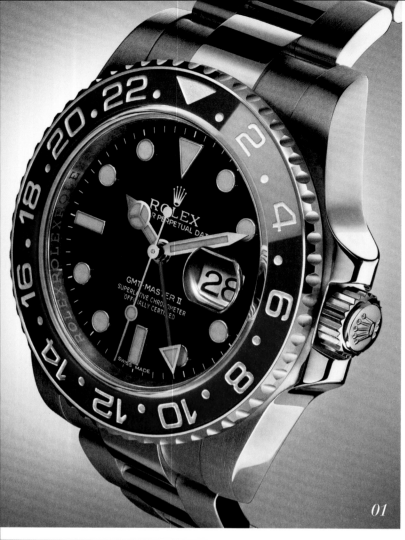

01

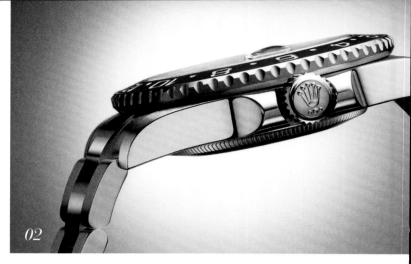

02

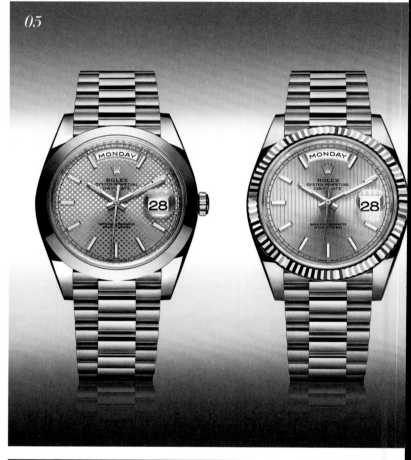

05

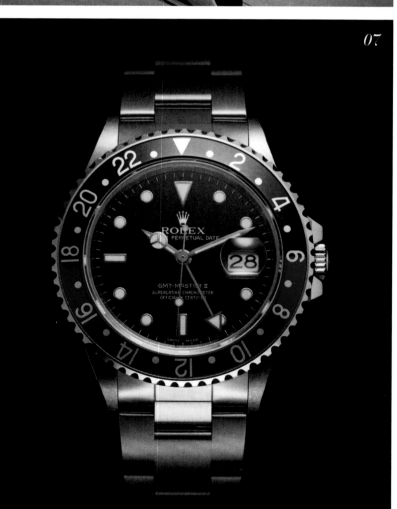

07

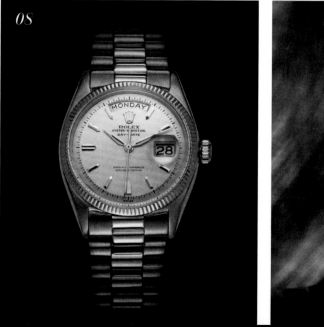

08

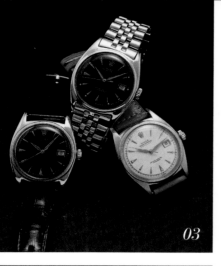

03

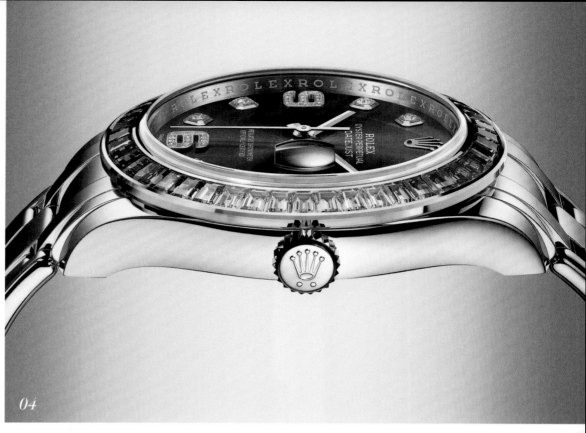

04

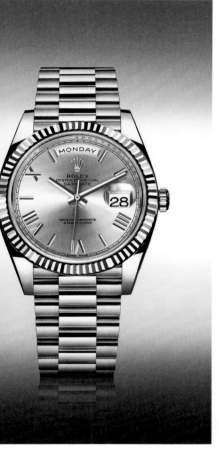

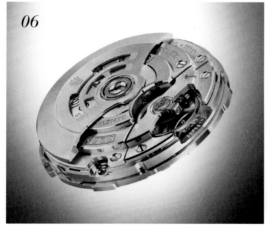

06

01./02. GMT-Master II, 2014
03. From left: Datejust, Ref. 6105, ca. 1951;
    Oyster Red Date, Ref. 6074, ca. 1950;
    Datejust, Ref. 6105, 1952
04. Datejust Pearlmaster 39, 2015
05. Day-Date 40, 2015
06. Calibre 3255, 2015
07. GMT-Master II, 1988
08. Day-Date, 1956
09. Day-Date II, 2010
10. Explorer II Orange Hand, ca. 1970
11. Explorer I, 1953

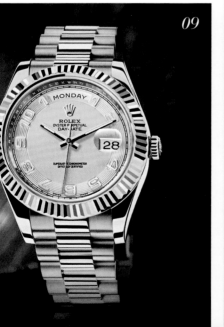

09

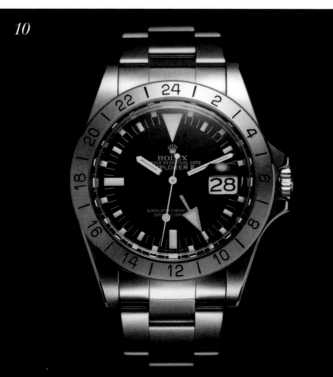

10

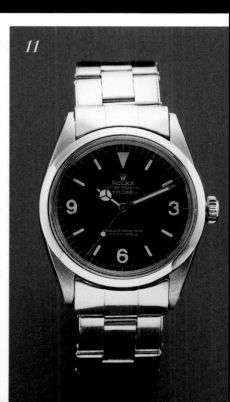

11

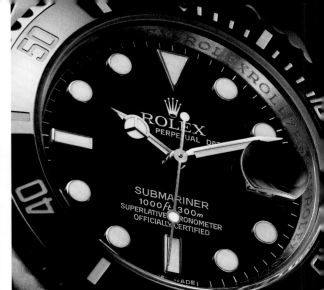

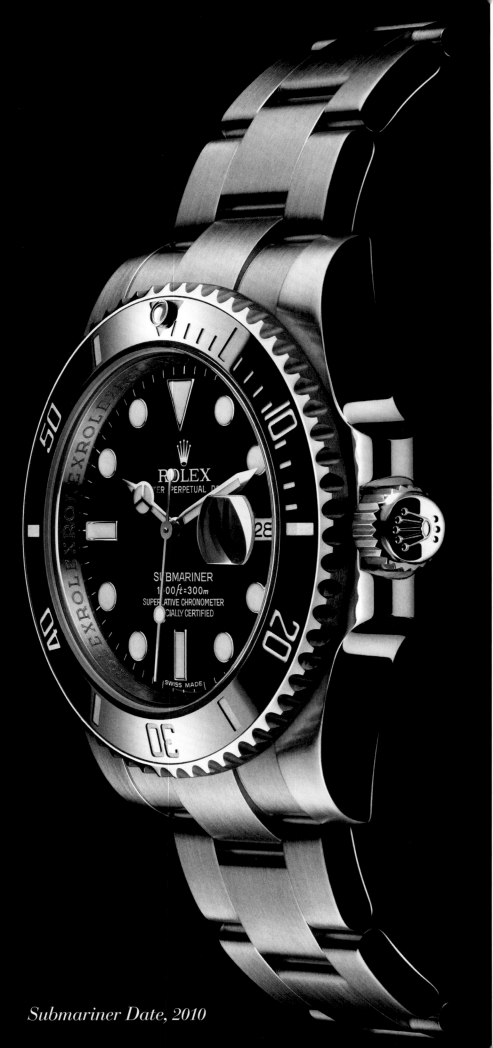

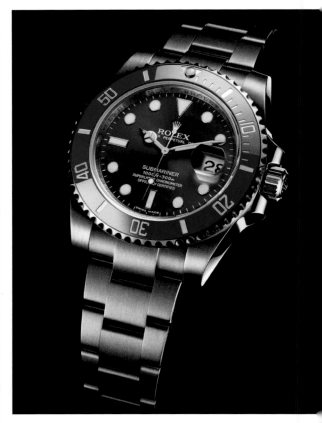

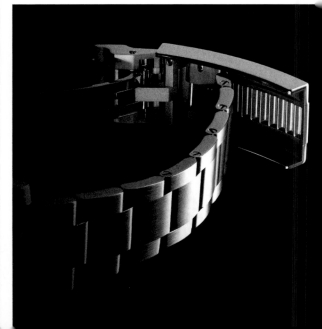

Submariner Date, 2010

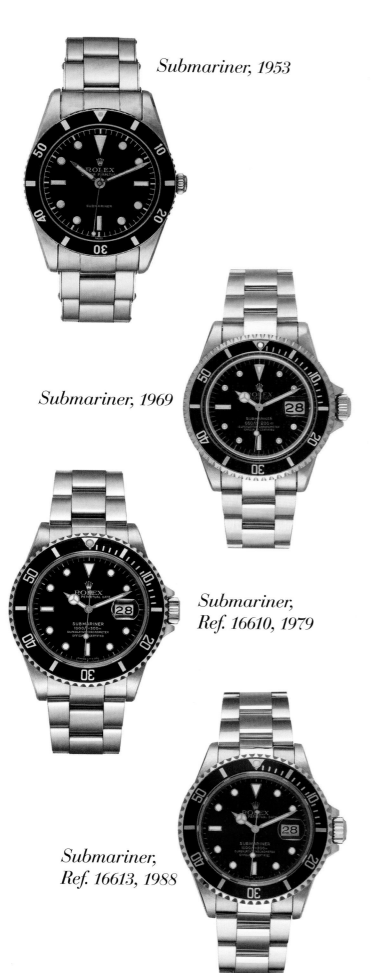

Submariner, 1953

Submariner, 1969

*Submariner,
Ref. 16610, 1979*

*Submariner,
Ref. 16613, 1988*

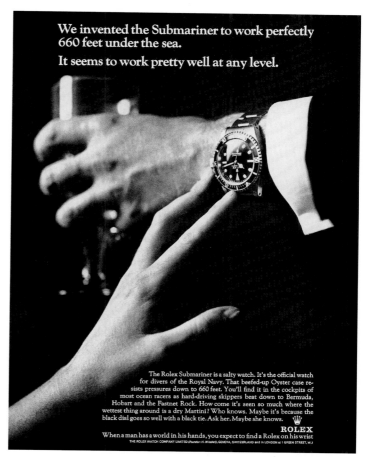

We invented the Submariner to work perfectly 660 feet under the sea.

It seems to work pretty well at any level.

The Rolex Submariner is a salty watch. It's the official watch for divers of the Royal Navy. That beefed-up Oyster case resists pressures down to 660 feet. You'll find it in the cockpits of most ocean racers as hard-driving skippers beat down to Bermuda, Hobart and the Fastnet Rock. How come it's seen so much where the wettest thing around is a dry Martini? Who knows. Maybe it's because the black dial goes so well with a black tie. Ask her. Maybe she knows.

ROLEX

When a man has a world in his hands, you expect to find a Rolex on his wrist

THE ROLEX WATCH COMPANY LIMITED (Founder: H. Wilsdorf), GENEVA, SWITZERLAND and in LONDON at 1 GREEN STREET, W.1

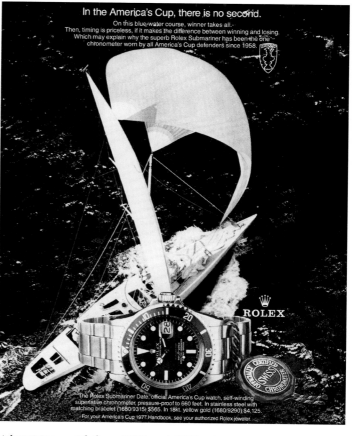

In the America's Cup, there is no second.

On this blue-water course, winner takes all.
Then, timing is priceless, if it makes the difference between winning and losing.
Which may explain why the superb Rolex Submariner has been the one chronometer worn by all America's Cup defenders since 1958.

ROLEX

The Rolex Submariner Date, official America's Cup watch, self-winding superlative chronometer, pressure-proof to 660 feet. In stainless steel with matching bracelet (1680/9315) $565. In 18kt. yellow gold (1680/9290) $4,125.
For your America's Cup 1977 Handbook, see your authorized Rolex jeweler.

Advertising motifs from 1956 and 1977 ◦ Anzeigenmotive von 1956 und 1977 ◦ Motifs de campagnes publicitaires en 1956 et 1977

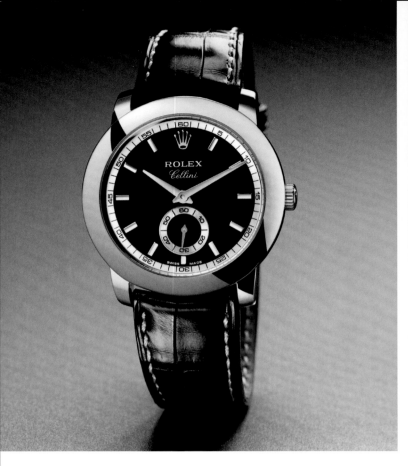

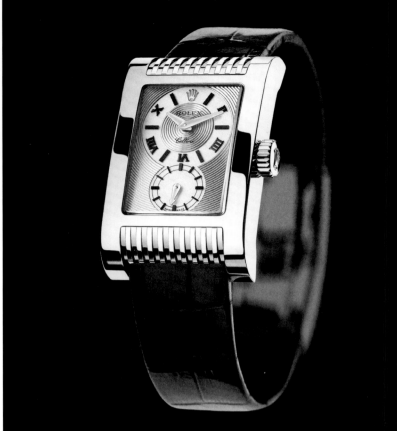

Mit der „Oyster" zu Weltruhm

Erstaunlich, aber wahr: Die Ursprünge von Rolex, einer der schweizerischsten Uhrenmarken überhaupt, haben mit der Eidgenossenschaft zunächst nur wenig zu tun. Sie gehen zurück auf den Kaufmann Hans Wilsdorf, geboren im nordbayerischen Kulmbach. Mit 19 Jahren kehrte er seiner deutschen Heimat den Rücken, um in der Schweiz Erfahrungen im Uhrenexport zu sammeln. 1903 packte er erneut seine Koffer und arbeitete in London als Angestellter eines Uhrenimporteurs.

Weil sich Hans Wilsdorf darüber ärgerte, dass der Handel die Uhrenqualität stark vernachlässigte, gründete er 1905 in London zusammen mit Alfred James Davis, der sich später als gerissener Filou entpuppen sollte, den Uhrengroßhandel Wilsdorf & Davis Ltd. Wegen massiver Konkurrenz brauchte es attraktive und vor allem nicht alltägliche Produkte. Da kamen die neuen Armbanduhren gerade recht. Als Bezugsquelle aktivierte er bereits vorhandene Kontakte in die Schweiz. Dort war ihm die 1878 gegründete Aegler S.A. in bester Erinnerung geblieben. In Biel stieß seine Offerte, England mit genau gehenden Armbanduhren zu versorgen, auf positive Resonanz. Erfolge ließen nicht auf sich warten: Die Importware verkaufte sich blendend.

1907 richtete das junge englische Unternehmen zur Vereinfachung der expandierenden Geschäftsbeziehungen ein eigenes Büro in der Schweiz ein. Hans Wilsdorf, der sich selbst fest in England etablierte und die britische Staatsbürgerschaft annahm, leistete mit seinem Handel einen beachtlichen Beitrag zum Siegeszug der Armbanduhr. Nun brauchte es nur noch einen zugkräftigen

Namen. Die Wahl fiel auf „Rolex" – in allen Weltsprachen gleichermaßen gut auszusprechen und, nach Aussagen der Witwe des Firmengründers, ein Kürzel aus jenem „rolling export", welchen der Schweizer Partner Aegler wegen des boomenden Geschäfts laufend praktizierte. Die Wirren des Ersten Weltkriegs führten zu einer Verlegung des Firmensitzes zunächst nach Biel und 1920 nach Genf, wo Wilsdorf am 17. Januar die Montres Rolex S.A. ins Handelsregister eintragen ließ.

Vier Aspekte der modernen Armbanduhr prägten Wilsdorf und Rolex entscheidend mit. Zunächst die Geschichte des offiziell geprüften Armbandchronometers: Schon 1914 konnte Hans Wilsdorf weltweit erstmals für eine Armbanduhr das begehrte Chronometerzeugnis „A" des Londoner Kew-Observatoriums entgegennehmen. Heute stellt die schweizerische Chronometerkontrolle COSC die meisten seiner Gangzeugnisse für das Haus Rolex aus. Insgesamt haben weit mehr als zehn Millionen Rolex-Uhrwerke das Adelsprädikat „Chronometer" erhalten.

Dann ist Rolex die Entwicklung der vollkommen wasserdichten Armbanduhr zuzuschreiben. 1926 erlangte Wilsdorf das erste Patent für ein mehrfach verschraubtes Gehäuse. Am 7. Oktober 1927 schwamm die englische Stenotypistin Mercedes Gleitze mit der neuen „Oyster" 15 Stunden und 15 Minuten lang durch den Ärmelkanal. Selbstverständlich drang kein Tropfen Wasser in die Schale. Seitdem wurde das geniale System von Rolex kontinuierlich optimiert, niemals jedoch entscheidend geändert.

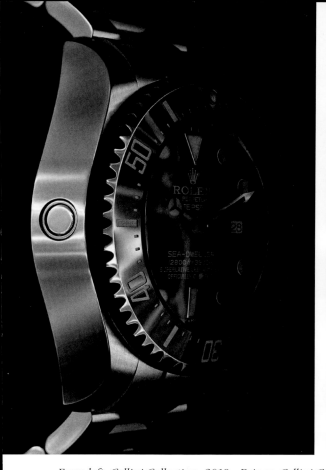
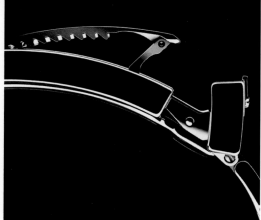
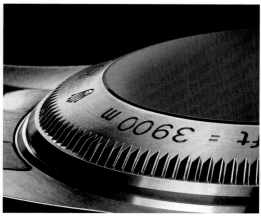
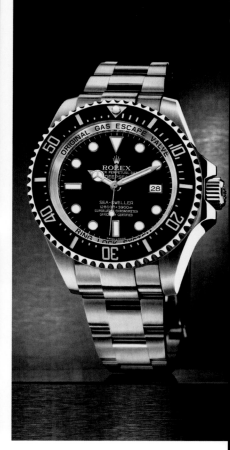

From left: Cellini Collection, 2010 ◦ Prince, Cellini Collection, 2007 ◦ Sea-Dweller Deepsea, 2014
Bottom: balance wheel with Parachrom blue hairspring ◦ Unten: Blaue Parachrom-Breguet-Spirale ◦ En bas : Spiral Parachrom bleu

Den dritten Meilenstein setzte Rolex mit dem Rotoraufzug bei Automatikarmbanduhren. Es war 1932, als Hans Wilsdorf die von Aegler entwickelte „Perpetual" als weltweit erste Armbanduhr mit automatischem Aufzug durch eine unbegrenzt drehende Schwungmasse präsentierte. Auch diesem revolutionären Selbstaufzugssystem ließ Rolex in den zurückliegenden Jahrzehnten eine stete Evolution angedeihen. Dabei legte die Manufaktur stets mehr Wert auf Zuverlässigkeit, Langlebigkeit und Präzision als auf minimale Dimensionen. Aktuell gilt Rolex als größte Schweizer Mechanikmanufaktur. Ausnahmslos alle Uhren werden mit eigenen Werken ausgestattet.

Schließlich entwickelten Hans Wilsdorf und Rolex zum 40. Firmenjubiläum im Jahr 1945 die „Datejust", eine wasserdichte Armbanduhr mit Automatikwerk und Fensterdatum. Datumsfenster im Zifferblatt hatte es schon vorher gegeben. Aber die Kombination aus wasserdichtem Gehäuse, Rotorautomatik und dieser Art der Datumsanzeige verkörperte eine echte Weltpremiere. In den USA wurde die „Datejust" als „Armbanduhr des 20. Jahrhunderts" bezeichnet. Krönend wirkte da noch die hilfreiche Datumslupe. Hierfür kam Wilsdorf, wie seine schlecht sehende Witwe erzählte, der Zufall zu Hilfe. Eines Morgens im Bad spritzte ein Wassertropfen exakt an der Stelle des Fensterdatums aufs Glas. Der Vergrößerungseffekt löste einen Heureka-Ruf aus.

Weitere Meilensteine des kinderlosen Firmengründers, der seinen umfangreichen Nachlass der Hans-Wilsdorf-Stiftung übertrug, heißen:

„Explorer", „GMT-Master", „Submariner" und „Day-Date". Alle sind sie auf ihre Weise einzigartig.

Obwohl die Schweiz dem weltmännischen Bayern und naturalisierten Engländer ihre Staatsbürgerschaft quasi auf dem goldenen Tablett anbot, lehnte Wilsdorf diese Offerte bis zu seinem Tod am 6. Juli 1960 ab. Bis zum Jahr 2004 gab es übrigens zwei rechtlich und finanziell völlig getrennte Rolex-Unternehmungen: In Biel fertigte die von der Familie Aegler und ihren Nachkommen beherrschte Manufacture des Montres Rolex S.A. ausschließlich Uhrwerke. Die 1920 von Hans Wilsdorf gegründete Montres Rolex S.A. zeichnete für die Herstellung der Fertiguhren verantwortlich. Zähe Verhandlungen führten in besagtem Jahr zur Vereinigung unter dem Dach des Genfer Unternehmensteils.

Nach dem milliardenschweren Kauf investierte die auf Perfektion und Präzision versessene Uhrenmanufaktur Rolex, die alle verbauten Kaliber und auch die winzigen Parachrom-Unruhspiralen selber fertigt, bis zu 500 Millionen Franken, um der Werkeproduktion am Bieler Stadtrand internationales Spitzenniveau zu verleihen. Die Einweihung des vorläufig letzten Bauabschnitts ging am 16. Oktober 2012 über die Bühne. Neben dem imposanten Baukomplex in Genf beherbergen auch die neuen Gebäude am Bieler Bözingenfeld Administration, Forschung und Entwicklung. Im Lastenheft für die Architekten rangierten Funktionalität, Energieeffizienz sowie das Wohlbefinden der rund 2 500 Mitarbeiterinnen und Mitarbeiter ganz oben. ◦

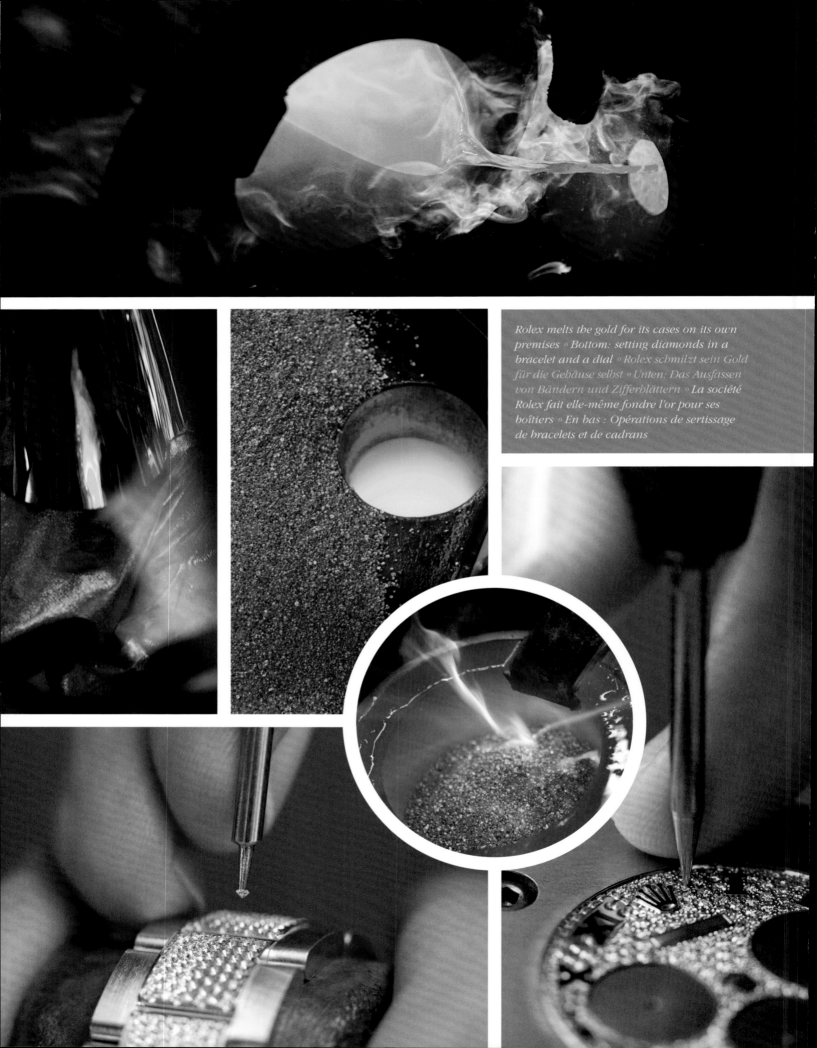

Rolex melts the gold for its cases on its own premises ◦ Bottom: setting diamonds in a bracelet and a dial ◦ Rolex schmilzt sein Gold für die Gehäuse selbst ◦ Unten: Das Ausfassen von Bändern und Zifferblättern ◦ La société Rolex fait elle-même fondre l'or pour ses boîtiers ◦ En bas : Opérations de sertissage de bracelets et de cadrans

La célébrité mondiale avec le modèle « Oyster »

Français

On a du mal à le croire, mais c'est la réalité : les débuts de Rolex, l'une des marques de montres suisses par excellence, n'ont pas grand-chose à voir avec la Confédération helvétique. Rolex doit ses débuts à Hans Wilsdorf, un commerçant natif de Kulmbach, une ville du nord de la Bavière. À 19 ans, il tourne le dos à son Allemagne natale pour se former en Suisse à l'exportation de montres. En 1903, il boucle à nouveau ses valises pour travailler à Londres chez un importateur de montres.

Mécontent du manque avéré de qualité des montres du commerce, Hans Wilsdorf s'associe en 1905 avec Alfred James Davis, lequel se révélera par la suite un fieffé filou, pour fonder à Londres Wilsdorf & Davis Ltd., une société spécialisée dans la vente de montres. La concurrence étant rude, il doit proposer des produits attirants et surtout sortant de l'ordinaire. Les nouvelles montres-bracelets, c'est justement ce qu'il faut. Pour se fournir, il active ses contacts en Suisse. Il a gardé un très bon souvenir d'Aegler SA, une manufacture créée en 1878 à Bienne. Sa proposition d'offrir sur le marché anglais des montres-bracelets d'une précision chronométrique trouve un écho favorable auprès d'Aegler. Le succès est immédiat : les montres importées se vendent à merveille.

En 1907, la jeune société anglaise ouvre une agence en Suisse pour simplifier les relations commerciales, toujours plus complexes. Hans Wilsdorf s'établit définitivement en Angleterre et prend la nationalité britannique, contribuant par son commerce grandement au succès triomphal des montres-bracelets. Désormais, il ne lui reste plus qu'à trouver un nom accrocheur. Le choix se porte sur « Rolex » — un nom simple à prononcer dans toutes les langues du monde et, selon les révélations de la veuve du fondateur de la société, l'abréviation de « rolling export », l'exportation à flux tendus pratiquée par son partenaire suisse Jean Aegler pour desservir un marché florissant. Les troubles de la Première Guerre mondiale amènent Rolex à transférer son siège à Bienne puis à Genève, où Wilsdorf fait inscrire la société Montres Rolex SA au registre du commerce le 17 janvier 1920.

Wilsdorf et Rolex marquent de leur empreinte quatre aspects de la montre-bracelet moderne. Tout d'abord, la certification officielle des chronomètres-bracelets : dès 1914, Hans Wilsdorf obtient pour la première fois au monde pour une montre-bracelet le certificat de précision de classe « A » délivré par l'observatoire de Kew. Aujourd'hui, la majorité des bulletins de marche établis par le Contrôle officiel suisse des chronomètres (COSC) le sont au bénéfice de la maison Rolex. Au total, plus de dix millions de mouvements Rolex ont obtenu l'éminente distinction « chronomètre ».

On doit aussi à Rolex le développement de la montre-bracelet totalement étanche. En 1926, Wilsdorf dépose le premier brevet de montre-bracelet à boîtier vissé hermétiquement. Le 7 octobre 1927, la sténotypiste anglaise Mercedes Gleitze traverse la Manche à la nage en 15 heures et 15 minutes avec au poignet une « Oyster ». Et pas une goutte d'eau ne pénètre dans le boîtier. Depuis, ce génial système Rolex n'a cessé d'être optimisé, sans jamais être modifié de manière déterminante.

La troisième avancée décisive de Rolex concerne le système de remontage à rotor pour montres-bracelets automatiques. En 1932, Hans Wilsdorf présente la « Perpetual » d'Aegler comme la première montre-bracelet au monde à remontage automatique, et ce, grâce à une masse oscillant en permanence sous l'effet de la gravité. Ces dernières décennies, Rolex n'a cessé d'améliorer ce système de remontage automatique révolutionnaire. Dans ce cadre, la manufacture privilégie toujours la fiabilité, la longévité et la précision plutôt que la réduction des dimensions. Aujourd'hui, Rolex est considérée comme la plus grande manufacture de précision suisse. Toutes les montres sont équipées des mouvements qu'elle fabrique.

Enfin, Hans Wilsdorf et Rolex mettent au point pour le 40e anniversaire de la société, en 1945, une montre-bracelet étanche à remontage automatique et affichage de la date dans un guichet. Le guichet n'est pas une nouveauté, mais plutôt la combinaison du boîtier étanche à l'eau, du remontage automatique par rotor et d'un affichage de la date de ce type qui constitue une vraie première mondiale. Aux États-Unis, la « Datejust » est baptisée « montre-bracelet du XXe siècle ». Mais le nec plus ultra, c'est la loupe pour la date. Pour cette invention, le destin vient en aide à Wilsdorf, selon le récit de sa femme malvoyante. Un matin dans la salle de bain, une goutte d'eau s'écrase à l'endroit du verre recouvrant le guichet. Voyant l'effet de grossissement obtenu, il s'exclame « Eureka ! ».

Sans enfants, le fondateur de la société lègue son impressionnante succession à la fondation Hans Wilsdorf, parmi laquelle des créations déterminantes : « Explorer », « GMT-Master », « Submariner » et « Day-Date ». Chacune d'entre elles est unique à sa manière.

La Suisse offre à ce Bavarois mondain et Anglais naturalisé la citoyenneté helvétique quasiment sur un plateau d'argent, mais Wilsdorf refusera cette offre jusqu'à sa disparition, le 6 juillet 1960. Jusqu'en 2004, il existera d'ailleurs deux sociétés Rolex complètement séparées sur un plan juridique et financier. À Bienne, la Manufacture des Montres Rolex SA, placée sous la direction de la famille Aegler et de ses descendants, fabrique exclusivement des mouvements. La société Montres Rolex SA, créée en 1920 par Hans Wilsdorf, se charge de fabriquer des montres terminées. Suite à d'âpres négociations, les deux branches se réuniront en 2004 sous l'égide de l'entité genevoise.

Après avoir déboursé plusieurs milliards, la manufacture de montres Rolex, éprise de perfection et de précision, qui fabrique elle-même tous les calibres qu'elle utilise et les minuscules spiraux Parachrom, investit pas moins de 500 millions de francs suisses pour élever la production de mouvements à un niveau de classe internationale, à la périphérie de la ville de Bienne. Le 16 octobre 2012, la dernière tranche de travaux en date est lancée. Venant compléter l'imposant complexe de Genève, les nouveaux bâtiments sis aux Champs-de-Boujean, à Bienne, abritent les départements de l'administration, de la recherche et du développement. Dans le cahier des charges soumis aux architectes, priorité absolue a été donnée à la fonctionnalité, à l'efficience énergétique et au bien-être des quelque 2 500 salariés. ○

Calibre 4130

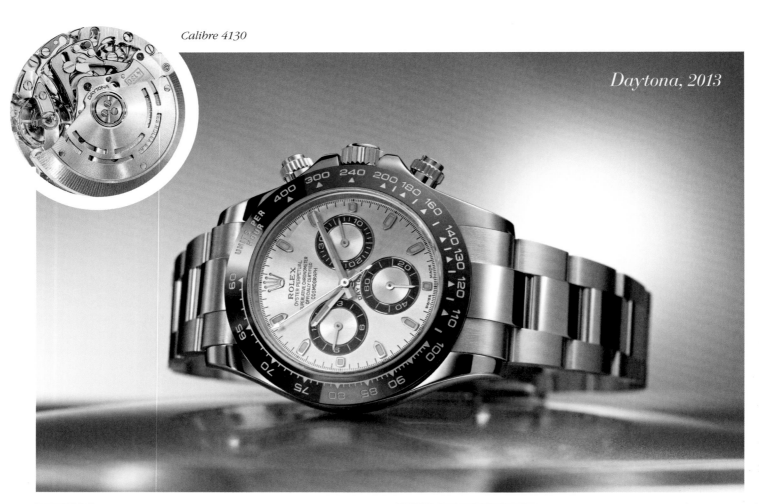

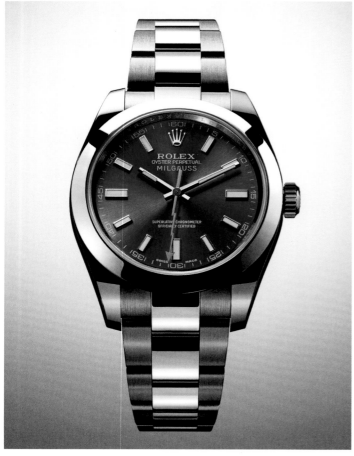

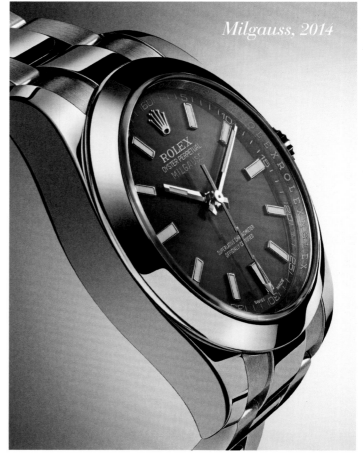

Milgauss, 2014

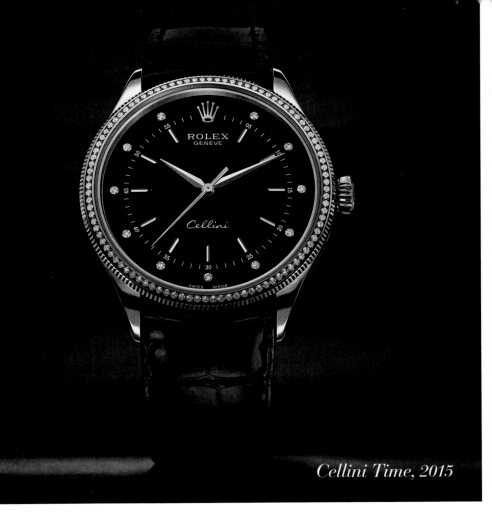

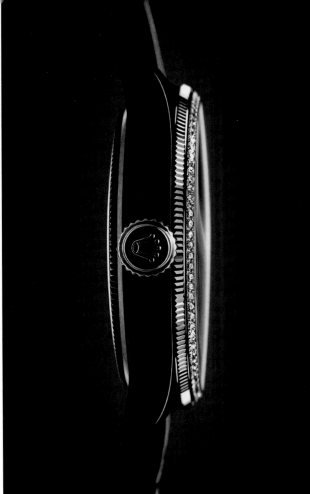

Cellini Time, 2015

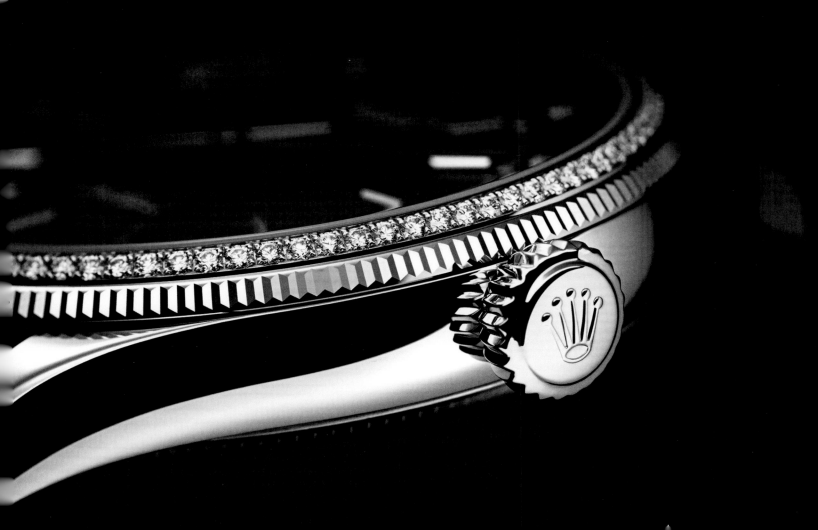

Seiko

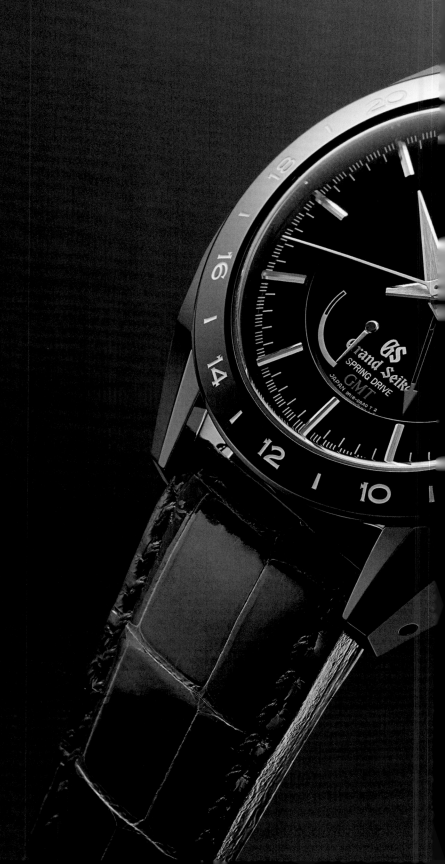

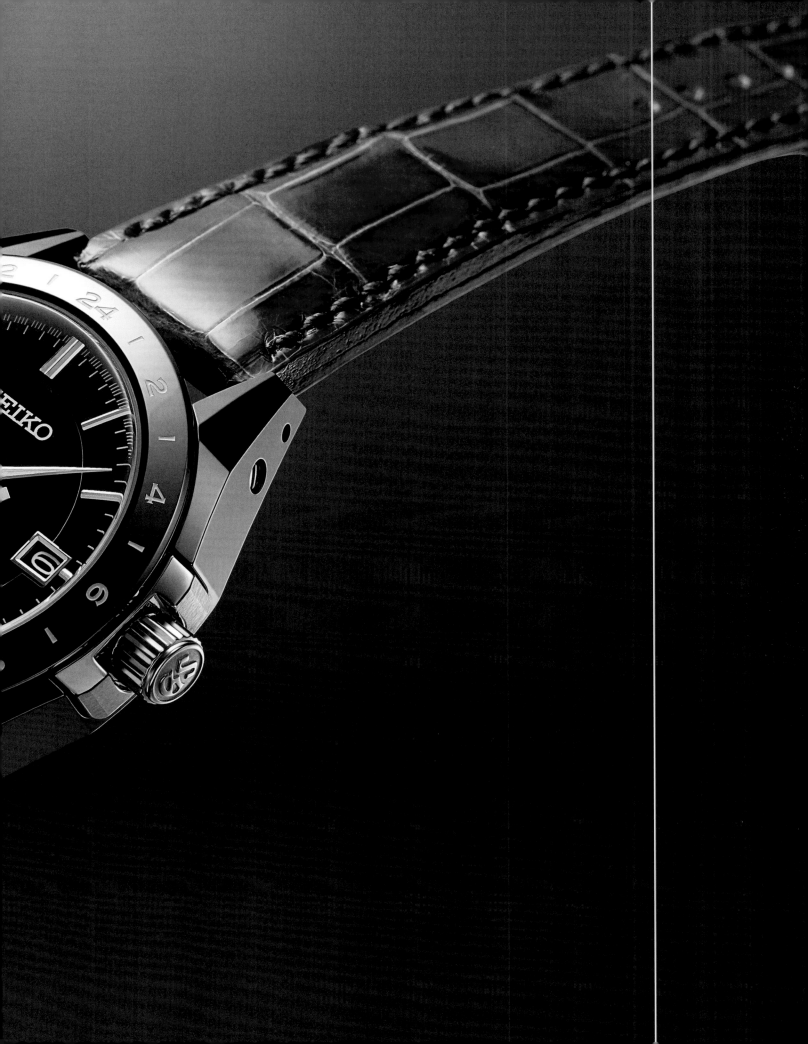

Everything from Japanese Manufactories

English

When Kintaro Hattori founded K. Hattori & Co. Ltd. in 1881, he probably never even dared to dream that his company would develop in the course of time into one of the world's largest watch manufacturers. Fully 43 years came and went before Seiko, the brand name that he had chosen, would adorn the dial of a timepiece. When the 22-year-old first started his business, he had nothing else in mind except to produce timepieces. He sold many different sorts of clocks and watches at his little shop in Tokyo. The lion's share of his merchandise was provided by importers based in the harbor city of Yokohama. Goods of Japanese provenance were in short supply in those years. He also occupied himself with repairs. His business flourished. In 1887, he was able to open a much more impressive store in the Ginza neighborhood, which was the up-and-coming business and shopping district in the Japanese capital city. He bought the building formerly occupied by the Choya newspaper soon afterwards. After thorough renovation, expansion and the installation of a new "Western-style" façade, Kintaro Hattori officially inaugurated the premises in 1895 as the new headquarters of a sizeable and impressive company. The building soon acquired landmark status, thanks in large measure to its circa 16-meter-tall clock tower, which could be seen from far and wide. Long before his Japanese competitors began to think likewise, its owner had already recognized the necessity of being able to fabricate his own timepieces. The starting shot for the

Seikosha factory was fired in 1892. ("Seiko" is the Japanese word for "exact.") The necessary machinery was scarce or unavailable in Japan, so the businessman embarked on a shopping trip to Europe and the United States. With the help of imported machinery, the number of units produced began to increase with astonishing speed. Hattori developed his own alarm clocks in 1899. By 1910, K. Hattori & Co. Ltd. occupied an outstanding position in the Land of the Rising Sun. Seikosha was the only Japanese factory that simultaneously manufactured wall clocks, table clocks and pocket watches, as well as hairsprings and mainsprings. The age of the wristwatch arrived with the little circular "Laurel" in 1913. Thereafter, between 30 and 50 of these models, which Hattori had equipped with his own enamel dial, were produced each day. By 1922, 60 percent of Seikosha's manufacturing capacity was devoted to pocket watches and wristwatches. A severe setback came in 1923, when the Great Kanto Earthquake, which destroyed his warehouse and factory, put a sudden end to production. Hattori was hard hit, but not discouraged. Despite some warnings, he presented the first wristwatches with the Seiko signature one year later—and their positive reception proved him right.

Rapid growth necessitated the construction of an additional factory in 1937. Given the logical name "Daini Seikosha," which means "Second Seikosha," it was the birthplace of diverse small timepieces. These included the pocket chronographs that the Japanese military urgently needed during the Second World War. It seems likely that Seiko relied on Swiss assistance to fill this order.

The "Marvel" began a new chapter, dedicated to precision, in Seiko's illustrious biography. In 1957, it was the first Japanese wristwatch to outperform its Swiss competitors and win an accuracy competition sponsored by the Asian subsidiary of the American Horological Society. A self-winding version debuted in 1959. Caliber 290 in the so-called "GyroMarvel" is the first Seiko movement with its own automatic subassembly. Just one year later, the grapevine was abuzz with talk about the legendary "Grand Seiko," the label's undisputed precision flagship. The staff at the manufacturing site in Suwa, west of Tokyo, had devoted intensive work toward optimizing this wristwatch. The results of their efforts were expressed in hand-wound Caliber 3180. Each watch encasing this caliber underwent extraordinarily strict tests before delivery. Seiko's self-imposed standards of precision were even more rigorous than their counterparts at the COSC, i.e., Switzerland's chronometer-testing authority. And this hasn't changed even one iota in the "Grand Seiko" to the present day. Further innovations, premieres, and superlative achievements followed in rapid succession. To coincide with the 1964 Summer Olympics, Seiko debuted the first hand-wound chronograph of Japanese provenance. One year later, the watch giant unveiled the first "made in Japan" professional diver's watch. A rotatable bezel was affixed atop its 38-millimeter-diameter steel case,

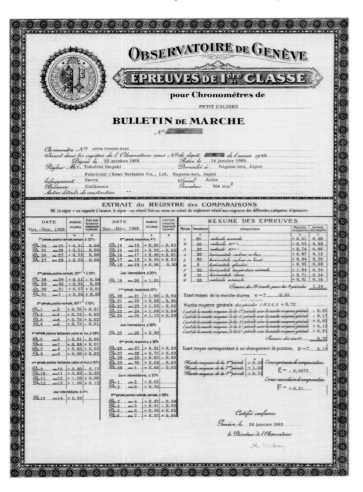

An observatory trial certificate from the Geneva Observatory ◦ Prüfzertifikat vom Genfer Observatorium ◦ Bulletin de marche de l'observatoire de Genève

which remained water-resistant to a pressure of 15 bar. Specially designed hand-wound Caliber 052 first won honors in chronometer competitions in 1966. It and high-frequency Caliber 5740, which debuted in 1967, both had balances paced at 36,000 semi-oscillations per hour. The "Bell-Matic," the first Japanese alarm wristwatch, made headlines the same year, when the starting shot was fired for an ambitious project designated by the numbers 6138 and 6139. Seiko joined the competition to create the world's first self-winding movement with built-in chronograph mechanism. After two years of intensive research in this complex topic, the first specimens of the "Seiko 5 Speed Timer" went on sale in May 1969. It was the world champion with regard to maturity for series manufacturing and marketing introduction. Ten years of research and development preceded the release of another world premiere: sales of the "Astron 35SQ," the world's first quartz wristwatch, began on December 25, 1969. Now let's fast-forward to 1999. After several strenuous attempts, Seiko unveiled a much-acclaimed synthesis of mechanical and electronic technologies. The "Spring Drive," a mechanical quartz watch without a battery, uses the power stored in its barrel to generate electrical energy for its quartz regulator, which oscillates at a pace of 32,768 hertz. The "Spring Drive Automatic" with rotor winding followed in 2005: one hour of motion on the wearer's wrist generates enough power to keep

this watch running for more than three hours. Things got really complicated in 2006, which saw the debut of the "Credor Sonnerie," a version of the "Spring Drive" with an elaborate automatic striking mechanism. A minute-repeater followed in 2010. One year before, in 2009, Seiko had celebrated 30 years of automatic chronographs with the debut of the "Ananta," which encased totally newly developed automatic Caliber 8R28, each of which concatenates no fewer than 292 components. The "Astron GPS Solar" has expressed electronic excellence since 2012: this wristwatch feels perfectly at home all over the globe because it "knows" a total of 39 time zones; when its wearer travels, it automatically resets itself to show the correct time in each new time zone. These are only a few of many examples which demonstrate the creativity and competence of this big brand from the Far East. Seiko's appeal is also due to its versatility. The spectrum of mechanical watches ranges from simple but reliable automatic movements to *grandes complications*, which are available in selected markets under the name "Credor." The vertical depth of manufacturing for the movements is very nearly 100 percent. And as far as the precision watchmaking that's built into every "Grand Seiko" is concerned, the clever Japanese are fully on a par with their Swiss colleagues. In the small but very fine Shizukuishi Watch Studio, where this exceptional timepiece is built, time really seems to stand still. ○

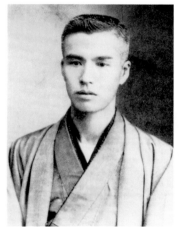

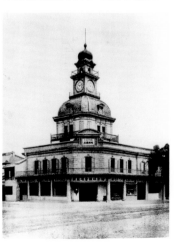

Clockwise, from top left: Kintaro Hattori, 1890 ○ Advertisement, 1881 ○ Clock shop, 1895 ○
Von oben links im Uhrzeigersinn: Kintaro Hattori, 1890 ○ Werbeanzeige, 1881 ○ Uhrengeschäft, 1895 ○
Dans le sens horaire en partant de la gauche : Kintaro Hattori, 1890 ○ Publicité, 1881 ○ Horlogerie, 1895

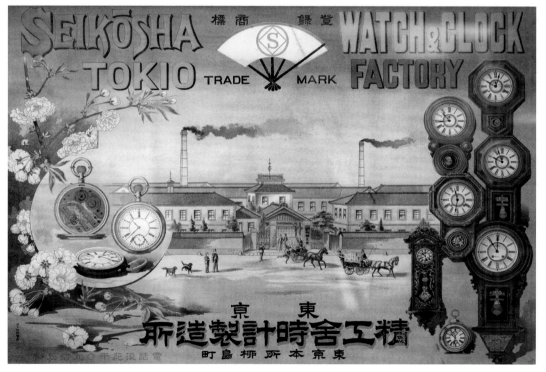

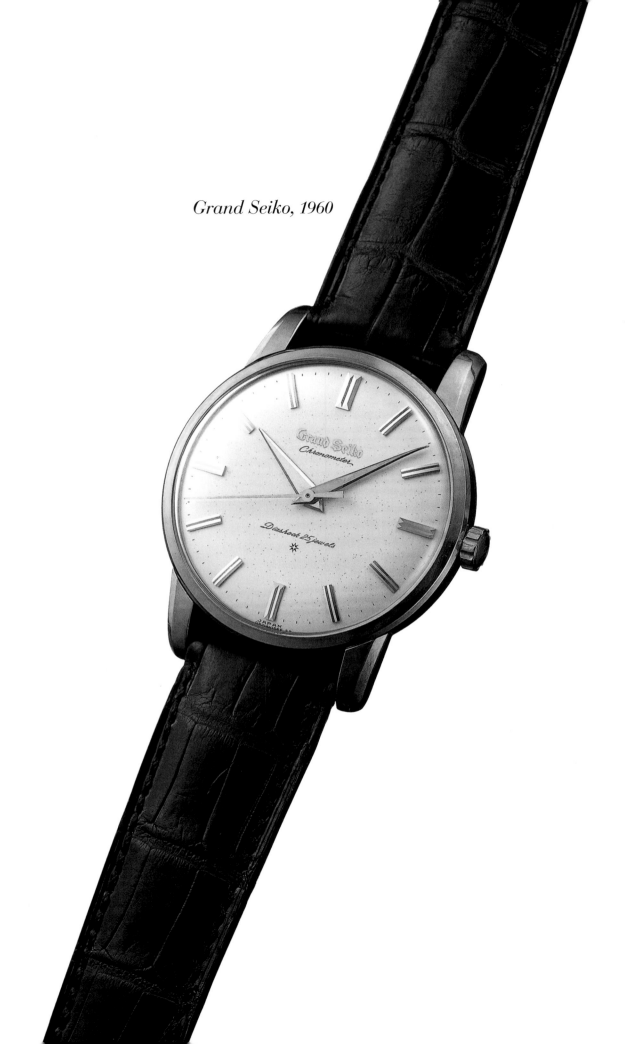

Grand Seiko, 1960

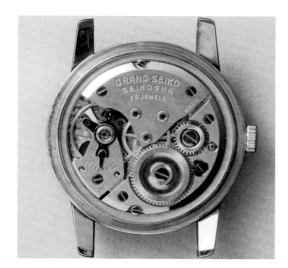

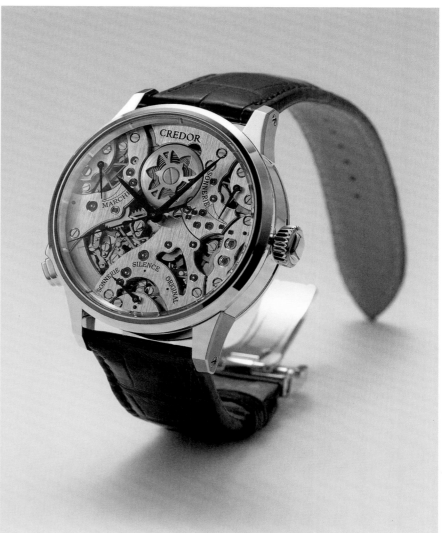

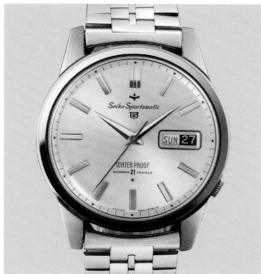

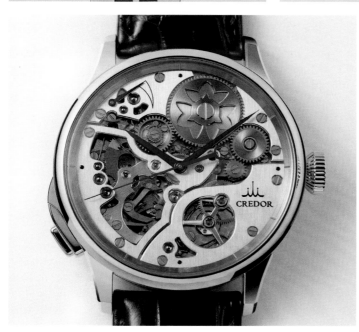

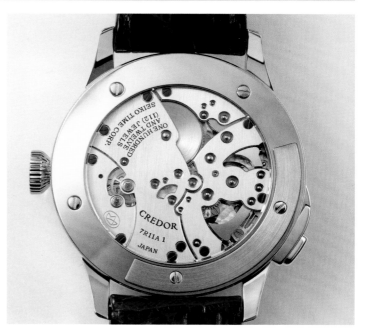

Clockwise from top left: First Grand Seiko, Cal. 3180, 1960 ◦ Credor Spring Drive Sonnerie, 2006 ◦
Credor Spring Drive Minute Repeater (back), 2010 ◦ Credor Spring Drive Minute Repeater, 2010 ◦ Sportsmatic 5, 1963

Alles aus japanischer Manufaktur

Deutsch

Daran, dass sich seine K. Hattori & Co. Ltd. im Laufe der Jahrzehnte zu einem der weltweit größten Uhrenhersteller entwickeln würde, wagte Kintaro Hattori 1881 vermutlich nicht einmal im Traum zu denken. Zunächst einmal dauerte es nach der Firmengründung schon ganze 43 Jahre, bis der von ihm erkorene Markenname Seiko erstmals das Zifferblatt einer Uhr zierte. Anfänglich hatte der 22-Jährige auch anderes im Sinn als eine Uhrenproduktion. In seinem kleinen Tokioter Laden verkaufte er Zeitmesser unterschiedlichster Art. Der Löwenanteil seiner Ware stammte von Importeuren aus der Hafenstadt Yokohama. Erzeugnisse japanischer Provenienz waren damals noch Mangelware. Daneben beschäftigte er sich auch mit Reparaturen. Das Geschäft florierte. Bereits 1887 konnte er ein deutlich repräsentativeres Lokal im Ginza-Viertel eröffnen, dem rapide aufstrebenden Geschäfts- und Einkaufsdistrikt der japanischen Hauptstadt. Wenig später erwarb er das Gebäude der Zeitung Choya. Nach gründlicher Renovierung, Erweiterung und der Fassadengestaltung im „Western-Style" konnte es Kintaro Hattori 1895 als neuen Stammsitz des mittlerweile stattlichen Unternehmens einweihen. Der entwickelte sich nicht zuletzt wegen des rund 16 Meter hohen Uhrturms zu einem weithin sichtbaren Wahrzeichen. Deutlich schneller als seine japanischen Mitbewerber hatte der Patron dann doch die Notwendigkeit eigener Uhrenfertigung erkannt. Das Startsignal für die Fabrik Seikosha war 1892 gefallen. „Seiko" bedeutet nichts anderes als „genau". Weil es in Japan noch an den nötigen Maschinen mangelte, begab sich der Unternehmer auf Einkaufstour nach Europa und in die Vereinigten Staaten von Amerika. Mit Hilfe importierter Anlagen kletterten die Stückzahlen erstaunlich schnell. 1899 lancierte Hattori selbst entwickelte Wecker. Gegen 1910 nahm die K. Hattori & Co. Ltd. im Land der aufgehenden Sonne schon eine herausragende Position ein. Als einzige Fabrik fertigte Seikosha Wand-, Tisch- und Taschenuhren gleichzeitig. Daneben entstanden auch Unruhspiralen und Zugfedern unter dem eigenen Dach. Das Zeitalter der Armbanduhr brach 1913 mit der kleinen runden „Laurel" an. Von diesem Modell, welches Hattori mit eigenem Emailzifferblatt ausstattete, entstanden fortan täglich 30 bis 50 Exemplare. 1922 lasteten Taschen- und Armbanduhren schon insgesamt 60 Prozent der Seikosha-Fertigungskapazitäten aus. Ein deutlicher Rückschlag war 1923 zu verzeichnen. Das große Kanto-Erdbeben zerstörte Lagerbestände und Fabrik, brachte die Produktion zum Erliegen. Doch Hattori ließ sich nicht entmutigen. Trotz mancher Warnungen präsentierte er schon ein Jahr später die ersten Armbanduhren mit der Signatur Seiko. Und die positive Resonanz gab ihm recht.

Ungestümes Wachstum verlangte 1937 nach Errichtung einer weiteren Fabrik. Logischerweise erhielt sie den Namen Daini Seikosha. In der „zweiten Seikosha" spielte sich fortan die Herstellung von Kleinuhren ab. Dazu gehörten auch Taschen-Chronographen, nach denen das japanische Militär während des Zweiten Weltkriegs dringend verlangte. Hierfür bediente sich Seiko aber vermutlich noch eidgenössischer Assistenz.

1956 leitete die „Marvel" ein neues, mit Präzision überschriebenes Kapitel der illustren Seiko-Biographie ein. Als erste japanische Armbanduhr gewann sie 1957 den Genauigkeitswettbewerb des fernöstlichen Ablegers der American Horological Society. Die Schweizer Konkurrenz musste sich geschlagen geben. Die Variante mit Selbstaufzugswerk kam 1959 in den Handel. Beim 290 in der sogenannten „GyroMarvel" handelte es sich um das erste Seiko-Kaliber mit eigener Automatik-Baugruppe. Danach dauerte es nur noch ein Jahr, bis das unangefochtene Präzisions-Flaggschiff in Gestalt der legendären „Grand Seiko" von sich reden machte. Bei der Kreation dieser Armbanduhr hatten sich Mitarbeiter in Suwa, einer Fertigungsstätte westlich von Tokio, intensiv mit Optimierungsmöglichkeiten beschäftigt. Das Resultat ihrer Bemühungen kam im Handaufzugskaliber 3180 zum Ausdruck. Jede der damit ausgestatteten Armbanduhren hatte sich vor der Lieferung außerordentlich strengen Tests zu unterziehen. Die selbst gesteckten Genauigkeitsanforderungen überstiegen sogar noch jene der amtlichen Schweizer Kontrollbehörde COSC. Und daran hat sich bei „Grand Seiko" bis heute kein Jota geändert. Weitere Innovationen, Premieren und Bestleistungen folgten Schlag auf Schlag. Anlässlich der Olympischen Sommerspiele 1964 stellte Seiko den ersten Handaufzugschronographen japanischer Provenienz vor. Nur ein Jahr später wartete der Uhrengigant mit der ersten professionellen Taucher-Armbanduhr „made in Japan" auf. Dem nassen Element widerstand das 38 Millimeter große Stahlgehäuse mit Drehlünette bis zu 15 Bar Druck. Ab 1966 punktete das speziell gestaltete Handaufzugskaliber 052 bei Chronometerwettbewerben. Stündlich 36 000 Unruh-Halbschwingungen waren auch eines der Kennzeichen des Hochfrequenz-Kalibers 5740 von 1967. Im gleichen Jahr machte der erste japanische Armbandwecker namens „Bell-Matic" von sich reden. Und es fiel der Startschuss für ein ehrgeiziges Projekt mit den Nummern 6138 und 6139. Seiko trat in den Wettstreit um das weltweit erste Automatikwerk mit integriertem Chronographen. Nach zwei Jahren intensiver Auseinandersetzung mit der komplexen Materie waren erste Exemplare des „Seiko 5 Speed Timer" im Mai 1969 käuflich zu erwerben. Er war der Weltmeister hinsichtlich Serienreife und Markteinführung. Ganze zehn Jahre Forschungs- und Entwicklungsarbeit waren einer weiteren Weltpremiere vorangegangen. Am 25. Dezember 1969 gelangte die „Astron 35SQ" als weltweit erste Quarz-Armbanduhr in den Handel. Wir springen ins Jahr 1999: Nach mehreren anstrengenden Anläufen wartete Seiko mit einer viel beachteten Synthese aus Mechanik und Elektronik auf. „Spring Drive", eine mechanische Quarzuhr ohne Batterie, nutzt die im Federhaus gespeicherte Antriebskraft zum Generieren elektrischer Energie für den mit 32 768 Hertz oszillierenden Quarz-Regulator. Die „Spring Drive Automatic" mit Rotoraufzug folgte 2005. Eine Stunde Bewegung am Handgelenk generiert Kraft für mehr als drei Stunden Funktion. Richtig kompliziert wurde es 2006 mit der „Credor Sonnerie", einer „Spring Drive" mit aufwendigem Selbstschlag-Mechanismus. Die Version mit Minutenrepetition gesellte sich 2010 hinzu. 30 Jahre Automatik-Chronograph zelebrierte Seiko 2009 mit „Ananta" und

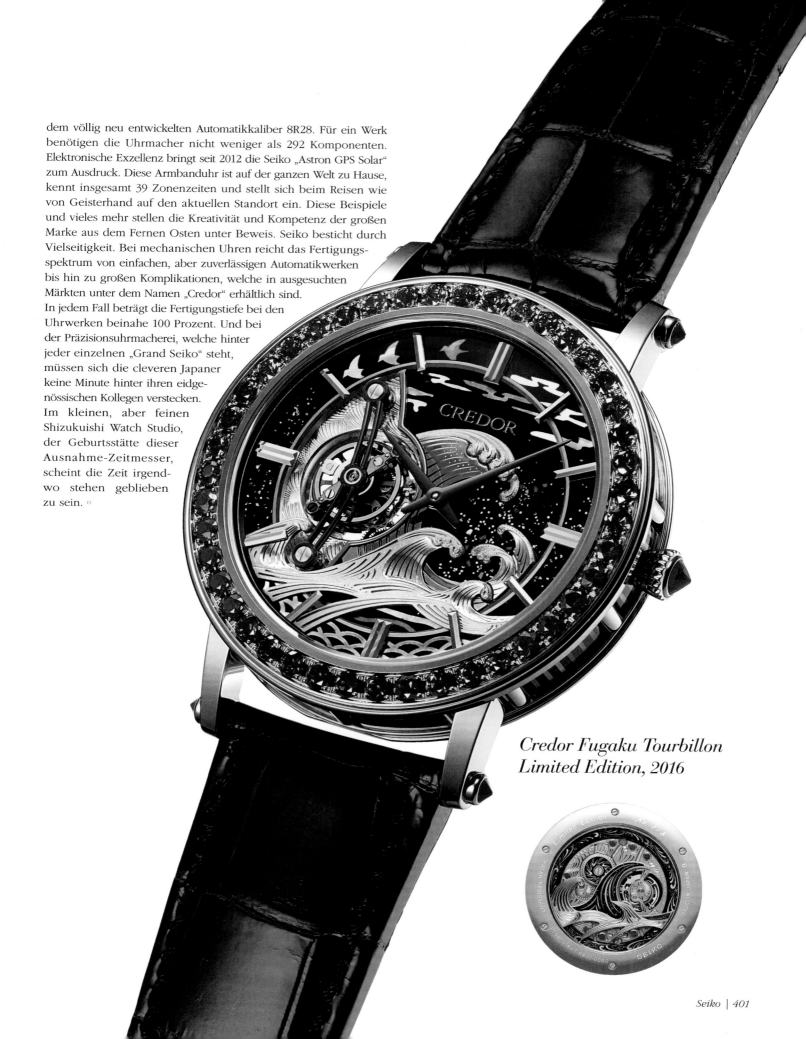

dem völlig neu entwickelten Automatikkaliber 8R28. Für ein Werk benötigen die Uhrmacher nicht weniger als 292 Komponenten. Elektronische Exzellenz bringt seit 2012 die Seiko „Astron GPS Solar" zum Ausdruck. Diese Armbanduhr ist auf der ganzen Welt zu Hause, kennt insgesamt 39 Zonenzeiten und stellt sich beim Reisen wie von Geisterhand auf den aktuellen Standort ein. Diese Beispiele und vieles mehr stellen die Kreativität und Kompetenz der großen Marke aus dem Fernen Osten unter Beweis. Seiko besticht durch Vielseitigkeit. Bei mechanischen Uhren reicht das Fertigungs-spektrum von einfachen, aber zuverlässigen Automatikwerken bis hin zu großen Komplikationen, welche in ausgesuchten Märkten unter dem Namen „Credor" erhältlich sind. In jedem Fall beträgt die Fertigungstiefe bei den Uhrwerken beinahe 100 Prozent. Und bei der Präzisionsuhrmacherei, welche hinter jeder einzelnen „Grand Seiko" steht, müssen sich die cleveren Japaner keine Minute hinter ihren eidge-nössischen Kollegen verstecken. Im kleinen, aber feinen Shizukuishi Watch Studio, der Geburtsstätte dieser Ausnahme-Zeitmesser, scheint die Zeit irgend-wo stehen geblieben zu sein. ◦

Credor Fugaku Tourbillon
Limited Edition, 2016

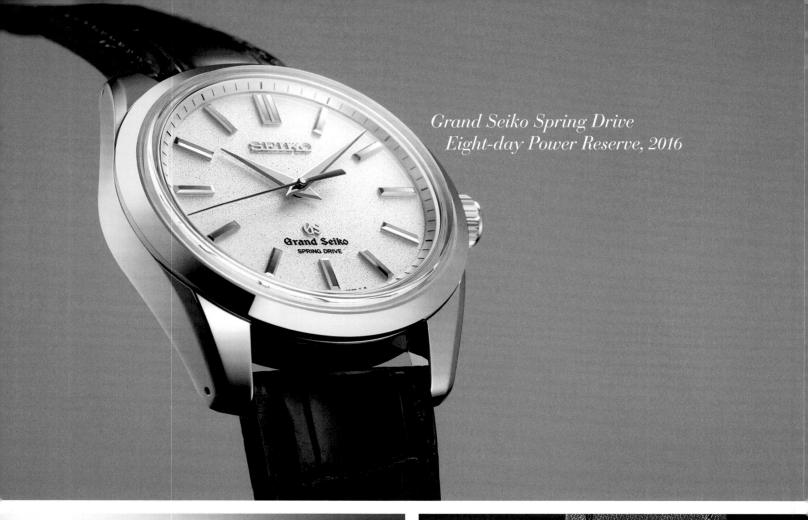

*Grand Seiko Spring Drive
Eight-day Power Reserve, 2016*

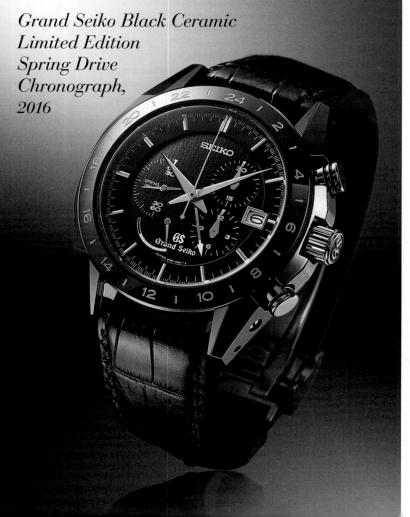

*Grand Seiko Black Ceramic
Limited Edition
Spring Drive
Chronograph,
2016*

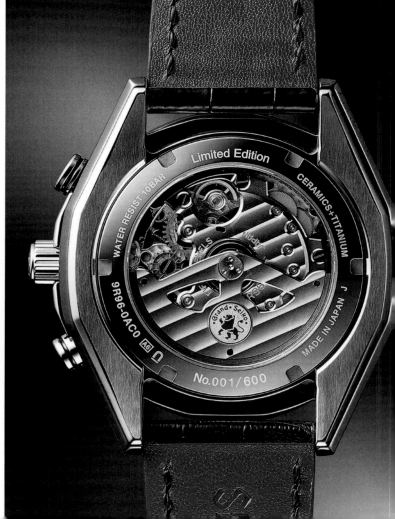

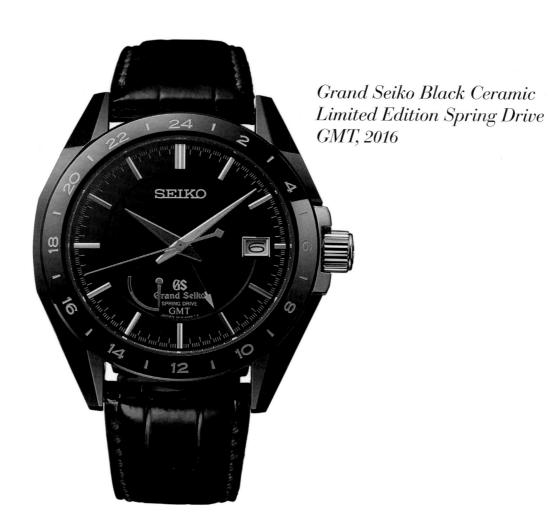

Grand Seiko Black Ceramic Limited Edition Spring Drive GMT, 2016

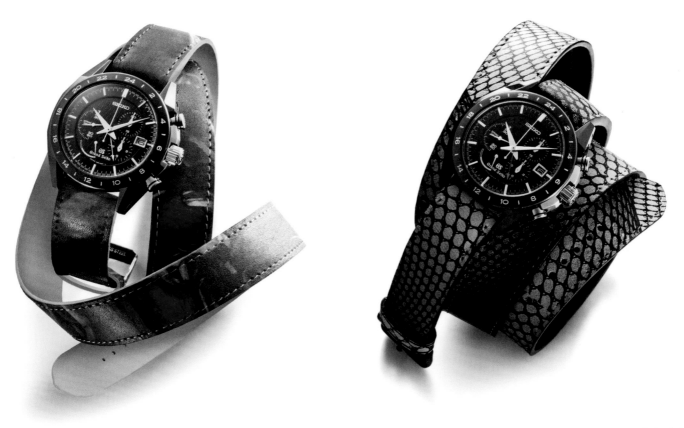

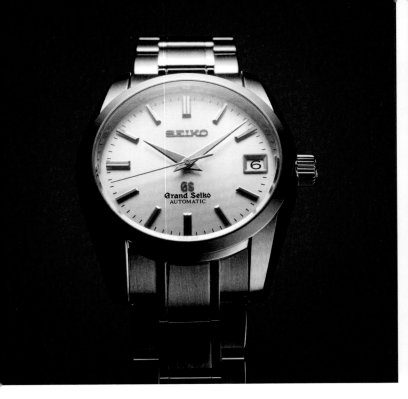

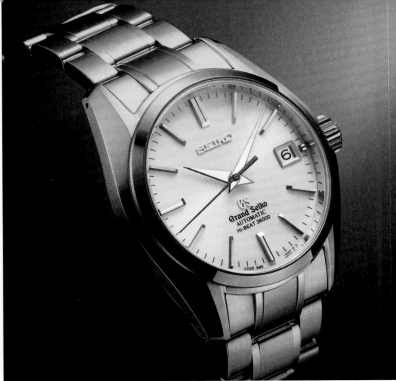

Fabrication japonaise sur toute la ligne

Français

En 1881, Kintaro Hattori est probablement à mille lieues d'imaginer que l'atelier d'horlogerie (K. Hattori & Co. Ltd.) qu'il vient de créer deviendra des décennies plus tard l'une des plus grandes manufactures de montres au monde. Dans un premier temps, 43 ans s'écouleront avant que Seiko, le nom de marque qu'il a choisi, n'orne le cadran d'une montre. Au début, ce jeune entrepreneur de 22 ans a d'autres projets que la fabrication de montres. Dans sa petite boutique à Tokyo, il vend des garde-temps de types très divers, provenant pour la plupart d'importateurs de la ville portuaire de Yokohama, car les produits d'origine japonaise sont encore rares à l'époque. Il effectue également des réparations. Ses affaires sont si florissantes que, dès 1887, il ouvre un local commercial nettement plus représentatif à Ginza, le quartier du commerce et des affaires en plein essor de la capitale japonaise. Peu de temps après, il acquiert le bâtiment du journal Chôya Shimbun. Celui-ci est entièrement rénové, agrandi et doté d'une façade dans le « style occidental » et, en 1895, Kintaro Hattori peut inaugurer le nouveau siège de son entreprise de taille désormais respectable. Le bâtiment devient un emblème local visible de loin, notamment grâce à son horloge culminant à environ 16 mètres. Kintaro Hattori a reconnu bien plus vite que ses concurrents japonais la nécessité de fabriquer également ses propres montres. Aussi, dès 1892, l'heure de créer la manufacture Seikosha (de « seiko », dont on associe la sonorité au succès et à la finesse) a-t-elle sonné. Comme il ne trouve pas les machines nécessaires au Japon, le jeune entrepreneur part s'approvisionner en Europe et aux États-Unis. Grâce aux machines importées, la production de montres augmente étonnamment vite. En 1899, K. Hattori lance des réveils de sa fabrication. Vers 1910, K. Hattori & Co. Ltd. occupe déjà une position prépondérante au pays du Soleil-Levant. Seikosha est la seule manufacture à fabriquer à la fois des horloges murales, des pendulettes et des montres de poche. Des spiraux et des ressorts de barillet sont également

réalisés en interne. L'ère des montres-bracelets débute en 1913 avec la « Laurel », une petite montre ronde que K. Hattori dote d'un cadran en émail de sa fabrication. La production journalière est de 30 à 50 exemplaires. En 1922, les montres de poche et les montres-bracelets absorbent déjà 60 pour cent des capacités de production de la manufacture. L'année 1923 est marquée par un net recul. En détruisant les stocks et la manufacture, le terrible séisme de Kanto porte un coup d'arrêt à la production. Mais K. Hattori ne se laisse pas décourager. Passant outre les avertissements de certains, il présente dès l'année suivante les premières montres-bracelets signées Seiko. L'écho positif reçu lui donne raison.

Une deuxième manufacture est construite en 1937 pour faire face à une croissance exponentielle. Elle est logiquement baptisée Daini Seikosha ou « deuxième Seikosha ». C'est là que seront dorénavant fabriquées les montres, ainsi que les chronographes de poche que l'armée japonaise réclame instamment durant la Seconde Guerre mondiale. Pour ces derniers toutefois, Seiko fait probablement encore appel aux services d'entreprises suisses.

En 1956, la « Marvel » marque dans l'illustre histoire de Seiko le début d'un nouveau chapitre caractérisé par la précision. Première montre-bracelet japonaise, elle remporte en 1957 le concours de précision organisé par la branche extrême-orientale de l'American Horological Society. La concurrence suisse doit s'avouer vaincue. C'est en 1959 qu'est commercialisée la version équipée d'un mouvement à remontage automatique. Sur la « GyroMarvel », le 290 est le premier calibre Seiko à mouvement automatique. Un an plus tard seulement, la légendaire « Grand Seiko », fleuron incontesté de la précision, fait sensation. Pour créer cette montre-bracelet, des collaborateurs de l'usine de Suwa, à l'ouest de Tokyo, ont beaucoup travaillé sur les possibilités d'optimisation. Le fruit de leurs

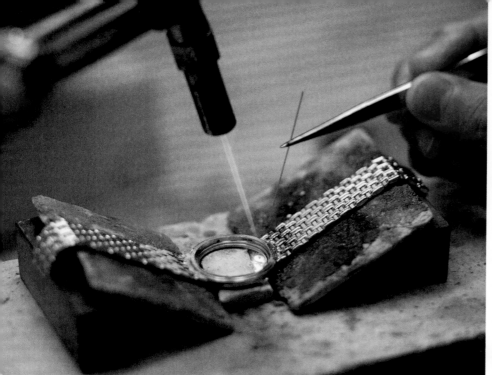
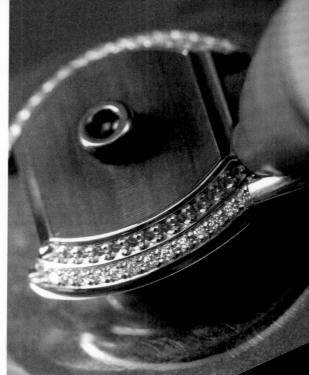

*Opposite page, from left: Grand Seiko Automatic, 2015 ◦ Grand Seiko Automatic Hi-Beat, 2009 ◦ This page, from left: Gold welding ◦ Diamond setting ◦ Diese Seite, von links: Goldschweißen ◦ Einsetzen der Diamanten ◦ **Cette page, de gauche à droite : Soudage de l'or ◦ Sertissage de diamants***

efforts est le calibre à remontage manuel 3180. Chaque montre-bracelet qui en est équipée doit subir des tests extrêmement sévères avant livraison. Les exigences que la marque s'impose en matière de précision dépassent même celles du Contrôle officiel suisse des chronomètres (COSC). Et jusqu'à aujourd'hui ces règles sont restées les mêmes, notamment pour la « Grand Seiko ». D'autres innovations, premières mondiales et performances record se succèdent à un rythme soutenu. Les Jeux olympiques de 1964 sont pour Seiko l'occasion de présenter le premier chronographe à remontage manuel de fabrication japonaise. Un an plus tard seulement, ce géant du monde horloger propose la première montre-bracelet de plongée professionnelle « made in Japan ». Le boîtier en acier de 38 millimètres à lunette tournante résiste en plongée à une pression de 15 bars. À partir de 1966, le calibre à remontage manuel 052, de conception spéciale, marque des points dans des concours de chronomètres. Le calibre de haute fréquence 5740, sorti en 1967, se caractérise entre autres par les 36 000 alternances horaires de son balancier. La même année, la « Bell-Matic », première montre-réveil japonaise, est un succès. C'est aussi la date de lancement d'un projet ambitieux, avec les calibres 6138 et 6139. Seiko affronte alors ses rivaux internationaux pour devenir la première manufacture au monde à créer un mouvement automatique à chronographe intégré. Au terme de deux ans d'études sur un sujet aussi complexe, les premiers exemplaires du « Seiko 5 Speed Timer » sont commercialisés en mai 1969. De tous les concurrents mondiaux, c'est le modèle le plus apte à être fabriqué en série et commercialisé. Dix bonnes années de recherche et développement ont précédé une autre première mondiale : le 25 décembre 1969, l'« Astron 35SQ » est la première montre-bracelet à quartz commercialisée dans le monde. Mais transportons-nous maintenant en 1999 : après plusieurs essais éprouvants, Seiko propose une synthèse très remarquée entre mécanique et électronique.

La « Spring Drive » est une montre à quartz mécanique sans pile qui utilise la force motrice emmagasinée dans le barillet pour générer l'énergie électrique nécessaire à son régulateur à quartz qui oscille à 32 768 Hertz. Elle est suivie en 2005 par la « Spring Drive Automatic », une montre à système de remontage par rotor. Portée une heure au poignet en mouvement, elle génère suffisamment de force motrice pour fonctionner plus de trois heures sans qu'il soit nécessaire de la remonter. La complexité technique atteint des sommets en 2006 avec la « Credor Sonnerie », une montre « Spring Drive » équipée d'un coûteux mécanisme de sonnerie à barillet propre. Une version à répétition minute viendra rejoindre en 2010 les deux modèles précédents. En 2009, Seiko célèbre 30 ans de chronographes automatiques avec la collection « Ananta » et le 8R28, un nouveau calibre automatique entièrement repensé. Rien moins que 292 composants sont nécessaires aux horlogers pour réaliser un mouvement. Depuis 2012, la Seiko « Astron GPS Solar » incarne l'excellence de la marque en matière électronique. Partout chez elle dans le monde, cette montre-bracelet connaît 39 fuseaux horaires et se règle en voyage comme par magie sur le lieu où se trouve son propriétaire. Ces exemples et bien d'autres caractéristiques encore démontrent la créativité et les compétences de la grande marque d'Extrême-Orient. Seiko séduit par sa diversité. Sa gamme de montres mécaniques s'étend de modèles à mouvements automatiques simples mais fiables à des modèles à grandes complications, proposés sous la dénomination « Credor » sur des marchés confidentiels. Dans chaque cas, la proportion de pièces fabriquées en interne sur les mouvements atteint quasiment 100 pour cent. Et dans l'horlogerie de précision, à la base de chaque modèle « Grand Seiko », les géniaux horlogers japonais n'ont absolument rien à envier à leurs collègues suisses. Le temps semble s'être arrêté dans le petit, mais non moins raffiné, atelier horloger de Shizukuishi, berceau de ce garde-temps d'exception. ◦

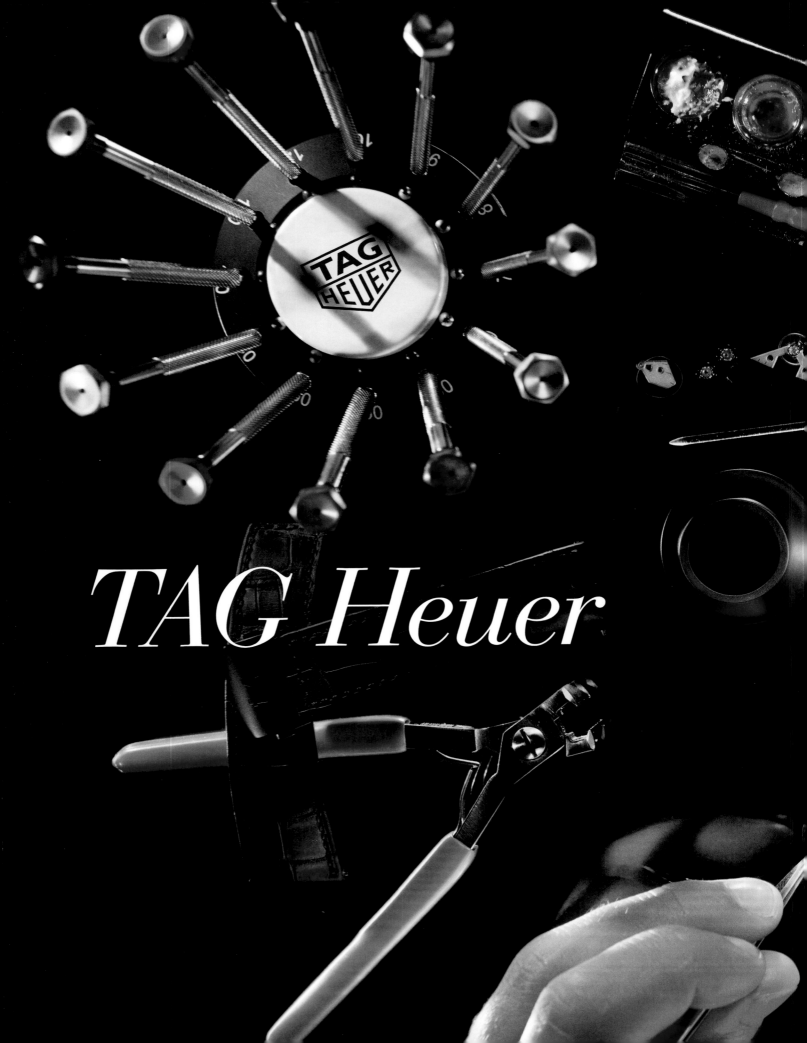

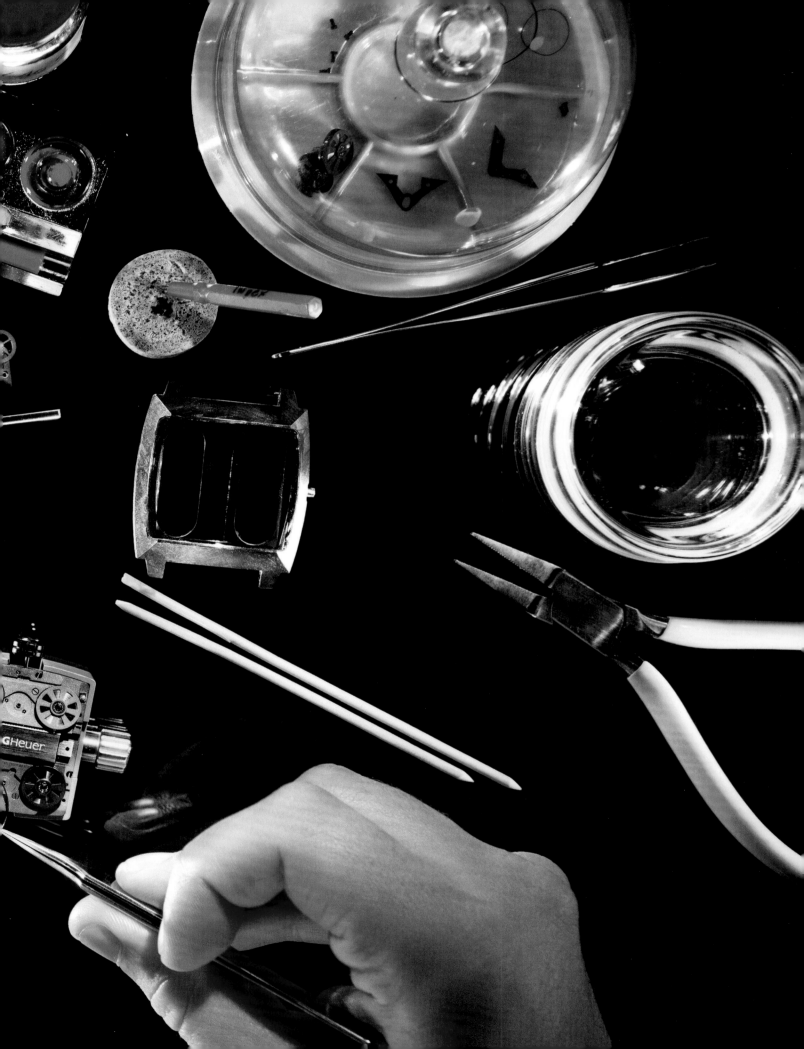

From Maker of Chronographs to Chronograph Manufacture

Heuer Becomes TAG Heuer

The chronograph runs like Ariadne's thread through the lives of Edouard Heuer (born in 1840) and his successors. Heuer became a watchmaker's apprentice at age 14. After acquiring his first professional experiences, the newly married young watchmaker went into business for himself at age twenty. His firm could be found in Bienne starting in 1865. It was here that he developed a crown winding for watches that made winding keys superfluous and culminated in his first patent in 1869. Various crises made it clear to him that success lay in specialization. That's why starting in 1882, his collection featured chronographs of his own provenance, including some which had been awarded patent protection. On Christmas Eve of the year 1886, Edouard Heuer applied for a patent to protect a modern oscillating pinion coupling which has not survived to the present day. Other relevant inventions followed.

Heuer was awarded a silver medal for his comprehensive spectrum of watches at the Exposition Universelle in Paris in 1889. When he died unexpectedly on April 30, 1892, he left a sizeable estate valued at 552,750 Swiss francs.

This financial basis enabled his sons Jules-Edouard and Charles-Auguste to continue the firm's activities. After the death of his brother on July 12, 1911, the responsibilities lay solely on the shoulders of Charles-A. Heuer, whose subsequent credits include a chronograph with a pulsometer scale and the "Mikrograph." The latter, which debuted in 1916, was a revolutionary stopwatch that achieved the unprecedented feat of accurately measuring elapsed intervals to the nearest 100th of a second. It and its derivatives convinced the timekeepers at the Olympic Games in 1920, 1924, and 1928. Wristwatches bearing the Heuer signature first became available in 1913. The next generation, embodied by Charles-Edouard II and Hubert-Bernard Heuer, took pains to systematically expand their activities on important markets. They acquired the Jules Jürgensen watch brand in 1919, which they kept until 1936. Heuer was obliged to cope with turbulent conditions after 1929 and during the ensuing global economic crisis. The label survived thanks to creative products, including stopwatches and chronographs for a wide variety of purposes. Prince Wilhelm of Sweden and designated US president Harry S. Truman adorned their wrists with golden wristwatches from Heuer in 1947.

The last generation is represented by Jack William Heuer, who was born in 1932. The son of Charles-Edouard Heuer II, Jack William Heuer studied engineering and afterwards acquired practical experience in the USA, thus qualifying for the executive post that we would later hold at Heuer, which he joined in 1958. As a shareholder in the family business, he merged Heuer with Leonidas to create Heuer-Leonidas in 1964. The ledgers record revenues of ten million Swiss francs in 1968; twice as much was earned a mere two years later in 1970. Urgent need for capital prompted the company to offer shares on the stock market.

Afterwards, Jack W. Heuer, who ranked among the undisputed pioneers of electronic chronographs, was obliged to generate dividends. But the generally poor situation and cheap Far Eastern competition repeatedly upset his plans. The consequences of the Quartz Crisis struck in 1976, followed by the aftermath of an unsuccessful stopwatch deal with China in 1981. The last official general meeting of Heuer-Leonidas SA convened on June 25, 1982. Jack W. Heuer, who refused to accept bankruptcy, along with the other owners of registered shares, had no choice but to agree to write off their share capital and dissolve existing reserves. Nouvelle Lemania and its associated shareholders, including the house of Piaget, became new majority shareholders. Starting in 1985, the majority of shares belonged to the internationally active TAG Group (Techniques d'Avant-Garde) and Heuer became TAG Heuer. Significantly more commercial watches, mostly encasing electronic quartz movements, along with skilful marketing (including the Formula One car race) and fresh advertising campaigns enabled TAG Heuer to resurrect itself like the proverbial phoenix from the ashes. Beginning in 1996, shares of TAG Heuer SA were again traded on the stock exchange. With annual sales of 715,000 watches and revenues of nearly 420 million Swiss francs in 1996, TAG Heuer advanced to fifth place among watch manufacturers in Switzerland. Jack W. Heuer returned as a consultant in 1997, the same year that witnessed a comeback of the legendary "Carrera" from 1963. TAG Heuer celebrated the renaissance of the avant-garde "Monaco" in 1998.

In September 1999, the French luxury group LVMH paid nearly 1.2 billion Swiss francs to acquire the traditional sport watch brand. Under LVMH's roof, TAG Heuer ascended to the status of a genuine watch manufactory with a large number of innovative calibers and a total of four production sites in the Jura region of western Switzerland. Jean-Christophe Babin, who is presently CEO of Bulgari, took the reins in November 2000. He initiated high-frequency chronographs such as the "360," "Mikrograph," "Mikrogirder", and "MikrotourbillonS." The "V4" concept watch, which was launched in 2004, brought the physicist and mathematician Guy Sémon on board. He can be thanked for numerous horological revolutions. A genuine sensation debuted in 2016. The big structural change began on March 1, 2014 with Jean-Claude Biver. The coordinating headman of Hublot, TAG Heuer and Zenith saw that action was urgently required. After becoming CEO at the beginning of 2015, he thoroughly revised the traditional Swiss brand. His credo: "TAG Heuer must offer watches that are priced appropriately for the buying power of its classical customers," i.e., at prices between 1,000 and 5,000 Swiss francs. Well-known brand ambassadors such as Cara Delevingne, Chris Hemsworth, David Guetta, or Tom Brady, the distinctive "Carrera Heuer 01," the successful "Connected" smartwatch, the aggressively priced "Carrera Heuer 02" tourbillon chronograph, and cooperation with the Red Bull Racing Team clearly show the direction in which this brand is headed. ○

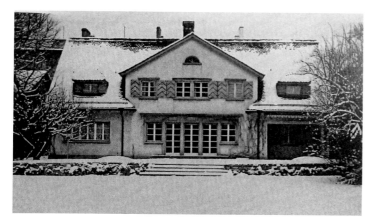

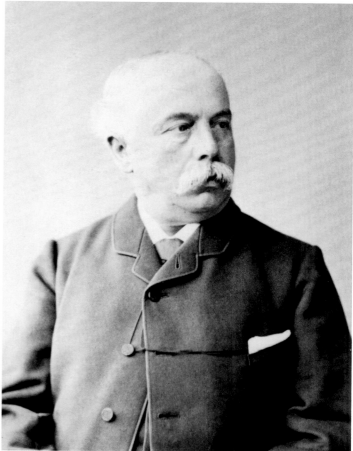

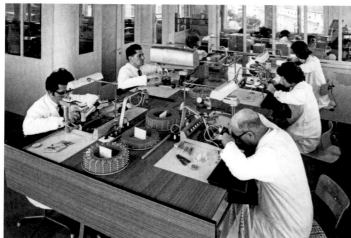

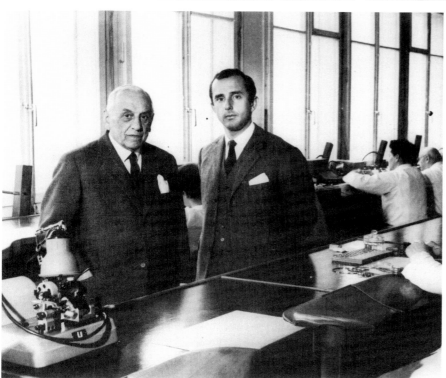

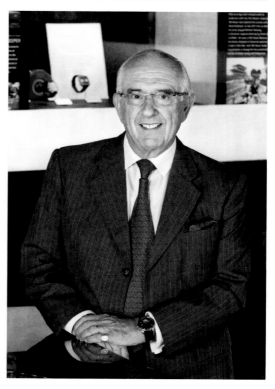

Clockwise from top left: Jack Heuer's childhood home, Bienne, c. 1940 ∘ Edouard Heuer, c. 1890 ∘ Jack Heuer, 2003 ∘ Jack Heuer and his father Charles-Edouard, 1958 ∘ Heuer in Bienne, 1968 ∘ Von oben links im Uhrzeigersinn: Jack Heuers Elternhaus, Biel, ca. 1940 ∘ Edouard Heuer, ca. 1890 ∘ Jack Heuer, 2003 ∘ Jack Heuer und sein Vater Charles-Edouard, 1958 ∘ Heuer in Biel, 1968 ∘ Dans le sens horaire en partant du haut, à gauche : Maison parentale de Jack Heuer, Bienne, vers 1940 ∘ Édouard Heuer, vers 1890 ∘ Jack Heuer, 2003 ∘ Jack Heuer et son père Charles-Édouard, 1958 ∘ Heuer à Bienne, 1968

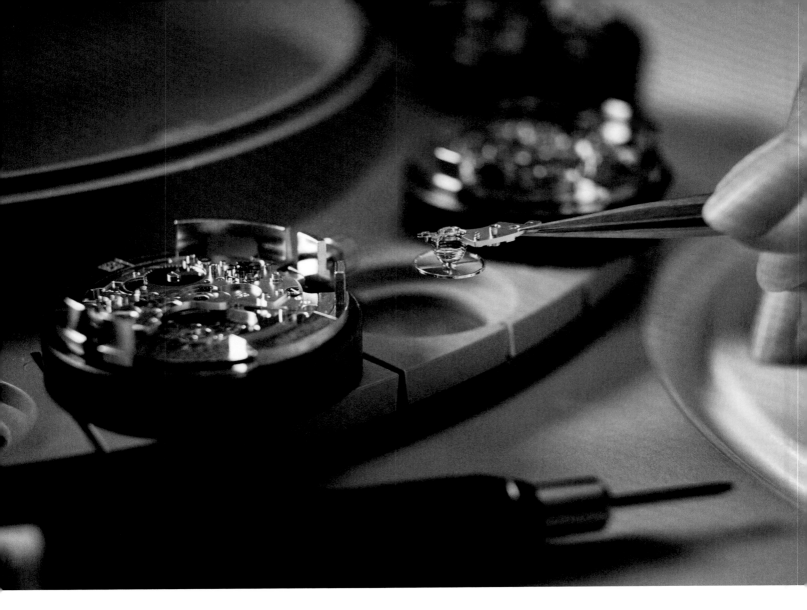

Movements, Cases and Dials

English

No legally unambiguous definition exists to define precisely what a watch manufactory is, but the company's products should include at least one caliber of its own. Furthermore, a firm that claims to be a manufacture should also fabricate the movement's bearing components, i.e., the plates, bridges, and cocks. The prestigious bell tolled for TAG Heuer with the debut of license Caliber 1887 in 2010. The reader should bear in mind that chronographs are especially demanding because of their complex functional inter-relationships. Swiss expertise expressed itself through differentiated analysis and optimization of the Japanese basic caliber. Achievements entirely of its own, on the other hand, are represented by self-winding Caliber Heuer 02 with chronograph and an innovative one-minute tourbillon, which has a rotating carriage made of light-weight and very sturdy carbon.

A mechanical timekeeping microcosm is strictly organized, and so too is a watch manufactory. Product developers, engineers, and designers collaborate at the brand's headquarters in La Chaux-de-Fonds. Whether the task involves a case, dial, hands, wristband, or movement, these specialists capably handle whatever comes their way. Movements fabricated in one's own manufactory are necessarily more costly than calibers purchased from third-party suppliers, so due to its capacity and to keep its prices competitive, TAG Heuer primarily also encases time-honored devices from Eta and Sellita. Prototype-makers likewise work at the brand's head-quarters. And the "torturers" occupy an extensive laboratory, where they impose the most extreme torments on prototypes destined for serial production.

The fabrication of T0 components takes place at Chevenez, a secluded village near the French border where workers can be found more easily. Milling and processing the plates, bridges, and cocks is entrusted to computer-guided and extremely adaptable manufacturing centers of the newest generation. Afterwards, the specialists in the neighboring T1 atelier require about 100 work steps to transform quality-tested components into finished chrono-graph movements. Here too, automation ensures optimally high quality. After thorough reorganization, the vertical range of manu-facture for the label's own movements now exceeds 50 percent.

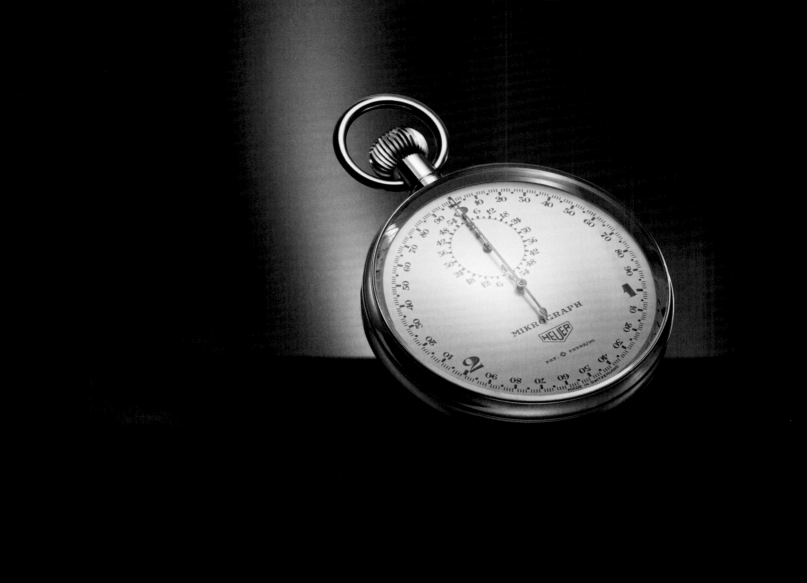

From left: Setting the hairspring ◦ Mikrograph, 1916 ◦ Von links: Einsetzen der Unruhspirale ◦ Mikrograph, 1916 ◦
De gauche à droite : Pose du spiral ◦ Mikrograph, 1916

Purchasers of wristwatches bearing the TAG Heuer signature also get protective cases from the brand's own manufactory. The label operates a highly modern case factory at Cornol near Chevenez with a workforce of circa 140 individuals who primarily produce steel cases for timepieces ranging from the "Aquaracer" to the "Monza." Most of the noise here is generated by menacing-looking stamping presses, but the computer-guided manufacturing centers are likewise impossible not to hear. Indefatigable robots and expert artisans share the tasks involved in the final polishing. After thorough quality control, which includes verification of watertight-ness, each case is sent to the assembly department.

The third member of the trio is ArteCad, a traditional dial factory. Nothing is too exotic for the specialists at this subsidiary of TAG Heuer. Their expertise also includes familiarity with tech-niques that were thought to have been forgotten. Complexity has drastically increased in recent years for new dials, e.g., for the face of the "Carrera Heuer 01" or the "Carrera Heuer 02T." These "faces of time" are veritable multipart three-dimensional artworks. Machines and manual craftsmanship go hand in hand in Tramelan.

The goal is absolute perfection because the dial and the hands that turn above it are usually the first details one notices when one scrutinizes at a watch.

The staff at the workshops in La Chaux-de-Fonds is responsible for the assembly and final control of the entire timepiece, which, of course, is considerably more than sum of its many parts. The watches that leave these ateliers are sent to destinations around the globe. TAG Heuer has traditionally been an international brand, and Jean-Claude Biver would like to significantly expand its presence throughout the world. ◦

Chronometric and Chronographic High Points Since 1887

The first highlight is undoubtedly the pocket chronograph with oscillating-pinion coupling, which was patented in 1887. The oscillating pinion performs the same function as a conventional gear coupling with swiveling lever, but significantly reduces the costs for a watch movement with a built-in chronograph.

Heuer's first pilot's watch won admirers thanks to the large ensemble formed by its dial and hands, which remains optimally legible even in unfavorable lighting conditions. The "Time of Trip" was first installed in the cockpits of aircraft and automobiles in 1911. The crew of the R34 rigid airship consulted this timepiece in 1919 during the first airborne crossing of the North Atlantic. Hugo Eckener followed their lead when he flew around the Earth aboard his "Graf Zeppelin" in 1929.

Development of the world's first stopwatch for extremely short intervals began in 1914. The patent was granted on October 2, 1916. To measure elapsed intervals to the nearest 100th of a second, the little balance oscillated at a pace of 50 hertz. According to the catalogue of models, the "Mikrograph" was suitable for "measuring the flights of projectiles and other short-term tests." Its price was 100 Swiss francs or 120 US dollars.

Collectors avidly covet Heuer's dashboard clocks. The "Autavia," a combined clock and stopwatch, first appeared in 1933 on instrument panels, where it reliably assisted drivers in races and rallies. Aircraft pilots and ships' captains likewise relied on timekeeping accessories from Bienne, which were known as "Rally-Master" starting in the late 1950s.

The 1930s also saw Heuer's specialists devoting intensive work to water-protected cases for chronographs. Wristwatch chronographs with water-resistant cases were exactly what the Swiss military wanted. Production of watertight military chronographs with black dials commenced in 1942.

Heuer's catalogue described the "Solunar" as a patented wristwatch "for fishermen, hunters, sailors, seamen, sportsmen, naturalists, scientists, physicians, farmers, bird breeders, gardeners, and foresters." In addition to the time of day or night, this watch also showed the moon's position, the lunisolar periods, and the tides. A push-piece could be pressed to synchronize the "Solunar" with published tables of local tides.

During the crucial prestart phases of a regatta, yachtsmen could rely on the "Yachting Timer," which was launched in 1962. After it had been switched on, the stopwatch reliably showed the current status of the countdown.

The "Autavia," which has an easily remembered name, celebrated a comeback in 1962, when it was released as a wristwatch chronograph with a rotatable bezel that clicked firmly into place atop the watertight steel case. This watch was available in various versions, e.g., for pilots, sport divers, or scientists.

Jack W. Heuer personally took care of the official timekeeping at the famous "12 Hours of Sebring" race, where he first heard about the legendary Carrera Panamericana racing event on open roads. That's why he chose the name "Carrera" for his chronograph creation, which debuted in 1963 with a spaciously planar and three-dimensional dial. The seconds scale was marked along a convex Plexiglas flange.

For the first time in the history of the chronograph, Heuer began marketing a specimen with a mono-window date display in 1965. Only hand-type or double-window date displays had been available prior to this debut. It was only logical that the chronographic world buzzed with gossip about the "Carrera 45 Dato."

Heuer's debut of the world's first modular chronograph with a microrotor for automatic winding, oscillating-pinion coupling and coulisse lever system coincided with the release of a new line of watches. The square case of the "Monaco" was also watertight. The Formula One pilot Jo Siffert was its optimal ambassador. Steve McQueen accordingly wore a "Monaco" on his wrist and an authentic pair of Jo Siffert racing coveralls when performed in the film *Le Mans*.

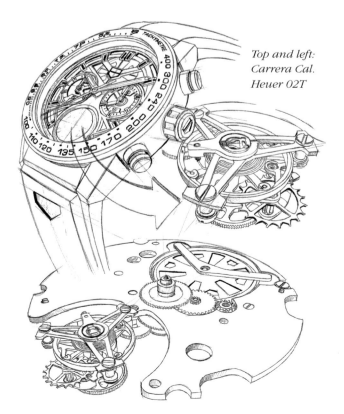

Top and left:
Carrera Cal.
Heuer 02T

Carrera Cal.
Heuer 01, 2016

Among the target group for the 1,000 series from 1992 were divers who had to work under extreme pressure conditions. Despite its remarkable watertightness (up to 100 bar), the case made do without a conventional helium-escape valve.

"Kirium Ti5," designed by Jörg Hysek, inaugurated a new design era. Its salient feature was the fluently organic transition between the case and the wristband. Furthermore, TAG Heuer introduced new high-performance materials: carbon was used for the dial, and titanium for the case and bracelet.

Mechanical or electronic? The "Monaco Sixty Nine" from 2003 answered this question with a simple "yes." Its innovative reversible case housed both a ticking mechanical caliber and a quartz-paced movement.

The "V4" debuted as a concept watch in 2004. It became a reality in 2009. Two micro-toothed belts with metal "souls" convey kinetic energy from the linear oscillating body to four barrels. A conventional balance steadily subdivides precious time into equal intervals.

The "Carrera Calibre 360 LE," a wristwatch with two movements, celebrated its world premiere in 2006. One caliber indicated the ordinary time of day, while the other could measure elapsed intervals to the nearest 100th of a second.

TAG Heuer returned to the status of a horological manufactory in 2010, which was also the year of its 150th anniversary. The engineers who designed automatic Caliber 1887 with oscillating-pinion chronograph relied on a Seiko movement. The 320 components that comprise this opus were fabricated entirely in Switzerland.

Born in 2011, the mechanical "Mikrotimer Flying 1000 Concept Chronograph" could measure elapsed intervals to the nearest 1,000th of a second. Analogously to its 100th-of-a-second predecessors, this

wristwatch has two separate power trains, each with its own oscillating and escapement system. Applications for ten patents testify to the ample inventiveness that culminated in its debut.

The mechanical "Mikrogirder" accurately measures brief intervals to the nearest 5/10,000th or 1/2,000th of a second. No conventional rate regulator can complete 7,200,000 semi-oscillations per hour, so three perpendicularly arranged metal blades collaborate in this construction, which was registered for ten patents and born in 2012.

This same year saw the "MikrotourbillonS2" add the crowning touch to the philosophy of ultra-fast oscillation. Each of the two tourbillons rotates at a different speed from its companion: the balance in the speedier tourbillon is paced at 50 hertz and serves as the pacemaker to precise measure elapsed intervals to the nearest 1/100th of a second. The more conventionally paced balance in the other tourbillon is responsible for the ordinary time display.

The "Carrera Heuer 01," which debuted in 2015, is based on Caliber 1887. Nothing is concealed by the specially designed front side. The transparent back likewise hides no details. Twelve parts create the modularly conceived "Carrera" case, which is watertight to ten bar. Its multipart architecture enables designers to combine diverse materials, surfaces, and colors.

Last but not least, we have the "Carrera" tourbillon chronograph from 2016. Jean-Claude Biver and his team demonstrate with Caliber Heuer 02T that complicated manufactory work, including an automatic chronograph and tourbillon, needn't be astronomically unaffordable. The unlimited version with a titanium case costs less than 15,000 euros. ○

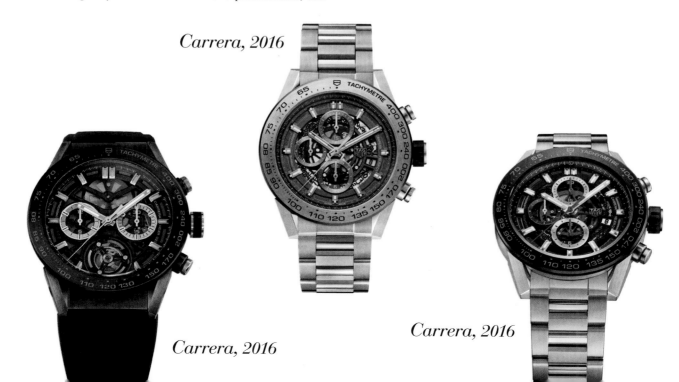

Carrera, 2016

Carrera, 2016

Carrera, 2016

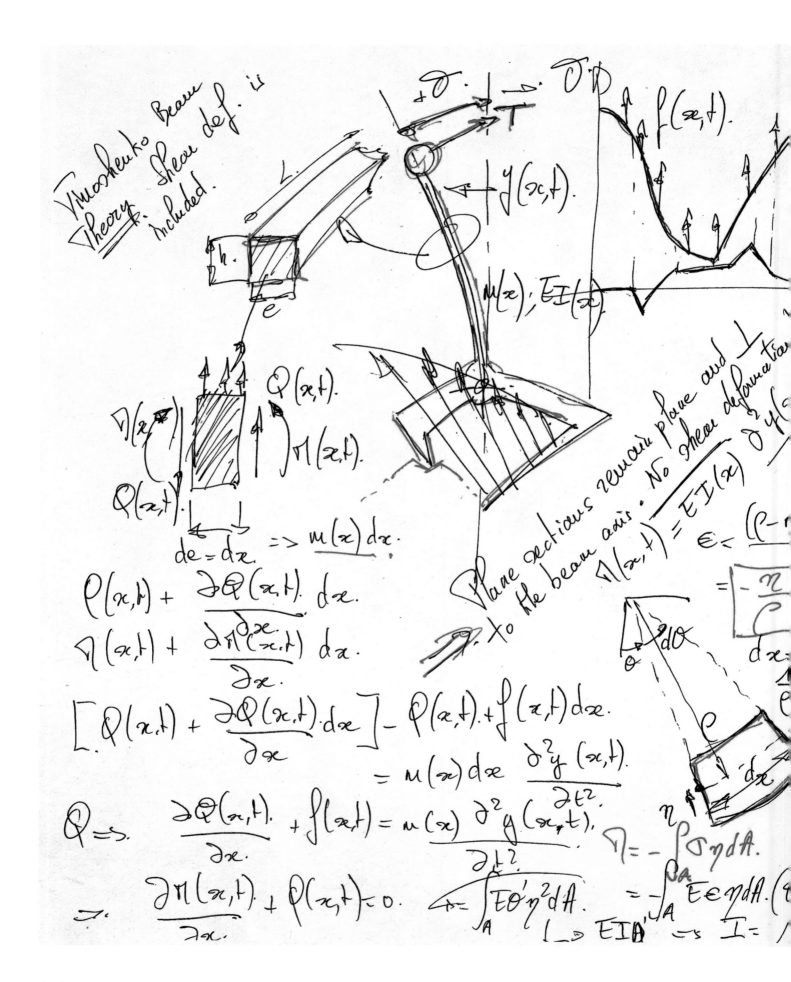

Timoshenko Beam Theory: shear def. is included.

$L.$

$h.$

e

$Q(x,t)$

$\mathbb{T}(x,t)$

$M(x,t)$

$Q(x,t).$

$de = dx.$ $\Rightarrow$ $m(x)dx.$

$Q(x,t) + \dfrac{\partial Q(x,t)}{\partial x} dx.$

$\mathbb{T}(x,t) + \dfrac{\partial \mathbb{T}(x,t)}{\partial x} dx.$

$\left[Q(x,t) + \dfrac{\partial Q(x,t)}{\partial x} dx \right] - Q(x,t) + f(x,t)dx.$

$= m(x)dx \dfrac{\partial^2 y(x,t)}{\partial t^2}.$

$Q \Rightarrow$ $\dfrac{\partial Q(x,t)}{\partial x} + f(x,t) = m(x) \dfrac{\partial^2 y(x,t)}{\partial t^2}.$

$\mathbb{T}.$ $\dfrac{\partial M(x,t)}{\partial x} + Q(x,t) = 0.$

$y(x,t).$

$m(x); EI(x).$

$f(x,t).$

Plane sections remain plane and $\perp$ to the beam axis. No shear deformation.

$M(x,t) = EI(x) \dfrac{\partial^2 y(x,t)}{\partial x^2}$

$\epsilon = \dfrac{(\rho - \eta)}{\rho}$

$= \dfrac{-\eta}{\rho}$

$d\alpha$

ρ

dx

$\eta = -\int_A \sigma \eta \, dA.$

$= -\int_A E\epsilon \eta \, dA.$

$M = \int_A E\theta' \eta^2 dA.$

$\rightarrow EIA'$ $\Rightarrow$ $I =$

$$\underline{\text{Free vibration}} - \frac{\partial^2}{\partial x^2}\left[EI(x)\,\frac{\partial^2 y(x,t)}{\partial x^2}\right] =$$

$$m(x)\cdot\frac{\partial^2 y(x,t)}{\partial t^2}\,; \quad -\frac{d^2}{dx^2}\left[EI(x)\,\frac{d^2 Y(x)}{dx^2}\right]\cdot F(t) =$$

$$m(x)\,Y(x)\,\frac{d^2 F(t)}{dt^2}; \quad \underline{\text{Where}}$$

$$Y(0) = 0; \quad \frac{dY(x)}{d(x)}\bigg|_{x=0} = 0; \quad \frac{d^2 Y(x)}{d(x)^2}\bigg|_{x=L} = 0.$$

$$x: \quad \frac{d}{dx}\left[EI(x)\,\frac{d^2 Y(x)}{dx^2}\right]\bigg|_{x=L} \quad \bigcirc \quad -\omega^2$$

$$\frac{1}{m(x)\,Y(x)}\,\frac{d^2}{dx^2}\left[EI(x)\,\frac{d^2 Y(x)}{dx^2}\right] = \frac{1}{F(t)}\,\frac{d^2 F(t)}{dt^2}$$

$$\Rightarrow \quad F(t) = A\sin\omega t + B\cos\omega t = C\cos(\omega t + \emptyset).$$

$$\underline{\text{Matrix Boundary Conditions}}$$

$$d\theta \quad \begin{bmatrix} 0 & 1 & 0 & 1 \\ 1 & 0 & 1 & 0 \\ -\sin\beta L & -\cos\beta L & \sinh\beta L & \cosh\beta L \\ -\cos\beta L & \sin\beta L & \cosh\beta L & \sinh\beta L \end{bmatrix} \begin{Bmatrix} Y(x) \\ Y'(x) \\ Y''(x) \\ Y'''(x) \end{Bmatrix} \begin{Bmatrix} A \\ B \\ C \\ D \end{Bmatrix}$$

$$-\eta\frac{d\theta}{dx} \quad \boxed{\text{Non-Trivial Solutions}}$$

$$\det.\begin{bmatrix} 0 & & \\ -\sin\beta L & & \\ -\cos\beta L & & \end{bmatrix} = 0.$$

$$\Rightarrow \quad 1 + \cos(\beta L)\cosh(\beta L) = 0.$$

$$= \beta^2 \sqrt{\frac{EI}{m}} \quad \boxed{\text{Solutions: Eigenvalues.}}$$

$$\omega_1 = 1.875^2\,\frac{EI}{mL^4} \quad \text{etc...} \quad \omega_2, \omega_3 \ldots$$

$$\text{response} \quad y(x,t) = \sum_{r=1}^{\infty} Y(x)\,\eta_r(t). \quad Y_r(x). \Rightarrow \text{Man normalized mode shapes.}$$

$$\ddot{\eta}_s(t) + \omega_s^2\,\eta_s(t) = - - -$$

$$\boxed{\eta = EI\,y''} \quad s(t) = \sum_0 - - -.$$

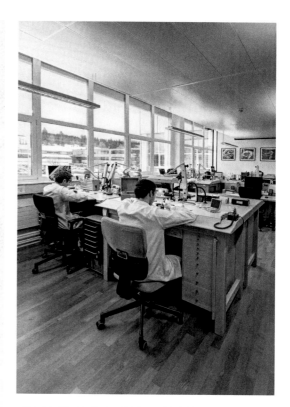

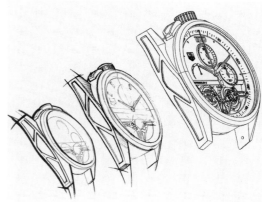

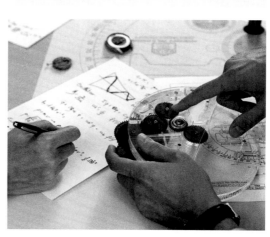

The art of manufacturing
Die Kunst der Manufakturarbeit
L'art de la manufacture

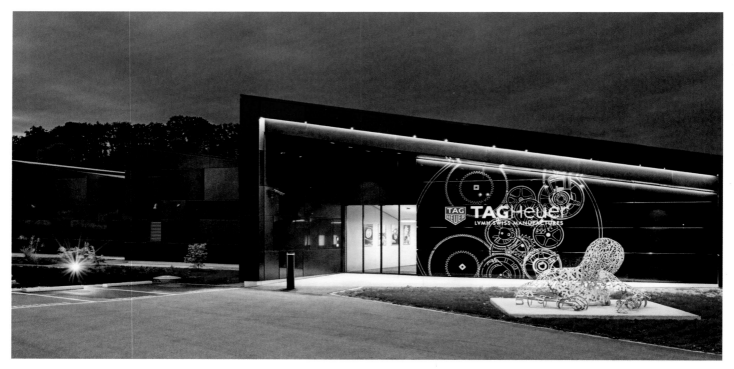

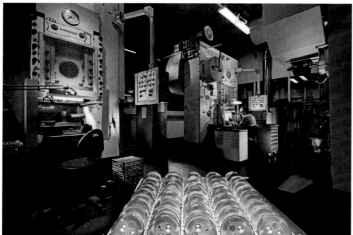

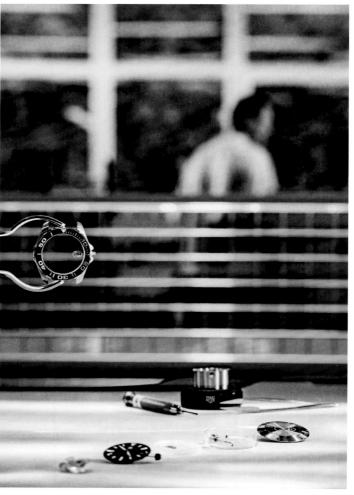

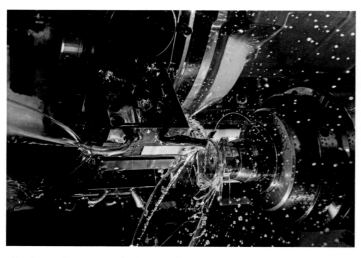

Clockwise from top: The home of TAG Heuer movements in Chevenz ○ Casing workshop ○ Machining and drilling holes in the cases ○ Dial creation ○ *Von oben im Uhrzeigersinn: Heimat der Uhrwerke von TAG Heuer in Chevenez ○ Gehäusefertigung ○ Bearbeitung der Gehäuse ○ Zifferblattherstellung* ○ *Dans le sens horaire en partant du haut : Manufacture des mouvements TAG Heuer à Chevenez ○ Emboîtage à La Chaux-de-Fonds ○ Usinage à Cornol ○ Frappe des cadrans chez ArteCad*

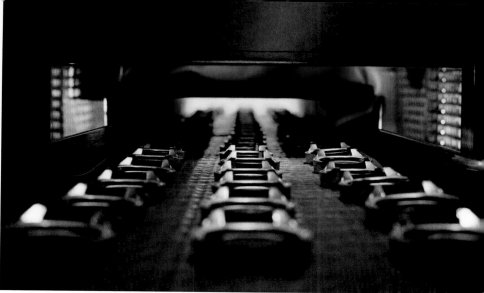
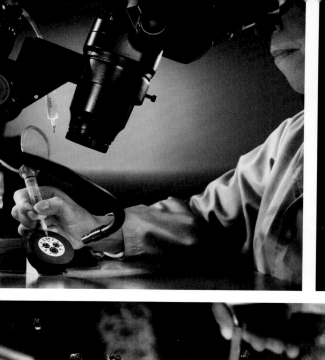
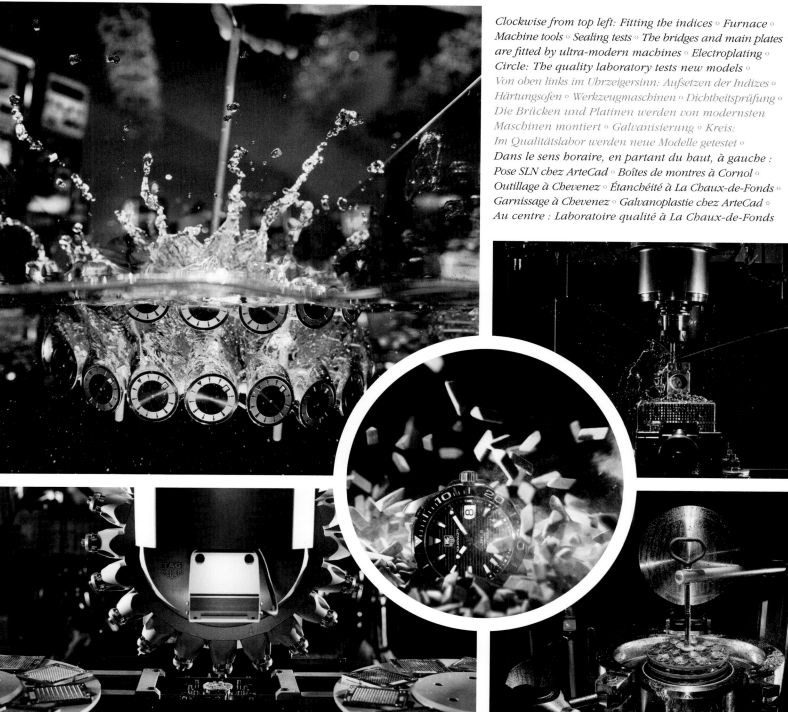

Clockwise from top left: Fitting the indices ○ Furnace ○
Machine tools ○ Sealing tests ○ The bridges and main plates
are fitted by ultra-modern machines ○ Electroplating ○
Circle: The quality laboratory tests new models ○

*Von oben links im Uhrzeigersinn: Aufsetzen der Indizes ○
Härtungsofen ○ Werkzeugmaschinen ○ Dichtheitsprüfung ○
Die Brücken und Platinen werden von modernsten
Maschinen montiert ○ Galvanisierung ○ Kreis:
Im Qualitätslabor werden neue Modelle getestet ○*

*Dans le sens horaire, en partant du haut, à gauche :
Pose SLN chez ArteCad ○ Boîtes de montres à Cornol ○
Outillage à Chevenez ○ Étanchéité à La Chaux-de-Fonds ○
Garnissage à Chevenez ○ Galvanoplastie chez ArteCad ○
Au centre : Laboratoire qualité à La Chaux-de-Fonds*

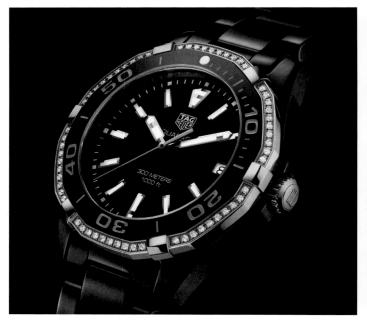
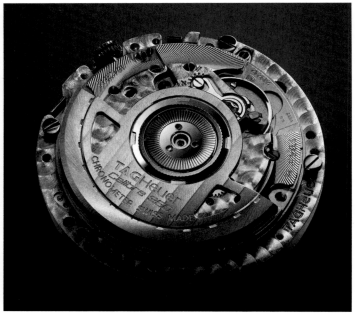

Von der Chronographenmanufaktur zur Chronographenmanufaktur

Aus Heuer wird TAG Heuer

Chronographen durchziehen das Leben von Edouard Heuer, Jahrgang 1840, und das seiner Nachfahren wie ein roter Faden. Mit 14 erhielt er seine Ausbildung zum Uhrmacher. Nach ersten beruflichen Erfahrungen und Eheschließung machte er sich mit 20 selbständig. Ab 1865 fand man seine Firma in Biel. Dort entwickelte er einen Kronenaufzug für Uhren, der die Schlüssel überflüssig machte und 1869 in sein erstes Patent mündete. Krisenzeiten führten ihm vor Augen, dass das Heil in der Spezialisierung zu suchen war. Ab 1882 beinhaltete die Kollektion deshalb auch – teilweise patentierte – Zeitschreiber eigener Provenienz. Am Heiligabend des Jahres 1886 beantragte Edouard Heuer Schutz für die moderne Schwingtriebkupplung, welche sich bis heute nicht überlebt hat. Weitere einschlägige Erfindungen folgten.

1889 konnte der Patron während der Weltausstellung in Paris eine Silbermedaille für sein umfassendes Uhrenspektrum entgegennehmen. Als er am 30. April 1892 überraschend verstarb, hinterließ er ein stattliches Nettovermögen von 552 750 Schweizer Franken.

Auf diesem finanziellen Fundament konnten die Söhne Jules-Edouard und Charles-Auguste I. die Aktivitäten fortführen. Nach dem Tod des Bruders am 12. Juli 1911 lastete die Verantwortung allein auf den Schultern von Charles-A. Heuer. Ihm sind u. a. ein Chronograph mit Pulsometer-Skala oder der „Mikrograph" von 1916 zu verdanken. Mit der revolutionären Stoppuhr konnte man erstmals

auf die Hundertstelsekunde genau stoppen. Sie und ihre Derivate überzeugten die Zeitnehmer bei den Olympischen Spielen 1920, 1924 und 1928. Seit 1913 gab es auch Armbanduhren mit der Signatur. Die nächste Generation in Person von Charles-Edouard II. und Hubert-Bernard kümmerte sich um den systematischen Ausbau wichtiger Märkte. 1919 legten sie sich (bis 1936) die Uhrenmarke Jules Jürgensen zu. Nach 1929 und der daraus resultierenden Weltwirtschaftskrise musste Heuer etliche Turbulenzen überstehen. Das Überleben sicherten kreative Produkte, darunter Stoppuhren und Chronographen für sehr unterschiedliche Zwecke. 1947 schmückten Prinz Wilhelm von Schweden sowie der designierte US-Präsident Harry S. Truman ihre Handgelenke mit goldenen Armbanduhren von Heuer.

Die letzte Generation repräsentierte der 1932 geborene Jack William Heuer. Durch sein Ingenieurstudium und Praxiserfahrung in den USA qualifizierte sich der Sohn von Charles-Edouard II. für die spätere Leitungsfunktion. 1958 trat er in die Fußstapfen seiner Vorfahren. Als Teilhaber am Familienunternehmen fusionierte er Heuer 1964 mit Leonidas zu Heuer-Leonidas. 1968 standen zehn Millionen Franken Umsatz in den Büchern, 1970 bereits das Doppelte. Dringender Kapitalbedarf führte zum Gang an die Börse. Fortan musste Jack W. Heuer, der zu den unangefochtenen Pionieren elektronischer Kurzzeitmesser gehörte, Dividenden erwirtschaften. Aber die allgemein schlechte Situation und die fernöstliche

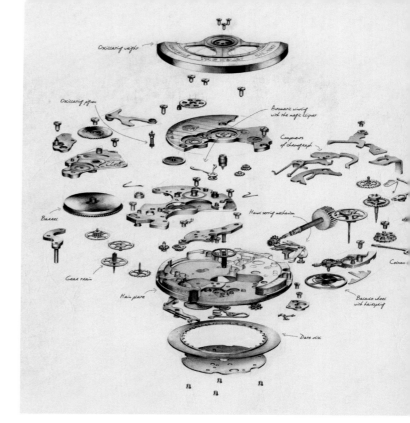

Billigkonkurrenz machten dicke Striche durch die Rechnung. Ab 1976 trafen ihn die Folgen der Quarz-Krise und 1981 die Auswirkungen eines misslungenen Stoppuhren-Geschäfts mit China. Die letzte öffentliche Generalversammlung der Heuer-Leonidas SA ging am 25. Juni 1982 über die Bühne. Jack W. Heuer, der einen Konkurs ablehnte, und die übrigen Inhaber von Namensaktien mussten der Abschreibung ihres Aktienkapitals und der Auflösung vorhandener Reserven zustimmen. Neue Hauptaktionäre wurden die Nouvelle Lemania und ihr nahestehende Anteilseigner, darunter das Haus Piaget. Ab 1985 gehörte die Aktienmehrheit der international tätigen TAG-Gruppe (Techniques d'Avant-Garde). Aus Heuer wurde TAG Heuer. Deutlich kommerziellere Uhren, meist mit elektronischen Quarzwerken, gekonntes Marketing unter Einbeziehung der Formel 1 und frische Werbekampagnen ließen TAG Heuer auferstehen wie Phoenix aus der Asche. Ab 1996 wurden die Aktien der TAG Heuer SA erneut an der Börse gehandelt. Mit jährlich über 715 000 verkauften Uhren und einem Umsatz von knapp 420 Millionen Schweizer Franken war TAG Heuer 1996 in der Schweiz zur Nummer fünf unter den Uhrenherstellern avanciert. 1997 brachte nicht nur Jack W. Heuer als Berater zurück, sondern auch die legendäre „Carrera" von 1963.

1998 zelebrierte TAG Heuer die Renaissance der avantgardistischen „Monaco".

Im September 1999 erwarb der französische Luxusmulti LVMH die traditionsreiche Sportuhrenmarke für knapp 1,2 Milliarden Schweizer Franken. Unter seinem Dach erlebte TAG Heuer den Aufstieg zur echten Uhrenmanufaktur mit einer Vielzahl innovativer Kaliber und insgesamt vier Produktionsstandorten im Westschweizer Jurabogen. Ab November 2000 lenkte Jean-Christophe Babin, aktuell Bulgari-CEO, die Geschicke. Er initiierte Hochfrequenz-Stopper wie „360", „Mikrograph", „Mikrogirder" und „MikrotourbillonS". Die 2004 lancierte Konzept-Armbanduhr „V4" brachte den studierten Physiker und Mathematiker Guy Sémon ins Haus. Ihm sind zahlreiche uhrmacherische Revolutionen zu verdanken. Eine echte Sensation steht noch 2016 ins Haus. Der große Strukturwandel begann am 1. März 2014 mit Jean-Claude Biver. Der koordinierende Chef von Hublot, TAG Heuer und Zenith entdeckte dringenden Handlungsbedarf, übernahm Anfang 2015 die Funktion des CEO und krempelte die Schweizer Traditionsmarke gründlich um. Sein Credo: „TAG Heuer muss Uhren anbieten, welche der Kaufkraft des klassischen Kunden gerecht werden." Und das ist eine Größenordnung zwischen 1 000 und 5 000 Schweizer Franken. Bekannte Markenbotschafter wie Cara Delevingne, Chris Hemsworth, David Guetta oder Tom Brady, die markante „Carrera Heuer 01", die erfolgreiche Smartwatch „Connected", der preisaggressive Tourbillon-Chronograph „Carrera Heuer 02" oder die Kooperation mit dem Red Bull Racing Team lassen erkennen, wohin die Reise künftig gehen wird. ○

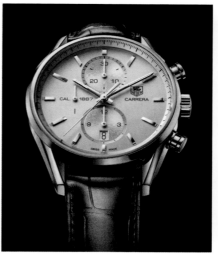

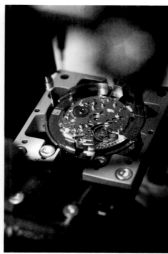

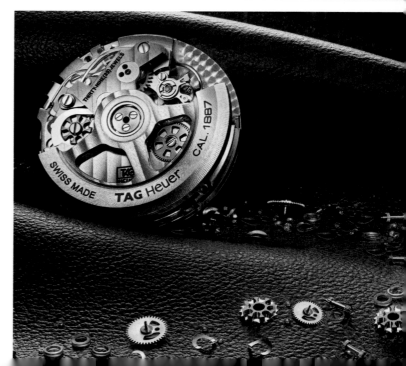

Cal. 1887

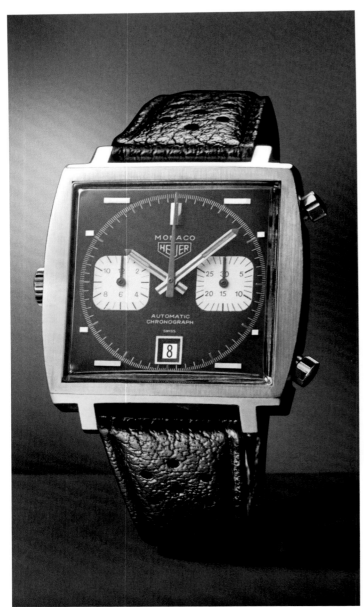

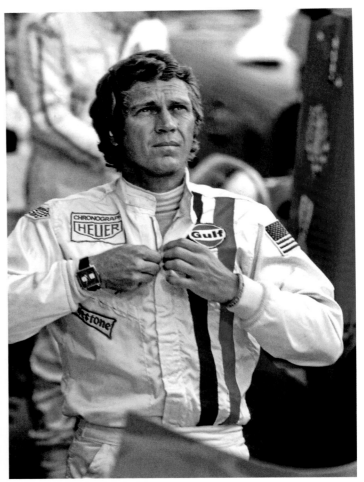

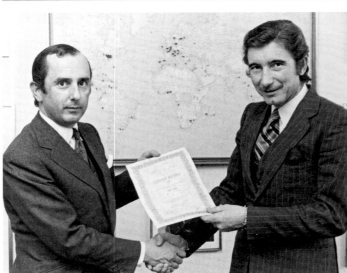

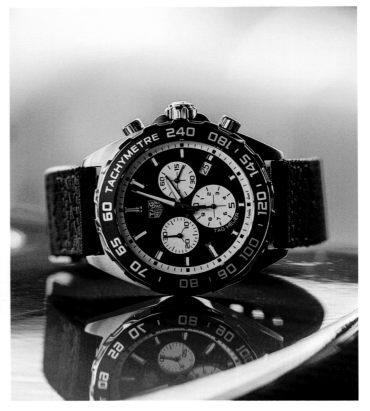

Clockwise from top left: Original Monaco, 1969 ◦ Steve McQueen, 1970 ◦ Formula 1 Red Bull Special Edition, 2016 ◦
Jack Heuer and Jo Siffert in Bienne, 1970

Werke, Gehäuse und Zifferblätter

Eine rechtlich eindeutige Definition dessen, was eine Uhrenmanufaktur ist, existiert nicht. Mindestens ein eigenes Kaliber sollte aber schon vorhanden sein. Darüber hinaus sollte eine Firma, welche diesen Anspruch für sich erhebt, mindestens auch die tragenden Teile, sprich die Platine, Brücken und Kloben des Uhrwerks fertigen. Die Stunde eigenen Tuns schlug bei TAG Heuer 2010 mit dem Lizenz-Kaliber 1887. Dabei gilt es zu wissen, dass Chronographen wegen der komplexen Funktionszusammenhänge besonders hohe Ansprüche stellen. Der eidgenössische Sachverstand zeigte sich in differenzierter Analyse und tiefgreifender Optimierung der japanischen Basis. Komplett eigene Leistung repräsentiert hingegen das Kaliber Heuer 02 mit Selbstaufzug, Chronograph und innovativem Minutentourbillon, dessen Drehgestell aus leichtem, hoch belastbarem Karbon besteht.

So geordnet, wie es in einem mechanischen Mikrokosmos zugeht, ist auch eine Manufaktur strukturiert. Bei TAG Heuer agieren die Produktentwickler, Konstrukteure und Designer am Hauptsitz in La Chaux-de-Fonds. Egal ob Gehäuse, Zifferblatt, Zeiger, Armband oder Uhrwerk: Sie kümmern sich um alles. Aus Gründen der Kapazität und kompetitiver Preise – Werke aus eigener Manufaktur sind grundsätzlich teurer als zugekaufte – verbaut TAG Heuer überwiegend auch Bewährtes von Eta und Sellita. Auch die Prototypisten erledigen ihren Job in der Zentrale. Ferner unterziehen die „Folterknechte" in einem ausgedehnten Labor alles, was in die Serienfertigung gehen soll, extremen Torturen.

In Chevenez, einem abgelegenen Dorf nahe an der französischen Grenze, wo sich Arbeitskräfte leichter finden lassen, geht die Teileproduktion T0 über die Bühne. Das Fräsen und Bearbeiten der Platinen, Brücken und Kloben obliegt computergesteuerten und extrem flexiblen Fertigungszentren der neuesten Generation. Anschließend benötigt das benachbarte Atelier T1 rund 100 Arbeitsschritte, damit aus qualitätsgeprüften Komponenten fertige Chronographenwerke werden. Automatisierung sorgt auch hier für

ein Optimum an Güte. Nach gründlicher Reorganisation erreicht die Fertigungstiefe bei den eigenen Werken inzwischen mehr als 50 Prozent.

Käufer einer Armbanduhr mit der Signatur TAG Heuer erhalten auch die schützende Schale aus eigener Manufaktur. In Cornol nahe Chevenez unterhält das Unternehmen eine hochmoderne Gehäusefabrik mit gegenwärtig circa 140 Beschäftigten. Sie produzieren in erster Linie Stahlgehäuse für Zeitmesser von „Aquaracer" bis „Monza". Martialisch anmutende Stanzen verursachen den meisten Lärm. Aber auch die computergesteuerten Fertigungszentren sind unüberhörbar. Die finale Politur teilen sich nimmermüde Roboter und Fachkräfte mit hoher Expertise. Nach gründlicher Kontrolle geht es an die Montage inklusive abschließendem Check auf Wasserdichte.

Dritte im Bunde ist ArteCad, eine traditionsreiche Zifferblattfabrik. Dieser Tochter von TAG Heuer ist nichts fremd. Das gilt auch für vergessen geglaubte Techniken. Bei neuen Zifferblättern, beispielsweise für die „Carrera Heuer 01" und „Carrera Heuer 02T", ist die Komplexität in den vergangenen Jahren sprunghaft angestiegen. Hier sind die „Gesichter der Zeit" vielteilige dreidimensionale Kunstwerke. Maschinelles und manuelles Tun gehen in Tramelan Hand in Hand. Ziel ist absolute Perfektion, denn jeder Blick auf die Uhr gilt in erster Linie dem Zifferblatt und den davor drehenden Zeigern.

Für die Assemblage und finale Kontrolle des Ganzen, welches bekanntlich mehr ist als die Summe seiner vielen Teile, sind schließlich die Werkstätten in La Chaux-de-Fonds zuständig. Von dort nehmen die fertigen Uhren ihren Weg in die ganze Welt. TAG Heuer ist traditionsgemäß eine internationale Marke. Aber Jean-Claude Biver möchte die Präsenz rund um den Globus noch deutlich ausbauen. ○

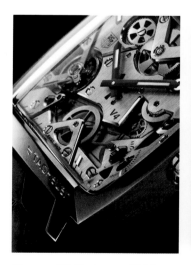
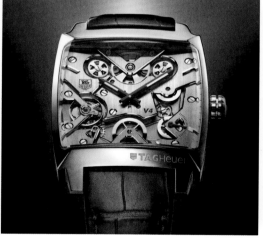
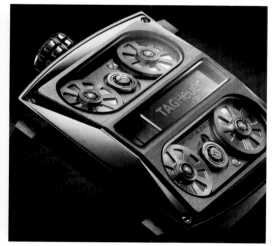

Monaco V4 Platinum, 2004

Chronometrische und chronographische Höhepunkte seit 1887

Als erstes Highlight kann zweifellos der 1887 patentierte Taschen-Chronograph mit Schwingtrieb-Kupplung gelten. Letztere erfüllte die gleiche Funktion wie die gängige Räderkupplung mit Schwenkhebel, minderte die Kosten für ein Uhrwerk mit Chronograph jedoch erheblich.

Die erste Borduhr von Heuer bestach durch ein großes, selbst bei ungünstigen Lichtverhältnissen bestens ablesbares Ensemble aus Zifferblatt und Zeigern. Ab 1911 fand man den „Time of Trip" in den Cockpits von Flugzeugen und Autos. 1919, bei der ersten Nordatlantiküberquerung eines Luftschiffs, blickte die Crew der R34 auf diese Uhr. Hugo Eckener tat Gleiches, als er die Erde 1929 mit dem Luftschiff „Graf Zeppelin" umrundete.

Die Entwicklung des weltweit ersten Tempo-Stoppers startete 1914. Das Patent datiert auf den 2. Oktober 1916. Für Hundertstelsekunden-Stoppgenauigkeit oszillierte die kleine Unruh mit 50 Hertz. Der „Mikrograph" eignete sich laut Modellkatalog „zum Erfassen der Flüge von Projektilen und andere Kurzzeittests". Sein Preis lag bei 100 Schweizer Franken oder 120 US-Dollar.

Hoch im Kurs bei Sammlern stehen die Dashboard-Zeitmesser von Heuer. Ab 1933 fand sich „Autavia", eine Kombination aus Uhr und Stopper, an vielen Armaturenbrettern. Dort unterstützte sie zum Beispiel Renn- und Rallyefahrer. Daneben vertrauten aber auch Flugzeugpiloten und Bootskapitäne auf die Zeit-Accessoires aus Biel, die ab den späten 1950er Jahren „Rally-Master" hießen.

Ebenfalls in den 1930ern beschäftigte sich Heuer sehr intensiv mit wassergeschützten Gehäusen für Chronographen. Der vorgestellte Armbandchronograph mit wasserabweisender Schale kam den Schweizer Militärs sehr entgegen. 1942 startete die Produktion wasserdichter Militär-Chronographen mit schwarzem Blatt.

1949 brachte mit der „Solunar" die patentierte Armbanduhr „für Fischer, Jäger, Segler, Seeleute, Sportsleute, Naturforscher, Wissenschaftler, Mediziner, Landwirte, Vogelzüchter, Gärtner, Forstmeister", wie im Modellkatalog zu lesen stand. Neben der Zeit stellte sie auch die Mondstellung, die Solunar-Perioden und die Gezeiten dar. Per Drücker erfolgte die Synchronisation mit lokal veröffentlichten Tabellen.

Regattaseglern in der schwierigen Vorstart-Phase half der 1962 lancierte „Yachting Timer". Nach dem Starten informierte der Stopper zuverlässig über den aktuellen Stand des Countdown.

Ein Comeback des einprägsamen Namens „Autavia" ist für 1962 zu verzeichnen. Und zwar in Gestalt eines Armbandchronographen. Die rastende Drehlünette am wasserdichten Stahlgehäuse war in unterschiedlichen Versionen beispielsweise für Piloten, Sporttaucher oder Wissenschaftler zu haben.

Bei den berühmten 12-Stunden-Rennen von Sebring kümmerte sich Jack W. Heuer höchstpersönlich um die offizielle Zeitnahme. Dort hörte er von dem legendären Straßenrennen Carrera Panamericana. Deshalb erhielt seine 1963 eingeführte Chronographen-Kreation mit großflächig und dreidimensional gestaltetem Zifferblatt den Namen „Carrera". Der Spannring im gewölbten Plexiglas trug die Sekundenskala.

Erstmals in der Geschichte des Chronographen brachte Heuer 1965 ein Exemplar mit Monofenster-Datumsanzeige auf den Markt. Bis dahin hatte es nur Zeiger- oder Doppelfensterdatum gegeben. Logischerweise machte auch die „Carrera 45 Dato" von sich reden.

Mit der Vorstellung des weltweit ersten Modul-Chronographen mit Mikrorotor-Selbstaufzug, Schwingtrieb-Kupplung und Kulissenschaltung ging bei Heuer auch eine neue Uhrenlinie einher. Die quadratische „Monaco"-Schale war zugleich auch wasserdicht. Als Botschafter empfahl sich der Formel-1-Pilot Jo Siffert. Folglich trug auch Steve McQueen im Film *Le Mans* zum Original-Jo-Siffert-Dress die „Monaco".

Die Serie 1000 von 1992 wandte sich u.a. an Taucher, welche unter extremen Druckverhältnissen arbeiten müssen. Trotz beachtlicher Wasserdichte bis 100 Bar Druck kam die Schale ohne das übliche Heliumventil aus.

„Kirium Ti5", gestaltet von Jörg Hysek, leitete 1998 eine neue Design-Ära ein. Signifikantes Charakteristikum war der fließende, organische Übergang vom Gehäuse zum Armband. Darüber hinaus präsentierte TAG Heuer neue Hochleistungs-Werkstoffe: Karbon für das Zifferblatt und Titan für Gehäuse und Band.

Mechanik oder Elektronik? Diese Frage stellte sich nicht bei der „Monaco Sixty Nine", Jahrgang 2003. Die innovative Armbanduhr vereinte im Wendegehäuse je ein tickendes und ein quarzgesteuertes Uhrwerk.

2004 debütierte die „V4" als Konzept-Uhr. 2009 war sie Realität. Die kinetische Energie des linearen Schwungkörpers wird über zwei Mikro-Zahnriemen mit Metallseele an gleich vier Federhäuser weitergeleitet. Den Zeittakt liefert eine konventionelle Unruh.

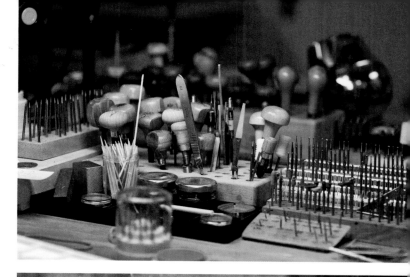

Eine Armbanduhr mit zwei Werken war 2006 die Weltpremiere namens „Carrera Calibre 360 LE". Eines indiziert die Uhrzeit, das andere stoppt Zeitintervalle auf die Hundertstelsekunde genau.

2010, im Jahr des 150. Firmenjubiläums, kehrte TAG Heuer zur uhrmacherischen Manufaktur zurück. Beim Automatikkaliber 1887 mit Schwingtrieb-Chronograph hatten die Konstrukteure auf ein Uhrwerk von Seiko zurückgegriffen. Die Fertigung des aus 320 Teilen zusammengefügten Œuvres erfolgt komplett in der Schweiz.

Auf die Tausendstelsekunde stoppt der mechanische „Mikrotimer Flying 1000 Concept Chronograph", dessen Geburtsstunde 2011 schlug. Analog zu den Hundertstelsekunden-Vorbildern besitzt die Armbanduhr zwei getrennte Getriebeketten mit je einem Schwing- und Hemmungssystem. Zehn angemeldete Patente künden vom vorangehenden Einfallsreichtum.

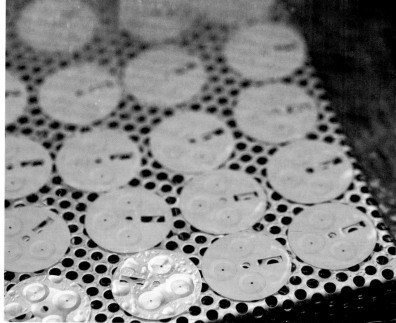

Kurzzeitmessung auf die 5/10000stel oder 1/2000stel Sekunde genau bietet der mechanische „Mikrogirder". Stündlich 7200000 Halbschwingungen leistet keiner der herkömmlichen Gangregler. Ergo kooperieren in dieser Konstruktion, für die gleich zehn Patente beantragt wurden, drei rechtwinklig angeordnete Metallklingen. Geburtsjahr war 2012.

Im gleichen Jahr krönte „MikrotourbillonS2" mit zwei unterschiedlich schnellen Tourbillons die Tempo-Philosophie. Das flottere, Unruhfrequenz 50 Hertz, liefert den Takt zum Stoppen auf die Hundertstelsekunde genau. Konventioneller Natur ist der für die Zeitanzeige zuständige Drehgang.

Auf dem Kaliber 1887 basiert die 2015 vorgestellte „Carrera Heuer 01". Die speziell gestaltete Vorderseite verbirgt nichts. Der Sichtboden ebenfalls nicht. Aus zwölf Teilen ist die bis zehn Bar wasserdichte „Carrera"-Schale modular konzipiert. Das eröffnet die Kombination unterschiedlicher Materialien, Oberflächen und Farben.

Bleibt last, but not least der „Carrera"-Tourbillon-Chronograph von 2016. Anhand des Kalibers Heuer 02T demonstrieren Jean-Claude Biver und sein Team, dass komplizierte Manufakturarbeit mit Automatikchronograph und Drehgang nicht unerschwinglich teuer sein muss. Die unlimitierte Ausführung mit Titangehäuse kostet weniger als 15000 Euro. ○

Opposite page: Time of Trip, 1911

ArteCad dial manufacturing, from top: Gemstone-setting
workshop ○ Surface treatment (sand blasting) ○ Index setting ○
Zifferblattfabrik ArteCad, von oben: Blick in die Werkstatt ○
Oberflächenbehandlung (mit Sandstrahl) ○ Zifferblattgestaltung ○
Manufacture de cadrans chez ArteCad, de haut en bas : Atelier
d'empierrage ○ Traitement de surface (sablage) ○ Pose des index

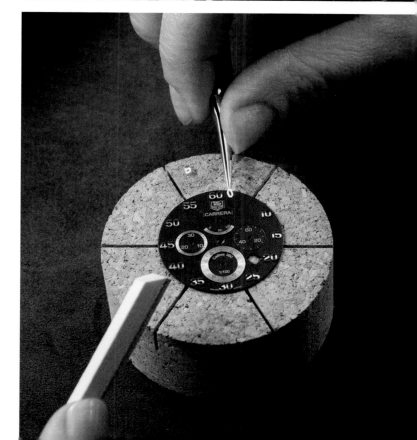

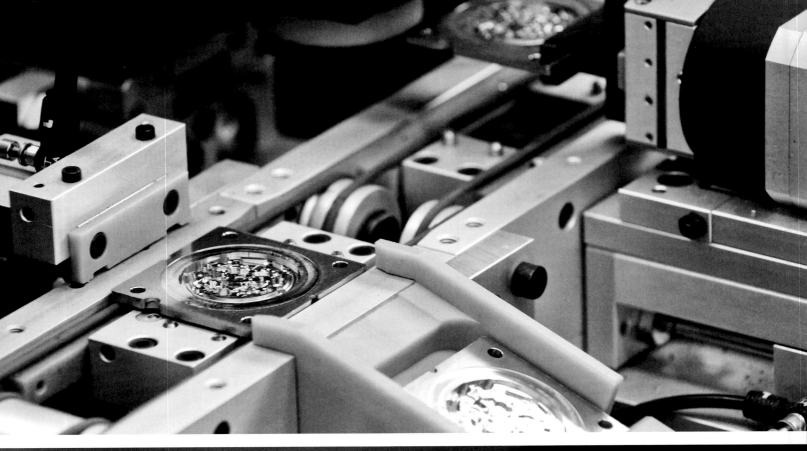

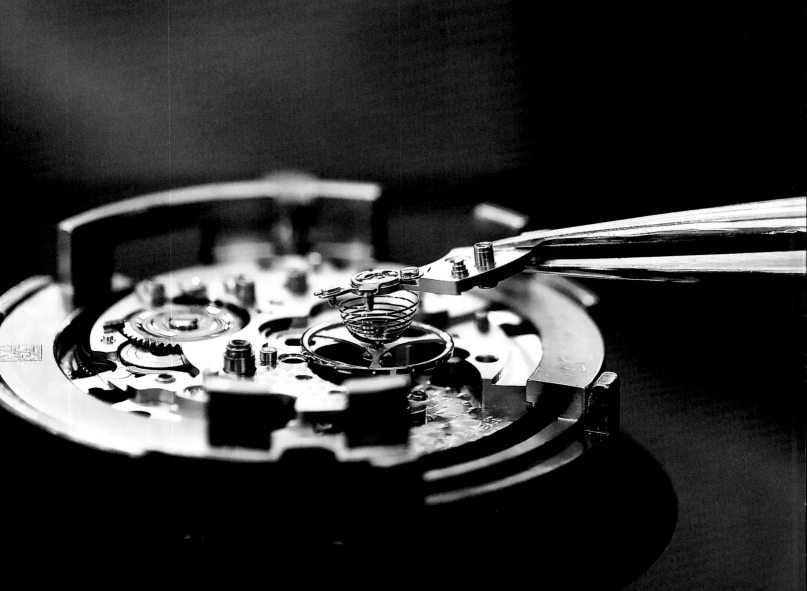

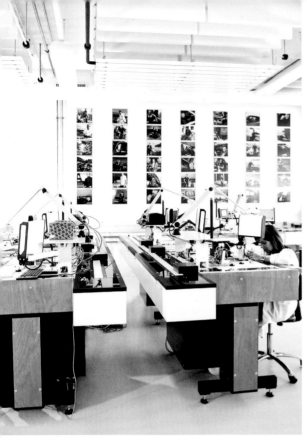

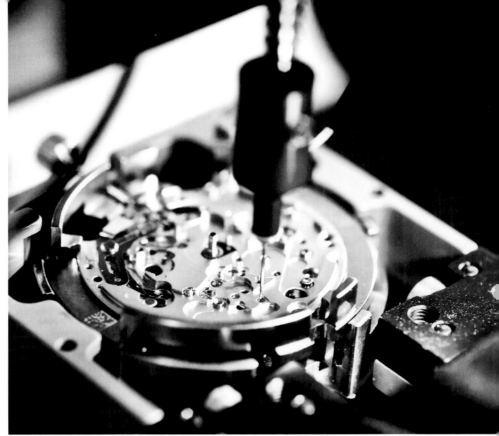

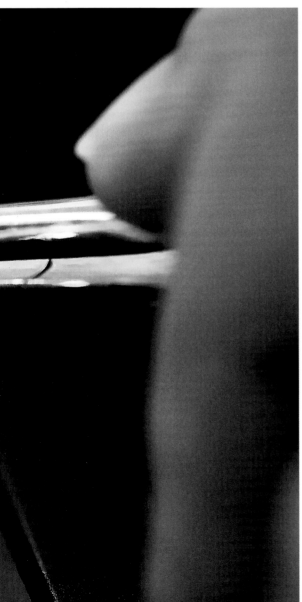

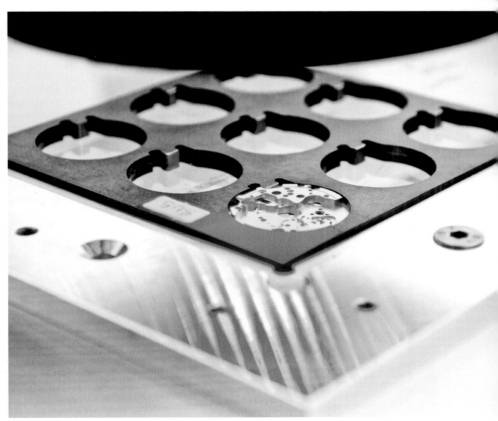

Clockwise from opposite page top left: Milling of the main plates of TAG Heuer movement ○ Movement assembly workshop ○ Drilling of the main plate, in-house movement ○ Laser controlling ○ Hairspring setting ○ Von oben links im Uhrzeigersinn: Fräsen der Hauptplatinen für die Uhrwerke von TAG Heuer ○ Uhrwerkmontage-Atelier ○ Bohren der Hauptplatine, Manufakturwerk ○ Kontrolle per Laser ○ Einsetzen der Unruhspirale ○ Dans le sens horaire, en partant du haut de la page ci-contre : Fraisage des platines des mouvements TAG Heuer ○ Atelier d'assemblage des mouvements ○ Perçage de la platine sur un mouvement de manufacture ○ Commande laser ○ Pose du spiral

D'une manufacture de chronographes à l'autre

Quand Heuer devient TAG Heuer

Français

Les chronographes sont un véritable fil rouge dans la vie d'Édouard Heuer (né en 1840) et de ses descendants. À l'âge de 14 ans, il entre en apprentissage chez un horloger. Après ses premières expériences professionnelles et son mariage, il s'installe à son compte. Il a alors 20 ans. En 1865, il s'établit à Bienne, où il élabore un remontoir à couronne qui rend toute clé inutile et lui vaudra son premier brevet, en 1869. En ces temps de crise, il prend conscience que le salut est dans la spécialisation. C'est pourquoi sa collection compte dès 1882 des chronographes – dont certains font l'objet de brevets – qu'il a lui-même fabriqués. La veille de la Noël 1886, Édouard Heuer dépose une demande de brevet pour le pignon oscillant moderne, invention restée d'actualité. D'autres inventions dans le domaine suivront.

En 1889, ce chef d'entreprise se voit décerner une médaille d'argent lors de l'Exposition universelle à Paris pour sa large gamme de montres. Lorsqu'il décède brusquement le 30 avril 1892, il laisse une confortable fortune nette de 552 750 francs suisses.

Cette base financière permet à ses fils Jules-Édouard et Charles-Auguste de poursuivre ses activités. Après le décès de Jules-Édouard le 12 juillet 1911, Charles-Auguste endosse seul la responsabilité des affaires. On lui doit entre autres un chronographe avec cadran pulsomètre et le « Mikrograph » en 1916. Cette montre stop révolutionnaire permet de chronométrer au 1/100e de seconde. Cette dernière et ses dérivées séduiront les chronométreurs des Jeux olympiques de 1920, 1924 et 1928. En 1913, Heuer signe ses premières montres-bracelets. La génération suivante, représentée par Charles-Édouard et Hubert-Bernard, s'attache à renforcer la présence de la marque sur des marchés clés. En 1919, ils rachètent la marque horlogère Jules Jürgensen, qu'ils revendront en 1936. Après 1929 et la crise économique mondiale qui s'ensuit, Heuer doit surmonter de nombreuses turbulences. La survie de la marque est assurée par des produits innovants, notamment des montres stop et des chronographes aux fonctions très variées. En 1947, le prince Guillaume de Suède et le président américain Harry S. Truman portent des montres-bracelets Heuer en or.

La dernière génération est représentée par Jack William Heuer. Né en 1932, le fils de Charles-Édouard acquiert les qualifications requises pour ses futures fonctions de direction grâce à ses études d'ingénierie et son expérience pratique aux États-Unis. En 1958, il suit les traces de ses aïeux. En qualité d'associé dans l'entreprise familiale, il opère en 1964 la fusion de Heuer avec Leonidas, donnant naissance à Heuer-Leonidas. En 1968, les livres comptables font apparaître dix millions de francs suisses de chiffre d'affaires, et le double dès 1970. Mais un besoin urgent en capitaux entraîne l'introduction en Bourse. Jack W. Heuer, qui figure parmi les pionniers incontestés des instruments de chronométrage électronique, doit désormais dégager des dividendes. Or, la conjoncture mondiale difficile et la concurrence asiatique à bas prix contrecarrent sérieusement ses plans. À dater de 1976, il est rattrapé par la crise du quartz et, en 1981, par les conséquences de l'annulation d'une commande de chronomètres par la Chine. La dernière assemblée générale publique des actionnaires de Heuer-Leonidas SA se tient le 25 juin 1982. Jack W. Heuer, qui s'oppose à la faillite, et les autres détenteurs d'actions nominatives doivent voter l'amortissement de leur capital social et la dissolution des réserves existantes.

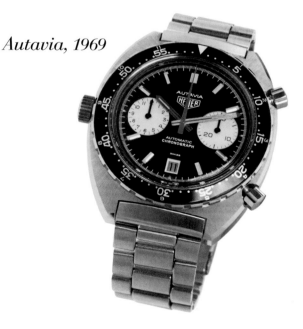

Autavia, 1969

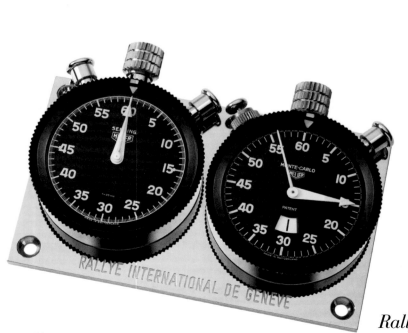

Rallye International de Genève, 1960

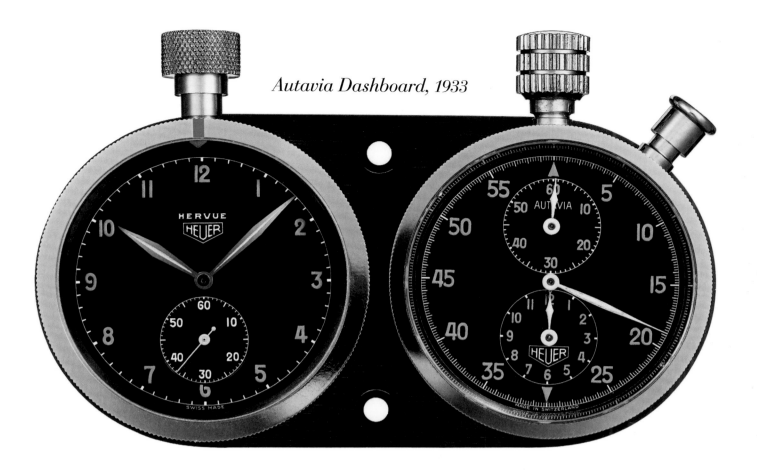

Autavia Dashboard, 1933

Les nouveaux actionnaires de contrôle sont la Nouvelle Lemania et ses propres actionnaires, dont la maison Piaget. À compter de 1985, la majorité des actions appartient au groupe d'envergure internationale TAG (Techniques d'Avant-Garde). Heuer devient TAG Heuer. Grâce à des montres nettement plus commerciales, le plus souvent dotées de mouvements électroniques à quartz, à un marketing réussi par l'association avec la Formule 1 et des campagnes publicitaires qui décoiffent, TAG Heuer renaît de ses cendres comme le Phénix. À partir de 1996, les actions TAG Heuer SA sont à nouveau négociées en Bourse. Avec plus de 715 000 montres vendues par an et un chiffre d'affaires de près de 420 millions de francs suisses, TAG Heuer devient en 1996 la cinquième société horlogère suisse. 1997 marque le retour de Jack W. Heuer comme conseiller, ainsi que de la légendaire « Carrera » de 1963.

En 1998, TAG Heuer célèbre la renaissance de l'avant-gardiste « Monaco ».

En septembre 1999, la multinationale française du luxe LVMH acquiert la marque de montres de sport riche d'une longue tradition pour près de 1,2 milliard de francs suisses. Sous son égide, TAG Heuer s'impose comme une véritable manufacture de montres, avec nombre de calibres innovants et quatre sites de production dans l'Arc jurassien en Suisse romande. À partir de novembre 2000, c'est Jean-Christophe Babin, l'actuel PDG de Bulgari, qui prend les rênes de la société. Il est à l'origine des chronomètres haute fréquence « 360 », « Mikrograph », « Mikrogirder » et « MikrotourbillonS ». L'ingénieur mathématicien Guy Sémon, à qui l'on devra de nombreuses révolutions dans le secteur de l'horlogerie, met un pied dans la maison à la faveur du développement de la concept watch révolutionnaire « Monaco V4 » lancée en 2004. La véritable sensation sera pour 2016. Une grande transformation structurelle débute le 1er mars 2014 avec Jean-Claude Biver, responsable de la coordination entre Hublot, TAG Heuer et Zenith. Découvrant qu'il est urgent de prendre des mesures, il endosse début 2015 la fonction de PDG et fait prendre un virage à 180 degrés à la marque suisse de tradition. Son crédo : « TAG Heuer doit proposer des montres à la mesure du pouvoir d'achat du client traditionnel. » Ce pouvoir d'achat est estimé entre 1 000 et 5 000 francs suisses. Les ambassadeurs célèbres de la marque, comme Cara Delevingne, Chris Hemsworth, David Guetta ou Tom Brady, ainsi que la remarquable « Carrera Heuer 01 », la montre connectée « Connected » plébiscitée, le chronographe tourbillon au prix concurrentiel « Carrera Heuer 02 » ou encore la coopération avec l'écurie Red Bull laissent deviner l'orientation qui sera prise à l'avenir. ○

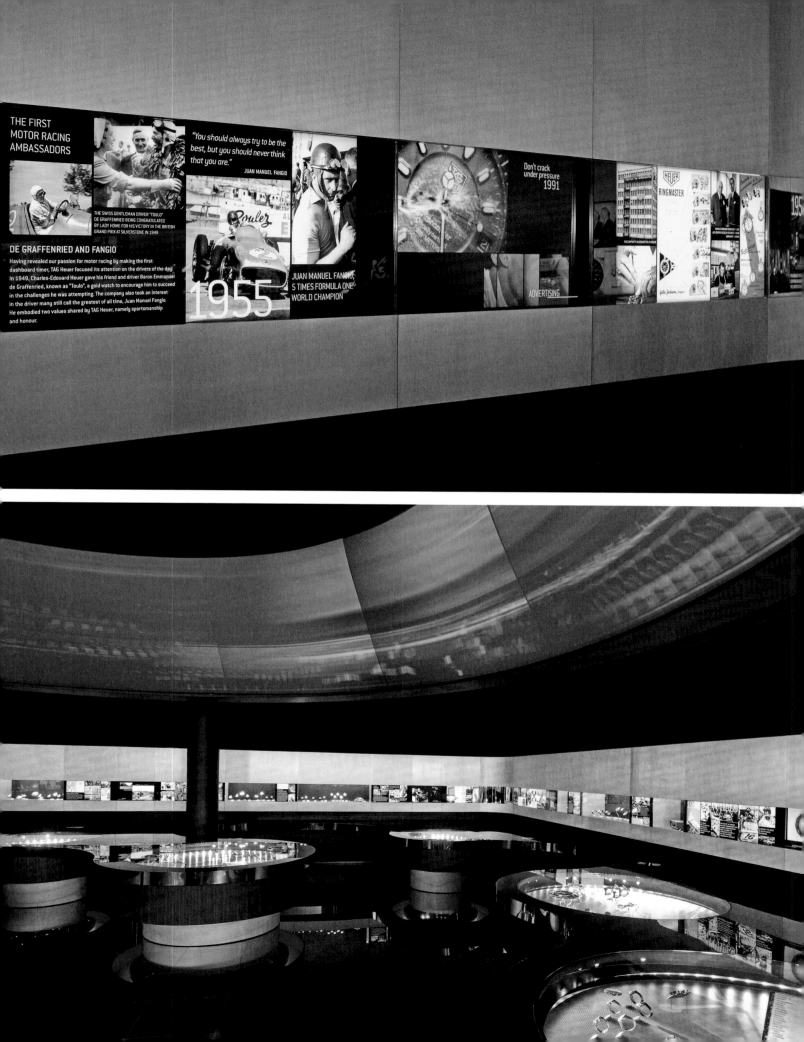

THE FIRST MOTOR RACING AMBASSADORS

"You should always try to be the best, but you should never think that you are."

JUAN MANUEL FANGIO

THE SWISS GENTLEMAN DRIVER "TOULO"
DE GRAFFENRIED BEING CONGRATULATED
BY LADY HOWE FOR HIS VICTORY IN THE BRITISH
GRAND PRIX AT SILVERSTONE IN 1949

1955

JUAN MANUEL FANGIO,
5 TIMES FORMULA ONE
WORLD CHAMPION

Don't crack
under pressure
1991

ADVERTISING

DE GRAFFENRIED AND FANGIO

Having revealed our passion for motor racing by making the first
dashboard timer, TAG Heuer focused its attention on the drivers of the day.
In 1949, Charles-Edouard Heuer gave his friend and driver Baron Emmanuel
de Graffenried, known as "Toulo", a gold watch to encourage him to succeed
in the challenges he was attempting. The company also took an interest
in the driver many still call the greatest of all time, Juan Manuel Fangio.
He embodied two values shared by TAG Heuer, namely sportsmanship
and honour.

Mouvements, boîtiers et cadrans

Juridiquement, il n'existe pas de définition claire de ce qu'est une manufacture de montres. On sait que le fabricant doit réaliser lui-même au moins un calibre. La société qui prétend à ce titre doit en outre fabriquer les éléments de soutien de l'ensemble des composants du mouvement, autrement dit la platine et les différents ponts. Pour TAG Heuer, l'heure de l'autonomie sonne en 2010, avec le calibre sous licence 1887 (Heuer 01). Il est important de savoir que les chronographes posent des défis très délicats, compte tenu des relations complexes entre les fonctions qu'ils proposent. Pour ce premier mouvement de manufacture, l'expertise suisse se manifestera dans l'analyse différenciée et l'optimisation très poussée du calibre japonais ayant servi de plateforme. Pour sa part, le calibre Heuer 02 à remontage automatique, avec son chronographe et son tourbillon minute novateur dans sa cage en carbone léger très résistant, sera entièrement réalisé en Suisse.

Une manufacture est structurée, comme tout microcosme mécanique. Chez TAG Heuer, les développeurs de produits, constructeurs et concepteurs travaillent au siège de La Chaux-de-Fonds : boîtiers, cadrans, aiguilles, bracelets et mouvements, ils s'occupent de tout. Pour des raisons de capacité et de compétitivité – les mouvements de manufacture sont foncièrement plus chers que les mouvements achetés –, TAG Heuer transforme aussi des calibres éprouvés des marques Eta et Sellita. Les prototypistes travaillent également au siège. Enfin, dans un immense laboratoire, des « tortionnaires » infligent les pires sévices aux produits destinés à la fabrication en série.

C'est à Chevenez, un village retiré près de la frontière française, où l'on trouve plus facilement la main-d'œuvre requise, qu'est créé l'atelier de fabrication (de pièces de mouvements) T0. Le fraisage et le façonnage des platines et des ponts sont confiés à des centres d'usinage flexibles de dernière génération pilotés par ordinateur. Dans l'atelier T1 voisin, une centaine d'opérations sont ensuite nécessaires pour obtenir des mouvements chronographes aboutis à partir d'éléments de qualité contrôlée. L'automatisation garantit ici aussi une qualité optimale. Suite à une réorganisation radicale, la part des composants fabriqués en interne sur les mouvements maison dépasse 50 pour cent.

Le client qui achète une montre-bracelet signée TAG Heuer bénéficie automatiquement de la protection inhérente à la production en interne. Dans la localité de Cornol, près de Chevenez, la société exploite une fabrique ultramoderne de boîtiers qui emploie actuellement environ 140 personnes. Celles-ci produisent essentiellement des boîtiers en acier pour les garde-temps de toutes les collections, de « Aquaracer » à « Monza », dans le bruit assourdissant des estampeuses à l'aspect guerrier, et dans une moindre mesure, des centres d'usinage pilotés par ordinateur. Le poli final est l'œuvre de robots infatigables et d'ouvriers hautement spécialisés. À un contrôle rigoureux succèdent l'assemblage et, pour finir, le test d'étanchéité.

Troisième membre de l'équipe, ArteCad est une manufacture de cadrans traditionnelle. Rien n'a de secret pour cette filiale de TAG Heuer installée à Tramelan, même les techniques réputées oubliées. La complexité s'est brusquement accrue ces dernières années pour les nouveaux cadrans, notamment ceux de la « Carrera Heuer 01 » et de la « Carrera Heuer 02T ». Ces « visages du temps » sont des œuvres d'art tridimensionnelles aux multiples composants. Les activités mécaniques et manuelles vont de pair chez ArteCad. L'objectif visé est la perfection absolue. Lorsqu'on jette un coup d'œil à sa montre, c'est en effet le cadran et les aiguilles tournant sur cet arrière-plan que l'on voit en premier.

L'assemblage et le contrôle final du produit, qui est comme on sait bien plus que la somme de ses multiples composants, sont assurés par les ateliers de La Chaux-de-Fonds. Les montres terminées se dispersent alors dans le monde entier. TAG Heuer a beau être déjà une marque internationale, Jean-Claude Biver souhaite encore renforcer cette présence tout autour du globe. ◦

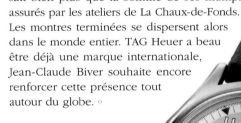

Solunar, 1949

Opposite page: TAG Heuer 360 museum, opened in 2008. More than 300 timepieces illustrating more than 150 years of history, from 1860 to the present day. Rich iconography. ◦ *Gegenüberliegende Seite: Reichhaltiges Bildmaterial versammelt das TAG Heuer 360 Museum. 2008 eröffnet, präsentiert es 300 Uhren aus über 150 Jahren Geschichte – von 1860 bis heute.* ◦ *Page ci-contre : Musée TAG Heuer 360, ouvert en 2008. Plus de 300 garde-temps illustrent plus de 150 ans d'histoire, de 1860 à nos jours. Riche iconographie.*

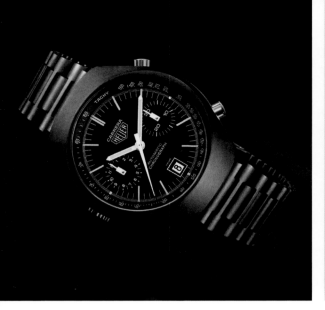

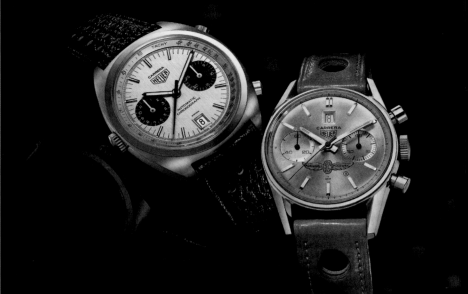

Jalons dans l'histoire des chronomètres et chronographes depuis 1887

Le premier temps fort est assurément le brevet du chronographe de poche à pignon oscillant en 1887. Ce pignon remplit la même fonction que les anciennes grandes roues des mouvements tout en réduisant nettement le coût d'un mouvement à chronographe.

Le « Time of Trip », premier compteur de bord Heuer, séduit par son grand ensemble cadran-aiguilles, parfaitement lisible même par mauvaise visibilité. Dès 1911, on le trouve dans les cockpits d'avions et les habitacles d'automobiles. En 1919, lors de la première traversée de l'Atlantique en dirigeable, l'équipage du R-34 s'en servira, tout comme Hugo Eckener lors de son tour du monde à bord du « Graf Zeppelin » en 1929.

Le développement du premier compteur mécanique au monde débute en 1914. Le brevet est délivré le 2 octobre 1916. Un petit balancier oscillant à 50 Hertz permet d'atteindre une précision de 1/100e de seconde. Comme l'indique le catalogue, le « Mikrograph » est fait « pour le chronométrage de projectiles et d'autres mesures de temps courts ». Il coûte 100 francs suisses, soit 120 dollars de l'époque.

Les compteurs de bord Heuer sont très prisés des collectionneurs. À compter de 1933, « Autavia », combinaison d'une montre et d'un chronomètre, figure sur de nombreux tableaux de bord, assistant par exemple des pilotes de course et de rallyes. Les pilotes d'avion et les capitaines de navire se fient eux aussi aux accessoires de mesure du temps fabriqués à Bienne, connus sous le nom de « Rally-Master » à partir de la fin des années 1950.

Toujours dans les années 1930, Heuer se consacre activement aux boîtiers étanches pour chronographes. Son chronographe de poignet à protection étanche plaît beaucoup à l'armée suisse, et en 1942 débute la production de chronographes étanches à cadran noir pour l'usage militaire.

L'année 1949 marque la sortie de la « Solunar », une montre-bracelet brevetée « pour les pêcheurs, chasseurs, marins, sportifs, naturalistes, scientifiques, médecins, agriculteurs, éleveurs d'oiseaux, jardiniers et gardes forestiers », comme le précise le catalogue de l'époque. Outre le temps, elle indique aussi la phase de lune, les heures où les poissons se nourrissent et les marées. Une pression sur un bouton-poussoir assure la synchronisation avec les tables de marées locales.

Lancé en 1962, le « Yachting Timer » aide les régatiers dans la phase délicate précédant le début de la course. Après le signal de départ, ce chronomètre renseigne de manière fiable sur l'état d'avancement du compte à rebours jusqu'au début effectif de la course.

« Autavia », nom facile à retenir, fait son retour en 1962, sous la forme d'un chronographe de poignet. La lunette tournante à l'intérieur du boîtier en acier est disponible en plusieurs versions, notamment pour pilotes, plongeurs sous-marins et scientifiques.

Alors qu'il assure en personne le chronométrage officiel d'une célèbre course automobile, les 12 Heures de Sebring, Jack W. Heuer entend parler de la Carrera Panamericana, une compétition automobile de légende. Aussi donne-t-il le nom « Carrera » au chronographe au grand cadran tridimensionnel qu'il présente en 1963. L'échelle des secondes est placée sur le réhaut, entre le cadran et le Plexiglas incurvé.

En 1965, pour la première fois dans l'histoire des chronographes, Heuer met sur le marché un modèle à guichet dateur remplaçant le quantième à aiguille ou à double guichet. Le « Carrera 45 Dato » fait donc logiquement parler de lui.

La présentation du premier chronographe au monde à mouvement modulaire et remontage automatique à micro-rotor, pignon oscillant et groupe commandes à came annonce l'arrivée d'une nouvelle ligne de montres chez Heuer. Le boîtier carré du chronographe « Monaco » est par ailleurs étanche. Le pilote de Formule 1 Jo Siffert s'avère être un excellent ambassadeur pour ce modèle. Steve McQueen, vêtu de la combinaison originale de Jo-Siffert, le portera également dans le film *Le Mans*.

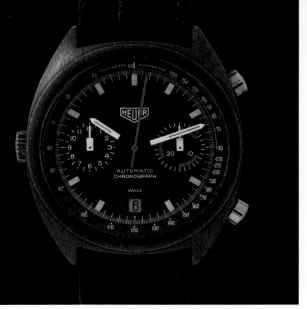
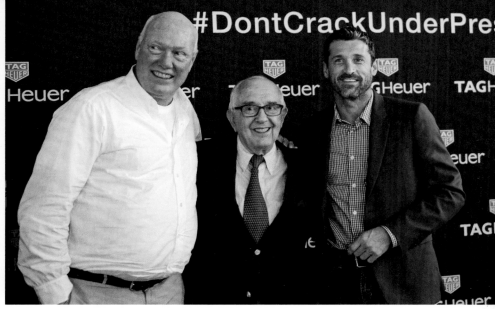

From left: Carrera, 1978 ◦ Carrera, 1969 and 1965 ◦ Heuer Monza, 1976 ◦ Jean-Claude Biver, Jack Heuer, and Patrick Dempsey, 2015

La série 1000 de 1992 s'adresse en particulier aux plongeurs sous-marins, qui doivent travailler par des pressions extrêmes. Le boîtier est d'une remarquable étanchéité jusqu'à 100 bars de pression, même sans l'habituelle valve à hélium.

Conçu par Jörg Hysek, la montre « Kirium Ti5 » marque en 1998 le début d'une nouvelle ère en matière de design. Elle se caractérise principalement par la transition organique fluide entre le boîtier et le bracelet. TAG Heuer propose en outre des matériaux ultra-performants : carbone pour le cadran, et titane pour le boîtier et le bracelet.

Mécanique ou électronique ? La question ne se pose pas pour la « Monaco Sixty Nine », sortie en 2003. Cette montre-bracelet innovante réunit dans son boîtier réversible un mouvement mécanique au recto et un mouvement à quartz au verso.

Débutant en 2004 comme concept watch, le « V4 » devient réalité en 2009. L'énergie cinétique du corps oscillant linéaire est transmise par deux courroies dentées microscopiques à âme métallique simultanément à quatre barillets. Le signal de rythme est fourni par un balancier traditionnel.

En 2006, la montre-bracelet à deux mouvements baptisée « Carrera Calibre 360 LE » est une première mondiale. Un mouvement affiche l'heure, l'autre chronomètre des intervalles de temps au 1/100ᵉ de seconde.

En 2010, année du 150ᵉ anniversaire de la société, TAG Heuer fait un retour à la fabrication en interne. Pour le calibre 1887 à remontage automatique et chronographe à pignon oscillant réalisé à la fin des années 1990, les constructeurs étaient partis d'une plateforme Seiko. Le nouveau mouvement constitué de 320 composants est cette fois entièrement fabriqué en Suisse.

Apparu en 2011, le concept chronographe mécanique « Mikrotimer Flying 1000 » mesure et affiche le 1/1000ᵉ de seconde. Comme ses prédécesseurs mesurant le 1/100ᵉ de seconde, cette montre-bracelet dispose pour l'embrayage de deux chaînes séparées, avec chacune un système d'oscillation et d'échappement. Les dix demandes de brevet déposées témoignent de la capacité d'innovation de la marque.

Le régulateur mécanique « Mikrogirder » découpe le temps au 5/10000ᵉ (1/2000ᵉ) de seconde. Aucun régulateur de marche existant n'avait encore atteint ses 7 200 000 alternances à l'heure. Cette construction de 2012, pour laquelle dix demandes de brevet sont en cours, fait appel à trois lamelles métalliques, dont une est perpendiculaire aux autres.

La même année, le chronographe « MikrotourbillonS », avec ses deux mécanismes de tourbillon de vitesses différentes, marque le couronnement de la philosophie de recherche de vitesse. Le tourbillon le plus rapide, qui affiche une fréquence de 50 Hertz, contrôle le chronographe au 1/100ᵉ de seconde. Le tourbillon responsable de l'affichage de l'heure se rapproche plus des tourbillons traditionnels.

Présenté en 2015, le chronographe « Carrera Heuer 01 » repose sur le calibre 1887. Côté cadran comme côté fond, le design spécialement épuré ne cache rien de la mécanique. La conception modulaire du boîtier « Carrera » étanche jusqu'à dix bars permet de jouer sur les matériaux, les surfaces et les couleurs de ses douze composants.

Le chronographe automatique avec tourbillon « Carrera » de 2016 est le dernier arrivé mais non le moindre. Avec le calibre Heuer 02T, Jean-Claude Biver et son équipe ont démontré que le travail complexe de fabrication en manufacture d'un chronographe à remontage automatique avec tourbillon ne devait pas forcément être inabordable. Les exemplaires de l'édition illimitée à boîtier en titane coûtent moins de 15 000 euros. ◦

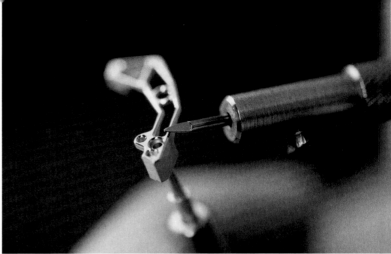
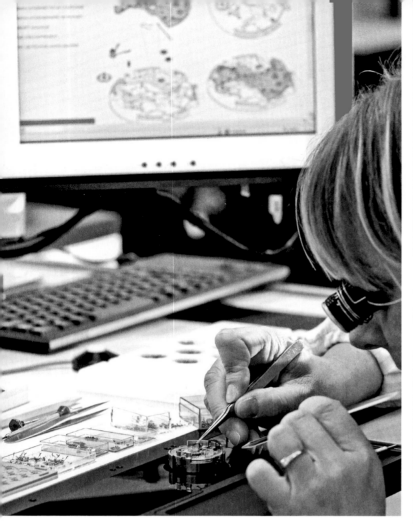
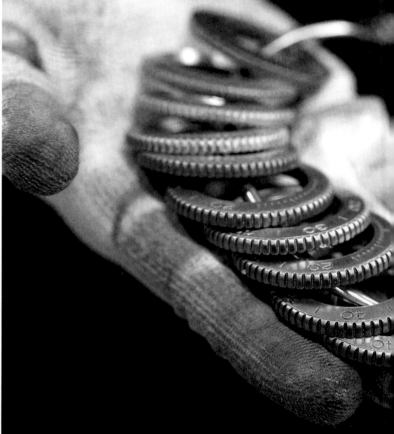
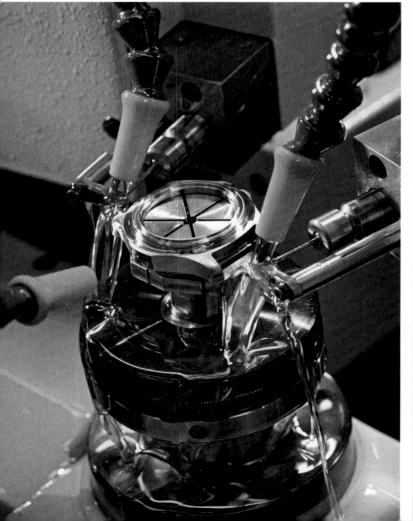
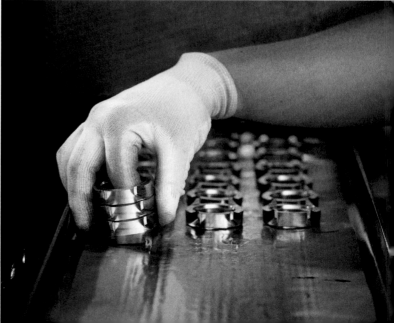

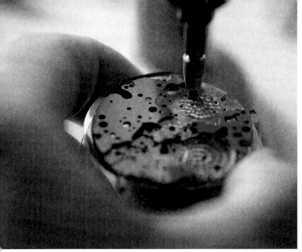
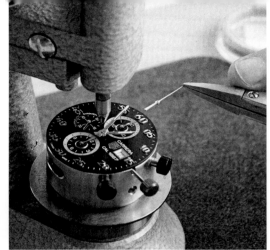
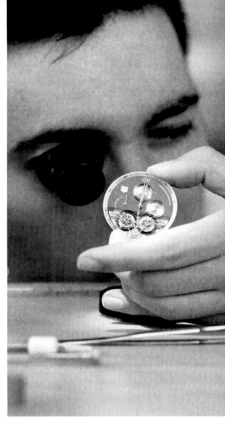
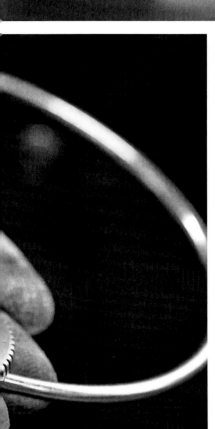
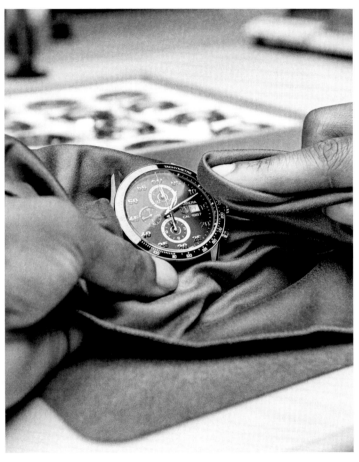

Movements, cases and dials:
TAG Heuer, as a true manufacture,
masters all the steps to produce a watch.
This extensive expertise contributes to
providing high-quality timepieces.
Uhrwerke, Gehäuse und Zifferblätter:
Als echte Manufaktur beherrscht
TAG Heuer alle Schritte der Uhren-
herstellung. Dank umfassender Expertise
entstehen hier Zeitmesser erster Güte.
Mouvements, boîtiers et cadrans :
TAG Heuer, en tant que véritable
manufacture, maîtrise toutes
les étapes de fabrication des montres.
Cette expertise étendue participe
à la production de garde-temps
d'exception.

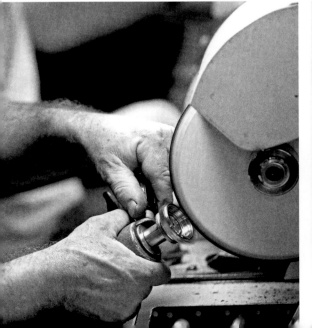
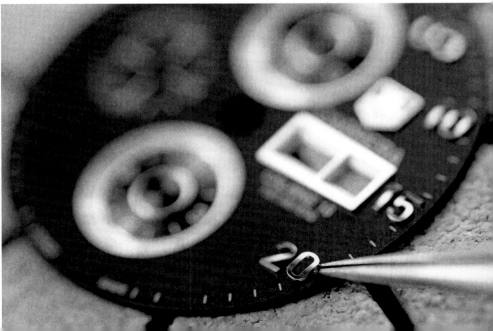

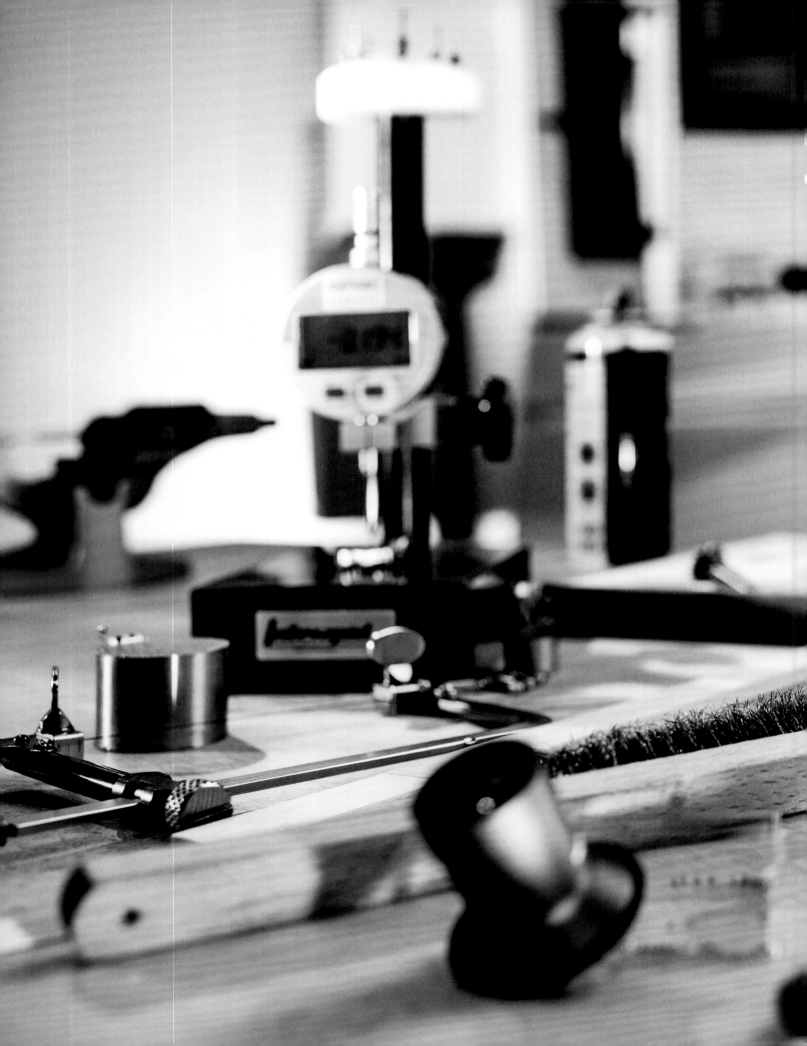

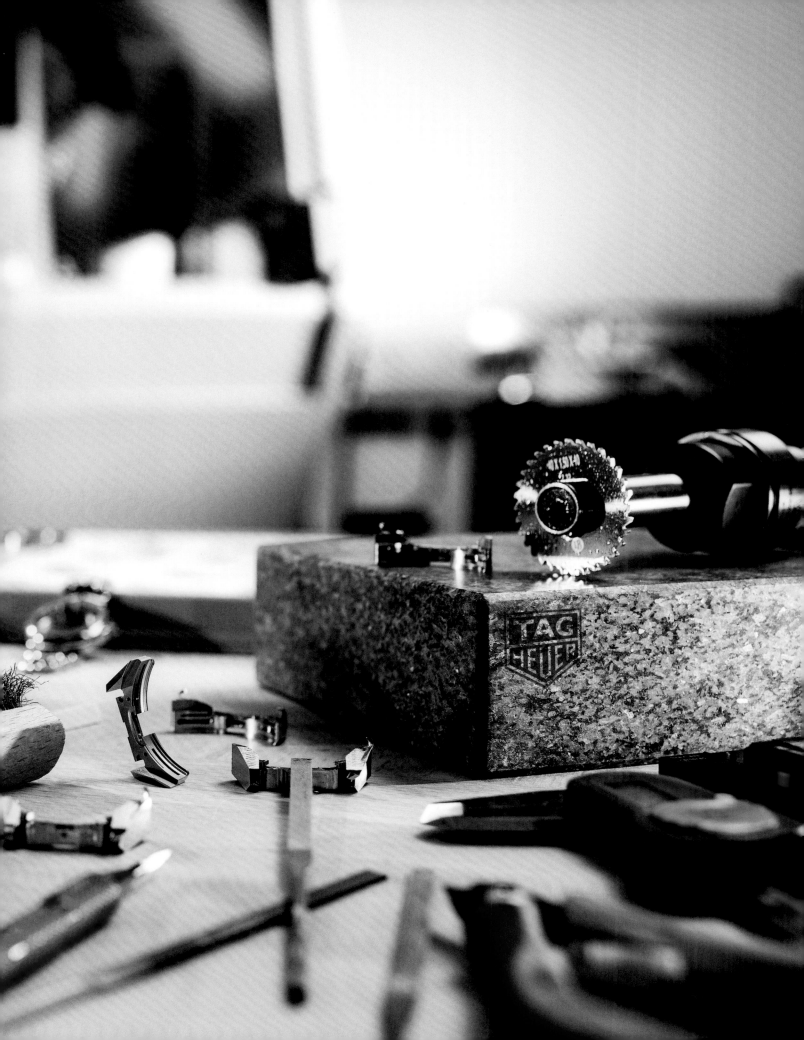

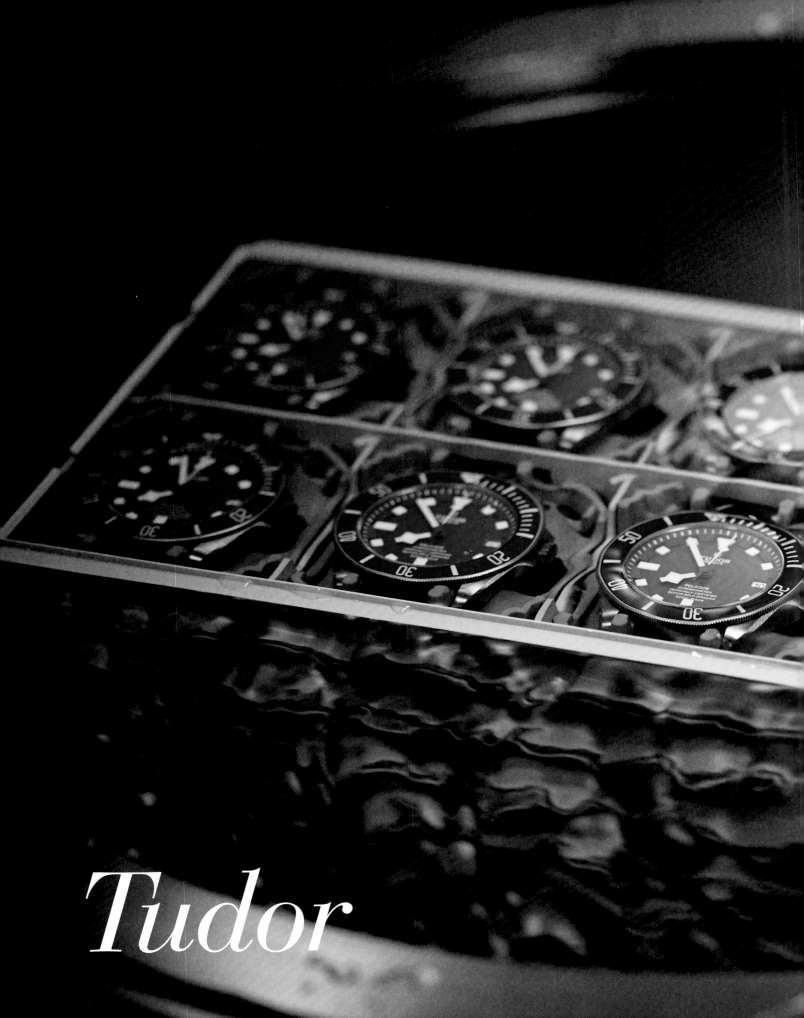

Tudor

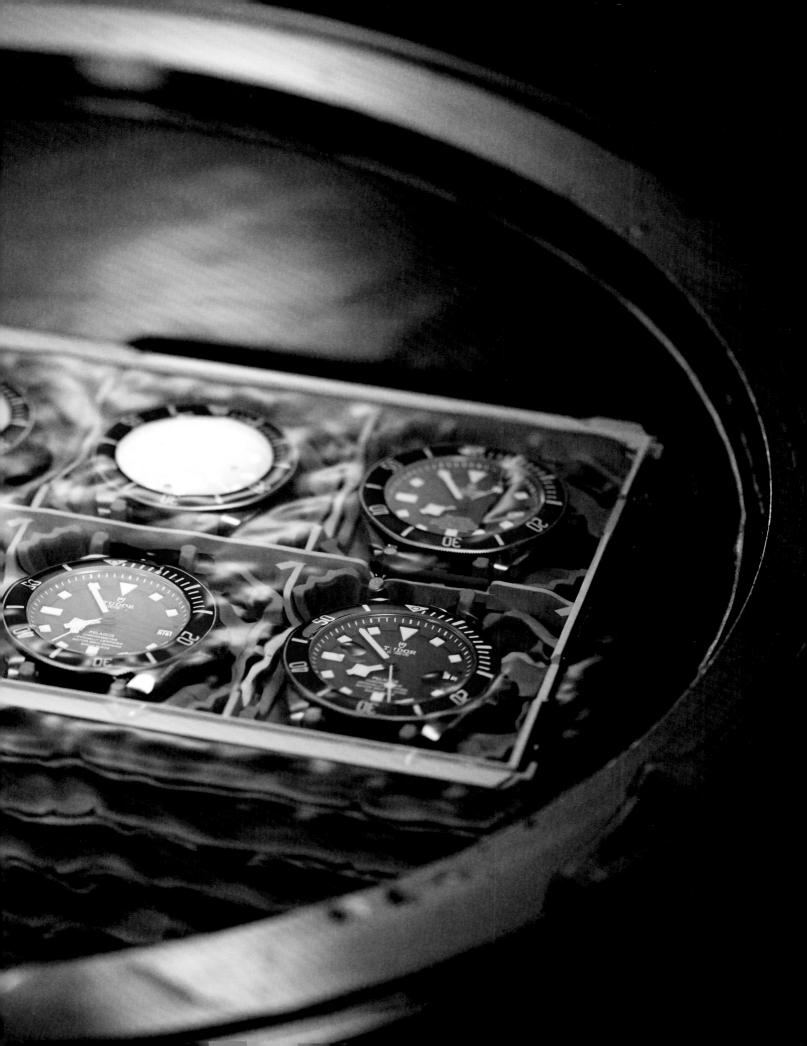

The Rolex for Beginners

Hans Wilsdorf wasn't only a marketing genius par excellence: he also cultivated a special affinity for Great Britain. It was in this island kingdom that he began his victory march into the world of precise time measurement in 1905. Three years later, this prompted him to establish the Rolex brand there. A native of Bavaria, Wilsdorf enthusiastically accepted British citizenship. It's well known that one cannot stand stably on only one leg, so Hans Wilsdorf registered a variety of brand names as protected trademarks. He ultimately used only a few of these monikers: for example, Tudor, which is the name of the English royal family that reigned from 1485 to 1603. When Hans Wilsdorf first took an interest in this name, the rights to its use where held by Isaac Blumenthal, a jewelry merchant in Geneva. A document on the letterhead of Horlogerie H. Wilsdorf and dated 1926 affirms that the Veuve de Philippe Hüther watch factory registered "The Tudor" brand for Wilsdorf's company to use on horological products of all varieties. The first timepieces bore this insignia without any additional logo. When Wilsdorf finally acquired unlimited rights in 1936, the stylized rose of the Tudor Dynasty was added as a trademark, initially with and later without a coat-of-arms. The name "Rolex SA" was engraved into the backs of the few pillow-shaped Tudor models from the 1940s. Hans Wilsdorf made no attempt to conceal the fact that watches bearing the Tudor signature were fabricated in his well-established manufactory. In 1952, Hans Wilsdorf provided a personal explanation about Montres Tudor SA, which he had founded in 1946. In this clarification, he explained that he had long studied the options for watches for ladies and gents that he could offer to his concessionaries at less costly prices than those charged for Rolex timepieces. This led him to create the new label. The "Tudor Oyster Prince" had two advantages that connoisseurs already appreciated in Rolex watches: the famous watertight Oyster case and the Perpetual rotor winding mechanism. Ladies could opt for self-winding and non-watertight models. To preserve the necessary distance between Tudor and Rolex, the product strategy was based on a synthesis of existing elements (i.e., time-honored Oyster cases) and movements from third-party suppliers (e.g., Eta). These affordably priced timepieces sold surprisingly well, especially in the Middle Kingdom. Frogmen in the French navy in the late 1960s consulted the dial of a Tudor model that looked remarkably similar to Rolex's "Submariner." The U.S. Navy Seals soon followed in the wake of their European colleagues. Collectors are eager to get their hands on the "Monte Carlo" chronograph (Reference 7159/0), which encases hand-wound column-wheel Caliber Valjoux 234. Tudor has repeatedly attracted admiring attention in recent years by recalling its illustrious historical diver's watches, which were revived in retro models like the "Pelagos" and the "Heritage Black Bay." Rolex's affiliated company acquired manufactory status in 2015 with the debut of caliber family MT56xx with rotor winding and silicon hairspring in the cases of the minimalistically styled watches of the "North Flag" line. Caliber MT5621 supports a date display and a power-reserve indicator; Caliber MT5612 offers only a date window; and Caliber MT5601, which debuted in 2016, embodies purism by animating only three hands. It ticks in various versions of the bestselling "Heritage Black Bay," including a much-acclaimed version with a bronze case. ◦

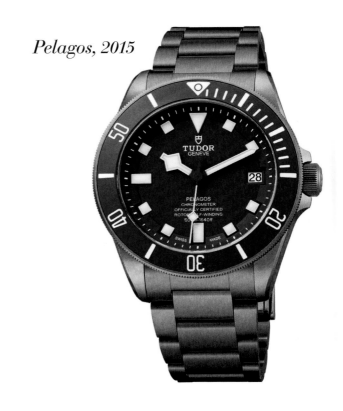

Pelagos, 2015

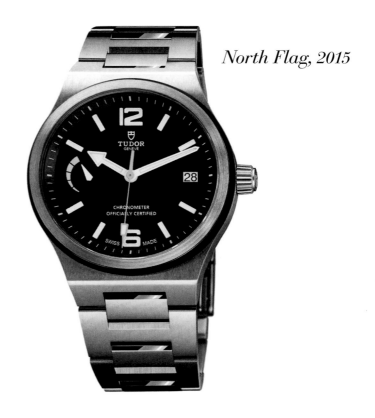

North Flag, 2015

Die Rolex für Einsteiger

Deutsch

Ohne jeden Zweifel besaß Hans Wilsdorf, Marketinggenie par excellence, eine besondere Affinität zum britischen Königreich. Hier nämlich hatte 1905 sein Siegeszug in die Welt der Präzisionszeitmessung begonnen, welcher drei Jahre später zur Gründung der Marke Rolex führte. Die englische Staatsbürgerschaft hatte der gebürtige Bayer ebenfalls aus innerster Überzeugung angenommen. Weil man auf einem Bein bekanntlich nicht gut steht, ließ sich Hans Wilsdorf ein beachtliches Spektrum unterschiedlicher Signaturen schützen. Genutzt hat er freilich nur wenige. So zum Beispiel Tudor, Name eines englischen Königsgeschlechts, welches von 1485 bis 1603 regierte. Als sich Hans Wilsdorf erstmals hierfür interessierte, lagen die Rechte noch beim Genfer Schmuckhändler Isaac Blumenthal. 1926 erklärte die Uhrenfabrik Veuve de Philippe Hüther auf einem Briefbogen der Horlogerie H. Wilsdorf, dass sie die Marke The Tudor für dieses Unternehmen habe registrieren lassen. Und zwar zur Nutzung auf uhrmacherischen Erzeugnissen aller Art. Die ersten Zeitmesser trugen denn auch nur diesen Schriftzug ohne weiteres Logo. Als Hans Wilsdorf 1936 endlich selbst über die uneingeschränkten Rechte verfügte, gesellte sich die stilisierte Rose der Tudor-Dynastie als Markenzeichen hinzu. Anfangs mit, später nur noch ohne Wappenschild. In den Gehäuseboden eines der wenigen kissenförmigen Tudor-Modelle aus den 1940er Jahren war noch unübersehbar der Name Rolex SA eingraviert. Hans Wilsdorf machte also keinen Hehl daraus, dass die Tudor signierten Uhren aus seiner längst etablierten Manufaktur stammten. 1952 gab Hans Wilsdorf eine persönliche Erklärung zur 1946 gegründeten Montres Tudor SA ab. Darin stand zu lesen, dass er über viele Jahre hinweg Möglichkeiten der Fabrikation einer Uhr studiert habe, welche seine Konzessionäre den Damen und Herren zu einem günstigeren Preis als die Rolex-Zeitmesser anbieten könnten. Genau aus diesem Grund sei das neue Label entstanden. Die „Tudor Oyster Prince" besitze zwei Vorteile, welche man schon von Rolex kenne: das berühmte wasserdichte Oyster-Gehäuse und den Perpetual-Rotor-Aufzug. Für Damen gebe es nicht wasserdichte und automatische Modelle. Zur Wahrung des gebotenen Abstands basierte die Produktstrategie auf einer Synthese aus Vorhandenem, also den bewährten Oyster-Gehäusen, und zugekauften Uhrwerken beispielsweise von Eta. Besonders im Reich der Mitte kamen diese preisgünstigen Armbanduhren erstaunlich gut an. In den späten 1960er Jahren blickten französische Marinetaucher auf eine Tudor, welche der „Submariner" von Rolex verblüffend ähnelte. Und die US-amerikanischen Navy Seals taten es ihren europäischen Kollegen gleich. Extrem hohen Stellenwert bei Sammlern genießt auch der Chronograph „Monte Carlo", Referenz 7159/0, mit dem Schaltrad-Handaufzugskaliber Valjoux 234. Durch die Rückbesinnung auf seine Taucheruhren-Geschichte, Retro-Modelle wie „Pelagos" und „Heritage Black Bay" erregte Tudor in den zurückliegenden Jahren immer wieder Aufmerksamkeit. 2015 begann schließlich auch bei der Rolex-Tochter das Manufaktur-Zeitalter. Die Kaliberfamilie MT56xx mit Rotoraufzug und Silizium-Unruhspirale debütierte in der reduziert gestalteten Uhrenlinie „North Flag". Das MT5621 besitzt Datums- und Gangreserveanzeige, das MT5612 lediglich ein Fensterdatum. Puristisch, das heißt nur mit drei Zeigern ausgestattet, tickt das 2016 vorgestellte MT5601 in verschiedenen Versionen des Bestsellers „Heritage Black Bay", darunter auch eine viel beachtete mit Bronzegehäuse. ○

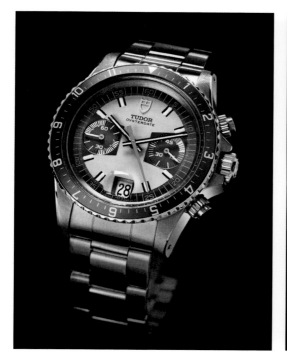
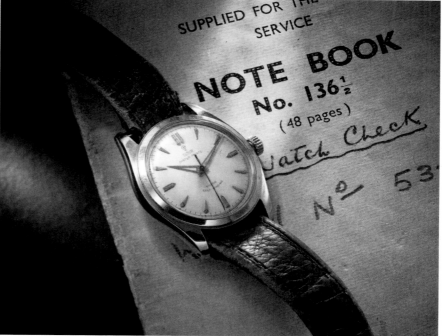

From left: Oysterdate Monte Carlo, 1973 ○ Oyster Prince, 1950s

La Rolex de l'homme de la rue

Français

Génie du marketing par excellence, Hans Wilsdorf a assurément une affinité particulière pour le royaume britannique. C'est en effet là qu'il débute en 1905 sa marche triomphale dans l'univers des garde-temps de précision, couronnée trois ans plus tard par la création de Rolex. C'est aussi une intime conviction qui conduit ce Bavarois d'origine à adopter la nationalité anglaise. Comme chacun sait, il ne faut jamais mettre tous ses œufs dans le même panier. Aussi Hans Wilsdorf dépose-t-il un large éventail de marques, dont il n'exploitera toutefois qu'un petit nombre. C'est le cas de la marque Tudor, qui évoque une dynastie ayant régné sur l'Angleterre de 1485 à 1603. Lorsque Hans Wilsdorf s'y intéresse pour la première fois, le joaillier genevois Isaac Blumenthal s'en est déjà assuré les droits. Mais, en 1926, la maison « Veuve de Philippe Hüther » déclare, sur une lettre à en-tête de l'Horlogerie H. Wilsdorf, avoir déposé, pour le compte de cette même entreprise, la marque « The Tudor » pour tous produits horlogers. Les premiers garde-temps Tudor porteront alors sur le cadran l'inscription Tudor, mais sans logotype. En 1936, lorsque Hans Wilsdorf acquiert enfin les droits exclusifs sur la marque, la rose stylisée de la dynastie des Tudors apparaît. Le blason dans lequel s'inscrit ce logotype disparaît ensuite. Sur l'un des rares modèles Tudor de forme coussin des années 1940, on voit nettement « Rolex SA » gravé sur le fond de boîte. Hans Wilsdorf ne fait donc aucun mystère de ce que les montres signées Tudor proviennent de sa manufacture depuis longtemps établie. En 1952, il fait une déclaration sur la société « Montres Tudor SA », qu'il a créée en 1946. On peut y lire qu'il a pendant de nombreuses années étudié les possibilités de fabriquer une montre que ses concessionnaires pourraient vendre à leur clientèle à un prix plus bas que les montres Rolex. C'est justement cela qui a présidé à la création du nouveau label Tudor. La « Tudor Oyster Prince » partage deux avantages exclusifs avec les Rolex : le fameux boîtier étanche Oyster et le mécanisme à rotor Perpetual. Des modèles étanches ou automatiques ne sont pas prévus pour les femmes. Pour conserver l'écart qui s'impose entre les deux marques, la stratégie produits de Tudor s'appuie sur une combinaison de l'existant, les boîtiers Oyster éprouvés, et des mouvements achetés, Eta surtout. Ces montres-bracelets abordables sont très appréciées en Chine. À la fin des années 1960, les plongeurs de la Marine nationale française sont équipés de Tudor qui ressemblent étrangement aux « Submariner » de Rolex. Plus tard, les Navy Seals de l'US Navy imiteront leurs homologues européens. Caractérisé par son calibre à remontage manuel Valjoux 234, le chronographe à roue à colonnes « Monte Carlo » (7159/0) est lui aussi très prisé des collectionneurs. Avec sa mise à jour des montres de plongée rétro « Pelagos » et « Heritage Black Bay », Tudor fait régulièrement sensation ces dernières années. Mais 2015 marque aussi pour la filiale de Rolex le passage à la fabrication maison. Les calibres MT56xx à système de remontage par rotor et spiral font leurs débuts dans la ligne de montres d'une grande sobriété « North Flag ». Le modèle MT5621 est doté de l'affichage de la date et de la réserve de marche, tandis que le MT5612 dispose simplement d'un guichet pour la date. Dans un style épuré, à savoir seulement équipé de trois aiguilles, le mouvement MT5601 présenté en 2016 anime différentes versions du best-seller « Heritage Black Bay », dont un modèle très prisé à boîtier en bronze. ○

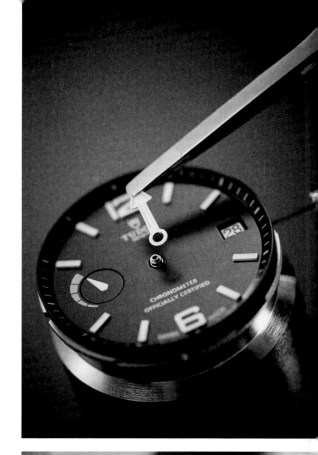

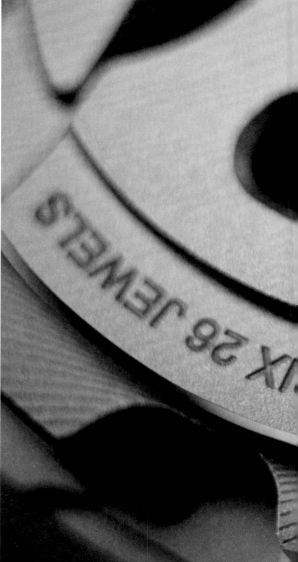

Self-winding manufactory Cal. MT5601

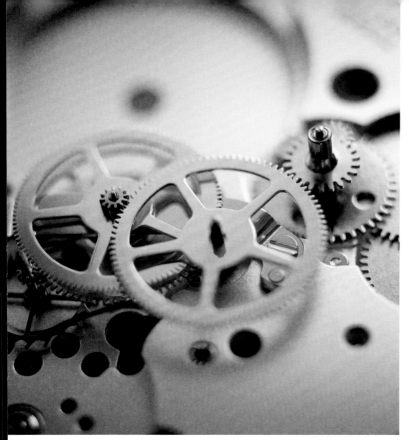

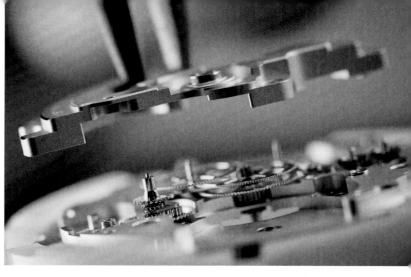

Clockwise, from top left: Installing the hour-hand on the North Flag with power-reserve display ◦ Installing the gear-train bridge of a manufactory caliber ◦ Detail of manufactory Caliber MT5612 ◦ Von oben links im Uhrzeigersinn: Montage des Stundezeigers der North Flag mit Gangreserveanzeige ◦ Räderwerk eines Manufakturkalibers ◦ Montage der Räderwerksbrücke eines Manufakturkalibers ◦ Detail des Manufakturkalibers MT5612 ◦ Dans le sens horaire, en partant du haut, à gauche : Montage de l'aiguille des heures de la North Flag avec affichage de la réserve de marche ◦ Rouage d'un calibre de manufacture ◦ Montage du pont rouage d'un calibre de manufacture ◦ Détail du calibre de manufacture MT5612

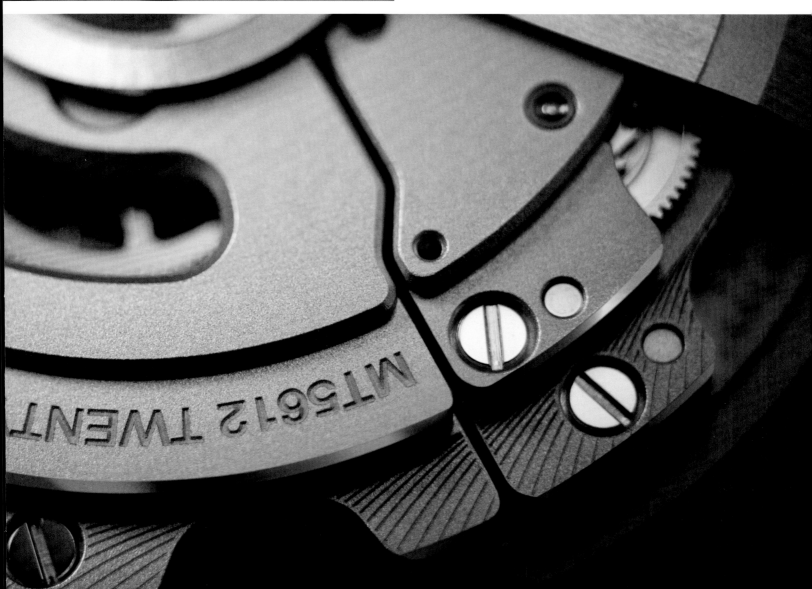

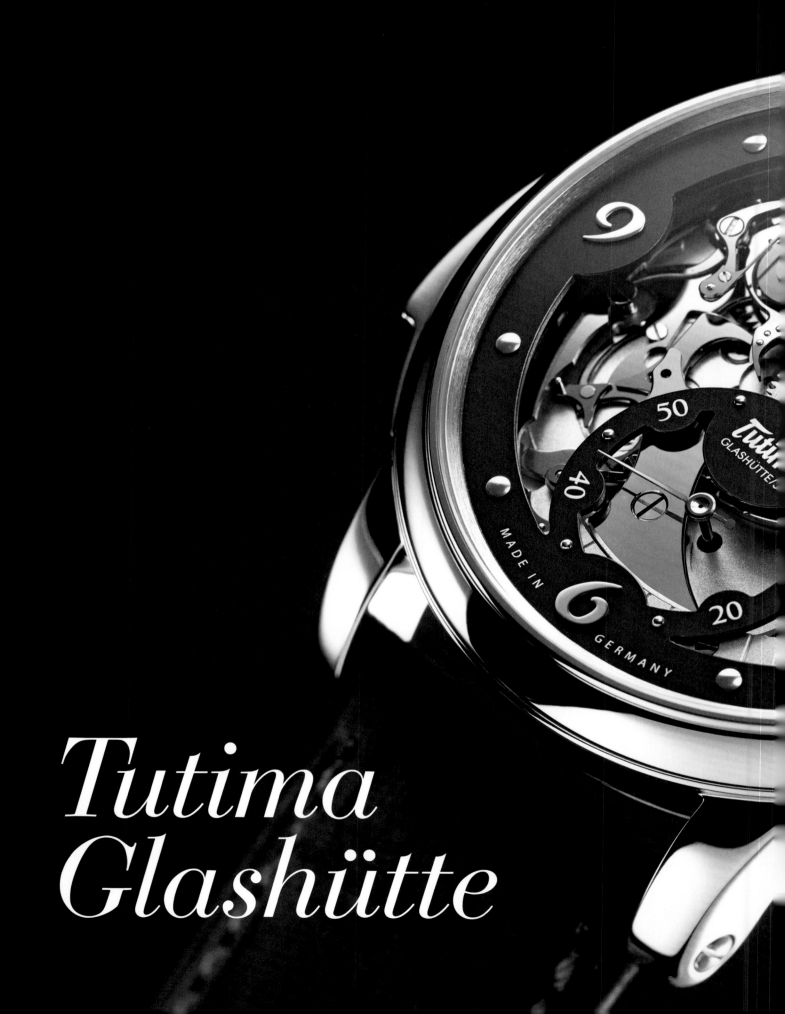

Tutima
Glashütte

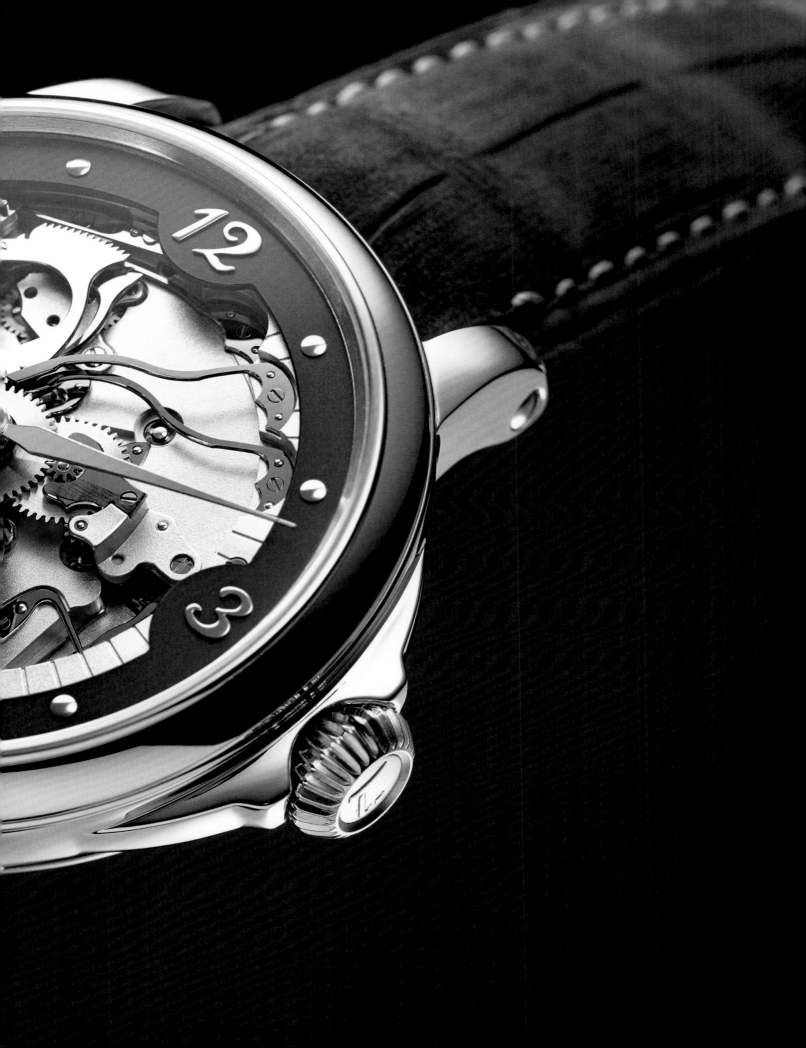

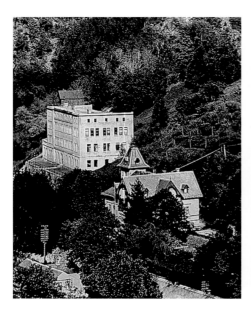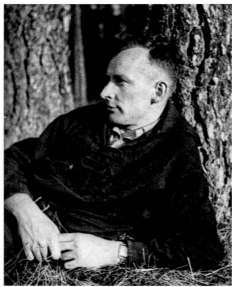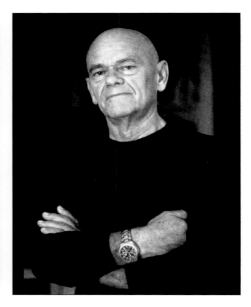

From left: Former UFAG Glashütte ◦ Dr. Ernst Kurtz ◦ Dieter Delecate ◦ Von links: Ehemalige UFAG Glashütte ◦ Dr. Ernst Kurtz ◦ Dieter Delecate ◦ De gauche à droite : Ancienne société UFAG Glashütte ◦ Dr. Ernst Kurtz ◦ Dieter Delecate

Back to Glashütte

English

Glashütte is undoubtedly the Mecca of German watch manufacturing, a distinction which the city has enjoyed not only since the fall of the Berlin Wall in 1989 and Germany's reunification one year later, but since 1845, when Ferdinand Adolph Lange relocated his watchmaking activities to the remote Müglitz Valley. Step by step, Lange initiated the establishment and promoted the growth of a branch of industry specializing primarily in luxurious products. Early in the 20th century, the industrialists began to realize that Glashütte's products should also appeal to broader classes of the population. The attorney Dr. Ernst Kurtz accordingly founded in 1926 a conglomerate of companies, which included among its members the Uhren-Rohwerke-Fabrik Glashütte AG (UROFA) and the Uhrenfabrik Glashütte AG (UFAG). The latter principally produced complete wristwatches. Top-quality products were launched with the signature "Tutima Glashütte." This catchy name was, of course, no mere coincidence: it was derived from the Latin adjective *tutus*, which means "secure" or "protected." The development and fabrication of affordably priced ébauches for wristwatches began under the direction of Ernst Kurtz and several competent colleagues. The purchase of a small Swiss watch factory and its transfer to Saxony compensated for lack of know-how in this field.

This led to the production of watch movements, including chronograph movements, which had no need to fear comparison with their dominant Swiss competitors. Dr. Kurtz and several of his colleagues were able to escape to the West on May 7, 1945, i.e., one day before an aerial bombardment decimated Glashütte. Beginning in 1951, the businessman and his team were again doing business at Ganderkesee in Lower Saxony. This is also where they created petite Caliber 570 for ladies' watches in 1956. Due to lack of sufficient liquidity for serial production, a former colleague from

Glashütte directed the business under the name "Norddeutsche Uhren-Rohwerkefabrik" (NUROFA). Sales and distribution of the watches were entrusted to Dieter Delecate, who was likewise one of Dr. Kurtz's former colleagues. The regime of the German Democratic Republic wasn't interested in the traditional name "Tutima," so the moniker could celebrate its glorious comeback at a new location. Dieter Delecate reintroduced the name: he registered it as a protected trademark on April 28, 1969 and afterwards established the Tutima Uhrenfabrik GmbH. Delecate deserves credit for building this business into a globally known brand. For several decades, a large portion of Glashütte's watchmaking tradition found its new home in picturesque Ganderkesee. As in the good old days, the company specialized in pilot's watches and especially aviator's watches with chronographs. Their production was based on Swiss movements from Lemania and naturally also from Eta. A successful retro model alluded to the distinctive chronograph pilot's watches that encased flyback Caliber Urofa 59, which was produced between 1941 and 1945. In the 1980s, the West German military needed an official pilot's chronograph. The military testing office drew up a precise list of specifications for the desired timepiece. After examining all submitted offers, the contract was awarded to Tutima in 1983. The opulent self-winding chronograph, equipped with robust Caliber Lemania 5100, was also impressive thanks to its ergonomially shaped buttons and its optimally legible counter for up to 60 elapsed minutes, which turned at the center of the dial.

With the acquisition of a prominent building near Glashütte's railway station in May 2011, Tutima gazed self-consistently "back to the future." After it underwent thorough renovations and was equipped with an ultramodern machine park, the building was ready to welcome a staff of technicians, engineers, watchmakers and other specialists. This new chronometric beginning in Saxony is also associated with a return to the erstwhile activities as a manufactory. To celebrate this renaissance, Dieter Delecate, his daughter Ute and his son Jörg wanted more than just the ordinary, e.g., a classical hand-wound or automatic movement. Instead, they envisioned a little mechanical jewel of a sort that had never before been exclusively "made in Glashütte." They set their sights on a wristwatch with a minute repeater. Due to its uncommon complexity, this additional function—which sequentially chimes the hours, the quarter hours, and the minutes on a pair of gongs—embodies the utmost in the art of mechanical watchmaking. The traditional excellence of Glashütte workmanship was handsomely embodied in gold-plated hand-wound Caliber T800, which is 32 millimeters in diameter and 7.2 millimeters tall. Each of this movement's 550 components is finely processed in accord with the rules and usages of horological craftsmanship. For example, all components in the cadrature for the striking mechanism have beveled edges. A tin plate is used to give Glashütte's characteristic black polish to planar surfaces. It's not surprising to learn that each of these ticking microcosms requires fully three months to assemble. The finest acoustic quality is assured by two gongs, anchored in the massive case and tuned a major third apart. To paraphrase Spinoza, "all things excellent are rare," so Tutima offers only 20 specimens of this timepiece in rose gold and five others in platinum cases. Of course, all of the responsible individuals were well aware that these watches would be reserved for the very few people whose discriminating standards insist on nothing less than the utmost exclusiveness. Tutima accordingly offers two additional movements of its own to aficionados and fans of fine Glashütte manufactory watchmaking. Purism par excellence radiates from 31-component Caliber T617 in the "Patria Small Second." This hand-wound movement has only three hands: one each for the hours, the minutes and the seconds. The solid gold trio is fabricated entirely by hand. Cosmopolites and frequent flyers appreciate the

"Patria Dual Time." This wristwatch encases upgraded Caliber T619, which amasses a 65-hour power reserve and offers a 12-hour display to indicate the time in a second time zone. After its relocation, Tutima naturally never lost sight of the traditional theme that it had cultivated for many decades. A new chronographic era began in 2013 with the "Saxon One." With regard to its functionality, this "time-writer" alludes to automatic Caliber Lemania 5100, production of which ceased long ago, and which offers a central counter for 60 elapsed minutes and an additional 24-hour hand. A special module had to be developed and manufactured to add these additional functions to time-honored Caliber Eta 7750. Both were created in the light-flooded manufacturing building in Glashütte. More than 50 percent of the value is added in Saxony, so the dials of these watches can bear both the word "Glashütte" and the phrase "made in Germany" as proof of their provenance. Traditional elements are reflected in the significantly bolder-looking version "M2," which is a successor of the identically named NATO chronograph that Tutima had developed for the German military. The color scheme for the hands aptly expresses military professionalism: eye-catching red on the three hands for the stopwatch function clearly contrasts with the white hue of the four other hands for the ordinary time of day or night.

Last but not least, the brand-new "M2 Seven Seas" feels perfectly at home in the silence of the underwater world. Its titanium case resists pressure to 50 bar, which means that this 44-millimeter-diameter titanium wristwatch can theoretically accompany its wearer to a depth of 500 meters. Tutima equips reliable automatic Caliber Eta 2836 with a specially designed rotor. On the one hand, this detail individualizes the timepiece; on the other hand, it contributes toward further increasing the amount of value added to the watch at Tutima's new home in Glashütte.

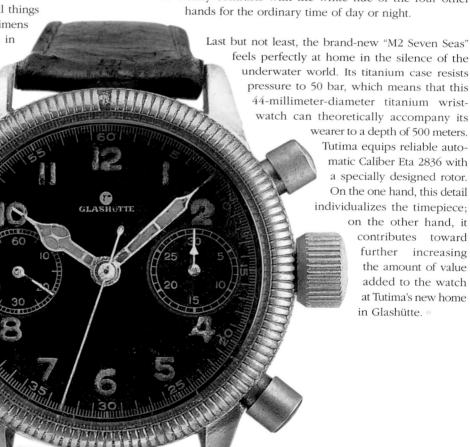

Pilot's chronograph, 1941

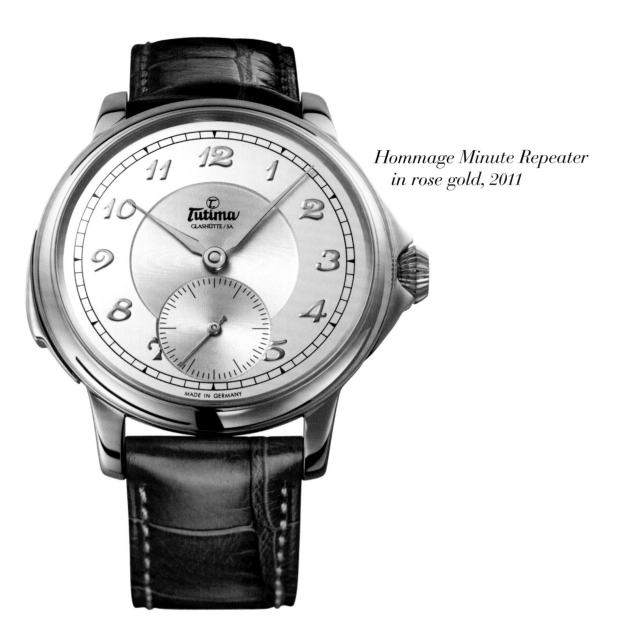

*Hommage Minute Repeater
in rose gold, 2011*

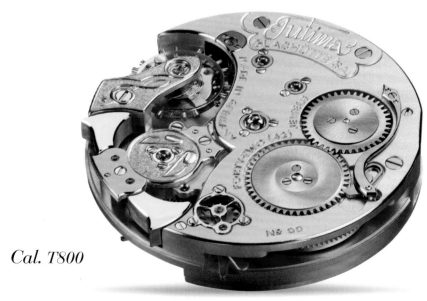

Cal. T800

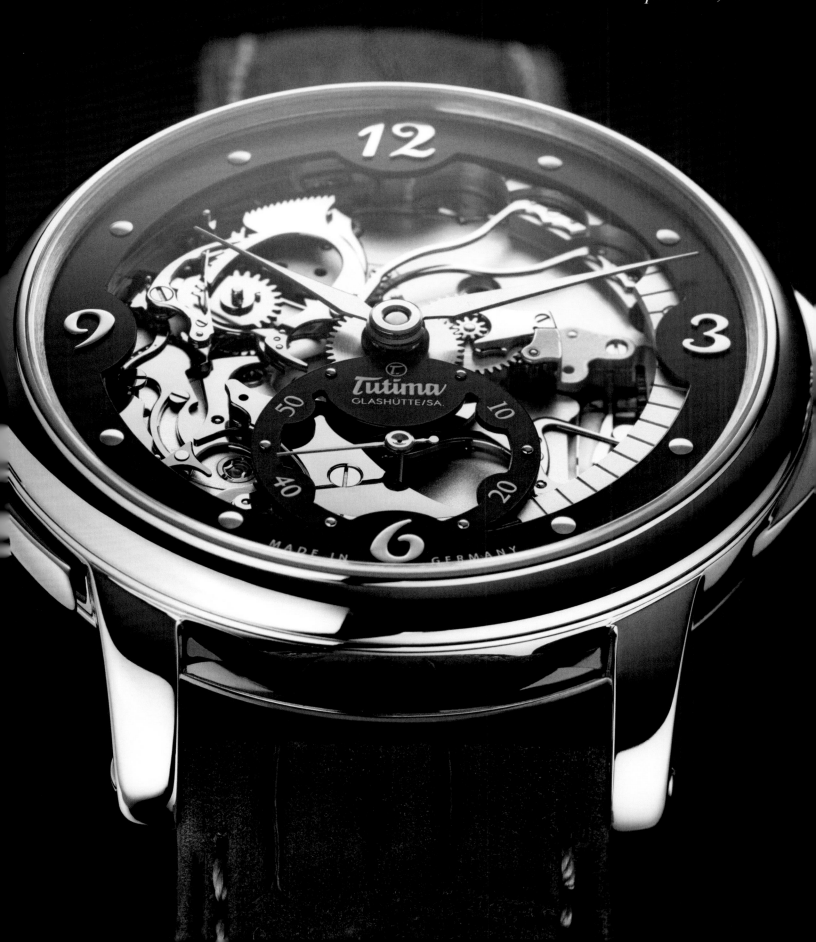

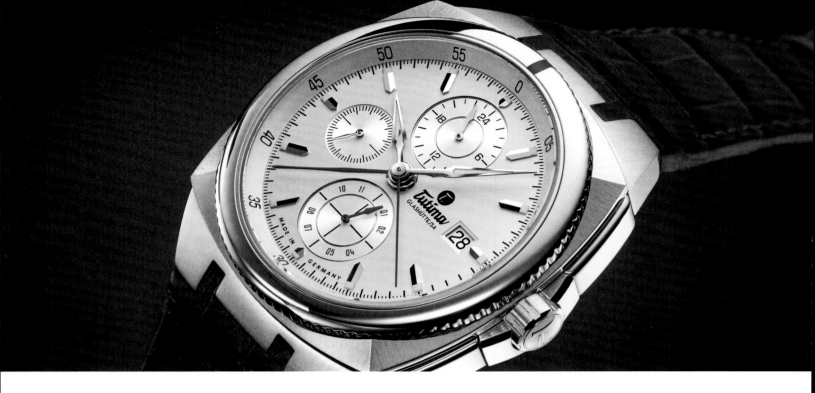

Zurück nach Glashütte

Deutsch

Ohne jeden Zweifel ist Glashütte das Mekka der deutschen Uhren-fertigung. Und zwar nicht erst seit dem Fall der Berliner Mauer im Jahr 1989 und der deutschen Wiedervereinigung ein Jahr später, sondern seit 1845. Damals verlegte Ferdinand Adolph Lange seine uhrmacherischen Aktivitäten ins abgeschiedene Müglitztal. Zug um Zug initiierte er den Aufbau des vornehmlich auf Luxusprodukte spezialisierten Industriezweigs. Anfang des 20. Jahrhunderts wuchs die Erkenntnis, dass Glashütter Produkte auch breitere Bevölkerungsschichten ansprechen sollten. In diesem Sinne erstellte der Jurist Dr. Ernst Kurtz 1926 ein Firmenkonglomerat, zu dem die Uhren-Rohwerke-Fabrik Glashütte AG (UROFA) sowie die Uhren-fabrik Glashütte AG (UFAG) gehörten. Letztere produzierte in erster Linie fertige Armbanduhren. Die Top-Qualitätsprodukte wurden mit der Signatur Tutima Glashütte lanciert. Der einprägsame Name kam natürlich nicht von ungefähr. Er leitet sich ab vom lateinischen Adjektiv „tutus". Und das steht für „sicher, geschützt". Unter Leitung von Ernst Kurtz und einigen kompetenten Mitstreitern begann die Entwicklung und Herstellung preiswerter Armbanduhr-Rohwerke. Der Kauf einer kleinen Schweizer Uhrenfabrik und ihr Transfer nach Sachsen kompensierten mangelndes Knowhow auf diesem Gebiet.

Auf diese Weise entstanden Uhrwerke bis hin zu Chronographen, welche den Vergleich mit der dominanten eidgenössischen Konkurrenz nicht fürchten mussten. Am 7. Mai 1945, einen Tag vor dem Bombenhagel auf Glashütte, gelang es Dr. Kurtz wie auch einigen Mitarbeitern, sich in den Westen zu retten. Ab 1951 begegnete man dem Unternehmer und seinem Team im niedersächsischen Ganderkesee. Dort entstand 1956 das kleine Kaliber 570 für Damenuhren. Mangels Liquidität zur Serienproduktion führte ein früherer Mitarbeiter aus Glashütte den Betrieb unter dem Namen Norddeutsche Uhren-Rohwerkefabrik (NUROFA) fort. Den Vertrieb der Uhren übernahm Dieter Delecate, ebenfalls früherer Mitarbeiter

von Dr. Kurtz. Da sich das Regime der Deutschen Demokratischen Republik nicht für den traditionsreichen Namen Tutima interes-sierte, feierte er am neuen Standort ein glorreiches Comeback. Dieter Delecate führte diesen Namen auf den Uhren wieder ein. Am 28. April 1969 meldete er den Namen auch zu seinem Schutz an. Anschließend rief er die Tutima Uhrenfabrik GmbH ins Leben. Der Aufstieg zur weltweit bekannten Marke ist ihm zu verdanken. Für mehrere Jahrzehnte fand ein bedeutender Teil der Glashütter Uhrentradition im malerischen Ganderkesee seine neue Heimat. Wie in guten alten Zeiten spezialisierte sich das Unternehmen auf Fliegeruhren und ganz besonders jene mit Chronograph. Die Produktion basierte auf Schweizer Uhrwerken von Lemania und natürlich Eta. An den markanten, zwischen 1941 und 1945 produzierten Piloten-Stopper mit dem Flyback-Kaliber Urofa 59 erinnerte ein erfolgreiches Retromodell. In den 1980er Jahren benötigte die deutsche Bundeswehr einen offiziellen Flieger-chronographen. Das Anforderungsprofil für diese Uhr hatte die militärische Erprobungsstelle exakt formuliert. Nach Prüfung der eingegangenen Offerten erhielt Tutima 1983 den Zuschlag. Der opulente Automatik-Chronograph, ausgestattet mit dem robusten Kaliber Lemania 5100, beeindruckte auch durch seine ergonomischen Bedientasten und den optimal ablesbaren, weil in der Mitte des Zifferblatts drehenden 60-Minuten-Zähler.

Mit dem Erwerb eines prominent nahe dem Glashütter Bahnhof gelegenen Gebäudes blickt Tutima seit Mai 2011 konsequent zurück in die Zukunft. Nach gründlicher Renovierung des Bauwerks sowie seiner Ausstattung mit einem hochmodernen Maschinenpark hielten Techniker, Konstrukteure, Uhrmacher und andere Facharbeiter ihren Einzug. Mit dem chronometrischen Neustart in Sachsen verknüpfte sich im weitesten Sinn auch die Rückkehr zu den einstigen Manufaktur-Aktivitäten. Etwas Normales, also ein klassisches

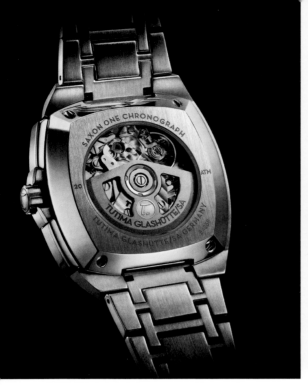
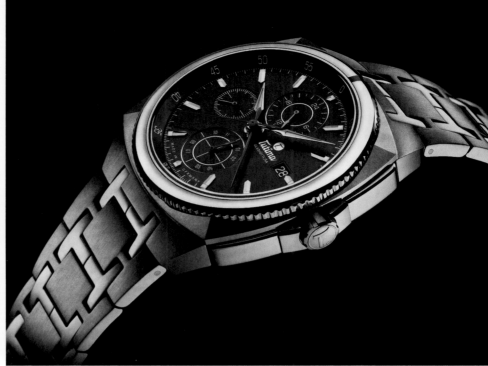

Saxon One Chronograph, 2013

Handaufzugs- oder Automatikwerk, war Dieter Delecate, Tochter Ute und Sohn Jörg anlässlich der Renaissance allerdings nicht aufregend genug. Ihnen schwebte ein mechanisches Kleinod vor, das es aus rein Glashütter Produktion noch nicht gegeben hatte. Gemeint ist eine Armbanduhr mit Minutenrepetition. Wegen ihrer ungemeinen Komplexität repräsentiert diese Zusatzfunktion, welche nacheinander Stunden, Viertelstunden und Minuten auf zwei Tonfedern schlägt, die allerhöchste Schule mechanischer Uhrmacherkunst. Das vergoldete Handaufzugswerk, Kaliber T800 mit 32 Millimetern Durchmesser und 7,2 Millimetern Bauhöhe, setzt die überlieferte Glashütter Wertarbeit ohne Einschränkung fort. Jedes der insgesamt 550 Bauteile ist nach den einschlägigen Regeln der Handwerkskunst feinbearbeitet. Zum Beispiel besitzen alle Komponenten der Schlagwerks-Kadratur anglierte Kanten. Die Oberflächen weisen eine Glashütter Schwarzpolitur mittels Zinnplatte auf. Kein Wunder, dass der Zusammenbau eines dieser tickenden Mikrokosmen drei Monate verschlingt. Für höchste Klangqualität bürgen zwei direkt im massiven Gehäuse verankerte Tonfedern mit Terz-Stimmung. Weil alles Vorzügliche selten ist, offeriert Tutima nur 20 Exemplare in Roségold und fünf in Platin. Natürlich waren sich alle Verantwortlichen der Tatsache bewusst, dass Derartiges nur ganz wenigen Menschen vorbehalten bleibt, welche das in jeder Hinsicht Besondere zum Maß ihrer hohen Ansprüche machen. Daher offeriert Tutima ausgewiesenen Kennern und Fans feiner Glashütter Manufaktur zwei weitere Uhrwerke eigener Provenienz. Purismus par excellence verstrahlt das 31 Millimeter große T617 in der „Patria Small Second". Dieses Handaufzugswerk besitzt nur drei Zeiger, je einen für die Stunden, Minuten und Sekunden. Das massivgoldene Trio ist komplett von Hand gefertigt. Kosmopoliten und Vielflieger kommen bei der „Patria Dual Time" auf ihre Kosten. In dieser Armbanduhr tickt das aufgewertete Kaliber T619 mit 65 Stunden Gangautonomie und zusätzlicher 12-Stunden-Anzeige zur Indikation einer zweiten Zonenzeit. Selbstverständlich hat Tutima nach dem Umzug auch das jahrzehntelang gepflegte Stammthema nicht aus den Augen verloren. Bereits 2013 begann mit der „Saxon One" eine neue chronographische Zeitrechnung. Hinsichtlich seiner Funktionalität rekurriert der zeitschreibende Bolide auf das längst nicht mehr produzierte Automatikkaliber Lemania 5100 mit zentralem 60-Minuten-Zähler und zusätzlichem 24-Stunden-Zeiger. Dieser erweiterte Nutzen des bewährten Basis-Uhrwerks Eta 7750 verlangte nach Entwicklung und Fertigung eines speziellen Moduls. Beides ging und geht im lichtdurchfluteten Glashütter Manufakturgebäude über die Bühne. Wegen des sächsischen Wertschöpfungsanteils jenseits von 50 Prozent dürfen alle Zifferblätter nicht nur die Herkunftsbezeichnung „Glashütte" tragen, sondern auch den Schriftzug „Made in Germany". Überlieferte Elemente in gleich mehrfacher Hinsicht spiegelt die deutlich markantere Version „M2" als Nachfolgerin des sogenannten Nato-Chronographen wider, den Tutima für die Bundeswehr entwickelt hatte. Die farbliche Gestaltung der Zeiger bringt militärische Professionalität zum Ausdruck: Das signifikante Rot an den drei Zeigern für die Stoppfunktion hebt sich deutlich ab von den anderen vier durchgängig weißen Zeit-Zeigern.

In der Stille der Unterwasserwelt fühlt sich schließlich die brandneue „M2 Seven Seas" zu Hause. Ihr Titangehäuse ist auf eine Druckdichte bis 50 Bar ausgelegt. Ergo können die Besitzerinnen und Besitzer der 44 Millimeter großen Titan-Armbanduhr rein theoretisch bis zu 500 Meter tief abtauchen. Tutima stattet das zuverlässige Automatikkaliber Eta 2836 mit einem speziell gestalteten Rotor aus. Dieses Merkmal dient zum einen der Individualisierung und zum anderen dem Wertschöpfungsanteil in der neuen Glashütter Heimat von Tutima. ○

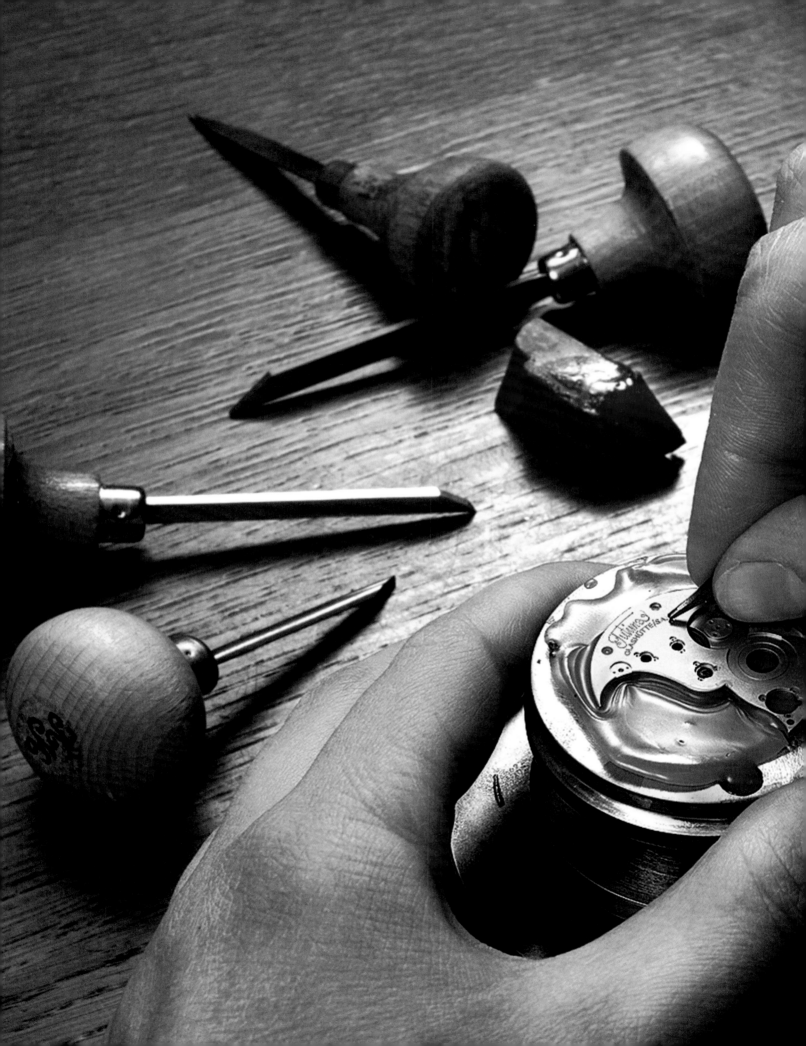

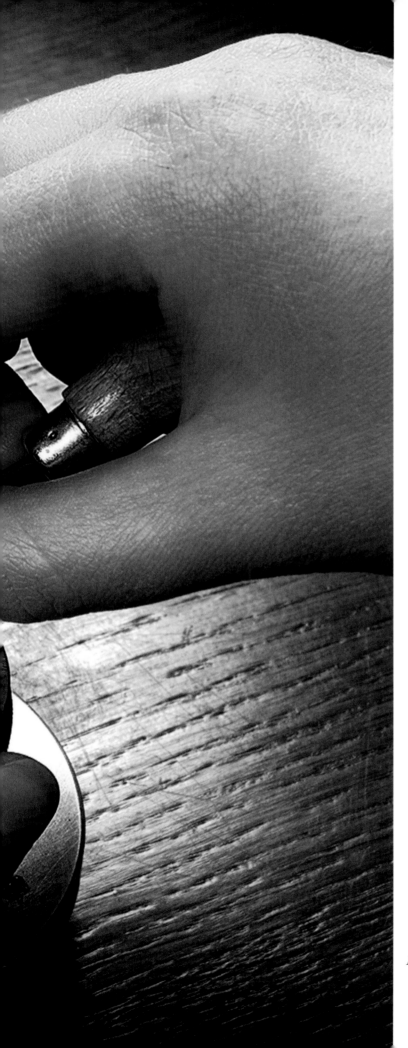

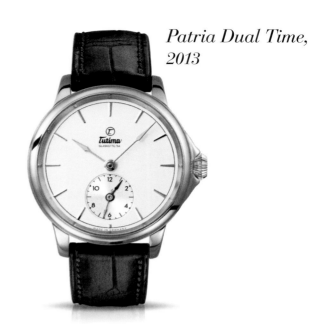

Patria Dual Time,
2013

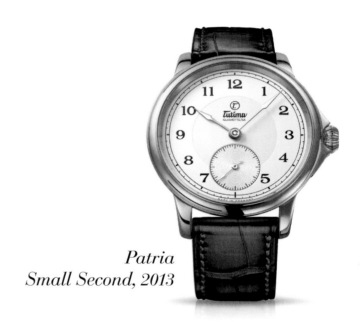

Patria
Small Second, 2013

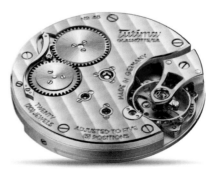

Cal. T617

Manual engraving ○ Handgravur ○ Gravure à la main

Le retour à Glashütte

Français

Glashütte est sans conteste le haut lieu de la fabrication horlogère allemande. Et ce pas uniquement depuis la chute du Mur en 1989 et la réunification un an plus tard, mais depuis 1845, date à laquelle Ferdinand Adolph Lange transfère ses activités horlogères dans cette petite ville de la vallée reculée de la Müglitz. Progressivement, il contribue à l'extension de ce secteur industriel essentiellement spécialisé dans les produits de luxe. Mais au début du XXᵉ siècle, on prend conscience que les produits de Glashütte doivent également s'adresser à de plus larges couches de la population. C'est dans cette optique que le juriste Ernst Kurtz crée en 1926 un conglomérat dans lequel figurent les sociétés UROFA (Uhren-Rohwerke-Fabrik Glashütte AG) et UFAG (Uhrenfabrik Glashütte AG). Cette dernière est spécialisée dans les montres-bracelets, des produits de première qualité lancés sous la signature Tutima Glashütte. Ce nom facile à retenir n'a bien sûr pas été choisi au hasard : il est dérivé de l'adjectif latin tutus, qui signifie « sûr, protégé ». Sous la direction d'Ernst Kurtz et de quelques collaborateurs compétents, la société commence à développer et à fabriquer des ébauches pour montres-bracelets à des prix abordables. L'achat et le transfert en Saxe d'une petite fabrique horlogère suisse compensent le manque de savoir-faire dans ce domaine.

On voit ainsi apparaître divers produits, allant des mouvements jusqu'aux chronographes, qui tiennent largement la comparaison avec la forte concurrence suisse. Le 7 mai 1945, un jour avant la pluie de bombes qui s'abat sur Glashütte, Ernst Kurtz et quelques-uns de ses employés parviennent à se réfugier à l'Ouest. À dater de 1951, on retrouve l'entrepreneur et son équipe à Ganderkesee, en Basse-Saxe. C'est là que le petit calibre 570 destiné à des montres pour femmes voit le jour en 1956. L'entreprise, qui manque de liquidités pour la production en série, est reprise par un ancien employé du temps de Glashütte sous le nom NUROFA (Norddeutsche Uhren-Rohwerkefabrik). La vente des montres est assurée par Dieter Delecate, également ancien employé d'Ernst Kurtz. Le régime de la RDA ne s'étant jadis pas intéressé à la célèbre signature Tutima, celle-ci fait un retour triomphal sur son nouveau site. Dieter Delecate réintroduit le nom sur le cadran. Le 28 avril 1969, il le dépose à titre de marque. Dans la foulée, il crée la société Tutima Uhren-fabrik GmbH. C'est à lui qu'elle doit d'être connue dans le monde entier. Durant des décennies, une grande partie des modèles signés Glashütter Tradition sont fabriqués dans la pittoresque ville de Ganderkesee. Comme à sa belle époque, l'entreprise se spécialise dans les montres d'aviateur et plus particulièrement celles dotées d'un chronographe. La production s'appuie sur des calibres Lemania et bien sûr aussi Eta. Un modèle rétro rappelle la célèbre montre

d'aviateur chronographe fabriquée entre 1941 et 1945 et son calibre retour-en-vol Urofa 59. Dans les années 1980, la Bundeswehr lance un appel d'offres pour une montre d'aviateur chronographe avec un cahier des charges précisément défini par le centre d'essai militaire. Après examen des offres, Tutima remporte le marché en 1983. Équipé du robuste calibre Lemania 5100, l'imposant chronographe automatique est impressionnant avec ses boutons poussoirs ergonomiques et son compteur 60 minutes bien en évidence parfaitement lisible.

Avec l'acquisition d'un imposant bâtiment proche de la gare de Glashütte, Tutima revient depuis mai 2011 vers son passé pour s'orienter résolument vers son avenir. Une fois soigneusement rénové et équipé d'un parc de machines ultramodernes, le bâtiment est investi par des techniciens, constructeurs, horlogers et autres spécialistes. Le renouveau des activités horlogères en Saxe est marqué dans une large mesure par le retour à l'ancienne production en manufacture. Pour marquer la renaissance de la marque, un modèle courant, à savoir un mouvement à remontage manuel ou automatique de base, n'est pas assez exaltant pour Dieter Delecate, sa fille Ute et son fils Jörg. Ils rêvent d'un bijou de mécanique, constitué, comme cela n'a encore jamais été fait, exclusivement de composants provenant de Glashütte, en l'occurrence une montre-bracelet à répétition minutes. Par sa complexité inouïe, cette fonction auxiliaire consistant à sonner les heures, les quarts d'heure puis les minutes sur deux marteaux représente le summum de l'art horloger mécanique. Le mouvement à remontage manuel doré, un calibre T800 de 32 millimètres de diamètre et 7,2 millimètres d'épaisseur, perpétue fidèlement la tradition de travail de qualité de Glashütte. Chacun des 550 composants est usiné selon les règles de l'art. Par exemple, tous les composants de la cadrature ont des arêtes anglées. Les surfaces arborent le poli miroir au bloc d'étain, typique de Glashütte. Rien d'étonnant à ce qu'il faille trois mois pour assembler l'un de ces microcosmes mécaniques. Deux timbres réson-nant en tierce majeure directement intégrés dans le boîtier massif assurent une sonorité parfaite. L'excellence étant synonyme de rareté, Tutima en propose seulement 25 exemplaires, 20 en or rose et cinq autres en platine. Tous les responsables sont bien sûr conscients du fait que ces modèles sont réservés à une élite très exigeante, en mal de singularité à tous les égards. Aussi, Tutima propose-t-elle à ces fins connaisseurs et adeptes de produits raffinés de Glashütte deux mouvements fabriqués en interne. Avec son calibre T617 de 31 millimètres, le « Patria Small Second » incarne le purisme par excellence. Ce garde-temps à remontage manuel

M2, 2013

possède trois aiguilles distinctes pour les heures, les minutes
et les secondes, un trio en or massif entièrement réalisé à la main.
Le « Patria Dual Time » fait le bonheur de ceux qui voyagent beaucoup, notamment en avion. Cette montre-bracelet est animée par
le calibre T619 revalorisé avec ses 65 heures de réserve de marche
et un cadran auxiliaire des heures pour un deuxième fuseau
horaire. Après son déménagement, Tutima n'oublie bien sûr pas
ce qui a été son cheval de bataille des décennies durant. Dès
2013, le « Saxon One » amorce une nouvelle ère pour les chronographes maison. D'un point de vue fonctionnel, ce garde-temps
d'exception s'appuie sur le calibre automatique Lemania 5100
à compteur 60 minutes et aiguille 24 heures. Mais ce dernier n'est
plus produit depuis longtemps et la marque part du calibre de base
éprouvé Eta 7750 pour développer et fabriquer un module spécial.
Les opérations nécessaires se déroulent alors, comme aujourd'hui
encore, dans les bâtiments inondés de lumière de la manufacture à Glashütte. La part de la valeur ajoutée en Saxe dépassant
50 pour cent, tous les cadrans peuvent non seulement porter
l'indication de provenance « Glashütte », mais aussi l'inscription
« Made in Germany ». Reprenant de nombreux éléments du
chronographe OTAN, jadis développé par Tutima pour l'Armée
allemande, la version « M2 » est beaucoup plus marquante. Les
couleurs retenues pour les aiguilles traduisent le professionnalisme militaire : le rouge significatif sur les trois aiguilles de la
fonction de chronométrage tranche nettement sur le blanc des
quatre autres aiguilles.

Enfin, la toute nouvelle « M2 Seven Seas », dont le boîtier en titane
est prévu pour résister à une pression de 50 bars, est faite pour
le monde du silence. Les propriétaires de cette montre-bracelet
de 44 millimètres de diamètre peuvent donc théoriquement plonger
avec elle jusqu'à 500 mètres de profondeur. Valeur sûre, le calibre
automatique Eta 2836 est équipé d'un rotor spécialement conçu
pour ce modèle, qui contribue à la fois à sa personnalisation et
à la valeur ajoutée que Tutima lui apporte sur son nouveau site. ◦

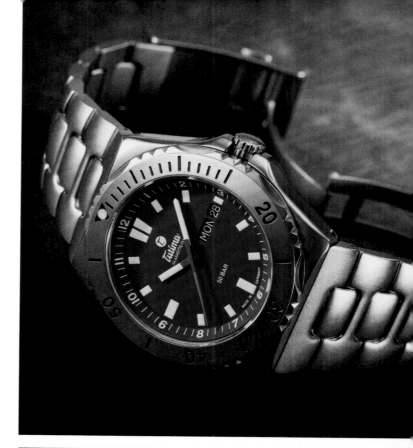

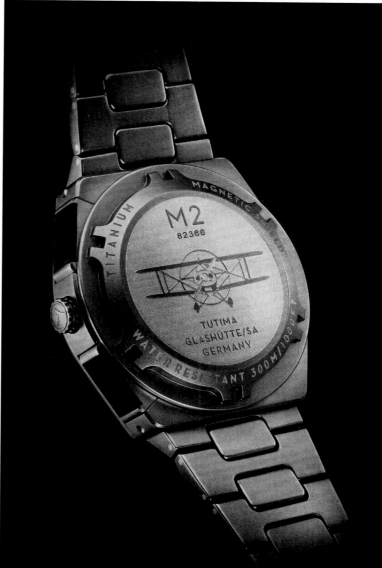

From top: M2 Seven Seas, tested to a pressure of 50 atmospheres, 2016 ◦
M2 back, 2013 ◦ Von oben: M2 Seven Seas, druckgeprüft bis 50 Bar, 2016 ◦
Rückseite der M2, 2013 ◦ De haut en bas : M2 Seven Seas, certifiée résistant
à une pression de 50 bars, 2016 ◦ Revers de la M2, 2013

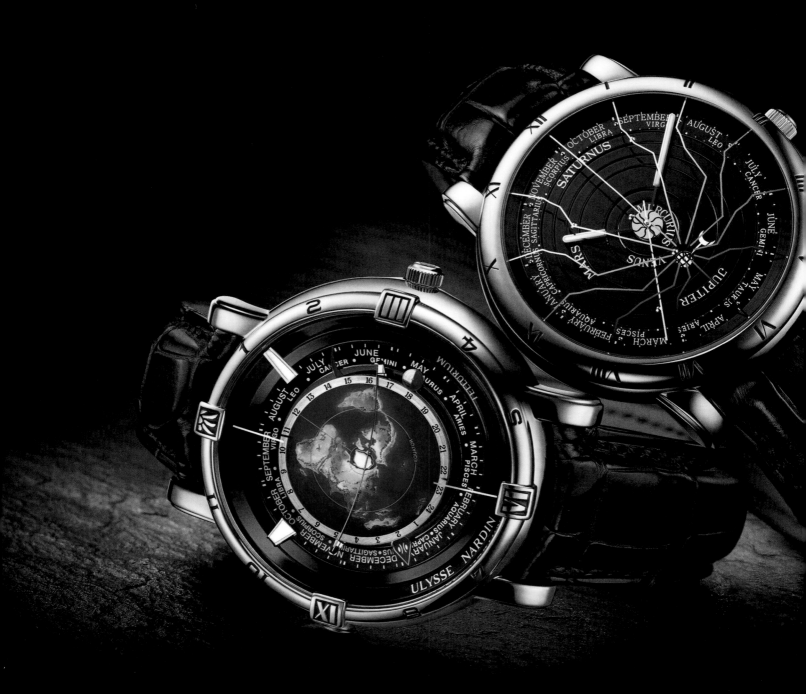

Ulysse Nardin

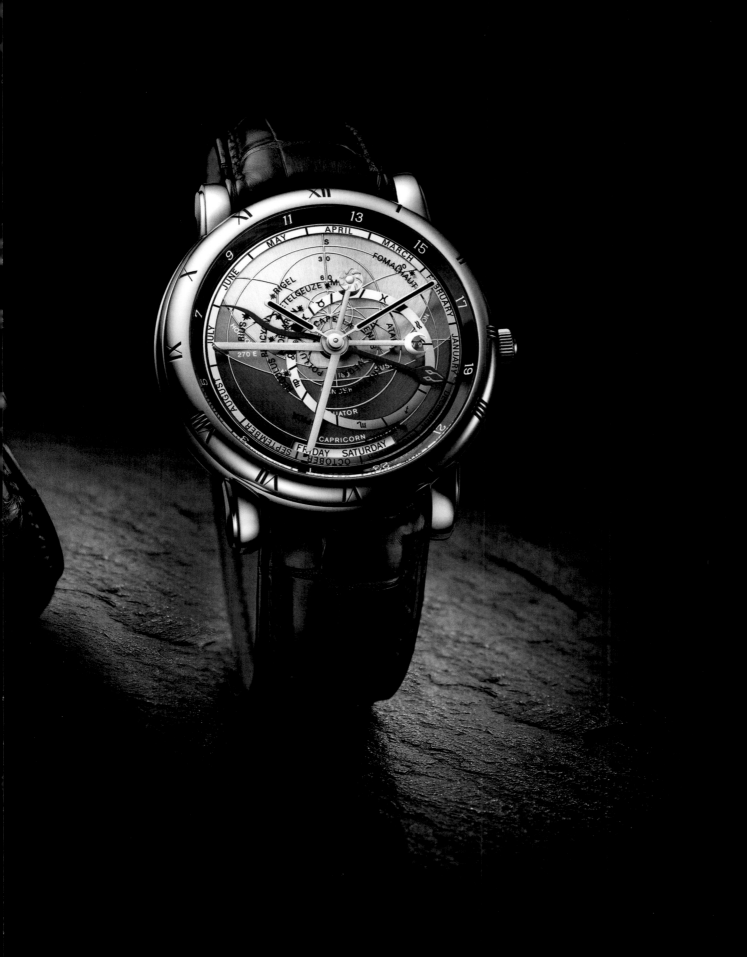

A Perfect Synthesis of Tradition and Innovation

Ulysse Nardin's history, which began in 1846, contains many high points, but the "Freak" carousel that debuted in 2001 revolutionized watchmaking in a truly unique way. Connoisseurs were particularly impressed by the innovative Dual-Direct escapement in thoroughly unconventional hand-wound Caliber UN-01, which can run for eight days between windings. Trailblazing technologies go hand in hand with unprecedented materials here. Classical components, e.g., lever and escape-wheel, can be sought in vain. Two drive wheels directly transfer energy to the rate regulator, which consists of a balance and its hairspring. For the first time ever, Ulysse Nardin fabricated the two wheels from pure silicon via the plasma method. Silicon is antimagnetic, combines the greatest possible hardness with low weight, and has very smooth surfaces which make oil superfluous. Two men stand behind this exceptional construction: Rolf Schnyder, a passionate watchmaker with a big heart, purchased the ailing firm in 1983 and transformed it into a recognized player in the art of sophisticated mechanical watchmaking; his sparring partner, the creative watchmaker and scientist Ludwig Oechslin, deserves credit for having enabled Ulysse Nardin to achieve numerous chronometric milestones.

Incidentally: the firm's founders were similarly characterized by utterly untamable inventiveness. New ideas matured in Ulysse Nardin's mind nearly every day. His creativity was commended by the jurors at the 1862 London Exhibition, who chose him to receive the highest award. The prize medal was followed by additional honors in ensuing years. This could well be why he was selected to submit one of Switzerland's contributions to the World's Columbian Exposition in Chicago in 1893. The richly decorated case of his watch recalled the Industrial Revolution of the second half of the 19th century. Its sculptor required more than 1,200 hours to complete this masterpiece. Ulysse Nardin had initially equipped his exceptional timepiece with a chronometer movement, but its later owner wanted a chronograph with a minute repeater. A nearly countless number of marine chronometers and navigator's watches earned global renown. For these genres alone, Ulysse Nardin received some 4,300 awards from various observatories and testing facilities. Wristwatches joined the collection after the turn from the 19th to the 20th century. A first chronograph for the wrist premiered at a very early date: namely, in 1912.

It seemed as though nothing could stop the series of successes—until the 1960s, when advances in electronic timekeeping gradually diminished the demand for nautical timepieces. Instead of responding appropriately, Ulysse Nardin's successors—most of whom were lawyers—languished in entrepreneurial lethargy.

Freak, 2001

To make matters worse, they totally misjudged the future of the wristwatch, so it looked as though the family business was doomed.

The proverbial knight in shining armor arrived in the guise of Rolf Schnyder, who invested a princely sum of 1.5 million Swiss francs. His intelligent revival of tradition and innovation, coupled with Ludwig Oechslin's ingenious engineering and designing, led to the creation of a trilogy of unique astronomical wristwatches in 1985. The first of these, the highly complex "Astrolabium Galileo Galilei," was depicted on the inside cover page of the 1988 edition of *The Guinness Book of Records*. This year also witnessed the debut of the "Planetarium Copernicus," which depicts the astronomical positions of five planets in relationship to the sun and the earth. The third member of the trio was the "Tellurium Johannes Kepler," which premiered in 1992.

Protected by numerous patents, the "Perpetual Ludwig" with perpetual calendar is meticulously conceived in every detail, convincingly user-friendly, and still remains state-of-the-art today. For the first time in the history of this horological complication, all indicators could be set and adjusted in both directions via the crown. Ulysse Nardin combined this mechanism and an intuitively operable time-zone display in the "GMT± Perpetual." The outsize date automatically synchronizes with trips through time zones, no matter whether the traveler journeys westward or eastward. "Sonata," which debuted in 2003, turned the history of the alarm wristwatch on its head: this watch's alarm is impossible not to hear and functions perfectly all around the globe. Needless to say, operating this technical marvel is mere child's play.

Ulysse Nardin celebrated its 160th anniversary by unveiling automatic Caliber UN-160, which the brand completely developed and fabricated in La Chaux-de-Fonds. Two of its finest features are the exclusive lubricant-free Dual-Ulysse escapement and the label's own Glucydur balance with variable moment of inertia.

This movement's debut catapulted Ulysse Nardin into the aristocratic circle of genuine manufactories. Another world premiere caused an appreciative stir in 2007: the silicon components of the Dual-Direct escapement in the "Freak DIAMonSIL" are coated with a layer of synthetic nano-crystalline diamond. This combination of materials offers the advantages of the world's hardest material, but keeps expenses at a very reasonable level. The "Moonstruck" followed in the astronomical footsteps of the "Trilogy of Time" in 2009. The utmost precision again characterized the accuracy of this watch's indicators. Approximately 100,000 years will come and go before the lunar display mistakenly shows a new moon rather than a full moon. Particularly impressive: the moon's current phase can be read exactly in relation to any desired location on the Earth's surface, and the timepiece also demonstrates the global dynamic of the tides, which are influenced by the moon and the sun. Prior to his unexpected death in 2011, Rolf Schnyder initiated the ambitious project to develop Caliber UN-118. DIAMonSIL was again the magic word in the escapement, and the balance cooperates with a freely oscillating silicon hairspring. Each of the 350 red gold specimens of this wristwatch, which sold out long ago, has an enamel dial crafted by Donzé, a traditional specialist and a member of the Ulysse Nardin Group. After the death of its charismatic owner in 2011, Ulysse Nardin came under the aegis of the French Kering Group. CEO Patrik Hoffmann, a longstanding associate of the deceased, logically continued his predecessor's successful strategy. The purchase of chronograph Caliber 137 also assured Ulysse Nardin's autonomy in this sophisticated field of mechanical timekeeping. But extensive optimizations were necessary, including the use of silicon components for the oscillating and escapement system, before it could be deployed as Caliber UN-150. The visionary Rolf Schnyder had financially participated in the production of these components several years previously. His goal: achieve independence from suppliers of indispensable key components for mechanical watches.

Patrik Hoffmann's unmistakable "handwriting" is evident in several debutantes in 2016. The undisputed highlight among them is the "Grand Deck Marine Tourbillon." Its inlaid dial and unconventional minute-hand recall this brand's traditional affiliation with all things nautical. The minute-hand, which resembles a miniature mainsail boom, is powered by a paper-thin wire made of high-tech fiber and sweeps from left to right throughout the course of an hour. Afterwards, the gear-train and time display take into account the quick but not instantaneous motion with which this unconventional hand returns to its starting position. The "Marine Chronograph Annual Calendar" obeys Ludwig Oechslin's principle of simple but highly functional cadratures. It needs only seven additional components for its practical calendar, which automatically knows the lengths of all the months except February. The distinctive "FreakWing" expresses horological high technology coupled with the fascination of the 35th America's Cup as represented by Ulysse Nardin's partnership with Artemis Racing. In this context, the movement's upper bridge for the minutes is inspired by the internal structure of the rigid sail, while the texture of the rotating hours disc evokes the mesh net of the multihull. The bezel and case back consist of carbon fiber, a material that's omnipresent in high-level sailing. All this proves once again that the centuries-old art of mechanical watchmaking and the high technologies of the 21st century complement one another marvelously well. ◦

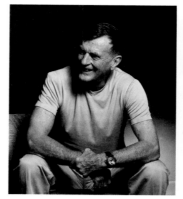
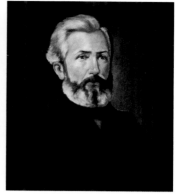
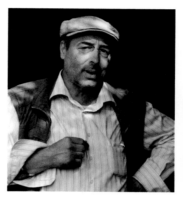
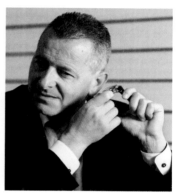

Clockwise from top left: Rolf Schnyder ◦ Ulysse Nardin ◦ Patrik Hoffmann ◦ Ludwig Oechslin

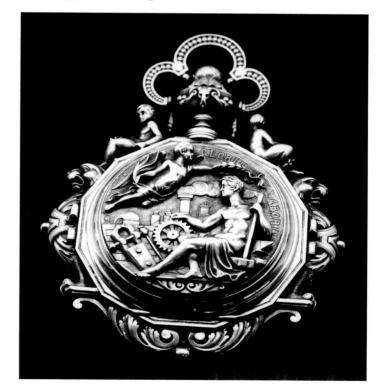

Top: Chicago, unique piece designed for the M2 World's Columbian Exposition in Chicago, 1893 ◦ Left: Dual-Ulysse escapement, Cal. UN-201 ◦ Oben: Chicago, Einzelstück, entworfen für die World's Columbian Exposition in Chicago, 1893 ◦ Links: Dual-Ulysse-Hemmung, Kaliber UN-201 ◦ Ci-dessus : Chicago, pièce unique conçue pour la World's Columbian Exposition à Chicago, 1893 ◦ À gauche : Échappement Dual-Ulysse, Cal. UN-201

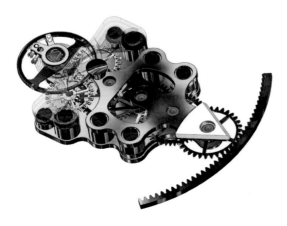

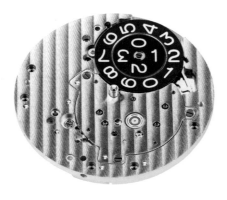
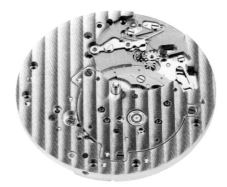
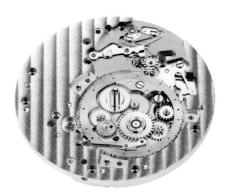

Top and middle: Automatic Cal. UN-160 ◦ Bottom: Freak DIAMonSIL, 2007

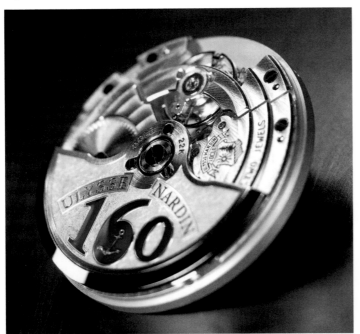
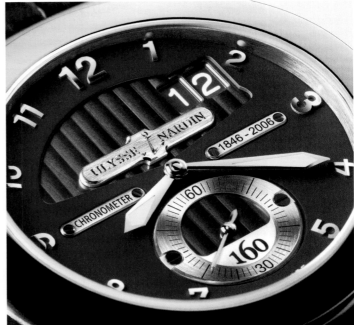
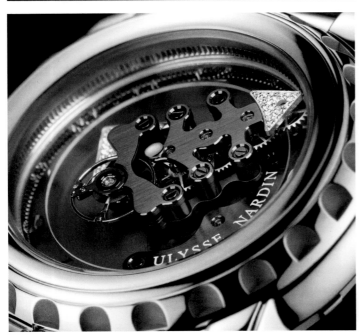
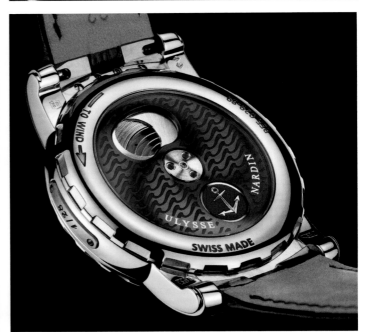

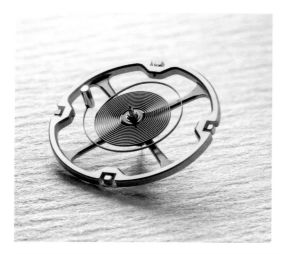

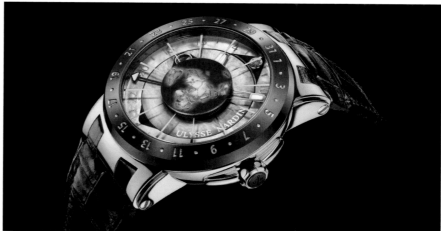

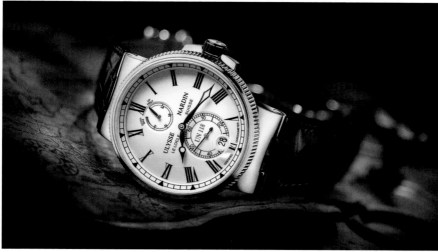

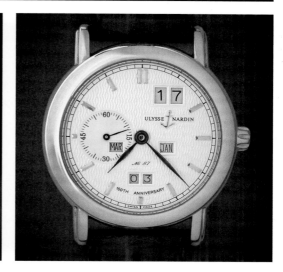

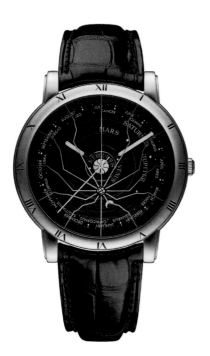

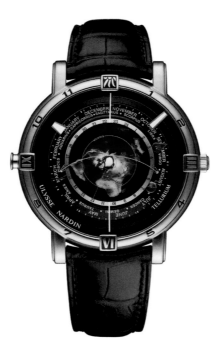

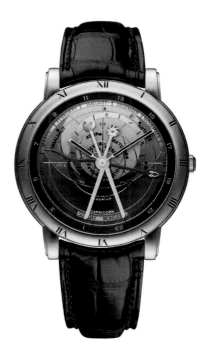

Top left: UN-118 caliber spiral balance spring ◦ Oben links: Unruhspirale, Kaliber UN-118 ◦ En haut à gauche: Spiral de balancier, calibre UN-118 ◦
Top right: Moonstruck, 2009 ◦ Middle, from left: Marine Chronometer Manufacture, 2012 ◦ Perpetual Ludwig, 1996 ◦ Bottom, from left:
Planetarium Copernicus, 1988 ◦ Tellurium Johannes Kepler, 1992 ◦ Astrolabium Galileo Galilei, 1985

Perfekte Synthese aus Tradition und Innovation

Höhepunkte gibt es definitiv viele in der bis 1846 zurückreichenden Biographie von Ulysse Nardin. Aber das 2001 vorgestellte „Freak"-Karussell hat die Uhrmacherei in wahrhaft einzigartiger Weise revolutioniert. Vor allem die innovative Dual-Direct-Hemmung des durch und durch ungewöhnlichen Handaufzugskalibers UN-01 mit acht Tagen Gangautonomie hat es in sich. Wegweisende Technologien gehen Hand in Hand mit bis dahin noch nie verwendeten Materialien. Klassische Bauteile wie beispielsweise Anker und Ankerrad sucht man vergebens. Zwei Antriebsräder übertragen die Energie direkt auf den Gangregler, bestehend aus Unruh und Unruhspirale. Erstmals überhaupt fertigt Ulysse Nardin besagtes Räderpaar im Plasmaverfahren aus reinem Silizium. Dieses amagnetische Material zeichnet sich aus durch höchstmögliche Härte bei geringem Gewicht: Besonders glatte Oberflächen machen Öl entbehrlich. Hinter dieser Ausnahme-Konstruktion stehen zwei Männer: Rolf Schnyder, ein passionierter Unternehmer mit Herz, erwarb die konkursreife Firma im Jahr 1983 und formte sie zu einer anerkannten Größe auf dem Gebiet anspruchsvoller mechanischer Uhrmacherkunst. Seinem Sparringspartner, dem kreativen Uhrmacher und Wissenschaftler Ludwig Oechslin, verdankt Ulysse Nardin zahlreiche chronometrische Meilensteine.

Schier unbändiger Einfallsreichtum hatte übrigens schon den Firmengründer ausgezeichnet. Beinahe jeden Tag reiften im Kopf von Ulysse Nardin neue Ideen. Dafür belohnte ihn die Jury der Londoner Weltausstellung 1862 mit der höchstmöglichen Auszeichnung in Gestalt einer Prize Medal. Weitere Ehrungen dieser Art folgten auf dem Fuße. Womöglich deshalb durfte er 1893 einen der Schweizer Beiträge zur Weltausstellung in Chicago, der World's Columbian Exposition, liefern. Das reich dekorierte Gehäuse seiner Uhr erinnerte an die industrielle Revolution in der zweiten Hälfte des 19. Jahrhunderts. Zur Vollendung benötigte der Skulpteur mehr als 1 200 Arbeitsstunden. Ursprünglich hatte Ulysse Nardin seinen Ausnahme-Zeitmesser mit einem Chronometerwerk ausgestattet. Der spätere Käufer dieser Uhr wünschte sich jedoch einen Chronographen mit Minutenrepetition. Weltgeltung erlangten beinahe unzählige Marinechronometer und Beobachtungsuhren. Allein hierfür konnte Ulysse Nardin sage und schreibe rund 4 300 Auszeichnungen verschiedener Observatorien und Prüfstellen entgegennehmen. Nach der Wende vom 19. zum 20. Jahrhundert fanden Armbanduhren in die Kollektion. Sehr früh, nämlich schon 1912, entstand ein erster Chronograph fürs Handgelenk.

Bis in die 1960er Jahre schien nichts die Erfolge trüben zu können. Dann aber ließen Fortschritte bei der elektronischen Zeitmessung den Bedarf an nautischen Zeitmessern sukzessive gegen null gehen. Anstatt adäquat gegenzusteuern, erstarrten die Nachfahren des Ulysse Nardin, meist Juristen von Beruf, in unternehmerischer Lethargie. Weil sie zu allem Überfluss auch noch die Zukunft der Armbanduhr völlig falsch einschätzten, schien das Ende des Familienunternehmens besiegelt.

Als sogenannter Weißer Ritter investierte Rolf Schnyder für damalige Verhältnisse stattliche 1,5 Millionen Schweizer Franken. Seine intelligente Verquickung von Tradition und Innovation sowie das konstruktive Genie von Ludwig Oechslin führten ab 1985 zu einer Trilogie einzigartiger astronomischer Armbanduhren. 1988 fand sich die Nummer eins, das hochkomplexe „Astrolabium Galileo Galilei", auf der Umschlagseite des Guinness-Buchs der Rekorde. Im gleichen Jahr debütierte das „Planetarium Copernicus", welches die astronomischen Positionen von fünf Planeten in Relation zu Sonne und Erde darstellt. Drittes im Bunde war das 1992 vorgestellte „Tellurium Johannes Kepler".

Durchdacht bis ins letzte Detail, überaus benutzerfreundlich und aktuell bis in die Gegenwart präsentierte sich der mehrfach patentierte „Perpetual Ludwig" mit ewigem Kalender. Erstmals in der Geschichte dieser uhrmacherischen Komplikation sind sämtliche Indikationen mit Hilfe der Krone in beide Richtungen einstell- und korrigierbar. Beim „GMT± Perpetual" kombinierte Ulysse Nardin diesen Mechanismus mit intuitiv handhabbarem Zeitzonen-Dispositiv. Das Großdatum folgt Zeit-Reisen ganz automatisch, egal ob in westlicher oder östlicher Richtung. „Sonata", vorgestellt 2003, stellte die Geschichte des Armbandweckers auf den Kopf. Ihr Alarm funktioniert unüberhörbar rund um den Globus. Kinderleichte Bedienung war und ist auch hier Ehrensache.

Sein 160. Firmenjubiläum zelebrierte Ulysse Nardin 2006 mit dem komplett selbst entwickelten und in La Chaux-de-Fonds gefertigten Automatikkaliber UN-160. Zwei seiner Filetstücke sind die exklusive, ohne Öl arbeitende Dual-Ulysse-Hemmung und die eigene Glucydur-Unruh mit variablem Trägheitsmoment.

Classico Schooner America, 2016

Mit diesem Uhrwerk katapultierte sich Ulysse Nardin in den aristokratischen Zirkel echter Manufakturen. Eine weitere Weltpremiere machte 2007 von sich reden. Im spektakulären „Freak DIAMonSIL" tragen die Silizium-Komponenten der Dual-Direct-Hemmung eine synthetische nano-kristalline Diamantschicht. Dieser Werkstoffmix bietet die Vorteile des härtesten Materials überhaupt, hält die Kosten aber in sehr überschaubarem Rahmen. An die astronomische „Trilogie der Zeit" knüpfte die „Moonstruck" von 2009. Oberste Präzision genoss einmal mehr die Anzeigegenauigkeit. Bis diese Armbanduhr anstelle des Neumonds einen Vollmond abbildet, verstreichen rund 100 000 Jahre. Besonders beeindruckend: das exakte Ablesen der aktuellen Mondphase in Relation zu einem beliebig bestimmbaren Punkt auf der Erdoberfläche sowie eine Demonstration der globalen Dynamik der von Mond und Sonne beeinflussten Gezeiten. Vor seinem überraschenden Tod im Jahr 2011, hatte Rolf Schnyder noch das ehrgeizige Kaliber-Projekt UN-118 in Angriff genommen. Bei der Hemmung lautete das Zauberwort erneut DIAMonSIL. Und die Unruh kooperiert mit einer völlig frei oszillierenden Siliziumspirale. Die 350 Rotgold-Exemplare der längst ausverkauften Armbanduhr besitzen ein Emailzifferblatt des traditionsreichen Spezialisten Donzé, einem Mitglied der Ulysse-Nardin-Gruppe. Nach dem Ableben des charismatischen Eigentümers im Jahr 2011 gelangte Ulysse Nardin unter das Dach der französischen Kering-Gruppe. Als CEO setzte Patrik Hoffmann, ein langjähriger Weggefährte des Verstorbenen, die erfolgreiche Strategie konsequent fort. Der Kauf des Chronographenkalibers 137 bescherte Ulysse Nardin Autonomie auch auf diesem anspruchsvollen Gebiet mechanischer Zeitmessung. Bevor es als Kaliber UN-150 zum Einsatz kam, waren jedoch tiefgreifende Optimierungen vonnöten, darunter die Verwendung von Silizium-Komponenten für das Schwing- und Hemmungssystem. An den

Produzenten dieser Bauteile hatte sich der visionäre Rolf Schnyder schon Jahre zuvor finanziell beteiligt. Sein Ziel: Unabhängigkeit von den Zulieferern unverzichtbarer Schlüsselkomponenten mechanischer Uhren.

Die unverkennbare Handschrift von Patrik Hoffmann tragen gleich mehrere Neuheiten des Jahres 2016: Als absolutes Highlight kann das „Grand Deck Marine Tourbillon" gelten. Sein Intarsien-Zifferblatt und der außergewöhnliche Minutenzeiger erinnern an die langen chronometrischen Verbindungen zum nassen Element. Angetrieben von einem hauchdünnen Draht aus Hightech-Faser bewegt sich der einem Großbaum nachempfundene Minutenzeiger innerhalb einer Stunde von links nach rechts. Logischerweise berücksichtigen Räderwerk und Zeitanzeige die Dauer der zügigen, aber nicht springenden Rückbewegung zum Ausgangspunkt. Dem von Ludwig Oechslin stets postulierten Prinzip einfacher, trotzdem jedoch höchst funktionaler Kadraturen folgt der „Marine Chronograph Annual Calendar". Nur sieben zusätzliche Teile braucht es für das hilfreiche Kalendarium, welches die Länge aller Monate mit Ausnahme des Februars kennt. Uhrmacherische Hochtechnologie, gepaart mit der durch den Partner Artemis Racing repräsentierten Faszination des 35. America's Cup, bringt die markante „FreakWing" zum Ausdruck. In diesem Sinn ist die obere Minutenbrücke des Uhrwerks von der inneren Struktur des starren Segels inspiriert. Den Flanken des Doppelrumpf-Boliden ähnelt die Beschaffenheit der rotierenden Stundenscheibe. Schließlich bestehen Glasrand und Boden aus Kohlenstofffaser, dem Werkstoff des Top-Segelsports schlechthin. Womit erneut bewiesen wäre, dass sich jahrhundertealte mechanische Uhrmacherkunst und die Hochtechnologien des 21. Jahrhunderts in wunderbarer Weise ergänzen können. ◦

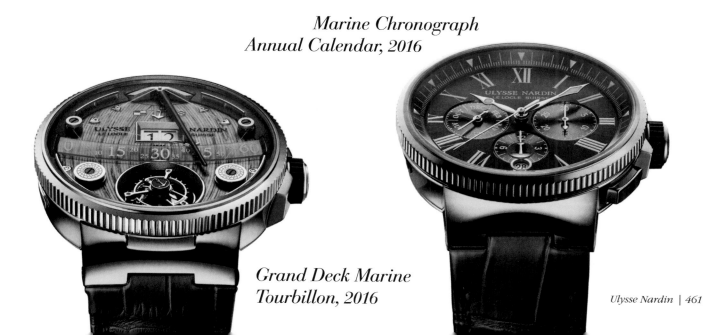

Marine Chronograph Annual Calendar, 2016

Grand Deck Marine Tourbillon, 2016

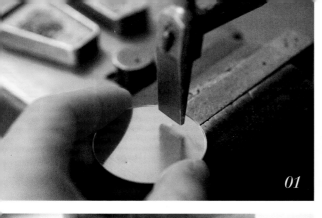

01

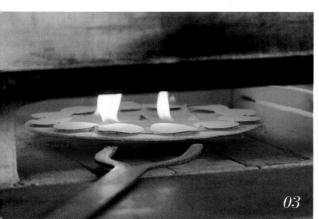

02

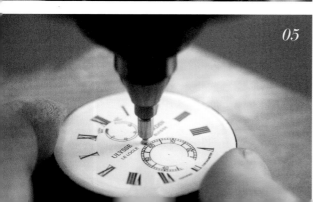

03

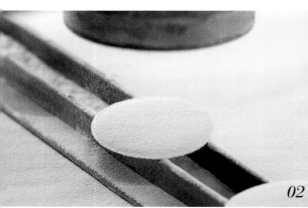

04

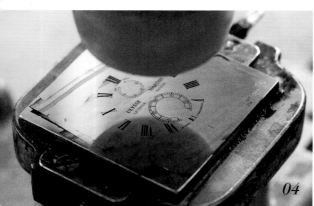

05

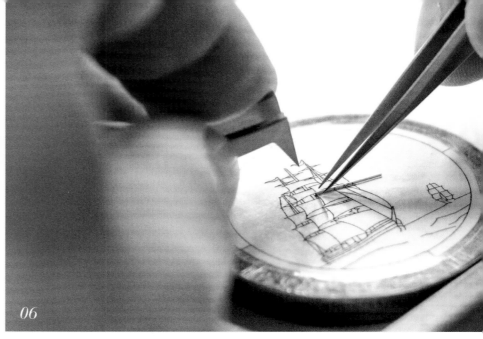

06

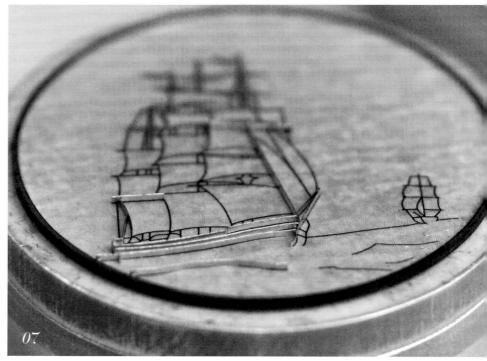

07

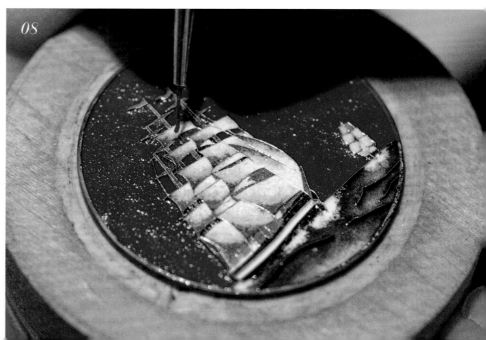

08

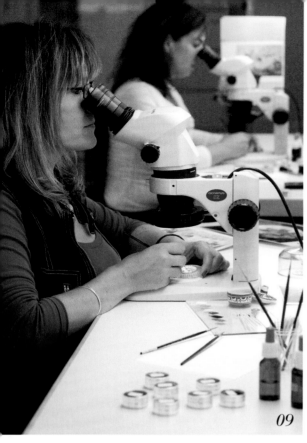

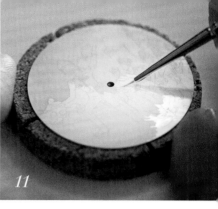

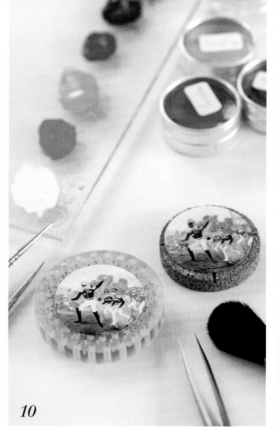

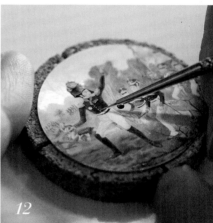

09

10

11

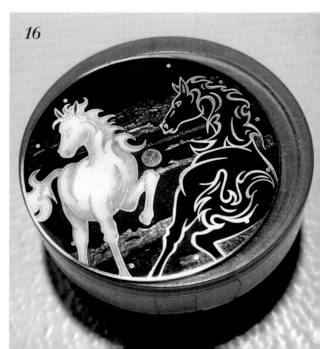

12

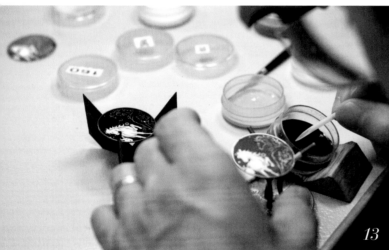

13

01–05: Enameling process: "Grand Feu" Marine Chronometer Manufacture ○ *Emaillierung, Grand-Feu-Technik: Marine Chronometer Manufacture* ○ **Émaillage grand feu : Marine Chronometer Manufacture**

06–08: Enameling process: "Cloisonné Enamel" Classico Kruzenshtern ○ *Emaillierung, Cloisonné-Technik: Classico Kruzenshtern* ○ **Émaillage cloisonné : Classico Kruzenshtern**

09–12: Miniature painting process: Classico Raevsky ○ *Miniaturmalerei: Classico Raevsky* ○ **Peinture de miniature : Classico Raevsky**

13–16: Enameling process: "Champlevé Enamel" Classico Horse ○ *Emaillierung, Champlevé-Technik: Classico Horse* ○ **Émaillage champlevé : Classico Horse**

14

15

16

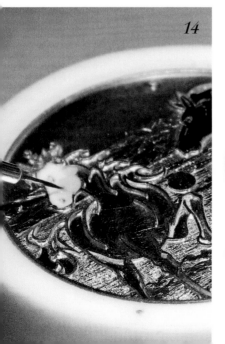

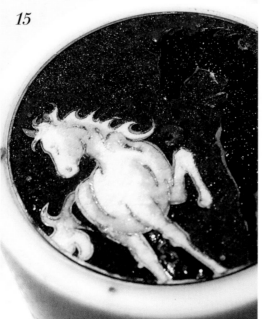

La synthèse parfaite entre tradition et innovation

Si les dates clés sont légion dans l'histoire d'Ulysse Nardin, qui débute en 1846, la sortie en 2001 du carrousel tourbillon « Freak » constitue une véritable révolution horlogère. On est particulièrement émerveillé par l'innovant double échappement direct Dual Direct du calibre à remontage manuel UN-01 totalement inédit, qui assure une réserve de marche de huit jours. Les technologies de pointe vont de pair avec l'emploi de matériaux jusqu'alors réservés à d'autres domaines. On cherchera en vain les composants classiques comme l'ancre et la roue d'échappement. Ce sont deux roues d'impulsion qui transmettent l'énergie directement à l'organe de régulation constitué par le balancier et le spiral. Ulysse Nardin réalise ces roues en silicium pur par traitement plasma, une véritable première. Ce matériau amagnétique se caractérise par une dureté extrêmement élevée pour un poids très faible, mais aussi par des surfaces lisses à tel point qu'aucun lubrifiant n'est nécessaire. Ce type de conception hors pair est l'œuvre de deux hommes. Rolf Schnyder, entrepreneur passionné et généreux, rachète en 1983 la manufacture au bord de la faillite et fait d'elle un acteur reconnu dans le segment de l'horlogerie mécanique de luxe. Son « sparring partner », le maître-horloger créatif et chercheur Ludwig Oechslin, est à l'origine de nombreuses avancées chronométriques pour le compte d'Ulysse Nardin.

Cette créativité débordante était déjà le propre du fondateur de l'entreprise : presque tous les jours, de nouvelles idées mûrissaient dans le cerveau d'Ulysse Nardin. Le jury de l'Exposition universelle de 1862 à Londres l'en récompense par la plus haute distinction, la « Prize Medal ». Et les honneurs comparables de se succéder. Cela vaut sans doute à la marque de faire partie en 1893 des contributeurs suisses à l'Exposition universelle de Chicago, également connue sous le nom de World's Columbian Exposition. Les riches ornementations du boîtier de sa montre, qui évoquent la Révolution industrielle de la seconde moitié du XIXe siècle, ont exigé plus de 1 200 heures de travail de la part du graveur. À l'origine, ce modèle inédit est équipé d'un mouvement de chronomètre. Le futur acquéreur de la montre la fera transformer en chronographe doté d'une répétition minutes. Un nombre pour ainsi dire incalculable de chronomètres de marine et de montres d'observation atteignent à la renommée internationale. Dans ce seul domaine, la marque reçoit pas moins de 4 300 récompenses de différents observatoires et organismes vérificateurs. Au début du XXe siècle, la collection admet en son sein des montres-bracelets. Très tôt, dès 1912, l'horloger met au point un chronographe de poignet.

Jusque dans les années 1960, rien ne semble pouvoir ternir cette réussite. Avec les progrès de la mesure électronique du temps, la demande en chronomètres de marine baisse alors progressivement jusqu'à devenir quasi nulle. Au lieu de redresser le cap, les descendants d'Ulysse Nardin, en majorité des juristes de formation, laissent l'entreprise sombrer dans la léthargie. Comme ils n'ont en outre pas su prendre la mesure de la saturation du marché de la montre-bracelet, ni anticiper son évolution, il semble que la fin de l'entreprise familiale soit scellée.

Rolf Schnyder, en véritable Chevalier blanc, investit 1,5 million de francs suisses, une coquette somme pour l'époque. Son judicieux mix de tradition et d'innovation ainsi que l'inventivité de Ludwig Oechslin sont à l'origine du développement, à partir de 1985, d'une trilogie de montres-bracelets astronomiques d'exception. En 1988, la première, le modèle « Astrolabium Galileo Galilei » à multiples complications, fait la couverture du *Livre Guinness des records*. La même année, on voit arriver sur le marché la « Planetarium Copernicus », qui affiche les positions astronomiques de cinq planètes par rapport au Soleil et à la Terre. Dernier modèle du trio, « Tellurium Johannes Kepler » sort en 1992.

Étudiée jusque dans le moindre détail, extrêmement conviviale, encore d'actualité et faisant l'objet de plusieurs brevets, la « Perpetual Ludwig » fait son entrée en scène avec un quantième perpétuel. Pour la première fois dans l'histoire de cette complication horlogère, la couronne permet l'ajustement et la rectification en avant ou en arrière de tous les indicateurs calendaires. Dans le modèle « GMT± Perpetual », Ulysse Nardin associe le quantième perpétuel à une fonction GMT activable de manière intuitive. La Grande Date s'ajuste automatiquement, que l'on voyage vers l'est ou vers l'ouest. En 2003, « Sonata » révolutionne l'histoire des montres réveils. Sa sonnerie cathédrale, qui est facilement perceptible, fonctionne indépendamment du fuseau horaire. Comme toujours, la marque se fait ici un point d'honneur à proposer des fonctions faciles à manier.

Sorti en 2006 pour célébrer le 160e anniversaire de la manufacture, le calibre automatique UN-160 est entièrement développé et fabriqué en interne à La Chaux-de-Fonds. Deux de ses « morceaux de choix » sont l'échappement Dual Ulysse non lubrifié exclusif et le balancier maison en glucidur à moment d'inertie variable.

Ce mouvement catapulte Ulysse Nardin dans le gotha des vraies manufactures. En 2007, une nouvelle première mondiale défraye la chronique. Dans le spectaculaire « Freak DIAMonSIL », les composants en silicium de l'échappement Dual Direct sont revêtus d'une couche de diamant nanocristallin de synthèse, une solution qui permet de bénéficier des propriétés du matériau le plus dur qui soit tout en maintenant les coûts à un niveau raisonnable. Dans la lignée des trois montres astronomiques composant la Trilogie du Temps, « Moonstruck » est encore un chef-d'œuvre de précision de l'affichage : il ne faudra pas moins de 100 000 ans pour voir s'afficher une pleine lune au lieu d'une nouvelle lune. Ce modèle sorti en 2009 présente entre autres caractéristiques remarquables : la détermination exacte de la phase de lune actuelle en un point quelconque à la surface de la Terre ainsi que la visualisation du mouvement des marées résultant de l'attraction combinée de la Lune et du Soleil. Avant son décès soudain en 2011, Rolf Snyder s'attelle à l'ambitieux projet de calibre UN-118. De nouveau, la formule magique pour l'échappement est DIAMonSIL. Le balancier est couplé à un spiral silicium oscillant en toute liberté. Les 350 exemplaires en or rouge de la montre-bracelet épuisée depuis longtemps sont dotés d'un cadran émail signé Donzé, une entreprise de tradition intégrée au groupe Ulysse Nardin. Après le décès de son charismatique propriétaire en 2011, Ulysse Nardin entre dans le giron du groupe français Kering. En sa qualité de PDG, Patrik Hoffmann, compagnon de route de longue date du défunt, poursuit cette stratégie gagnante. Avec l'achat du calibre 137 Ebel pour chronographes, Nardin devient également autonome dans ce secteur très pointu de l'horlogerie mécanique. Après de nécessaires optimisations majeures, notamment l'insertion de composants en silicium dans l'organe oscillant et l'échappement, ce calibre fait son entrée en scène sous le nom de UN-150. De nombreuses années auparavant, le visionnaire Rolf Schnyder a pris une participation financière dans les fabricants de ces composants. Son objectif : l'indépendance à l'égard des fournisseurs des incontournables composants clés de montres mécaniques.

La signature reconnaissable entre mille de Patrik Hoffmann se retrouve sur plusieurs nouveautés de l'année 2016, parmi lesquelles une véritable prouesse technologique, le « Grand Deck Marine Tourbillon ». Son cadran en marqueterie et son aiguille des minutes inédite rappellent la longue tradition des chronomètres de marine. Mue par un nanofil dans une fibre de haute technologie, l'aiguille des minutes aux allures de bôme de voilier traverse le cadran de gauche à droite en l'espace d'une heure. Les engrenages et l'affichage ne tiennent bien sûr pas compte de la durée du retour à zéro, qui est rapide mais non sautant. Le « Marine Chronograph Annual Calendar » est l'expression de la volonté constamment réaffirmée par Ludwig Oechslin de créer des cadratures à la fois simples et extrêmement performantes. Il suffit de sept composants supplémentaires pour obtenir l'utile complication du calendrier, qui reconnaît chaque mois à son nombre de jours, sauf février. L'alliance de la haute technologie horlogère et de la fascination pour la 35e Coupe de l'America, qui se traduit par un partenariat avec Artemis Racing, trouve son expression dans la très originale « FreakWing ». En l'occurrence, le pont des minutes est inspiré de la structure interne de l'aile rigide de certains catamarans. Le disque des heures rotatif rappelle par sa constitution les flancs de ces bolides à deux coques. Enfin, les bords du verre et le fond de boîte sont en fibre de carbone, le matériau par excellence des voiliers de compétition. N'est-ce pas là une fois de plus la preuve que la haute horlogerie mécanique pluricentenaire et les hautes technologies du XXIe siècle se complètent à merveille ?

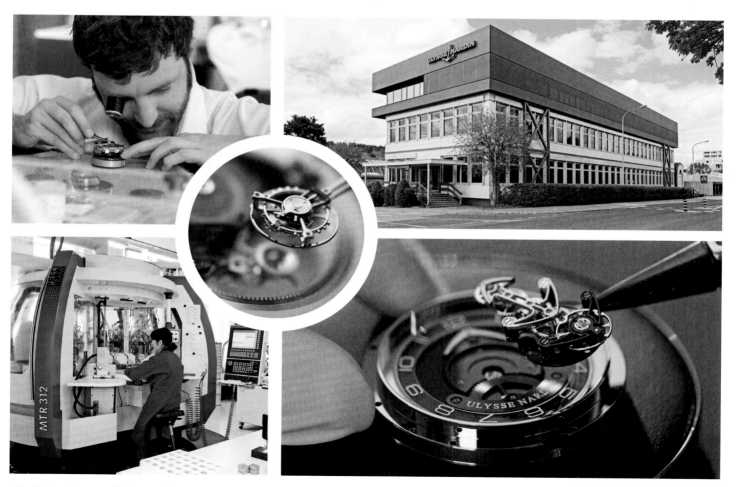

Clockwise, from top left: Complication workshop ○ La Chaux-de-Fonds ○ Freak movement assembly ○ Movements construction ○ Circle: Tourbillon carriage assembly ○ Von oben links im Uhrzeigersinn: Komplikationen-Atelier ○ La Chaux-de-Fonds ○ Montage eines Uhrwerks, Modell Freak ○ Fertigung der Uhrwerke ○ Kreis: Montage des Tourbillonkäfigs ○ Dans le sens horaire, en partant du haut, à gauche : Atelier de complications ○ La Chaux-de-Fonds ○ Assemblage du mouvement de la Freak ○ Assemblage de mouvements ○ Au centre : Assemblage d'une cage de tourbillon

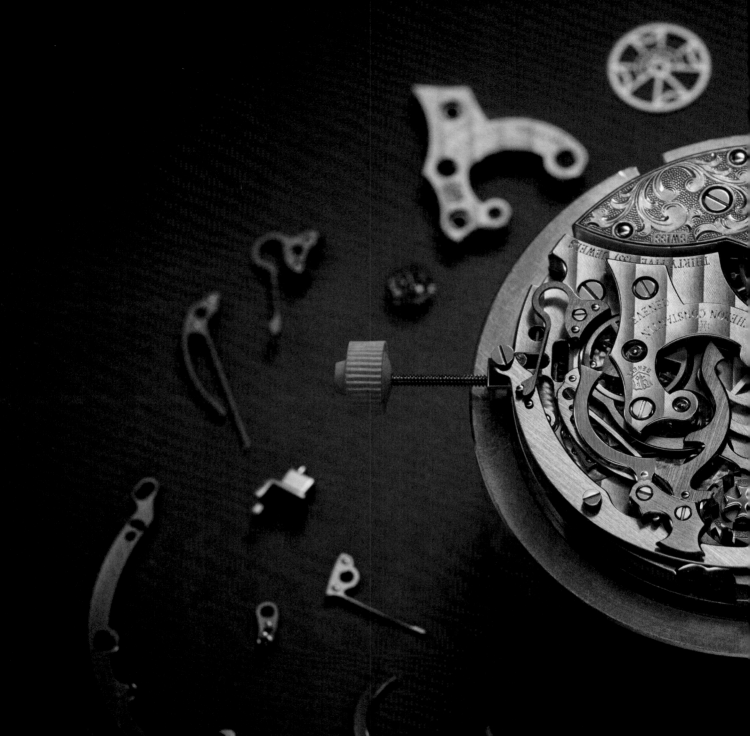

Vacheron Constantin

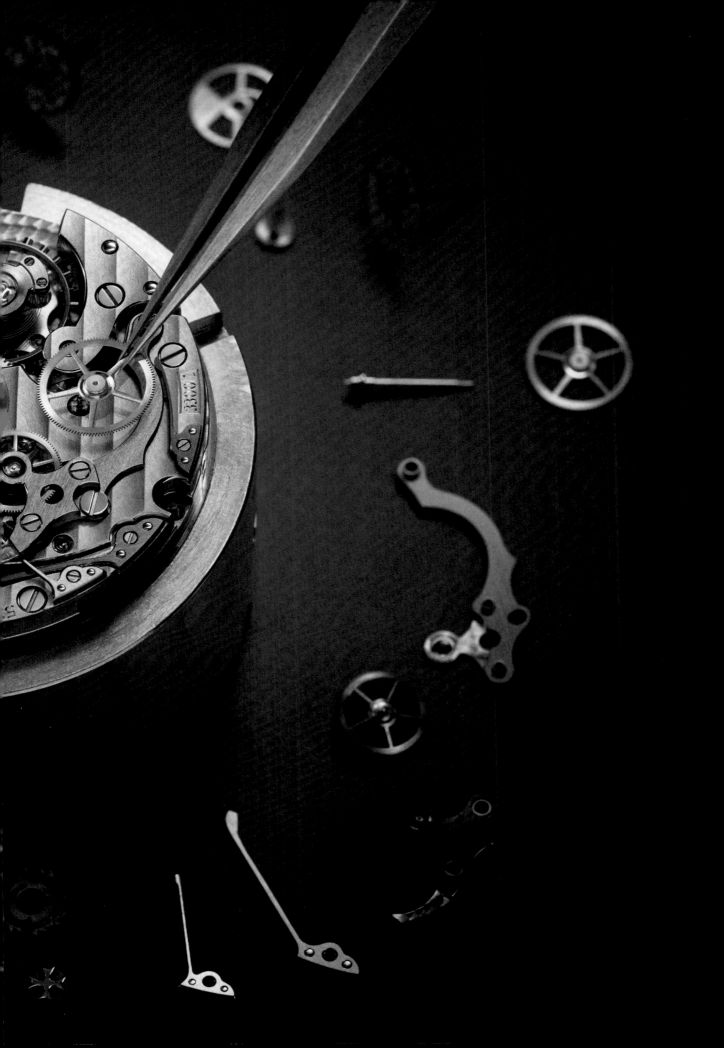

In Geneva since 1755

Everyone agrees that Vacheron Constantin ranks among the oldest Swiss watch manufactories, but opinions differ about the brand's origins. Some historians claim that Jean-Marc Vacheron, who was born in 1731, completed his training as a watchmaker at age 20 and was accepted into the elite circle of Geneva's "cabinotiers." These artisans worked in garrets in Geneva's Saint-Gervais district, where they performed tasks assigned to them by others, who made complete watches. But the impassioned craftsman wasn't satisfied with this arrangement, so he opened his own watch atelier in the old city of Geneva in 1755. Another version asserts that Jean-Marc Vacheron began training as a watchmaker in 1755 and went into business for himself in 1785. According to a third story, he set up his watch-making shop in the center of Geneva in 1775.

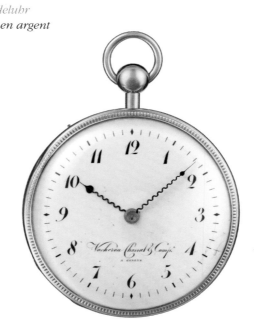

No matter which of these tales is closest to the truth, we can be sure that Jean-Marc Vacheron lived through turbulent times because of the contemporary political situation. The father of five children, it is unlikely that the thought of "luxury" ever crossed his mind. The Vacheron household eked out a very modest living, but the family stuck together. Jean-Marc's son Jacques-Barthélemy supported his ailing father as best he could. Hefty loans were unavoidable on several occasions. Long-lasting improvements weren't forthcoming until 1819, when a partner joined the business: François Constantin, the prosperous son of a textile and grain merchant, was also a talented salesman. Vacheron & Constantin was born. Its motto: "Faire mieux si possible, ce qui est toujours possible," i.e., "Do better if possible, and it's always possible." With remarkable and sometimes rather unorthodox methods, the new partner breathed fresh life into the sickly company.

The business had the good fortune to take Georges-Auguste Leschot on board in 1839. An ingenious watchmaker and inventor, Leschot had quietly developed various machines to manufacture watch movements with greater precision. With their help, the production of highly precise calibers could begin two years later. By 1860, the company could be regarded as a genuine manufacture: this means that all components were made in the firm's own workshops, which were located on the Rhône Island in central Geneva. The firm's name was changed to César Vacheron & Co. when co-owner Jean-François Constantin, a nephew of François Constantin, left the company in 1867. The name became Charles Vacheron & Cie. after César's death in 1869. The well-known name Vacheron & Constantin was restored in 1875, when Jean-François Constantin returned to the company, which registered the Maltese cross as its trademark in 1880. The brand added the first wristwatches to its collection around 1910.

Top: the bezel sets the hands and winds the mainspring of this early bangle watch, 1889 ○ *Oben: Frühe Spangenarmbanduhr mit Lünetten-Zeigerstellung und Aufzug, 1889* ○ *En haut : Sur cette ancienne montre-agrafe, la lunette met à l'heure les aiguilles et remonte le ressort de barillet, 1889*

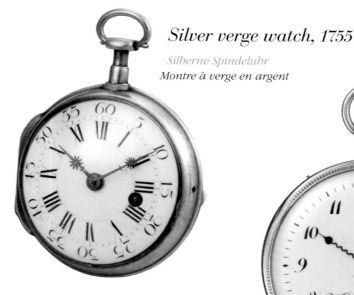

Silver verge watch, 1755

Silberne Spindeluhr
Montre à verge en argent

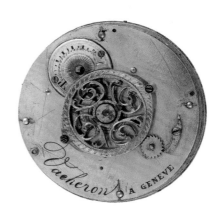

Golden verge watch, 1812

Goldene Spindeluhr
Montre à verge en or

Charles Constantin joined the management of the family business on June 1, 1914, but the company suffered hard times due to war and economic crisis. Vacheron & Constantin manufactured a mere 211 movements in 1932. In the face of this disheartening situation, the family tradition was continued by Charles' nephew Léon Constantin rather than by his children. Lack of sufficient liquidity compelled Léon Constantin to negotiate with Jaeger-LeCoultre in 1938, with the result that Vacheron & Constantin became a member of the S.A.P.I.C. Group (Société anonyme de produits industriels et commerciaux), whose aegis spanned the Jaeger-LeCoultre watch factory, the Jaeger-LeCoultre sales organization, the Ed. Jaeger factories for watches and aircraft instruments in Paris, and other business fields in Paris, London, and New York. Charles Constantin sold his block of shares in 1940 to the Genevan industrialist Georges Ketterer, who restored Vacheron & Constantin's health by revising its collection, which encased very meticulously finished ébauches supplied by LeCoultre. Vacheron & Constantin's name was changed to Vacheron Constantin by Georges' son Jacques Ketterer: as majority shareholder and director, he eliminated the "&" from the appellation in 1974.

Turbulence soon loomed on the horizon. In the wake of the Quartz Crisis, the former Saudi oil minister Yamani took over 85 percent of Vacheron Constantin's shares in 1987. The new owner didn't want to invest, so the noble and traditional company was again offered for sale in 1996. The Vendôme Luxury Group, which would later become the Richemont Group, bought the brand for the comparatively low price of 45 million euros. Under its wings,

Vacheron Constantin could resume showing its true strengths. Impressive proofs of this are the firm's gigantic building, which stands in the Genevan suburb of Plan-les-Ouates, where its neighbors include numerous other luxury watch manufactories. Vacheron Constantin also operates a factory in Vallée de Joux.

The number of third-party movements has successively declined and the label's own manufactory calibers have gained greater significance. The assortment includes manually wound and self-winding calibers with every conceivable complication, e.g., tourbillon, calendar, or repeater. For the first time in its long history, Vacheron Constantin can proudly show three of its own chronograph movements in 2015: hand-wound caliber 3200 is a monopusher chronograph with a tourbillon; manually wound caliber 3300 is a classical monopusher chronograph movement; and self-winding caliber 3500, with an uncommonly elaborate split-seconds function, holds a world record thanks to its extreme slimness of 5.2 millimeters.

All of Vacheron Constantin's movements satisfy the requirements of the Geneva Seal, which became even stricter in 2011. To earn this coveted hallmark, the assembly, fine adjustment, and encasement of mechanical calibers and additional modules (e.g., calendar cadratures) must occur in the Canton of Geneva. The movement and the case belong inseparably together. Everything revolves around the completed watch, its watertightness, functionality, rate autonomy, and, above all, its precision. After seven consecutive days in various positions, the watch must not deviate from perfect timekeeping by more than one minute. ◦

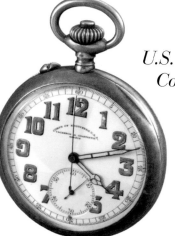

U.S. Army
Corps of Engineers, 1918

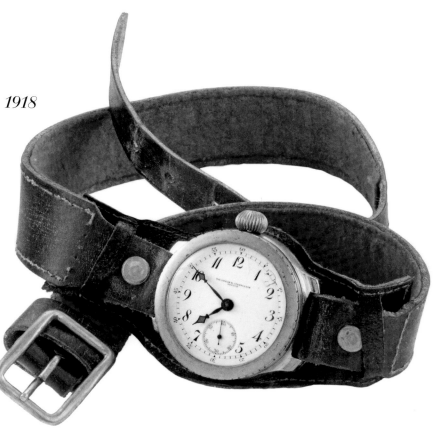

Pilot's Watch, 1903

This watch is worn outside the flight overalls strapped to the pilot's thigh ◦ *Die Uhr wurde über der Fliegermontur am Oberschenkel getragen* ◦ *Montre portée sur la combinaison du pilote au niveau de la cuisse*

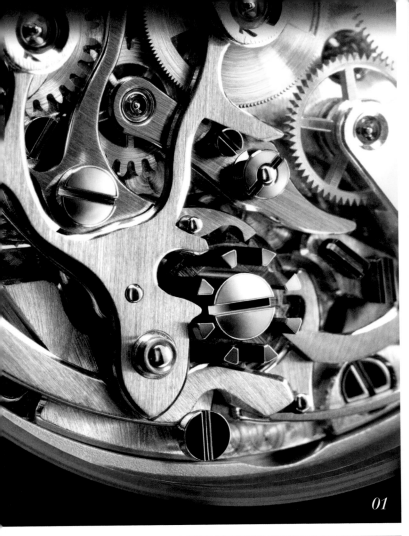

01

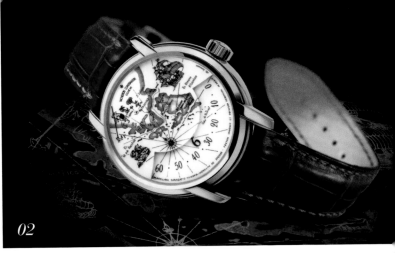

02

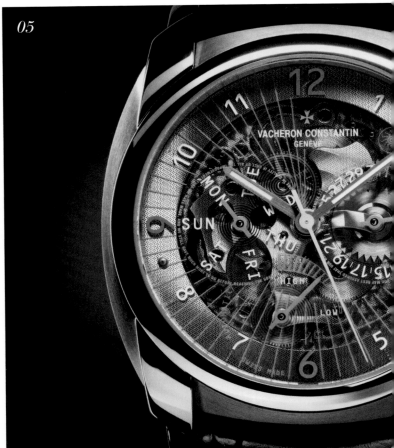

05

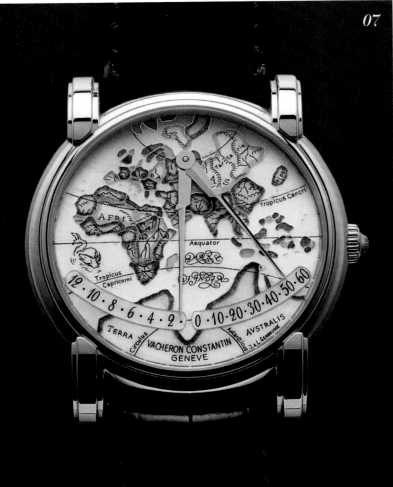

07

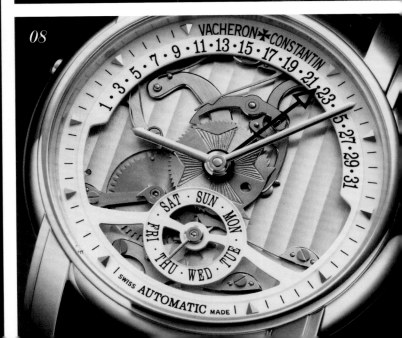

08

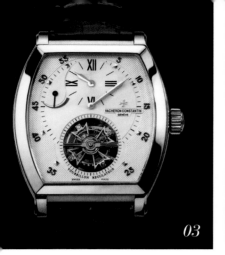

03

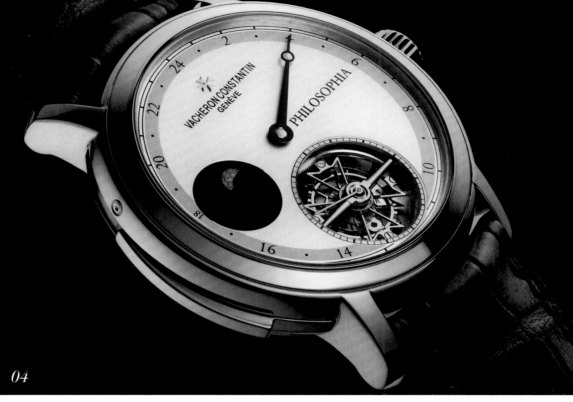

04

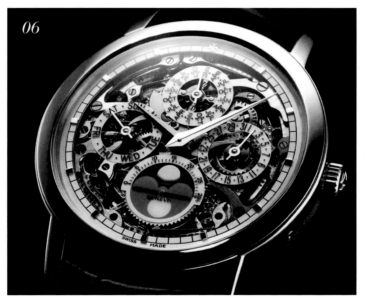

06

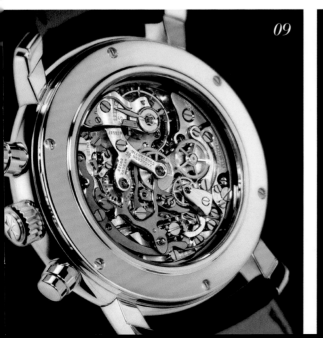

09

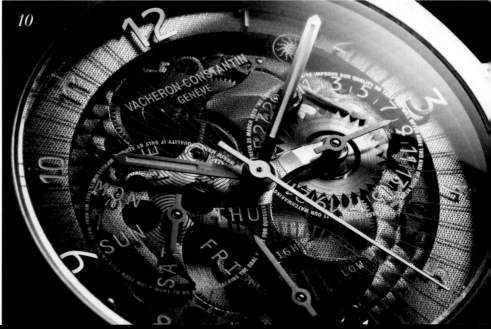

10

Seit 1755 in Genf

Deutsch

Über die Anfänge von Vacheron Constantin, eine der ältesten Schweizer Uhrenmanufakturen, herrschen geteilte Meinungen. Einmal könnte es so gewesen sein, dass der 1731 geborene Jean-Marc Vacheron nach Beendigung seiner Ausbildung zum Uhrmacher als 20-Jähriger in den elitären Zirkel der Genfer Cabinotiers aufgenommen wurde. Diese führten in ihren „Kabinetten" unter den Dächern des Genfer Stadtteils Saint-Gervais kontemplativ jene Arbeiten aus, welche ihnen die Fertigsteller von Uhren vermittelten. Weil dem passionierten Handwerker das nicht reichte, eröffnete er 1755 in der Genfer Altstadt sein eigenes Uhrenatelier. Eine andere Version besagt, dass Jean-Marc Vacheron erst 1755 mit seiner Ausbildung zum Uhrmacher begann und sich 1785 selbstständig machte. An dritter Stelle steht zu lesen, dass er sich 1775 im Zentrum der Genfer Uhrmacherei eingerichtet habe.

So oder so erlebte Jean-Marc Vacheron nicht zuletzt aus politischen Gründen ausgesprochen stürmische Zeiten. Das Wort Luxus dürfte beim Vater von fünf Kindern selten gefallen sein. Stattdessen ging es bei den Vacherons recht bescheiden zu. Aber die Familie hielt zusammen. Jacques-Barthélemy, der Sohn, unterstützte seinen gesundheitlich angeschlagenen Vater Jean-Marc nach Kräften. Zeitweise musste er zur Sicherung der Existenz größere Darlehen aufnehmen. Nachhaltige Besserung trat erst 1819 ein, als sich François Constantin, begüterter Sohn eines Stoff- und Getreidehändlers sowie begnadeter Verkäufer, hinzugesellte. Vacheron & Constantin war geboren. Der Leitspruch: „Faire mieux si possible, ce qui est toujours possible." (Besser machen, wenn möglich – und das ist immer möglich!) Durch extravagante und teilweise recht unorthodoxe Methoden gelang es dem neuen Partner, die infolge der Französischen Revolution heftig kränkelnde Firma mit neuem Leben zu erfüllen.

1839 kam, ein Glücksfall, Georges-Auguste Leschot ins Haus. Der geniale Uhrmacher und Erfinder hatte in aller Stille verschiedene Maschinen für eine präzisere Herstellung von Uhrwerken entwickelt. Schon zwei Jahre später lief mit ihrer Hilfe die Produktion neuer Präzisionskaliber an. Ab 1860 konnte die Firma als wirkliche Manufaktur gelten, das heißt, alle Teile entstanden in den Werkstätten auf der zentral gelegenen Genfer Rhône-Insel. 1867 schied Jean-François Constantin, der Neffe von François Constantin, als Mitinhaber aus. Der Firmenname wurde zu César Vacheron & Co, nach dessen Tod 1869 zu Charles Vacheron & Cie. Um den weithin bekannten Namen Vacheron & Constantin erneut aufleben zu lassen, kehrte Jean-François Constantin 1875 wieder in die Firma zurück. 1880 ließ man sich das Malteserkreuz als Markenzeichen rechtlich schützen. Erste Armbanduhren fanden gegen 1910 in die Kollektion.

Ab dem 1. Juni 1914 sah man Charles Constantin im Management des Familienunternehmens. Doch einmal mehr brachten Krieg und Wirtschaftskrise Wechselbäder mit sich. 1932 stellte Vacheron & Constantin gerade einmal 211 Uhrwerke fertig. Angesichts dieser misslichen Situation setzten nicht die Kinder von Charles, sondern sein Neffe Léon Constantin die familiäre

Tradition fort. Der musste 1938 mangels ausreichender Liquidität mit Jaeger-LeCoultre verhandeln. So wurde Vacheron & Constantin ein Mitglied der Gruppe S.A.P.I.C. (Société anonyme de produits industriels et commerciaux), unter deren Dach sich neben der Uhrenfabrik Jaeger-LeCoultre auch die Jaeger-LeCoultre Vertriebsorganisation, die Ed. Jaeger Fabriken für Uhren und Flugzeuginstrumente in Paris sowie andere Geschäftsfelder in Paris, London und New York befanden. 1940 verkaufte Charles Constantin sein Aktienpaket an den Genfer Industriellen Georges Ketterer. Unter seiner Ägide gesundete Vacheron & Constantin durch eine geänderte Kollektionspolitik unter Verwendung sorgfältig feinbearbeiteter Rohwerke von LeCoultre. Dass Vacheron & Constantin zu Vacheron Constantin mutierte, ist dem Sohn Jacques Ketterer zu verdanken. 1974 verfügte er den Wegfall des „&" in seiner Eigenschaft als Mehrheitsaktionär und Direktor.

Auch danach ging es turbulent weiter. Infolge der Quarzuhren-Krise übernahm 1987 der ehemalige saudische Erdölminister Yamani 85 Prozent der Vacheron-Constantin-Aktien. Investieren wollte der neue Eigentümer jedoch nicht. Deshalb stand das noble Traditionsunternehmen 1996 abermals zum Verkauf. Nun stieg die Vendôme Luxury Group, heute Richemont-Gruppe, für vergleichsweise schlappe 45 Millionen Euro ein. Unter deren Fittichen konnte und durfte Vacheron Constantin endlich wieder seine Stärken zeigen. Ausdrucksstarke Beweise sind das riesige Firmengebäude im Genfer Vorort Plan-les-Ouates, dicht neben etlichen anderen Luxusuhren-Manufakturen, und eine moderne Fabrikationsstätte im Vallée de Joux.

Inzwischen ist die Zahl zugekaufter Uhrwerke sukzessive gesunken. Die eigene Werkemanufaktur gewinnt zunehmend an Bedeutung. Im Programm sind Kaliber mit manuellem und automatischem Aufzug, mit allen erdenklichen Komplikationen wie Tourbillon, Kalendarium oder Repetitionsschlagwerk. Seit 2015 kann Vacheron erstmals in der langen Unternehmensgeschichte auch drei hauseigene Chronographenwerke vorweisen: das 3200 mit Handaufzug, Eindrücker-Steuerung und Tourbillon, das 3300 als klassisches Chronographenwerk ebenfalls nur mit einem Bediendrücker und manuellem Aufzug sowie schließlich das ultraflache 3500. Trotz seiner minimalen Bauhöhe von nur 5,2 Millimetern besitzt dieser Weltrekord Selbstaufzug und die besonders aufwendige Schleppzeigerfunktion.

Ausnahmslos alle Uhrwerke von Vacheron Constantin genügen den 2011 deutlich verschärften Qualitätsanforderungen des Genfer Siegels. Ihm zufolge müssen die mechanischen Uhrwerke und zusätzliche Module, beispielsweise Kalenderwerke, im Kanton Genf montiert, justiert und eingeschalt worden sein. Werk und Gehäuse gehören untrennbar zusammen. Hier drehen sich die Dinge um die fertige Uhr, und zwar ihre Wasserdichtigkeit, Funktion, Gangautonomie sowie vor allem auch die Präzision. Nach sieben aufeinanderfolgenden Tagen in unterschiedlichen Positionen darf die Abweichung von der Norm nicht mehr als eine Minute betragen.

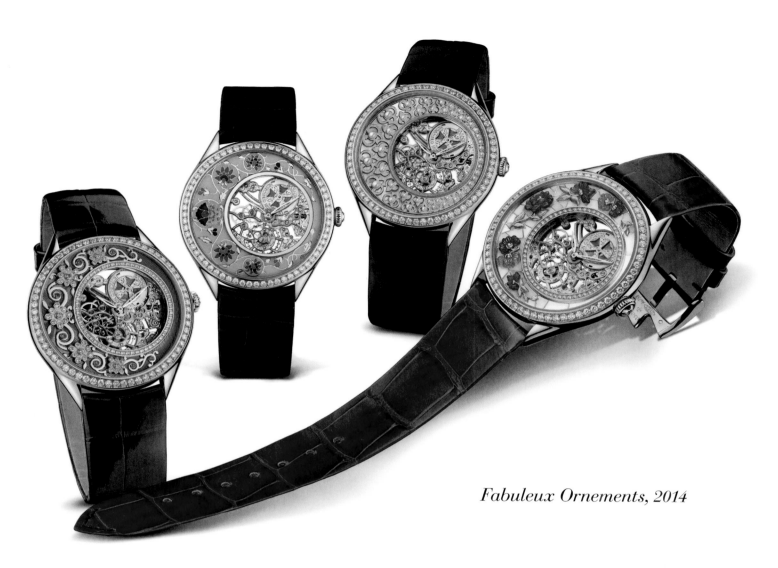

Fabuleux Ornements, 2014

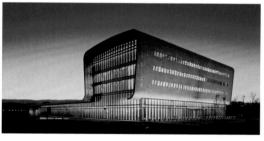

The futuristic factory building
in Plan-les-Ouates, 2005
*Das futuristische Fabrikgebäude
in Plan-les-Ouates, 2005*
*Le bâtiment futuriste de la
manufacture à Plan-les-Ouates,
2005*

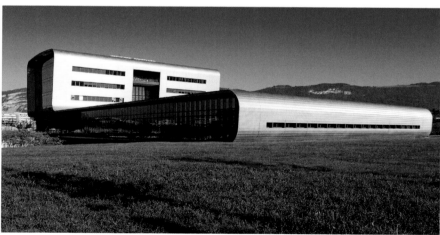

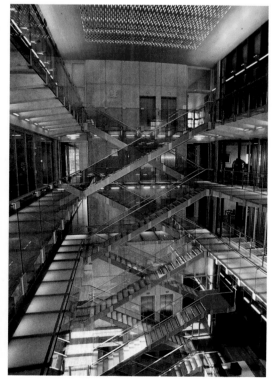

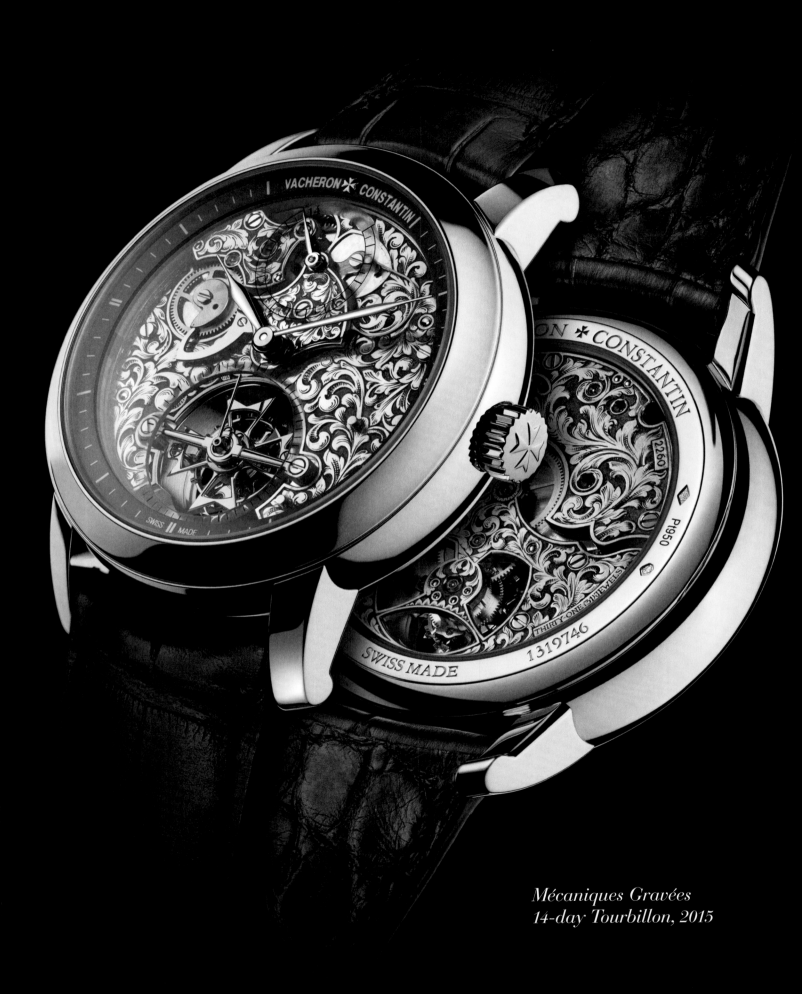

Mécaniques Gravées
14-day Tourbillon, 2015

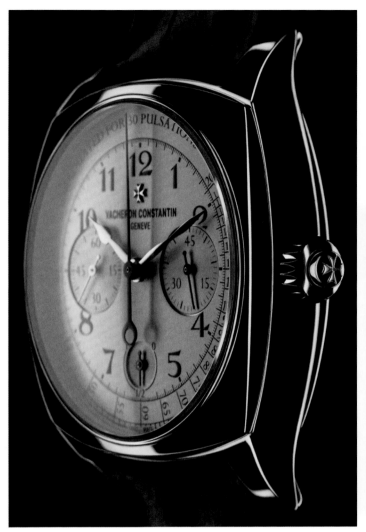

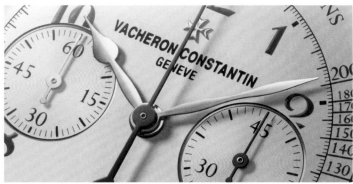

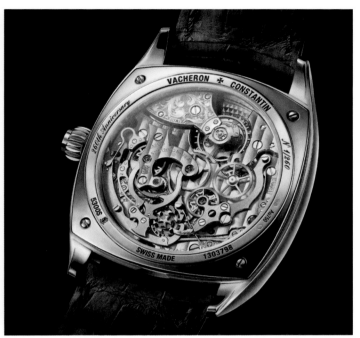

Harmony Chronograph, 2015

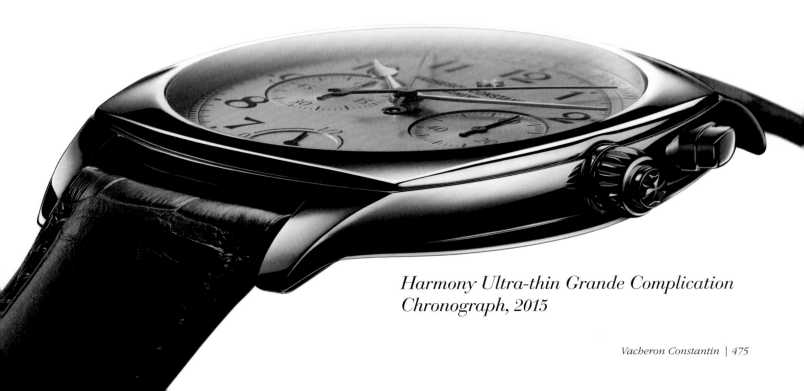

Harmony Ultra-thin Grande Complication
Chronograph, 2015

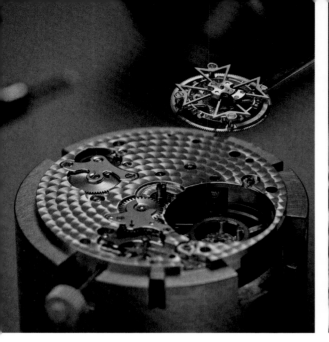
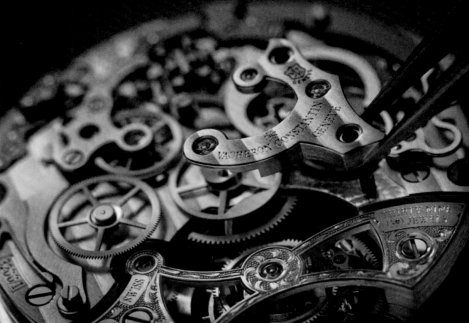

Établi à Genève depuis 1755

Français

Les avis divergent sur les débuts de Vacheron Constantin, l'une des plus grandes manufactures horlogères suisses. Selon certains, une fois sa formation d'horloger terminée, Jean-Marc Vacheron aurait été admis en 1751, à l'âge de 20 ans, dans le cercle fermé des cabinotiers genevois. Ces derniers réalisaient religieusement les travaux que leur confiaient les établisseurs dans leurs « cabinets » situés dans les mansardes du faubourg genevois de Saint-Gervais. Cet artisan passionné ne pouvant s'en contenter, il aurait ouvert en 1755 son atelier d'horlogerie dans la vieille ville de Genève. Pour d'autres, Jean-Marc Vacheron n'aurait débuté sa formation d'horloger qu'en 1755 et se serait établi à son compte en 1785. Pour d'autres encore, il n'aurait emménagé qu'en 1775 au cœur du quartier horloger genevois.

Quoi qu'il en soit, Jean-Marc Vacheron a traversé des temps très tumultueux, notamment pour des raisons politiques. Père de cinq enfants, le mot luxe lui est étranger. Les Vacheron vivent au contraire très modestement, mais la famille est soudée. Jacques-Barthélemy, son fils, soutient comme il peut son père à la santé fragile. Par moments, ce dernier doit souscrire des emprunts assez importants pour assurer leur subsistance. La situation s'améliore enfin durablement en 1819, lorsqu'il s'associe à François Constantin, fils fortuné d'un marchand de tissus et céréales, et vendeur surdoué. La société Vacheron & Constantin est née. Elle se donne pour devise : « Faire mieux si possible, ce qui est toujours possible. » Par des méthodes insolites et parfois vraiment pas orthodoxes, le nouveau partenaire parvient à donner un nouveau souffle à la société sérieusement affaiblie au sortir de la Révolution française.

En 1839, par un heureux hasard, Georges-Auguste Leschot rejoint la société. Ce génial horloger et inventeur a mis au point en toute discrétion des machines pour fabriquer des mouvements encore plus précis. Grâce à son aide, la production de nouveaux calibres de précision débute deux ans plus tard seulement. Dès 1860, la société peut être considérée comme une vraie manufacture, tous les composants étant réalisés dans les ateliers de l'île du Rhône,

au centre de Genève. En 1867, Jean-François Constantin, neveu de François Constantin, se démet de ses fonctions d'associé. La société devient César Vacheron & Co. puis, à la mort de César en 1869, Charles Vacheron & Cie. En 1875, Jean-François Constantin réintègre la société pour faire revivre le nom Vacheron & Constantin, qui jouissait d'une certaine notoriété. En 1880, la croix de Malte est déposée à titre de marque. La collection s'enrichit de ses premières montres-bracelets vers 1910.

Le 1er juin 1914, Charles Constantin rejoint l'équipe dirigeante de l'entreprise familiale. Puis la guerre et la crise économique font à nouveau souffler le chaud et le froid. En 1932, Vacheron & Constantin fabrique tout juste 211 mouvements. Face à cette situation fâcheuse, ce ne sont pas les enfants de Charles, mais son neveu Léon Constantin qui reprend le flambeau pour perpétuer la tradition familiale. En 1938, la société manquant de liquidités, il doit composer avec Jaeger-LeCoultre. Vacheron & Constantin devient alors membre de la SAPIC (Société anonyme de produits industriels et commerciaux), société regroupant en son sein, outre la manufacture horlogère Jaeger-LeCoultre et la société de vente des produits Jaeger-LeCoultre, les établissements Ed. Jaeger qui fabriquent à Paris des montres et instruments de mesure pour l'aviation mais aussi d'autres produits dans d'autres spécialités à Paris, Londres et New York. En 1940, Charles Constantin vend ses actions à l'industriel genevois Georges Ketterer. Sa nouvelle politique en matière de collection, qui repose sur l'utilisation de mouvements LeCoultre de précision usinés avec soin, assure le rétablissement de Vacheron & Constantin. C'est à Jacques Ketterer, le fils de Georges, que l'on doit la transformation du nom Vacheron & Constantin en Vacheron Constantin. En 1974, il décrète, en qualité d'actionnaire majoritaire et de directeur, la suppression du « & ».

La période qui suit reste troublée. La crise du quartz veut que Yamani, l'ancien ministre saoudien du Pétrole, rachète 85 pour cent des actions Vacheron Constantin. Or ce nouveau propriétaire ne veut

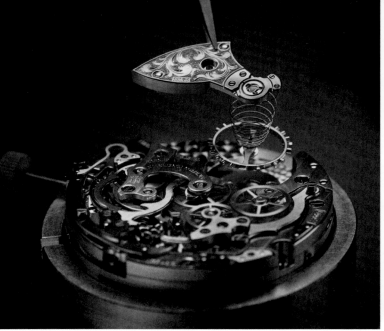

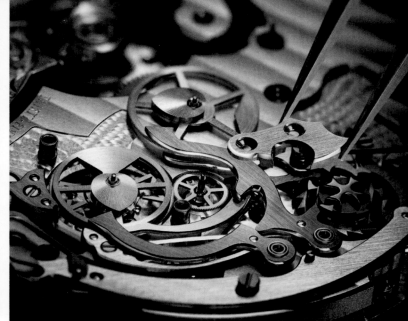

Left page: Caliber 3200, 2015 ○ Right page: Caliber 3300, 2015

pas investir. La noble entreprise traditionnelle est à nouveau en vente en 1996. C'est Vendôme Luxury Group, actuel groupe Richemont, qui entre dans l'affaire pour 45 millions d'euros, une somme relativement modeste. Sous l'égide du nouvel acquéreur, Vacheron Constantin peut enfin exprimer à nouveau son potentiel. Témoins marquants de sa puissance : l'immense bâtiment de la société dans le faubourg genevois de Plan-les-Ouates, qui jouxte nombre de manufactures de montres de luxe, et l'atelier de fabrication moderne dans la vallée de Joux.

Entre-temps, le nombre de mouvements achetés a progressivement diminué. Ils sont fabriqués à la manufacture, dont l'importance ne cesse de grandir. Au programme figurent des calibres à remontage manuel ou automatique, avec toutes les complications imaginables, telles le tourbillon, le calendrier ou la répétition minutes. En 2015, Vacheron est pour la première fois de sa longue histoire en mesure de proposer trois mouvements de chronographes : le 3200 à remontage manuel, déclenchement dynamique et tourbillon, le 3300, comme tout mouvement de chronographe classique, simplement doté d'un remontage manuel et d'un seul poussoir, et enfin, le 3500, un mouvement ultraplat. Malgré un record mondial de minceur avec seulement 5,2 millimètres, ce mouvement dispose du remontage automatique et de la rattrapante, une fonction d'une exceptionnelle complexité.

Les mouvements Vacheron Constantin remplissent tous les exigences qualité du Poinçon de Genève, qui ont été nettement durcies en 2011. Selon ces critères, les mouvements mécaniques et les modules supplémentaires, tels les modules de calendrier, doivent être assemblés, réglés et emboîtés dans le canton de Genève. Mouvement et boîtier sont indissociables. Cela est déterminant pour les caractéristiques de la montre terminée, à savoir son étanchéité à l'eau, sa fonction, sa réserve de marche et avant tout sa précision. Après sept jours successifs dans différentes positions, l'écart par rapport à la norme ne doit pas dépasser la minute. ○

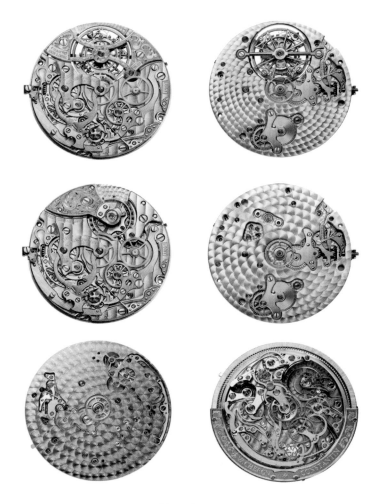

Calibers from 2015:
3200 ○ 3300 ○ 3500

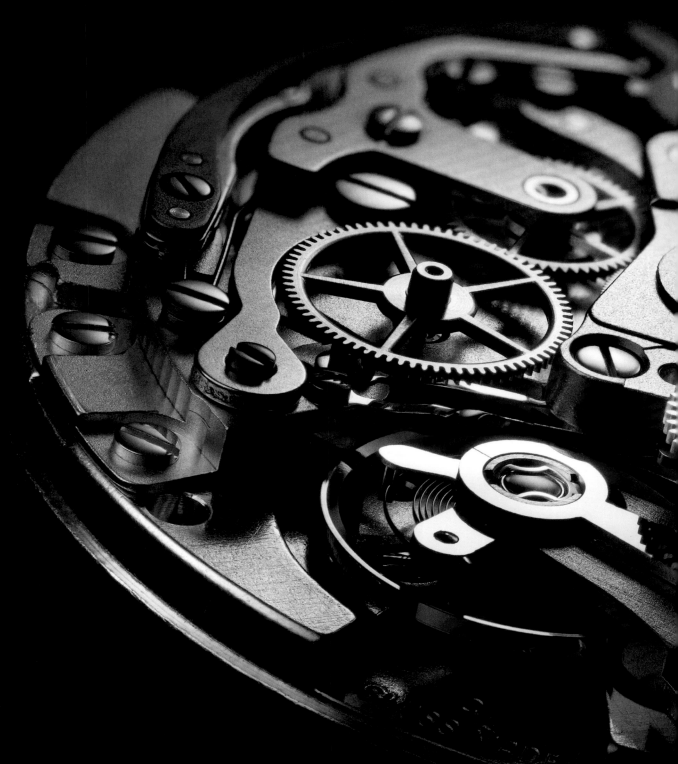

Zenith

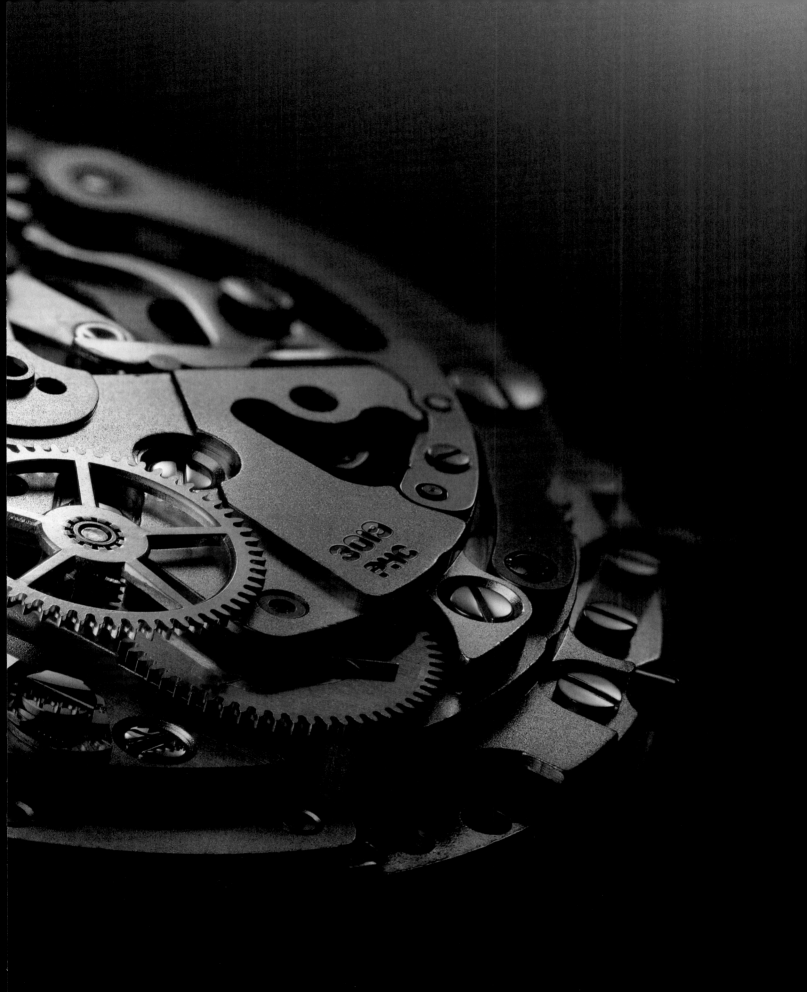

Zenith Means the Highest Point

The number 1,500 says more than many words. It stands for the numerous prizes, medals, and other international honors for outstandingly precise watchmaking that Zenith has received in the course of its 150-year history. Even in his wildest dreams, Georges Favre-Jacot surely hadn't imagined such an abundance of commendations when he started his chronometric activities in the little town of Le Locle in Switzerland's Jura region in 1865. The highest precision likewise didn't play an important role at this time. The 22-year-old Favre-Jacot simply wanted to use efficient manufacturing methods to produce affordably priced and satisfactory accurate pocketwatches. He instituted the so-called "American" production method, which involved the use of innovative machinery to fabricate components so accurately that they needed nearly no post-processing and could readily be replaced by identical parts whenever a watch needed repair.

In the early 20th century, the owner realized that his company could only continue to grow if it gained access to markets outside Europe. He accordingly asked his nephew James Favre to journey to the USA, Russia, and China. The traveling salesman returned with a briefcase full of valuable orders. An important milestone was laid in 1911, when Favre's company was incorporated as Zenith SA. This also gave its ticking products a catchy name, which soon adorned the dials, movement, and cases. Wristwatches of various constructions, including some with helpful additional functions, were added to the collection during the First World War: built-in alarms woke the watch's owner or reminded him of imminent appointments, and chronographs could measure elapsed intervals. In the years between the two World Wars, Zenith's timepieces repeatedly ranked among the winners of the chronometer competitions sponsored by the Swiss observatories in Geneva and Neuchâtel. The wristwatch chronometer encasing extremely precise hand-wound caliber 135, which was manufactured from 1948 to 1962, is avidly sought by collectors today.

With ample experience in the relevant disciplines, Zenith participated in the international race to develop the first self-winding wristwatch chronographs in the 1960s. Several years of developmental work bore fruit on January 10, 1969, when Zenith convened a press conference in Geneva and proudly presented "El Primero,"

the world's first wristwatch with a chronograph, a ball-borne rotor, and a balance with a frequency of five hertz to measure elapsed intervals to the nearest tenth of a second. But the thrill of victory didn't last very long. By the mid 1970s, the brand's new owners had grown disenchanted with mechanical timekeeping. They ordered the staff to destroy both the existing stock and the tools that had produced the calibers. Fortunately, a veteran watchmaker refused to obey the order: Charly Vermot concealed in a rambling attic everything he could get his hands on. Thanks to his foresighted disobedience, Zenith was optimally equipped when the mechanical wristwatch celebrated its renaissance in the mid 1980s—and "El Primero" was able to return in 1986.

There's another reason why 1969 was important in Zenith's biography: it marked the founding of the Mondia-Zenith-Movado holding company, of which Zenith Radio Corporation in the USA became a majority shareholder in 1971. When the majority of shares were sold to the DIXI financial group, Swiss shareholders again took Zenith's helm in 1978. They relinquished it in 1999, when the French luxury corporation LVMH (Moët Hennessy – Louis Vuitton) took the manufactory under its wings, where it remains in 2015, which is its 150th anniversary year.

The portfolio's flagship remains "El Primero," which is currently available in diverse variants, for example, classical, sporty, with a tourbillon, or an "open heart," i.e., with a large aperture in the dial and plate to offer horological voyeurs deep views of the timekeeping mechanisms. Zenith has fabricated a classical self-winding movement named "Elite" since 1994. This model welcomed a sister with a central second-hand in 2015. The "Academy Christophe Colomb," which debuted in 2008 and afterwards underwent several improvements, is unique in its construction. Amassing a 50-hour power reserve, automatic caliber 8800 has an off-center platinum rotor and a cardanically suspended container to house its oscillating and escapement system. Thanks to this ingenious device, the rate-regulating organs always remain in a horizontal position, as they do in classical marine chronometers, so that gravity, one of the arch-enemies of mechanical watches, cannot interfere with the regularity of the balance's oscillations. ◦

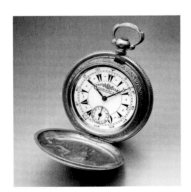
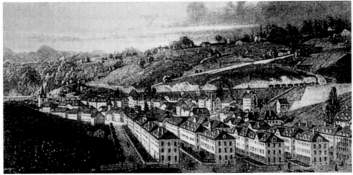
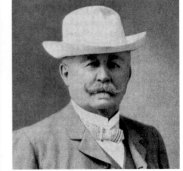

From left: typical Savonnette pocket-watch for the Ottoman market, ca. 1880 ◦ Historical view of Le Locle ◦ Von links: Typische Savonnette-Taschenuhr für den osmanischen Markt, ca. 1880 ◦ Historische Stadtansicht von Le Locle ◦ Depuis la gauche : Montre de poche savonnette pour le marché ottoman, vers 1880 ◦ Vue historique du Locle

Georges Favre-Jacot (1843–1917)

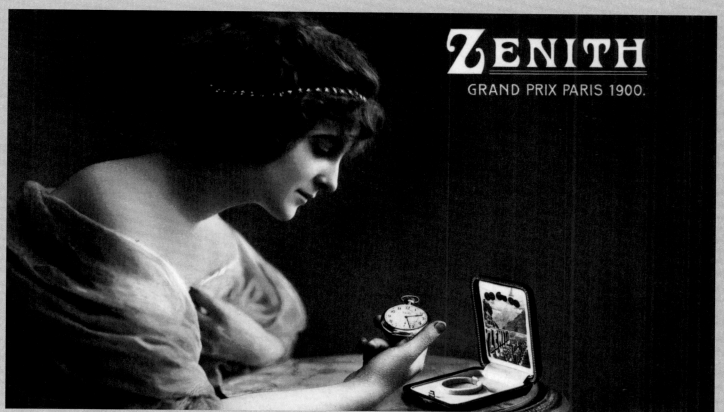

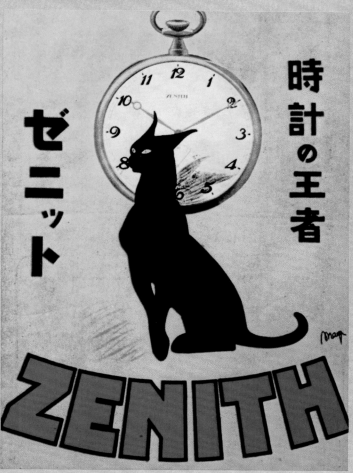

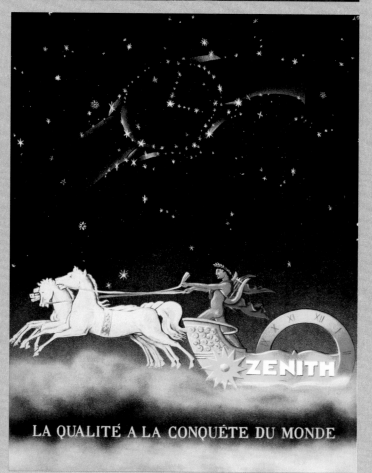

Clockwise from top: advertising motif after 1900 ◦ Advertising from the 1920s ◦ Advertising motif for the Japanese market ◦ Von oben im Uhrzeigersinn: Anzeigenmotiv nach 1900 ◦ Anzeige der 1920er Jahre ◦ Werbemotiv für den japanischen Markt ◦ Dans le sens horaire : Motif publicitaire après 1900 ◦ Publicité des années 1920 ◦ Motif publicitaire pour le marché japonais

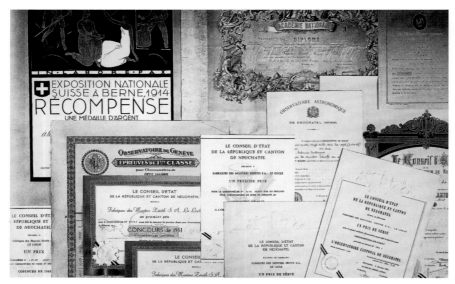

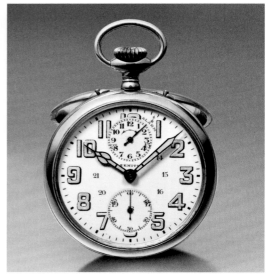

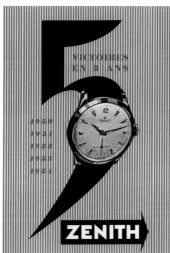

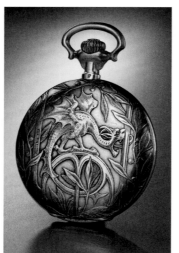

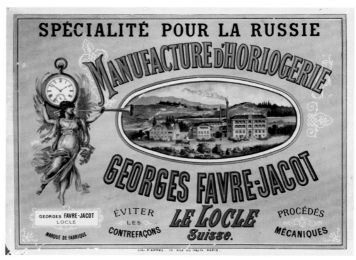

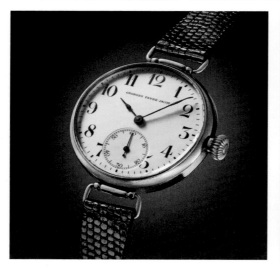

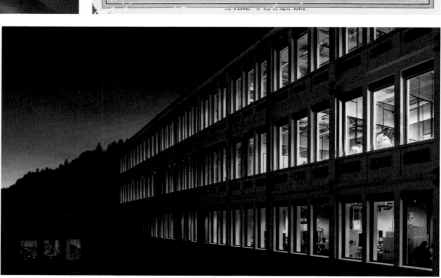

Top: observatory awards from the 1950s ○ Alarm, 1920 ○ Middle from left: advertisement from the 1950s ○ Case for a pocket-watch for the Italian market, 1911 ○ Enamel sign for the Russian market, ca. 1880 ○ Bottom: wristwatch, 1911 ○ Façade of the factory in Le Locle ○
Oben: Observatoriumspreise der 1950er Jahre ○ Wecker, 1920 ○ Mitte von links: Anzeige aus den 1950er Jahren ○ Taschenuhrgehäuse für den italienischen Markt, 1911 ○ Emailschild für den russischen Markt, ca. 1880 ○ Unten: Armbanduhr, 1911 ○ Fabrikfassade in Le Locle ○
En haut : Prix décernés par divers observatoires dans les années 1950 ○ Réveil, 1920 ○ Au milieu depuis la gauche : Publicité des années 1950 ○ Boîtier de montre de poche pour le marché italien, 1911 ○ Ornement émaillé pour le marché russe, vers 1880 ○ En bas : Montre-bracelet, 1911 ○ Façade de la manufacture au Locle

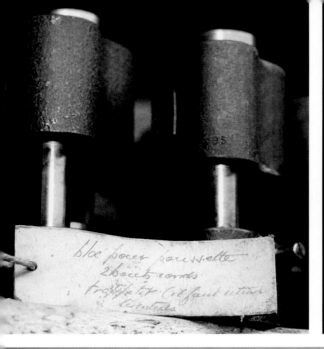

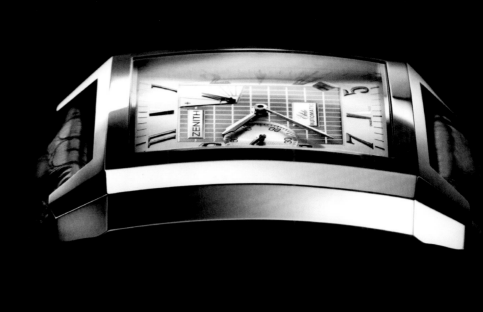

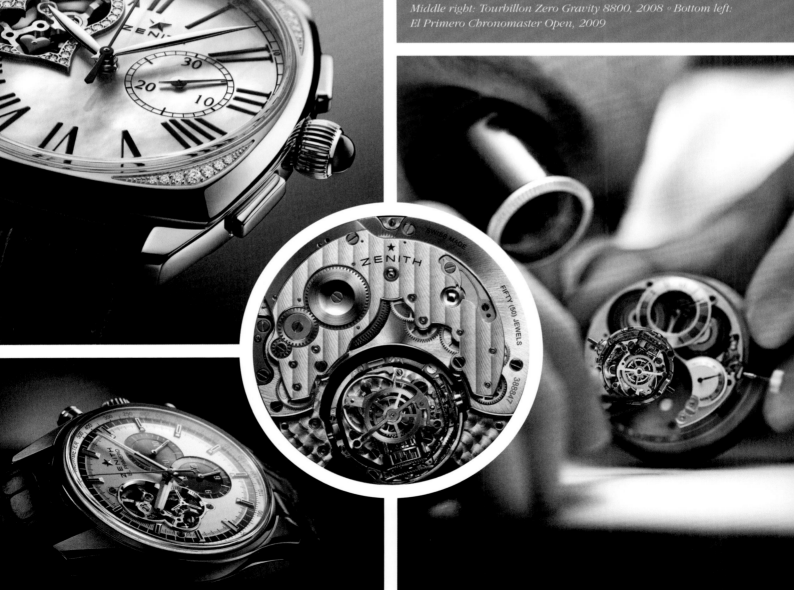

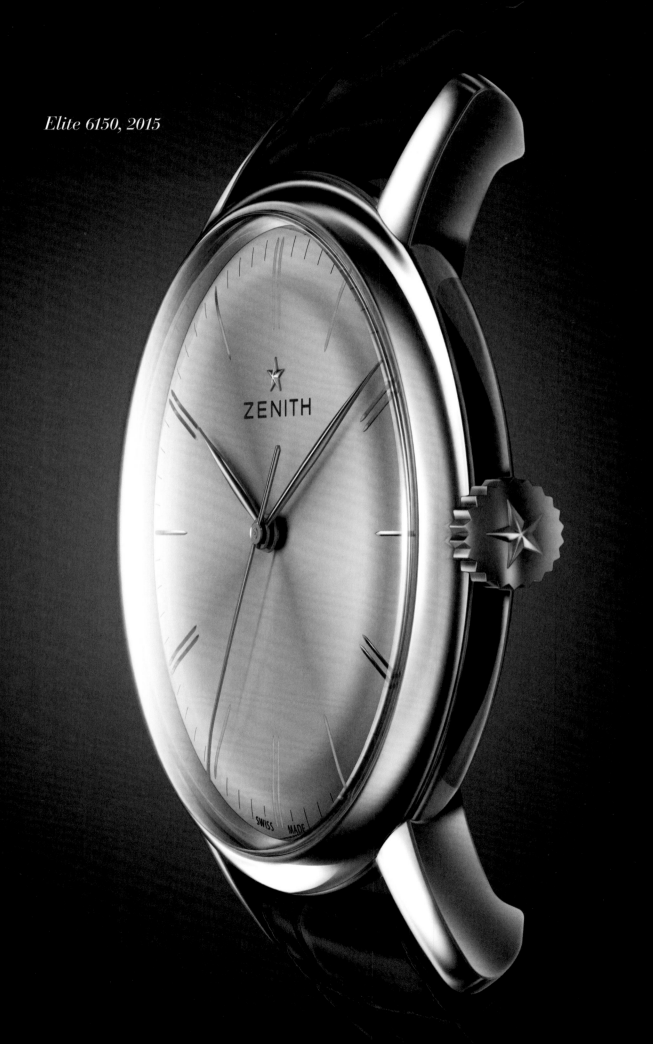

Elite 6150, 2015

Zenith heißt ganz oben

Deutsch

1500, diese Zahl sagt mehr als viele Worte. Sie steht für die zugesprochenen Preise, Medaillen sowie sonstigen internationalen Ehrungen für herausragend präzise Uhrmacherei, welche die Uhrenmanufaktur Zenith im Laufe ihrer 150-jährigen Firmengeschichte erhalten hat. An eine derartige Fülle von Auszeichnungen hätte Georges Favre-Jacot nicht einmal im Traum zu denken gewagt, als er 1865 im Jurastädtchen Le Locle seine chronometrischen Aktivitäten startete. Auch höchste Präzision spielte zu diesem Zeitpunkt noch keine übergeordnete Rolle. Dem gerade einmal 22-Jährigen ging es allein um eine rationelle und deshalb preisgünstige Produktion zufriedenstellend genauer Taschenuhren. Als bewährtes Vehikel nutzte er die sogenannte amerikanische Fertigungsmethode. Das bedeutet: die Nutzung neuartiger Maschinen zur Produktion passgenauer Komponenten, welche sich später ohne große Nacharbeit zusammenbauen und im Servicefall problemlos austauschen ließen.

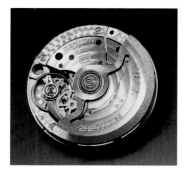

Im frühen 20. Jahrhundert erkannte der Patron die Bedeutung außereuropäischer Märkte für weiteres Wachstum. Neffe James Favre reiste auf Bitten des Onkels in die Vereinigten Staaten von Amerika, nach Russland und China. Nach jeder Rückkehr hatte er lohnende Aufträge in seiner Aktentasche. Dann kam das Jahr 1911 und mit ihm ein wichtiger Meilenstein. Die Favre'sche Unternehmung mutierte zur Aktiengesellschaft Zenith S.A. Damit bekamen auch die tickenden Kinder endlich einen einprägsamen Namen, der bald schon die Zifferblätter, Werke und Gehäuse zierte. Armbanduhren unterschiedlichster Bauart, darunter auch solche mit hilfreichen Zusatzfunktionen, fanden während des Ersten Weltkriegs in die Kollektion. Wecker erinnerten ans Aufstehen oder an Termine, Chronographen gestatteten das Stoppen von Zeitintervallen. In den Jahren zwischen beiden Weltkriegen fand sich der Name Zenith regelmäßig ganz oben in den Ergebnislisten der von den eidgenössischen Observatorien Genf und Neuenburg ausgeschriebenen Chronometerwettbewerbe. Hoch in der Sammlergunst rangieren heute die Armbandchronometer mit dem äußerst präzisen, von 1948 bis 1962 hergestellten Handaufzugskaliber 135.

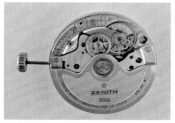

In den 60er Jahren beteiligte sich die einschlägig erfahrene Manufaktur am internationalen Wettstreit um die Einführung des ersten Armbandchronographen mit Selbstaufzug. Die jahrelangen Bemühungen fruchteten. Am 10. Januar 1969 konnte Zenith während einer Pressekonferenz in Genf nicht ganz ohne Stolz „El Primero" als weltweit erste Armbanduhr mit Chronograph, Kugellagerrotor und fünf Hertz Unruhfrequenz für Zehntelsekunden-Stoppungen vorstellen. Die Freude am Erreichten währte zunächst einmal nicht sonderlich lange. Mitte der 70er Jahre verloren die neuen Markeneigentümer den Spaß an derartiger Mechanik. Sie ordneten die Vernichtung aller Lagerbestände und obendrein auch noch sämtlicher Werkzeuge an. Zum Glück widersetzte sich ein altgedienter Uhrmacher dem Diktat. Charly Vermot versteckte alles, dessen er habhaft werden konnte, klammheimlich auf einem der weitläufigen Dachböden. So war Zenith bestens gerüstet, als die mechanische Armbanduhr Mitte der 80er Jahre ihre Renaissance feierte: 1986 kehrte „El Primero" zurück.

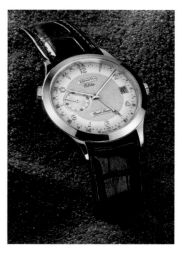

Für die Firmenbiographie ist 1969 auch noch aus einem anderen Grund wichtig: Das Jahr brachte die Gründung der Holding Mondia-Zenith-Movado, an der 1971 die amerikanische Zenith Radio Corporation eine Mehrheitsbeteiligung erwarb. Mit Übernahme der Aktienmehrheit durch die Finanzgruppe DIXI geboten von 1978 bis 1999 wieder Schweizer Aktionäre über Zenith. Dann nahm der französische Luxusmulti LVMH Moët Hennessy – Louis Vuitton die Manufaktur unter seine Fittiche. Und dort befindet sich das Traditionslabel auch noch 2015, im Jahr des 150. Geburtstags.

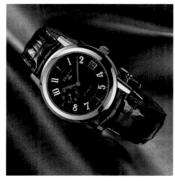

Zugpferd des Portfolios ist und bleibt „El Primero" in höchst unterschiedlichen Ausführungen. Klassisch, sportlich, mit Tourbillon oder auch „offenem Herzen". Bei dieser Variante eröffnet ein großes Guckloch in Zifferblatt und Platine Mechanik-Voyeuren tiefe Einblicke. Seit 1994 produziert Zenith auch ein klassisches Automatikwerk namens „Elite", das 2015 eine Schwester mit Zentralsekunde bekam. Weltspitze, weil einzigartig in ihrer Konstruktion, ist die 2008 vorgestellte und seitdem mehrfach verbesserte „Academy Christophe Colomb". Beim Automatikkaliber 8800 mit dezentralem Platinrotor und 50 Stunden Gangautonomie haben die Konstrukteure das Schwing- und Hemmungssystem in einen kardanisch aufgehängten Container ausgelagert. Durch diesen genialen Schachzug befindet es sich, wie auch die klassischen Marinechronometer, immer in waagrechter Position. Die Schwerkraft, einer der größten Genauigkeitsfeinde mechanischer Uhren, hat so beim besten Willen keinen Angriffspunkt mehr. ◦

Top to bottom:
Automatic calibre 670
Automatic calibre 685
Elite Class Dual Time, 2002
Elite Port Royal, 2003

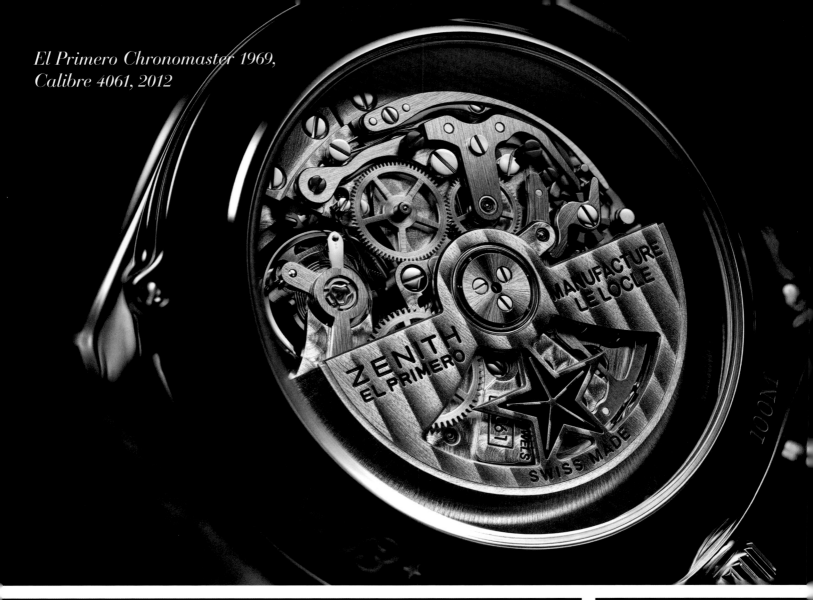

El Primero Chronomaster 1969,
Calibre 4061, 2012

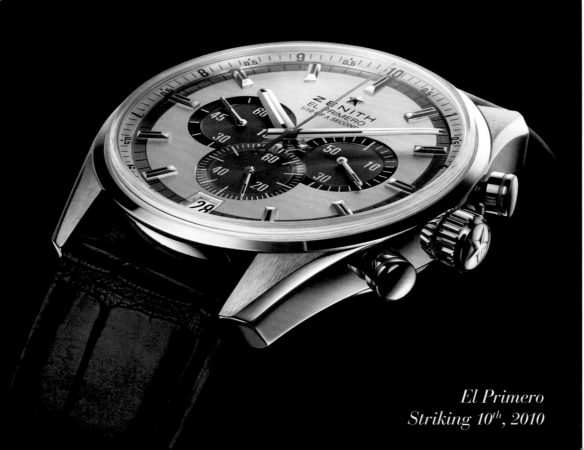

El Primero
Striking 10ᵗʰ, 2010

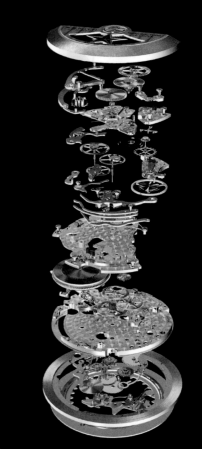

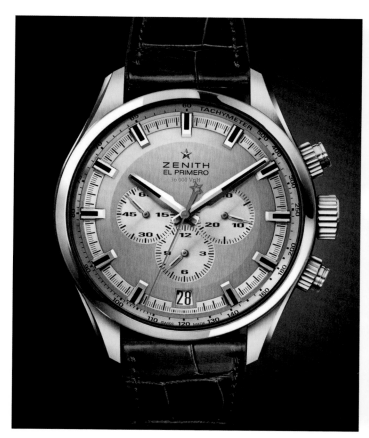

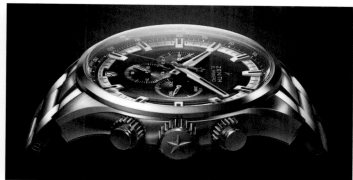

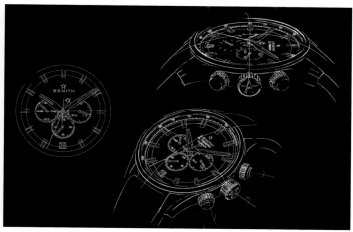

El Primero Sport, 2015

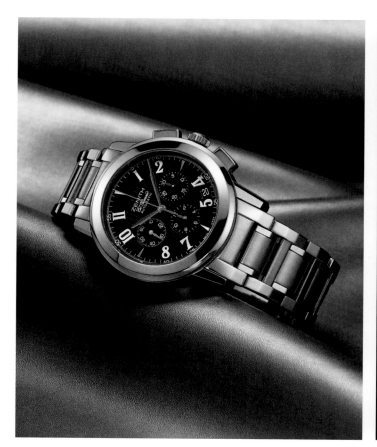

El Primero Port Royal, 2003

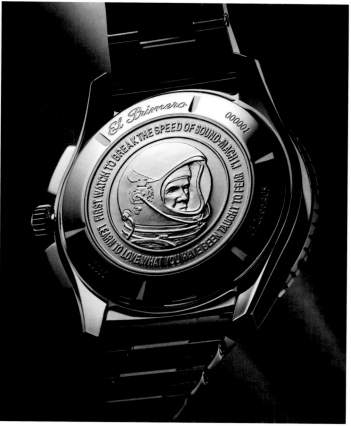

El Primero Stratos Flyback Striking 10ᵗʰ Felix Baumgartner, 2013

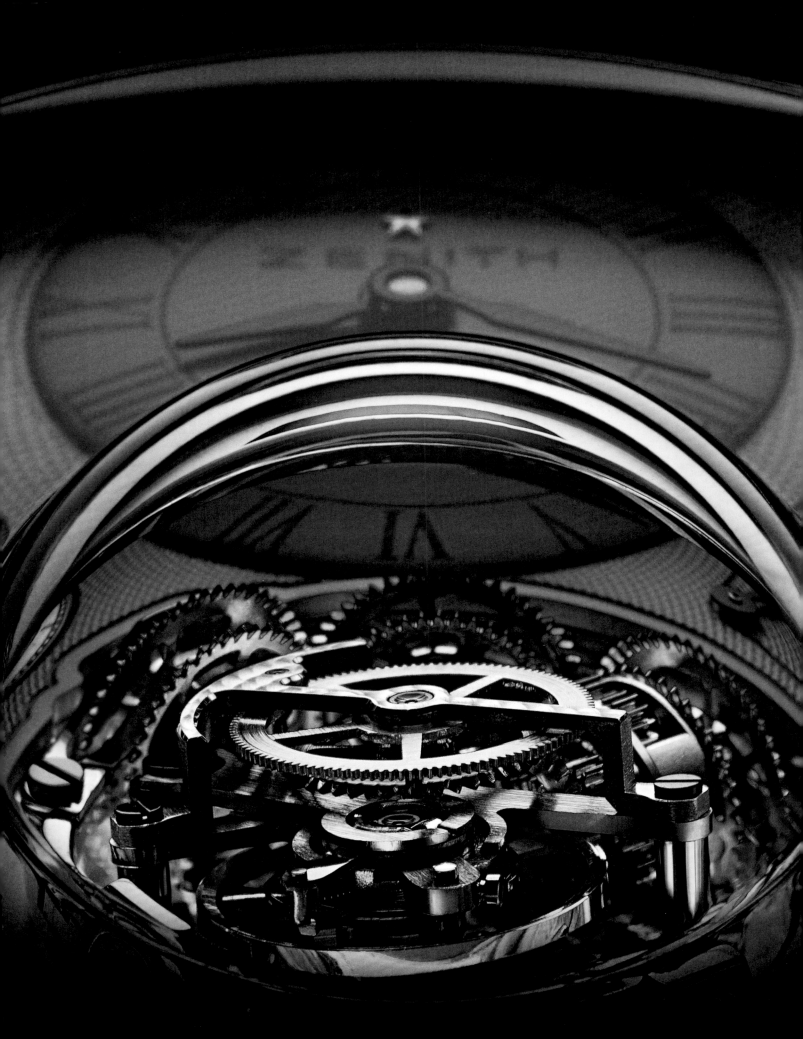

Zenith ou les sommets de l'horlogerie

1 500, ce nombre en dit plus qu'un long discours. Il correspond aux prix, médailles et autres distinctions internationales attribués à la manufacture horlogère Zenith au cours de 150 ans d'existence pour une horlogerie d'une précision hors pair. Lorsqu'il débute ses activités horlogères au Locle, bourgade du Jura suisse, Georges Favre-Jacot n'aurait jamais osé espérer, même en rêve, une telle profusion de récompenses. D'autant plus que la très haute précision ne joue pas encore à l'époque un rôle prépondérant. À tout juste 22 ans, il cherche simplement à fabriquer de façon rationnelle et donc à un prix avantageux des montres de poche d'une précision raisonnablement supérieure. Il recourt pour ce faire à un système éprouvé, la méthode américaine de production, c'est-à-dire l'utilisation de machines d'un nouveau type pour produire des composants mieux ajustés, que l'on puisse assembler sans trop de rectifications et échanger facilement à des fins d'entretien.

Au début du XXe siècle, le patron de l'entreprise découvre l'importance des marchés extra-européens pour continuer à se développer. Georges charge son neveu James Favre de se rendre aux États-Unis, en Russie et en Chine. De chaque voyage, il rapporte des commandes rémunératrices dans sa serviette. Puis vient l'année 1911 et avec elle une étape clé : l'entreprise de Favre devient la société par actions Zenith SA. Les créations horlogères reçoivent ainsi enfin un nom facile à retenir, qui ne tardera pas à orner cadrans, mouvements et boîtiers. Des montres-bracelets des types les plus divers, dont certaines avec d'intéressantes fonctions supplémentaires, viennent enrichir la collection durant la Première Guerre mondiale, notamment des réveils qui aident à se lever ou honorer un rendez-vous, et des chronographes qui permettent de mesurer des durées. Dans l'entre-deux guerres, le nom Zenith figure régulièrement en tête des listes de résultats des concours de chronométrie organisés par les observatoires fédéraux de Genève et Neuchâtel. Aujourd'hui, les collectionneurs sont friands de chronomètres-bracelets équipés du calibre 135 à remontage manuel extrêmement précis, qui seront fabriqués de 1948 à 1962.

Disposant de toutes les compétences appropriées, la manufacture participe dans les années 1960 à la compétition qui fait rage au niveau international concernant l'introduction du premier chronographe-bracelet à remontoir automatique. Les années d'efforts portent leurs fruits. Le 10 janvier 1969, lors d'une conférence de presse à Genève, Zenith peut non sans fierté présenter « El Primero », la première montre-bracelet à chronographe, rotor à roulements à billes et fréquence de balancier cinq hertz pour une précision au dixième de seconde. Dans un premier temps, la société n'a pas le temps de se réjouir d'un tel exploit. Au milieu des années 1970 en effet, les nouveaux propriétaires de la marque n'apprécient plus ce type de mécanique et décident de détruire les stocks existants ainsi que tous les outils. Heureusement, un horloger blanchi sous le harnais s'oppose à ce dictat. Charly Vermot dissimule secrètement dans l'un des vastes greniers tout ce dont il peut s'emparer. Zenith est donc parfaitement outillé au retour en grâce de la montre-bracelet mécanique au milieu des années 1980 : en 1986, « El Primero » réapparaît.

Dans l'histoire de la marque, 1969 est aussi une date importante à un autre titre : c'est l'année où se forme la holding Mondia-Zenith-Movado, dont Zenith Radio Corporation aux États-Unis devient actionnaire majoritaire en 1971. Avec le rachat de la majorité des actions par le groupe financier DIXI, des actionnaires suisses règnent à nouveau sur Zenith de 1978 à 1999. Ensuite, c'est le groupe français du luxe LVMH Moët Hennessy – Louis Vuitton qui prend la manufacture sous son aile. Et c'est dans ce même groupe que se trouve cette marque de tradition en 2015, l'année de son 150e anniversaire.

Le fer de lance du portefeuille est depuis toujours « El Primero », dans les versions les plus diverses : classique, sportive, à tourbillon ou à « cœur battant à fleur de peau ». Dans cette dernière, une généreuse ouverture ménagée dans le cadran et la platine offre aux amoureux de mécanique une vue dégagée sur le cœur. Depuis 1994, Zenith produit également « Elite », un mouvement automatique classique qui a hérité en 2015 d'une petite sœur avec aiguille des secondes au centre. Présentée en 2008 et depuis plusieurs fois perfectionnée, l'« Academy Christophe Colomb » est reine de sa catégorie par son architecture exceptionnelle. Sur le calibre automatique 8800 à rotor décentré en platine et 50 heures de réserve de marche, les constructeurs ont déporté la masse oscillante et l'échappement dans une cage montée sur cardans. Grâce à ce coup de génie, l'échappement est toujours horizontal, comme sur les traditionnels chronomètres de marine. La gravité, l'une des plus grandes ennemies de la précision des montres mécaniques, ne peut en aucun cas s'exercer. ∘

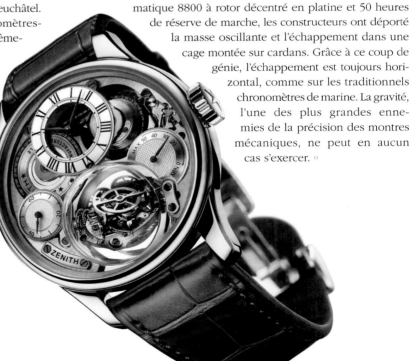

Academy Christophe Colomb Hurricane, 2014

Credits

The images related to the individual companies

Alpina
Ateliers deMonaco
Audemars Piguet
Baume & Mercier
Blancpain
Breguet
Breitling
Carl F. Bucherer
Bulgari
Cartier
Chopard
Chronoswiss
Corum
Eterna
Frédérique Constant
Girard-Perregaux
Glashütte Original
de Grisogono
Hanhart
Hublot
IWC Schaffhausen
Jaeger-LeCoultre
Junghans
A. Lange & Söhne
Longines
Montblanc
Nomos Glashütte
Omega
Oris
Panerai
Parmigiani Fleurier
Patek Philippe
Piaget
Porsche Design
Rolex
Seiko
TAG Heuer
Tudor
Tutima Glashütte
Ulysse Nardin
Vacheron Constantin
Zenith

come from each company's own museum or collection.

Cover © Audemars Piguet
Back Cover © Vacheron Constantin

The images related to the modern wristwatches were made available by the companies' press offices in Switzerland and Germany.
Some additional photos were provided by the cpb archive in Munich.
Pictures from other sources (e.g., auction houses or private collections) are listed below:

Breguet
(pp 60–63) all images Collection Breguet, Paris except (p 61 top) Christie's Geneva;
(p 62 bottom middle) Louvre, Paris, loan of Château de Fontainebleau;
(p 63 top right) Louvre, Paris;
(p 65 middle) Antiquorum, Geneva

Cartier
(p 102 top, middle, bottom left)
Collection Cartier, Paris

Hublot
(pp 204–205) cpb archive, Munich

Junghans
(p 253) images of Max Bill watches courtesy of Max, Binia + Jakob Bill Stiftung, © VG Bild-Kunst, Bonn 2022

A. Lange & Söhne
(p 259 bottom) cpb archive, Munich;
(p 259 top) A. Lange & Söhne Collection, Glashütte

Parmigiani Fleurier
(p 334) image of Bugatti Chiron courtesy of Bugatti Automobiles

Patek Philippe
(p 340 bottom middle, p 342 middle)
Patek Philippe Museum, Geneva;
(p 344 bottom middle) Christie's, Geneva

Piaget
(p 363) Collection of the Piaget Museum, Geneva

Vacheron Constantin
(pp 468–469) collection of the Vacheron Constantin Museum, Geneva

The publishers thank the brands Alpina, Ateliers deMonaco, Baume & Mercier, Bulgari, Frédérique Constant, Girard-Perregaux, de Grisogono, Hanhart, Junghans, Oris, Parmigiani Fleurier, Porsche Design, Seiko, TAG Heuer, Tutima Glashütte, and Ulysse Nardin, whose financial support helped to make this book's publication possible.

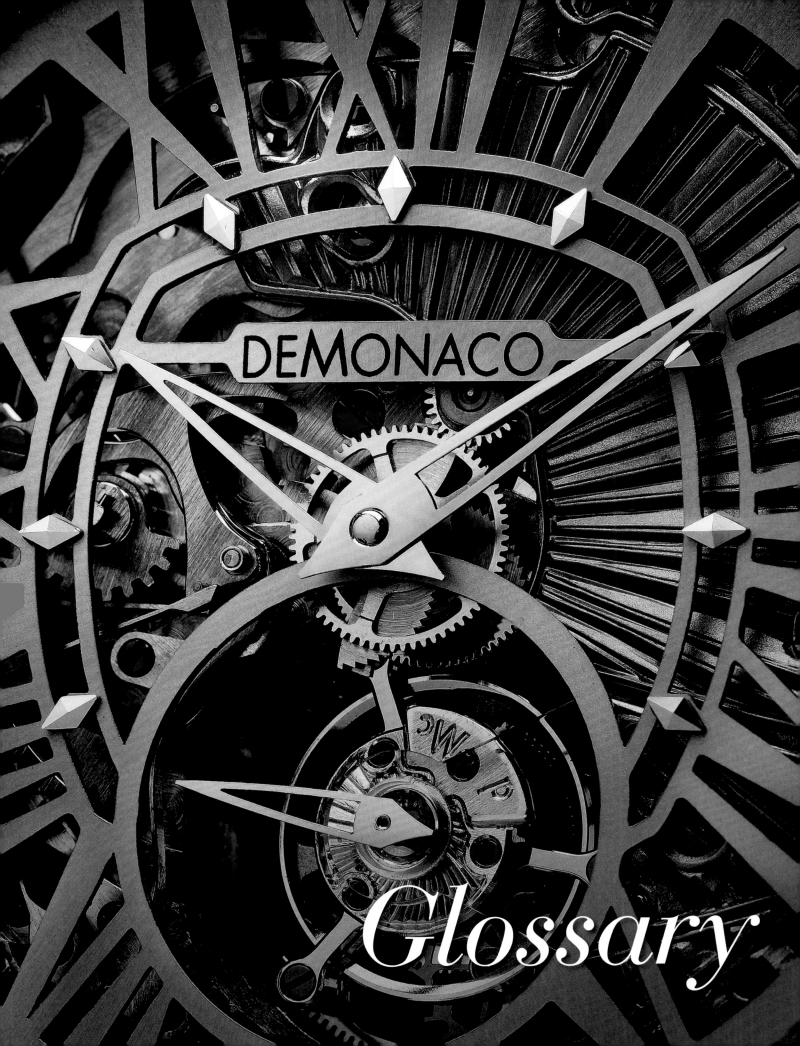

Glossary

Glossary

Age of the moon
An → indication showing the number of days that have elapsed since the last new moon. In a synodic month, the interval from one new moon to the next lasts exactly 29 days, 12 hours, 44 minutes, and 3 seconds.

Analog time display
The time is indicated by a pair of hands. The current time can be told by comparing the relative positions of the hour and minute hand.

Antimagnetic (correctly: nonmagnetic)
A watch is described as antimagnetic if it's protected against the negative influences of magnetic fields. It can be termed "nonmagnetic" if it continues to run in a magnetic field of 4,800 A/m (amperes per meter) and afterwards deviates from correct timekeeping by no more than 30 seconds per day.

Automatic winding
An additional mechanism that uses energy derived from the motions of the wearer's forearm to tighten the → mainspring of a mechanical watch.

Balance
A circular metal hoop that oscillates together with the → balance-spring in portable timepieces. Their oscillations subdivide the steady passage of time into brief segments, which are ideally of identical duration.

Balance-spring
The balance-spring can be described as the "soul" of a mechanical watch. The inner end of this spring is attached to the balance-staff and the outer end is affixed to the balance-cock. The balance-spring's elasticity assures that the → balance oscillates regularly. The duration of each swing is determined by the active length of the balance-spring and by the moment of inertia of the balance's rim.

Beveling
Beveled edges on steel parts are a distinguishing feature of fine watches. The edges can be beveled mechanically with the aid of a pantograph or, in the finest luxury watches, the beveling is performed manually with a file. Ideally, each angle on the edge should be precisely 45 degrees. Beveling has no effect on the movement's function.

Bezel
This term can have various meanings in watchmaking. Strictly speaking, a bezel is an annulus surrounding the crystal of a watch's → case. The crystal is pressed into the bezel, which is then mounted on the middle part of the case. The term is often used, however, to denote rotatable rings on the fronts of watch cases.

Breguet balance-spring
A balance-spring with a high upward curvature at its outer end to assure uniform "breathing." This type of spring is costly to manufacture, so nowadays it is installed only in very high-quality → calibers.

Cadrature
An additional switching mechanism in complicated watches, e.g., a mechanism for a → chronograph, → repeater, or calendar. The term is derived from the French word *cadran* (dial). Cadratures accordingly denote traditional switching mechanisms installed on the dial side of the movement and thus under the dial. Depending on its construction, a cadrature may be integrated into a movement or mounted on a separate → plate and connected to the movement as an additional module.

Caliber
The dimensions and shape of a watch movement and its parts. The caliber designation facilitates precise identification, e.g., when ordering spare parts. Manufactory calibers should be distinguished from movement-blanks furnished by suppliers of → ébauches. The former are movements that → manufactories produce for their own use. Aficionados sometimes talk about exclusive or reserved calibers: these are movement-blanks that ébauche suppliers develop and/or produce solely for individual customers. These calibers are not available for other → établisseurs.

Case
The protective exterior housing of a watch. Cases are made in diverse versions and materials. Two common varieties of cases for pocket watches are the open (Lépine) and closed (Savonnette) case. Water-resistant or → watertight cases are commonly used for wristwatches. Watch cases are made in many different shapes (round, square, oval, rectangular, tonneau) and materials (platinum, gold, silver, steel, titanium, carbon, aluminum, plastic).

Chaton (setting)
A circular metal disc with a hole to accept a bearing jewel. Chatons are either pressed or screwed into place in watch movements. Because of their attractive appearance, screwed chatons are again used today, especially by manufactories in Glashütte.

Chronograph
A chronograph—or more correctly "chronoscope"—is a watch with an hour hand, a minute hand, and a special additional mechanism which, at the push of a button, alternately starts a (usually) centrally positioned elapsed-seconds hand, stops it, and returns it to its zero position. The ordinary time display remains unaffected. Depending on the particular version, a chronograph may also have counters to tally the minutes and hours that have elapsed since the chronograph was switched on. When the zero-return button is pressed, the counters' hands likewise return to their starting positions. Two-button chronographs have predominated since the 1930s: one button starts and stops the stopwatch; the other returns the elapsed-time hands to zero. Such chronographs facilitate additive stopping, i.e., the elapsed-second hand can be halted and restarted from its last position as often as desired. Depending on its balance's frequency, a mechanical wristwatch chronograph can measure elapsed intervals to the nearest tenth of a second.

Chronograph rattrapante
This type of chronograph can simultaneously time two or more events that began at the same moment. The split-seconds mechanism is useful for measuring intermediate times at races.

Chronometer
A chronometer is a precise timepiece which has proven the accuracy of its rate during a 15-day test at one of the official watch-testing authorities, e.g., the COSC (Contrôle Officiel Suisse des Chronomètres) in Switzerland. In the five → positions "crown left," "crown up," "crown down," "dial up," and "dial down," the average daily rate must remain within −4 and +6 seconds; the average daily deviation may not exceed two seconds; and the greatest deviation of rate must not be greater than five seconds. All watches are tested at 23°, 8°, and 38° Celsius. Only after a watch has passed the chronometer test does it earn the right to bear the word "chronometer" on its dial and to be marketed together with a corresponding certificate.

Column-wheel chronograph
A rotatable component which can have five, six, seven, eight, or nine columns controls the start, stop, and zero-return functions in a classical chronograph. With each switching process, the column-wheel rotates clockwise through an exactly defined angle. If the end of the switch yoke comes to rest atop a column, the latter keeps the former lifted; if the switch yoke's end falls between two columns, gentle pressure from the spring holds it down.

Complication
An additional mechanism in a mechanical watch, e.g., → chronograph, → perpetual calendar, or → tourbillon.

Coulisse switching

A switching mechanism to control a → chronograph. A moveable cam, differently shaped depending on the particular → caliber, embodies the "program" to start the chronograph, stop it, and return its elapsed-time hands to zero. Chronographs with coulisse or cam switching are technically less complex, but no less reliable than their counterparts with column-wheels.

Counterfeit

A plagiarized imitation of a popular and usually high-quality watch. Certificates, invoices, and etuis no longer adequately protect purchasers against counterfeiters because these supplementary items can likewise be counterfeited.

Crown

A button or knob to wind a watch, to set its hands and/or to adjust its date display. The crown was formerly also sometimes used to control the → chronograph.

Crown winding

Until the late 19th century, many pocket watches were wound, and their hands were often set with the aid of a little key. With modern crown winding, both tasks are performed via a small → crown equipped with a mechanism to switch between the winding and setting functions.

Crystals

Four kinds of crystals are used on wristwatches. Glass crystals are found primarily on early wristwatches: they resist scratches, but are very fragile and prone to breakage. Plexiglas progressively replaced glass for watch crystals beginning in the early 1940s: Plexiglas crystals are unbreakable, but become scratched relatively easily. Crystals made of mineral glass have a hardness of 5 on the Mohs scale and are therefore much more robust than plastic crystals. Nowadays, high-quality watches usually use → sapphire for their crystals: with a hardness of 9 on the Mohs scale, sapphire is extremely resistant to scratches and breakage, but can only be processed with special diamond-tipped tools.

Date display

→ Indicator of the date, either in analog form via a hand (hand-type date display) or digitally via a printed ring (window date). The watch industry uses two separate digits (one in the "tens" and another in the "ones" column) for the outsize date display. Date displays can be divided into three types: creeping, half-creeping, and jumping. The first two types are gradually advanced by the movement. Jumping indicators successively amass power from the movement throughout the day, store the energy in a → spring, and suddenly release the stored → energy exactly at midnight to power the switching mechanism.

Digital time display

The time of day or night is shown by numerals.

Diver's watches

Wristwatches manufactured in compliance with clearly defined standards and intended for underwater use. Professional diver's watches must be watertight to a minimum depth of 200 meters (660 feet). For the diver's safety, a jeweler should annually test the watch's watertightness and functionality.

Ébauche

French term for "sketch." In specialized French watchmaking vocabulary, the movement-blank for a watch is also known as an ébauche or blanc.

Enamel

A colored vitreous coating applied to a metal substrate. The enameling technique has been used on watches (e.g., the dial and the → case) for more than 350 years. An enamel dial was almost a standard feature on fine watches in the early 20th century. Enamel dials have become extremely rare nowadays, partly because of the high cost of manufacturing them.

Energy

The ability to do work. Potential energy is needed to power watches. In mechanical timepieces, this potential energy can be stored in a tightened → mainspring (spring power) or in a manually raised weight (gravity power).

Escapement

A mechanism that conveys the power of the → mainspring in small impulses to the watch's oscillating system (→ balance and → balance-spring) and also prevents the movement from racing ahead and exhausting its stored energy. In a watch with a balance paced at 28,800 semi-oscillations per hour, the escapement allows the gear-train to advance 691,200 times each day. This adds up to more than a billion impulses in the course of four years—approximately six times the performance of a human heart.

Établisseur

A watchmaker who purchases components (e.g., movement, dial, hands, and → case) from specialized manufacturers and then assembles these parts to make complete watches.

Fabrication number

Many watch manufacturers use fabrication numbers to identify their products. The makers number the movement, the → case, or both. An officially tested → chronometer must always bear a movement number. This number is also printed on the accompanying test certificate.

Faceted

Steel and brass components in very fine watches have faceted edges (→ beveling). The bevel should ideally have an angle of 45 degrees.

Finishing

The final decorations and processing given to a watch.

Frequency

Oscillations per unit of time, measured in hertz (Hz). Pendulums are most commonly used to regulate the rate of stationary clocks, while mobile timepieces rely on → balances. Both components move to and fro at a particular frequency. The pendulum of a one-second pendulum clock requires exactly one second to swing from one extreme point of its arc to the other, so it has a frequency of 0.5 Hz or 1,800 semi-oscillations per hour (A/h). Early balances were paced at 7,200 or 9,000 A/h. The frequency of balances in pocket watches was increased first to 12,600 and later to the now common standard of 18,000 A/h (2.5 Hz). This frequency originally developed into the most common standard for wristwatches. To increase precision, watch manufacturers later raised the frequency to 21,600 A/h (3 Hz), 28,800 A/h (4 Hz), 36,000 A/h (5 Hz), 57,600 (8 Hz), and even 72,000 (10 Hz). A speedier balance frequency is associated with greater energy consumption. Higher rotational speeds must also be taken into account when lubricating the → escapement.

Full calendar

A complete calendar showing the current day, date, and month.

Geneva waves

An undulating pattern commonly used to decorate the bridges and cocks in fine → calibers. Generally found only on high-quality movements, Geneva waves are applied prior to galvanic ennoblement but remain visible afterwards.

GMT

Greenwich Mean Time, also known as World or Universal Time Coordinate (UTC), is the time at zero degrees longitude, i.e., the meridian that passes through Greenwich, England. GMT is currently used as the standard time for navigation and international radio communication.

Gregorian calendar

After lengthy preparations and the abrupt elimination of ten full days, the calendar reform instituted by Pope Gregory XIII took effect in Rome on October 15, 1582. This new calendar rectified the tiny residual error that had been accumulating since the introduction of the → Julian calendar in 45 BC, which had erroneously added 0.0078th of a day to each year.

The special feature of the Gregorian calendar, which corrects this inaccuracy, is that it eliminates three leap days from the leap-year cycle every 400 years. Leap days are not included in secular years (i.e., years that begin a new century) unless the year is evenly divisible by 400. There will accordingly not be a 29th day in February in the years 2100, 2200, and 2300.

Guilloche
The engraving of fine and sometimes artistically interwoven patterns on the → cases or dials of watches. Traditional guilloche is performed manually with the aid of antique machines.

Hallmarks
Stamps punched into → cases specifying the type and fineness of the → precious metal, the country and sometimes also the city of origin, the year of manufacture, and the case-maker. Other punched hallmarks may represent the trademark of the manufacturing or delivering watch company, a → reference or a serial number.

Hand-wound movement
Depending on its construction, a hand-wound movement may contain 80 or more components, which can be divided into eight main functional groups:
1. the power system as the energy-providing organ;
2. the transfer system to convey → energy to
3. the subdividing system, also known as the → escapement;
4. the regulating system;
5. the motion-work (dial-train);
6. the organs for indicating the time;
7. the hand-setting system; and
8. the winding system.

Hand-wound watch
A timepiece in which the → mainspring must be wound by hand.

Hour counter
A constructive feature of some → chronographs to tally the number of hours that have elapsed since the stopwatch function was triggered. Most hour counters can tally up to twelve hours. Pressing the zero-return button triggers the hand on the hour counter to return to its zero position.

Indication
A display, e.g., of the time, date, day of the week, month, equation of time, → power reserve, or second time zone.

Jewels
An internationally used term denoting the → stones in a watch movement.

Julian calendar
The cycle of three 365-day years and one 366-day leap year can be traced back to Gaius Julius Caesar. But the Julian year is 0.00078th of a day longer than the actual astronomical year. This error accumulated over the centuries, prompting Pope Gregory XIII to correct the Julian calendar in 1582.

Lever
The lever is one of the most complicated parts of a mechanical watch. Its twofold task is to transfer power from the gear-train to the → balance, thus keeping the latter in oscillation, and to prevent the gear-train from racing ahead and quickly exhausting the energy stored in the mainspring.

Lever escapement
Currently the most commonly used type of → escapement in mechanical watches. The Swiss lever escapement predominates among today's high-quality wristwatches.

Ligne
A unit of length traditionally used to express the dimensions of watch. The ligne is derived from the *pied du roi*, i.e., the French foot. For example: a circular movement can have a diameter of 11 lignes and a rectangular movement may measure 8¾ by 12 lignes. One ligne is equivalent to 2.2558 millimeters.

Luminous dial
A dial from which the time can also be read in the dark.

Mainspring
An elastic and spirally-coiled strip of steel which stores the energy needed to power a portable mechanical watch.

Manufactory (French: manufacture)
According to the unwritten laws of watchmaking, a watch manufacturer may only describe itself as a manufactory if it makes at least one → ébauche or movement-blank of its own. The industry uses the term → établisseur to denote a manufacturer who assembles watches from pre-made movement-blanks.

Mechanical geared clocks
Timepieces powered by a → mainspring. The rate is regulated either by a → balance with a balance-spring or by a pendulum. The development of the geared clock probably resulted from the mechanism used to propel planetariums, which have been known since the late 13th century. The oldest mechanical geared clock in German-speaking Europe is probably the clock in Strasbourg's cathedral. This clock was completed in 1352. Functional geared clocks probably first appeared in England toward the end of the 13th century.

Moon's phases
Depending on the relative positions of the sun, the earth, and the moon, the moon waxes through various phases from new moon through waxing half moon to full moon, then wanes through waning half moon to new moon. One lunation is equal to circa 29.5 days.

Perpetual calendar
A complex calendar mechanism, consisting of approximately 100 parts, that automatically takes into account the different lengths of the months. Most perpetual calendars won't require manual correction until February 28, 2100.

Plate
Also known as movement plate. A metal plate bearing the bridges, cocks, and other parts of a watch movement.

Position
Unlike pocket watches, wristwatches are worn in many different orientations, which can be divided into the following positions: "crown up," "crown down," "crown right," "dial up," and "dial down." A precise wristwatch accordingly undergoes fine adjustment in each of these five positions.

Power reserve
Energy potential beyond the normal winding interval of a wristwatch (24 hours). The power reserve usually ranges between 10 and 16 hours. However, the propulsive force of the → mainspring declines during this remaining interval, which causes a reduction in the watch's rate performance.

Power-reserve display
Indication of the currently remaining → power reserve in a mechanical movement.

Precious metals
The precious metals gold, platinum, and silver are usually used for the → cases of wristwatches. Gold alloys are available in various finenesses: 333/1,000 (8 karat), 375/1,000 (9 karat), 575/1,000 (14 karat), or 750/1,000 (18 karat). Admixture of other metals (e.g., copper) determines the hue. Winding rotors are often made of 21-, 23-, or 24-karat gold. The fineness of platinum is usually 950/1,000.

Pulsometer
A time- and work-saving scale on the dial (usually on a → chronograph) to help the wearer measure pulse rates. Depending on the calibrations, the user starts the chronograph, counts either 20 or 30 pulse beats and then stops the chronograph: the tip of the halted elapsed-seconds hand will point to the calculated pulse rate per minute.

Rate autonomy

The entire running interval of a mechanical movement, i.e., the time that elapses from complete winding of the → mainspring to motionlessness due to its slackening.

Reference

A manufacturer-specific combination of letters and numbers to classify various watch models. The reference number often also includes information about the type, case material, movement, dial, hands, wristband, and embellishment with gemstones.

Regulator dial

Off-center → indication of the hours and seconds.

Repeater striking-train

An elaborate additional function which enables a movement to audibly chime the current time with greater or lesser accuracy. Depending on the specific repeater, the timepiece may be triggered to chime every quarter hour, every eighth of an hour (i.e., every 7½ minutes), every five minutes, or every minute.

Retrograde display

A retrograde hand advances stepwise along an arc of a circle to indicate the time, date, or day of the week and then suddenly jumps back to its starting position when it reaches the end of the arc.

Rotor:

An unlimitedly rotatable oscillating weight in watches with → automatic winding. Depending on the construction of the self-winding mechanism, the rotor acts to tighten the → mainspring in either one or both of the rotor's directions of rotation. Central rotors rotate above the entire movement; microrotors are inset into the plane of the movement.

Sapphire crystal

A scratch-resistant crystal with a hardness of 9 on the Mohs scale. Only diamond is harder.

Satin finish

A fine, silky, matt finish on metal surfaces.

Shock absorption

A system to prevent breakage of the slender and thus very delicate pivots of the balance-staff. For this purpose, the ring jewel and cap jewel of the balance-staff's bearing are elastically affixed in the → plate and balance-cock. If the watch suffers a strong impact, the jewels can shift laterally and/or axially. A wristwatch with a shock-absorption system should be able to survive a plunge from a height of one meter (3.3 feet) onto an oak floor without suffering damage and without showing significant rate deviations afterwards.

Silicon

Brass, steel, and synthetic rubies predominated for centuries in the fabrication of mechanical movements. But silicon, the basic material of electronic microchips, became acceptable in the world of haute horlogerie in 2001. In its monocrystalline form, silicon has the same crystalline structure as diamond. Silicon is 60 percent harder and 70 percent lighter than steel, → nonmagnetic, and resistant to corrosion. Even without post-processing, silicon has an extremely smooth surface which makes lubricant oil unnecessary. Silicon is elastic, but not plastically formable. Specially treated "Silinvar" expands only very minimally when warmed, so it's well suited as a material for → balance-springs. A neologism derived from the words "invariable silicon," Silinvar was co-developed by Patek Philippe, Rolex, and the Swatch Group in collaboration with the Centre Suisse d'Electronique et de Microtechnique (CSEM) and the Institut de Microtechnique of the University of Neuchâtel.

Skeletonized movement

A watch movement in which the → plate, bridges, cocks, barrel, and sometimes also the → rotor are pierced and filed away to leave only the material that is indispensable for each component's proper function. The openwork enables connoisseurs to peer into the depths of the movement. Wristwatches with skeletonized movements first appeared in the 1930s.

Spring

Many different types of springs are used in watch movements. Alongside the → balance-spring and → mainspring, locking springs and retainer springs are also commonly used.

Stainless steel

A popular alloy containing the metals steel, nickel, and chrome with admixed molybdenum or tungsten. Stainless steel does not rust and is extremely resistant and → antimagnetic, but is comparatively difficult to process.

Stones

To reduce friction in precise watches, jewels are inserted into the most important bearings, as pallet stones on the lever and on the impulse-pin.

Stop-seconds function

A mechanism to momentarily halt the movement and/or the second-hand so the hands can be set with to-the-second accuracy.

Stopwatches

Unlike → chronographs, stopwatches do not display the ordinary time of day. In simply constructed stopwatches, the movement is halted to stop the second-hand.

Tourbillon

Invented by Abraham-Louis Breguet in 1795 and patented by him in 1801, the tourbillon compensates for the center-of-gravity error in the oscillating system (→ balance and → balance-spring) of a mechanical watch. The French word *tourbillon* means "whirlwind." In horological contexts, it refers to a construction in which the entire oscillating and escapement system is mounted inside a cage of the lightest possible weight. The cage continually and uniformly rotates around its own axis, usually completing one rotation per minute. These constant rotations compensate for the negative influences exerted by the earth's gravity on the accuracy of a watch's rate when the timepiece is in a vertical → position. The tourbillon's rotations also improve the watch's rate performance. A tourbillon has no effect on the accuracy of the rate when the timepiece is in a horizontal position.

Tourbillon cage

Made of steel or (in modern wristwatches) also of titanium or aluminum, this delicate cage contains the oscillating and escapement system (→ balance, balance-staff, → balance-spring, → lever, escape-wheel) in a watch with a → tourbillon. The tourbillon cage usually completes one rotation around its own axis every minute. A tourbillon cage should be rigid, filigreed and as lightweight as possible. Fabricating it is one of the most challenging horological tasks.

Watertight wristwatches

Watches can be labeled with the word "watertight" if they are resistant against perspiration, sprayed water, and rain. Furthermore, they must not allow any moisture to penetrate their → cases while submerged to a depth of one meter (3.3 feet) for 30 minutes. The additional phrase "50 meters" or "5 bar" means that these watches have been tested at the corresponding pressure by their manufacturers. Nevertheless, it is not recommendable to swim—and it is even less recommendable to dive—while wearing such watches.

World-time indication

Watches with world-time indication normally show 24 time zones on their dials. Exceptional world-time watches indicate the time in as many as 37 different time zones.

Glossar

Analoge Zeitanzeige

Zeitanzeige per Zeigerpaar. Die aktuelle Uhrzeit ergibt sich aus der Stellung von Stunden- und Minutenzeiger zueinander.

Anglierung

Merkmal feiner Uhren sind u. a. die gebrochenen Kanten der Stahlteile. Die Anglierung wird entweder maschinell mit Pantographen (Storchenschnabel) angebracht oder bei Luxusuhren auf höchstem Niveau in überlieferter Form per Feile von Hand ausgeführt. Idealerweise beträgt der Kantenwinkel exakt 45 Grad. Die Anglierung besitzt keine Auswirkungen auf die Funktion des Uhrwerks.

Anker

Eines der kompliziertesten Teile mechanischer Uhren. Seine Aufgabe besteht zum einen darin, die Kraft vom Räderwerk auf die → Unruh zu übertragen, um deren Schwingungen aufrechtzuerhalten. Andererseits verhindert er das ungebremste Ablaufen des aufgezogenen Räderwerks.

Ankerhemmung

Heute die am meisten verbreitete → Hemmung bei mechanischen Uhren. Bei hochwertigen Armbanduhren beherrscht mittlerweile die Schweizer Ankerhemmung das Feld.

Antimagnetisch, korrekt amagnetisch

Eine Uhr ist amagnetisch, wenn sie gegen die negativen Einflüsse magnetischer Felder geschützt ist. Sie darf dann als amagnetisch bezeichnet werden, wenn sie in einem Magnetfeld von 4800 A/m (Ampere pro Meter) weiterläuft und anschließend eine Gangabweichung von höchstens 30 Sekunden/Tag aufweist.

Automatischer Aufzug

Zusatzmechanismus, welcher die (Arm-)Bewegungen zum Spannen der → Zugfeder einer mechanischen Uhr nutzt.

Breguet-Spirale

→ Unruhspirale mit hochgebogener Endkurve für gleichförmigeres „Atmen". Aufgrund beträchtlicher Herstellungskosten ist sie heute nur noch in sehr hochwertigen → Kalibern zu finden.

Chaton

Kreisrundes, gebohrtes Metallstück zur Aufnahme eines Lagersteins, durch Einpressen oder Verschrauben im Uhrwerk befestigt. Vor allem Glashütter Uhrenmanufakturen verwenden heute aus optischen Gründen wieder verschraubte Chatons.

Chronograph

Unter Chronograph oder – sprachlich korrekter – Chronoskop versteht man eine Uhr mit Stunden- und Minutenzeiger, deren spezieller Zusatzmechanismus das Starten, Stoppen und Nullstellen eines (meist) zentral positionierten Sekundenzeigers per Knopfdruck ermöglicht. Die Zeitanzeige bleibt davon unberührt. Je nach Ausführung besitzen Chronographen zudem Zählzeiger für die seit Beginn der Stoppung verstrichenen Minuten und Stunden. Nach Betätigung des Nullstelldrückers springen auch die Zählzeiger in ihre Ausgangsposition zurück. Seit den 1930er Jahren dominierte der Zwei-Drücker-Chronograph. Ein Drücker dient dem Starten und Anhalten, der andere dem Nullstellen. Diese Chronographen ermöglichen Additionsstoppungen, d. h. der Chronographenzeiger kann beliebig oft angehalten und aus der zuletzt eingenommenen Position heraus erneut gestartet werden. Abhängig von der Unruhfrequenz können mechanische Armbandchronographen bis auf die Zehntelsekunde genau stoppen.

Chronograph-Rattrapante

Mit seiner Hilfe lassen sich zwei oder mehr Vorgänge simultan stoppen, sofern sie gleichzeitig beginnen. Der Schleppzeiger-Mechanismus eignet sich bei Wettrennen auch zum Erfassen von Zwischenzeiten.

Chronometer

Präzisionsuhr, welche ihre Ganggenauigkeit im Rahmen einer 15-tägigen Kontrolle bei einer offiziellen Uhrenprüfstelle (z. B. der COSC Contrôle Officiel Suisse des Chronomètres in der Schweiz) unter Beweis gestellt hat. In den fünf → Lagen „Krone links", „Krone oben", „Krone unten", „Zifferblatt oben" und „Zifferblatt unten" muss der mittlere tägliche Gang zwischen −4 und +6 Sekunden liegen; die mittlere tägliche Gangabweichung darf 2 Sekunden, die größte Gangabweichung 5 Sekunden nicht überschreiten. Alle Uhren werden bei Temperaturen von 23,8 und 38 °C geprüft. Erst nach dem Bestehen der Chronometerprüfung darf eine Uhr auf dem Zifferblatt die Bezeichnung Chronometer tragen und mit einem entsprechenden Zertifikat vermarktet werden.

Datumsanzeige

→ Indikation des Datums entweder analog durch einen Zeiger (Zeigerdatum) oder digital durch einen bedruckten Ring (Fensterdatum). Beim sogenannten Großdatum verwendet die Uhrenindustrie zwei separate Ziffern für die Zehner und Einer. Grundsätzlich zu unterscheiden sind schleichende, halbspringende und springende Datumsanzeigen. Erstere werden vom Werk allmählich fortgeschaltet. Springende Indikationen entziehen dem Uhrwerk im Laufe des Tages sukzessive Kraft und speichern sie in einer → Feder. Die → Energie wird exakt um Mitternacht für den Schaltvorgang freigesetzt.

Digitale Zeitanzeige

Darstellung der Uhrzeit in Ziffern.

Drehgang

Andere Bezeichnung für → Tourbillon.

Drehgestell

Feiner Käfig aus Stahl oder – bei modernen Armbanduhren – auch aus Titan oder Aluminium, welcher bei Uhren mit → Tourbillon zur Aufnahme des Schwing- und Hemmungssystems (→ Unruh, Unruhwelle, → Unruhspirale, → Anker, Ankerrad) dient. Das Drehgestell rotiert in der Regel einmal pro Minute um seine Achse. Es soll fest, filigran und möglichst leicht sein. Seine Herstellung gehört zu den uhrmacherischen Herausforderungen.

Ebauche

Französischer Begriff für Entwurf. In der französischen Uhrmachersprache wird das Rohwerk einer Uhr als blanc oder ébauche bezeichnet.

Edelmetalle

Für die → Gehäuse von Armbanduhren werden in aller Regel die Edelmetalle Gold, Platin und Silber verwendet. Gold gibt es mit einem Feingehalt von 333/1000 (8 Karat), 375/1000 (9 Karat), 575/1000 (14 Karat) oder 750/1000 (18 Karat). Die Legierung mit anderen Metallen (z. B. Kupfer) bestimmt den Farbton. 21-, 23- oder 24-karätiges Gold findet man z. B. bei Aufzugsrotoren. Bei Platin beträgt der Feingehalt 950/1000.

Edelstahl

Populäre Legierung aus den Metallen Stahl, Nickel und Chrom unter Beifügung von Molybdän oder Wolfram. Sie ist rostfrei, extrem widerstandsfähig und → amagnetisch, jedoch vergleichsweise schwer zu bearbeiten.

Email

Französisches Wort für farbigen Glasfluss auf Metall. Bei Uhren (Zifferblätter und → Gehäuse) findet die Emailtechnik seit mehr als 350 Jahren Anwendung. Nach der Wende vom 19. zum 20. Jahrhundert gehörte das Email-Zifferblatt beinahe zum Standard für feine Uhren. Inzwischen ist es – nicht zuletzt auch aus Kostengründen – extrem rar geworden.

Energie

Gespeicherte Arbeitsfähigkeit. Zum Antrieb von Uhren ist ein Energiepotenzial erforderlich. Dies kann bei mechanischen Uhren eine gespannte → Zugfeder (Federkraftantrieb) oder ein hochgezogenes Gewicht (Schwerkraftantrieb) sein.

Etablisseur

Uhrenhersteller, welcher Komponenten (Werk, Zifferblatt, Zeiger, → Gehäuse) bei spezialisierten Fabrikanten einkauft und zu fertigen Zeitmessern verarbeitet.

Ewiger Kalender

Komplexes Kalenderwerk, bestehend aus etwa 100 Teilen, welches die unterschiedlichen Monatslängen in aller Regel bis zum 28. Februar 2100 ohne manuelle Korrektur berücksichtigt.

Fabrikationsnummer

Zur Identifikation ihrer Produkte greifen viele Uhrenhersteller auf Fabrikationsnummern zurück. Dabei nummerieren sie Werk oder → Gehäuse oder beides. Offiziell geprüfte → Chronometer müssen in jedem Fall eine Werksnummer tragen. Diese findet sich auf dem Prüfzertifikat wieder.

Facette

Bei sehr feinen Uhren besitzen die Stahl- und Messingteile facettierte Kanten (→ Anglierung). Diese Kantenflächen sollten idealerweise im Winkel von 45 Grad angebracht sein.

Fälschungen

Plagiate beliebter, in der Regel sehr hochwertiger Uhren. Zertifikate, Rechnungen und Etuis schützen vor Fehlkäufen schon lange nicht mehr. Sie werden ebenfalls massenhaft gefälscht.

Feder

In Uhrwerken kommen Federn unterschiedlichster Natur zum Einsatz. Neben → Unruhspirale und → Zugfeder sind dies vor allem Sperr- und Haltefedern.

Finissage

Fein- oder Fertigbearbeitung einer Uhr.

Frequenz

Schwingungen pro Zeiteinheit, gemessen in Hertz (Hz). Bei ortsfesten Uhren findet man als gangregelndes Organ primär das Pendel, während mobile Uhren eine → Unruh besitzen. Beide bewegen sich mit einer bestimmten Frequenz hin und her. Das Pendel einer Sekundenpenduluhr benötigt von Umkehrpunkt zu Umkehrpunkt exakt eine Sekunde. Es verfügt also über eine Frequenz von 0,5 Hz oder 1800 Halbschwingungen/Stunde (A/h). Frühe Unruhschwinger brachten es auf 7200 bis 9000 A/h. Bei Taschenuhren wurde die Frequenz zunächst auf 12600 und später auf den allgemein üblichen Standard von 18000 A/h (2,5 Hz) gesteigert. Auch bei den Armbanduhren entwickelte sich diese Unruhfrequenz anfänglich zur gängigen Norm. Zur Steigerung der Präzision erhöhten die Uhrenfabrikanten die Schlagzahl auf 21600 A/h (3 Hz), 28800 A/h (4 Hz), 36000 A/h (5 Hz), 57600 (8 Hz) und sogar 72000 (10 Hz). Mit höherer Unruhfrequenz verknüpft sich steigender Energiebedarf. Außerdem verlangen zunehmende Rotationsgeschwindigkeiten Berücksichtigung beim Schmieren der → Hemmung.

Gangautonomie

Gesamte Laufzeit eines mechanischen Uhrwerks, also der Zeitraum zwischen dem vollständigen Aufzug und dem Stehenbleiben wegen Entspannung der → Zugfeder.

Gangreserve

Über das normale Aufzugsintervall einer Armbanduhr (24 Stunden) hinausreichendes Energiepotenzial. Üblicherweise bewegt sich die Gangreserve in einer Größenordnung zwischen 10 und 16 Stunden. Allerdings lässt die Antriebskraft der → Zugfeder in dieser verbleibenden Zeitspanne nach, was zu einer Reduzierung der Gangleistungen führt.

Gangreserveanzeige

→ Indikation der aktuell verbleibenden → Gangreserve bei mechanischen Uhrwerken.

Gehäuse

Schützende Hülle einer Uhr. Gehäuse gibt es in den unterschiedlichsten Ausführungen und Materialien. So unterscheidet man bei Taschenuhren u. a. zwischen offenen (Lépine) und geschlossenen Gehäusen (Savonnette). Für Armbanduhren werden z. B. wassergeschützte oder → wasserdichte Gehäuse verwendet. Außerdem gibt es eine Vielzahl unterschiedlicher Gehäuseformen (rund, quadratisch, oval, rechteckig, tonneauförmig) und -materialien (Platin, Gold, Silber, Stahl, Titan, Carbon, Aluminium, Kunststoff).

Genfer Streifen

Häufig verwendete rippenförmige Dekoration auf den Brücken und Kloben feiner → Kaliber. Sie wird vor der galvanischen Veredelung aufgebracht, bleibt aber dennoch erkennbar. Genfer Streifen finden sich im Allgemeinen nur bei hochwertigen Werken.

Gläser

Für Armbanduhren gibt es vier verschiedene Sorten von Gläsern: Kristallgläser kommen hauptsächlich bei frühen Armbanduhren vor. Sie sind zwar kratzfest, dafür jedoch sehr bruchempfindlich. Zu Beginn der 1940er Jahre wurden die Kristallgläser mehr und mehr von Kunststoffgläsern (Plexiglas) abgelöst. Die sind zwar unzerbrechlich, verkratzen aber relativ leicht. Mineralgläser besitzen eine Härte von 5 Mohs und sind daher wesentlich robuster als Kunststoffgläser. Heute wird in hochwertigen Uhren zumeist → Saphirglas verwendet. Bei einer Härte von 9 Mohs ist es extrem kratz- und bruchfest, jedoch nur mit speziellen Diamantwerkzeugen zu bearbeiten.

GMT

Greenwich Mean Time; Welt- oder Universalzeit (UTC) am Greenwich-Nullmeridian. Die mittlere Zeit von Greenwich gilt heute als Standard im Navigationswesen und im internationalen Funkverkehr.

Gregorianischer Kalender

Am 15. Oktober 1582 trat in Rom nach langen Vorarbeiten und der Eliminierung von zehn ganzen Tagen eine von Papst Gregor XIII. verfügte Kalenderreform in Kraft. Sie beseitigte den winzigen Restfehler des 45 v. Chr. eingeführten Julianischen Kalenders. Nach diesem war das Jahr um 0,0078 Tage zu lang. Das Spezifikum des Gregorianischen Kalenders, das diesen Fehler beseitigt, besteht darin, innerhalb von 400 Jahren drei Schalttage ausfallen zu lassen – und zwar in allen nicht durch 400 teilbaren Säkularjahren (Jahre des vollen Jahrhunderts). Demnach werden die Jahre 2100, 2200 und 2300 ohne den 29. Februar auskommen müssen.

Guillochieren

Gravieren feiner, teilweise kunstvoll verschlungener Muster in Uhrengehäuse oder Zifferblätter. Traditionelles Guillochieren geschieht von Hand mit Hilfe überlieferter Maschinen.

Handaufzugsuhr

Zeitmesser, bei dem die → Zugfeder von Hand gespannt werden muss.

Handaufzugswerk

Ein Handaufzugswerk besteht, je nach Konstruktion, aus 80 und mehr Teilen. Es lässt sich in acht wichtige Funktionsgruppen einteilen:
1. das Antriebssystem als energiespendendes Organ
2. das Übertragungssystem zur Weiterleitung der → Energie an
3. das Verteilungssystem, auch → Hemmung genannt
4. das Reguliersystem
5. das Zeigerwerk
6. die Organe zur Zeitanzeige
7. das Zeigerstellsystem
8. das Aufzugssystem

Hemmung

Mechanismus, der die Kraft der → Zugfeder in kleinen Stößen an das Schwingsystem (→ Unruh und → Unruhspirale) einer Uhr weiterleitet und das ungebremste Ablaufen des Uhrwerks verhindert. Bei einer Unruhfrequenz von 28800 Halbschwingungen/Stunde lässt sie das Räderwerk täglich 691 200 Mal vorrücken. Im Laufe von vier Jahren ergibt dies mehr als eine Milliarde Kraftstöße. Das entspricht etwa der sechsfachen Leistung eines menschlichen Herzens.

Indikation
Anzeige z. B. von Zeit, Datum, Wochentag, Monat, Äquation, → Gangreserve, zweiter Zonenzeit.

Jewels
Internationale Bezeichnung für die → Steine eines Uhrwerks.

Julianischer Kalender
Auf Gaius Julius Cäsar geht der Rhythmus von jeweils drei Normaljahren mit 365 Tagen und einem 366-tägigen Schaltjahr zurück. Allerdings ist das Julianische Jahr gegenüber den wahren astronomischen Gegebenheiten um 0,0078 Tage zu lang. Deshalb musste Papst Gregor XIII. den Julianischen Kalender im Jahr 1582 korrigieren.

Kadratur
Begriff für ein zusätzliches Schaltwerk komplizierter Uhren, z. B. der Mechanismus für einen → Chronographen, ein → Repetitionsschlagwerk oder ein Kalendarium. Der Name leitet sich ab vom französischen Wort cadran (Zifferblatt). Deshalb meinen Kadraturen traditionsgemäß Schaltwerke auf der Vorderseite des Uhrwerks und damit unter dem Zifferblatt. Je nach Konstruktion kann die Kadratur ins Uhrwerk integriert oder auf einer separaten → Platine montiert und als Modul additiv mit dem Uhrwerk verbunden sein.

Kaliber
Dimension und Gestalt eines Uhrwerks und seiner Teile. Die Kaliberbezeichnung ermöglicht eine exakte Identifikation u. a. zum Zweck der Ersatzteilbestellung. Von den konfektionierten Rohwerken der → Ebauches-Lieferanten sind Manufakturkaliber zu unterscheiden. Letztere sind Uhrwerke, die → Manufakturen für ihren eigenen Bedarf produzieren. Gelegentlich ist auch von exklusiven oder reservierten Kalibern die Rede. Dahinter verbergen sich Rohwerke, welche Ebauches-Fabrikanten ausschließlich für einzelne Kunden entwickeln und/oder produzieren. Auf diese Kaliber haben andere → Etablisseure keinen Zugriff.

Komplikation
Zusatzmechanismus bei mechanischen Uhren wie → Chronograph, → ewiger Kalender oder → Tourbillon.

Krone
Knopf zum Aufziehen einer Uhr, zum Zeigerstellen und/oder Korrigieren von → Datumsanzeigen sowie früher auch zum Steuern von → Chronographen.

Kronenaufzug
Bis ins späte 19. Jahrhundert erfolgten Aufzug und/oder Zeigerstellung vieler Taschenuhren mit Hilfe eines kleinen Schlüssels. Beim modernen Kronenaufzug geschieht beides über eine kleine → Krone mit Umschaltvorrichtung.

Kulissenschaltung
Schaltwerk zur Steuerung eines → Chronographen. Ein beweglicher, je nach → Kaliber unterschiedlich ausgeformter Schaltnocken liefert das „Programm" für die Funktionen Start, Stopp und Nullstellung. Chronographen mit Kulissen- oder Nockenschaltung sind technisch weniger aufwendig, aber nicht minder zuverlässig als die Pendants mit Schalt- oder Säulenrad.

Lagen
Armbanduhren werden im Gegensatz zu Taschenuhren in vielen verschiedenen Lagen getragen. Es sind dies die Positionen „Krone oben", „Krone unten", „Krone rechts", „Zifferblatt oben" und „Zifferblatt unten". Bei Präzisionsuhren erfolgt daher eine Regulierung in diesen fünf Lagen.

Leuchtzifferblatt
Zifferblatt, von dem sich auch bei Dunkelheit die Zeit ablesen lässt.

Linie
Überlieferte Maßeinheit für Uhrwerksdimensionen, abgeleitet vom „Pied du Roi", dem französischen Fuß. Beispielsweise 11‴ bei runden Werken oder 8¾ x 12‴ bei Formwerken. Eine Linie entspricht 2,2558 Millimetern.

Lünette
In der Uhrmacherei ein vielschichtiger Begriff. Streng genommen versteht man darunter den Glasreif eines → Gehäuses. Dieser wird mit dem eingepressten Glas auf dem Mittelteil des Gehäuses montiert. Oft werden heute auch auf der Vorderseite von Uhrengehäusen befestigte Drehringe als Lünetten bezeichnet.

Manufaktur
Nach den ungeschriebenen Gesetzen der Uhrmacherei darf sich ein Uhrenhersteller nur dann Manufaktur nennen, wenn er mindestens ein → Ebauche oder Rohwerk selbst fertigt. Die Fertigsteller von Uhren unter Verwendung konfektionierter Rohwerke heißen branchenintern → Etablisseure.

Mechanische Räderuhren
Uhren, die über eine → Zugfeder angetrieben werden. Die Regulierung des Gangs erfolgt durch eine → Unruh mit Spiralfeder oder ein Pendel. Die Entwicklung der Räderuhr resultiert vermutlich aus dem Antriebsmechanismus von Planetarien, die seit dem späten 13. Jahrhundert bekannt sind. Als älteste mechanische Räderuhr des deutschsprachigen Raums kann wohl die Uhr des Straßburger Münsters gelten. Sie wurde 1352 fertiggestellt. In England gab es aber vermutlich schon gegen Ende des 13. Jahrhunderts funktionsfähige Räderuhren.

Mondalteranzeige
→ Indikation zum Ablesen der Anzahl von Tagen, die seit dem letzten Neumond vergangen sind. Im synodischen Monat beträgt das Zeitintervall von Neumond zu Neumond exakt 29 Tage, 12 Stunden, 44 Minuten und 3 Sekunden.

Mondphasen
Der Mond durchläuft seine von der Stellung Sonne, Mond, Erde abhängigen Licht- bzw. Mondphasen (Neumond – erstes Viertel – Vollmond – letztes Viertel – Neumond) innerhalb einer Lunation von etwa 29,5 Tagen.

Platine
Auch Werkplatte. Eine Metallplatte, welche die Brücken, Kloben und sonstigen Bestandteile eines Uhrwerks trägt.

Pulsometer
Arbeitserleichternde Zifferblattskalierung (meist bei → Chronographen) zum Messen der Pulsfrequenz. Abhängig von der Graduierung sind nach dem Starten des Chronographen 20 oder 30 Pulsschläge zu zählen. Die Spitze des gestoppten Chronographenzeigers zeigt dann auf die hochgerechnete Pulsfrequenz pro Minute.

Punzen
Stempel in → Gehäusen, die Auskunft geben über die Art und den Feingehalt des verwendeten → Edelmetalls, über Herkunftsland und teilweise Herkunftsort, über das Herstellungsjahr sowie über den Gehäusemacher. Hinzu kommen häufig das Markenzeichen der herstellenden oder liefernden Uhrenfirma, eine → Referenz- und eine Seriennummer.

Referenz
Herstellerspezifische Kombination von Buchstaben und Ziffern zur Klassifizierung seiner verschiedenen Uhrenmodelle. Die Referenznummer beinhaltet oft auch Informationen über Typ, Gehäusematerial, Werk, Zifferblatt, Zeiger, Armband und Ausstattung mit Edelsteinen.

Regulatorzifferblatt
Außermittige → Indikation von Stunden und Sekunden.

Repetitionsschlagwerk
Aufwendige Zusatzfunktion eines Uhrwerks, die es gestattet, die aktuelle Zeit mehr oder minder genau akustisch wiederzugeben. Je nach Ausführung des Schlagwerkes unterscheidet man zwischen Uhren mit

Viertelstunden-, Achtelstunden- (bzw. 7½-Minuten-), 5-Minuten- oder Minutenrepetition.

Retrograde Anzeige
Zeiger, der sich zur → Indikation von Zeit, Datum oder auch Wochentag schrittweise über ein Kreissegment bewegt und, am Ende der Skala angekommen, ruckartig in seine Ausgangsposition zurückspringt.

Rotor
Unbegrenzt drehende Schwungmasse bei Uhren mit → automatischem Aufzug. Je nach Konstruktion des Selbstaufzugs wird die → Zugfeder in eine oder beide Drehrichtungen gespannt. Zu unterscheiden sind Zentral- und Mikrorotoren. Erstere drehen sich über dem ganzen Werk, letztere sind in die Werksebene integriert.

Saphirglas
Kratzfestes Uhrenglas der Härte 9 Mohs. Eine noch größere Härte weist nur der Diamant auf.

Satinierung
Feiner, seidiger und matter Schliff auf Metalloberflächen.

Schaltradchronograph
Bei klassischen Chronographenkalibern erfolgt die Steuerung der Funktionen Start, Stopp und Nullstellung über ein drehbar gelagertes Bauteil mit – je nach Konstruktion – fünf, sechs, sieben, acht oder neun Säulen. Bei jedem Schaltvorgang bewegt es sich um einen exakt definierten Winkel im Uhrzeigersinn weiter. Kommt das Ende einer Schaltwippe auf einer Säule zu liegen, wird es durch diese angehoben. Fällt es hingegen zwischen zwei Säulen, sorgt leichter Federdruck für eine Absenkung.

Silizium
Jahrhundertelang beherrschten Messing, Stahl und synthetischer Rubin die Fertigung mechanischer Uhrwerke. Seit 2001 ist auch Silizium, das Basismaterial elektronischer Mikrochips, salonfähig. In monokristalliner Form besitzt es die gleiche Kristallstruktur wie Diamant. Es ist 60 Prozent härter und 70 Prozent leichter als Stahl, → amagnetisch und korrosionsfest. Auch ohne Nachbearbeitung verfügt es über eine extrem glatte Oberfläche, die Öl entbehrlich macht. Silizium ist elastisch, aber nicht plastisch verformbar. Speziell behandeltes „Silinvar" dehnt sich bei Erwärmung so gut wie nicht aus. Somit eignet es sich auch zur Herstellung von → Unruhspiralen. Invariables Silizium „Silinvar" ist eine Gemeinschaftsentwicklung von Patek Philippe, Rolex und der Swatch Group mit dem Zentrum für Elektronik und Mikrotechnik (CSEM) und dem Institut für Mikrotechnik der Universität Neuchâtel.

Skelettwerk
Uhrwerk, bei dem → Platine, Brücken, Kloben, Federhaus und ggf. → Rotor so weit durchbrochen werden, dass nur noch das für die Funktion unabdingbar notwendige Material übrigbleibt. Auf diese Weise kann man durch das Uhrwerk schauen. Armbanduhren mit skelettiertem Uhrwerk gibt es seit Mitte der 1930er Jahre.

Steine
Bei Präzisionsuhren werden zur Verminderung der Reibung in den wichtigsten Lagern, an den Ankerpaletten und der Ellipse Steine eingesetzt.

Stoppsekunde
Vorrichtung zum Anhalten von Uhrwerk und/oder Sekundenzeiger, um die Uhrzeit sekundengenau einstellen zu können.

Stoppuhren
Im Gegensatz zu → Chronographen besitzen Stoppuhren keine Zeitanzeige. Bei einfachen Konstruktionen wird das Werk angehalten, wenn der Sekundenzeiger stoppen soll.

Stoßsicherung
System zum Schutz der feinen und deshalb sehr empfindlichen Zapfen der Unruhwelle vor Bruch. Zu diesem Zweck sind die Loch- und Decksteine der Unruhwellenlager federnd in → Platine und Unruhkloben befestigt. Bei harten Stößen geben sie entweder lateral und/oder axial nach.

Eine stoßgesicherte Armbanduhr soll einen Sturz aus einem Meter Höhe auf einen Eichenholzboden unbeschadet überstehen. Außerdem darf sie danach keine wesentlichen Gangabweichungen aufweisen.

Stundenzähler
Konstruktionsmerkmal mancher → Chronographen zur Erfassung der seit Beginn des Stoppvorgangs verstrichenen Stunden. Meist werden bis zu zwölf Stunden gezählt. Die Betätigung des Nullstelldrückers bewirkt auch eine Rückstellung des Stundenzählers.

Taucheruhren
Nach klar definierten Normen hergestellte Armbanduhren für Einsätze unter Wasser. Professionelle Taucheruhren müssen dem nassen Element bis mindestens 200 Meter Tiefe widerstehen. Zur eigenen Sicherheit sollte jede Taucheruhr einmal jährlich vom Juwelier auf ihre Dichtigkeit hin überprüft werden.

Tourbillon
Von Abraham-Louis Breguet 1795 erfundene und im Jahr 1801 patentierte Konstruktion zur Kompensation von Schwerpunktfehlern im Schwingsystem (→ Unruh und → Unruhspirale) mechanischer Uhren. Beim Tourbillon (dt.: Wirbelwind) ist das komplette Schwing- und Hemmungssystem in einem möglichst leichten Käfig untergebracht. Dieser dreht sich innerhalb einer bestimmten Zeitspanne (meist eine Minute) einmal um seine Achse. Auf diese Weise können die negativen Einflüsse der Erdanziehung in den senkrechten → Lagen einer Uhr ausgeglichen und die Gangleistungen gesteigert werden. In waagrechter Lage hat das Tourbillon dagegen keinen Einfluss auf die Ganggenauigkeit.

Unruh
Kreisrunder Metallreif, welcher in tragbaren Uhren durch seine Oszillationen gemeinsam mit der → Unruhspirale für das Unterteilen der kontinuierlich verstreichenden Zeit in möglichst gleichlange Abschnitte sorgt.

Unruhspirale
Die Unruhspirale kann als Seele einer mechanischen Uhr bezeichnet werden. Mit ihrem inneren Ende ist sie an der Unruhwelle, mit dem äußeren am Unruhkloben befestigt. Durch ihre Elastizität sorgt sie dafür, dass die → Unruh gleichmäßig hin und her schwingt. Ihre aktive Länge bestimmt neben dem Trägheitsmoment des Unruhreifens dessen Schwingungsdauer.

Vollkalendarium
Komplettes Kalendarium mit Anzeige von Tag, Datum und Monat.

Wasserdichte Armbanduhren
Uhren dürfen die Bezeichnung „wasserdicht" tragen, wenn sie gegen Schweiß, Spritzwasser und Regen resistent sind. Zudem darf in einem Meter Tiefe mindestens 30 Minuten lang keine Feuchtigkeit eindringen. Der Zusatz „50 Meter" oder „5 Bar" besagt, dass diese Uhren vom Fabrikanten einem entsprechenden Prüfdruck ausgesetzt wurden. Dennoch empfiehlt es sich nicht, mit solchen Uhren zu schwimmen oder gar zu tauchen.

Weltzeitindikation
Uhren mit Weltzeitindikation zeigen normalerweise 24, in Ausnahmefällen bis zu 37 Zonenzeiten auf einem Zifferblatt an.

Zugfeder
Elastisches und spiralförmig aufgewickeltes Stahlband zur Speicherung der Antriebsenergie für tragbare mechanische Uhren.

Glossaire

Acier inoxydable
Alliage répandu d'acier, de nickel et de chrome additionné de molybdène ou de tungstène, il est inoxydable, extrêmement résistant et → amagnétique, mais relativement difficile à travailler.

Affichage analogique de l'heure
Affichage de l'heure à l'aide de deux aiguilles. La position relative de l'aiguille de l'heure et de celle des minutes donne l'heure qu'il est à un moment donné.

Affichage de l'heure universelle
Les montres équipées de cette fonction affichent normalement 24, et, dans certains cas particuliers, jusqu'à 37 fuseaux horaires.

Affichage de la date
→ Indication de la date par voie analogique à l'aide d'une aiguille (date à aiguille) ou numérique par un disque portant des chiffres apparaissant dans un guichet. Dans la « grande date », les chiffres du quantième sont portés par deux disques. On distingue essentiellement entre les affichages à disque traînant, semi-sautant et sautant. Dans le premier cas, la date est progressivement avancée par le mouvement. Dans le dernier cas, le disque soustrait à plusieurs reprises dans la journée de l'énergie qu'il emmagasine dans un → ressort. Cette → énergie est libérée à minuit précis pour le changement de date.

Affichage des phases de Lune
→ Indication donnant le nombre de jours depuis la dernière nouvelle lune. Dans le mois synodique, l'intervalle séparant deux nouvelles lunes est de 29 jours, 12 heures, 44 minutes et 3 secondes.

Affichage numérique de l'heure
Indication de l'heure par des chiffres.

Affichage rétrograde
Aiguille servant à → l'indication de l'heure, de la date ou encore de la semaine, qui se déplace graduellement sur un segment de cercle et, une fois qu'elle a atteint la fin de l'échelle graduée, revient brusquement dans sa position initiale.

Amagnétique
Une montre amagnétique, couramment qualifiée d'antimagnétique, ne subit pas les influences négatives des champs magnétiques. Une montre ne peut être qualifiée d'amagnétique que si, après exposition à un champ magnétique de 4800 Ampère/mètre, elle continue à fonctionner avec un écart de marche d'au plus 30 secondes/jour.

Ancre
L'une des pièces les plus compliquées des montres mécaniques. Elle sert d'une part à transférer la force motrice du train de roues au → balancier, pour entretenir ses oscillations. Elle empêche d'autre part le déroulement incontrôlé du train de roues une fois celui-ci remonté.

Anglage
Les montres de luxe se caractérisent entre autres par des chanfreins sur les arêtes des pièces métalliques. L'anglage est réalisé par usinage au pantographe ou à la main à l'aide d'une lime suivant la tradition, pour une finition haut de gamme. Dans l'idéal, l'angle du chanfrein est de 45 degrés. L'anglage n'influe en rien sur le fonctionnement du mouvement.

Arrêt seconde
Dispositif permettant d'arrêter le mouvement et/ou l'aiguille des secondes pour effectuer une mise à l'heure à la seconde près.

Autonomie
Durée de fonctionnement totale d'un mouvement mécanique, c'est-à-dire le temps s'écoulant entre le remontage complet et l'arrêt suite au relâchement du → ressort de barillet.

Balancier
Anneau métallique qui, sur les montres de poche et les montres-bracelets, garantit par ses oscillations, avec le → spiral de balancier, que le temps qui s'écoule soit divisé en périodes de durées égales avec la plus grande précision possible.

Boîtier (ou boîte)
Écrin protecteur d'un mouvement. On trouve des boîtiers des types les plus divers. On distingue par exemple souvent pour les montres de poche entre boîtiers ouverts (Lépine) et boîtiers fermés (Savonnette). Pour les montres-bracelets, on utilise des boîtiers imperméables ou → étanches. Les boîtiers existent dans des formes très diverses (forme ronde, carrée, ovale, rectangulaire ou tonneau) et dans différents matériaux (platine, or, argent, acier, titane, carbone, aluminium ou matériau synthétique).

Cadran de type régulateur
→ Indication décentrée des heures et des secondes.

Cadran luminescent
Cadran sur lequel on peut lire l'heure même dans l'obscurité.

Cadrature
Terme désignant un mécanisme de déclenchement auxiliaire des montres à complications, notamment le mécanisme de déclenchement d'un → chronographe, d'une → répétition ou d'un calendrier. Le terme vient du mot cadran. En effet, les cadratures proprement dites sont traditionnellement sous le cadran, et par conséquent sur le dessus du mouvement. Suivant son architecture, la cadrature peut être intégrée dans le mouvement ou montée sur une → platine distincte en tant que module complémentaire du mouvement.

Cage de tourbillon
Fine cage d'acier ou – sur les montres-bracelets modernes – également en titane ou en aluminium qui, sur les montres à → tourbillon, est destinée à recevoir la masse oscillante et l'échappement (→ balancier, axe de balancier, → spiral de balancier, → ancre, roue d'ancre). La cage pivote en règle générale une fois par minute sur son axe. Elle doit être solide, ajourée et la plus légère possible. Sa fabrication fait partie des défis posés aux horlogers.

Calendrier complet
Calendrier de montre affichant le jour de la semaine, le quantième (date) et le mois.

Calendrier grégorien
Entrée en vigueur le 15 octobre 1582 à Rome après de longs travaux préparatoires, la réforme du calendrier décrétée par le pape Grégoire XIII entraîne la suppression de 10 jours du calendrier. Elle supprime ainsi la faible erreur résiduelle du → calendrier julien, introduit en 45 av. J.-C. Dans celui-ci en effet, la date de l'équinoxe avançait de 0,0078 jour chaque année. L'astuce du calendrier grégorien pour corriger cette erreur est de supprimer sur une période de 400 ans trois jours intercalaires – à savoir ceux des années séculaires dont le millésime n'est pas divisible par 400. Les années 2100, 2200 et 2300 devront ainsi se passer de 29 février.

Calendrier julien
C'est Jules César qui institue ce rythme de périodes de quatre années comportant chacune trois années communes de 365 jours et une année bissextile de 366 jours. Mais l'année julienne est plus longue de 0,0078 jour par rapport à la réalité astronomique, ce qui conduit le pape Grégoire XIII à réformer le calendrier julien en 1582.

Calibre

Dimension et forme d'un mouvement et de ses pièces. La désignation d'un calibre permet une identification exacte, notamment pour la commande de pièces de rechange. Il faut distinguer entre ébauches confectionnées par les fournisseurs → d'ébauches et calibres de manufacture. Ces derniers sont des mouvements que les → manufactures produisent pour leurs propres besoins. On parle parfois aussi de calibres exclusifs ou réservés. Ce sont des mouvements que les fabricants d'ébauches mettent au point et/ou produisent pour certains clients en particulier. Les autres → établisseurs ne peuvent se procurer ces mouvements.

Chanfrein

Sur les montres haut de gamme, les composants en acier et en laiton sont chanfreinés (→ anglage). Dans l'idéal, l'angle du chanfrein sur chaque composant est de 45 degrés.

Chaton

Bague métallique sertissant une pierre servant de coussinet ensuite fixée dans le mouvement par sertissage ou vissage. Aujourd'hui, les manufactures horlogères de Glashütte plus particulièrement recourent de nouveau à des chatons vissés pour des raisons esthétiques.

Chronographe

Un chronographe, ou plus exactement un chronoscope, est une montre dotée d'aiguilles des heures et des minutes, et sur laquelle un mécanisme auxiliaire permet, par une action sur des poussoirs, de mettre en marche, arrêter et remettre à zéro une aiguille des secondes placée, le plus souvent, au centre du cadran. L'affichage de l'heure n'est pas modifié. Selon les versions/modèles, les chronographes peuvent disposer d'aiguilles de compteurs des minutes et heures écoulées depuis la mise en marche du chronométrage. En activant le poussoir de remise à zéro, les aiguilles des compteurs reviennent dans leur position initiale. Depuis les années 1930, le chronographe à deux poussoirs prédomine. Un poussoir sert à mettre en marche et arrêter les aiguilles des compteurs, l'autre à les remettre à zéro. Ces chronographes permettent d'effectuer des chronométrages successifs, autrement dit d'arrêter les aiguilles des compteurs et de les remettre en marche à partir de la position précédente autant de fois qu'on le souhaite. Suivant la fréquence du balancier, les chronographes mécaniques permettent des chronométrages jusqu'au dixième de seconde près.

Chronographe à roue à colonnes

Sur les calibres de chronographes classiques, la commande des fonctions Mise en marche, Arrêt et Remise à zéro est assurée par une roue pivotante comportant – selon sa conception – cinq, six, sept, huit ou neuf colonnes. À chaque mise en marche, la roue se déplace d'un angle précis dans le sens horaire. Lorsque l'extrémité d'une bascule d'interrupteur vient se poser sur une colonne, cette dernière la fait se relever. Lorsqu'elle se pose entre deux colonnes, une légère pression de ressort la fait s'abaisser.

Chronomètre

Contrairement aux → chronographes, les chronomètres n'ont pas d'affichage de l'heure. Sur les modèles de conception simple, le mouvement se met en marche ou s'arrête d'une pression sur un bouton lors d'un chronométrage.

Commande (ou enclenchement) à came

Mécanisme de déclenchement pour la commande d'un → chronographe. Une came (navette) mobile de forme différente selon le → calibre fournit le « programme » des fonctions Mise en marche, Arrêt et Remise à zéro. Les chronographes commandés par came sont techniquement moins complexes, mais non moins fiables que leurs homologues à roue à colonnes.

Complication

Mécanisme auxiliaire des montres mécaniques, du type → chronographe, → quantième (ou calendrier) perpétuel ou → tourbillon.

Compteur d'heures

Dispositif particulier de certains → chronographes permettant de totaliser le nombre d'heures écoulées depuis le moment où l'on met en marche la trotteuse. Le plus souvent, ce mécanisme permet de comptabiliser jusqu'à 12 heures. Une pression sur le poussoir de Remise à zéro fait revenir l'aiguille du compteur d'heures à son point de départ.

Contrefaçons

Copies de montres prisées, en général de très grande valeur. Certificats, factures et écrins étant aussi très souvent contrefaits, ils ne protègent plus depuis longtemps de mauvais achats.

Côtes de Genève

Décoration en forme de nervures sur les ponts, notamment de balancier, de → calibres haut de gamme. Appliquée avant la finition par galvanoplastie, elle reste néanmoins visible. En général, les côtes de Genève n'ornent que les mouvements de grande valeur.

Couronne

Bouton que l'on pousse pour remonter une montre, régler les aiguilles et/ou corriger les → indications de date et autrefois aussi pour commander le → chronographe.

Ébauche

Mouvement de montre incomplet. Autrefois, les horlogers genevois appelaient le mouvement d'une montre blanc ou ébauche.

Échappement

Mécanisme qui distribue par à-coups l'énergie du → ressort de barillet à la masse oscillante (→ balancier et → spiral de balancier) d'une montre et empêche le déroulement incontrôlé du mouvement. Pour une fréquence de balancier de 28800 alternances/heure, l'échappement fait avancer le train de roues 691 200 fois par jour, ce qui représente plus d'un milliard d'impulsions en quatre ans. Cela correspond à peu près au quadruple des performances d'un cœur humain.

Échappement à ancre

→ L'échappement actuellement le plus répandu sur les montres mécaniques. Sur les montres-bracelets haut de gamme, c'est l'échappement à ancre suisse qui domine.

Émail

Substance vitrifiable servant à décorer le métal. La technique de l'émaillage est utilisée depuis plus de 350 ans sur les montres (cadrans et → boîtiers). Au début du XXe siècle, les cadrans émaillés étaient presque la norme pour les montres haut de gamme. Depuis – notamment pour des raisons de coûts – cette pratique est devenue extrêmement rare.

Énergie

Capacité motrice stockée. Pour l'entraînement des montres, des réserves énergétiques sont nécessaires. Sur les montres et pendules mécaniques, elles sont fournies respectivement par un → ressort de barillet (entraînement à ressort) tendu et par un poids que l'on remonte (entraînement par gravité).

Établisseur

Fabricant d'horlogerie qui achète chez des fabricants spécialisés des éléments (mouvements, cadrans, aiguilles, → boîtiers) qu'il assemble ensuite pour réaliser des garde-temps terminés.

Étanche (montre-bracelet)

Pour pouvoir porter la mention « étanche » ou « waterproof », une montre-bracelet doit résister à la transpiration, aux projections d'eau et à la pluie. De plus, elle ne doit laisser pénétrer aucune humidité en cas d'immersion d'au moins 30 minutes à un mètre de profondeur. La mention supplémentaire « 50 mètres » ou « 5 bars » indique que ces montres ont été soumises par le fabricant à une pression d'essai correspondante. Toutefois, il n'est pas recommandé de nager, voire de plonger, avec ce type de montres.

Finissage

Dernières opérations d'usinage et de finition d'une montre.

Fréquence

Nombre d'oscillations par unité de temps, mesuré en hertz (Hz). Sur les pendules, le principal organe régulateur de marche est le pendule, tandis que les montres disposent d'un → balancier. Ces deux organes se balancent d'un côté à l'autre à une fréquence donnée. L'organe oscillant d'une pendule à secondes met exactement une seconde pour aller d'un point extrême à l'autre. Sa fréquence est donc de 0,5 Hz ou 1800 A/h (alternances/heure). Les oscillateurs à balancier anciens atteignaient de 7 200 à 9 000 A/h. Sur les montres de poche, la fréquence a d'abord été réglée sur 12 600 puis portée à 18 000 A/h (2,5 Hz), valeur la plus courante. Sur les montres-bracelets également, cette fréquence de balancier est devenue la norme dans un premier temps. Pour augmenter la précision, les fabricants de montres ont porté le nombre d'alternances à 21 600 A/h (3 Hz), 28 800 A/h (4 Hz), 36 000 A/h (5 Hz), 57 600 A/h (8 Hz) et jusqu'à 72 000 A/h (10 Hz). Une fréquence de balancier plus élevée s'accompagne d'une augmentation des besoins en énergie. Par ailleurs, des vitesses de rotation supérieures exigent plus de précautions lors du graissage de → l'échappement.

GMT
Greenwich Mean Time ; temps universel coordonné (TUC) au méridien de Greenwich. Le temps moyen de Greenwich est aujourd'hui la norme dans la navigation et les communications radio internationales.

Guillochage
Gravure de fins motifs, souvent entrelacés, sur le → boîtier ou le cadran d'une montre. Le guillochage traditionnel est effectué à la main, au moyen d'une lime et d'un burin.

Indication
Affichage de l'heure, du quantième, du jour de la semaine, du mois, de l'équation du temps, → de la réserve de marche ou d'un second fuseau horaire.

Indication de réserve de marche
→ Indication, sur les montres mécaniques, de la → réserve de marche restante.

Jewels
Désignation internationale du nombre de rubis (→ pierres précieuses) d'un mouvement de montre.

Ligne
Mesure traditionnelle utilisée pour indiquer la taille d'un mouvement, elle dérive du « pied du Roi » ou pied français. Exemples : 11''' pour un calibre de forme ronde et 8¾ x 12''' pour les calibres de toutes les autres formes. Une ligne équivaut à 2,2558 millimètres.

Lunette
Terme qui en horlogerie désigne des pièces très différentes. Au sens strict, c'est l'anneau qui porte la glace d'un(e) → boîtier (boîte) de montre. Celui-ci est ajusté à cran avec la glace enchâssée sur la partie médiane (carrure) de la boîte. Aujourd'hui, on appelle aussi très souvent lunette (lunette tournante), le cercle rotatif placé sur l'avant du cadran.

Manufacture
D'après les lois non écrites de l'horlogerie, un fabricant de montres ne peut prétendre à l'appellation manufacture que s'il fabrique lui-même au moins une → ébauche. Les fabricants de montres utilisant des ébauches produites en externe sont appelés dans le métier des → établisseurs.

Mécanisme à remontage automatique
Mécanisme auxiliaire qui utilise les mouvements (du poignet) pour armer le → ressort de barillet d'une montre mécanique.

Métaux précieux
Pour les → boîtiers des montres-bracelets, on utilise en général de l'or, du platine et de l'argent. On recourt à de l'or de différentes finesses : 333/1 000 (8 carats), 375/1 000 (9 carats), 575/1 000 (14 carats) ou 750/1 000 (18 carats). L'alliage avec d'autres métaux (cuivre, par exemple) détermine la teinte. On trouve de l'or titrant 21, 23 ou 24 carats dans les rotors de remontage automatique. La finesse du platine utilisé est de 950/1 000.

Montre à remontage manuel
Garde-temps dont le → ressort de barillet doit être armé à la main.

Montre-chronomètre
Montre dont la précision de marche a été attestée par des contrôles d'une durée de 15 jours effectués par un organisme officiel (comme le Contrôle officiel suisse des chronomètres (COSC) en Suisse). Dans les cinq → positions « couronne à gauche », « couronne en haut », « couronne en bas », « cadran dessus » et « cadran dessous », la moyenne de marche quotidienne doit rester comprise entre −4 et +6 secondes, l'écart de marche journalier moyen ne doit pas dépasser 2 secondes, et le plus grand écart de marche 5 secondes. Toutes les montres sont contrôlées à des températures de 8, 23 et 38 °C. Seules celles ayant passé avec succès les tests de précision peuvent porter sur leur cadran la mention « chronomètre » et être commercialisées avec le certificat officiel correspondant.

Montre et horloge (à mouvement) mécanique
Montre entraînée par un → ressort de barillet. Le réglage de la marche est assuré par un oscillateur, comme l'ensemble → balancier-spiral. L'origine de la montre mécanique remonterait au mécanisme d'entraînement des planétaires, connus depuis la fin du XIIIᵉ siècle. Le plus ancien mouvement mécanique dans la zone germanophone est assurément celui de l'horloge de la cathédrale de Strasbourg, réalisé en 1352. En Angleterre, il y aurait eu des horloges à mouvement mécanique dès la fin du XIIIᵉ siècle.

Montres de plongée
Montres-bracelets fabriquées suivant des normes clairement établies pour des interventions sous l'eau. Les montres de plongée professionnelles doivent rester étanches au minimum jusqu'à 200 mètres de profondeur. Pour des raisons de sécurité, il faut faire vérifier l'étanchéité d'une montre de plongée au moins une fois par an par un horloger.

Mouvement à remontage manuel
Un mouvement à remontage manuel se compose, selon sa construction, de 80 pièces ou plus. On peut le subdiviser en huit grands groupes fonctionnels :
1. système d'entraînement, organe qui fournit l'énergie
2. système de transmission qui achemine → l'énergie au
3. système de distribution, également appelé → échappement
4. système de régulation
5. minuterie
6. organes d'affichage de l'heure
7. système de réglage des aiguilles
8. système de remontage

Mouvement squelette
Mouvement sur lequel → la platine, les ponts, le barillet et éventuellement le → rotor sont tellement ajourés qu'il ne reste plus que le matériau absolument indispensable pour le bon fonctionnement des pièces. On peut ainsi voir à travers le mouvement. On trouve des montres-bracelets à mouvement squelette depuis le milieu des années 1930.

Numéro de fabrication
Pour identifier leurs produits, la plupart des fabricants de montres numérotent leurs mouvements ou leurs → boîtiers, voire les deux. Le numéro de mouvement doit impérativement apparaître sur les → chronomètres officiellement testés. Ce numéro est repris dans le rapport d'essai.

Pare-chocs
Système permettant d'éviter que les fins et donc très fragiles pivots du balancier ne se cassent. Pour ce faire, les pierres glace et contre-pivots des paliers de l'axe de balancier sont maintenus dans des → platines et des coqs servant d'amortisseurs. En cas de chocs importants, ils cèdent dans un plan latéral et/ou axial. Une montre-bracelet protégée contre les chocs doit résister sans dommage à une chute d'une hauteur d'un mètre sur un parquet en chêne. Elle ne doit pas non plus accuser un écart de marche significatif.

Phases de la Lune
Durant une lunaison, qui dure environ 29,5 jours, la Lune traverse différentes phases (d'illumination) dépendant des positions relatives du Soleil, de la Lune et de la Terre : Nouvelle Lune – Premier Quartier – Pleine Lune – Dernier Quartier – Nouvelle Lune.

Pierres précieuses

Sur les montres de précision, des pierres précieuses sont utilisées pour réduire les frottements au niveau des principaux paliers, des palettes d'ancre et des ellipses.

Platine

Plaque métallique qui soutient les ponts et les divers autres organes d'un mouvement.

Poinçons

Marques dans les → boîtiers qui renseignent sur la nature et la finesse du → métal précieux utilisé, sur le pays d'origine et parfois sur le lieu d'origine, la date de fabrication et aussi sur l'identité du fabricant du boîtier. À cela s'ajoutent souvent le logo de la société qui a fabriqué ou livré la montre, un numéro de → référence et un numéro de série.

Positions

Contrairement aux montres de poche, les montres-bracelets sont portées dans cinq positions différentes : « couronne à gauche », « couronne en haut », « couronne en bas », « cadran dessus » et « cadran dessous ». Les montres de précision sont donc vérifiées dans ces cinq positions.

Pulsomètre

Cadran dont les divisions facilitent (généralement sur les → chronographes) la prise du pouls. Selon le type de graduation, il faut compter 20 ou 30 pulsations après la mise en marche du chronographe. L'aiguille de chronographe activée pointe sur la fréquence du pouls extrapolée sur une minute.

Quantième (ou calendrier) perpétuel

Calendrier complexe constitué d'une centaine d'éléments qui prend généralement en compte les durées différentes des mois jusqu'au 28 février 2100 sans exiger de correction manuelle.

Rattrapante de chronographe

Ce mécanisme permet de chronométrer simultanément deux événements ou plus, à condition qu'ils débutent au même instant. Il permet aussi de prendre les temps intermédiaires dans des compétitions.

Référence

Combinaison de lettres et de chiffres utilisée par un fabricant de montres pour classer ses différents modèles. Le numéro de référence contient parfois aussi des informations sur le type de modèle, le matériau du boîtier, le mouvement, le cadran, les aiguilles, le bracelet et les pierres précieuses incrustées.

Remontoir

Jusqu'à la fin du XIXᵉ siècle, le remontage et/ou le réglage des aiguilles de nombreuses montres de poche se faisaient à l'aide d'une petite clef. Sur le remontoir moderne, ces deux opérations s'effectuent grâce à une petite → couronne à inverseur.

Réserve de marche

Réserve d'énergie suffisante entre deux remontages d'une montre-bracelet (24 heures). Habituellement, la réserve de marche est de l'ordre de 10 à 16 heures. Toutefois, la force d'entraînement du → ressort de barillet diminuant au fil du temps, une baisse des performances est inévitable.

Ressort

Des ressorts des types les plus divers sont utilisés dans les mouvements. Outre les → spiraux et → ressorts de barillet, ce sont principalement des ressorts de blocage et de maintien.

Ressort de barillet

Ruban d'acier élastique enroulé en spirale servant à emmagasiner la force motrice pour les montres de poche et les montres-bracelets.

Rotor

Masse oscillante tournant librement sur les montres à → mécanisme de remontage automatique. Suivant la conception de ce dernier, le → ressort de barillet est armé dans un seul ou dans les deux sens de rotation.

On distingue les rotors centraux et les microrotors. Les premiers pivotent sur l'ensemble du mouvement, les derniers sont intégrés au niveau du mouvement.

Satinage

Opération matifiant et adoucissant la surface d'un composant métallique au moyen d'abrasifs fins.

Silicium

Pendant des siècles, le laiton, l'acier ou le rubis synthétique ont dominé la fabrication des mouvements. Depuis 2001, le silicium, matériau de base des micropuces électroniques, est également admis. À l'état de monocristal, il a la même structure que le diamant. Il est 60 pour cent plus dur et 70 pour cent plus léger que l'acier, → amagnétique et résistant à la corrosion. Même sans finition particulière, il possède une surface extrêmement lisse, qui rend tout lubrifiant inutile. Le silicium est élastique, mais non malléable. Soumis à un traitement spécial, le « Silinvar » ne se dilate quasiment pas sous l'effet de la chaleur. Il se prête ainsi à la fabrication de → spiraux de balancier. Le silicium invariable « Silinvar » résulte d'un projet de développement commun entre Patek Philippe, Rolex et Swatch Group ainsi que le Centre suisse d'électronique et de microtechnique (CSEM) et l'Institut de microtechnique de l'université de Neuchâtel.

Sonnerie à répétition

Fonction auxiliaire complexe d'un mouvement, qui produit des sons correspondant à différentes divisions de l'heure. Selon le modèle de sonnerie, on différencie entre montres répétitions à quarts, demi-quarts (soit 7½ minutes), à 5 minutes ou à minutes.

Spiral Breguet

→ Spiral à spire extérieure relevée pour permettre au ressort de se développer régulièrement durant son expansion et sa contraction. Au vu des coûts de fabrication considérables, il n'est plus utilisé aujourd'hui que dans les → calibres de grande précision.

Spiral de balancier

On peut dire que c'est l'âme de toute montre mécanique. Ce ressort est fixé par son extrémité intérieure au balancier et par son extrémité extérieure au coq. Par son élasticité, il assure la régularité des oscillations du → balancier. La longueur active du spiral et le moment d'inertie du balancier déterminent la durée des oscillations (ou période).

Tourbillon

Inventé en 1795 par Abraham-Louis Breguet et breveté en 1801, ce dispositif sert à compenser les perturbations infligées par la gravité à l'organe d'oscillation (→ balancier et → spiral de balancier) des montres mécaniques. L'organe d'oscillation et l'échappement sont logés dans une cage aussi légère que possible qui fait un tour sur son axe en une durée déterminée (généralement une minute). Cela permet de compenser les écarts de marche liés à la gravité terrestre dans les → positions verticales et d'améliorer les performances de marche. En position horizontale, le tourbillon n'a en revanche aucune influence sur l'exactitude de marche.

Verre saphir

Verre de montre résistant aux rayures et d'une dureté de 9 sur l'échelle de Mohs. Seul le diamant présente une dureté plus élevée.

Verres

Pour les montres-bracelets, il existe quatre types différents de verres. Les verres en cristal sont surtout utilisés sur les montres-bracelets anciennes. Résistant bien aux rayures, ils sont toutefois très cassants. Au début des années 1940, les verres en cristal sont progressivement évincés par des verres en matériaux synthétiques (Plexiglas). Certes incassables, ils se rayent par contre très facilement. Avec une dureté de 5 sur l'échelle de Mohs, les verres minéraux sont nettement plus robustes que les verres synthétiques. Aujourd'hui, on monte sur les montres de grande valeur le plus souvent du → verre saphir. D'une dureté de 9 sur l'échelle de Mohs, il est extrêmement résistant aux rayures et aux cassures, mais doit être travaillé avec des outils diamantés spéciaux.

Imprint

© 2022 teNeues Verlag GmbH

Introduction & texts by Gisbert L. Brunner

Photo editing & captions (Audemars Piguet, Breguet, Breitling, Cartier, Chopard, Glashütte Original, Hublot, IWC Schaffhausen, Jaeger-LeCoultre, A. Lange & Söhne, Montblanc, Omega, Panerai, Patek Philippe, Piaget, Rolex, Vacheron Constantin, Zenith) by Christian Pfeiffer-Belli
Editorial management by Regine Freyberg
Design & layout by Sophie Franke
Translation by Howard Fine (English),
Hanna Lemke (German),
Claude Checconi & Christèle Jany (French)
Copy editing by Howard Fine (English),
Dr. Simone Bischoff & Hanna Lemke (German),
Claude Checconi & Christèle Jany (French)
Production by Nele Jansen
Photo editing (Alpina, Ateliers deMonaco, Baume & Mercier, Blancpain, Carl F. Bucherer, Bulgari, Chronoswiss, Corum, Eterna, Frédérique Constant, Girard-Perregaux, de Grisogono, Hanhart, Junghans, Longines, Nomos Glashütte, Oris, Parmigiani Fleurier, Porsche Design, Seiko, TAG Heuer, Tudor, Tutima Glashütte, Ulysse Nardin) & color separation by David Burghardt/db-photo.de

ISBN 978-3-96171-185-7

Library of Congress Control Number: 2018962299
Printed in Slovakia by Neografia a.s.

Bibliographic information published by the Deutsche Nationalbibliothek. The Deutsche Nationalbibliothek lists this publication in the Deutsche Nationalbibliografie; detailed bibliographic data are available on the Internet at dnb.dnb.de.

Published by teNeues Publishing Group

teNeues Verlag GmbH
Ohmstraße 8a
86199 Augsburg, Germany

Düsseldorf Office
Waldenburger Straße 13
41564 Kaarst, Germany
e-mail: books@teneues.com

Augsburg/München Office
Ohmstraße 8a
86199 Augsburg, Germany
e-mail: books@teneues.com

Berlin Office
Lietzenburger Straße 53
10719 Berlin, Germany
e-mail: books@teneues.com

Press Department Stefan Becht
Phone: +49-152-2874-9508 / +49-6321-97067-97
e-mail: sbecht@teneues.com

teNeues Publishing Company
350 Seventh Avenue, Suite 301
New York, NY 10001, USA
Phone: +1-212-627-9090
Fax: +1-212-627-9511

www.teneues.com

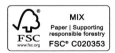

teNeues Publishing Group
Augsburg / München
Berlin
Düsseldorf
London
New York

teNeues